T0214880

Lecture Notes in Computer Science 9226

Commenced Publication in 1973
Founding and Former Series Editors:
Gerhard Goos, Juris Hartmanis, and Jan van Leeuwen

More information about this series at http://www.springer.com/series/7409

De-Shuang Huang · Kang-Hyun Jo
Abir Hussain (Eds.)

Intelligent Computing Theories and Methodologies

11th International Conference, ICIC 2015
Fuzhou, China, August 20–23, 2015
Proceedings, Part II

 Springer

Editors
De-Shuang Huang
Tongji University
Shanghai
China

Abir Hussain
Liverpool John Moores University
Liverpool
UK

Kang-Hyun Jo
University of Ulsan
Ulsan
Korea, Republic of (South Korea)

ISSN 0302-9743   ISSN 1611-3349   (electronic)
Lecture Notes in Computer Science
ISBN 978-3-319-22185-4   ISBN 978-3-319-22186-1   (eBook)
DOI 10.1007/978-3-319-22186-1

Library of Congress Control Number: 2015945123

LNCS Sublibrary: SL3 – Information Systems and Applications, incl. Internet/Web and HCI

Printed on acid-free paper

Springer International Publishing AG Switzerland is part of Springer Science+Business Media
(www.springer.com)

Preface

The International Conference on Intelligent Computing (ICIC) was started to provide an annual forum dedicated to the emerging and challenging topics in artificial intelligence, machine learning, pattern recognition, bioinformatics, and computational biology. It aims to bring together researchers and practitioners from both academia and industry to share ideas, problems, and solutions related to the multifaceted aspects of intelligent computing.

ICIC 2015, held in Fuzhou, China, August 20–23, 2015, constituted the 11th International Conference on Intelligent Computing. It built upon the success of ICIC 2014, ICIC 2013, ICIC 2012, ICIC 2011, ICIC 2010, ICIC 2009, ICIC 2008, ICIC 2007, ICIC 2006, and ICIC 2005 that were held in Taiyuan, Nanning, Huangshan, Zhengzhou, Changsha, China, Ulsan, Korea, Shanghai, Qingdao, Kunming, and Hefei, China, respectively.

This year, the conference concentrated mainly on the theories and methodologies as well as the emerging applications of intelligent computing. Its aim was to unify the picture of contemporary intelligent computing techniques as an integral concept that highlights the trends in advanced computational intelligence and bridges theoretical research with applications. Therefore, the theme for this conference was "Advanced Intelligent Computing Theories and Applications." Papers focusing on this theme were solicited, addressing theories, methodologies, and applications in science and technology.

ICIC 2015 received 671 submissions from 25 countries and regions. All papers went through a rigorous peer-review procedure and each paper received at least three review reports. On the basis of the review reports, the Program Committee finally selected 233 high-quality papers for presentation at ICIC 2015, included in three volumes of proceedings published by Springer: two volumes of *Lecture Notes in Computer Science* (LNCS) and one volume of *Lecture Notes in Artificial Intelligence* (LNAI).

This volume of *Lecture Notes in Computer Science* (LNCS) 9226 includes 74 papers.

The organizers of ICIC 2015, including Tongji University and Fujian Normal University, China, made an enormous effort to ensure the success of the conference. We hereby would like to thank the members of the Program Committee and the reviewers for their collective effort in reviewing and soliciting the papers. We would like to thank Alfred Hofmann, executive editor at Springer, for his frank and helpful advice and guidance throughout and for his continuous support in publishing the proceedings. In particular, we would like to thank all the authors for contributing their papers. Without the high-quality submissions from the authors, the success of the conference would not have been possible. Finally, we are especially grateful to the International Neural Network Society and the National Science Foundation of China for their sponsorship.

June 2015
De-Shuang Huang
Kang-Hyun Jo
Abir Hussain

ICIC 2015 Organization

General Co-chairs

De-Shuang Huang   China
Changping Wang   China

Program Committee Co-chairs

Kang-Hyun Jo   Korea
Abir Hussain   UK

Organizing Committee Co-chairs

Hui Li   China
Yi Wu   China

Award Committee Chair

Laurent Heutte   France

Publication Co-chairs

Valeriya Gribova   Russia
Zhi-Gang Zeng   China

Special Session Co-chairs

Phalguni Gupta   India
Henry Han   USA

Special Issue Co-chairs

Vitoantonio Bevilacqua   Italy
Mike Gashler   USA

Tutorial Co-chairs

Kyungsook Han   Korea
M. Michael Gromiha   India

International Liaison

Prashan Premaratne   Australia

Publicity Co-chairs

Juan Carlos Figueroa   Colombia
Ling Wang   China
Evi Syukur   Australia
Chun-Hou Zheng   China

Exhibition Chair

Bing Wang   China

Organizing Committee Members

Jianyong Cai   China
Qingxiang Wu   China
Suping Deng   China
Lin Zhu   China

Program Committee Members

Andrea Francesco Abate, Italy
Waqas Haider Khan Bangyal, Pakistan
Shuhui Bi, China
Qiao Cai, USA
Jair Cervantes, Mexico
Chin-Chih Chang, Taiwan, China
Chen Chen, USA
Huanhuan Chen, China
Shih-Hsin Chen, Taiwan, China
Weidong Chen, China
Wen-Sheng Chen, China
Xiyuan Chen, China
Yang Chen, China
Cheng Cheng, China
Ho-Jin Choi, Korea
Angelo Ciaramella, Italy
Salvatore Distefano, Italy
Jianbo Fan, China
Minrui Fei, China
Juan Carlos Figueroa, Colombia

Huijun Gao, China
Shan Gao, China
Dunwei Gong, China
M. Michael Gromiha, India
Zhi-Hong Guan, China
Kayhan Gulez, Turkey
Phalguni Gupta, India
Fei Han, China
Kyungsook Han, Korea
Laurent Heutte, France
Wei-Chiang Hong, Taiwan, China
Yuexian Hou, China
Peter Hung, Ireland
Saiful Islam, India
Li Jia, China
Zhenran Jiang, China
Xin Jin, USA
Joaquín Torres-Sospedra, Spain
Dah-Jing Jwo, Taiwan, China
Vandana Dixit Kaushik, India

Sungshin Kim, Korea
Yoshinori Kuno, Japan
Takashi Kuremoto, Japan
Xiujuan Lei, China
Bo Li, China
Guo-Zheng Li, China
Kang Li, UK
Peihua Li, China
Shi-Hua Li, China
Shuai Li, Hong Kong, China
Shuwei Li, USA
Xiaodi Li, China
Xiaoou Li, Mexico
Chengzhi Liang, China
Bingqiang Liu, China
Ju Liu, China
Shuo Liu, USA
Weixiang Liu, China
Xiwei Liu, China
Yunxia Liu, China
Chu Kiong Loo, Malaya
Geyu Lu, China
Ke Lu, China
Yingqin Luo, USA
Pabitra Mitra, India
Tarik Veli Mumcu, Turkey
Roman Neruda, Czech Republic
Ben Niu, China
Sim-Heng Ong, Singapore
Seiichi Ozawa, Japan
Vincenzo Pacelli, Italy
Shaoning Pang, New Zealand
Francesco Pappalardo, Italy
Surya Prakash, India
Prashan Premaratne, Australia
Yuhua Qian, China
Daowen Qiu, China
Ivan Vladimir Meza Ruiz, Mexico
Seeja K.R., India
Fanhuai Shi, China

Wilbert Sibanda, South Africa
Jiatao Song, China
Stefano Squartini, Italy
Wu-Chen Su, Taiwan, China
Zhan-Li Sun, China
Antonio E. Uva, Italy
Mohd Helmy Abd Wahab, Malaysia
Bing Wang, China
Jingyan Wang, USA
Ling Wang, China
Shitong Wang, China
Xuesong Wang, China
Yi Wang, USA
Yong Wang, China
Yuanzhe Wang, USA
Yufeng Wang, China
Yunji Wang, USA
Wei Wei, Norway
Zhi Wei, China
Ka-Chun Wong, Canada
Hongjie Wu, China
Qingxiang Wu, China
Yan Wu, China
Junfeng Xia, USA
Shunren Xia, China
Bingji Xu, China
Xin Yin, USA
Xiao-Hua Yu, USA
Shihua Zhang, China
Eng. Primiano Di Nauta, Italy
Hongyong Zhao, China
Jieyi Zhao, USA
Xiaoguang Zhao, China
Xing-Ming Zhao, China
Zhongming Zhao, USA
Bojin Zheng, China
Chun-Hou Zheng, China
Fengfeng Zhou, China
Yong-Quan Zhou, China

Additional Reviewers

Haiqing Li
JinLing Niu
Li Wei
Peipei Xiao
Dong Xianguang
Shasha Tian
Li Liu
Xuan Wang
Sen Xia
Jun Li
Xiao Wang
Anqi Bi
Min Jiang
Wu Yang
Yu Sun
Bo Wang
Wujian Fang
Zehao Chen
Chen Zheng
Hong Zhang
Huihui Wang
Liu Si
Sheng Zou
Xiaoming Liu
Meiyue Song
Bing Jiang
Min-Ru Zhao
Xiaoyong Bian
Chuang Ma
Yujin Wang
Wenlong Hang
Chong He
Rui Wang
Zhu Linhe
Xin Tian
Mou Chen
F.F. Zhang
Gabriela Ramírez
Ahmed Aljaaf
Neng Wan
Abhineet Anand
Antonio Celesti
Aditya Nigam
Aftab Yaseen

Weiyuan Zhang
Shamsul Zulkifli
Alfredo Liverani
Siti Amely Jumaat
Muhammad Amjad
Angelo Ciaramella
Aniza Mohamed Din
En-Shiun Annie Lee
Anoosha Paruchuri
Antonino Staiano
Antony Lam
Alfonso Panarello
Alfredo Pulvirenti
Asdrubal Lopez-Chau
Chun Kit Au
Tao Li
Mohd Ayyub Khan
Azizi Ab Aziz
Azizul Azhar Ramli
Baoyuan Wu
Oscar Belmonte
Biao Li
Jian Gao
Lin Yong
Boyu Zhang
Edwin C. Shi
Caleb Rascon
Yu Zhou
Fuqiang Chen
Fanshu Chen
Chenguang Zhu
Chin-Chih Chang
Chang-Chih Chen
Changhui Lin
Cheng-Hsiung Chiang
Linchuan Chen
Bo Chen
Chen Chen
Chengbin Peng
Cheng Cheng
Jian Chen
Qiaoling Chen
Zengqiang Chen
Ching-Hua Shih

Chua King Lee
Quistina Cuci
Yanfeng Chen
Meirong Chen
Cristian Rodriguez Rivero
Aiguo Chen
Aquil Mirza Mohammed
Sam Kwong
Yang Liu
Yizhang Jiang
Bingbo Cui
Yu Wu
Chuan Wang
Wenbin Chen
Cheng Zhang
Dhiya Al-Jumeily
Hang Dai
Yinglong Dai
Danish Jasnaik
Kui Liu
Dapeng Li
Dan Yang
Davide Nardone
Dawen Xu
Dongbo Bu
Liya Ding
Shifei Ding
Donal O'Regan
Dong Li
Li Kuang
Zaynab Ahmed
David Shultis
Zhihua Du
Chang-Chih Chen
Erkan İMal
Ekram Khan
Eric Wei
Gang Wang
Fadzilah Siraj
Shaojing Fan
Mohammad Farhad
  Bulbul
Mohamad Farhan
    Mohamad Mohsin

Fengfeng Zhou
Feng Jiqiang
Liangbing Feng
Farid García-Lamont
Chien-Yuan Lai
Filipe de O. Saraiva
Francesco Longo
Fabio Narducci
Francesca Nardone
Francesco Camastra
Nhat Linh Bui
Hashim Abdellah
Hashim Moham
Gao Wang
Jungwon Yu
Ge Dingfei
Geethan Mendiz
Na Geng
Fangda Guo
Gülsüm Gezer
Guanghua Sun
Guanghui Wang
Rosalba Giugno
Giovanni Merlino
Xiaoqiang Zhang
Guangchun Cheng
Guanglan Zhang
Tiantai Guo
Weili Guo
Yanhui Guo
Xiaoqing Gu
Hafizul Fahri Hanafi
Haifeng Wang
Mohamad Hairol Jabbar
H.K. Lam
Khalid Isa
Hang Su
Guangjie Han
Hao Chu
Ben Ma
Hao Men
Yonggang Chen
Haza Nuzly Abdull
   Hamed
Mohd Helmy Abd Wahab
   Pang
Hei Man Herbert

Hironobu Fujiyoshi
Guo-Sheng Hao
Huajuan Huang
Hiram Calvo
Hongjie Wu
Hongjun Su
Hitesh Kumar Sharma
Haiguang Li
Hongkai Chen
Qiang Huang
Tengfei Zhang
Zheng Huai
Joe Huang
Jin Huang
Wan Hussain Wan Ishak
Xu Huang
Ying Hu
Ho Yin Sze-To
Haitao Zhu
Ibrahim Venkat
Jooyoung Lee
Josef Moudířk
Li Xu
Jakub Smid
Jianhung Chen
Le Li
Jianbo Lu
Jair Cervantes
Junfeng Xia
Jinhai Li
Hongmei Jiang
Jiaan Zeng
Jian Lu
Jian Wang
Jian Zhang
Jianhua Zhang
Jie Wu
Jim Jing-Yan Wang
Jing Sun
Jingbin Wang
Wu Qi
Jose Sergio Ruiz Castilla
Joaquín Torres
Gustavo Eduardo Juarez
Jun Chen
Junjiang Lin
Junlin Chang

Juntao Liu
Justin Liu
Jianzhong Guo
Jiayin Zhou
Abd Kadir Mahamad
K. Steinhofel
Ka-Chun Wong
Ke Li
Kazuhiro Fukui
Sungshin Kim
Klara Peskova
Kunikazu Kobayashi
Konstantinos Tsirigos
Seeja K.R.
Kwong Sak Leung
Kamlesh Tiwari
Li Kuang
K.V. Arya
Zhenjiang Lan
Ke Liao
Liang-Tsung Huang
Le Yang
Erchao Li
Haitao Li
Wei Li
Meng Lei
Guoqi Luo
Huihui Li
Jing Liang
Liangliang Zhang
Liang Liang
Bo Li
Bing Li
Dingshi LI
Lijiao Liu
Leida Li
Lvzhou Li
Min Li
Kui Lin
Ping Li
Liqi Yi
Lijun Quan
Bingwen Liu
Haizhou Liu
Jing Liu
Li Liu
Ying Liu

Zhaoqi Liu	Musheer Ahmad	Wei Cui
Zhe Liu	Monika Verma	Rozaida Ghazali
Yang Li	Naeem Radi	Raghuraj Singh
Lin Zhu	Aditya Nigam	Rey-Sern Lin
Jungang Lou	Aditya Nigam	S.M. Zakariya
Lenka Kovářová	Nagarajan Raju	Sabooh Ajaz
Shaoke Lou	Mohd Najib Mohd Salleh	Alexander Tchitchigin
Li Qingfeng	Yinan Guo	Kuo-Feng Huang
Qinghua Li	Zhu Nanli	Shao-Lun Lee
Lu Huang	Mohammad Naved	Wei-Chiang Hong
Liu Liangxu	Qureshi	Toshikazu Samura
Lili Ayu Wulandhari	Su Rina	Sandhya Pundhir
Xudong Lu	Sanders Liu	Jin-Xing Liu
Yiping Liu	Chang Liu	Shahzad Alam
Yutong Li	Patricio Nebot	Mohd Shamrie Sainin
Junming Zhang	Nistor Grozavu	Shanye Yin
Mohammed Khalaf	Zhixuan Wei	Shasha Tian
Maria Musumeci	Nobuyuki Nezu	Hao Shen
Shingo Mabu	Nooraini Yusoff	Chong Shen
Yasushi Mae	Nureize Arbaiy	Jingsong Shi
Manzoor Lone	Kazunori Onoguchi	Nobutaka Shimada
Liang Ma	R.B. Pachori	Lu Xingjia
Manabu Hashimoto	Yulei Pang	Shun Chen
Md. Abdul Mannan	Xian Pan	Silvio Barra
Qi Ma	Binbin Pan	Simone Scardapane
Martin Pilat	Peng Chen	Salvador Juarez Lopez
Asad Khan	Klara Peskova	Shenshen Liang
Maurizio Fiasché	Petra Vidnerová	Sheng Liu
Max Talanov	Qi Liu	Somnath Dey
Mohd Razali MD Tomari	Prashan Premaratne	Rui Song
Mengxing Cheng	Puneet Gupta	Seongpyo Cheon
Meng Xu	Prabhat Verma	Sergio Trilles-Oliver
Tianyu Cao	Peng Zhang	Subir Kumar Nandy
Minfeng Wang	Haoqian Huang	Hung-Chi Su
Muhammad Fahad	Qiang Fu	Sun Jie
Michele Fiorentino	Qiao Cai	QiYan Sun
Michele Scarpiniti	Haiyan Qiao	Shiying Sun
Zhenmin Zhang	Qingnan Zhou	Sushil Kumar
Ming Liu	Rabiah Ahmad	Yu Su
Miguel Mora-Gonzalez	Rabiah Abdul Kadir	Hu Zhang
Aul Montoliu	R. Rakkiyappan	Yan Qi
Yuanbin Mo	Ramakrishnan	Hotaka Takizawa
Marzio Pennisi	Chandrasekaran	Tanggis Bohnuud
Binh P. Nguyen	Yunpeng Wang	Yang Tang
Mingqiang Zhang	Hao Zheng	Lirong Tan
Muhammad Rashid	Radhakrishnan Delhibabu	Yao Tuozhong
Shenglin Mu	Rohit Katiyar	Tian Tian

Tianyi Wang
Toshiaki Kondo
Tofik Ali
Tomáš Ken
Peng Xia
Gurkan Tuna
Tutut Herawan
Zhongneng Xu
Danny Wu
Umarani Jayaraman
Zhichen Gong
Vibha Patel
Vikash Yadav
Esau Villatoro-Tello
Prashan Premaratne
Vishnu Priya Kanakaveti
Vivek Srivastava
Victor Manuel
  Landassuri-Moreno
Hang Su
Yi Wang
Chao Wang
Cheng Wang
Jiahai Wang
Jing Wang
Junxiao Wang
Linshan Wang
Waqas Haider Bangyal
Herdawatie Abdul Kadir
Yi Wang
Wen Zhang
Widodo Budiharto
Wenjun Deng
Wen Wei
Wenbin Chen
Wenzhe Jiao
Yufeng Wang
Haifeng Wang
Deng Weilin
Wei Jiang
Weimin Huang
Wufeng Tian
Wu-Chen Su
Daiyong Wu
Yang Wu

Zhonghua Wu
Jun Zhang
Xiao Wang
Wei Xiong
Weixiang Liu
Wenxi Zhang
Wenye Li
Wenlin Zhang
Li-Xin He
Ming Tan
Yi-Gang Zhang
Xiangliang Zhang
Nan Xiang
Xiaobo Zhang
Xiaohu Wang
Xiaolei Wang
Xiaomo Liu
Lei Wang
Yan Cui
Xiaoyong Zhang
Xiaozhen Xue
Jianming Xie
Xin Gao
Xinhua Xiao
Xin Lu
Xinyu Zhang
Liguang Xu
Hao Wu
Xun Li
Jin Xu
Xin Xu
Yuan Xu
Xiaoyin Xu
Yuan Xu
Shiping Chen
Xiaoyan Sun
Xiaopin Zhong
Atsushi Yamashita
Yanen Guo
Xiaozhan Yang
Zhanlei Yang
Yaqiang Yao
Bei Ye
Yan Fu
Yehu Shen

Yu-Yen Ou
Yingyou Wen
Ying Yang
Yingtao Zhao
Yanjun Zhao
Yong Zhang
Yoshinori Kobayashi
Yesu Feng
Yuan Lin
Lin Yuan
Yugandhar Kumar
Yujia Li
Yupeng Li
Yuting Yang
Yuyan Han
Yong Wang
Yingying Wei
Yingxin Guo
Xiangjuan Yao
Guodong Zhao
Huan Zhang
Bo Zhang
Gongjie Zhang
Yu Zhang
Zhao Yan
Wenrui Zhao
Zhao Yan
Tian Zheng
Zhengxing Hang
Xibei Yang
Zhenxin Zhan
Zhipeng Cai
Lingyun Zhu
Yanbang Zhang
Zheng-Ling Yang
Juan Li
Zongxiao He
Zhengyu Ouyang
Will Zhu
Xuebing Zhang
Zhile Yang
Yi Zhang

Contents – Part II

Contents – Part I

Efficiency and Effectiveness Metrics in Evolutionary Algorithms and Their Application

Guo-Sheng Hao[1(✉)], Chang-Shuai Chen[1], Gai-Ge Wang[1],
Yong-Qing Huang[1,3], De-Xuan Zhou[2], and Zhao-Jun Zhang[2]

[1] School of Computer Science and Technology,
Jiangsu Normal University, Xuzhou 221116, Jiangsu, China
guoshenghaoxz@tom.com
[2] School of Electrical Engineering and Automation,
Jiangsu Normal University, Xuzhou 221116, Jiangsu, China
zzj921@163.com
[3] School of Mathematics and Computer, Tongling University,
Tongling 244000, Anhui, China
yongqinghuangbs@sina.com

Abstract. Efficiency and effectiveness are two important metrics for the evaluation of evolutionary algorithms (EAs). Firstly, there exist a number of efficiency metrics in EA, such as population size, number of termination generation, space complexity, and time complexity and so on. But the relationship of these metrics is left untouched. And evaluating or comparing EAs with one of these metrics or using them separately is unfair. Therefore it is necessary to consider their relationship and give proper metrics combination. We conclude that the product of population size and number of generation should be less than the value of search space size, and the product of time complexity and space complexity should also be less than a constant. Secondly, we study the relationship between efficiency and effectiveness. Based on these two metrics, we conclude that not only EAs can be compared, but also problems hardness can be measured. The results reveal important insights of EAs and problems hardness.

Keywords: Evolutionary algorithms · Performance · Efficiency · Effectiveness · Problems hardness · Population size · Stop generation

1 Introduction

Evolutionary algorithms (EAs) [1] are problem independent and have been reported to perform relatively well on problems such as: search, optimization, and artificial intelligence.

The analysis of problem hardness (difficulty) has been studied for decades [2]. Measuring hardness of an instance for a particular algorithm is typically done by comparing the optimization precision reached after a certain number of iterations compared to other algorithms, and/or by comparing the number of iterations taken to reach the best solution [3]. There have been numerous efforts identifying challenges

© Springer International Publishing Switzerland 2015
D.-S. Huang et al. (Eds.): ICIC 2015, Part II, LNCS 9226, pp. 1–12, 2015.
DOI: 10.1007/978-3-319-22186-1_1

such as isolation, deception, multi-modality, the size of basins of attraction [4], as well as landscape metrics [5], autocorrelation structures and distributions of local minima [6].

In optimization algorithms, performance measurement is an important topic and effectiveness and efficiency are two aspects of the performance of algorithms. Firstly, effectiveness refers to the quality of the obtained solution and most of its indicators are directly related to the optimized objective(s) [7], and there is also effective metric about the solution (population) diversity [8]. Secondly, efficiency is usually characterized by its runtime behavior, i.e., the order of its computation time and its memory require-ments [9] and its indicators are mainly about the algorithm. Efficiency can be increased drastically with the use of domain knowledge which can be embedded in algorithms, while effectiveness highly depends on the properties of the problem [10]. To our knowledge, the issues that relationship between efficiency and effectiveness, the rela-tionship among metrics of efficiency are still not analyzed deeply. And this is the purpose of this paper.

2   Efficiency Metrics of Evolutionary Algorithms

2.1   Population Size and Termination Generation Number

To calculate the efficiency, for the population size, only the number of individuals who get evaluation from environment is considered, but not from surrogate, in other words, who get real evaluation but not virtual evaluation. As shown in Fig. 1, there are two kinds of individuals' fitness evaluation: directly from environment and from the sur-rogate. Generally surrogate is developed for cheaper fitness functions, but at the cost of accuracy of fitness estimation [11].

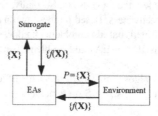

Fig. 1. Individuals evaluation structure in EAs.

The product of population size and the number of generation is the number of individuals whose information is used by the algorithm. In other words, the product is the number of individuals who get their evaluation from environment or whose eval-uation is the real one but not the estimated one by surrogate. With the same obtained best solution X_{best}^{real}, the less the product is, the less resources the algorithm costs, and the higher the efficiency is.

Label the number of those individuals who have the real evaluation till the current evolutionary generation t as $N_{ind}(t)$, which can be simply explained as the product of the population size $|P|$ and t, and there should be:

$$N_{ind}(t) = |P| \times t \tag{1}$$

$N_{ind}(t)$ should not be more than $|S|$. It can be taken as granted that EAs are better than the traversal methods on searching optima, and the latter will have to visit all the individuals in search space. Therefore, for the termination generation T of each algorithm, there should be:

$$|S| \geq N_{ind}(T) \tag{2}$$

The relationship among $|P|$, T, $N_{Ind}(T)$ and $|S|$ is illustrated in Fig. 2, in which, the outer curve of the shadow area is $|P| \times T = |S|$, which is also the upper bound of $N_{ind}(T)$. The lower bounds for T and $|P|$ are labeled as T_0 and $|P|_0$ respectively.

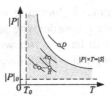

Fig. 2. Relationship of population size and generation number

According to (3), it is easy to know that the points A, B and C, in Fig. 2, are the examples of $N_{Ind}(T)$, while point D is not. The product value of $|P| \times T$ corresponding to points decreases with the order of D, A, B and C.

In fact, $N_{ind}(T)$ as a metrics of efficiency has been used many times [12, 13], but the relationship between $|P|$ and T was not given.

2.2 Space Complexity and Time Complexity

The definitions of $N_{ind}(T)$ does not include enough elements. For example, although $N_{ind}(T)$ consider the time-complexity with T and space-complexity with $|P|$, they are only the basic complexity.

Space-complexity of the algorithms is related with population size, but it can't be simply represented with it. There are many memory-based algorithms, that use the same population size as other algorithms, but their space complexity is more than others.

Similarly, time-complexity of the algorithm is related with the termination generation number, but it can't be simply represented with it. There are many data-mining embedded algorithms that spend more time to extract rules to guide the algorithms to explorer/exploit promising search space. In addition to the search process, establishing a high quality model in EAs might require longer time [14]. In a competing heuristics situation [15], time-complexity is also heavy, because the solutions are repeated generated and evaluated until it dominates the worst one in current population.

In order to compare the space-complexity, the memory cost per individual is considered. Label the memory cost of an algorithm during the evolutionary process as M_C, and define the memory cost per individual as:

$$m_C = \frac{M_C}{|S|} \qquad (3)$$

Similarly, in order to compare the time-complexity, the time cost of an algorithm per individual is considered. Label the time cost during the evolutionary process as T_C, and define the time cost per individual as:

$$t_C = \frac{T_C}{|S|} \qquad (4)$$

It has been mentioned many times in literature that under certain efficiency condition, time complexity is inversely proportion the space complexity. Sometimes, in order to save time, an algorithm will have to run with high space complexity, or in order to save memory to run in long time. Therefore, the relationship of m_C and t_C is shown in Fig. 3.

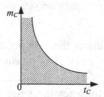

Fig. 3. Relationship between m_C and t_C

For a certain problem, algorithm with high m_C will be with the low t_C. Therefore, the product of power of m_C and power of t_C also has a upper bound. With this idea, a complexity metric can be defined as:

$$p_C = \frac{|S|}{|S| + \omega_{P1} M_C \times T_C} \qquad (5)$$

where $\omega_{P1} = e^{\lfloor ln|S| \rfloor - \lfloor ln(M_C \times T_C) \rfloor}$, and it is for the balance of magnitudes. Therefore, there is $p_C \in (0, 1)$.

2.3   Efficiency of Evolutionary Algorithms

The algorithm should not visit every individual. It is assumed that the increasing of M_C and T_C will result in the decreasing of $|P| \times T$. Therefore, m_C and t_C are inversely proportion to $N_{ind}(T)$ with the same algorithm and the same problem. Therefore, there is:

$$\frac{\partial N_{ind}(T)}{\partial M_C} < 0, \frac{\partial N_{ind}(T)}{\partial T_C} < 0 \tag{6}$$

Based on the relationship of $|P|$ and T, one of the metrics of the efficiency of algorithm can be defined as:

$$p_{N1} = 1 - \omega_{P2}\frac{N_{ind}(T)}{|S|} = 1 - \omega_{P2}\frac{|P| \times T}{|S|} \tag{7}$$

where $\omega_{P1} = e^{\lfloor ln|S| \rfloor - \lfloor ln(|P| \times T) \rfloor}$, which is for the balance of magnitudes. It is easy to know $p_{N1} \in [0, 1]$. The bigger p_{N1} is, the better the efficiency is.

Based on the analysis of M_C, T_C and $N_{ind}(T)$, the following basic logistics can be given:

$$M_C \uparrow \Rightarrow T_C \downarrow, T_C \uparrow \Rightarrow M_C \downarrow, M_C \uparrow \Rightarrow N_{Ind}(T) \downarrow,$$
$$T_C \uparrow \Rightarrow N_{Ind}(T) \downarrow, N_{Ind}(T) \downarrow \Rightarrow P_{N1} \uparrow, M_C \uparrow \Rightarrow P_C \downarrow, T_C \uparrow \Rightarrow P_C \downarrow \tag{8}$$

Based on the above logistics, the efficiency metric can be defined as:

$$p_{e1} = \omega_1 p_{N1} + \omega_2 p_C \tag{9}$$

where $\omega_1 + \omega_2 = 1$.

This efficiency metric includes parameters of population size, number of termination generation, time-complexity and memory-complexity. All the values of these parameters can be calculated prior to the running of the algorithm. Compared with effectiveness, this is the advantage of efficiency metrics.

3 Relationship of Efficiency and Effectiveness of Evolutionary Algorithms

3.1 Effectiveness in EAs

It is sure that, for two algorithms, the one, which obtains a solution nearer from the real optimum than that of another algorithm, has higher effectiveness. Therefore, a measurement of the effectiveness of the algorithm based on distance of obtained solution from the real optimum can be given. Generally, it is difficult to know the real optimum, but a solution better than or equal to the best obtained one can be given. Label this solution as $X_{best}^{virtaul}$, the precision as δ. Label $d_b = \left| X_{best}^{real} - X_{best}^{virtual} \right|$, then d_b can be regarded as the metric for the improvement of population. According to Deb's necessary properties for metrics [8], the effectiveness p_{e2} of the performance should satisfy:

$$p_{e2} \in [0, 1], \frac{\partial p_{e2}}{\partial d_b} < 0 \tag{10}$$

This means that with the decreasing of the distance, the effectiveness increases. The *progress rate* [16] is one of the metrics that satisfy this property, but it is complex to calculate.

For simplicity, the expression is as following:

$$p_{e2} = \frac{\delta}{||X_{best}^{real} - X_{best}^{virtual}|| + \delta} = \frac{\delta}{d_b + \delta} \tag{11}$$

3.2 Relationship Between Efficiency and Effectiveness

Usually, the longer time an EA run (with a sufficient population size), the higher the solution quality will be [14]. However, in real-world scenarios, the computational resources are often limited, which leads to a tradeoff between efficiency and the effectiveness. This means that although efficiency p_{e1} and effectiveness p_{e2} are two elements of the performance **p**, they are not independent. If an algorithm is assigned with big value for c_P or T, it will be expected to have higher effectiveness, while large value of c_P or T also means low efficiency, in another word, an algorithm with low p_{e1} should have high p_{e2}.

Assuming that for a certain algorithm, p_{e1} could be inversely proportion to p_{e2}. Intuitively, for a certain algorithm, when the efficiency is improved, the effectiveness will be decreased, and vice versa. This means that for a certain algorithm, the improvement of efficiency will bring the low effectiveness and the improvement of effectiveness will bring the low efficiency. Therefore, there should be a tradeoff between p_{e1} and p_{e2}.

But for different algorithm variants, the performance will be different. Some of the variants can increase the efficiency and some of them can increase the effectiveness and some of them can increase both of the effectiveness and efficiency. The last kind of algorithm is the one that is preferred.

The relationship that p_{e1} is inversely proportional to p_{e2} can be seen from the aspect of space precision δ. Denote the dimension or number of variants as d, which means that $X = \{x^{(i)}|i = 1, 2,..., d\}$, and D_i as the scope of i-th dimension. Label the upper and low bound of every dimension as u_{Di} and l_{Di}, and $d_D = \prod_{i=1}^{d}(u_{Di} - l_{Di})$, then there is:

$$|S| = \prod_{i=1}^{d} |D_i| = \prod_{i=1}^{d} \frac{U_{Di} - L_{Di}}{\delta} = \frac{d_D}{\delta^d} \tag{12}$$

where d_D is a constant value and $\delta > 0$. Therefore, according to (8) and (12), there is:

$$\frac{\partial p_{e1}}{\partial \delta} > 0, \frac{\partial p_{e2}}{\partial \delta} < 0 \tag{13}$$

Therefore, from the viewpoint of decision variable δ, p_{e1} is approximately inversely proportional to p_{e2}, and they can be seen as conflict objectives. For a certain algorithm, the relationship of p_{e1} and p_{e2} is illustrated in Fig. 4, in which, the intersection point

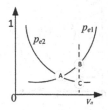

Fig. 4. Relationship between pe1 and pe2 for a certain algorithm

"A" of p_{e1} and p_{e2} is the tradeoff. And for the point of "B", p_{e1} is increased while p_{e2} is decreased. Similarly, for the point of "C", p_{e2} is increased while p_{e1} is decreased.

The relationship that p_{e1} is inversely proportional to p_{e2} also can be seen from the aspect of complexity metric p_C. It is easy to know that:

$$\frac{\partial p_C}{\partial M_C} < 0, \frac{\partial p_C}{\partial T_C} < 0 \tag{14}$$

And there is

$$\frac{\partial p_{e1}}{\partial M_C} < 0, \frac{\partial p_{e1}}{\partial T_C} < 0, \frac{\partial p_{e2}}{\partial M_C} > 0, \frac{\partial p_{e2}}{\partial T_C} > 0 \tag{15}$$

Based on the above analysis, the logistics relationship among the parameters and the algorithm's performance is given in Fig. 5.

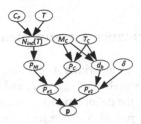

Fig. 5. Logistic relationships among the parameters and algorithm performance

Algorithm variants are always proposed in order to improve the performance. For example, in Fig. 6, it can be taken as granted that the original performance tradeoff is at the point ⓪. Then the tradeoff is improved to points "A", "B" and "C", during which "A" is the best tradeoff, which has both high p_{e1} and p_{e2}.

3.3 Definition of Performance

Based on the above analysis, a more sophisticated performance indicator is proposed as:

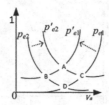

Fig. 6. Relationships among algorithms' efficiency and effectiveness

$$p = (p_{e1}, p_{e2}) \tag{16}$$

Based on this indicator, the definition of dominance relationship on performance can be given.

Definition 1 (Dominance Relationship on Performance). For two performance vector $p_1 = (p_{e1,1}, p_{e2,1})$ and $p_2 = (p_{e1,2}, p_{e2,2})$, The preference orders on the set of performance vector can be defined.

- $p_1 \succ p_2(p_1$ dominates $p_2)$, if $\mathbf{p}_1$ is not worse than $\mathbf{p}_2$ in any element, and is better in at least one of the elements;
- $p_1 \succeq p_2(p_1$ weakly dominates $p_2)$, if $\mathbf{p}_1$ is not worse than $\mathbf{p}_2$ in any elements;
- $p_1 = p_2(p_1$ equals $p_2)$, if $\mathbf{p}_1 = \mathbf{p}_2$;
- $p_1 \| p_2(p_1$ and p_2 are incomparable to each other), if neither $\mathbf{p}_1$ weakly dominates $\mathbf{p}_2$ nor $\mathbf{p}_2$ weakly dominates $\mathbf{p}_1$.

The relationship $\prec, \preceq$ are defined accordingly, i.e., $p_1 \succ p_2$ is equivalent to $p_2 \prec p_1$, etc.

4   Application of Performance Measurement

Performance **p** is determined by an algorithm A and the optimization problem pl. This from the definition of p_{e1} is the variables of c_p and T, and both c_p and T are related with the algorithm A. Similarly, from the definition of p_{e2}, it can be seen that the variables of δ and d_b are also related with pl and A. Therefore, for distinguishing performance, $\mathbf{p}(pl, A)$ can be used to explicitly illustrate that the performance is for the concrete problem pl and concrete algorithm A.

4.1   Algorithm Comparison

For the same problem pl with the same space size $|S|$, all EAs can be compared with performance **p**. the algorithms domination definition is given as following.

Definition 2 (Algorithms Domination Relationship on Certain Problem). For two EAs A_1 and A_2, suppose their performance is $\mathbf{p}_1(pl, A_1)$ and $p_2(pl_2, A\neg_2)$ respectively. The performance orders can be defined as:

- $A_1 \succ_{pl} A_2 (A_1$ dominates $A_2)$, if $\mathbf{p}_1$ is not worse than $\mathbf{p}_2$ in any element, and is better in at least one of the elements;
- $A_1 \succeq_{pl} A_2 (A_1$ weakly dominates $A_2)$, if $\mathbf{p}_1$ is not worse than $\mathbf{p}_2$ in any elements;
- $A_1 =_{pl} A_2 (A_1$ equals $A_2)$, if $\mathbf{p}_1 = \mathbf{p}_2$;
- $A_1 \|_{pl} A_2 (A_1$ and A_2 are incomparable to each other), if neither $\mathbf{p}_1$ weakly dominates $\mathbf{p}_2$ nor $\mathbf{p}_2$ weakly dominates $\mathbf{p}_1$.

The relationship $\prec, \preceq$ are defined accordingly, i.e., $A_1 \succ_{pl} A_2$ is equivalent to $A_2 \prec_{pl} A_1$, etc.

Suppose that all the problems compose of a set $PL = \{pl_1, pl_2, \ldots\}$. Based on this set, the theoretic case for the algorithms comparison can be defined as $A_1 \succ_{PL} A_2$ and so on. For the same problem with the same precision δ, if three EAs: a_1, a_2 and a_3 obtain the same optima and the efficiency are corresponding to the points C, B and A respectively in Fig. 1, then there is $\mathrm{p}_1(A_1, pl) \succ \mathrm{p}_2(A_2, pl) \succ \mathrm{p}_3(A_3, pl)$ and $A_1 \succ_{pl} A_2 \succ_{pl} A_3$.

4.2 Problem Hardness Measurement

For different problems pl_1, pl_2 with the same space size $|S|$, if the same algorithm performs differently on them with $\mathrm{p}_1(pl_1, A\neg)$ and $\mathrm{p}_2(pl_2, A\neg)$, then the definition about problem hardness can be given in the following definition.

Definition 3 (Hardness Domination Relationship of Problems on Certain Algorithm). For a concrete EAs A, label the performance for two problems pl_1 and pl_2 as $\mathrm{p}_1(pl_1, A\neg)$ and $\mathrm{p}_2(pl_2, A)$ respectively. The preference orders can be defined as:

- $pl_1 \prec_A pl_2 (pl_1$ dominates $pl_2)$, if $\mathbf{p}_1$ is not worse than $\mathbf{p}_2$ in any element, and is better in at least one of the elements;
- $pl_1 \prec_A pl_2 (pl_1$ weakly dominates $pl_2)$, if $\mathbf{p}_1$ is not worse than $\mathbf{p}_2$ in any elements;
- $pl_1 =_A pl_2 (pl_1$ equals $pl_2)$, if $\mathbf{p}_1 = \mathbf{p}_2$;
- $pl_1 \|_A pl_2 (pl_1$ and pl_2 are incomparable to each other), if neither $\mathbf{p}_1$ weakly dominates $\mathbf{p}_2$ nor $\mathbf{p}_2$ weakly dominates $\mathbf{p}_1$.

The relationship $\prec, \preceq$ are defined accordingly, i.e., $pl_1 \succ_A pl_2$ is equivalent to $pl_2 \prec_A pl_1$, etc.

Suppose that all the algorithms compose of a set $Al = \{A_1, A_2, \ldots\}$. The theoretic case for the problems comparison can be defined as $pl_1 \succ_{Al} pl_2$ and so on.

Generally speaking, there are two kinds of measurements for the problems difficulty: prior estimation and post calculation. One can see that the above definition is based on the post calculation.

4.3 Comparison Design

Efficiency and effectiveness are metrics for the comparison of algorithms (and its parameters' value) and problems. These two metrics are related with at least three elements: algorithm, value of parameters of the algorithm and the optimized problem.

Based on the comparison of these three elements, experiments can be carried out to validate the relationship between efficiency and effectiveness. The comparison of these elements is shown in Table 1, in which, the last column gives the suitability to compare for the corresponding case. And in this table, 0 means the same case, 1 means the difference case.

Table 1. Comparison of elements

Number	Algorithms	Values of parameters	Optimized problems	Comparison
1	0	0	0	None
2	0	0	1	Hardness
3	0	1	0	Performance
4	0	1	1	Suitability
5	1	0	0	Performance/None
6	1	1	0	Performance
7	1	0	1	None
8	1	1	1	None

- For the case of No. 1, the three elements, algorithms, values of parameters and optimized problems, are all with the same situations; therefore, it is unnecessary to compare the performance. But generally, it is always to be utilized to calculate the mean and variance of algorithms' performance.
- For the case of No. 2, two elements, algorithms and its parameters' values are same and only the optimized problems are different. Therefore, by the performance of the algorithm, the hardness of the optimized problem can be compared.
- Similar to the case of No. 2, No. 3 can be used to compare the influence of parameters' values on the performance of algorithms. And the case of No. 4 can be used to compare the performance for different problems with different parameters' values.
- For the case of No. 5, when the compared algorithms have the same parameters, it can be used to compare the performance for different algorithms with the same parameters' value and for the same optimized problems. But generally speaking, since parameters are part of an algorithm, different algorithm may have parameters with different meaning resulting in their incomparability. Therefore, the value in the last column is performance/none. Similarly, for the case of No. 6, the performance can be compared.
- For the case of No. 7 and 8, both the algorithms and optimized problems are different; therefore, both of these two elements are incomparable.

Based on the analysis in Table 1, the experiments to compare performance can be as the cases of No. 3, 5 and 6, and the experiments to compare hardness can be as the case of No. 2.

5 Conclusion

As two important metrics for the evaluation of evolutionary algorithms (EAs), efficiency and effectiveness are studied in this paper. For efficiency metrics, population size, number of termination generation, space complexity, and time complexity and their relationship were studied. We conclude that the product of population size and number of generation should less than the search space size, and the product of time complexity and space complexity should also less than a constant. Secondly, we study the relationship between efficiency and effectiveness. Based on these two metrics, we conclude that not only EAs can be compared, but also problems hardness can be measured. The results reveal important insights of EAs and problems hardness.

Acknowledgement. This work is supported by the National Natural Science Foundation of China under Grant No. 61305149, 61403174, by Jiangsu Provincial Natural Science Foundation under Grant No. BK20131130, by China Ministry of Education, Humanities and Social Sciences Youth Foundation under Grant No. 11YJC630074 and by Jiangsu Overseas Research & Training Program for University Prominent Young & Middle-aged Teachers and Presidents.

References

1. Yang, X.-S.: Nature-Inspired Optimization Algorithms. Elsevier, Oxford (2014)
2. Malan, K.M., Engelbrecht, A.P.: Ruggedness, funnels and gradients in fitness landscapes and the effect on PSO performance. Paper presented at the 2013 IEEE Congress on Evolutionary Computation (CEC), 20–23 Jun 2013
3. Xin, B., Chen, J., Pan, F.: Problem difficulty analysis for particle swarm optimization: deception and modality. In: Proceedings of the First ACM/SIGEVO Summit on Genetic and Evolutionary Computation, Shanghai, China, pp. 629–630 (2009)
4. Gras, R.: How efficient are genetic algorithms to solve high epistasis deceptive problems? Paper presented at the IEEE World Congress on Computational Intelligence, 1–6 Jun 2008
5. Smith-Miles, K., Lopes, L.: Measuring instance difficulty for combinatorial optimization problems. Comput. Oper. Res. **39**(5), 875–889 (2012). Elsevier B.V
6. Malan, K.M., Engelbrecht, A.P.: A survey of techniques for characterising fitness landscapes and some possible ways forward. Inf. Sci. **241**, 148–163 (2013). doi:10.1016/j.ins.2013.04.015
7. Jin, Y., Sendhoff, B.: Trade-off between performance and robustness: an evolutionary multiobjective approach. In: Fonseca, C.M., Fleming, P.J., Zitzler, E., Deb, K., Thiele, L. (eds.) EMO 2003. LNCS, vol. 2632, pp. 237–251. Springer, Heidelberg (2003)
8. Deb, K., Jain, S.: Running performance metrics for evolutionary multi-objective optimizations. Paper presented at the Proceedings of the Fourth Asia-Pacific Conference on Simulated Evolution and Learning (SEAL 2002), Singapore (2002)
9. Merz, P.: Advanced fitness landscape analysis and the performance of memetic algorithms. Evol. Comput. **12**(3), 303–325 (2004). doi:10.1162/1063656041774956
10. Lu, C.-C., Yu, V.F.: Data envelopment analysis for evaluating the efficiency of genetic algorithms on solving the vehicle routing problem with soft time windows. Comput. Ind. Eng. **63**(2), 520–529 (2012). doi:10.1016/j.cie.2012.04.005

11. Jin, Y.: Surrogate-assisted evolutionary computation: recent advances and future challenges. Swarm Evol. Comput. **1**(2), 61–70 (2011)
12. Cooper, J., Hinde, C.: Improving genetic algorithms' efficiency using intelligent fitness functions. In: Chung, P.H., Hinde, C., Ali, M. (eds.) IEA/AIE 2003. LNCS, vol. 2718, pp. 636–643. Springer, Heidelberg (2003)
13. Sastry, K., Goldberg, D.E., Pelikan, M.: Efficiency enhancement of probabilistic model building genetic algorithms. In: Illinois Genetic Algorithms Laboratory (2004)
14. Sastry, K., Pelikan, M., Goldberg, D.: Efficiency enhancement of estimation of distribution algorithms. In: Pelikan, M., Sastry, K., CantúPaz, E. (eds.) Scalable Optimization via Probabilistic Modeling. Studies in Computational Intelligence, vol. 33, pp. 161–185. Springer, Heidelberg (2006)
15. Tvrdík, J., Misik, L., Krivy, I.: Competing heuristics in evolutionary algorithms. In: Intelligent Technologies-Theory and Applications, pp. 159–165 (2002)
16. Beyer, H.-G., Schwefel, H.-P., Wegener, I.: How to analyse evolutionary algorithms. Theoret. Comput. Sci. **287**(1), 101–130 (2002). doi:10.1016/S0304-3975(02)00137-8

A Genetic Local Search Algorithm for Optimal Testing Resource Allocation in Module Software Systems

Ruimin Gao[✉] and Siyan Xiong

School of Computer Science and Technology,
University of Science and Technology of China (USTC), Hefei, China
gaorm@mail.ustc.edu.cn

Abstract. As modern software systems have expanded continuously, the problem of how to optimally allocate the limited testing resource during the software testing phase attracted lots of attention. The Optimal Testing Resource Allocation Problems (OTRAPs) involve seeking for an optimal allocation of limited testing resource. There are two major objectives in the OTRAPs: reliability and cost. Since the designers pay more and more attention to reducing the cost, in this paper, we studied OTRAPs with the latter objective. In previous work, approaches based on genetic algorithms have been claimed to be strong alternatives in solving the problem. Hence, in this paper we proposed a new algorithm based on genetic algorithm and local search strategy (GLSA) to solve the OTRAPs. Experimental results show that the algorithm proposed can obtain better performance than some existing approaches for solving the software testing resource problem.

Keywords: Genetic algorithm · Software testing · Optimal allocation · Modular software system

1 Introduction

A software development process consists of four phases [1]: specification, designing, coding and testing. The testing phase consists of several stages including module testing, integration testing, system testing and acceptance testing [2]. During the testing phase the software system is tested to detect and correct faults. Researches show that the testing phase is the most costly phase and approximately 40–50 % of the resources consumed during the software development are testing resources [3]. Each software module is tested independently during the module phase and not all the modules are equally important, therefore how to allocate the limited resource is an important issue.

There are numerous publications on OTRAPs [1–19] which are proposing new formulations of OTRAPs or utilizing problem-solving approaches. To formulate an OTRAP, the relationship of reliability, cost and resources should be defined firstly. The relationship between reliability is described by Reliability Growth Models (SRGMs). The reliability can be described by failure severity, the number of failures, time occurrence and so on [4]. Researchers use different theories, like non-homogeneous

© Springer International Publishing Switzerland 2015
D.-S. Huang et al. (Eds.): ICIC 2015, Part II, LNCS 9226, pp. 13–23, 2015.
DOI: 10.1007/978-3-319-22186-1_2

Poisson process, chaos theory and uncertainty theory, to build a SRGM. Therefore there are publications of maximizing the reliability with constrained testing resource to ensure that the reliability of software systems will be maximized [5–8] or the remaining faults can be minimized [1, 3]. In practical applications, cost means profit. Hence in recent years, researchers pay more attention on the testing cost by minimizing the testing cost with constrained testing resource and reliability [9, 10]. In practical problems designers want to minimum the cost under conditions of achieving a certain level of software reliability. Therefore, in this paper, we studied OTRAPs with the goal to minimize the testing cost with the constraint of reliability and testing resource.

There many optimization approaches have been proposed to solve OTRAPs, like dynamic programming [11], and nonlinear programming [12–14]. When the solution space of the problem is multimodal, these methods may be easily trapped in a local optimal. Genetic algorithms (GA) are inspired by Darwin's theory of evolution. GA has a strong capability of global search which will escape from local optima. In previous work, the best way to solve OTRAPs of the goal to minimize the testing cost with the constraint of reliability and testing resource is GA [9, 10, 19]. To accelerate the genetic algorithm in converging on optimal solutions, we add a local search operator in GA.

The rest of the paper is organized as follows. In Sect. 2, the problem formulation is given. After that, Sect. 3 introduces the genetic local search algorithm in detail. In Sect. 4, empirical studies and discussions are presented to compare the proposed algorithm with other approaches. Finally, in Sect. 5, conclusions are presented.

2   Problem Formulations

Because a software system typically consists of multiple modules, the model that formulates the relationship between the testing resource and reliability (or cost) needs to be addressed. We address this problem with SRGMs, and the SRGMs defined based on the following notations and assumptions [9, 10, 15].

Notations:

W: Total testing resource
C: Total testing cost
$W(t)$: Cumulative testing effort in the time interval $(0, t]$.
$\omega(t)$: Current testing-effort expenditure rate at time t.
$m(t)$: Mean value function in NHPP (Non-Homogeneous Poisson Process).
p: Probability of perfect debugging of a fault.
a: The number of dormant faults in the software at the beginning of testing.
b: Fault detection rate.
α: Error generation rate.
β: Constant in logistic learning function.

Assumptions

(1) The software failure phenomenon can be described by a NHPP.
(2) The software is subject to failures at random times during execution caused by faults remaining in the software.

(3) Each time a failure is observed, an immediate effort takes place to decide the cause of the failure in order to remove it.
(4) During the fault removal process, the fault is reduced by one with probability p or remains unchanged with probability $1-p$.
(5) During the fault removal process, new faults can be generated. The fault generation rate is proportional to the rate of fault removal.
(6) Fault removal rate per remaining fault assumed to be non-decreasing inflection S-shaped logistic function.

Under the above assumptions, the removal process can be described with testing resource as follows [10]:

$$\frac{dm(t)}{dt} = pb(W(t))(a(t) - m(t))\frac{dW(t)}{dt} \qquad (1)$$

Where

$$b(W(t)) = \frac{b}{1 + \beta e^{-bW(t)}} \qquad (2)$$

$$a(t) = (a + \alpha m(t)) \qquad (3)$$

Solving Eq. (1) in under the condition $m(t = 0) = 0$ and $W(t = 0) = 0$ we can get:

$$m(t) = \frac{a}{1 - \alpha}\left[1 - \left(\frac{(1 + \beta)e^{-bW(t)}}{1 + \beta e^{-bW(t)}}\right)^{p(1-\alpha)}\right] \qquad (4)$$

Based on SRGMs, the mean value function of fault removal process for i-th module can be calculated as:

$$m_i(t) = \frac{a_i}{1 - \alpha_i}\left[1 - \left(\frac{(1 + \beta_i)e^{-b_i W_i(t)}}{1 + \beta_i e^{-b_i W_i(t)}}\right)^{p_i(1-\alpha_i)}\right] \quad i = 1, 2, \ldots\ldots N \qquad (5)$$

Where N is the number of modules in a software system.

W_i is cumulative testing resource in the interval $(0, t]$ for i-th module. Based on the above equation, the mean value function of a software system with N modules is calculated as:

$$m(W) = \sum_{i=1}^{N} v_i \frac{a_i}{1 - \alpha_i}\left[1 - \left(\frac{(1 + \beta_i)e^{-b_i W_i}}{1 + \beta_i e^{-b_i W_i}}\right)^{p_i(1-\alpha_i)}\right] \quad i = 1, 2, \ldots\ldots N \qquad (6)$$

Where v_i is the weight attached to i-th module.

The measure of defining software reliability at time t as given by Huang et al. [16] is, "the ratio of the cumulative number of detected faults at time t to the expected number of initial fault content of the software." The initial fault content of the proposed model is $\frac{a}{1-\alpha}$. Then the reliability of i-th module is calculated as:

$$R_i = \frac{m_i(t)}{\frac{a_i}{1-\alpha_i}} = 1 - \left(\frac{(1+\beta_i)e^{-b_iW_i}}{1+\beta_ie^{-b_iW_i}}\right)^{p_i(1-\alpha_i)} \quad i = 1,2\ldots\ldots N \tag{7}$$

Most commonly cost model for each module is defined as [20]:

$$C_i(T) = C_{1i}m_i(T) + C_{2i}(m_i(\infty) - m_i(T)) + C_{3i}(T) \ i = 1,2\ldots\ldots N \tag{8}$$

Where the first and second term is the cost of testing and debugging during the testing phase and operational phase respectively and the last one is the cost of testing up to the release time.

For SRGMs with testing resource the above cost function can be modified to describe the costs per unit testing effort expenditure as:

$$C_i(W(T)) = C_{1i}m_i(W(T)) + C_{2i}(m_i(\infty) - m_i(W(T))) + C_{3i}(W(T)) \ i = 1,2\ldots\ldots N \tag{9}$$

The total cost of the whole system can be calculated as:

$$C = \sum_{i=1}^{N}\left\{(C_{1i} - C_{2i})\frac{v_ia_i}{1-\alpha_i}\left[1 - \left(\frac{(1+\beta_i)e^{-b_iW_i}}{1+\beta_ie^{-b_iW_i}}\right)^{p_i(1-\alpha_i)}\right] + \frac{C_{2i}a_i}{1-\alpha_i} + C_3W_i\right\} \tag{10}$$

Given an optimal testing-resource allocation problem, we aimed at minimizing the testing cost when the reliability of each module is at least R_0. Then the optimal testing-resource allocation problem can be formulated as:

Minimize

$$C = \sum_{i=1}^{N}\left\{(C_{1i} - C_{2i})\frac{v_ia_i}{1-\alpha_i}\left[1 - \left(\frac{(1+\beta_i)e^{-b_iW_i}}{1+\beta_ie^{-b_iW_i}}\right)^{p_i(1-\alpha_i)}\right] + \frac{C_{2i}a_i}{1-\alpha_i} + C_3W_i\right\}$$

Subject to

$$\sum_{i=1}^{N}W_i \le W$$

$$W_i \ge 0, i = 1,\ldots\ldots,N$$

$$R_i \ge R_0, i = 1,\ldots\ldots,N \tag{11}$$

3   GLSA for Testing-Resource Allocation

GA is inspired by Darwin's theory of evolution. Godberg [21] gave the introduction of GA. To accelerate the genetic algorithm in converging on optimal solutions, we add a local search operator in GA. The pseudo-code of GLSA is presented in Algorithm 1.

Algorithm 1. Pseudo-Code of GLSA

1:   Randomly generate an initial population with N individuals $P = \{\mathbf{x}_1, \mathbf{x}_2 \ldots \ldots \mathbf{x}_N\}$
2:   The generation counter $t=0$
3:   while $t < t_{max}$ (maximum generation number)
4:       Evaluate the fitness of all the individuals in the population, the fitness of individual $\mathbf{x}_i$ is denoted as $f(\mathbf{x}_i)$
5:       Select parent individuals to do the crossover operation
6:       if f(offspring)<f(parent)
7:              Replace parent by offspring
8:       end if
9:       for i=1:N
10:              Perform mutation on individual $\mathbf{x}_i$ with probability P_1 to generate an offspring $\mathbf{x}_i'$
11:                  if $f(\mathbf{x}_i') < f(\mathbf{x}_i)$
12:                      $\mathbf{x}_i = \mathbf{x}_i'$
13:                  end if
14:              Perform local search around each individual with probability P_2 to generate an offspring $\mathbf{x}_i''$
15:                  if $f(\mathbf{x}_i'') < f(\mathbf{x}_i)$
16:                      $\mathbf{x}_i = \mathbf{x}_i''$
17:                  end if
18:       end for
19:   end while

3.1   Fitness Function

How to evaluate the quality of a solution is an important issue in a GA. In our testing resource allocation problem, the fitness function is given by:

$$Fitness = \sum_{i=1}^{N} \left\{ (C_{1i} - C_{2i}) \frac{v_i a_i}{1 - \alpha_i} \left[1 - \left(\frac{(1 + \beta_i) e^{-b_i W_i}}{1 + \beta_i e^{-b_i W_i}} \right)^{p_i(1 - \alpha_i)} \right] + \frac{C_{2i} a_i}{1 - \alpha_i} + C_3 W_i \right\} \quad (12)$$

3.2   The Selection, Crossover and Mutation Operator

Population diversity and selective pressure are two important issues in the evolution process of a genetic search. Strong selective pressure may lead to early convergence to a local optima. If the selective pressure is weak and the population is too diverse, the search will be ineffective. Hence we use tournament selection here.

In the crossover process, two chromosomes are formed by swapping sets of genes of two parent chromosomes, hoping that at least one child will have genes that improve its fitness. In the proposed algorithm, we use arithmetical crossover operator.

In the mutation process, the algorithm may escape from a local optima. We use Gaussian mutation here.

3.3 Local Search Strategy

GA possesses the strong capability of global search and usually not very sensitive to initial solutions. The effectiveness of GA has been demonstrated on OTRAPs, but GA also has shortcomings like poor capability of local search. To accelerate the genetic algorithm in converging on optimal solutions, we combine GA and local search operator to solve OTRAPs.

In this paper, the local search operator is applied to the selected individual. Assume the selected individual x corresponds to a system consisting of N components ($N \geq 2$) and the fitness of individual x is $f(x)$. The pseudo-code of the local search is presented in Algorithm 2.

Algorithm 2. Pseudo-Code of the local search procedure on an individual x

1: Randomly select two components of x, denoted as x_1, x_2
2: Initialize an archive $S = \emptyset$. This archive will be used to store the feasible solutions obtained by local search
3: Decrease the resource of the two components by $m * \min(x_1, x_2)$,(ie, $x_1' = x_1 - m * \min(x_1, x_2)$ and $x_2' = x_2 - m * \min(x_1, x_2)$) to get a new individual x_{i1}
4: Increase the resource of the two components to get x_{i2}
5: Increase the resource of one component and Decrease the resource of the other component at the same time to get two new individuals x_{i3} and x_{i4}
6: Decrease the resource of one component to get two new individual x_{i5} and x_{i6}
7: Decrease the resource of one component to get two new individual x_{i7} and x_{i8}
8:    Include all the new individuals into S
9:    i=1;
10: while i<9
11:        Randomly select an individual x' from S
12:        if x' violate all the constraints and f (x')<f(x)
13:            x=x'
14:            break
15:        else
16:              i=i+1
17:        if i=9
18:            get the one which has the smallest fitness among the new individuals and parent individual to replace x
19:        end if
20: end while

The local search operator randomly selects two components. The resources of the two components are modified to generate 8 new individuals. A newly generated individual is reserved in an archive S. Randomly select an individual from S. If the new individual violate all the constraints and the fitness of the new individual is less than $f(\mathbf{x})$, the new individual will replace $\mathbf{x}$. Otherwise, if all the individuals in S do not violate the constraints, get the one which has the smallest fitness among the new individuals and parent individual to replace $\mathbf{x}$.

4 Numerical Example

In this section, the performance of GLSA is evaluated on a numerical example. The simulation is carried out in MATLAB R2014b software under a 64 bit windows 8.1 environment. The processor of my PC is core(TM) i5-4200 M with a CPU frequency at 2.5 GHZ, as well as 8 GB memory.

The results are compared with two algorithms. The first one is Simulated Annealing (SA). The other algorithm, denoted as K-A algorithm, was proposed by Kapur and Aggarwal [10]. K-A algorithm used GA to solve OTRAPs. For K-A, the population size is 50, the crossover probability is 0.8 and the mutation probability is 0.05. For GLSA, the population size is 50, the crossover probability is 0.8, the mutation probability is 0.05 and the local search probability is 0.05. For SA, the initial temperature is 10000 and the temperature update factor is 0.975.

Consider a software system consisting of six modules proposed in [10]. The parameters estimate for each module is shown in Table 1. The total testing resource is 50000. The cost parameters $C_{1i} = C_1 = 2, C_{2i} = C_2 = 10$ and $C_3 = 0.5$.

Table 1. Parameters used for OTRAPs

Module	a_i	b_i	α_i	β_i	v_i	p_i
M1	218	0.00147	0.0344	0.9736	0.1543909	0.95
M2	105	0.00955	0.6513	0.5321	0.0743626	0.91
M3	231	0.00174	0.2959	0.4129	0.1635977	0.9
M4	286	0.00252	0.1148	0.7276	0.2025496	0.92
M5	260	0.00429	0.1433	0.3697	0.184136	0.91
M6	312	0.00267	0.0123	0.6321	0.2209632	0.99

4.1 The Convergence Behavior Study

In this experiment, the aim is to evaluate the convergence behavior of GLSA, K-A and SA on a case study. The reliability value is set to 0.93. Each algorithm runs 15 times and the result in the table is the average of 15 times. The last row provides results of the Wilcoxon Test, while "w-d-l" indicates GLSA is superior, not significantly different or inferior to the corresponding compared algorithms (Table 2).

The convergence curve of SA, K-A and GLSA are plotted in Fig. 1, where the x-axis represents the number of generations and the y-axis represents the cost. As we

Table 2. Average results of SA, K-A and GLSA with differerent generations

Generation	SA	K-A	GLSA
100	38033.51	35764.42	32765.23
200	37565.20	33604.63	28446.26
300	37270.25	30217.77	25949.91
400	36867.54	28321.97	23937.33
500	35825.89	27293.51	23131.70
600	35227.39	26277.05	22500.39
700	34750.99	25088.98	22494.36
800	33257.69	25009.90	21454.55
900	32904.24	23873.14	21194.72
1000	33034.07	23043.93	21277.61
2000	29648.00	21392.55	20614.45
3000	26162.72	21005.52	20686.10
4000	24757.10	21095.74	20532.21
5000	24317.54	21010.47	20529.96
6000	24100.22	20871.98	20528.93
7000	23246.10	21214.30	20527.14
8000	22630.71	20869.81	20527.81
9000	22162.83	20919.29	20525.88
10000	21959.27	20898.93	20525.88
Wilcoxon Test	18-1-0	19-0-0	–

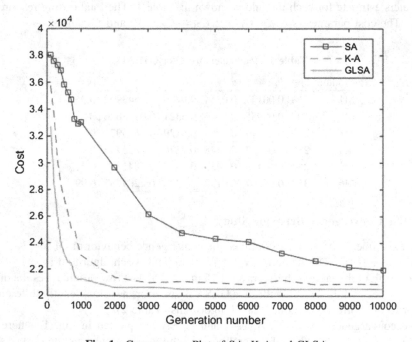

Fig. 1. Convergence Plot of SA, K-A and GLSA

can see, all the methods converged rather fast and the solution obtained by GLSA was significantly better than other two algorithms. Furthermore, the Wilcoxon test (with a significance level 0.05) has been employed to compare the overall performance of the three methods. For most generation numbers, the results of GLSA are significantly than others.

Assume that For K-A, the population size is P, the crossover probability is p_c and the mutation probability is p_m. For GLSA, the population size is P', the crossover probability is p'_c, the mutation probability is p'_m and the local search probability is p_l. For SA, the initial circulation number is u. The generation number of the three algorithms is t_{max}. Therefore, we can calculate that the number of fitness evaluations of K-A is $P + (p_c \times P + p_m \times P) \times t_{max}$, the number of fitness evaluations of GLSA is $P' + (p'_c \times P' + p'_m \times P' + p_l \times 8 \times P') \times t_{max}$, and the number of fitness evaluations of SA is $1 + u \times t_{max}$. In this experiment, the number of fitness evaluations of K-A is $50 + 43\, t_{max}$, the number of fitness evaluations of GLSA is $50 + 63\, t_{max}$ and the number of fitness evaluations of SA is $1 + 200\, t_{max}$.

4.2 Performance Over Different Reliability Values

The experiment above-mentioned only provides a case study on which the GLSA outperformed the other two algorithms, but it is not sufficient to draw a conclusion. Therefore, we take another experiment to compare the algorithms by varying the reliability constraint. Each algorithm runs 15 times and the result in the table is the average of 15 times (Table 3).

Table 3. Average results of SA, K-A and GLSA with differerent reliability constraint values

Reliability constraint value	SA	K-A	GLSA
0.81	19157.31	19290.78	19176.89
0.84	19919.86	19531.35	19397.61
0.87	20144.25	19738.36	19662.84
0.9	20580.31	20075.07	20022.95
0.93	20877.67	20711.46	20526.40
0.96	21564.95	21456.13	21353.93
Wilcoxon Test	4-2-0	5-1-0	–

The results are plotted in Fig. 2, where the x-axis represents reliability constraint value and the y-axis represents the cost. As we can see, the solution obtained by GLSA was significantly better than other two algorithms. Furthermore, the Wilcoxon test (with a significance level 0.05) has been employed to compare the overall performance of the three methods. For most reliability constraint value, the results of GLSA are significantly than others.

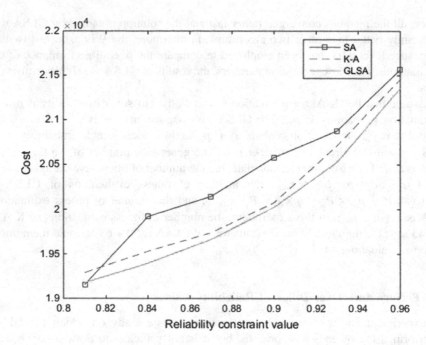

Fig. 2. Comparison among SA, K-A and GLSA under different reliability constraint values

5 Conclusion

In this paper, we proposed a genetic local search algorithm to solve the optimal testing resource problem by minimizing the total software testing cost with the constraints of limited testing resource and to achieve the desired level of reliability for each module. Through the experiments of a numerical example, we can draw a conclusion that the GLSA converged significantly better than the compared algorithms and it can find a better solution.

References

1. Ohtera, H., Yamada, S.: Optimal allocation and control problems for software-testing resources. IEEE Trans. Reliab. **39**, 171–176 (1990)
2. Dai, Y.-S., Xie, M., Poh, K.-L., Yang, B.: Optimal testing-resource allocation with genetic algorithm for modular software systems. J. Syst. Softw. **66**, 47–55 (2003)
3. Yamada, S., Ichimori, T., Nishiwaki, M.: Optimal allocation policies for testing-resource based on a software reliability growth model. Math. Comput. Model. **22**, 295–301 (1995)
4. Lyu, M.R., Rangarajan, S., Van Moorsel, A.P.: Optimal allocation of test resources for software reliability growth modeling in software development. IEEE Trans. Reliab. **51**, 183–192 (2002)
5. Coit, D.W.: Economic allocation of test times for subsystem-level reliability growth testing. IIE Trans. **30**, 1143–1151 (1998)

6. Coit, D.W., Smith, A.E.: Reliability optimization of series-parallel systems using a genetic algorithm. IEEE Trans. Reliab. **45**, 254–260, 266 (1996)
7. Yang, B., Xie, M.: Testing-resource allocation for redundant software systems. In: 1999 Pacific Rim International Symposium on Dependable Computing, Proceedings, pp. 78–83, IEEE (1999)
8. Yang, B., Xie, M.: A study of operational and testing reliability in software reliability analysis. Reliab. Eng. Syst. Saf. **70**, 323–329 (2000)
9. Jha, P., Gupta, D., Yang, B., Kapur, P.: Optimal testing resource allocation during module testing considering cost, testing effort and reliability. Comput. Ind. Eng. **57**, 1122–1130 (2009)
10. Kapur, P., Aggarwal, A.G., Kapoor, K., Kaur, G.: Optimal testing resource allocation for modular software considering cost, testing effort and reliability using genetic algorithm. Int. J. Reliab. Qual. Saf. Eng. **16**, 495–508 (2009)
11. Leung, Y.-W.: Dynamic resource-allocation for software-module testing. J. Syst. Softw. **37**, 129–139 (1997)
12. Huang, C.-Y., Lo, J.-H.: Optimal resource allocation for cost and reliability of modular software systems in the testing phase. J. Syst. Softw. **79**, 653–664 (2006)
13. Hou, R.-H., Kuo, S.-Y., Chang, Y.-P.: Efficient allocation of testing resources for software module testing based on the hyper-geometric distribution software reliability growth model. In: Seventh International Symposium on Software Reliability Engineering, 1996 Proceedings, pp. 289–298, IEEE (1996)
14. Berman, O., Ashrafi, N.: Optimization models for reliability of modular software systems. IEEE Trans. Softw. Eng. **19**, 1119–1123 (1993)
15. Ohba, M.: Software reliability analysis models. IBM J. Res. Dev. **28**, 428–443 (1984)
16. Huang, C.-Y., Lo, J.-H., Kuo, S.-Y., Lyu, M.R.: Optimal allocation of testing-resource considering cost, reliability, and testing-effort. In: 10th IEEE Pacific Rim International Symposium on Dependable Computing, 2004 Proceedings, pp. 103–112, IEEE (2004)
17. Espadas, J., Molina, A., Jiménez, G., Molina, M., Ramírez, R., Concha, D.: A tenant-based resource allocation model for scaling software-as-a-service applications over cloud computing infrastructures. Futur. Gener. Comput. Syst. **29**, 273–286 (2013)
18. Kang, D., Jung, J., Bae, D.H.: Constraint-based human resource allocation in software projects. Softw. Pract. Exp. **41**, 551–577 (2011)
19. Johri, P., Nasar, M., Chanda, U.: A genetic algorithm approach for optimal allocation of software testing effort. Int. J. Comput. Appl. **68**, 21–25 (2013)
20. Kapur, P., Garg, R., Kumar, S.: Contributions to Hardwave and Software Reliability. World Scientific, Singapore (1999)
21. Golberg, D.E.: Genetic Algorithms in Search, Optimization, and Machine Learning. Addion wesley, Boston (1989)

An Improved Artificial Bee Colony Algorithm Based on Particle Swarm Optimization and Differential Evolution

Fengli Zhou$^{(\boxtimes)}$ and Yanxia Yang

Faculty of Information Engineering,
Wuhan University of Science and Technology City College,
Wuhan 430083, China
thinkview@163.com, 34355363@qq.com

Abstract. In order to improve the problem that Artificial Bee Colony (ABC) is good at exploring but lack of exploitation, two new solution search strategies named PSO-DE-PABC and PSO-DE-GABC are proposed based on Particle Swarm Optimization (PSO) and Differential Evolution (DE). PSO-DE-PABC generates new candidate position around the random particle to improve divergence. PSO-DE-GABC generates new candidate position around the global best solution to accelerate the convergence, and differential vectors are also used to increase the divergence. Besides, Dimension Factor (DF) is introduced to control the search rate of the algorithms. A new scout strategy considering current swarm state is used to replace the original random scout strategy to enhance the local search ability. Comparison with basic ABC, GABC (Gbest-guided ABC) and ABC/best algorithm is given on 10 groups of standard benchmark function. The results show that PSO-DE-GABC and PSO-DE-PABC have better convergence rate and accuracy.

Keywords: Hybrid optimization · Artificial bee colony (ABC) · Particle swarm optimization (PSO) · Differential evolution (DE) · Solution search strategy · Scout strategy

1 Introduction

Artificial Bee Colony (ABC) is proposed by Karaboga in 2005 is a heuristic search algorithm for numerical optimization, and experiments show that ABC is a very effective optimization algorithm [1, 2]. Compared with other optimization algorithms, ABC algorithm has many advantages such as simple structure, easy to implement and less control parameters, etc. [3, 4] Therefore, ABC algorithm has been used and studied widely. For example, Karaboga et al. used ABC to train the weight of feedforward neural network [5]; Rao et al. optimized distributed network to reduce losses in network transmission by using ABC [6]; Szeto et al. used ABC algorithm to optimize and schedule the vehicle routing in urban road [7]. However, ABC algorithm also exists some problems like other search algorithms, such as premature convergence, easy to fall into local minimum, stagnation at the late searching stage and so on. Based on these problems, many researchers have proposed more effective algorithm by combining

© Springer International Publishing Switzerland 2015
D.-S. Huang et al. (Eds.): ICIC 2015, Part II, LNCS 9226, pp. 24–35, 2015.
DOI: 10.1007/978-3-319-22186-1_3

ABC and other algorithms' optimization theory. For example, Alatas et al. who were inspired by chaos theory, introduced chaotic sequence into initialization phase and search strategies, improved global convergence abilities of the algorithm [8]; Zhu et al. proposed GABC (Gbest-guided ABC) algorithm guided by global best solution based on particle swarm principles [9]; Gao et al. were inspired by the idea of differential evolution and proposed ABC/best algorithm [10]. A series of test experiments show that the search strategies of basic ABC algorithm are more inclined to explore in search space, which resulted the decline of convergence speed and accuracy.

In order to better balance exploration and exploitation ability of ABC algorithm, two new ABC search strategies named PSO-DE-PABC and PSO-DE-GABC are proposed based on Particle Swarm Optimization (PSO) and Differential Evolution (DE); then dimension factor is introduced to control the search strategy and improve the convergence speed. Besides, the paper presents a scout strategy by using current population information, and it has better convergence ability as compared with random scout strategy in basic ABC. Through comparing with basic ABC, GABC (Gbest-guided ABC) and ABC/best algorithm based on 10 groups of standard benchmark function, the simulation experiments show that PSO-DE-GABC and PSO-DE-PABC have higher convergence rate and accuracy for most functions.

2 Artificial Bee Colony Algorithm

Artificial bee colony (ABC) Algorithm is a swarm intelligence optimization algorithm that can imitate the behavior of honey bees in nature. Leader, Scouter and Follower are three different types of artificial bee in artificial bee colony. Leader is responsible for excavating nectar source and corresponds with the food source; Follower selects nectar source based on information provided by Leader, and the number of Leader, Follower and nectar source is the same. If the quality of nectar source is still not improved after Leader and Follower exploit it for *limit* times, corresponding Leader will abandon here and transform itself for Scouter to exploit other nectar source.

The position of nectar source represents a potential solution of the problem to optimize. The following parameters need to be set artificially before executing ABC algorithm: D is the dimension of search space; SN is the number of nectar sources, Leaders and Followers; *limit* is the limit times for continuously exploiting without any increase; MCN is the maximum number of iterations. The position of nectar source is recorded as $X_i = (x_{i1}, x_{i2}, \ldots, x_{iD})$, where $i = 1, 2, \ldots, SN$. First, Leader is randomly initialized to search space and calculate the fitness value of its corresponding nectar source's position. In the basic ABC, the fitness function is defined as Eq. (1):

$$fit_i = \begin{cases} \frac{1}{1+f_i}, & f_i \geq 0 \\ 1 + |f_i|, & f_i < 0 \end{cases} \tag{1}$$

Where f_i is the function value of i-th nectar source; fit_i is the fitness value of i-th nectar source's function value after conversion.

Then Followers will use round-robin method to choose their own nectar source position according to the nectar source information provided by Leaders, which is based on the Eq. (2):

$$P_i = fit_i / \sum_{j=1}^{SN} fit_j \tag{2}$$

The search strategy of Leaders and Followers is shown in Eq. (3):

$$v_{i,j} = x_{i,j} + \varphi_{i,j}(x_{i,j} - x_{r,j}) \tag{3}$$

Where $v_{i,j}$ is the position of new generated nectar source; $x_{i,j}$ is the position of original i-th nectar source; $r \in \{1, 2, \ldots, SN\}$ and $r \neq i, j \in \{1, 2, \ldots, D\}$, i and j are randomly generated integer; $\varphi_{i,j} \sim U[-1,1]$ and U is uniformly distributed.

According to the Eq. (1), the fitness value of new nectar source can be calculated and compared it with itself, if the new solution has better fitness value, it will replace the old solution; otherwise it will be discarded. Then checking times not updated continuously of its nectar source, if the number is greater than *limit*, executing scout strategy that discarding the nectar source here and finding new nectar source according to the following Eq. (4):

$$x_{i,j} = lb_j + \psi_{i,j}(ub_j - lb_j) \tag{4}$$

Where lb_j and ub_j are the lower and upper bounds of x_{ij}, $\psi_{i,j} \sim U[0,1]$.

3   Improved Artificial Bee Colony Algorithm

3.1   New Search Strategy

The search strategy equation of basic ABC algorithm randomly selects one individual and one dimension at each update, thus resulting in slow convergence speed and convergence stagnation at the later stage [11]. Referring to the update equation of particle swarm algorithm, literature [9] proposed search strategy guide Eq. (5) based on the global optimal solution:
"GABC":

$$v_{i,j} = x_{i,j} + \varphi_{i,j}(x_{i,j} - x_{r,j}) + \psi_{i,j}(g_j - x_{i,j}) \tag{5}$$

Where $r \in \{1, 2, \ldots, SN\}$ and $r \neq i$; $\varphi_{i,j} \sim U(-1, 1)$; $\psi_{i,j} \sim U[0, C]$ and C is a non-negative constant; g is the global optimal solution of population.

C in Eq. (5) is similar to the acceleration factor in PSO algorithm, which has very important effects for the exploring and exploitation abilities of ABC algorithm. For example, Eq. (5) will degrade into Eq. (3) when $C = 0$, therefore basic ABC can be seen as a special case of GABC; exploitation ability of GABC will strengthen when C increases, then the convergence speed will accelerate but the exploring ability will decline, so the algorithm is easy to fall local minimal point. Experimental results show that $C = 1.5$ is a relatively good choice for most test functions.

Inspired by differential evolution, literature [10] adopted differential evolution equations to replace search strategy equations in basic ABC and proposed ABC/best algorithm, its update equations are shown in Eqs. (6) and (7):

"ABC/best/1":

$$v_{i,j} = g_j + \varphi_{i,j}(x_{r1,j} - x_{r2,j}) \tag{6}$$

"ABC/best/2":

$$v_{i,j} = g_j + \varphi_{i,j}(x_{r1,j} - x_{r2,j}) + \varphi_{i,j}(x_{r3,j} - x_{r4,j}) \tag{7}$$

Where $r1, r2, r3, r4, \in \{1, 2, \ldots, SN\}$ and $r1, r2, r3, r4 \neq i$.

This paper combines updating formulas of PSO and differential evolution and proposes two new ABC search strategy formulas after considering the global convergence ability and convergence speed. Equations (8) and (9) are as follows:

"PSO-DE-PABC":

$$v_{ij} = x_{rj} + \psi_{ij}(g_j - x_{rj}) + \varphi_{ij}(x_{pj} - x_{qj}) \tag{8}$$

"PSO-DE-GABC":

$$v_{ij} = g_j + \psi_{ij}(g_j - x_{rj}) + \varphi_{ij}(x_{pj} - x_{qj}) \tag{9}$$

Where $r, p, q, \in \{1, 2, \ldots, SN\}$ and $r \neq i, p \neq q; \psi_{ij} \sim U[0, 1]$.

By the comparison Eq. (8) with Eq. (5), the biggest difference between them is that Eq. (8) has introduced idea of random individual choice in differential evolution to each item, and improves diversity in population search process by making full use of population's individual information. The experimental results in literature [12] have showed that the exploring ability of Eq. (6) is much higher than Eq. (7). Therefore this paper integrates elitist strategy in differential evolution and incentive Strategy in PSO, proposes Eq. (9) to balance the exploring and exploitation abilities in search process, and adds a disturbance that is guided by global optimal solution based on Eq. (6).

3.2 Dimension Factor

In order to improve the convergence speed of PSO-DE-PABC and PSO-DE-GABC, this paper proposes selection operator based on dimensions to control the convergence performance. Unlike one-dimensional search strategy in general ABC and inspired by index-crossover operator in differential evolution, this paper presents a dimension factor parameter named DF to control the convergence speed. The algorithm is described as follows: after updating a dimension selected randomly, DF dimensions behind itself will be updated at the same time.

3.3　New Scout Strategy

The random scout strategy in basic ABC has given up previous search information of population and thrown the artificial bees into search space randomly. These will weaken local search ability of ABC and slower the convergence speed. Therefore, a more effective scout strategy is proposed and shown in Eq. (10):

$$x_{ij} = x_{ij} + \psi_{ij}(g_j - x_{ij}) + \varphi_{ij}\left[\omega_{max} - \frac{iteration}{MCN}(\omega_{max} - \omega_{min})\right] \bullet x_{ij} \qquad (10)$$

Where $\omega_{max} = 1$; $\omega_{min} = 0.2$; $\varphi_{ij} \sim$ U(-1, 1); $\psi_{ij} \sim$ U[0, 1]; *iteration* is current iteration's algebra, and *MCN* is max iterations' algebra.

Unlike random scout strategy, the paper has searched around the Scouters themselves, its search range will reduce continually with iterations and guided by global best position. So it has strong global search ability in the early and has strong local search ability in the later.

Specific steps of improved algorithm are as follows:

(1) Set parameters named *SN*, *DF*, *MCN* and *limit*, initialize Leader population randomly.
(2) Calculate fitness value fit_i of each Leader.
(3) Evaluate global optimal solution g.
(4) Leaders update nectar source v_{ij} according to Eq. (8) or (9) and execute greedy selection.
(5) Calculate the probability p_i of Followers' choice to nectar source based on Eq. (2).
(6) Followers choose nectar source according to the probability p_i, update nectar source v_{ij} by Eq. (8) or (9) and execute greedy selection.
(7) Evaluate global optimal solution g.
(8) Check whether the number of continuously update is greater than *limit* of the nectar source or not, if so, transform Leaders in this area into Scouters and execute scout strategy in formula (10).
(9) Determine whether the termination condition is satisfied or not, if so, output the result and end the algorithm; otherwise, go to step (4).

4　Experiment Design and Analysis

In order to verify the efficiency of PSO-DE-PABC and PSO-DE-GABC proposed in this paper, 10 different test functions are used for simulation, and comparing the experimental results with basic ABC [2], GABC [9] and ABC/best [10]. The simulation is realized on the computer with Intel Core i5-2400, memory of 4 GB, main frequency of 3.09 GHz, the development environment is Visual Studio 2010 and using standard C language.

This paper uses 10 commonly benchmark functions [12, 13], Table 1 gives the name, search space and optimal solution position of test functions. $f_1 \sim f_4$ are unimodal functions, around the optimal solution of Schwefel 1.2 has a very small descent

Table 1. Name, search space and optimal solution of test functions

Function	Function name	Search space	Optimal solution
f_1	Shifted Sphere	$[-100,100]$	$(0,0.1,\ldots,0.1)$
f_2	Schwefel 1.2	$[-100,100]$	$(0,0,\ldots,0)$
f_3	Shifted Schwefel 2.22	$[-100,100]$	$(0,0.1,\ldots,0.1)$
f_4	Shifted Schwefel 2.21	$[-100,100]$	$(0,0.1,\ldots,0.1)$
f_5	Rosenbrock	$[-30,30]$	$(1,1,\ldots,1)$
f_6	Shifted Rastrigin	$[-5.12,5.12]$	$(0,0.1,\ldots,0.1)$
f_7	Shifted Griewangk	$[-600,600]$	$(0,0.1,\ldots,0.1)$
f_8	Shifted Ackley	$[-32,32]$	$(0,0.1,\ldots,0.1)$
f_9	Penalized	$[-50,50]$	$(-1,-1,\ldots,-1)$
f_{10}	Penalized2	$[-50,50]$	$(1,1,\ldots,1)$

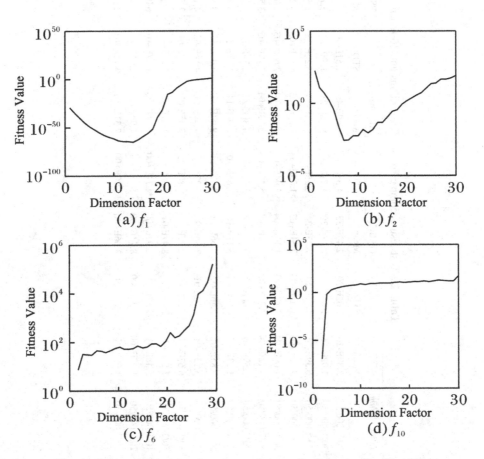

Fig. 1. Effect of dimension factor's number on simulation results ($D = 30$)

Table 2. Test results comparison for 5 algorithms and 10 test functions

Function	D/MCN/Num	ABC		GABC		ABC/best/1		PSO-DE-GABC		PSO-DE-PABC	
		Mean	STD	Mean	STD	Mean	STD	Mean	STD	Mean	STD
f_1	20/1 000/50	3.39E-20	2.62E-20	2.40E-33	4.38E-33	4.69E-46	7.51E-46	1.31E-54	1.32E-54	1.22E-80	2.88E-80
	30/1 000/50	1.92E-11	1.67E-11	2.41E-20	1.71E-20	9.22E-28	1.48E-27	5.73E-31	6.65E-31	5.36E-49	6.59E-49
	50/2 000/50	6.39E-14	4.60E-14	7.23E-24	6.95E-24	9.91E-33	6.22E-33	1.57E-33	1.91E-33	4.57E-56	6.48E-56
f_2	20/1 000/50	2.55E+03	1.02E+03	2.51E+03	1.05E+03	3.13E+03	1.46E+03	3.21E-07	6.79E-07	3.77E-04	1.11E-03
	30/1 000/50	1.09E+04	2.17E+03	1.16E+04	2.94E+03	1.14E+04	2.61E+03	2.04E-01	1.61E-01	2.85E-01	7.15E-01
	50/2 000/50	3.12E+04	4.03E+03	3.13E+04	5.40E+03	3.73E+04	7.79E+03	3.51E+00	1.80E+00	6.30E-01	4.12E-01
f_3	20/1 000/50	3.59E-10	1.80E-10	5.24E-17	2.70E-17	4.30E-24	2.74E-24	4.34E-30	4.17E-30	1.13E-48	1.02E-43
	30/1 000/50	1.13E-05	4.68E-06	3.11E-10	1.63E-10	2.07E-14	1.27E-14	5.06E-17	4.63E-17	2.57E-26	2.20E-26
	50/2 000/50	6.85E-07	2.46E-07	3.06E-12	9.57E-13	9.07E-17	4.07E-17	8.77E-19	4.70E-19	3.15E-30	1.75E-30
f_4	20/1 000/50	3.13E-01	1.12E-01	7.88E-02	3.22E-02	7.99E-02	4.18E-02	4.14E-18	2.23E-17	1.29E-53	6.95E-53
	30/1 000/50	6.67E-01	1.72E-01	2.97E-01	9.20E-02	3.45E-01	1.07E-01	2.71E-10	1.45E-09	1.54E-26	8.24E-26
	50/2 000/50	1.51E+00	2.68E-01	8.47E-01	1.38E-01	1.01E+00	1.84E-01	3.04E-17	1.10E-16	7.08E-19	3.77E-18
f_5	10/1 000/50	5.48E-01	5.58E-01	2.93E+00	4.74E+00	3.71E+00	1.27E+01	6.00E+00	6.42E+00	2.22E+00	3.83E+00
	20/1 000/50	1.97E+00	1.97E+00	8.74E+00	1.87E+01	1.96E+01	2.91E+01	1.62E-01	2.20E-01	5.18E+00	9.83E+00
	30/2 000/50	1.99E-01	2.01E-01	3.06E+00	1.10E+00	2.31E+01	3.10E+01	2.84E+00	4.90E+00	4.57E+00	9.59E+00

(Continued)

Table 2. (Continued)

Function	D/MCN/Num	ABC		GABC		ABC/best/1		PSO-DE-GABC		PSO-DE-PABC	
		Mean	STD	Mean	STD	Mean	STD	Mean	STD	Mean	STD
f_6	20/1 000/50	1.10E-12	2.84E-12	0.00E+00	0.00E+00	0.00E+00	0.00E+00	0.00E+00	0.00E+00	0.00E+00	0.00E+00
	30/1 000/80	1.66E-01	4.52E-01	2.00E-14	6.49E-14	0.00E+00	0.00E+00	0.00E+00	0.00E+00	0.00E+00	0.00E+00
	30/2 000/100	2.03E-01	3.97E-01	0.00E+00	0.00E+00	0.00E+00	0.00E+00	0.00E+00	0.00E+00	0.00E+00	0.00E+00
f_7	20/1 000/50	1.44E-16	5.09E-17	1.37E-16	4.7E-17	1.47E-16	1.99E-17	6.29E-17	5.50E-17	1.48E-17	3.77E-17
	30/1 000/50	1.46E-13	1.19E-13	2.63E-16	5.35E-17	2.81E-16	5.54E-17	1.18E-16	2.77E-17	1.11E-16	0.00E+00
	50/2 000/50	1.45E-15	1.37E-15	5.18E-16	6.60E-17	5.18E-16	5.97E-17	2.52E-16	4.91E-17	2.07E-16	3.77E-17
f_8	20/1 000/50	1.16E-09	7.37E-10	2.23E-14	3.86E-15	1.94E-14	4.27E-15	7.43E-15	1.71E-15	6.25E-15	1.71E-15
	20/1 000/50	1.78E-05	1.03E-05	3.55E-10	9.15E-11	8.67E-14	1.84E-14	1.48E-14	1.12E-15	9.92E-15	2.95E-15
	20/1 000/50	9.01E-07	4.34E-07	4.48E-12	1.54E-12	6.57E-14	4.35E-15	2.73E-14	2.70E-15	2.32E-14	2.69E-15
f_9	20/1 000/50	3.08E-12	6.67E-12	3.46E-14	1.39E-13	6.28E-32	6.81E-32	5.65E-32	5.69E-32	6.86E-32	6.71E-32
	30/1 000/50	9.72E-10	2.21E-09	7.59E-10	3.85E-09	3.79E-30	5.45E-30	6.09E-32	5.48E-32	4.29E-32	4.57E-32
	50/2 000/50	6.16E-11	2.30E-10	3.47E-11	1.87E-10	2.83E-32	2.86E-32	2.89E-32	2.87E-32	2.31E-32	2.53E-32
f_{10}	20/1 000/50	1.52E-19	4.06E-19	1.35E-32	2.21E-34	1.35E-32	2.60E-40	2.18E-32	4.36E-32	1.35E-32	2.60E-40
	30/1 000/50	6.62E-11	2.00E-10	1.13E-20	1.23E-20	1.23E-28	2.34E-28	1.05E-31	1.15E-32	1.37E-32	8.85E-34
	50/2 000/50	4.68E-14	4.58E-14	1.08E-24	9.52E-25	1.36E-32	3.70E-34	1.35E-32	2.21E-34	2.00E-32	3.41E-32

Table 3. Average results comparison for 5 scout strategies and 3 test functions

Function	Scout strategy	ABC	GABC	PSO-DE-GABC	PSO-DE-PABC
f_2	None	1.10E+04	1.11E+04	1.31E+00	5.40E-01
	Random	1.29E+04	1.27E+04	2.81E+02	1.37E+01
	Literature [19]	1.11E+04	9.03E+03	6.65E-01	3.10E-01
	Literature [20]	1.06E+04	8.52E+03	9.68E-01	3.69E-01
	The Article	7.08E+03	4.13E+03	2.34E-01	2.58E-01
f_4	None	7.33E-01	2.14E-01	9.61E-03	1.65E-06
	Random	6.61E-01	2.61E-01	2.02E-01	6.69E-02
	Literature [19]	1.53E-02	2.28E-03	1.47E-03	2.56E-05
	Literature [20]	2.99E-02	3.65E-03	1.78E-03	2.96E-06
	The Article	1.38E-09	3.54E-11	8.24E-11	3.90E-20
f_8	None	1.74E-05	3.16E-10	1.47E-14	8.73E-15
	Random	1.85E-05	4.10E-10	1.90E-14	1.44E-14
	Literature [19]	1.71E-05	3.98E-10	1.79E-14	1.39E-14
	Literature [20]	1.82E-05	3.81E-10	1.71E-14	1.32E-14
	The Article	1.46E-05	3.71E-10	1.90E-14	1.37E-14

gradient, Schwefel 2.21 is a badly optimized unimodal function for ABC algorithm; Rosenbrock is a very complex function, it is a unimodal function when its dimension is less than 4 and its global optimal solution will be surrounded by narrow valley, but it will be transformed into a multi-modal function when its dimension increases to 4, Shang et al. mathematically proved that it had two minimal points in 4 to 30 dimensions [14]. $f_6 \sim f_{10}$ are multi-modal functions, Griewangk's number of local minimal points will be reduced with the increasing of dimensions.

Comparing PSO-DE-GABC and PSO-DE-PABC with basic ABC, GABC and ABC/best/1, they initialize population randomly. D is the dimension of search space, MCN is the max iteration algebra in optimization process, Num is the individual size of population ($SN = Num/2$), $limit = 0.6 * SN * D$. Figure 1 shows that different DF influences on different function's performance, the setting is as follows: the value of DF in $f_1 \sim f_4$ and $f_7 \sim f_{10}$ is 5, and in $f_5 \sim f_6$ is 1.

Run 5 algorithms 30 times independently and the test results are shown in Table 2. Mean is the average value of computational results that can reflect convergence precision; STD is the standard deviation of results that can reflect convergent stability.

It can be seen from Table 2 that the proposed algorithm has higher convergence precision for unimodal functions $f_1 \sim f_4$, especially for f_2 which declines slowly; the proposed scout strategies have better effect for f_4 and make it converge to the global optimal solution fast.

Table 3 shows the comparison of optimal results with non-scout strategy, random scout strategy, literature [15], literature [16] and the proposed scout strategy for functions f_2, f_4 and f_8. The experimental parameters can refer to parameter setting when dimension's value is 30.

It can be seen from Table 3 that non-scout strategy has better effect than random scout strategy; the proposed scout strategy has better local exploitation ability for unimodal functions f_2 and f_4 that decline slowly; in the typical multi-modal function f_8, the effect of proposed scout strategy is better than the other three except for non-scout strategy. Therefore, the scout strategy in this paper can balance local exploitation ability and global search ability, and has better convergence performance.

For multi-modal functions $f_5 \sim f_{10}$, the proposed algorithm balances exploring and exploitation abilities well in the search process in addition to f_5, and make it converge to the global optimal solution quickly. In order to reflect the algorithm's search process more directly, Fig. 2 shows the convergence curve of following typical test functions named f_1, f_2, f_4, f_7, f_8 and f_9. From the curves in Fig. 2 it can be seen that PSO-DE-GABC and PSO-DE-PABC have faster convergence speed and better convergence accuracy.

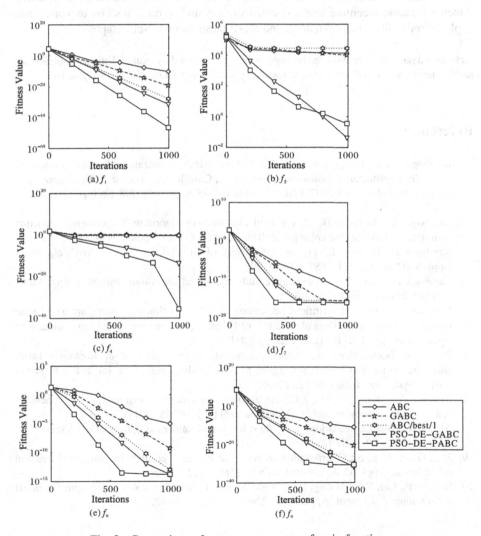

Fig. 2. Comparison of convergence process for six functions

5 Conclusion

Inspired by particle swarm optimization and differential evolution, this paper proposes PSO-DE-GABC and PSO-DE-PABC algorithms on the basis of global convergence ability and convergence speed, and introduces a simple and effective parameter (DF, Dimension Factor) to control the convergence speed of these two search strategies. In addition, this paper proposes a new scout strategy to improve the local search capability of algorithm, which can search effectively for functions with low gradient. The simulation results by 10 typical test functions show that PSO-DE-GABC and PSO-DE-PABC fully use individual information of population and effectively improve the convergence speed of ABC. However, the choice of dimension factor value in this paper is based on artificial experience, the next step we can study that factors including dimension factor, incentive strategy, elitist strategy and so on impact on exploring and exploitation abilities of population, and make these factors "self-adaptive".

Acknowledgement. The work in this paper is in part supported by Wuhan University of Science and Technology city college scientific research project under Grant No. 2014cyybky011.

References

1. Karaboga, D., Basturk, B.: Artificial Bee Colony (ABC) optimization algorithm for solving constrained optimization problems. In: Melin, P., Castillo, O., Aguilar, L.T., Kacprzyk, J., Pedrycz, W. (eds.) IFSA 2007. LNCS (LNAI), vol. 4529, pp. 789–798. Springer, Heidelberg (2007)
2. Karaboga, D., Basturk, B.: A powerful and efficient algorithm for numerical function optimization: artificial bee colony (ABC) algorithm. J. Global Optim. **39**(3), 459–471 (2007)
3. Karaboga, D., Basturk, B.: On the performance of artificial bee colony (ABC) algorithm. Appl. Soft Comput. **8**(1), 687–697 (2008)
4. Karaboga, D., Akay, B.: A comparative study of artificial bee colony algorithm. Appl. Math. Comput. **214**(1), 108–132 (2009)
5. Karaboga, D., Akay, B.: Artificial bee colony (ABC) algorithm on training artificial neural networks[C]. In: Proceedings of the 2007 IEEE 15th Signal Processing and Communications Applications, pp. 1–4. IEEE, Piscataway (2007)
6. Rao, R.S., Narasimham, S., Ramalingaraju, M.: Optimization of distribution network configuration for loss reduction using artificial bee colony algorithm. Int. J. Electr. Power Energy Syst. Eng. **1**(2), 116–122 (2008)
7. Szeto, W.Y., Wu, Y.Z., Ho, S.C.: An artificial bee colony algorithm for the capacitated vehicle routing problem. Eur. J. Oper. Res. **215**(1), 126–135 (2011)
8. Alatas, B.: Chaotic bee colony algorithms for global numerical optimazation. Expert Syst. Appl. **37**(8), 5682–5687 (2010)
9. Zhu, G., Kwong, S.: Gbest-guided artificial bee colony algorithm for numerical function optimization. Appl. Math. Comput. **217**(7), 3166–3173 (2010)
10. Gao, W.F., Liu, S.Y., Huang, L.L.: A global best artificial bee colony algorithm for global optimization. J. Comput. Appl. Math. **236**(11), 2741–2753 (2012)

11. Gao, W.F., Liu, S.Y., Huang, L.L.: A novel artificial bee colony algorithm based on modified search equation and orthogonal learning. IEEE Trans. Syst. Man Cybern. Part B Cybern. **43**(3), 1011–1024 (2013)
12. Gao, W.F., Liu, S.Y.: A modified artificial bee colony algorithm. Comput. Oper. Res. **39**(3), 687–697 (2012)
13. Yao, X., Liu, Y., Lin, G.M.: Evolutionary programming made faster. IEEE Trans. Evol. Comput. **3**(2), 82–102 (1999)
14. Shang, Y.W., Qiu, Y.H.: A note on the extended rosenbrock function. Evol. Comput. **14**(1), 119–126 (2006)
15. Barnharnsakun, A., Achalakul, T., Sirinaovakul, B.: The best-so-far selection in artificial bee colony algorithm. Appl. Soft Comput. **11**(2), 2888–2901 (2011)
16. Zhang, Y.X., Tian, X.M., Cao, Y.P.: Artificial bee colony algorithm with modified search strategy. J. Comput. Appl. **32**(12), 3326–3330 (2012)

New Approaches to the Problems of Symmetric and Asymmetric Continuous Minimax Optimizations

Siyan Xiong[✉] and Ruimin Gao

School of Computer Science and Technology,
University of Science and Technology of China (USTC), Hefei, China
shane@mail.ustc.edu.cn

Abstract. Depending on whether or not satisfying the symmetric condition, minimax problems could be considered either symmetric or asymmetric. In this paper, firstly we propose a new method for efficiently solving symmetric minimax problem using coevolutionary particle swarm optimization, in which a two-way alpha-beta pruning evaluation method and a simulated annealing based replacement strategy are employed. Secondly, in order to address asymmetric situations, a relaxation-based PSO is also introduced. Numerical simulations based on a series of symmetric and asymmetric minimax test functions are performed. Results of the proposed approaches and other state-of-the-art methods demonstrate that our methods could achieve considerably higher accuracy with less or equal number of function evaluations and CPU time.

Keywords: Continuous minimax optimization · Coevolution · Relaxation method · Particle swarm optimization

1 Introduction

Continuous minimax problem is searching optimal real-valued solutions with the best worst-case performance. We are often faced with a problem that how to make decisions under uncertainties or adversaries. The conventional approach like expected value optimization has a main issue that it neglects the worst-case effects. This is particularly important if the worst-case scenario can cause drastic results or risk-averse strategies are preferred. Minimax optimization, defines an optimality that considers the most robust decision under all possible scenarios. Minimax problems thus occur in many applications, such as robust control [1, 2], robust design [3, 4], game theory and gaming [5, 6], constraint optimization [7–10], etc.

The nature of the minimax problem inherently involves two variable spaces, one refers to decision variable (the optimal solution to be sought), and another relates to scenario (uncertain or adversarial variable). The continuous minimax problem can thus be formulated as

© Springer International Publishing Switzerland 2015
D.-S. Huang et al. (Eds.): ICIC 2015, Part II, LNCS 9226, pp. 36–46, 2015.
DOI: 10.1007/978-3-319-22186-1_4

$$\min_{x \in X} \max_{y \in Y} f(x, y). \tag{1}$$

where $x = [x_1, \ldots, x_{N_x}]^T$ is a N_x-dimensional decision variable and $y = [y_1, \ldots, y_{N_y}]^T$ is a N_y-dimensional scenario variable. $X \subseteq \mathbb{R}^{N_x}$ and $Y \subset \mathbb{R}^{N_y}$ are closed and bounded subsets in continuous spaces. $f : \mathbb{R}^{N_x + N_y} \to \mathbb{R}$ is a real-valued objective function on $X \times Y$.

As continuous minimax optimization possess many obstacles and limitations on both classical programming and normal evolutionary computation methods, several two-population coevolutionary algorithms (CEAs) have been proposed [8, 9, 11–14], etc. Based on a game point of view, a minimax problem is treated as a zero-sum game, and the main idea of CEA is discretizing the strategy space by populations and thereby obtaining a static matrix game. During a coevolution process, the original problem is approximated by iteratively generated static matrix game that has a denser population around the solution.

There also exists several non-coevolutionary methods such as SAEA [15], CrDE [16], etc. Techniques like Nash ascendancy or surrogate model are applied to transform the minimax problem into a normal optimization problem, but CEAs are still the most promising paradigm based on the existing empirical studies.

However, CEAs are only suitable for minimax problems that satisfy (2),

$$\min_{x \in X} \max_{y \in Y} f(x, y) = \max_{y \in Y} \min_{x \in X} f(x, y). \tag{2}$$

i.e. the symmetric problems. Equation (2) does not hold in all cases, for asymmetric minimax which not meet the condition, CEAs are easy to fail [17]. To our best knowledge, there is no existing method capable of efficiently solving both symmetric and asymmetric minimax problem.

In this paper, two novel approaches are proposed, including a simulated annealing coevolutionary particle swarm optimization with alpha-beta pruning (SACPP) for solving symmetric minimax problem, and another PSO algorithm based on the relaxation method (RelaxPSO) aim at the asymmetric cases.

The remainder of this paper is organized as follows. In Sect. 2, related properties and work of symmetric minimax are reviewed, then SACPP is proposed and explained in details. Then, a relaxation method for asymmetric minimax is introduced, and RelaxPSO is described in Sect. 3. Next, in Sect. 4, a series of test functions are presented, and experimental results are compared. Finally, conclusions and future work of this paper are drawn in Sect. 5.

2 Symmetric Minimax Problem

2.1 Coevolutionary Algorithms

Above all, for solving minimax problem, we are aim to approximate the maximized $\Psi(x) = \max_{y \in Y} f(x, y)$, and find out the global minimum of $\Psi(x)$ with respect to x. The symmetric condition implies that there exists a $y^* \in Y$ such that

$$\min_{x \in X} \max_{y \in Y} f(x, y) = \min_{x \in X} f(x, y^*). \tag{3}$$

If we could find out or closely approximate this single y^*, symmetric minimax problem would degenerate into a normal and relatively simple optimization problem. Moreover, this condition also implies that there exists a x^* such that

$$f(x, y^*) \le f(x^*, y^*) \le f(x^*, y), \forall x \in X, \forall y \in Y. \tag{4}$$

(4) means that the saddle point solution (x^*, y^*) is also a Nash equilibrium of the two-player zero-sum continuous simultaneous game. At a Nash equilibrium no player can benefit by unilaterally deviating from the current chosen strategy. Therefore, when (x^*, y^*) is reached, a stable basin of attraction is thereby established.

Based on these properties, several CEAs have been proposed and try to accelerate the solving process. CEAs analogously maintain two populations in X and Y space respectively. In population P_X, the fitness evaluation is according to $\max_{y \in P_Y} f(x, y)$, while in population P_Y, the fitness assignment is via $\min_{x \in P_X} -f(x, y)$ (smaller is better). Using this evaluation method based on security strategy, CEAs can concurrently approach x^* in P_X and y^* in P_Y.

Most of the subsequent CEAs have been focused on changing the ways of generating the next populations, such as using other optimization algorithms like PSO[9, 14], DE [18], etc., alternating the methods for evaluating fitness values, such as [2, 14, 17], etc., and/or modifying the mechanisms of individual replacement, such as [4], etc.

However, in each iteration, most of these works still required $N \times M$ fitness evaluations. Considerable population size and number of iterations are needed to ensure the accuracy and robustness of the algorithm when dealing with practical problems, hence the accumulated cost is expensive. Moreover, so far these methods lack a reasonable method for individual replacement, most of them are based on complete replacement or random replacement.

2.2 Proposed Method

Evaluation with Two-Way Alpha-Beta Pruning. Based on above discussion, we first propose an alpha-beta pruning method to reduce the redundant fitness evaluations, then present a replacement mechanism based on simulated annealing. A two-population coevolutionary particle swarm optimization together with these two improvements are bundled into a fast algorithm called SACPP. PSO is a famous population-based optimization algorithm [19], we choose it as the backbone mainly because of its simplicity and previous success in black-box real-valued optimization. Due to limited space here, we are not going to review the detail of PSO here.

Having considered that we just want to find out the global minimum of $\Psi(x)$, it does not require approximating $\Psi(x_i)$ for all x_i in the population. Similarly, we do not need to approximate $\Phi(y_i)$ entirely. Given population $P_X = (x_1, \ldots, x_{N_{P_X}})$, $P_Y = (y_1, \ldots, y_{N_{P_Y}})$, we want to compute the following with the minimum cost

$$x^* = x_{i^*}, i^* = \arg\min_{N_{P_X}} \Psi(x_i)$$
$$y^* = x_{j^*}, j^* = \arg\max_{N_{P_Y}} \Phi(x_i)$$

(5)

Consider a simple example, $P_X = [P_{X_1}, P_{X_2}, P_{X_3}]$, $P_Y = [P_{Y_1}, P_{Y_2}, P_{Y_3}]$, its utilities matrix is shown in Table 1.

The original continuous strategy space has been transformed in to discrete set by population-based optimization algorithms. Therefore (5) can be considered a discrete minimax tree search problem.

By using an alpha-beta pruning algorithm, it enables us to reduce considerable function evaluations, and ensure the correct solution to (5). Consider we want to calculate the value of the root, an ordinary alpha-beta pruning algorithm will be

$$\begin{aligned} Minimax(root) &= \min(\max(6,5,4), \max(1,9,2), \max(8,7,3)) \\ &= \min(6, \max(1,9,x), \max(8,y,z)) \\ &= \min(6, a, b) \, where \, a \geq 9, b \geq 8 \\ &= 6. \end{aligned}$$

(6)

However, we need to compute both the minimax value of X and the maximin value of Y. Therefore, our proposed algorithm is two-way pruning. Moreover, the effectiveness of alpha-beta pruning is highly dependent on the order in which the nodes are expanded [20]. We use a dynamic move-ordering heuristic by trying (i^*, j^*) first, the detailed pseudo code can be found in Algorithm 1. Through this algorithm, we can save the vast majority of the computation without affecting the correctness which is not guaranteed by the iterative method used in [14] even in this simple example.

Table 1. Simple example

$f(x,y)$	y_1	y_2	y_3
x_1	6	5	4
x_2	1	9	2
x_3	8	7	3

Algorithm 1. Evaluation via two-way alpha-beta pruning

```
1: loop
2:        i* = arg min_N F_X
3:        j* = arg max_M F_Y
4:        if P_{X_i*} and P_{X_j*} all evaluated then
6:              x* = P_{X_i*}
7:              y* = P_{Y_j*}
8:              return F_X, F_Y, x*, y*
9:        if f(P_{X_i*}, P_{Y_j*}) not evaluated then
10:             F' = f(P_{X_i*}, P_{Y_j*})
11:             F_{X_i*} = max(F', F_{X_i})
12:             F_{Y_j*} = min(F', F_{Y_i})
13:       else
14:             if P_{X_i*} not evaluated then
15:                   j ← j if f(P_{X_i*}, P_{Y_j}) not evaluated
16:                   F' = f(P_{X_i*}, P_{Y_j})
17:                   F_{X_i*} = max(F', F_{X_i})
18:                   F_{Y_j} = min(F', F_{Y_i})
19:             else
20:                   i ← i if f(P_{X_i}, P_{Y_j*}) not evaluated
21:                   F' = f(P_{X_i}, P_{Y_j*})
22:                   F_{Y_j*} = min(F', F_{Y_i})
23:                   F_{X_i} = max(F', F_{X_i})
```

Replacement by Simulated Annealing. Apart from fitness evaluation in CEAs, replacement mechanism is also critical. Previously, in most of CEAs, the individual replacements between the generations are determined by randomly selection or fitness comparisons. Neither of two seems reasonable, in the former case, good information may be randomly destroyed every generation, while in the latter situation, fitness in different generations is not transitive due to the lack of objective fitness function. In fact, the fundamental problem of CEAs is: how to build an effective algorithm when at time t_1, one believes $A \prec B$ and perhaps later one will believe $B \prec A$ at time t_2.

To address this inherent defect of coevolution, a replacement strategy based on simulated annealing is introduced. At generation k, simulated annealing takes current state s, generated candidate state s_n, and the temperature schedule t_k that governs the Metropolis acceptance probability as input, and evaluates to the following probability

$$P(e, e_n, t_k) = \begin{cases} \exp^{-\frac{e_n - e}{t_k}} & \text{if } e_n < e, \\ 1 & \text{otherwise.} \end{cases} \qquad (7)$$

Where $e = fit(s)$ return the function value of the state.

In SACPP, this replacement method is used for updating $pbest_i$, the previous best position for each particle. The motivation is the fitness assignment may not be accurate in the early evolutionary process, the initial x^* and y^* may actually be away from the actual optimal. High temperature t_k can avoid getting stuck in the local optima caused by previous over-optimistic evaluation. As the coevolution continues, temperature t_k slowly decreases as the evaluation getting more and more accurate.

3 Asymmetric Minimax Problem

3.1 Difficulties in Asymmetric Minimax

Since the problems do not satisfy the symmetric condition, the symmetric evaluation method seems not so reasonable. Several subsequent works try to tackle asymmetric problems by using asymmetric evaluation methods. In [17], a rank-based CEA was proposed, fitness assignment is based on the maximum rank of each y_i achieves against P_X. The main purpose of doing so is to reward any y_i that makes at least one x_i worse. In addition, several heuristic evaluation methods for evaluating scenario population can be found in [21], but the empirical results in [4] and [21] have shown that no one strictly performs well.

In fact, asymmetric minimax problem implies that there is no single y^* that satisfies (3), instead, it requires a set S_Y that consists of multiple y. When we use CEAs on asymmetric problems, each population itself as a single-objective EA with finite population size will inevitably drift and converge due to three factors: (i) selection errors, (ii) selection noises, and (iii) the disruptions of stochastic operators.

For asymmetric problems, this property is not acceptable. Consequently, some artificial diversity control methods have been proposed. In {Cramer, 2009 #61}, a non-coevolutionary model is proposed for the asymmetric problem. It claimed that using random sampling can introduce ultimate diversity thus can avoid the pitfalls in CEAs. It uses a uniform randomly generated scenario population at each generation, although the difficulty of determining the scenario fitness is removed, a considerable sample size is required to ensure the accuracy of this semi-random search.

In [22], there is a different perspective that minimax problem can be reformulated as an equivalent optimization problem with infinite constraints.

$$\begin{aligned} &\underset{x}{\text{minimize }} \sigma \\ &\text{subject to } f(x, y) \le \sigma, \forall y \in Y. \end{aligned} \qquad (8)$$

Apparently the problem has infinite constraints is still intractable.

An iterative relaxation procedure is able to find an approximate solution by relaxing the problem (8) into a series of finite constraints problems as described in Algorithm 2.

Algorithm 2. Relaxation Framework

1: **Pick** $y_0 \in Y$ **and set** $S_Y = \{y_0\}$

2: $(x^*, f_m) = \arg\min_{x \in X} \left\{ \max_{y \in S_Y} f(x, y) \right\}$

3: $(y^*, f_M) = \arg\max_{y \in Y} f(x^*, y)$

4: **if** $f_M - f_m < \grave{o}_R$ **then**

5:      **return** (x^*, y^*)

6: **else**

7:      **insert** y^* **to** S_Y, **goto step 2**

The threshold $\in_R$ specifies the accuracy of the desired solution. Under reasonable assumptions, the main loop has been proven to terminate after a finite number of iterations. The returned result would be the solution to (8) because it met or very slightly violated the constraints. Hence it is also the solution to the original problem (1).

3.2 Proposed Method

If the number of outer iterations k becomes medium or large, the whole relaxation procedure would be expensive as well. Generally, if we directly combine relaxation and EAs, the total time complexity would be $O(k^2 NP)$, where P is the population size, and N is the average number of inner generation at each iteration. In this case, how to reduce the total cost is critically important.

To reduce the potential computational effort of the relaxation method, two approaches can be considered. (i) Reducing or limiting the number of outer iterations k, or (ii) decreasing the number of function evaluations within each iteration, particularly in step 2 of Algorithm 2. For (i), it is hard to reduce the necessary iterations. For example, assume we have a problem which has two different maximizers, thus at least two iterations are required to populate the minimum set S_Y. However, limiting the maximum value of k, and replacing the most irrelevant element in S_Y when the set is filled is still possible. While for (ii), at each iteration, about $P \times k$ times function evaluations are required but some of them are not necessary. An ordinary alpha-beta evaluation method shown in Algorithm 3 is applied to prune redundant function evaluations.

Algorithm 3. Evaluate with Set

1: **for** $i \leftarrow 1, N$ **do**

2:      $fitness(X_i) = f(X_i, S_{Y_i})$

3: **loop**

4:      $i^* = \arg\min_N fitness(X)\$$

5:      **if** X_i for all $y \in S_Y$ have been evaluated **then**

6:          **return** $fitness(X)$

7:      $j = successor(S_Y, i^*)$

8:      $fitness(X_{i^*}) = max(fitness(X_{i^*}), f(X_{i^*}, S_{Y_j})$

---

4 Numerical Experiments

In this section, we simulate and compare the performances of our proposed approaches and several other state-of-the-art approaches. Specifically, we investigate the performance of CrDE [16] and IterPSO [14] with our SACPP with pruning evaluation on symmetric problems; ASGA [4], CrDE [16] with our RelaxPSO on asymmetric problems respectively. Eleven symmetric test cases and eight asymmetric test cases with known reference values are selected to study the behaviors of the algorithms. In all benchmarks, each algorithm is executed 50 times on each problem, and the mean, median, and standard deviation of the mean square error (MSE) of x and y, together with the average number of function evaluations(FE, rounded to integer) are reported.

In all experiments, the parameters of proposed algorithms are as follows, some of them, such as acceleration coefficients are set through early experience with particle swarm optimization,

- population size $= max(10 * N_x, 10 * N_y)$,
- inertia weight $w = 0.4$, acceleration coefficients $c_1 = c_2 = 2.0$,
- relaxation threshold $\in_R = 10^{-3}$,
- initial temperature $= 50$, linearly decrease to 0,
- stop criterion: max iterations $= 20 + 10(N_x + N_y)$ exceeded, or relative change in the global best function value over the last 20 iterations is less than 10^{-6}.

4.1 Overall Performances

Table 2 summarizes the results of symmetric benchmark. First of all, IterPSO and SACPP as CEAs required significantly less amount of function evaluations and CPU time. In order to compare the results of different algorithms, a Wilcoxon rank-sum test with level $P < 0.05$ is conducted, then the "w-d-l" results are recorded. For CrDE, the test result is 11-0-0, it implies that SACPP performs better consistently for all test functions than CrDE. For IterPSO, SACPP performs 10-1-0, only on B1, the result has not reached the significant level.

The results of asymmetric benchmark are listed in Table 2. Similarly, the number of fitness evaluations of RelaxPSO is much smaller than ASGA and CrDE. The accuracy of RelaxPSO outperformed ASGA and CrDE on 7 of 8 test functions other than function A2. The two plane function A2 can lead to high performance variance of relaxation method because in A2, the worst case $max_y f(x, y)$ just equals to the counterpart x, predefined threshold $\in_R$ is too coarse then caused the premature breaking of the loop. In addition, note that there is another advantage of RelaxPSO that it can obtain the precise y in set_y, while other two approaches are incapable of.

4.2 Effectiveness of the Evaluation Methods

In order to analyze the effectiveness of the two-way alpha-beta pruning, SACPP and SACP (without pruning) are tested on the same benchmark respectively with other

Table 2. Results of symmetric and asymmetric benchmarks

Problem		CrDE	IterPSO	SACPP	Problem		ASGA	CrDE	RelaxPSO
A1	mean	2.69×10^{-14}	3.45×10^{-13}	5.08×10^{-15}	A2	mean	8.68×10^{-3}	5.10×10^{-4}	5.36×10^{-3}
	median	6.71×10^{-16}	2.41×10^{-15}	3.74×10^{-16}		median	0.00	0.00	0.00
	std	2.19×10^{-14}	1.85×10^{-12}	1.01×10^{-14}		std	3.32×10^{-2}	2.18×10^{-3}	8.90×10^{-3}
	FE	100000	1773	1480		FE	100000	100000	9212
A5	mean	6.24×10^{-8}	1.63×10^{-13}	0.00	A3	mean	9.21×10^{-2}	4.58×10^{1}	5.36×10^{-3}
	median	1.38×10^{-8}	6.84×10^{-15}	0.00		median	0.00	6.47	0.00
	std	1.05×10^{-7}	3.76×10^{-13}	0.00		std	3.28×10^{-1}	4.91×10^{1}	8.90×10^{-3}
	FE	100000	4525	3679		FE	100000	100000	9654
B1	mean	2.62×10^{-4}	5.29×10^{-10}	5.29×10^{-10}	A4	mean	5.55×10^{-2}	4.49	5.78×10^{-2}
	median	1.81×10^{-4}	5.29×10^{-10}	5.29×10^{-10}		median	6.31×10^{-3}	4.35	2.88×10^{-11}
	std	2.92×10^{-4}	5.72×10^{-13}	1.22×10^{-13}		std	2.12×10^{-1}	2.07×10^{8}	2.27×10^{-1}
	FE	100000	5003	4429		FE	100000	100000	13597
B2	mean	6.14×10^{-3}	2.59×10^{-5}	5.10×10^{-6}	C8	mean	2.20×10^{-5}	5.94×10^{-3}	4.87×10^{-18}
	median	1.36×10^{-3}	9.91×10^{-6}	5.03×10^{-6}		median	8.49×10^{-6}	1.61×10^{-3}	6.13×10^{-20}
	std	1.08×10^{-2}	5.22×10^{-5}	4.31×10^{-7}		std	3.90×10^{-5}	1.09×10^{-2}	1.25×10^{-17}
	FE	100000	4996	4575		FE	100000	100000	1636
B3	mean	1.02×10^{-2}	1.51	3.88×10^{-7}	C9	mean	8.02×10^{-2}	2.38×10^{-2}	5.13×10^{-8}
	median	5.40×10^{-3}	2.44×10^{-7}	1.81×10^{-10}		median	3.55×10^{-2}	1.55×10^{-2}	2.83×10^{-9}
	std	1.28×10^{-2}	3.67	1.78×10^{-6}		std	9.50×10^{-2}	2.71×10^{-2}	1.27×10^{-7}
	FE	100000	5130	5123		FE	100000	100000	8116
B4	mean	1.51×10^{-2}	4.21×10^{-9}	8.84×10^{-11}	C10	mean	2.89×10^{-2}	1.08×10^{-2}	2.29×10^{-10}
	median	8.70×10^{-3}	7.12×10^{-10}	8.84×10^{-11}		median	1.42×10^{-2}	5.89×10^{-3}	2.96×10^{-11}
	std	1.87×10^{-2}	1.20×10^{-8}	8.16×10^{-14}		std	3.11×10^{-2}	1.29×10^{-2}	5.02×10^{-10}
	FE	100000	8843	6632		FE	100000	100000	8033
B5	mean	8.41×10^{-5}	2.09×10^{-3}	6.80×10^{-7}	C14	mean	5.15×10^{-1}	4.86×10^{-2}	3.56×10^{-5}
	median	6.20×10^{-5}	5.46×10^{-5}	3.71×10^{-7}		median	4.89×10^{-1}	3.38×10^{-2}	2.34×10^{-5}
	std	7.13×10^{-5}	7.16×10^{-3}	1.71×10^{-6}		std	3.64×10^{-1}	3.38×10^{-2}	3.64×10^{-5}
	FE	100000	9906	8825		FE	100000	100000	33799
B6	mean	2.22×10^{-2}	6.70×10^{-2}	4.12×10^{-4}	C15	mean	0.44×10^{-1}	4.17×10^{-2}	5.42×10^{-5}
	median	1.82×10^{-2}	1.02×10^{-2}	8.16×10^{-12}		median	3.77×10^{-1}	3.58×10^{-2}	1.95×10^{-5}
	std	1.82×10^{-2}	1.66×10^{-1}	2.02×10^{-3}		std	3.04×10^{-1}	2.67×10^{-1}	6.96×10^{-5}
	FE	100000	14555	13328		FE	100000	100000	20949
B7	mean	2.55×10^{-1}	7.92×10^{-2}	1.33×10^{-7}					
	median	1.97×10^{-1}	7.83×10^{-4}	1.40×10^{-11}					
	std	2.00×10^{-1}	1.59×10^{-1}	8.67×10^{-7}					
	FE	100000	25170	26423					
C12	mean	1.49×10^{-2}	4.19×10^{-2}	4.01×10^{-13}					
	median	9.55×10^{-3}	4.73×10^{-8}	4.00×10^{-13}					
	std	1.38×10^{-2}	2.05×10^{-1}	1.49×10^{-14}					
	FE	100000	14759	10068					
C13	mean	9.20×10^{-3}	1.37×10^{-3}	8.21×10^{-16}					
	median	7.97×10^{-3}	6.82×10^{-9}	7.35×10^{-16}					
	std	5.79×10^{-3}	9.64×10^{-3}	2.86×10^{-16}					
	FE	100000	16532	12169					
Wilcoxon Test		11-0-0	10-1-0	-			7-1-0	7-1-0	-

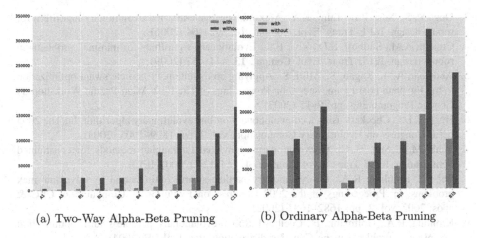

(a) Two-Way Alpha-Beta Pruning   (b) Ordinary Alpha-Beta Pruning

Fig. 1. Average number of function evaluations with or without pruning methods

parameters fixed. The average numbers of fitness evaluations of SACPP and SACP are shown in Fig. 1a. From the figure we could see that the two-way alpha-beta pruning is highly efficacious, for problem B7, SACPP used only 25000 in average while SACP required more than 3×10^5.

Distinct from two-way alpha-beta pruning method, an ordinary pruning method is used in RelaxPSO. From Fig. 1b, we can see that due to the absence of killer move heuristic, the effectiveness is weaker than 1a. However, as the complexity of the function grows, this method can save up to one half number of FEs.

5 Conclusions and Future Work

This paper demonstrates two novel methods, namely SACPP for symmetric and RelaxPSO for asymmetric minimax problems respectively. The proposed methods have shown to compare favorably with existing approaches to solving these problems.

SACPP and RelaxPSO also left a lot to be improved. The PSO version we used is ordinary because of its simplicity and clearance, in practical applications, our proposed techniques are suitable for any population-based continuous optimization algorithms. On the other hand, when we have no knowledge about symmetry of the problem to be solved, a reasonable hybrid strategy will be very useful.

References

1. Marzat, J., Walter, E., Piet-Lahanier, H.: Worst-case global optimization of black-box functions through Kriging and relaxation. J. Glob. Optim. **55**, 707–727 (2013)
2. Hur, J., Lee, H., Tahk, M.-J.: Parameter robust control design using bimatrix co-evolution algorithms. Eng. Optim. **35**, 417–426 (2003). Taylor and Francis

3. Ong, Y.-S., Nair, P.B., Lum, K.Y.: Max-min surrogate-assisted evolutionary algorithm for robust design. IEEE Trans. Evol. Comput. **10**, 392–404 (2006)
4. Cramer, A.M., Sudhoff, S.D., Zivi, E.L.: Evolutionary algorithms for minimax problems in robust design. IEEE Trans. Evol. Comput. **13**, 444–453 (2009)
5. Abdelbar, A.M., Ragab, S., Mitri, S.: Applying co-evolutionary particle swarm optimization to the Egyptian board game seega. In: Proceedings of The First Asian-Pacific Workshop on Genetic Programming, pp. 9–15 (2003)
6. Hughes, E.J.: Checkers using a co-evolutionary on-line evolutionary algorithm. In: The 2005 IEEE Congress on Evolutionary Computation, vol. 2, pp. 1899–1905 (2005)
7. Tahk, M.-J., Sun, B.-C.: Coevolutionary augmented Lagrangian methods for constrained optimization. IEEE Trans. Evol. Comput. **4**, 114–124 (2000). IEEE
8. Shi, Y., Krohling, R.A.: Co-evolutionary particle swarm optimization to solve min-max problems. In: Proceedings of the 2002 Congress on Evolutionary Computation, CEC & apos; 2002, vol. 2, pp. 1682–1687 (2002)
9. Krohling, R.A., Hoffmann, F., Coelho, L.S.: Co-evolutionary particle swarm optimization for min-max problems using Gaussian distribution, vol. 1, IEEE (2004)
10. Segundo, G.A., Krohling, R.A., Cosme, R.C.: A differential evolution approach for solving constrained min–max optimization problems. Expert Syst. Appl. **39**, 13440–13450 (2012). Elsevier
11. Barbosa, H.J.: A genetic algorithm for min-max problems. In: Goodman, E. (ed.) Proceedings of the First International Conference on Evolutionary Computation and Its Applications, pp. 99–109 (1996)
12. Herrmann, J.W.: A genetic algorithm for minimax optimization problems. In: Proceedings of the 1999 Congress on Evolutionary Computation, CEC 1999, vol. 2 (1999)
13. Krohling, R.A., dos Santos Coelho, L.: Coevolutionary particle swarm optimization using Gaussian distribution for solving constrained optimization problems. IEEE Trans. Syst. Man Cybern. Part B Cybern. **36**, 1407–1416 (2006). IEEE
14. Masuda, K., Kurihara, K., Aiyoshi, E.: A novel method for solving min-max problems by using a modified particle swarm optimization. In: 2011 IEEE International Conference on Systems, Man, and Cybernetics (SMC), pp. 2113–2120 (2011)
15. Zhou, A., Zhang, Q.: A surrogate-assisted evolutionary algorithm for minimax optimization. In: 2010 IEEE Congress on Evolutionary Computation (CEC), pp. 1–7 (2010)
16. Lung, R.I., Dumitrescu, D.: A new evolutionary approach to minimax problems. In: 2011 IEEE Congress on Evolutionary Computation (CEC), pp. 1902–1905 (2011)
17. Jensen, M.T.: Robust and flexible scheduling with evolutionary computation. BRICS, Computer Science Department, University of Aarhus (2001)
18. Huang, F.-Z., Wang, L., He, Q.: An effective co-evolutionary differential evolution for constrained optimization. Appl. Math. Comput. **186**, 340–356 (2007). Elsevier
19. Eberhart, R.C., Kennedy, J.: A new optimizer using particle swarm theory. In: Proceedings of the Sixth International Symposium on Micro Machine and Human Science, vol. 1, pp. 39–43 (1995)
20. Russell, S., Norvig, P.: Artificial Intelligence: A Modern Approach, vol. 25. Prentice-Hall, Englewood Cliffs (1995). Citeseer
21. Branke, J., Rosenbusch, J.: New approaches to coevolutionary worst-case optimization. In: Rudolph, G., Jansen, T., Lucas, S., Poloni, C., Beume, N. (eds.) PPSN 2008. LNCS, vol. 5199, pp. 144–153. Springer, Heidelberg (2008)
22. Parpas, P., Rustem, B.: An algorithm for the global optimization of a class of continuous minimax problems. J. Optim. Theor. Appl. **141**, 461–473 (2009). Springer

A Video Steganalysis Algorithm
for H.264/AVC Based on the Markov Features

Ting Da[1], ZhiTang Li[1,2(✉)], and Bing Feng[1]

[1] Department of Computer Science and Technology,
Huazhong University of Science and Technology, Wuhan, China
dpting1222@163.com, lzt@hust.edu.cn
[2] Network and Computing Center,
Huazhong University of Science and Technology, Wuhan, China

Abstract. A new steganalysis algorithm based on the features of the original domain was proposed combining spatial correlation with temporal correlation among video frames. First, divide video frames into a new kind of special frame structure of P_0-I-P_1, calculate the luminance component for every frame of the macroblocks. And then we use the Eulerian-distance of vectors to express the correlation between macroblocks to get the most similar macroblock of adjacent frames. Thus, we combine the co-occurrence matrix model with Markov model into the P_0-I and I-P_1 frame type. We try to determine whether any data hiding is embedded in the video according to the changes of correlation between video frames. The experimental results show that the algorithm proposed in this paper has a high detection probability, and it calls the promising result and deserves further studies.

Keywords: Steganalysis · Co-occurrence matrix · Markov · Correlation

1 Introduction

With the rapid development of network communication and information processing technologies, steganalysis of video has been rapidly developed as a new information security technology in recent years. Since the shocking 9.11 incident, illegal activities, which undermine social stability and social security by taking advantage of public information network and various of digital media to carry data for communicating, have attracted wide attention of the government and the public. As the countermeasure, steganalysis is analytical technology, focus on steganography, which identifies the existence of embedded data and evaluates the size of embedded data by analyzing the digital media, and then try to extract, restore or destruct the hidden data. The development of video compression techniques and network streaming media service make the range of applications of video more and more widely. Therefore, video, as a common form of digital media, will surely become an important carrier of steganography due to its characteristics such as large amount of data, rich content and complex statistical characteristics. In conclusion, it is necessary to study the video steganalysis [1].

Foundation Item: The National Natural Science Foundation of China (61272407).

D.-S. Huang et al. (Eds.): ICIC 2015, Part II, LNCS 9226, pp. 47–59, 2015.
DOI: 10.1007/978-3-319-22186-1_5

Steganalysis can be divided into original domain steganalysis and transform domain steganalysis according to its scope: or be divided into steganalysis based on the identification and steganalysis based on statistics, or be divided into specific steganalysis and universal steganalysis. Present studies of video steganalysis mainly aim at specific steganographic algorithm, but still no mature methods for universal steganalysis. In view of the influence of spread-spectrum embedding for video based on temporal relativity, Xu et al. [2] proposed a steganalysis method based on temporal relativity. To detect the existence of embedded data in spread-spectrum stego video clips effectively, Liu et al. [3] proposed a steganalysis approach based on inter-frame collusion strategy according to temporal characteristics of video sequence. Sun et al. [4] proposed a steganalysis method based on motion estimation, which extracts the motion vector field as features that reflect the time-varying characteristics, and then uses the supported vector machine to detect the embedded data.

General blind detection is a kind of atomic detection method, it can detect any steganographic methods by training the original video and the stego video. But the characteristics of video are complex, so there is no mature universal steganalysis method. Most existing steganalysis methods ignore two main characteristics of the video: temporal correlation and spatial correlation. Correlation between frames and the pixels of each frame can reflect the differences between videos. Because the hidden data is independent of carrier, the spatiotemporal correlation of the carrier will change after being embedded [5]. This paper presents an idea combining spatial correlation with temporal correlation, co-occurrence matrix of feature vector is obtained by Gray-level Co-occurrence Matrix as the correlation between macroblocks, and then establish Markov model of inter-frames, we try to determine whether any data is embedded in the video according to the changes of correlation between video frames using SVM [6] classifier.

2 Steganographic Method of H.264/AVC

2.1 Steganographic Method

On one hand, with the development of Internet video services, it's urgent to upgrade the Internet content security supervision, on the other hand, H.264/AVC (H.264 for short), as a kind of video compression standard with efficient compression and good network adaptability, has been widely used in Internet. Therefore, researches on steganography and steganalysis for H.264 video have become increasingly hot at home and abroad. Compared to early MPEG and H.263, H.264 is more excellent, and it can provide high coding efficiency than MPEG and H.263, the video capture chip output stream are generally in the form of YUV data stream. At present, about eighty percent of videos are H.264 format. In addition, the traditional video steganalysis method is difficult to detect the steganographic algorithm since the format for H.264 is complex [7, 8]. Compared with static digital image, video has large amount of information and complex structure, so it must be encoded and compressed before transmission in order to save network bandwidth and storage space, for video encoding can reduce spatial and temporal redundancy [9, 10].

In original domain (YUV data stream) steganography method, data is embedded into the pixels of video, and then compress the video with H.264/AVC standard, this method is not conducive to information extraction because it will disturb the embedded data during video compression process. So most existing steganographic method embed data into quantized DCT coefficients (especially quantized DCT coefficients in I frames) to meet the need of covert communication [11].

Generally, steganographic algorithm only changes middle frequency coefficients since the changes of high frequency coefficients are easy to be detected and the changes of low frequency coefficients will damage the integrity of the original data. And the steganographic algorithm only modifies little in DCT coefficient in order to achieve the requirement of data hiding. And now many steganographic methods anti DCT domain steganalysis is proposed. The modification of any DCT coefficient will cause all pixel values of a macroblock change in different degree. The transformation process of coefficient is shown in Fig. 1. So the change of pixel value in original domain is much higher than the change of coefficients in DCT domain. And this paper proposes a steganalysis scheme based on video original domain. Video Steganalysis in original domain can eliminate the need for video encoding and decoding process. So we can quickly improve the efficiency, reduce the running time of the algorithm and improve the performance of the algorithm [12, 13].

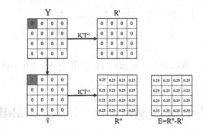

Fig. 1. The transformation process of coefficient in the DCT domain

2.2 Frame Types of H.264/AVC

H.264/AVC video is usually composed of a plurality of continuous GOP (Group of Pictures) sequences, each GOP comprises a plurality of frames, frame type usually consists of I-frame, P-frame and B-frame. I-frame uses intra coding, P-frame uses forward frequency estimation and B-frame uses bi-directional estimation. The first frame of GOP is I-frame, the other frames are P-frames and B-frames. I-frame is a special frame which can be fully reconstructed without referring other frames. P-frame using motion compensation to send the difference of motion vector between I-frame and P-frame which are before it. B-frame is predicted by P-frame or I-frame before and after it. In general, the most frequently-used frame structure is IPPIPP or IBPIBP. This paper mainly discusses the frame structure of IPPIPP (in a similar way in the frame structure of IBPIBP). The following is a typical sequence of frames: $I_1P_2P_3I_4P_5$. $P_6I_7P_8P_9I_{10}P_{11}P_{12}I_{13}\ldots\ldots$, in which $I_1P_2P_3$ is a GOP.

Usually we can load hidden data in I-frame. I-frame is the key frame of each GOP and it has the following characteristics [14, 15]. I-frame is the reference frame of P-frame and B-frame, so it directly affects the quality of other frames in the same group. So I-frame has a very important position in video, it can directly affect the P-frames of the same GOP, but have no effect on the P-frames in the other GOP. In the original video (not data hiding), there is much correlation between frames, that means each I-frame has high correlation with the P-frame after it. So if the hidden data is embedded in I-frame, P-frame will be affected in the same GOP. According to the characteristics of GOP, if the hidden data is embedded in the I_4-frame, it will directly affect the $I_4P_5P_6$ in this GOP. And P_5-frame references to I_4-frame. Although P_3-frame is before I_4-frame, it is not existed in the same GOP. I_4-frame has no influence on P_3-frame. Taking advantage of this features to construct a new kind of special frame structure named P_0-I-P_1 as shown in the following Fig. 2. Because of the particularity of frame type, we begin with P_3-frame. For example: $P_3I_4P_5$ $P_6I_7P_8$ $P_9I_{10}P_{11}$ $P_{12}I_{13}P_{14}$...... And then we calculate the feature vectors which will change if any data is embeded into the P_0-I frame and I-P_1 frame.

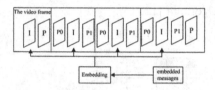

Fig. 2. The video frame is constructed of a new kind of special frame structure named P_0-I-P_1

3   Co-Occurrence Matrix – Markov Model Building

3.1   Co-Occurrence Matrix Model Building

In general, a video can be seen as a series of images. Natural images are usually continuous and smooth, and there is a strong correlation between adjacent pixels in an image. However, once secret data is embedded into it, the correlation between adjacent pixels will reduce since the embedded data is independent of the carrier image. In this paper, the co-occurrence matrix will be applied to excavate the correlation between pixels in the luminance component (Y) of the video frame.

The Gray-level Co-occurrence Matrix (GLCM) is a common method to describe the texture feature through the study of spatial correlation in gray level. The traditional histogram method can only reflect the first-order statistics of data but beyond its correlation. Owing to the texture is formed as grayscale distribution appearing repeatedly in space, therefore, there is a gray relation between two pixels which separated by a distance in the image space (frame), namely spatial correlation characteristics of gray in the image (frame). Haralick et al. [16] have extracted 14 kinds of eigenvalue by using the GLCM, and eight kinds of them are the parameters with strong ability, namely average (a_1) and standard deviation (b_1) of energy, average (a_2) and standard deviation (b_2) of entropy, average (a_3) and standard deviation (b_3) of inertia moment, average (a_4) and standard deviation (b_4) of correlation [17].

Research shows that the sensitiveness of human eyes to the luminance is inferior to the chrominance, thus scholars often choose to embed secret data into luminance and the co-occurrence matrix of this paper is applied in the luminance component (Y) of the video frame.

We make up the co-occurrence matrix model between the 16*16 macroblock of P_0-I frame and the 16*16 macroblock of I-P_1 frame. The specific steps are as follows:

(1) Compress the grayscale of original image in order to reduce calculation amount and conduct histogram specification after quantizing the image level as 16 (the general provision grayscale of image is 16);
(2) Calculate four co-occurrences matrixes with 16*16 macroblock whose distance is 1 and the angles are respectively $0°$, $45°$, $90°$ and $135°$, namely Q_1, Q_2, Q_3, Q_4;
(3) Normalization of co-occurrence matrixes;
(4) Calculate the four texture feature parameters of co-occurrence, including energy, entropy, inertia moment and correlation;
(5) Calculate the average and standard deviation of energy, entropy, inertia moment and correlation as 8 dimensional (8-D for short) texture feature (a_1, b_1, a_2, b_2, a_3, b_3, a_4, b_4).

A lot of mathematical methods are used to describe variables relationship. But those methods widely used in general can be divided into two kinds. One is the similar or related coefficient and the another is distance [18]. This paper will use the Eulerian-distance to describe the correlation between macroblocks.

The Eulerian-norm (Also known as 2-norm or Euclid-norm) is defined as:

$$\|x\|_2 = \left(\sum_{i=1}^{n} |x_i|^2 \right)^{\frac{1}{2}} \quad x = (x_1, x_2, \cdots, x_n)^T \in C^n \tag{1}$$

We assume that a frame of video has M * N pixels, and the pixel value belongs to a collection P. P = {0, 1,..., 255}. First, we divide each frame into 16*16 macroblocks and get 8-D texture characteristics of adjacent 16*16 macroblocks to constitute two vectors ($\mathbf{P_1}$ and $\mathbf{P_2}$). Then we use the Eulerian-distance of vectors to express the correlation between macroblocks. We use the Eulerian-distance to express the correlation of two macroblocks which will reflect the similarity of their comprehensive feature commendably.

The two vectors $\mathbf{P_1}$ and $\mathbf{P_2}$ mentioned above are defined as:

$$P_1 = [a_1^1 \, b_1^1 \, a_2^1 \, b_2^1 \, a_3^1 \, b_3^1 \, a_4^1 \, b_4^1] \tag{2}$$

$$P_2 = [a_1^2 \, b_1^2 \, a_2^2 \, b_2^2 \, a_3^2 \, b_3^2 \, a_4^2 \, b_4^2] \tag{3}$$

Then we calculate the 2-norm of difference between $\mathbf{P_1}$ and $\mathbf{P_2}$, which is the Eulerian-distance, namely $\mathbf{P}$, between macroblocks. P is defined as:

$$\|P\|_2 = \left(\sum |P_1 - P_2|^2 \right)^{\frac{1}{2}} \tag{4}$$

This paper uses the Eulerian-distance of 8-D feature vector to describe the correlation of adjacent 16*16 macroblocks, which means the smaller the Eulerian-distance is, the stronger the correlation of adjacent 16*16 macroblocks will be.

The specific steps to judge the similarity of macroblocks between two adjacent frames are as follows:

(1) Divide the two frames into 16*16 macroblocks;
(2) Set the two adjacent frames as A_0 and B_0 (A_0 is the former and B_0 is the latter). For example of b_0 is the macroblock in B_0-frame which has a same position like a_0-macroblock in A_0-frame. Then we choose the macroblocks at eight directions (right side, upper right, right above, upper left, left side, bottom left, right below and bottom right) around b_0, namely b_1, b_2, b_3, b_4, b_5, b_6, b_7 and b_8. As shown in the Fig. 3.

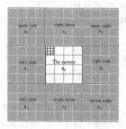

Fig. 3. The position relationship of b_0-macroblock between adjacent macroblocks

(3) We calculate the Eulerian-distances of 8-D feature vector among a_0-macroblock in A_0-frame and b_0-macroblock to b_8-macroblock in B_0-frame. Then we elect a macroblock with minimum Eulerian-distance as the most similar one to a_0-macroblock.

$$P_{min} = \sum_{i=1}^{n=8} \min(P_0, P_1, P_2, P_3, P_4, P_5, P_6, P_7, P_8) \tag{5}$$

As shown in the Fig. 4. Grey b_0-macroblock and the current a_0-macroblock is the same position. Black b_4-macroblock is most relevant to the current a_0-macroblock.

In general, there is a great deal of correlation between adjacent frames and macroblocks which have the same position in different frames which change slowly. Because the objects in the video frames will happen a certain displacement offset which is depend on the moving speed, the correlations among eight adjacent macroblocks in two frames are stronger than macroblocks at the same position when the moving speed is fast. And the specific direction is related to the moving direction of object. Generally speaking, the position direction of the macroblocks which have a stronger correlation between adjacent frames is the moving direction of object (if the correlation in the same position between macroblocks is the strongest, which means the objects in two frames are in a state of rest).

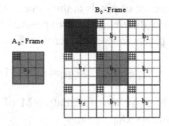

Fig. 4. The position relationship between adjacent frames

3.2 Markov Model Building

Due to the temporal domain characteristic of video frames, there is a strong correlation between pixels and correlation between frames is also one of the most important features in the video. The video can be seen as a frame structure which has a strong correlation between images, adjacent macroblocks and pixels, because it will change after data hiding which is independent of carrier. When we combine these two correlations (between pixels or frames), it makes for revealing macro differences between videos. WEI et al. [19] analyses the correlation between adjacent pixels differences in image and the pixels itself based on mutual data. The function $f(x_1, x_2, \cdots x_n)$ describes the smoothness of the pixel group $G = (x_1, x_2, \cdots x_n)$, functions are defined as follows:

$$f(x_1, x_2, \cdots x_n) = \sum_{i-1}^{n-1} |x_{i-1} - x_i| \tag{6}$$

This discriminant function is representing the spatial correlation of pixels in group G. The value of x_i is different from the concrete method to determine. The more noise in G, the greater the value of the function $f(x)$ will be.

Its result shows that adjacent pixels and differents of adjacent pixels are independent and the adjacent pixels in natural images have similar distribution model. By the same token, there is similar feature between frames in the video or macroblocks at the same position. When we embed secret data into a video, this model will be destroyed, which means we can test the steganography by extracting some sensitive features related to this feature. In this paper, we extract some statistical features, which are the correlation between frames will be changed when we embed secret data, to apply to blind detection and use the pixel difference matrix to calculate the Markov transfer matrix. Finally, we have got good detection effects.

For a given video, each section is consists of GOP. Each GOP is consists of several fames and every frame is also made up by a lot of 16*16 macroblocks. Thus, we combine the co-occurrence matrix model with Markov model which are applied in a special frame structure named P_0-I-P_1. That means, this method will be applied in P_0-I and I-P_1 frame type structure. And then, feature vector is extracted from this two frame structures as the training features.

(1) When we get the B_i-macroblock which is the most similar to A_0-macroblock, we should calculate the difference of pixels at the same position in the A_0-macroblock and B_i-macroblock. These differences are expressed by D_{0j}, and the m(m = 1, ...,16) represent the 16 pixels in every macroblock.

$$D_{0,j} = A_m^0 - B_m^j \qquad (m = 1, \cdots, 16) \tag{7}$$

(2) Then we calculate the Markov transfer matrix M of the same position in two macroblocks using the matrix D_{0j}.

$$M_{n,n} = P_r\left(D_{i+1,j+1} = n | D_{i+1,j+1} = n\right) \tag{8}$$

(3) Next, we will transfer the diagonal values in matrix M into a one-dimensional M^1 and calculate its variance $S(x)$. $E(x)$ is the mean value of M^1.

$$E(x) = \bar{x} = \frac{1}{N}\sum_{i=1}^{N} x_i \quad x_i \in M^1 \tag{9}$$

$$S(x) = \frac{1}{N-1}\sum_{i=1}^{N} (x_i - \bar{x})^2 \quad x_i \in M^1 \tag{10}$$

(4) The combination of this two feature vectors are extracted from the P_0-I frame structure and I-P_1 frame structure as the SVM training features.

3.3 Co-Occurrence Matrix - Markov Model Analysis

The co-occurrence matrix-Markov model is mainly composed of two parts: P_0-I frame type and I-P_1 frame type. We will analyze these two frame types as follows.

1. P_0-I frame type

P_1-frame and I-frame belong to the different GOP. Before the data hiding, P_0-frame and I-frame are correlated strongly, which is expressed the number of same elements in pixel values more than different elements between P_0-frame and I-frame. The bigger values of matrix M^1 are focus on the former elements and the latter elements value is smaller. This uneven distribution of M^1 elements means the return of vector M is volatile, namely the variance is larger.

After embedding the secret data, the correlation between P_0-frame and I-frame turns smaller, which is expressed the number of same elements in pixel value decreases and the different elements increases. Thus, the bigger value will also decrease and the smaller value will increase in the matrix M^1, and it will be more uniform which means the undulation of returned vector M will become smaller, namely the variance is smaller.

2. I-P_1 frame type

In the I-P_1 frame type, the P_1-frame and I-frame belong to the same GOP and the I-frame is a reference for P_1-frame. That means each I-frame has a high correlation with the P_1-frame after it. Thus, after embedding the secret data, the same elements in I-frame and P-frame will increase and the different one will decrease. Similarly we can get the returned vector M^1 which will become more volatile.

4 The Experimental Results and Experimental Analysis

4.1 Experimental Procedure

In the experiment, We used the reference software JM of H.264/AVC video standard system for test video files as the platform, and we maked use of the standard test sequences as the video test sequences, such as akiyo, deadline and yuv video, etc. JM is a free open source project to provide complete solutions of recording, transferring and streaming audio and video. The digital (1,2,3,...) represents the frame sequence number of the each sequence image. The frame structure is discussed in this paper: $I_1P_2P_3I_4P_5P_6I_7P_8P_9I_{10}P_{11}P_{12}I_{13}\ldots$

This paper selects 30 video sequences from the different types of online download. It contains the video content which changes quickly and slowly, such as football test video. Each video interval is divided into a plurality of video clips by the 30 frames for an interval. A total of more than 8000 video frames, of which more than 5000 frames as training frames, 3000 frames as the test frame. The test video is embed by the most common steganography algorithm which pixels are embedded about 1 bit secret data according to the parity in the compressed domain. The embedding rate is 10 %, 15 % which avoids the loss of confidential data. And then the video sequence is obtained as the vector sequence and makes an classification experiments. In this paper, SVM is used as the classifier, the kernel function is RBF function.

The embedding rate is 10 % (Embed 633 bits of secret data per frame):

$$((352 * 288)/16) * 10\% = 633.6 \tag{11}$$

The embedding rate is 15 % (Embed 950 bits of secret data per frame):

$$((352 * 288)/16) * 15\% = 950.4 \tag{12}$$

4.2 The Experimental Result

The experimental results which the video with P_0-I and I-P_1 frame type is the video before and after embedding, for example, are shown in the following Fig. 5.

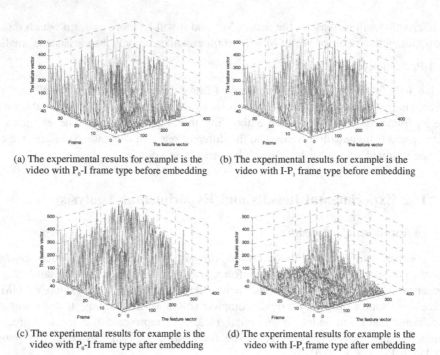

(a) The experimental results for example is the
video with P_0-I frame type before embedding

(b) The experimental results for example is the
video with I-P_1 frame type before embedding

(c) The experimental results for example is the
video with P_0-I frame type after embedding

(d) The experimental results for example is the
video with I-P_1 frame type after embedding

Fig. 5. The experimental results for example is the video with P_0-I frame type and I-P_1 frame type before and after embedding

The experimental result, for example, is the rate of correct detection which the stego video and original video in different embedding ratios. In this case, the features in this paper which consider both spatial redundancy and temporal redundancy perform better than the Rana et al. [20] features which just consider spatial redundancy between adjacent pixels. We compared the two algorithms as shown in the following Table 1.

In this paper, experiments show that the proposed steganalysis method in the 0.15 embedding rate, the correct probabilities for detection of stego video is close to 100 %. With the increase of embedding rate, the correct detection probability is increasing gradually. We compared the two algorithms, the one in this paper and the other Rana et al. [20] features for test video. The final result as shown in the following Table 2. Compared with the image, the video has a large amount of data, a lot of image methods cannot be directly applied to the video detection methods. Therefore, we should combine with intra and inter frame correlations are the most effective way. The algorithm proposed in this paper has a high detection probability, and then it calls the promising effect and deserves additional studies.

Table 1. The detection probability which the stego video and original video in different embedding ratios using different algorithms

Video classification	Embedding ratio	Stego in proposed	Stego in Rana et al.	Original in proposed	Original in Rana et al.
Harbour	0.15	100 %	86.35 %	91.57 %	73.64 %
	0.1	94.84 %	76.81 %	87.62 %	62.24 %
Container	0.15	98.97 %	84.16 %	90.52 %	67.28 %
	0.1	95.91 %	78.06 %	89.79 %	66.98 %
Football	0.15	95.92 %	77.23 %	82.47 %	61.59 %
	0.1	84.53 %	76.75 %	76.38 %	55.36 %
Husky	0.15	100 %	82.57 %	91.75 %	72.61 %
	0.1	97.97 %	80.49 %	89.69 %	71.35 %
Coastguard	0.15	100 %	87.33 %	93.81 %	70.84 %
	0.1	95.95 %	78.63 %	89.89 %	65.55 %
Hall-monitor	0.15	97.96 %	83.65 %	86.41 %	67.32 %
	0.1	92.63 %	75.82 %	83.15 %	63.49 %

Table 2. The final result of the detection probability which the stego video and original video in different embedding ratios using different algorithms

	Embedding ratio	Proposed	Rana et al.
Stego video	0.15	98.80 %	72.62 %
	0.1	94.63 %	70.02 %
Original video	0.15	89.07 %	65.53 %
	0.1	86.43 %	62.79 %

5 Conclusion

In this paper, a new steganalysis scheme based on features of the original domain was proposed utilizing the temporal and the spatial correlation among video frames. This paper presents an idea combined correlation between the temporal and the spatial between adjacent frames and pixels. Then co-occurrence feature vector matrix is obtained by the Gray-level Co-occurrence Matrix as the correlation between macroblocks to establish Markov model of inter-frames, we try to determine whether any secret data is embedded in the video according to the changes of correlation between video frames. Compared with the image, a lot of image steganalysis cannot be directly applied to the video algorithms. Combining with intra and inter frame correlation is the most effective way in video steganalysis. The experimental results show that the algorithm proposed in this paper has a high detection probability, and then it can conduct additional studies. It is necessary to make mainly further improvement on the correct detection rate and optimize this algorithm mainly focused on the low bit rate of videos.

References

1. XU, C.Y.: Research on video steganography and video steganalysis. The PLA Information Engineering University (2009)
2. Xu, C.Y., Ping, X.J.: Video steganalysis based on spatial-temporal correlation. J. Image Graph. **1006–8961**, 1331–1337 (2010)
3. Liu, B., Liu, F.L., Yang, C.F.: Video steganalysis scheme based on inter-frame collusion. J. Commun. **1000-436X**, 41–49 (2009)
4. Sun, Y.F., LIU, F.L.: Digital video steganalysis algorithm based on motion estimation. Pattern Recogn. Artif. Intell. **1003–6059**, 759–765 (2010)
5. Deng, Q.L., Lin, J.J.: Image steganalysis based on co-ocurrence matrix. Microcomput. Inf. **1008–0570**, 6–8 (2009)
6. Burges, C.: A tutorial on support vector machines for pattern recognition. Data Min. Knowl. Disc. **2**, 121–167 (1998)
7. Tian, L.H., Zheng, N.N., Xue, J.R., et al.: A CAVLC-based blind watermarking method for H.264/AVC compressed video. In: Asia-Pacific Services Computing Conference. APSCC, Yilan, Taiwan, pp. 1295–1299 (2008)
8. Wang, B., Feng, J.C.: A chaos-based steganography algorithm for H.264 standard video sequences. In: International Conference on Communications, Circuits and Systems, ICCCAS 2008, Xiamen, China, pp. 750–753 (2008)
9. Su, P., Li, M., Chen, I.: A content-adaptive digital watermarking scheme in H.264/AVC compressed videos. In: International Conference on Intelligent Information Hiding and Multimedia Signal Processing, IIHMSP, Harbin, China, pp. 849–852 (2008)
10. Bhattacharya, S., Chattopadhyay, T., Pal, A.: A survey on different video watermarking techniques and comparative analysis with reference to H.264/AVC. In: 2006 ISCE International Symposium on Consumer Electronics, Samos Island, Greece, pp. 1–6 (2006)
11. Ma, X.J., Li, Zh.T., Tu, H., Zhang, B.C.: A data hiding algorithm for H.264/AVC video streams without intra-frame distortion drift. IEEE Trans. Circ. Syst. Video Technol. **20**(4), 1320–1330 (2010)
12. Fridrich, J., Goljan, M., Hogea, D.: Steganalysis of JPEG images breaking the F5 algorithm. In: Petitcolas, F.A.P. (ed.) IH 2002. LNCS, vol. 2578, pp. 310–323. Springer, Heidelberg (2002)
13. Fridrich, J., Goljan, M., Hogea, D.: Attacking the outguess. In: Proceeding of ACM Workshop on Multimedia and Security, pp. 967–982 (2002)
14. Wang, M., Fan, K., Yue, B., et al.: A content protection scheme for H.264-based video sequence. In: 8th ACIS International Conference on Software Engineering, Artificial Intelligence, Networking and Parallel/Distributed Computing, SNPD 2007, Qingdao, China, pp. 388–393 (2007)
15. Hartung, F., Girod, B.: Digital watermarking of MPEG-2 coded video in the bitstream domain. In: IEEE International Conference on the Acoustics, Speech, and Signal Processing, ICASSP, Munich, Germany, pp. 2621–2624 (1997)
16. Lu, C.S., Chung, P.C., Chen, C.F.: Unsupervised texture segmentation via wavelet transform. Pattern Recogn. **30**(0031–3203), 729–742 (1997)
17. Guo, D.J., Song, Z.C.: A study on texture image classifying based on gray-level co-ocurrence matrix. For. Mach. Woodwork. Equip. **1001–4462**, 21–23 (2005)
18. Cao, J.N., Li, D.R., Guan, Z.Q.: Study on approach of detection for video image basedon decomposable markov network. Acta Optica Sinica **0253–2239**, 312–318 (2005)

19. WEI, S.Y., NING, C., GAO, Y.X.: Biomimetic gait recognition based on motion contours wavelets analysis and mutual information. In: 2010 3rd International Congress on Image and Signal Processing, 1, pp. 404–408. IEEE, Piscataway (2010)
20. Rana, V., Mishra, R., Bora, P.K., Kashyap, S.: Novel scheme of video steganalysis for detecting antipodal watermarks. In: Proceedings IEEE Region 10 Conference-TENCON, pp. 1–5 (2008)

Multi AUV Intelligent Autonomous Learning Mechanism Based on QPSO Algorithm

Jian-Jun Li[1,3(✉)], Ru-Bo Zhang[1,2], and Yu Yang[3]

[1] College of Computer Science and Technology, Harbin Engineering University,
Harbin 150001, China
[2] College of Electromechanical and Information Engineering,
Dalian Nationalities University, Dalian 116600, Liaoning, China
[3] School of Computer and Information Engineering,
Harbin University of Commerce, Harbin 150028, China
517718768@qq.com

Abstract. Intelligent autonomous learning is a hotspot of research on multi-agent system research, because of the PSO algorithm and multiple AUV system has loose coupling intelligent group structure, multiple AUV system flexibility and openness, multiple AUV system can be combined with swarm intelligence algorithm. So, using the swarm intelligence research of particle swarm optimization (pso) autonomous intelligent AUV underwater robot autonomous learning mechanism, can greatly improve the performance of many of AUV system. Quantum PSO algorithm and PSO algorithm based on the experimental verification, to prove QPSO algorithm has the certain superiority in the AUV more autonomous learning, in the process of each iteration self-learning optimal state, not only improve the efficiency of the algorithm, but also for the current to search for the optimal state, improve the accuracy of search algorithm.

Keywords: QPSO · Multi autonomous underwater vehicle · Intelligent agent · Autonomous learning mechanism

1 Introduction

Intelligent system from the beginning of American spacecraft, NASA's deep space 1 [1, 2], the control of vehicle autonomous spacecraft [3], the Mars rover [4]. Intelligent system has been advanced applications in aerospace field, at the same time, in recent years, autonomous underwater robot (Autonomous Underwater Vehicle, AUV) of intelligent system has been the development of the [5, 6].

Autonomous underwater robot is a kind of AUV used for a wide range of underwater robot, can participate in various tasks including underwater survey, check the deep structure of the underwater detection of marine characteristics, seabed mapping, the laying of submarine cables, searching for the downed aircraft, and naval mine etc. AUV autonomous behavior can be based on intelligent autonomous learning to do. AUV intelligent autonomous learning through intelligent frame of corresponding consideration of intelligent behavior reasoning. The use of intelligent autonomous agent framework, through the sensors to perceive the environment and take action to help the AUV intelligent decision-making, implementation of task planning and task planning.

© Springer International Publishing Switzerland 2015
D.-S. Huang et al. (Eds.): ICIC 2015, Part II, LNCS 9226, pp. 60–67, 2015.
DOI: 10.1007/978-3-319-22186-1_6

The flexibility and openness of the multi AUV system, the multi AUV system combination and swarm intelligence algorithm. The intelligent community structure, PSO algorithm and multi AUV system are loosely coupled and can be a AUV as one or a group of particles. AUV as an agent not only can exchange information with the best in the current population of AUV, can also study independently in several iterations, to complete the accumulation of knowledge, so as to improve the ability to solve problems, implementation of swarm intelligent optimization [7, 8].

2 PSO Algorithm

2.1 The Definition of PSO

Particle swarm optimization algorithm, called particle swarm optimization algorithm (Partical Swarm Optimization), abbreviated as PSO, is an evolutionary computation technique (Evolutionary Computation), proposed in 1995 by Dr. Eberhart and Dr. Kennedy from the behavior of birds of prey. Compared with the genetic algorithm, PSO has the advantage of simple and easy to realize and there is no need to adjust many parameters. Has been widely used in function optimization, neural network training, fuzzy system control and other genetic algorithm applications [9–12].

Find the two best values, the particle updates its velocity and position according to the following formula:

$$V_{id}^{k+1} = \omega V_{id}^k + c_1 r_1 (P_{id}^k - X_{id}^k) + c_2 r_2 (P_{gd}^k - X_{id}^k) \tag{1}$$

$$X_{id}^{k+1} = X_{id}^k + V_{id}^{k+1} \tag{2}$$

Among them ω as the inertia weight $d = 1, 2, \cdots, D; i = 1, 2, \cdots, n; k$ for the current iteration number; V_{id} for the particle velocity;

c_1 and c_2 For non negative constant, known as the learning factor; r_1 and r_2 For the random distribution between the [0,1]. To prevent blind search particles, generally recommend its position and speed is limited to a certain range of $[-X_{max}, X_{max}]$, $[-V_{max}, V_{max}]$.

2.2 Optimization of Multi AUV System PSO

From Eqs. (1) and (2) analysis, each dimension of the search space is independent of each other, PSO can be simplified to one dimension. In order to simplify the calculation, in addition to the assumption that population of I particles, the particles remain intact, individual particle best position and the whole group optimal position remains unchanged, as p_α and g_α So, $\beta_1 = c_1 r_1$, $\beta_2 = c_2 r_2$. Then (1) and (2) can be simplified to

$$V(k + 1) = \omega V(k) + \beta_1(p_d - x(k)) + \beta_2(g_d - x(k)) \tag{3}$$

$$X(k + 1) = X(k) + V(k + 1) \tag{4}$$

Through the establishment of a AUV micro grid operation environment, AUV can dynamically select the appropriate. In the system of AUV AUV through the mechanism of autonomous learning, improving the ability of control and autonomy of each AUV.

Micro grid environment of independent study for mS, Autonomous Learning AUV $a_{i,j}$ Be located (i,j), $a_{i,j} = (a_1, a_2,, \cdots, a_n)(i,j = 1, 2, \cdots, mS_{size})$. Micro grid size is $mS_{size} \times mS_{size}$. For the neighborhood to help AUV learning

$$S_{i,j} = \{a_{ij1}, a_{ij2}, a_{i1j}, a_{i2j}, a_{i1j1}, a_{i1j2}, a_{i2j1}, a_{i2j2}\} \tag{5}$$

To adapt to the value of the comparison according to the self-learning mechanism between neighborhoods, have the best fitness value AUV.

3 QPSO Algorithm

3.1 QPSO Algorithm Definition

In the basic evolutionary formula particles $X_{i,j}(t+1) = P_{i,j}(t) \pm \frac{L_{i,j}(t)}{2} \cdot \ln\left[1/u_{i,j}(t)\right]$ $u_{i,j}(t) \sim U(0,1)$, How to effectively control $L_{i,j}(t)$, to make it converge to 0, so that the individual particle best position P_i, to get the global best position of G group. Through the following two ways [13–15].

The first method of converting $L_{i,j}(t)$.

$$L_{i,j}(t) = 2a \cdot |P_{i,j}(t) - X_{i,j}(t)| \tag{6}$$

The evolution equation is transformed into particle

$$X_{i,j}(t+1) = P_{i,j}(t) \pm a \cdot |P_{i,j}(t) - X_{i,j}(t)| \cdot \ln\left[1/u_{i,j}(t)\right] u_{i,j}(t) \sim U(0,1) \tag{7}$$

The second kind of method is introduced into the algorithm mean best position $C(t)$.

$$C(t) = (C_1(t), C_2(t), \cdots, C_n(t))$$
$$= \frac{1}{N}\sum_{i=1}^{N} P_i(t) = \left(\frac{1}{N}\sum_{i=1}^{N} P_{i,1}(t), \frac{1}{N}\sum_{i=1}^{N} P_{i,2}(t), \cdots, \frac{1}{N}\sum_{i=1}^{N} P_{i,1}(t)\right) \tag{8}$$

$L_{i,j}(t)$ conversion

$$L_{i,j}(t) = 2a \cdot |C_j(t) - X_{i,j}(t)| \tag{9}$$

The evolution equation is transformed into particle

$$X_{i,j}(t+1) = P_{i,j}(t) \pm a \cdot \left| C_j(t) - X_{i,j}(t) \right| \cdot \ln\left[{1}/{u_{i,j}(t)} \right] u_{i,j}(t) \sim U(0,1) \tag{10}$$

In the formula a is the expansion - contraction factor. The QPSO algorithm is applied to the above formula of the particle swarm algorithm, called quantum behaved particle swarm optimization algorithm.

3.2   QPSO Algorithm Flow

In the M dimension of the search space, QPSO algorithm is composed by a solution to the problem of particle group $X = \{X_1, X_2, \ldots, X_N\}$, At the time of t, the i particle position $X_i(t) = [X_{i,1}(t), X_{i,2}(t), \ldots, X_{i,M}(t)]$, $i = 1, 2, \ldots, n$, Particle velocity vector. The individual best position $P_i(t) = [P_{i,1}(t), P_{i,2}(t), \ldots, P_{i,M}(t)]$, The global best position for the group $G_i(t) = [G_1(t), G_2(t), \ldots, G_M(t)]$, and $P_g(t) = G(t)$, Where g is the best position of the particle. In the global search, $g \in \{1, 2, \ldots, N\}$.

The best position of individual particle i $pbest$ by the formula (11) to determine the:

$$P_i(t) = \begin{cases} X_i(t) & \text{if} \quad f[X_i(t)] < f[P_i(t-1)] \\ P_i(t-1) & \text{if} \quad f[X_i(t)] \geq f[P_i(t_i-1)] \end{cases} \tag{11}$$

The group of global best position a By the formulas (12) and (13) to determine the:

$$g = \arg \min_{1 \leq i \leq N} \{f[p_i(t)]\} \tag{12}$$

$$G(t) = P_g(t) \tag{13}$$

Updating formula of QPSO algorithm for particle position:

$$P_{i,j}(t) = \varphi_j(t) \cdot p_{i,j}(t) + \left[1 - \varphi_j(t)\right] \cdot G_j(t) \, \varphi_j(t) \sim U(0,1) \tag{14}$$

$$X_{i,j}(t+1) = P_{i,j}(t) \pm a \cdot \left| C_j(t) - X_{i,j}(t) \right| \cdot \ln\left[{1}/{u_{i,j}(t)} \right] u_{i,j}(t) \sim U(0,1) \tag{15}$$

4   Simulation Experiment

4.1   The Experimental Parameters

The main parameters of the QPSO algorithm is as follows: the initial population is evenly distributed in the $[-100, 100]^M$, M said the dimension of test functions, the number of iterations for the L = 50, the results of each integer value. Learning factor

Table 1. Test function.

Test functio	Range	Explain
$f_1(x) = \sin^6(5\pi x)$	[0,1]	5 the global optimal value equivalence
$f_2(x) = \sin^6\left(5\pi\left(x^{3/4} - 0.05\right)\right)$	[0,1]	5 uneven distribution of the global optimal value
$f_3(x, y) = \sum_{i=1}^{4} f_i(x)^2$	[0,1]	1 the global optimal values and 6 local optimal value

QPSO algorithm $C_1 = C_2 = 2$, ω is set to 0.8, Expansion —contraction factor α in three different interval [1.0,0.6], [0.8,0.6], [0.6,0.6] with the increase in the number of iterations decreases linearly. Through the simulation of MATLAB PSO algorithm and QPSO algorithm in the optimization results of different test function (Table 1).

5 Experimental Verification

Different groups formed in the QPSO algorithm, the optimal value of the interaction between each other in different range of particles, after several iterations of autonomous learning, to complete the updating of the information and the accumulation of knowledge, so as to improve the AUV autonomous learning ability, to achieve global optimization AUV. Figures 1, 2 and 3 shows the 3 test functions optimization estimation of the convergence speed of PSO algorithm QPSO algorithm. So we can see that the QPSO algorithm is better than PSO algorithm in autonomous learning advantage.

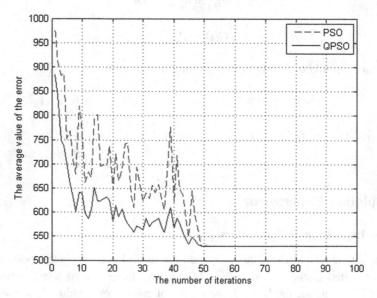

Fig. 1. Function $f_1(x) = \sin^6(5\pi x)$ the test results.

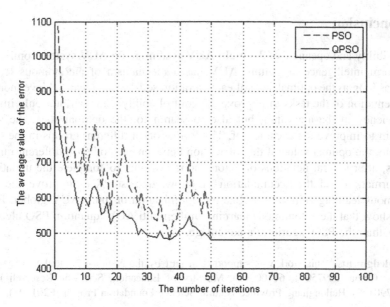

Fig. 2. Function $f_2(x) = \sin^6(5\pi(x^{3/4} - 0.05))$ the test results.

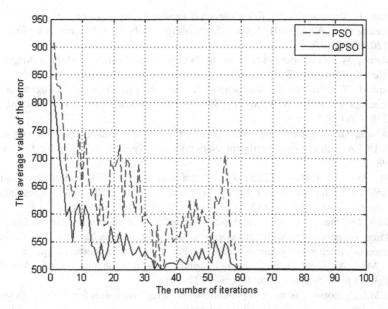

Fig. 3. Function $f_3(x, y) = \sum_{i=1}^{4} f_i(x)^2$ the test results.

6   Conclusion

The flexibility and openness of the multi AUV system, multi AUV system combination and swarm intelligence algorithm. AUV has a mechanism of autonomous learning ability, as long as the optimal state of each iteration self-learning, can achieve their own implementation of the tasks to improve the control ability. This will not only improve the efficiency of the algorithm, but also to search for the optimal state, the search algorithm to improve the accuracy of. The formation of different groups in the QPSO algorithm, the optimal value of the interaction between each other in different range of particles, after several iterations of autonomous learning, to complete the updating of the information and the accumulation of knowledge, so as to improve the AUV autonomous learning ability, to realize the optimization of multi AUV global. Experiments show that the autonomous learning in the multi AUV, quantum PSO algorithm is better than the PSO algorithm's advantage.

Acknowledgments. This work was supported in part by the National Natural Science Foundation of China (60975071, 61100005), Ministry of Education, Scientific Research Project (13YJA790123), Heilongjiang Province Natural Science Foundation Project (F201425).

References

1. Bresina, J., Washington, R.: Robustness via run-time adaptation of contingent plans. In: Proceedings of AAAI-2001 Spring Symposium: Robust Autonomy, Stanford, USA, pp. 120–128 (2010)
2. Matthews, W.: Computer takes over to land stricken aircraft. C4ISR – Magazine of Net-Centric Warfare, 225–231 (2008)
3. Bresina, J., Washington, R.: Robustness via run-time adaptation of contingent plans. In: Proceedings of AAAI-2001 Spring Symposium: Robust Autonomy, Stanford, USA, pp. 248–256 (2001)
4. Blackburn, M.: Autonomy and intelligence – a question of definitions. In: Proceedings of ONR-UCLA Autonomous Intelligent Networks and Systems Symposium, Los Angeles, pp. 98–105 (2002)
5. Li, H., Popa, A., Thibault, C., Seto,M.: A software framework for multi-agent control of multiple AUVs for underwater MCM. In: Proceedings of IEEE Autonomous Intelligent Systems Conference, Povoa de Varzim, Portugal, pp. 189–196 (2010)
6. Hudson, J.: Three-dimensional path-planning for a communications and navigations aid working cooperatively with autonomous underwater vehicles. In: Proceedings of Autonomous Intelligent Systems Conference, Povoa de Varzim, Portugal, pp. 12–20 (2011)
7. Seto, M.L.: An agent to optimally re-distribute control in an under actuated AUV. Int. J. Intell. Defence Support Syst. **4**(1), 3–19 (2010)
8. Seto, M.L.: Autonomous mission-planning with energy constraints for AUVs with side scan sonars. In: Proceedings of IEEE International Conference Machine Learning and Applications, pp. 156–165 (2011)
9. Phong, P.D., Ho, N.C., Thuy, N.T.: Multiobjective particle swarm optimization algorithm and its application to the fuzzy rule based classifier design problem with the Order Based semantics of linguistic terms. In: Proceeding of the 10th IEEE RIVF International Conference on Computing and Communication Technologies, pp. 12–17 (2013)

10. Sanjeevi, S.G., Nikhila, A.N., Khan, T., Sumathi, G.: Comparison of Hybrid PSO-SA Algorithm and Genetic Algorithm for Classification. Comput. Eng. Intell. Syst. **2**(3), 37–45 (2012)
11. Sun, J., Feng, B., Xu, W.B.: Particle swarm optimization with particles having quantum behavior. In: Proceedings of Congress on Evolutionary Computation, pp. 325–331 (2004)
12. Sun, J., Xu, W.B., Feng, B.: A global search strategy of quantum behaved particle swarm optimization. In: Proceedings of IEEE Conference on Cybernetics and Intelligent Systems, pp. 111–116 (2004)
13. Ravi, K., Rajaram, M.: Optimal location of FACTS devices using improved particle swarm optimization. Electrical Power and Energy Syst. **49**(2), 333–338 (2013)
14. dos Santos Coelho, L.: A quantum particle swarm optimizer with chaotic mutation operator. Chaos, Solitons Fractals **37**(5), 1409–1418 (2008)
15. Kundu, R., Das, S., Mukherjee, R., Debchoudhury, S.: An improved particle swarm optimizer with difference mean based perturbation. NeuroComput. **129**(3), 315–333 (2014)

A Directional Evolution Control Model for Network

Gang Luo and Yawei Zhao[(✉)]

College of Engineer and Information Technology,
University of Chinese Academy of Sciences,
Yu Quan Road 19A, Shi Jing Shan District, Beijing 100049, China
zhaoyw@ucas.ac.cn

Abstract. In order to control the evolution direction of network to achieve certain goals, we proposed Directional Evolution Control Model for Network Based on Percolation Coefficient (PC-DECMN). Firstly, after the analysis of the spreading behavior based on SI model, we found that the influence scope was only related to the topology structure of network and initial source of infection. So the definition of percolation coefficient to measure the influence scope was proposed based on the transfer closure of adjacency matrix. Then, PC-DECMN was established based on the definition of percolation coefficient. Simulation experiment using the guarantee data of certain financial institution shows that, PC-DECMN can effectively control the evolution of guarantee network so that it can resist the risk better, compared with the natural evolved guarantee network.

Keywords: Network evolution · Percolation coefficient · Spreading · Topology structure

1 Introduction

The structure of network in reality is usually changing over time, a new node or edge may be joined or removed in different time. In many scenarios, we need network to evolve towards a particular direction to achieve certain goals. For example, when a new infectious disease comes, the government would take measures to prevent further spreading, in fact, the government's isolation measure is exactly one kind of human intervention to network evolution; Another example, in the financial sector, what banks worry about most is that, if a company cannot repay the loan, the guarantee cannot repay for it neither, and this guarantee itself also have applied loans, at this time, the risks would spread along the chain of guarantee relationships, eventually it may result in a large number of enterprises bankruptcy, and the banks will suffer huge losses. Therefore, in addition to do good credit work before making loans, banks should also consider the structural problems of guarantee network formed by guarantee relationships; a good guarantee network structure itself is able to inhibit the rapid propagation of risks.

This paper is organized as follows: Sect. 2 describes the study of the basic theory of complex networks, network evolution and network spreading; Sect. 3 firstly describes the concept of percolation coefficient, and then the directional evolution control model

© Springer International Publishing Switzerland 2015
D.-S. Huang et al. (Eds.): ICIC 2015, Part II, LNCS 9226, pp. 68–79, 2015.
DOI: 10.1007/978-3-319-22186-1_7

for network was proposed, after that we analyzed a typical case to show how to control the evolution direction; Sect. 4 verifies the control model through performing simulation experiment of guarantee network in the financial field; Sect. 5 is the conclusion.

2 Related Works

The random graph theory raised by Erdős and Rényi was regarded as the first systematic research of complex network in mathematics [1]. After that Watts and Strogatz revealed the small world character of complex network in their paper, which corrected the mistake that the structure of complex network totally random [2]. Barabási and Albert revealed another important characteristic—scale free [3] in 1999. These two papers laid the foundation of complex network.

The main task of network evolution research is to study the change regulation of topological structure over time. Leskovec and Faloutsos studied a large quantity of networks and found that, the density of edges increased faster than linear increment, while the average distance between nodes decrease over time [4]. Li and Chen proposed a novel evolving network model with the new concept of local-world connectivity, which exists in many physical complex networks. They found that this local-world evolving network model can maintain the robustness of scale-free networks and can improve the network reliance against intentional attacks, which is the inherent fragility of most scale-free networks [5]. Barrat et al. presented a general model for the growth of weighted networks in which the structural growth is coupled with the edges' weight dynamical evolution. The model generates weighted graphs exhibiting the statistical properties observed in several real-world systems [6]. There are still a lot of researches related with community structure. Palla et al. tracked the community in different moment and divided the possible events of community evolution into 6 categories: growth, contraction, merging, splitting, birth and death [7]. Beyond the studies above, visualization of network evolution is also popular. Ahn et al. focused on the visualization of network evolution and described a taxonomy tasks for visual analysis of network evolution [8].

SI (Susceptible-Infected) Model [9] is the basic epidemic model. In SI model, the probability of individual being infected when it contacts with an infectious source is β, the ratio of susceptible individuals at time t is $s(t)$, while the infected individuals' is $i(t)$. They satisfy $s(t) + i(t) = 1$ and the expression of $i(t)$ is as follows:

$$i(t) = \frac{i_0 e^{\beta t}}{1 - i_0 + i_0 e^{\beta t}}, i_0 = (0) \tag{1}$$

which corresponds to Logistic curve.

Epidemic spreading has been well studied some time. Lü et al. proposed a model to emphasize the essential difference between information spreading and epidemic spreading, found that the spreading effectiveness can be sharply enhanced by introducing a little randomness into the regular structure [10]. Wang et al. proposed a general epidemic threshold condition that applies to arbitrary graphs and proved that, under reasonable approximations, the epidemic threshold for a network is closely related to the largest eigenvalue of its adjacency matrix [11]. Ganesh et al. identified topological properties of the graph that determine the persistence of epidemics. They showed that if the ratio of cure to infection rates is smaller than the spectral radius of the graph, then the mean epidemic lifetime is with the same order of $log(n)$, where n is the number of nodes [12]. Garas et al. modeled the spreading of a crisis by constructing a global economic network and applying the Susceptible-Infected-Recovered (SIR) epidemic model with a variable probability of infection, surprisingly found that countries with much lower GDP like Belgium are able to initiate a global crisis. Using the k-shell decomposition method, they obtained a measure of "centrality" as a spreader of each country in the economic network and found the 12 most central countries that are the most likely to spread a crisis globally according to the measure [13]. Wang et al. proposed an improved CSR model for rumor spreading in mobile social networks. In this model, people's accepting threshold for rumor was taken into account. The multi-agent simulation results indicated that the information spreading process is sensitively dependent on initial conditions [14]. The optimization problem of selecting the most influential nodes is NP-hard. Kempe and Tardos provided a provable approximation guarantees based on greedy strategy for efficient algorithms. Computational experiments on large collabo-ration networks showed that the approximation algorithms significantly out-perform node selection heuristics based on the well-studied notions of degree centrality and distance centrality from the field of social networks [15].

The statistical properties of SIR epidemics in random networks have been under-stood for some time, but the explicit dynamics have been understood mainly through simulation. To address this shortcoming, Volz proposed a system of nonlinear ordinary differential equations to model SIR dynamics in random networks. He also provided a simple means of tracking the evolution of the degree distribution among susceptible or infected [16].

It is known that almost all the studies focused on fields like social network, epi-demic spreading, communication network, economic network and Internet, according to the paragraphs above. These networks share something in common—the evolution of network can rarely be controlled by human beings. And to the best of our knowl-edge, there is scarcely any researches try to control the evolution process of network according to the topological structure or spreading characteristics of network. However, in some cases, we indeed need to control the evolution of network, which is exactly what this paper aims to research.

3   Directional Evolution Control Model for Network

3.1   The Representation of Network

Networks are usually represented as $N = (V, E)$, where V represents the node-sets of network, $V = \{a_1, a_2, ..., a_n\}$, while E represents the edge-sets of network, $E = \{< a_i, aj > | a_i, a_j \in V\}$. $< a_i, a_j >$ means that there exists an edge pointed from node a_i to a_j, and it can be shortened as a_{ij}. All the edges formed an n by n matrix M, $M = (a_{ij})_{n \times n}$, which is the adjacency matrix of network N.

The adjacency network defines the directed connections between nodes in network, while sometimes we need to get all the connected relationships no matter nodes are directly connected or not, so the reachability matrix is needed. We can get the reachability matrix by evaluating the transfer closure of adjacency matrix:

$$M_t = M + M^2 + M^3 + \ldots \tag{2}$$

The operator "+" in formula (2) is logical plus, which follows the operation law: $0 + 0 = 1, 0 + 1 = 1, 1 + 0 = 1$, and $1 + 1 = 1$. There is a large quantity of algorithms to evaluating the transfer closure of adjacency matrix, whereas the Warshall algorithm [17] is the classical one between them.

When the information spread to another node, it is equivalent to the change of state for the node, so as diseases and energy. So for the convenience to demonstrate, we describe the spreading of states in network to represent the spreading of information, disease and energy etc. We say a node been infected when another node spread its state to it.

3.2 Percolation Coefficient

In reality, problems like how to evaluate the spreading scope when one node has been infected is always being concerned. For example, in the social network, rumors spread from some influential people compared with ordinaries usually result in different spreading scope. In order to leverage the spreading scope, we introduce a new notation: percolation density $\rho_{i,\infty}$, which represents the ratio of infected nodes when the ith node is the source of infection. It is easy to know that, nodes never fail to be infected as long as the source of infection is reachable to them according to SI model. So we can demonstrate $\rho_{i,\infty}$ with the help of transfer closure of adjacency matrix:

$$\rho_{i,\infty} = \frac{1}{n}(\sigma_i[M_t] + E_i) \cdot (\sigma_i[M_t] + E_i)^T \tag{3}$$

where $\sigma_i[M_t]$ denotes the ith row of matrix M_t, the plus operation of corresponding elements in $\sigma_i[M_t] + E_i$ is logical plus. The following is the definition of E_i:

$$E_i = [e_1 \quad e_2 \quad \cdots \quad e_n] \tag{4}$$

There is only one element is none zero in E_i, which means

$$e_j = \begin{cases} 0, j \neq i \\ 1, j = i \end{cases}, j = 1, 2, \cdots, n \tag{5}$$

We define the percolation coefficient of network N as follows based on the definition of percolation density:

$$c(N) = \sum_{i=1}^{n} \rho_{i,\infty} \cdot P(S_{i,0} = 1) \tag{6}$$

where $P(S_{i,\infty} = 1)$ represent the probability that the state of the ith node is 1 at time 0, in other words, the probability of the ith node being the infectious source.

The algorithm description is given below according to the definition of percolation coefficient:

Input: The transfer closure of network adjacency matrix M_t, and the probability distribution P for every node becoming the infectious source.

Output: the percolation coefficient of network.

Pseudo-code:

```
Calculate_Percolation_Coefficient (Mt, P)
  result := 0
  for i := 1 to n do
    density := 0
    for j := 1 to n do
      if j = i or Mt[i, j] != 0 then
        density := density + 1
    result := result + density * P[i] / n
  return result
```

In general, the probability of different node becoming the infectious source is different. For example, in SNS, a rigorous man may not be the spreading source of rumor. $P(S_{i,0} = 1)$ is exactly the variable to measure the probability of the initial infectious source. But for convenience of study, we made the probability of every node becoming infectious source equal, thus we got $P(S_{i,0} = 1) = 1/n$. Plugging it along with the definition of $\rho_{i,\infty}$ into formula (6), we got the percolation coefficient of network as follows:

$$c(N) = \frac{1}{n^2} \sum_{i=1}^{n} (\sigma_i[M_t] + E_i) \cdot (\sigma_i[M_t] + E_i)^T \tag{7}$$

From the formula (7), we know that, the percolation coefficient only depends on the topological structure of the whole network when the probability of every node being the source of infection is equal. The network in reality is usually complex, but when we dive into the local, there are only a few basic structures. We will analyze five typical structures below and for convenience to calculate, formula (7) will be used.

The structures shown in Fig. 1 are islands structure (Fig. 1a), star divergent structure (Fig. 1b), star convergent structure (Fig. 1c), ring structure (Fig. 1d), and chain structure (Fig. 1d), respectively. We assume there are n nodes in all of the structures. Every circle represents a node and filled circles represent nodes have been infected; the spreading direction is identical to the direction of edge. Considering that the undirected network is the special case of directed network, so we talk about the directed one for more universality.

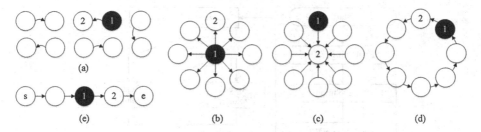

Fig. 1. Five local network structures

Most of the nodes are not connected to each other in the islands structure (Fig. 1a), so states can only spread in a very small connected sub-network. It is easy to know that the percolation coefficient of this structure is $c(N) = 1/n^2 \cdot (n/2 + (n/2) \cdot 2) = 3/2n$.

In the star divergent structure (Fig. 1b), the periphery nodes are certain to be infected when the central node becomes the source of infection, while the central node will not when the periphery nodes become the source of infection. So the percolation coefficient is $c(N) = 1/n^2 \cdot ((n-1) \cdot 1 + 1 \cdot n) = (2n - 1) / n^2$.

It is just the opposite situation in case of star convergent structure (Fig. 1c) in contrast with the previous one. The percolation coefficient is $c(N) = 1/n^2 \cdot ((n - 1) \cdot 2 + 1 \cdot 1) = (2n - 1) / n^2$.

In the ring structure (Fig. 1d), the transfer closure of adjacency matrix will be all one matrix owing to the strong connectivity of the network, which means all of the nodes will be infected so as any one become the source of infection. The percolation coefficient is $c(N) = 1/n^2 \cdot (n \cdot n) = 1$.

In the chain structure (Fig. 1e), the closer position of infected node is to s, smaller the dissemination scope will be. The percolation coefficient is $c(N) = 1/n \cdot (1/n + 2/n + \ldots + 1) = (n + 1) / 2n$.

According to the percolation coefficient, we can easily draw a conclusion that the ring structure is the best one to spread state, the chain structure takes the second place, while the islands structure is the best one to block the spreading in network.

3.3 Directional Evolution Control Model for Network

The percolation coefficient usually changes when the topology structure of network changes. The increment of percolation is defined as follows:

$$\Delta c = c(N') - c(N) \tag{8}$$

where $c(N')$ is the percolation coefficient after the evolution of network, while $c(N)$ is the percolation coefficient of original network.

From the definition of percolation coefficient in Sect. 3.2, we found that it can represent the spreading scope in network well. In real world, we expect the dissemination scope of some state like computer virus, risk smaller, while some others like energy, information larger. In fact, the evolution process of many networks can interfered by human beings, so we can try to control the evolution process of network, expecting the directional evolution to fulfill certain goals to our advantage.

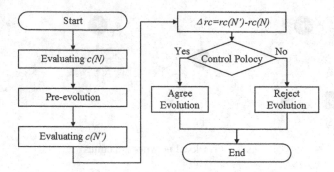

Fig. 2. Flow chart of directional evolution control model

Figure 2 describes the flow chart of directional evolution control model for network. The pre-evolution step means just adjusting the adjacency matrix of network according to the upcoming evolution, rather than let the network evolving really. This model is inspired by the theory of evolution in biology. In nature, there are always some living beings eliminated because of the selection of nature, while the survivals is more fit to nature. The core idea of this model is to let the network evolving in certain direction by constantly selecting. So in the network evolution, we can take measures similar to nature. The control policy is designed to select the evolution by letting Δc fulfill certain constrains like less than a threshold or more than it. For example, in guarantee network, what banks don't want to see most is the widespread of risk when some companies can't pay a debt. So Δc should be restricted to be less than some threshold. While in computer networks, we want to connect more computers in every evolution if only coverage is considered, and in this case we should limit Δc larger than some threshold.

The algorithm of PC-DECM is described below:

Input: The adjacency matrix of original network M, the adjacency network of pre-evolved network M', the probability distribution of nodes becoming initial infectious source P in the original network, and the probability distribution of nodes becoming initial infectious source P' in the pre-evolved network.

Output: Accept the evolution or not.

Pseudo-code:

```
Directional_Evolution_Control_Model (M, M', P, P')
  Mt := Warshall (M)
  Mt' := Warshall(M')
  c := Calculate_Percolation_Coefficient (Mt)
  c' := Calculate_Percolation_Coefficient (Mt')
  delta := c' - c
  if Control_Policy (delta) satisfied then
    return AGREE
  else
    return REJECT
```

where Warshall denotes the algorithm to get the transfer closure of adjacency matrix. Other algorithms can also be applied. Control_Polcy is a function to judge according to Δc, for example if we limit Δc no more than certain value THRESHOLD, we can implement it like the pseudo-code following:

```
Control_Policy (delta)
   if delta > THRESHOLD then
      return false
   else
      return true
```

Next we will analyze one of the basic local structures to illustrate how to control the direction of evolving network. Taking the chain structure as an example, the solid line in Fig. 3(a), (b) and (c) described the structure of original network, while the dashed line in Fig. 3(a), (b) and (c) and the removed 4th node in Fig. 3(d) described the pre-evolution of the network.

For case in Fig. 3(a), it is obvious that, $\Delta c = 1/n \cdot (\sum i/n) - (n + 1)/2n = 0$, which means the evolution has no contribution to the spreading. For case in Fig. 3(b), $\Delta c = 1/n \cdot (\sum i/n + 2/n + 1/n) - (n + 1)/2n = 3/n^2$ the percolation coefficient increased, which means the newly added edge can improve the spreading. For case Fig. 3(c), the newly add edge made the formation of ring in the network, $\Delta c = 1 - (n + 1)/2n = (n - 1)/2n \approx 1/2$. It is clearly that the evolution in this case greatly promoted the spreading. For case Fig. 3 (d), the 4th node disappeared (we need $n \geq 4$) along with the edged connected with it, $\Delta c = 1/n^2 \cdot (3 + 2 + 1 + (n - 4) + (n - 3) + \dots + 1) - (n + 1)/2n = (-4n + 12)/n^2 < 0$, which means this evolution can restrain the spreading. By taking different steps to Δc above, we can control the direction of network evolution. In case of guarantee network, for evolution in Fig. 3(a), new loan didn't increase the risk, so we should grant the application. Although evolution in Fig. 3(b) will make contribution to the spreading of risk, the percolation coefficient is $3/n^2$, which is with the same order of n^{-2}, that is to say, the effect can be ignored with the growth of n in a large network. So this application should also be granted. For case in Fig. 3(c), because of the formation of ring, the nodes at the end of the chain can affect the nodes in the head like the 1th node currently when risk occurred while formerly not. So this application should be rejected. Banks should let the applicant choose another guarantor. Case in Fig. 3(d) means the loan has been paid off and the company no longer guarantee for others.

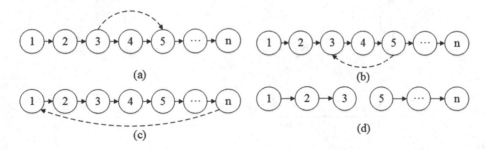

(a)

(b)

(c)

(d)

Fig. 3. Example of network evolution

From the example above we can find that the probability which causes great change of Δc is very low (Fig. 3c). So there is no possible to continuous reject evolution.

4 Experiments

In order to verify the proposed model, the following is the simulation based on the real data of a financial institution.

The data sets we use are the monthly list of guarantee relationships spanning more than 60 months from 2006 to 2012.

Every company as a node and every guarantee relationship as an edge, a directed guarantee network can be built. Use the data of October 2006 as experimental data, and measure the basic parameters of the guarantee network, main parameters can be found in Table 1. The argument of the "Maximal connected sub-network" is the number of nodes it contains, "Average clustering coefficient" and "Average path length" are measures of the maximal connected sub-network after converting to non-directed network. Although there is bias between the "Average path length" with the theoretical value in the WS small-world network, it is in the acceptable range. The "Average clustering coefficient" is relatively large, consistent with the theory well.

What Fig. 4 shown is the double logarithm plot of the degree distribution P (k) versus the out-degree k, we can see the distribution of out-degree of nodes meets the

Table 1. Basic measurement of guarantee network

Parameter name	Parameter value
Number of nodes	32878
Number of edges	31486
Maximal connected sub-network	8721
Average clustering coefficient	0.1146
Average path length	16.2124

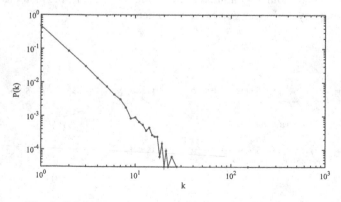

Fig. 4. Double logarithm plot of the out-degree distribution

power-law distribution, it is a scale-free network, and the power-law coefficient is −3.0278. The distribution of in-degree is similar, not repeated.

As the guarantee network meet the characteristics of small-world and scale-free, we got a complex network.

4.1 Experimental Method

The network evolution is what the topology of network changes in the time dimension, and the topology is the relationship between nodes and edges, so we takes a random way to add edges to guarantee network to simulate network evolution, select one from the point-edge changing set randomly every moment for pre-evolution, simulate 1000 moments. In order to test the proposed model, set up two groups, which take similar manner to simulate network evolution. The natural evolution group evolves without any control while the controlled one refuses the pre-evolution when the increment of percolation coefficient exceeds the preset threshold 1/8721, and randomly select another node as a starting point as a newly-add edge (simulating reselect guarantor in guarantee loan). From Table 1 we know that, the number of nodes and edges in the guarantee network are in the same order, indicating that the guarantee network is very sparse and there are a lot of potential disconnected edges in the network, therefore, the probability that after adding an edge randomly it leads to dramatic change is very small. In order to make the experiment effect more obviously, mix some changes which may lead to dramatic increment of percolation coefficient into optional edge changing set in advance.

After the evolution of 1000 times, apply the SI model to the two guarantee networks to simulate the process of risk spreading, every simulation select a point as the initial risk infection source randomly. Repeat 1000 times, take the average of ratios of companies infected by risks at every time and compare them.

4.2 Result and Analysis of Experiment

Figure 5 shows the changing process of percolation coefficient in 1000 moments of natural evolved guarantee network and controlled one. In order to make the percolation coefficient change of neighboring time clearer, intercept the moment in 0–99.

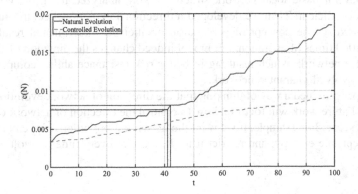

Fig. 5. Percolation coefficient changing trends

As can be seen from the figure, because the natural evolution group accepts all pre-evolution of network, resulting in its percolation coefficient increasing very fast, e.g. at 42, 43 moment in the figure, percolation coefficient greatly increases; however the controlled evolution group would reject pre-evolution when its percolation coefficient increases too large, which makes the percolation coefficient of evolution control group increase gently.

Figure 6 displays the ratio of risk infected companies versus time t for the two networks after 1000 moment evolution using SI model. As can be seen from the figure, at every moment, the ratio of infected companies in evolution control group is lower than the natural evolved one, indicating that if insolvency happens, the network of controlled evolution group has stronger ability to resist risks.

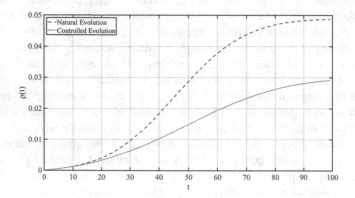

Fig. 6. The proportion of infected nodes changing trends

5   Conclusions

In this paper, we introduced the concept of percolation coefficient which can be used to represent the spreading scope of network, after analyzing the spreading behavior of network. Several basic local network structures were analyzed using the concept of percolation coefficient. Then we developed a directional evolutional control model for network based on the concept of percolation coefficient. Experimental results show that, after 1000 times of evolution, our model indeed changes the direction of evolution for guarantee network, which resulting in better risk resistance ability compared with the natural evolved guarantee network.

Our paper provided a way of thinking that the direction of network evolution can be controlled. Future work will focus on how to control the direction of network evolution in different scene. For example, if we want the network more orderly, we can introduce some concept like entropy, and restrict it to decrease at every time of evolution.

References

1. Erdős, P., Rényi, A.: On the evolution of random graphs. Sel. Pap. Alfréd Rényi **2**, 482–525 (1976)
2. Watts, D.J., Strogatz, S.H.: Collective dynamics of 'small-world'networks. Nature **393** (6684), 440–442 (1998)
3. Barabási, A.L., Albert, R.: Emergence of scaling in random networks. Science **286**(5439), 509–512 (1999)
4. Leskovec, J., Kleinberg, J., Faloutsos, C.: Graph evolution: densification and shrinking diameters. ACM Trans. Knowl. Disc. Data (TKDD) **1**(1), 2 (2007)
5. Li, X., Chen, G.: A local-world evolving network model. Physica A: Stat. Mech. Appl. **328** (1), 274–286 (2003)
6. Barrat, A., Barthélemy, M., Vespignani, A.: Modeling the evolution of weighted networks. Phys. Rev. E: Stat., Nonlin., Soft Matter Phys., 70(6 Pt 2), 066149 (2004)
7. Palla, G., Barabási, A.L., Vicsek, T.: Quantifying social group evolution. Nature **446**(7136), 664–667 (2007)
8. Ahn, J.W., Plaisant, C., Shneiderman, B.: A task taxonomy for network evolution analysis. IEEE Trans. Vis. Comput. Graph. **20**(3), 365–376 (2014)
9. Anderson, R.M., May, R.M.: Infectious Diseases of Humans, vol. 1. Oxford University Press, Oxford (1991)
10. Lü, L., Chen, D.B., Zhou, T.: The small world yields the most effective information spreading. New J. Phys. **13**(12), 123005 (2011)
11. Wang, Y., Chakrabarti, D., Wang, C., Faloutsos, C: Epidemic spreading in real networks: an eigenvalue viewpoint. In: Proceedings 22nd International Symposium on Reliable Distributed Systems, 2003, pp. 25–34. IEEE (2003)
12. Ganesh, A., Massoulié, L., Towsley, D: The effect of network topology on the spread of epidemics. In: Proceedings IEEE INFOCOM 2005. 24th Annual Joint Conference of the IEEE Computer and Communications Societies, vol. 2, pp. 1455–1466. IEEE (2005)
13. Garas, A., Argyrakis, P., Rozenblat, C., Tomassini, M., Havlin, S.: Worldwide spreading of economic crisis. New J. Phys. **12**(11), 113043 (2010)
14. Hui, W.,Jiang-Hong, H., Lin, D., et al.: Dynamics of rumor spreading in mobile social networks. Acta. Phys. Sin. 2013, 62(11), 110505 (2003)
15. Kempe, D., Kleinberg, J., Tardos, É.: Maximizing the spread of influence through a social network. In: Proceedings of the Ninth ACM SIGKDD International Conference on Knowledge Discovery and Data Mining, pp. 137–146. ACM (2003)
16. Volz, E.: SIR dynamics in random networks with heterogeneous connectivity. J. Math. Biol. **56**(3), 293–310 (2008)
17. Warshall, S.: A theorem on boolean matrices. J. ACM **9**(1), 11–12 (1962)

SRSP-PMF: A Novel Probabilistic Matrix Factorization Recommendation Algorithm Using Social Reliable Similarity Propagation

Ruliang Xiao[1,2(✉)], Yinuo Li[1,2], Hongtao Chen[1,2], Youcong Ni[1,2], and Xin Du[1,2]

[1] Faculty of Software, Fujian Normal University, Fuzhou 350117, China
xiaoruliang@fjnu.edu.cn
[2] Fujian Provincial University Engineering Research Center of Big Data Analysis and Application, Fuzhou 350117, China

Abstract. Recommendation systems have received great attention for their commercial value in today's online business world. Although matrix factorization is one of the most popular and most effective recommendation methods in recent years, it also encounters the data sparsity problem and the cold-start problem, which leads it is very difficult problem to further improve recommendation accuracy. In this paper, we propose a novel factor analysis approach to solve this hard problem by incorporating additional sources of information about the users and items into recommendation systems. Firstly, it introduces some unreasonable prior hypothesises to the features while using probabilistic matrix factorization algorithm (PMF). Then, it points out that it is neccesary to give two new hypothesises about conditional probability distribution of user and item feature and buliding some concepts such as social relation, social reliable similarity propagation metrics, and social reliable similarity propagation algorithm (SRSP). Finally, a kind of a novel recommendation algorithm is proposed based on SRSP and probabilistic matrix factorization (SRSP-PMF). The experimental results show that our method performs much better than the state-of-the-art approaches to long tail recommendation.

Keywords: Recommendation system · Social relation · Reliable similarity propagation metrics · Probabilistic matrix factorization

1 Introduction

Matrix factorization (MF) is one of the most popular and most effective recommendation methods in recent years. Probabilistic matrix factorization (PMF) is one of the important methods to achieve matrix factorization. However, there is a priori hypothesis in the probabilistic matrix factorization that feature vectors about user and item subject to the spherical Gauss prior distribution, and mean is zero. This assumption internals does not take into account the innate relation between the user and the user, item and item, such as social relations of users, and item similarity relation. Thus, this priori assumption is not reasonable.

© Springer International Publishing Switzerland 2015
D.-S. Huang et al. (Eds.): ICIC 2015, Part II, LNCS 9226, pp. 80–91, 2015.
DOI: 10.1007/978-3-319-22186-1_8

Further, if the similarity relation is putted into probabilistic matrix factorization model, we will not accurately calculate similarity too. As the number of Internet users and items more and more rapidly increasing, score density to item also is becoming more separated and diversified. When the user score density is large, the traditional similarity measure methods can get more reliable results, however. When the user score density is too small, user behavior data is too sparse, and traditional similarity calculation method will lacks calculated basis so that results will be not gotten reliable. For example,cosine similarity and modified cosine similarity for the user both assume that the unrated items should be rated as 0. When these rated data are sparser, influence to the result will be greater. Facing the same problem, when people compute the relevant similarity, these common rating item set may be very small, even for 0. Assumption for too much score, and too little real score data, can not support the similarity computation reliablely.

To solve the above problems, we put forward a novel social similarity propagation based on probability matrix decomposition model (SRSP-PMF). We modify the constraints on prior assumptions of feature vectors by integrating the similarity between users and the similarity between items. For the problem of cold start, using the mothod of social relation to propagate similarity, we can calculate the similarity of users while score density is lower so as to improve the accuracy of the similarity metrics and their effectiveness.

2 Related Work

In the face of unreasonable prior hypothesises to the feature vectors of probabilistic matrix factorization, most approach is used to modify this prior hypothesis with user's social relation and trust relationship, and to improve the effect of the recommendation model. They assume that all the users are not independent and not identically distributed; Their assumption emphasizes the social interactions or connections among users.

The literature [1] proposes a method integrating social network framework and the user-item rating matrix, based on probabilistic factor analysis. The literature [2] assumes that latent vectors in different domains are generated from a common Gaussian distribution with an unknown mean vector and full covariance matrix. By inferring the shared mean and covariance of the Gaussian from given cross-domain rating matrices, the latent factors are aligned across different domains, which enables us to predict ratings in different domains. Literature [3] aims at providing more realistic and accurate recommendations, and propose a factor analysis-based PMF optimization framework to incorporate the user trust and distrust relationships into the recommender systems. Literature [4] investigates the social recommendation problem on the basis of psychology and sociology studies, which exhibit two important factors: individual preference and interpersonal influence. The authors propose a social contextual recommendation framework. This framework is based on a probabilistic matrix factorization method to incorporate individual preference and interpersonal influence to improve the accuracy of social recommendation. Literature [5] constrains probabilistic matrix factorization with deriving the Gibbs updates for the side feature vectors for items, and think that the application of PMF in the recommendation was confined to simply integrated with external information. However, it is difficult to obtain social relations and trust relations and other external information under normal circumstances [6].

When using the similarity constraint of feature vectors to correct the prior hypothesises, it is necessary to accurately measure similarities between users and items. But, how to accurately calculate the degree of similarity between the users is a challenge. In recent years, there has been a lot of many researches on different similarity calculation methods.

These methods are mainly divided into two classes that based on memory [7] and based on model [4–9]. According to the information sources, and the former can be divided into two kinds: implicit and explicit information. Literature [7] proposes a method of user similarity calculation model based on the matrix decomposition model to get feature vector, and makes the feature vector cosine similarity between users as user similarity. The method has higher efficiency and better accuracy that is compared with the traditional method, but it can not fundamentally solve the problem of the cold start users and data sparseness effects. Literature [8, 9] compute user similarity with the external information, such as geographic information and semantic information to track the trajectory, to mine the degree of similarity between the users, and achieve to good effect. But, because of the limitation of technology, it is often difficult to get accurately.

In term of recommendation method based on memory, some researchers adopt similarity propagation method to calculate the similarity. Literature [10] find more reliable neighbor through similarity propagation, however, the costs with time and space of the algorithm is too large, and too complicated. Literature [11] constructs an new idea to make user similarity transmit along a limited path only those greater than the threshold, to increase the nearest neighbors quantity, and to reduce the impact of the sparse data problem. So, it improves the quality of recommendation. This method ignores that too little score data to calculate user similarity accurately.

Social relationship is another relationship which differs from the above between users. Social relation and user similarity fuse with probability matrix factorization model has greatly improved the recommendation accuracy [12, 13]. Literature [14] refines the classification of social relation and further improve the the accuracy of recommendation on the basis of [12]. Trust, as a kind of social relationship, is proved to be positively correlated with the similarity [15]. Therefore, social relations can also be used to help measure of correlation between users. Literature [16] proposes a model of social relationship through user ratings and other information, and then combines similarity into recommendation system. In a certain extent, the method can improve the accuracy of recommendation. However, the method is only a linear weighted simple fusion with the similarity relation.

Our method differs from the previous practice, we construct social relation, social reliable similarity propagation algorithm, and also modify feature vector constraint of probability matrix factorization, the proposed recommendation algorithm is a novel probabilistic matrix factorization recommendation algorithm using social reliable rimilarity propagation (SRSP-PMF).

3   Constructed New Factor Constraints

The traditional calculating similarity method is effected by data sparsity. User rate data is more dense, the similarity calculation is more reliable; on the contrary, and similarity calculation is more unreliable. On the other hand, due to the user's level of activity and registration time is uneven, the number of different user's scores vary widely. Therefore, we can calculate similarity between users which ratings are more, and compute the similarity between users rating less which can obtain from the user's social relationships. So, we can avoid the users who rating less among other users' unreliable similarity computation, improve the overall reliability of the similarity. For convenience of description next, give following definitions.

Definition 1: Cold start users and active users. When users ratings less than the threshold T, these users are called cold start users, cold start users are denoted as the set LU; Users ratings more than threshold T are active users, usually, they are denoted as set AU.

Traditional similarity computation method are not accurately when the data is sparse. But, when the data is dense, it is reliable. The active users' ratings are more dense, the more reliable of similarity computation.

Give the definition of reliable similarity as follows.

Definition 2: Reliable similarity. When we compute similarity, the number of rating scores is bigger than the threshold T, the similarity becomes more reliable, it is called as reliable similarity.

The results of similarity computation are all reliable similarity between active users by exploring the Definition 1 and Definition 2.

Definition 3: Social relation. In the social network, the degree of association of users is the social relation; it can be measured by the path number among users, and the length of the path.

3.1   Social Relation Metrics

In this paper, similarity propagation is based on social relation. Therefore, the first step must construct a model of the social relation; try to quantify the users' social relation. Random walk is the common method to measure the relation between vertexes in the graph, which has the advantage of computing propagation. The method holds the social relation's characteristic [13, 14], can be applicable for measure the degree of users' social relation. So, this method is usually used to compute the social relation.

First, we will construct the social network graph based on the users' ratings and the user demographic information. If we want to calculate a user u with other users' social relation, starting from the node of the user to random walk from the bipartite graph. When random walk to other user's node, we first decide if continue to random walk based on the probability α, if decide stop random walk, restarting random walk from the node of user u, if decide to continue random walk that select one node equal probability to continue random walk which is connect with the current node. After some times of restart iteration L, every user node is visited several times are convergence to a value and now stop random walk. The more times visiting on a user, these users are closer, and stronger social relation. When stopping random walk, we can use the following formula to calculate the social relation between user u and other user v.

$$Relation(u, v) = \frac{|in(v)|}{|out(u)|} \tag{1}$$

$In(v)$ denotes the visit times of user v, and $out(u)$ denotes the restart random walk times of user u, respectively.

In the social network graph, the bigger social relation has the characteristic: there are more paths between two nodes, there is shorter path.

3.2   SRSP: Social Reliable Similarity Propagation Algorithm

All users can be divided into cold start users and active users, according to the Definition 1. Firstly, the users are sorted according to ratings of users from more to less, and a new similarity matrix can be generated. Then, based on Definition 1, the similarity matrix can be divided into four areas according to the score threshold T, as shown in Fig. 1.

In Fig. 1, Area1 and Area4 stand for active user similarity matrix and cold start user similarity matrix respectively. And Area2, Area3 are the matrixes of similarity between active user and cold start user. Due to the symmetry of similarity matrix, Area1 and Area4 are symmetric matrix, and Area 2, Area3 are transposed matrix.

By Definition 2, the user similarities of Area1 are reliable similarity, and others are not reliable similarity. The main ideas of the propagation of reliable similarity as follows, firstly, calculate the reliable similarity of Area1; and then, propagate the reliable similarity of Area 1 to Area2, Area3 and Area4 according to the social

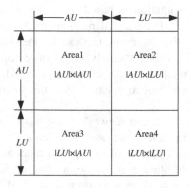

Fig. 1. Area division of similarity matrix

Algorithm 1 SRSP
Input:
Ratinginformation,
Demographicinformation,
Scorethreshold T,
Restartprobability α.
Input:
User similarity matrix.
Step1:calculate the reliable similarity of Area1;
Step2:according to the social corelation of users, propagate the reliable similarity of Area 1 to Area2;
Step3:in the same way,propagate the reliable similarity to Area3;
Step4:in the same way,propagate the reliable similarity to Area4;
Step5: Return the similarity matrix composed of the above four areas.

Fig. 2. SRSP: similarity propagation algorithm using social relation

correlation of users. Similarity propagation algorithm using social relation shows as following algorithm SRSP (also see Fig. 2).

Step 1. Constructs social network diagram according to the ratings and demographic information,and measure social correlation between users using restart random walk method on social network diagram, get social relationship matrix *Relation*; then, determine the active users set *AU* and cold start users set *LU* according to the number of user ratings threshold *T*

Step 2. Measures the similarity between active users using traditional similarity calculation method, and then the similarity matrix of Area1 is finished

Step 3. Calculates similarity matrix of Area2 and Area3 depends on social relation matrix *Relation*. According to the literature [12, 14], user social relation and user similarity is positive correlation, therefore, when the social relation between user A and user B is stronger, the similarity between A and user C is closer to the similarity between B and C. The similarities of Area2 and Area3 are similarities between active users and cold start users, and for each pair of active user *i* and cold start user *j*, the similarity *Sim* (*i, j*) can be calculated by formula (2).

$$Sim(i,j) = \frac{\sum\limits_{u \in K(j)} Sim(u,i)Relation(j,u)}{\sum\limits_{u \in K(j)} Relation(j,u)} \tag{2}$$

In formula (2), *K(j)* is the set of social neighbors of user *j*, which includes *Top-K* active users who have strongest social relation with *j*, *Relation(j, u)* is the social relation between user *j* and user *u*. The similarities of Area2 can be calculated by formula (2), and for the symmetry of similarity matrix, matrix of Area3 is the transpose of Area2.

Step 4. Calculates Area4 according to user social relation and similarity matrixes of Area2 and Area3, the similarities of Area4 are between cold start users, and the calculation of Area4 is similar to Step3, such as formula (3).

$$Sim(i,j) = \begin{cases} \dfrac{\sum\limits_{u\in K(j)} Sim(u,i)Relation(j,u)}{2\sum\limits_{u\in K(j)} Relation(j,u)} + \dfrac{\sum\limits_{u\in K(i)} Sim(u,j)Relation(i,u)}{2\sum\limits_{u\in K(i)} Relation(i,u)} & (i \neq j) \\ \\ 1 & (i = j) \end{cases} \qquad (3)$$

$K(i)$ and $K(j)$ are social neighbors of user i and user j respectively, and each of them includes *Top-K* active users who have strongest social relation with corresponding user. *Relation* (i, u) and *Relation* (j, u) are the social relation between i and u, j and u respectively. As for the symmetry of similarity matrix of Area4, $Sim(i, j)$ is equal to $Sim(j, i)$.

Step 5. Returns the similarity matrix composed of the above four areas.Step 1 has higher complexity, so it can be calculated with off line. In Step 2, if the complexity that calculates the similarity between two users is N, the complexity of Step 2 is $N*|AU|^2$. Step 3 and Step 4 is the core of the algorithm, each cold start user needs to find out the social neighbors in active users, and the complexity of this work is $|Top\text{-}K|*|AU|*|LU|$, the complexity of similarity propagation based on (2) and (3) is $|Top\text{-}K|*|AU|*|LU|$ and $2*|Top\text{-}K|*|LU|^2$ respectively. From what has been discussed above, after the offline calculation of social relation matrix, the complexity of SRSP is $N*|AU|^2 + |Top\text{-}K|*|AU| *|LU| + 2*|Top\text{-}K|*|LU|^2$.

4 SRSP-PMF: Novel Probabilistic Matrix Factorization Recommendation Modal

Supposing that recommendation system, there are M items, N users, R denoted score matrix, R_{ui} is the score of user u on item i, which users and items map the association to the feature space of the F dimension, and then $U \in \Re^{F\times N}$ and $V \in \Re^{F\times M}$ respectively represents as users and items feature matrix, U_u as feature vectors of user u, V_i as feature vectors of objects i. There are two hypotheses following within probability matrix factorization.

(1) All rates obey independent Gauss mixture distribution, as shown in formula (4).

$$p(R|U,V,\sigma^2) = \prod_{u=1}^{N}\prod_{i=1}^{M} [N(R_{ui}|g(U_u^T V_i), \sigma_R^2)]^{I_{u,i}} \qquad (4)$$

Which $I_{u,i}$ is the indicator function, if there is a score of user u on item i, its value is 1 else 0, the magnitude of each user of items are subject to mean value of the relevant user and item feature vector. In the meantime, σ is denoted as variance of the Gauss distribution, andσis the parameters of the model, and used to control the fluctuation.

(2) Each user or item feature vector obeys spherical Gauss prior distribution that its mean value is 0, as shown in the formula (5).

$$p(U|\sigma_U^2) = \prod_{i=1}^{N} N(U_i|\ 0, \sigma_U^2 I),$$

$$p(V|\sigma_V^2) = \prod_{j=1}^{M} N(V_j|\ 0, \sigma_V^2 I).$$

(5)

Where, i is the unit vector. However, similar users or similar items have similar feature vector. Then, under the assumption (2), we have a conclusion as follows.

Theorem 1: Users or items feature vector is constrained by two factors, and subjected to zero mean Gauss prior distribution, and approach to the similar neighbor feature vector.

According to the Theorem 1, the conditional probability distribution of user and item feature vector is shown as following.

$$p(U|\ T, \sigma_U^2, \sigma_T^2) \propto p(U|\ \sigma_U^2) p(U|\ T, \sigma_T^2)$$

$$= \prod_{u=1}^{N} N(U_u|\ 0, \sigma_U^2 I) \times \prod_{u=1}^{N} N\left(U_u|\ \sum_{v \in N_u} T_{u,v} U_v, \sigma_U^2 I\right)$$

(6)

$$p(V|\ T, \sigma_V^2, \sigma_T^2) \propto p(V|\ \sigma_V^2) p(V|\ T, \sigma_T^2)$$

$$= \prod_{i=1}^{M} N(V_i|\ 0, \sigma_V^2 I) \times \prod_{i=1}^{M} N\left(V_i|\ \sum_{j \in N_i} T_{i,j} V_j, \sigma_V^2 I\right)$$

(7)

Where, $T_{u,v}$ is the weight of the similarity neighborhood between user u and v. $T_{i,j}$ is the same, N_u acts as similar neighborhood of user u.

After combining formula (6) and (7) into the PMF (probabilistic matrix factorization) model, using user item similarity with SRSP metric method, and similar weight T under neighborhood of conventional similarity item, we may get a novel probabilistic matrix factorization recommendation algorithm using social reliable similarity propagation. This model is denoted as following (Fig. 3).

According to Bayes reasoning, posterior probability of user and item feature vector can get as following (8).

$$p(U, V|\ R, T_U, T_V, \sigma_R^2, \sigma_{TU}^2, \sigma_{TV}^2, \sigma_U^2, \sigma_V^2) \propto$$
$$p(R|\ U, V, \sigma_R^2) p(U|\ T_U, \sigma_U^2, \sigma_{TU}^2) p(V|\ T_V, \sigma_V^2, \sigma_{TV}^2)$$

(8)

We substitute formula (8) with (5), (6) and (7), then, a conditional probability function can be given as followed.

$$p(U,V\mid R,T_U,T_V,\sigma_R^2,\sigma_{TU}^2,\sigma_{TV}^2,\sigma_U^2,\sigma_V^2) \propto \prod_{u=1}^{N}\prod_{i=1}^{M}\left[N(R_{ui}\mid g(U_u^T V_i),\sigma_R^2)\right]^{I_{u,i}}$$

$$\times \prod_{u=1}^{N} N\left(U_u\mid \sum_{v\in N_u} T_{u,v}U_v,\sigma_{TU}^2 I\right) \times \prod_{u=1}^{N} N(U_u\mid 0,\sigma_U^2 I) \qquad (9)$$

$$\times \prod_{i=1}^{M} N\left(V_i\mid \sum_{j\in N_i} T_{i,j}V_j,\sigma_{TV}^2 I\right) \times \prod_{i=1}^{M} N(V_i\mid 0,\sigma_V^2 I)$$

Taking the logarithm, we further get the formula

$$\ln p(U,V\mid R,T_U,T_V,\sigma_R^2,\sigma_{TU}^2,\sigma_{TV}^2,\sigma_U^2,\sigma_V^2) \propto$$

$$-\frac{1}{2\sigma_R^2}\sum_{u=1}^{N}\sum_{i=1}^{M} I_{u,i}(R_{u,i}-U_u^T V_i)^2 - \frac{1}{2\sigma_U^2}\sum_{u=1}^{N} U_u^T U_u - \frac{1}{2\sigma_V^2}\sum_{i=1}^{M} V_i^T V_i$$

$$-\frac{1}{2\sigma_{TU}^2}\sum_{u=1}^{N}\left(\left(U_u-\sum_{v\in N_u} T_{U_{u,v}}U_v\right)^T \left(U_u-\sum_{v\in N_u} T_{U_{u,v}}U_v\right)\right) \qquad (10)$$

$$-\frac{1}{2\sigma_{TV}^2}\sum_{i=1}^{M}\left(\left(V_i-\sum_{j\in N_i} T_{V_{i,j}}V_j\right)^T \left(V_i-\sum_{j\in N_i} T_{V_{i,j}}V_j\right)\right)$$

$$+\eta(\sigma_R,\sigma_{TU},\sigma_{TV},\sigma_U,\sigma_V) + C$$

The function $\eta(\sigma_R,\sigma_{TU},\sigma_{TV},\sigma_U,\sigma_V)$ does not include some items of parameter matrix U, V; C is a constant. The maximum of P is equivalent to minimizing the loss function (11).

$$C = \frac{1}{2}\sum_{u=1}^{N}\sum_{i=1}^{M} I_{u,i}(R_{u,i}-U_u^T V_i)^2 + \frac{\lambda_1}{2}\sum_{u=1}^{N} U_u^T U_u + \frac{\lambda_2}{2}\sum_{i=1}^{M} V_i^T V_i$$

$$+\frac{\lambda_3}{2}\sum_{u=1}^{N}\left(\left(U_u-\sum_{v\in N_u} T_{U_{u,v}}U_v\right)^T \left(U_u-\sum_{v\in N_u} T_{U_{u,v}}U_v\right)\right) \qquad (11)$$

$$+\frac{\lambda_4}{2}\sum_{i=1}^{M}\left(\left(V_i-\sum_{j\in N_i} T_{V_{i,j}}V_j\right)^T \left(V_i-\sum_{j\in N_i} T_{V_{i,j}}V_j\right)\right)$$

To learn the parameters with stochastic gradient descent method, we construct gradient direction as shown (12) and (13).

$$\frac{\partial C}{\partial U_u} = \sum_{i=1}^{M} I_{u,i}V_i\left(U_u^T V_i - R_{u,i}\right) + \lambda_1 U_u + \lambda_3\left(U_u-\sum_{v\in N_u} T_{U_{u,v}}U_v\right) \qquad (12)$$

$$\frac{\partial C}{\partial V_i} = \sum_{u=1}^{N} I_{u,i}U_u\left(U_u^T V_i - R_{u,i}\right) + \lambda_2 V_i + \lambda_4\left(V_i-\sum_{j\in N_i} T_{V_{i,j}}V_j\right) \qquad (13)$$

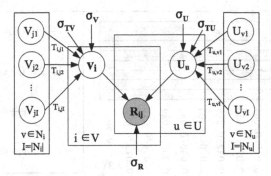

Fig. 3. SRSP-PMF: probabilistic matrix factorization based on SRSP

5 Experiments and Result Analysises

The data set gets from the MovieLens and Netflix website. The size of the former is 1 M, which includes 6040 users, 3952 movies and 1000209 rating records. The last includes 6000 users, 4000 movies and 987019 rating records, which is extracted from the open Netflix competition data set according to mean distribution.

We employ the Root Mean Square Error (RMSE) to measure the prediction quality of our proposed approaches in comparison with other recommendation methods. The metrics *RMSE* is defined as:

$$RMSE = \frac{\sqrt{\sum_{i=1}^{N}(p_i - q_i)^2}}{N} \tag{14}$$

where p_i denotes the rating of user i, p_j denotes the rating of user i as predicted by a method, and N denotes the number of tested ratings.

We adopted the conventional Pearson similarity method, and comparative experiments between SRSP-PMF and conventional PMF.

In the experiment, set $\lambda_1 = \lambda_2 = 0.08$ in the model PMF, $\lambda_1 = \lambda_2 = 0.04$ in the model SRSP-PMF, $N = 100$ to neighborhood size. In the model of gradient descent learning, we set learning step $r = 0.0005$, at least iterations $K = 10$, the $F = 40$ dimension of the feature space is mapped to the users and items. In the SRSP algorithm based on experience, we set random walk restart probability $\alpha = 0.7$, restart the iterative parameter $L = 12800$, similar neighbor number *Top-N* = 100, social neighbor number *Top-K* = 20, threshold $T = 50$.

Firstly, according to the experience, adding constriction $\lambda_3 + \lambda_4 = 0.08$, and they are all positive. Then these two parameters were adjusted respectively in the experiment, MovieLens 1 M data set and Netflix data set. The experimental results are shown in Fig. 4.

From Fig. 4, it can be seen that the experimental results have the same changing trend in the MovieLens data set and Netflix data set. According to optimal results of each model, the *RMSE* effect of the control model is 0.84558, 0.86929, respectively.

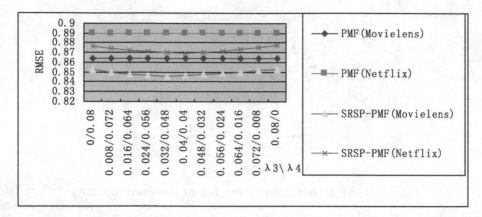

Fig. 4. SRSP-PMF and PMF *RMSE* changing with $\lambda_3\backslash\lambda_4$ on two data set

SRSP-PMF recommended result is better than PMF. We can have a conclusion that feature vector of user and item is close to the feature vector of similar neighbor. So, it is also proved that the above theorem is correct.

6 Conclusion

By analyzing defects of the similarity computation method and similarity which used in the traditional social relationship and users' social relation, proposed using social relationships of reliable similarity propagation PMF algorithm (SRSP-PMF). Experiments show that the SRSP-PMF algorithm can better measure the ratings sparse of cold start user and the similarity between the user and other users, and also improve the accuracy of PMF recommendation.

Acknowledgment. This work is supported by the Humanities and Social Science Research Projects of the Ministry of Education of P.R.C (11YJA860028).

References

1. Ma, H., Yang, H., Lyu, M.R., et al.: Sorec: social recommendation using probabilistic matrix factorization. In: Proceedings of the 17th ACM Conference on Information and Knowledge Management, pp. 931–940. ACM (2008)
2. Iwata, T., Takeuchi, K.: Cross-domain recommendation without shared users or items by sharing latent vector distributions. In: Proceedings of the Eighteenth International Conference on Artificial Intelligence and Statistics, pp. 379–387 (2015)
3. Ma, H., Lyu, M.R., King, I.: Learning to recommend with trust and distrust relationships. In: Proceedings of the Third ACM Conference on Recommender Systems, pp. 189–196. ACM (2009)
4. Jiang, M., Cui, P., Wang, F., et al.: Scalable recommendation with social contextual information. IEEE Trans. Knowl. Data Eng. **26**(11), 2789–2802 (2014)

5. Severinski, C., Salakhutdinov, R.: Bayesian Probabilistic Matrix Factorization: A User Frequency Analysis. arXiv preprint arXiv:1407.7840 (2014)
6. O'Donovan, P., Agarwala, A., Hertzmann, A.: Collaborative filtering of color aesthetics. In: Proceedings of the Workshop on Computational Aesthetics, pp. 33–40. ACM (2014)
7. Yu, L., Pan, R., Li, Z.: Adaptive social similarities for recommender systems. In: Proceedings of the fifth ACM Conference on Recommender Systems, pp. 257–260. ACM (2011)
8. Li, Q., Zheng, Y., Xie, X., et al.: Mining user similarity based on location history. In: Proceedings of the 16th ACM SIGSPATIAL International Conference on Advances in Geographic Information Systems, p. 34. ACM (2008)
9. Ying, J.J.C., Lu E.H.C., Lee, W.C., et al.: Mining user similarity from semantic trajectories. In: Proceedings of the 2nd ACM SIGSPATIAL International Workshop on Location Based Social Networks, pp. 19–26. ACM (2010)
10. Zhao, Q.-Q., Lu, K., Wang, B.: SPCF: a memory based collaborative filtering algorithm via propagation. Chin. J. Comput. 36(3), 671–676 (2013)
11. Hu, F.-H., Zheng, X.-L., Gan, H.-H.: Collaborative filtering algorithm based on similarity propagation. Comput. Eng. 37(10), 50–51(2011)
12. Ma, H., Yang, H., Lyu, M.R., et al.: Sorec: social recommendation using probabilistic matrix factorization. In: Proceedings of the 17th ACM Conference on Information and Knowledge Management, pp. 931–940. ACM (2008)
13. Yang, X., Steck, H., Liu, Y.: Circle-based recommendation in online social networks. In: Proceedings of the 18th ACM SIGKDD International Conference on Knowledge Discovery and Data Mining, pp. 1267–1275. ACM (2012)
14. Guo, L., Ma, J., Chen, Z., et al.: Learning to recommend with social relation ensemble. In: Proceedings of the 21st ACM international conference on Information and knowledge management, 2599–2602. ACM (2012)
15. Bhuiyan, T.: A survey on the relationship between trust and interest similarity in online social networks. J. Emerg. Technol. Web Intell. 2(4), 291–299 (2010)
16. Yu, Y., Qiu, G.: Research on collaborative filtering recommendation algorithm by fusing social network. New Technol. Libr. Inf. Serv. 006, 54–59 (2012)

Adaptive Fuzzy PID Sliding Mode Controller of Uncertain Robotic Manipulator

Minh-Duc Tran[1] and Hee-Jun Kang[2(✉)]

[1] Graduate School of Electrical Engineering, University of Ulsan,
Ulsan 680-749, South Korea
ductm.ctme@gmail.com
[2] School of Electrical Engineering, University of Ulsan,
Ulsan 680-749, South Korea
hjkang@ulsan.ac.kr

Abstract. This paper presents a novel adaptive fuzzy sliding mode controller (AFSMC) that combines a PID sliding mode control (SMC) and adaptive fuzzy control (AFC) for robotic manipulator in the presence of uncertainties and external disturbances. First, the controller is developed based on PID sliding mode framework. Then, two fuzzy logic systems are applied, a fuzzy logic system is used to replace the discontinuous control term, and an adaptive fuzzy inference engine is utilized to approximate the unknown term in the equivalent control part of SMC. Thus, the system not only can handle the large uncertainties and does not require the knowledge of their upper bounds to be known in advance, but can also attenuate the chattering phenomenon in conventional SMC without degrading the system robustness. Finally, the simulation results of a two-link robot manipulator are presented to demonstrate the effectiveness of the proposed control method.

Keywords: Fuzzy sliding mode control · Adaptive fuzzy system · Adaptive tracking control · Robot manipulator

1 Introduction

General robotic manipulators involve many uncertainties and external disturbances in their dynamics, such as payload variations, friction, and external disturbances. Due to these uncertainties, the control system cause poor performance and may make the closed-loop system unstable. Thus, control of the practical robot manipulator systems are challenging problems in the control field to achieve high performance and small tracking errors. Many advanced control schemes have been proposed to reduce the effects of robotic uncertainties such as proportional-integral-derivative (PID) control [1], adaptive control [2], robust control [3], sliding mode control [4–9], and intelligent control [10–12].

Among several control methods, sliding mode control (SMC) is well known as one of the most powerful way to overcome system parameter uncertainties and external disturbance [4–6]. The main advantages of SMC are strong robustness with respect to the system uncertainties, fast transient response, good transient performance, and simple design for both linear and nonlinear systems. However, the upper bounds of the

© Springer International Publishing Switzerland 2015
D.-S. Huang et al. (Eds.): ICIC 2015, Part II, LNCS 9226, pp. 92–103, 2015.
DOI: 10.1007/978-3-319-22186-1_9

uncertainties have to be known in advance in order to guarantee the stability of the closed-loop system. Moreover, the major drawback of SMC in practical application is the chattering phenomenon which is caused by discontinuous control action when the system state operates near the sliding surface. To deal with this problem, the most common methods replaced the sign function in the switching control by a saturating approximation [21] or boundary layer technique [9]. The boundary layer method was proposed to eliminate the chattering by defining a boundary layer around the sliding surface and then approximate the discontinuous control by continuous function within this boundary layer. As the result, the chattering elimination is achieved, however, if system uncertainties are large, a thicker boundary layer must be applied to eliminate the chattering problem. Thus, it makes the tracking error bigger, reduces the feedback of the system and vice versa.

Fuzzy logic system has been widely used as a successful approach to many real applications system with uncertainties [12–14]. The most important feature of fuzzy logic system is to use the expert knowledge directly; therefore, it has ability to express the amount of ambiguity in human thinking, and deal with the experience effectively. However, the main drawback of fuzzy system is lack of adequate analysis and precise mathematical models. Because of these drawbacks, fuzzy sliding mode control which combine sliding mode control and fuzzy control have been proposed and it has become a popular topic [13–21]. From authors' point of view, the fuzzy sliding mode controller can be classified into three categories:

(1) Using adaptive fuzzy logic to enhance the SMC system by cope with modeling inaccuracies and external disturbances [15, 16]. However, these methods used saturation to replace sign function in the reaching phase of SMC, therefore, the tracking error will be increased and the feedback of the system will be reduced.
(2) Using fuzzy logic system to adaptively tune the sliding mode gain in the discontinuous part of conventional SMC [17, 18]. As a result, the chattering phenomenon is attenuated.
(3) Using fuzzy logic system instead of the sign function in the reaching phase to eliminate the chattering in the conventional SMC [19–21].

All things considered, in this paper, an adaptive fuzzy PID sliding mode control is proposed for uncertain robotic manipulators. At first, a PID sliding mode controller which using the integral term in the sliding surface is considered for trajectory tracking of robot motion [7, 8]. Then, in order to improve the system performance, two fuzzy logic systems are applied. To reduce the chattering and improve the robustness of the system, a fuzzy logic system is used to replace the switching control term, and to enhance the SMC system as well as cope with modeling inaccuracies and external disturbances, an adaptive fuzzy inference engine is utilized to approximate the unknown term in the equivalent control part of SMC. The stability of the closed-loop system is guaranteed by using the Lyapunov theory and the tracking error asymptotically converge to zero in the presence of large uncertainties and external disturbances.

2 System Dynamics and Preliminaries

2.1 Robot System Model

Consider the dynamics equation of a serial n-links robot manipulator [22]

$$M(q)\ddot{q} + C(q,\dot{q})\dot{q} + G(q) + F(q,\dot{q}) = \tau \tag{1}$$

where $q(t), \dot{q}(t), \ddot{q}(t) \in R^n$ are the vector of joint accelerations, velocities and positions, respectively. $M(q) \in R^{n \times n}$ is the inertial matrix, $C(q,\dot{q}) \in R^{n \times n}$ expresses the centripetal and Coriolis matrix, $G(q) \in R^n$ represents the gravity torques vector, $\tau \in R^n$ is the control torque, and $F(q,\dot{q}) \in R^n$ is the unstructured uncertainties including friction and others disturbances.

For convenience, the inertial matrix $M(q)$ and the Coriolis matrix $C(q,\dot{q})$ in the dynamic model have the following useful structural properties;

Property 1: $M(q)$ is a symmetric positive definite matrix

$$m_1 \|x\|^2 \leq x^T M(q)x \leq m_2 \|x\|^2; \forall x \in R^n \tag{2}$$

where m_1, m_2 are known positive scalar constants, $x \in R^n$ is a vector, $\|\|$ denotes the Euclidean vector norm.

Property 2: $M(q) - 2C(q,\dot{q})$ is a skew symmetric matrix

$$x^T[\dot{M}(q) - 2C(q,\dot{q})]x = 0 \tag{3}$$

for any vector $x \in R^n$.

Assumption 1: The term $F(q,\dot{q})$ is a bounded function satisfying

$$|F(q,\dot{q})| \leq \bar{F} \tag{4}$$

The control objective of this paper is to design a stable control law to ensure that the real joint position vector q can track the desired joint position vector q_d, that is, the tracking error asymptotically converge to zero in spite of uncertainties and external disturbances.

3 Controller Design

In this section, an adaptive fuzzy PID sliding mode controller is proposed for the nonlinear robot dynamic system (3) in the presence of large uncertainties and external disturbances. First, the controller structure is developed based on PID sliding mode control. Then, two fuzzy systems for reaching and equivalent parts are proposed to handle uncertainties and to solve the chattering problem in conventional SMC.

3.1   Traditional SMC

Lets $e(t) = q_d(t) - q(t)$ be the tracking error vector of robot manipulators. In order to apply the SMC, the first step is designed the sliding surface function which contains the integral part in addition to the derivative term

$$s = \dot{e} + \lambda_1 e + \lambda_2 \int_0^t e \, dt \tag{5}$$

where $\lambda_1 = diag(\lambda_{11}, \lambda_{12}, \ldots, \lambda_{1n})$, $\lambda_2 = diag(\lambda_{21}, \lambda_{22}, \ldots, \lambda_{2n})$, with $\lambda_{1i}, \lambda_{2i} > 0$ for every $i = 1, 2, \ldots n$, $s = [s_1, s_2, \ldots, s_n]^T$, $e = [e_1, e_2, \ldots, e_n]^T$. It can be easily seen that, $s = 0$ is a stable sliding surface and $e \to 0$ as $t \to \infty$.

The robot dynamic equations can be rewritten based on the sliding surfaces as follow

$$M\dot{s} = -Cs + f + F(q, \dot{q}) - \tau \tag{6}$$

where

$$f = M(\ddot{q}_d + \lambda_1 \dot{e} + \lambda_2 e) + C(\dot{q}_d + \lambda_1 e + \lambda_2 \int_0^t e \, dt) + G$$

According to the sliding mode design procedure, the control input u consists of the components

$$\tau = u_{eq} + u_{SW} = u_{eq} + K_{SW} sign(s) \tag{7}$$

where $K_{SW} = diag(k_{SW1}, k_{SW2}, \ldots, k_{SWn})$, $k_{SW1}, k_{SW2}, \ldots k_{SWn}$ are positive constants. When the system operates on the sliding mode, the SMC theory gave us $\dot{s} = 0$, the equivalent control can be obtained by equation $\dot{s} = 0$ for nominal system in the absence of the uncertainties and external disturbances

$$u_{eq} = f + Ks \tag{8}$$

where $K = diag(k_1, k_2, \ldots, k_n)$, $k_1, k_2, \ldots k_n$ are positive constants.

The stability of the close loop system (7) can be easily proved by Lyapunov theory if the gains of the switching controller are bigger than the upper bounds of uncertainties. However, for handling large uncertainties and external disturbances, a large gain of the switching controls K_{SW} must be applied; this leads to large chattering appear in the control input signal, it is not appropriate to be applied in robotics in real applications. In our proposed controller, first, to reducing the effect of uncertainties and external disturbances in the equivalent part, an adaptive fuzzy system is applied to compensate the unknown $F(q, \dot{q})$. Then, fuzzy logic system is designed to attenuate the chattering problem which cause by sign function in reaching phase of conventional SMC. In the following subsections, these above components are described in detail.

3.2 Fuzzy Design for Reaching Phase

In order to eliminate the chattering phenomenon, fuzzy logic system is used for reaching phase. By using this method, the robust behavior of the system not only can guarantee but also can eliminate the chattering problem in conventional SMC.

The reaching law for our proposed control is selected as

$$u_{SW} = k_{fs}u_{fs} \tag{9}$$

where k_{fs} is the normalization factor of the output variable, u_{fs} is the output of the fuzzy SMC (FSMC), which is determined by the normalized values of s and $\dot{s}$.

$$u_{fs} = FSMC(s, \dot{s}) \tag{10}$$

where $FSMC(s, \dot{s})$ is the functional characteristics of the fuzzy linguistic decision schemes. The membership function of the linguistic variable can be defined as the following fuzzy subsets:

$$s = \dot{s} = \{NB, NS, Z, PS, PB\} \tag{11}$$

$$u_{fs} = \{NVH, NH, NVB, NB, NM, NS, Z, PS, PM, PB, PVB, PH, PVH\} \tag{12}$$

where NVH: negative very huge, NH: negative huge, NVB: negative very big, NB: negative big, NM: negative medium, NS: negative small, Z: zero, PS: positive small, PM: positive medium, PB: positive big, PVB: positive very big, PH: positive huge, PVH: positive very huge. The fuzzy logic system rule is design as Table 1.

Table 1. Rule-base of FSMC

u_{fs}		S				
		NB	NS	Z	PS	PB
$\dot{s}$	NB	PVH	PH	PVB	PB	PM
	NS	PH	PVB	PB	PM	PS
	Z	PM	PS	Z	NS	NM
	PS	NS	NM	NB	NVB	NH
	PB	NM	NB	NVB	NH	NVH

3.3 Adaptive Fuzzy Approximation in Equivalent Control Part

In order to handle the large unknown uncertainties and external disturbances in the equivalent control, an adaptive fuzzy logic system (AFLS) as show in Fig. 1 is adopted to compensate the unknown function $F(q, \dot{q})$ in each joint of robotic manipulator. Then, adaptive learning algorithms are derived to adjust on-line the output weights in the AFLS.

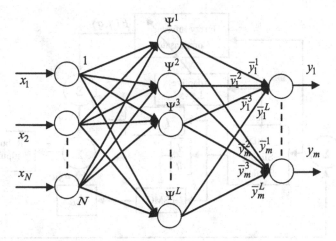

Fig. 1. Structure of adaptive fuzzy system

Thus, the equivalent control (8) is replaced by following equation

$$u_{eq} = f + \hat{F}(q, \dot{q}) + Ks \tag{13}$$

where

$$\hat{F}(q, \dot{q}) = \begin{bmatrix} \hat{F}_1 \\ \hat{F}_2 \\ \vdots \\ \hat{F}_n \end{bmatrix} = \begin{bmatrix} W_{F1}^T \Psi(e, \dot{e}) \\ W_{F2}^T \Psi(e, \dot{e}) \\ \vdots \\ W_{Fn}^T \Psi(e, \dot{e}) \end{bmatrix} = W_F^T \Psi(e, \dot{e}) \tag{14}$$

Letting the optimal parameter matrix of the AFLS is W_F^*, the approximating error is

$$\tilde{F}(q, \dot{q}) = \begin{bmatrix} W_{F1}^{*T} \Psi(e, \dot{e}) \\ W_{F2}^{*T} \Psi(e, \dot{e}) \\ \vdots \\ W_{Fn}^{*T} \Psi(e, \dot{e}) \end{bmatrix} - \begin{bmatrix} W_{F1}^T \Psi(e, \dot{e}) \\ W_{F2}^T \Psi(e, \dot{e}) \\ \vdots \\ W_{Fn}^T \Psi(e, \dot{e}) \end{bmatrix} + \begin{bmatrix} \varepsilon_1 \\ \varepsilon_2 \\ \vdots \\ \varepsilon_n \end{bmatrix} = \tilde{W}_F^T \Psi(e, \dot{e}) + \varepsilon \tag{15}$$

where $\tilde{W}_F = W_F^* - W_F$. It is assumed that the term ε is a bounded function satisfying

$$\varepsilon \le \bar{\varepsilon} \tag{16}$$

The overall control τ becomes

$$\tau = f + \hat{F}(q, \dot{q}) + Ks + k_{fs} u_{fs} \tag{17}$$

The diagram of the proposed controller is shown in below figure (Fig. 2).

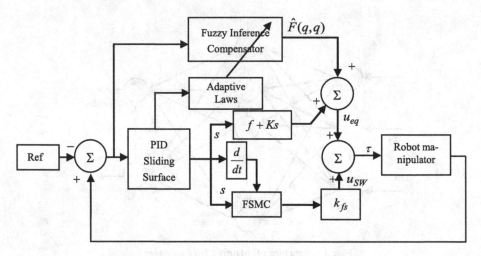

Fig. 2. Block diagram of the proposed controller

Theorem 1: Consider the n–link robot manipulator (1), if the sliding mode surface is chosen as (5), the controller is (17), and the adaptive update laws of the fuzzy logic system are designed as (18), then the tracking error asymptotically converge to zero.

$$\dot{W}_{Fi} = \Gamma_i^{-1} s_i \Psi(e, \dot{e}) \tag{18}$$

where $i = 1, 2, \ldots, n$, Γ_i are constant symmetric positive definite matrices.

4   Simulation Results

In order to verify the effectiveness of the adaptive fuzzy PID sliding mode control method, its overall procedure is applied for a two-link planar robotic manipulator. The simulations are performed in the MATLAB-Simulink environment using ODE 4 solver with a fixed-step size of 10^{-4} s.

The dynamic equation of the two-link robot is described as follows

$$M(q)\,\ddot{q} + C(q, \dot{q}) + G(q) + F(q, \dot{q}) = \tau(t) \tag{19}$$

where

$$M(q) = \begin{bmatrix} l_2^2 m_2 + 2l_1 l_2 m_2 c_2 + l_1^2 (m_1 + m_2) & l_2^2 m_2 + l_1 l_2 m_2 c_2 \\ l_2^2 m_2 + l_1 l_2 m_2 c_2 & l_2^2 m_2 \end{bmatrix}$$

$$C(q, \dot{q}) = \begin{bmatrix} -m_2 l_1 l_2 s_2 \dot{q}_2^2 - 2m_2 l_1 l_2 s_2 \dot{q}_1 \dot{q}_2 \\ m_2 l_1 l_2 s_2 \dot{q}_1^2 \end{bmatrix}$$

$$G(q) = \begin{bmatrix} m_2 l_2 g c_{12} + (m_1 + m_2) l_1 g c_1 \\ m_2 l_2 g c_{12} \end{bmatrix}$$

where $q = (q_1, q_2)^T$ is the joint variable vector, m_1 and m_2 are the link masses, l_1 and l_2 are the link lengths, gravity $g = 9.81$. The symbols s_1, s_2, s_{12} and c_1, c_2, c_{12} are respectively defined as: $s_1 = \sin(q_1)$, $s_2 = \sin(q_2)$, $s_{12} = \sin(q_{12})$, $c_1 = \cos(q_1)$, $c_2 = \cos(q_2)$, $c_{12} = \cos(q_{12})$.

The friction modeling for the dynamic system, including the viscous friction and the Coulomb friction torques, are selected as

$$F(q, \dot{q}) = \begin{bmatrix} \tau_{d_1} \\ \tau_{d_2} \end{bmatrix} = \begin{bmatrix} 8\sin(\dot{q}_1) + 1.3 sign(\dot{q}_1) \\ 8\sin(\dot{q}_2) + 1.3 sign(\dot{q}_2) \end{bmatrix} \tag{20}$$

The external disturbances are selected as

$$\tau_d = \begin{bmatrix} \tau_{d_1} \\ \tau_{d_2} \end{bmatrix} = \begin{bmatrix} 0.5\sin(2t) \\ 0.8\cos(2t) \end{bmatrix} \tag{21}$$

The parameters values employed to simulate the robot are given as of $l_1 = 1$ m, $l_2 = 0.8$ m, $m_1 = 1$ kg and $m_2 = 1.5$ kg. The reference trajectories are given by

$$q_{1d} = 0.3\sin(t)$$
$$q_{2d} = 0.5\sin(t) \tag{22}$$

The initial states are chosen as

$$q_1(0) = 0.1, \; q_2(0) = -0.2, \; \dot{q}_1(0) = 0, \; \dot{q}_2(0) = 0 \tag{23}$$

To illustrate the effectiveness of the proposed control method (NAF-PIDSMC) (17), in simulation we compared with two other controllers:

- A conventional PID sliding mode control (PIDSMC)

$$\tau = f + Ks + K_{SW} sign(s) \tag{24}$$

where K and K_{SW} are diagonal positive matrixes chosen as $\lambda_1 = diag(10, 10)$, $\lambda_2 = diag(0.1, 0.1)$, $K = diag(10, 10)$ and $K_{SW} = diag(10, 15)$.

- Adaptive fuzzy PID sliding mode control (AF-PIDSMC) which used only an adaptive fuzzy inference engine to approximate the unknown term in the equivalent control part of SMC

$$\tau = f + \hat{F}(q, \dot{q}) + Ks + K_{SW} sign(s) \tag{25}$$

where K and K_{SW} are diagonal positive matrixes chosen as $\lambda_1 = diag(10, 10)$, $\lambda_2 = diag(0.1, 0.1)$, $K = diag(10, 10)$ and $K_{SW} = diag(1, 0.5)$.

The parameter in the proposed controller are selected as $\lambda_1 = diag(10, 10)$, $\lambda_2 = diag(0.1, 0.1)$, $K = diag(10, 10)$ and $K_{fs} = diag(1, 0.5)$.

The simulation results for these control methods are shown in Figs. 3, 4 and 5. The tracking performance and tracking error are depicted in Figs. 3 and 4. As can be seen from these figures, the proposed controller (NAF-PIDSMC) delivers the smallest tracking errors compared with the PIDSMC and AF-PICSMC. In addition, it can be seen that the uncertainties and external disturbances are greatly compensated by using AFLS, therefore, the errors from NAF-PID-SMC and AF-PICSMC are highly reduced compared with the PICSMC method.

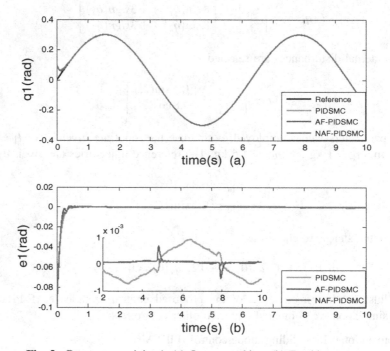

Fig. 3. Responses at joint 1: (a) Output tracking, (b) Tracking error.

Figure 5 shows the time histories of the applied control torque input. From the figure it can be observed that, in the PIDSMC scheme, the big gain has to be applied in the switching control term to make the system stable, hence a lot of chattering is appeared in the control input signal. Meanwhile, AFLC can greatly compensate for uncertainties and external disturbances in the AF-PIDSMC scheme, therefore the chattering phenomenon is significantly reduced because the small gain is applied in the discontinuous term. Especially, in the proposed control scheme, the smaller control effort is applied, the higher precise tracking error is achieved, and the chattering phenomenon is eliminated.

In summary, the simulation results demonstrate that the proposed controller is highly efficient in the control of the robotic manipulators in the presence of large uncertainties and external disturbances.

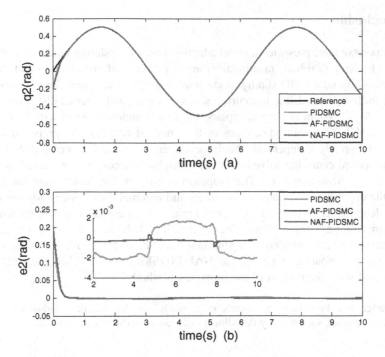

Fig. 4. Responses at joint 2: (a) Output tracking, (b) Tracking error.

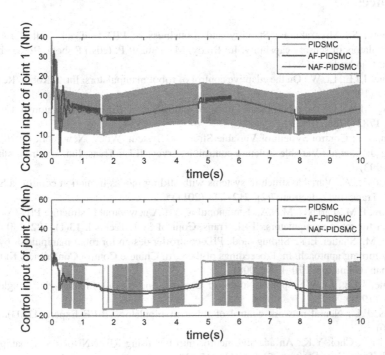

Fig. 5. Control input at joints 1 and 2.

5 Conclusion

In this paper, we have presented a novel adaptive fuzzy PID sliding mode controller for trajectory tracking of robotic manipulators in the presence of uncertainties and external disturbances. Based on PID sliding mode framework, one adaptive fuzzy system was used to handle with modeling inaccuracies, uncertainties and external disturbances and one fuzzy logic system is used to replace the discontinuous control term in the conventional SMC. The main advantages of this method are: (1) The proposed control strategy can greatly compensate the large uncertainties and external disturbances. (2) The proposed controller solved the major drawback of conventional SMC, which is the chattering phenomenon. (3) The proposed controller does not require the knowledge of the upper bounds of any uncertainties and external disturbances. Moreover, the adaptive learning algorithms of the fuzzy logic system are derived to adjust on-line without any offline training phase. The strict proof of the stability of the closed-loop system has been accomplished. The simulation results of the two-link robotic manipulator system comparing the proposed NAF-PIDSMC, AF-PIDSMC, and PIDSMC demonstrated the effectiveness of the proposed method.

Acknowledgment. Following are results of a study on the "Leaders Industry-university Cooperation" Project, supported by the Ministry of Education(MOE).

References

1. Arimoto, S., Miyazaki, F.: Stability and robustness of PID feedback control for robot manipulators of sensory capability. In: Brady, M., Paul, R.P. (eds.) Robotic Research. MIT Press, Cambridge (1984)
2. Slotine, J.J.E., Li, W.: On the adaptive control of robot manipulators. Int. J. Robot. Res. **6**(3), 49–59 (1987)
3. Spong, M.W.: On the robust control of robot manipulators. IEEE Trans. Automat. Contr. **37** (11), 1782–1786 (1992)
4. Utkin, V.I.: Control System of Variable Structure System. Wiley, New York (1976)
5. Hung, J.Y., et al.: Variable structure control: a survey. IEEE Trans. Ind. Electron. **40**(1), 2–22 (1993)
6. Sabanovic, A.: Variable structure systems with sliding modes in motion control- a Survey. IEEE Trans. Ind. Electron. **7**(2), 212–223 (2011)
7. Jafarov, E.M., Parlakci, M.N.A., Istefanopulos, Y.: A new variable structure PID-Controller design for robot manipulators. IEEE Trans. Control Syst. Technol. **13**(1), 122–130 (2005)
8. Ataei, M., Shafiei, E.Y.: Sliding mode PID-controller design for robot manipulator by using fuzzy turning approach. In: Proceedings of the 27th Chinese Control Conference, Kunming, Yunnan, China, pp. 170–174 (2008)
9. Slotine, J.J.E., Li, W.: Applied Nonlinear Control, pp. 41–190. Prentice-Hall, EngleWood Cliffs (1991)
10. Lewis, F.L.: Neural network control of robot manipulators. IEEE Expert. **11**(03), 64–75 (1996)
11. Lee, M.J., Choi, Y.K.: An adaptive neurocontroller using RBFNN for robot manipulator. IEEE. Trans. Ind. Electron. **51**(3), 711–717 (2005)

12. Hwang, G.C., Lin, S.C.: A stability approach to fuzzy control design for nonlinear systems. Fuzzy Set. Syst. **48**, 179–287 (1992)
13. Kim, S.W., Lee, J.J.: Design of a fuzzy controller with fuzzy sliding surface. Fuzzy Set Syst. **71**, 359–367 (1995)
14. Yu, X., Man, Z., Wu, B.: Design of fuzzy sliding mode control system. Prentice Fuzzy Set Syst. **95**(3), 295–306 (1998)
15. Yoo, B., Ham, W.: Adaptive fuzzy sliding mode control of uncertain nonlinear systems. IEEE Trans. Fuzzy Syst. **6**, 315–321 (1998)
16. Wu, T.Z., Juang, Y.T.: Adaptive fuzzy sliding mode controller of uncertain nonlinear systems. ISA Trans. **47**, 279–285 (2008)
17. Roopaei, M., Jahromi, M.Z., Jafari, S.: Adaptive gain fuzzy sliding mode control for the synchronization of nonlinear chaotic gyros. Chaos **19**, 013125 (2009)
18. Abdelhameed, M.N.: Enhancement of sliding mode controller by fuzzy logic with application to robotic manipulator. Mechatronics **15**, 439–458 (2005)
19. Roopaei, M., Jahromi, M.Z.: Chattering-free fuzzy sliding mode control in MIMO uncertain systems. Nonlinear Anal. **71**, 4430–4437 (2009)
20. Yau, H.T., Chen, C.L.: Chattering free fuzzy sliding mode control strategy for uncertain chaotic systems. Chaos Solitons Fract. **30**, 709–718 (2006)
21. Li, T.H.S., Huang, Y.C.: MIMO adaptive fuzzy terminal sliding mode controller for robotic manipulators. Inf. Sci. **180**(23), 4641–4660 (2010)
22. Craig, J.J.: Introduction to Robotics Mechanics and Control, 3rd edn. Pearson Prentice Hall, Upper Saddle River (2005)
23. Khalil, H.K.: Nonlinear Systems, 2nd edn. Prentice Hall, Englewood Cliffs (1996)

Multiple Convolutional Neural Network
for Feature Extraction

Guo-Wei Yang[1,2] and Hui-Fang Jing[1(✉)]

[1] College of Information Engineering,
Nanchang Hangkong University, Nanchang 330063, China
{ygw_ustb, jacelly1314}@163.com
[2] School of Technology, Nanjing Audit University, Nanjing 211815, China

Abstract. Recent theoretical studies indicate that Deep Neural Network has been applied to many image processing tasks. However, learning in deep architectures is still difficult. One of the neural network, Convolutional Neural Network (CNN) has gained great success in image recognition and it builds features by automatic learning. More importantly, CNN can operate directly on the gray image, so it can be directly used for processing classification of the image. In order to utilize CNN to recognize plant leaf, a hierarchical model based on CNN is proposed in this paper. We firstly do some pre-processing, such as illumination changes, rotation and leaf distortion. After that, we applied the method of CNN to extract the features of leaves pictures. One focus on our network is about the depth of CNN, which affects the ability of capability of convolution. Thus, we try our best to choose the best depth of CNN with several experiments. Moreover, in order to destroy the symmetry of networks, the strategies used in this paper is to add a mathematical formula for feature map connection between convolutional layer and sampling layer. The experimental results show that the proposed method is quite effective and feasible. And we also applied other classification methods to the ICL dataset. By contrast, our classification is much better than other methods.

Keywords: Classification · CNN · Image recognition · Plant leaf classification · Feature extraction

1 Introduction

Image Classification is an active research topic in computer vision and pattern recognition. In addition, it has been used in many applications, such as face recognition, biomedical image analysis. Traditional method depends largely on the choice of the characteristics, and the researchers tend to choose the characteristics, which is blind in some extent. In order to solve the above problems, this paper proposes a Deep Learning (DL) method that can learn features automatically on two-dimensional image.

The concept stems from the *Deep Learning* of artificial networks, including multi-layer, multi-hidden Layer Perceptron (MLP) is a kind of *Deep Learning* structure. Deep structure of local (non-linear processing involves multiple cell layers) non-convex objective minimum cost function is prevalent mainly training difficult.

© Springer International Publishing Switzerland 2015
D.-S. Huang et al. (Eds.): ICIC 2015, Part II, LNCS 9226, pp. 104–114, 2015.
DOI: 10.1007/978-3-319-22186-1_10

Before 2006, verities of machine learning method has been applied to all kinds of classification tasks. Such as Support Vector Machine (SVM), Boosting and Decision Tree have gained great success in the classification task [1, 2]. Many scholars focus on SVM and have got state-of-the-art results [3]. However, deep neural network re-boomed, when Hinton and Salakhutdinov proposed an efficient way to train deep architecture through layer-wised pre-training [4] for solving optimization problems associated with the deep structure of hope, and then proposed a multi-layer automatic encoder deep structure. *Deep learning* have great advantage in feature extracting, because they have better designs for modeling and training and get much more complex features than shallow architectures [5, 6]. In 1980, CNN was first proposed by Japanese scholars Fukushima, then Yann LeCun, et al. got further improve and perfect [7], which uses spatial relationship relative reduction in the number of training parameters to improve the performance of BP. In 2003, Sven Behnke conducted generalization, the same year, Patrice Simard, et al. made further simplified [8]. After Dan Ciresan, et al. further optimizing in 2011 and 2012, they created the best results at the time of recognition in multiple image databases (MNIST, NORB, HWDB1.0, CIFAR10) with CNN [9, 10]. Moreover, depth study also appeared in many deformed structures such as DAE, DCN and so on.

In this paper, we propose a Convolutional Neural Network (CNN) to solve image recognition problem, and this is a difficult problem because of lighting conditions, and the images are usually noisy or broken. We applied to extract plant leaf features and then classify these features. Then we analyzed the results of this process and the results prove the performance surpass pure other classification method.

The rest of the article is organized as follows. First, we describe the classical architecture about CNN and other approaches on Sect. 2. We then detail our CNN architecture, followed by experiments that evaluate our method. Finally, we conclude the whole paper with several discussions of future works.

2   Related Works

2.1   Convolutional Neural Network Architecture

Convolutional Neural Network is specifically designed to deal with a class of two-dimensional data of the multi-layer neural network. CNN is one of the important algorithm for *Deep Learning*, because it can be reached prior to the dissemination of improving network efficiency BP by reducing the number of network-related spatial data mining on the training parameters and requires very little data pre-processing. And CNN shows excellent results in many applications. In [7], it contrasts multiple recognition algorithm, the conclusion is superior to all over CNN algorithms. CNN is improved by BP neural network and similar to BP. Moreover, they both have adopted a forward propagation to calculate the output value, back propagation adjust the weights and bias. The biggest difference between CNN and standard is that the neurons of CNN between adjacent layers is not fully connected, but part of the connection, which is the perception of a nerve cell area from the upper part of the nerve cells, rather than BP, as connected with all the neurons.

CNN has three important ideological framework: local perception of the area, weights sharing, down-sampling space or time domain, which makes the network to achieve a relative displacement, scaling and distortion characteristics have remained unchanged. To reflect these three basic concepts, CNN will integrate feature extraction capabilities into a Multilayer Perceptron (MLPs) through restructuring or reducing weights. From [11], it is easy to understand that the right values and the required number of training data in close contract. With weights sharing, it is more similar to biological network, which can reduce the complexity of the network model [12].

CNN in close contact and spatial information makes it particularly suitable for image processing and understanding, and can automatically extracted the wealth of relevant characteristics from the image. In image and voice recognition tasks, multi-layer back-propagation neural network training network can learn complex, multi-dimensional, non-linear classification of surfaces from a large number of training examples [13].

2.2   Basic Framework for CNN

CNN is a multi-layer neural network, which is composed of multiple convolution layer and sub-sampling layer. Moreover, each layer is composed of multiple independent neurons. It basic framework is shown in Fig. 1. Basic principle and derivation can also be learned [14] simply. CNN consists of three different types of layers, convolution layer, sub-sampling (optional), and a fully connected layer. Convolution layer is responsible for feature extraction, which uses two key concepts: weights sharing and field accepting. The sub-sampling layer performs local averaging and sub-sampling, reducing the sensitivity of the offset and distortion. What full connectivity layer do is to perform a normal classification just like a traditional MLP networks.

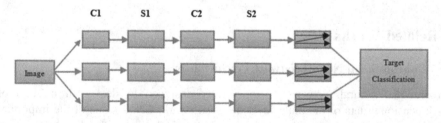

Fig. 1. Basic framework for CNN

Figure 1 shows that CNN consists of two layers, composed of two sub-sampling alternately. C layer is identified as Convolutional layer, also referred to as feature extraction layer. The input of each neuron is connected with a local receptive field of previous layer and it extracts the local feature. C layer has many different two-dimensional feature maps, a characteristic diagram extracting a characteristic, which means that can extract various features. When extracting features, the weight of the same feature map are shared. Even if the same convolution kernel, different feature maps use different convolution kernel. C layer preserved different local features, so that the extracted features have a rotation and translation invariance.

S layer identifies the sub-sampling level, also called characteristic mapping layer, which is responsible for the characteristic C layer obtained sub-sampling, so that the extracted features having scaling invariance. S layer only do a simple scaling layer mapping, relatively few neurons weights, which is also simply calculation. In general, convolution and sub-sampling layer appear assembly. But [8] indicates that convolution and sampling layer can be produced through a convolution layer embodiment convolution step size which is equal to embodiment 1 of the sub-sampling layer removing. Convolution is equivalent to one layer of the two functions previously without affecting the performance of CNN, so sub-sampling layer is optional layer.

In the CNN end layer is generally connected to one or several full connection, the final output of the number of nodes is classified target number. Moreover, the purpose of training is to make the output of CNN as close as possible to the original label. The layer is always in use in CNN, its role is similar to the standard MLP. After all the connections in the hidden layer and output layer information conversion and calculation process is complete forward propagation of a learning process, the final result output to the outside by the output layer.

In the learning process, the update of the CNN weights is based on back-propagation algorithm (B-P) [15], which update the weights ϖ of neurons by the time t to t + 1.

$$\varpi(t+1) = \varpi(t) + \eta\delta(t)\chi(t) \tag{1}$$

Here, η is the learning rate, $\chi(t)$ is the input to neurons. And $\delta(t)$ is the error term of neurons, the error term for the output neurons and hidden neurons have different expression. It should be noted, though generally increasing with the number of iterations the learning rate decreasing, but usually, the learning rate is the same for all weights η.

Results of sampling layer generates down-sampling enter after. If there have N input maps, they also have N output maps, although output map is smaller than input map.

$$x_j^l = f(\beta_j^l \, down(X_j^{l-1}) + b_j^l) \tag{2}$$

Here, down() represents down-sampling function. Difficulty here is that the calculation of the error signal in FIG. Assume down-sampling layer and the next layer is a layer of convolutional layer. If you fall behind all sampling layer fully connected network, then the error signal diagram can be obtained directly by the back-propagation algorithm.

2.3 Compared with Other Approaches

The easiest way of Deep Learning is to take advantage of the characteristics of artificial neural networks. Artificial Neural Network (ANN) have a system hierarchy itself, if given a neural network, we assume that the output and the input are the same. Then we train adjust its parameters, obtaining in each layer weights. Bourlard and Kamp [16] first proposed auto-association in 1988, to achieve this reproduction, the automatic encoder is necessary to capture the most input of the input data, so as PCA, to find the main component that may represent the original information. AE can be simply learned by error back propagation algorithm [17]. Auto-encoder (AE) is a neural network as much as possible to reproduce the input signal. AE neural is an unsupervised learning algorithm that applies back propagation, setting the output equal to the input. It directly encodes the input layer and then decodes the hidden layer. The aim of an AE is to learn a compressed or sparse representation (encoding) for a set of data.

Auto-encoder has some variants, here briefly the next two. If we add the L1 Regularity Auto-encoder limit on the basis (L1 mainly constrained nodes in each layer as 0, most have only a few non-zero, this is the Sparse source of the name), we can get Sparse Auto-encoder. In fact, every expression was to limit as much as possible sparse code each time.

Automatic noise reduction encoder DA is based on automatic encoder, the training data is added noise, so automatic encoder must learn to eliminate this noise and get real noise pollution had not been entered [18, 19]. Therefore, it forced the encoder to learn more robust expression of the input signal, which is than encoder.

There are some other methods of Deep Neural Network, here is not outlined. By other means, we summarize some of the advantages of CNN. CNN is mainly used to identify displacement, scaling and other forms of distortion invariant two-dimensional graphics. Because of CNN's feature detection layer lean through training data, when using CNN, avoiding explicit feature extraction, and implicitly learn from training data. Also due to the same neuron weights, the same surface feature mapping can be a parallel study, which is the convolution of the network with respect to a big advantage neuronal networks connected to each other. Convolutional Neural Network with its special structure has a local weights sharing in speech recognition and image processing unique advantages, its layout closer to the actual biological neural networks, the weights sharing reduces the complexity of the network, in particular multi-dimensional input vector images can be directly input network feature that avoids the process of feature extraction and classification data reconstruction complexity.

Convolutional neural network, it avoids the explicit characteristics of sampling, implicitly learning from the training data. This makes it significantly different from other convolution neural network classifier based on neural networks, through restructuring and reducing weights will feature extraction capabilities integrated into a multi-layer perceptron. It can directly handle grayscale images, can be directly used for processing image-based classification.

CNN more general neural network has the following advantages in image processing: First, inputting image and the network topology can be a good fit. Second, feature extraction and pattern classification is at the same time, and at the same time producing in training. Third, weight sharing can reduce training parameters of the network, so that the neural network structure becomes more simple and adaptable.

The close relationship between this layer and the airspace contact information of CNN is making it suitable for image processing and understanding. Moreover, its salient features automatic extraction of images side also showed a relatively better performance.

3 The Architecture Used in This Paper

First, we assume that the input gray-scale image is an image of the blades 64 × 64. Plant leaves frame identification of CNN based on this design is shown in Fig. 2.

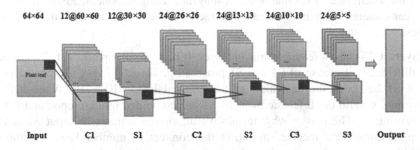

Fig. 2. Plant leaf identification framework

Input. We first do pre-processing for our leaves image as the input, it includes gradation processing, filtering process and gray-scale. When the image size is not 64 × 64, we use bilinear difference computation to scale image. Thus, it can meet the requirements of a standard input.

C1 Layer. C1 is a feature extraction layer, and its input is either from the input layer or sampling layer. In our method, C1 gets 12 two-dimensional feature maps for 60 size. In fact, it is obtained by a 5 × 5 convolution kernel. Each map of a convolution layer has the same size as the convolution kernel. When the convolution kernel is too small, it cannot be effectively extracted local feature, and when the convolution kernel is too large, the complexity extraction features may far exceed the ability of the convolution kernel represents. Therefore, it is important to set the appropriate convolution kernel to improve the performance of CNN, and it is also the difficulty to tune parameters for CNN. We use a 5 × 5 convolution kernel in the image. When rolling again, we finally get a (64–5 + 1) × (64–5 + 1) feature map. Convolution kernel move one step every time. Each features of convolution layer is a layer different from the convolution kernel on the front of all map elements and corresponding accumulated a bias, and then it obtained by seeking Sigmoid. In practice, it plus an offset term when convolution. For image χ, it uses convolution kernel ϖ convolution, offset term b. The output y convolution is:

$$y = Sigmoid(\varpi\chi + b) \tag{3}$$

S1 Layer. S1 is a sub-sampling layer, it gets 12 feature maps for 30 × 30 size. Sub-sampling layer is a layer of a map of the sampling handling, what sampling model here is a small area on one of the adjacent map of the statistical aggregation. Moreover, it is get maximum value of the small region. In this paper, it sums by all non-overlapping sub-block χ of 2 × 2 in C1. Then it multiplies by weights ϖ and pluses an offset b. The calculation of sub-sampling is:

$$y = Sigmoid(\varpi \cdot \underset{x_i \in x}{sum}(\chi_i) + b) \tag{4}$$

Because the feature map size of C1 is 60×60, the final result for sub-sampling is 30×30 feature map. In CNN, the general scaling factor is 2. The reason for this is to control the zoom rate of decline, since scaling is scaling exponent. Reduced too fast also means more rough image feature extraction, and image detail will lose more features.

C2 Layer. C2 is also a feature extraction layer. It has a similar place with C1, but also some differences. Because of C1 and S1 processing, virtually every neuron receptive field of S1 has been the equivalent of the original image 10×10. When making convolution, several or all feature maps by S1 are combined into an input, and the do the convolution. The reason why not always all of the features as input S1 is that incomplete connection mechanism allows the connection number keeping within a reasonable range. At the same time, the most important is destroying symmetry networks. Strategies used in this paper is that the feature map of S1 is connected to all that can be divisible by 1 feature maps. That is to say all of the feature maps of C2. The feature map 2 of S1 is connected to all even-numbered feature maps. The following is connected through this method (Fig. 3). Here, 'o' indicates a connection.

	0	1	2	3	4	5	6	7
0	o							
1	o	o						
2	o		o					
3	o	o		o				
4	o				o			
5	o	o	o			o		
6	o						o	
7	o	o		o				o
8	o		o					
9	o	o			o			
10	o							
11	o	o	o	o		o		

Fig. 3. The connection between C2 layer and S1 layer

The remaining layers of convolution and sub-sampling do the same thing as previous layers. Only as the depth increases, the extracted features is more abstract and more expressive capability. That is the depth meaning of Deep Neural Network. By C1, S1, C2, S2, the extracted feature have the ability to express. But the correct accuracy rates is low. It shows that the depth affects the ability of CNN.

4 Experiments

4.1 Data Set and Experiment Setup

The first step in identifying the image processing matter blade, blade pretreated by contour information calculation, obtaining the identification of the blade eigenvectors.

Here, we used the leaves images from ICL dataset. For more information to this database, please infer to [20].

First, we pick and choose those typical leaf shape and full leaf (Fig. 4), then do pre-processing for leaves image. Our image processing including gradation processing, filtering process. First, in order to remove the interference color, we go to the image color processing, the color difference in the background by drinking blade, converting to the grayscale image. Second, the filtering processing, the image of the blade after filtering operation to remove the blade internal hole and the surrounding noise such as small debris, making it more suitable for contour extraction.

Fig. 4. The typical leaves of ICL

The original data set contains some leaf petiole, which may affect the robustness of classification algorithm. We transform those pictures into 64 × 64 gray pixels. Because the acquisition of the blade may result in the acquisition of the operator petioles of varying lengths, the information it provides cannot be used as basis for classification. So before making classification, we first remove the petiole.

Because convolutional neural network in Fig. 2 only accept gray-scale images for 64 × 64 size, the ICL database is high-definition color images. After above pre-processing, we then change the picture size with bilinear interpolation geometric. Finally, we put the processing data to our Convolutional Neural Network.

The choice of CNN layers is also a problem. We know the number of parameters is the convolution kernel and offset + sampling layer scale factor and offset + full connectivity layer weights and bias, and the number of parameters entire link layer is more than two orders of magnitude than the number of arguments before the four-story.

Another innovation in this paper is to destroy the symmetry of networks, our strategies here used is adding a mathematical formula for the feature map connection between convolutional layer and the sampling layer. We can see the connection between C2 and S1 in Fig. 3.

In our experiment, our method calls CNN-1, and the most important parameter is the size of mini-batch. In this paper, the mini batch setting as 1 seem to take the fastest time of convergence. Random samples of training sequence, then, each gradient produced is random. There is also an argument learning rate. In the experiment, increasing the situation does not move mini-batch training will appear appropriate to increase the learning rate that can make training cost to decline. The parameters used in our experiment are in Table 1.

During this experiment, our experimental environment is shown in Table 2.

Table 1. The parameters of the architecture

Method	Activation function	Batch size	Learning rate	Alpha	Kernel size	Scale
CNN-1	Sigmoid	100	0.01	1	5	2×2

Table 2. The experimental environment

Hardware environment	Software environment
CPU: Intel Core i3-2370	Operating System: Windows 64-bit
RAM: 8G	Development Tool: Matlab 2012b

4.2 Results on the ICL Database

In order to compare the performance of CNN-1 algorithm proposed in this paper, 10 classes of Table 3 report the resulting classification on the best set for new model. We also took another 15 kinds of random leaf data from ICL with the same CNN-1. As can be seen in the table, our training works remarkably well, and in most cases yield significantly better classification performance than SVM.

Table 3. The comparison of the accuracy

Method	10 Classes	15 Classes
SVM	80 %	77 %
CNN-1	95 %	92 %

In these experiments, we can observe three different topics, which is showed by feature extraction obviously. First, the input layer of the data is the better the more normalized. For image processing, such as scaling and graying, is on the back of a great influence classification results. Second, right convolutional neural network values shared not only significantly reducing the number of required training of weights, greatly reducing the amount of training data required, but also reduced the complexity of neural convolutional neural network, reducing the over-fitting problem. Network layer choice is also important, great depth on CNN performance, inadequate depth will weaken CNN feature extraction capability. In order to break the symmetry, from the first layer to the second convolution layer is not full link layer. They can automatically learn the combination coefficient or human requirements. Finally, the choice of the classifier is not discussed as an important point of this article.

In conclusion, Convolutional Neural Network gives a better performance than the state-of-the-art classification method in our experimental data.

5   Conclusions and Future Works

For the experiment above, we have introduced a very simple training principle for CNN, and we find that CNN have better performance on feature extraction. Traditional artificial leaf extracting feature recognition method based on experience from the blade, and then on the classification features, blindness and low defect classification accuracy. This principle can be used to train and stack features to initialize a deep neural network. A series of image classification experiments were performed to evaluate this new training principle.

Moreover, on a conceptual prospective, CNN is just a Deep Neural Network method to classify images. Other classification method can also achieve classification task. The experimental results with SVM suggest that SVM may also encapsulate a form of robustness in the representations they learn, possibly because of their stochastic nature, which introduces noise in the representation during training.

Although CNN give us a better resolution on feature extraction, there are some problems about CNN architecture we should notice. So for the future work, we will try to identify the blade-based parallelization framework of CNN in order to accelerate the training speed. We also want to play some other more interesting things, such as dropout, max-out, max pooling. At the same time, we want to collect more leaves data to train CNN, so that it can identify more plants. Finally, we hope to get a simple method, widely used, the speedup satisfactory results.

Acknowledgments. The authors would like to sincerely thank the Institute of Machine Learning and Systems Biology of Tongji University and Professor Guo-Wei Yang (No.61272077).

References

1. Cai, C.Z., et al.: SVM-Prot: web-based support vector machine software for functional classification of a protein from its primary sequence. Nucleic Acids Res. **31**(13), 3692–3697 (2003)
2. Christian, S., Laptev, I., Caputo, B.: Recognizing human actions: a local SVM approach. In: Proceedings of the 17th International Conference on Pattern Recognition, 2004, ICPR 2004. vol. 3, IEEE (2004)
3. Mavroforakis, M.E., Theodoridis, S.: A geometric approach to support vector machine (SVM) classification. IEEE Trans. Neural Netw. **17**(3), 671–682 (2006)
4. Hinton, G.E., Salakhutdinov, R.R.: Reducing the dimensionality of data with neural networks. Science **313**(5786), 504–507 (2006)
5. Bengio, Y.: Learning deep architectures for AI. Found. Trends Mach. Learn. **2**(1), 1–127 (2009)
6. Bengio, Y., Courville, A., Vincent, P.: Representation learning: a review and new perspectives, 1–1 (2013)
7. Lecun, Y., Bottou, L., Bengio, Y.: Gradient-based learning applied to document recognition. Proc. IEEE **86**(11), 2278–2324 (1998)
8. Simard, P.Y., Steinkraus, D., Platt, J.C.: Best practice for convolutional neural networks applied to visual document analysis. In: ICDAR, pp. 958–962. IEEE, Los Alamitos (2003)

9. Cires, D.C., Meier, U., Masci, J.: Flexible, high performance convolutional neural networks for image classification. In: Proceedings of the Twenty-Second International Joint Conference on Artificial Intelligence, vol. 2, pp. 1237–1242 (2011)

10. Cires, D.C., Meier, U., Schmidhuber, et al.: Multi-column deep neural networks for image classification. In: IEEE Conference on Computer Vision and Pattern Recognition, pp. 3642–3649, New York (2012)

11. Baum, E., Haussler, D.: What size net gives valid generalization (J). Neural Comput. **1**, 151–160 (1989)

12. Krizhevsky, A., Sutskever, I., Hinton, G.E.: ImageNet Classification with Deep Convolutional Neural Networks. In: NIPS (2012)

13. Duda, R., Hard, P., Stork, D.: Pattern Recognition, 2nd edn. Wiley-Interscience, New York (2000)

14. Behnke, S.: Hierarchical Neural Networks for Image Interpretation [M], 1(4), 541–551 (1989)

15. Lippmann, R.: An introduction to computing with neural nets [J]. IEEE ASSP Magazine **4**, 22 (1987)

16. Bourlard, H., Kamp, Y.: Auto-association by multilayer perceptrons and singular value decomposition. Biol. Cybern. **59**(4–5), 291–294 (1988)

17. Rumelhart, D.E., Hinton, G.E., Williams, R.J.: Learning representations by back-propagating errors. Nature **323**(6088), 533–536 (1986)

18. Vincent, P., Larochelle, H., Bengio, Y., Manzagol, P.-A.: Extracting and composing robust features with denoising AEs. In: ICML (2008)

19. Vincent, P., Larochelle, H., Lajoie, I., Bengio, Y., Manzagol, P.-A.: Stackeddenoising AEs: learning useful representations in a deep network with a local denoising criterion. J. Mach. Learn. Res. **11**, 3371–3408 (2010)

20. http://mlsbl.tongji.edu.cn/chinese/file-database.asp

21. Li, B., Huang, D.S.: Locally linear discriminant embedding: an efficient method for face recognition. Pattern Recogn. **41**(12), 3813–3821 (2008)

22. Huang, D.S., Du, J.-X.: A constructive hybrid structure optimization methodology for radial basis probabilistic neural networks. IEEE Trans. Neural Netw. **19**(12), 2099–2115 (2008)

23. Huang, D.S., Horace, H.S.Ip, Chi, Z.-R.: A neural root finder of polynomials based on root moments. Neural Comput. **16**(8), 1721–1762 (2004)

24. Huang, D.S., Jiang, W.: A general CPL-AdS methodology for fixing dynamic parameters in dual environments. IEEE Trans. Syst. Man Cybern. Part B **42**(5), 1489–1500 (2012)

25. Huang, D.S.: Systematic Theory of Neural Networks for Pattern Recognition (in Chinese). Publishing House of Electronic Industry of China, Beijing (1996)

26. Huang, D.S.: Radial basis probabilistic neural networks: model and application. Int. J. Pattern Recogn. Artif. Intell. **13**(7), 1083–1101 (1999)

27. Wang, X.-F., Huang, D.S.: A novel density-based clustering framework by using level set method. IEEE Trans. Knowl. Data Eng. **21**(11), 1515–1531 (2009)

Hybrid Deep Learning for Plant Leaves Classification

Zhiyu Liu[1]([⊠]), Lin Zhu[1], Xiao-Ping Zhang[1], Xiaobo Zhou[1],
Li Shang[2], Zhi-Kai Huang[3], and Yong Gan[4]

[1] Institute of Machine Learning and Systems Biology, College of Electronics
and Information Engineering, Tongji University, Shanghai, China
liuzhiyu429@163.com
[2] Department of Communication Technology, College of Electronic Information
Engineering, Suzhou Vocational University, Suzhou 215104, Jiangsu, China
[3] College of Mechanical and Electrical Engineering,
Nanchang Institute of Technology, Nanchang 330099, Jiangxi, China
[4] College of Computer and Communication Engineering,
Zhengzhou University of Light Industry, Zhengzhou, China

Abstract. Recently, deep learning is very popular, it has been applied into
many applications, In this paper, a new neural network, hybrid deep learning is
introduced, which included AutoEncoder(AE) and convolutional neural network
(CNN). This neural network is applied for extracting the features of the plant
leaves. In this paper, we proved that hybrid deep learning can extract better
features for classification task. We apply the hybrid deep learning to extract
features of leaf pictures, and then we classify leaves using those features with
SVM, the result suggests that this method is not only better than pure SVM, but
also better than pure AE and pure CNN.

Keywords: Hybrid deep learning · CNN · AE · Classification · Feature
extracting

1 Introduction

There are approximately fifty thousand plants on the earth, classification and identification of plants species is very important to botanical research and agricultural production. It is foundational for people to understand the basic characteristics of plants.
To traditional research of plant classification, traditional plant classification is based on
artificial extract features. It has strong subjective, and person is easy to make mistake,
what is more, it is very difficult for people to extract features. Currently, deep learning
is very popular. Deep learning has the advantage of extract features automatically, the
method take great convenience to plant classification, and it also generate better result
on plant classification. We also find hybrid deep learning performs better in leaf
classification. This paper is combined with a new method which is called hybrid deep
learning consist of AE and CNN, in this way, we get better experimental result. In the
years before 2006, Support Vector Machine (SVM) has gained great success in the
classification task [1–3]. The method of SVM performs very good [2, 4], and it really

© Springer International Publishing Switzerland 2015
D.-S. Huang et al. (Eds.): ICIC 2015, Part II, LNCS 9226, pp. 115–123, 2015.
DOI: 10.1007/978-3-319-22186-1_11

can get state-of-the-art results. Many people focus on SVM and have done a lot of interesting work. In this paper, we extract features through hybrid deep learning consisting of AE and CNN, and then we do classification using these features with SVM. The results prove that the performance surpass pure SVM and pure Autoencoder.

2 Prerequisites

2.1 Preprocessing

Each pixel in the RGB color image has three channel datum, respectively R (red), G (green), B (blue), each channel uses an integer of 0 to 255 to represent the luminance of the channel. So for each pixel, RGB pictures need to use 24 -bit data to represent itself, each pixel represents a grayscale image of only the luminance channel, each pixel use the 8-bit data is represent itself.

In this way, by operation with image gray, 24-bit date RGB image changes into 8-bit data gray image. We can reduce the amount of data to 1/3 of the original amount of data. Common gray algorithms are as follow.

Component method:

$$I = R \quad or \quad I = G \quad or \quad I = B \tag{1}$$

Average method:

$$I = \frac{R + G + B}{3} \tag{2}$$

Maximum method:

$$I = \max(R, G, B) \tag{3}$$

In this paper, we use the maximum method.

Data normalization is very important in the training process of the neural network data, if we do not use data normalization, it may leads trained model accuracy rate to be very poor. Data normalization is to change the range of data into a certain fixed area, such as $[-11]$, among them, x' is the data which after normalization, x is the data which before normalization. μ is the mean of the data, σ is the variance of the data.

2.2 Basic AutoEncoder

Bourlard and Kamp first proposed autoassociation in 1988, a kind of neural network which replace the output layer with the data same as the input layer. The aim of an autoencoder is to learn a compressed, distributed representation (encoding) for a set of data, Autoencoder is based on the concept of sparse coding that is proposed in a seminal paper by Olshausen in 1996. While in the generic neural network the output layer is set with the labels of the corresponding input.

AE neural is an unsupervised learning algorithm using back propagation, set the output most equal to the input. Firstly, it directly encodes the input layer, it decodes the hidden layer. AE neural tries to transform inputs into outputs with the least possible amount of distortion, and the purpose is to learn a compressed or sparse representation (encoding) for a set of data. The architecture of AE is shown in Fig. 1:

In the AE, we should define a feature extracting function called encoder in a special parameter form firstly. With the encoder, we can straightforward compute the feature vector efficiently. The formula is:

$$h = f_e(x) \tag{4}$$

Where h is the feature vector or sometimes called representation, f is the encoder function, x is the input vector and theta is the parameter vector. To get the output, we should define another function called decoder:

$$r = g_\theta(h) \tag{5}$$

Where r is the output, g is the decoder function, h is the feature vector from (4) and theta is the parameter vector. So its encoder and decoder define the AE, and many different kind of training principles exit. Theta is earned on the task of reconstructing as well as possible the original input at the same time. One of most common training principle is attempting to lowest the reconstruct error defined as:

$$\min_\theta L(r, x) \tag{(6)}$$

Note how the main objective is to make reconstruction error low on the training examples, and by generalization, where the probability is high under the unknown data-generating distribution. For the minimization of reconstruction error to capture the structure of the data-generating distribution, it is therefore important that something in the training criterion or the parameterization prevents the AE from learning the identity function, which would yield zero reconstruction error everywhere.

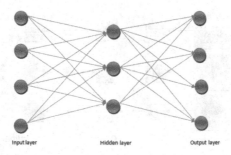

Fig. 1. The architecture of AE.

Basic AE training consists of finding a value of parameter vector theta minimizing reconstruction error

$$v(\theta) = \sum_t L(x^{(t)}, g_\theta(f_\theta(x^{(t)}))) \tag{7}$$

and x is the training example. This minimization is usually carried out by stochastic gradient descent as in the training mulit-layer-perceptrons (MLP). The most commonly used encoder and decoder are:

$$\begin{aligned} f_\theta(x) &= s_f(b + Wx) \\ g_\theta(h) &= s_g(d + W'h) \end{aligned} \tag{8}$$

Where s_f and s_g are non-linear.

2.3   Convolutional Neural Network

Convolution neural network is a multilayer neural network design inspired by biological visual perception net, it shows its great advantage in the field of image recognition [5–7]. Convolution neural network is combined by A series of simple and complex unit cell. Convolution neural network has invariance to rotation of the input pattern. It is a special multilayer perception designed to identify two-dimensional shapes, the network structure used for translation, scaling, tilting or other forms of deformation has a invariant height. It has become a hot topic in speech analysis and image recognition. TO sum up, convolution neural network is very fit for dealing with image [8–10].

Convolution neural network is a multilayer neural network, each floor is composed of many two-dimensional planes. The architecture of CNN is shown in Fig. 2:

In the CNN, the CNN convolutions through the input image and three trainable filter and bias. The filtering process is shown in Fig. 1. it produce three feature map layer in c1 after convolution, then summed with characterized in the map for each group of four pixels, add the weights, add the bias. We can get three s2 layer feature map by a sigmoid function. These Maps get into the filter layer wave, in this way, we

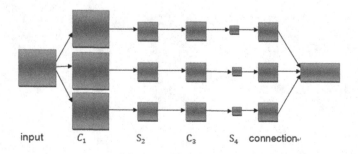

input C_1 S_2 C_3 S_4 connection

Fig. 2. The architecture of CNN

can get c3 layer, with the same method with s2 layer, we can get s4 layer. At last, these pixels are changed to an input vector to the traditional neural network.

Generally, C layer is a feature extraction layer, the input layer is connected with local receptive field in the previous layer, we can extract local feature. S layer is a feature map layer, each compute layer network is composed by some feature maps. Each feature is mapped to a surface. The weights of neurons in the surface are the same. Feature mapping structure use sigmoid function to be activating function, so that feature map has displacement invariance.

At last, because all the neurons in each mapping surface share weights, we can reduce the number of free parameters. Then we can reduce the complexity of the network preferences. Each feature extraction layer in CNN follows a Computing layer, we will get higher distortion tolerance.

2.4 A Brief Introduction to SVM

Support Vector Machines (SVMs) are supervised learning models with associated learning algorithms that analyze data and recognize patterns, used for classification and regression analysis. SVM can perform both linear classification and non-linear classification. The SVM perform classification through kernel trick, implicitly mapping their inputs into high-dimensional feature spaces. Considering our experiment, we will introduce SVM used for no-linear classification.

The SVM uses no-linear kernel for non-linear classification. This non-linear kernel allows the algorithm to fit the maximum-margin plane in a transformed feature space. The transformation may be nonlinear and the transformed space is high dimensional; thus though the classifier is a plane in the high-dimensional feature space, it may be nonlinear in the original input space. There are many no-linear kernel functions however the Gaussian kernel is most popular:

$$K(x,y) = \exp\left(\frac{\|x - y\|^2}{2\sigma^2}\right) \tag{9}$$

And we will adopt Gaussian kernel in our experiment. For more information about non-linear kernel, please refer to [11].

3 The Architecture Used in This Paper

In this paper, the author combined a hybrid deep learning consisting of Autoencoder and CNN with SVM. For clarity, the architecture is shown in Fig. 3:

The algorithm used to train hybrid deep learning is as follows: firstly, we trained a shallow AE, after this process, we can obtain a optimized weight as w1, then we deal with the input with the optimized AE, in this way, we get the hidden layer, we define it with a1, with the process, we get w1, a1, now, we can look a1 as input, then we train another high AE, which use a1 as input, in this way, we can get the right w2 and a2, we will keep on doing this operation until the process can achieve a proper layer.

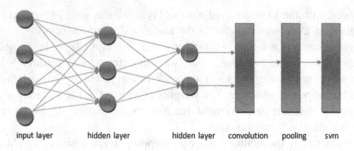

Fig. 3. The architecture of the author design

Secondly, we set a2 as input, we deal the input with convolution, the kernel can not neither too large nor too small, in this convolution, we choose the kernel as 3×3, after this process, then we will get 11 feature maps, each map has the size of $(32 - 3 + 1) \times (32 - 3 + 1)$, convolution kernel move one step each time, then, we train these features maps to mapping pass a sub-sampling layer s2, which is also called pooling, then we can get 11 feature maps for 15×15 size. Sub-sampling layer is a layer of a map of the sampling handling, after these processes, we can get 11 feature maps for 15×15 size, then we fine tune the hybrid deep learning with the data, at last, we unfold the feature maps, At last, we trained with SVM. The algorithm is in Table 1:

Table 1. Training algorithm

1. Change the image data into 64×64 matrix.
2. Activation[0] = input data.
3. For i=1, 2, 3...n, Train the AE[i], use activation[i-1] as the input, and get the weight w[i].
4. Unfold these AE in a deep learning network.
5. Adjust and fine the deep network, choose the proper layer to combine with CNN.
6. Trained input by convolution to c1.
7. Deal c1 with feature map (pooling), change into s2, then unfold s2 to one-dimensional.
8. Add a SVM to the last level for classification task, to train the SVM.

4 Experiment

In the experiment, we chose sigmoid activation function for the AE, the sigmoid function is:

$$sigmoid(x) = \frac{1}{1 + e^{-x}} \tag{10}$$

in the AE, our target is to get the optimize object:

$$minimize_{W,b,d} L_{DAE} = \frac{1}{2N} \sum_N \left(x^{(i)} - r^i \right)^2 \tag{11}$$

In our experiment, we used the leave pictures from ICL lab. For more information about this database, please infer to [12]. Here we choose 10 classes in the database, the experiment is implemented on MATLAB 2014b, Windows 7 Pro 64-bit, Intel(R) Core (TM) i3-2370, 6.00 GB RAM.

For all the samples, the proportion of training samples and testing samples is 2:1. We first transform those pictures to 64 * 64 gray pixels. Because the size of the picture is not the same, and the noise in the picture can affect our experiment result, after the Preprocessing, we can a fine 64 * 64 gray pixels.

In our experiment, the hybrid deep learning include the AE and CNN, our AE architecture is 4096-2048-1024-2048- 4096, and we chose the third layer to transport to next structure, our CNN model constitute one such class of models [2, 10], we chose the last layer to perform classify. We first convolute input with 13 kernels of size 3 * 3; the second is pooling over 4 non-overlapping regions of the image. For clarity, we show the process in Fig. 4, and the parameters used in our experience are in Table 2.

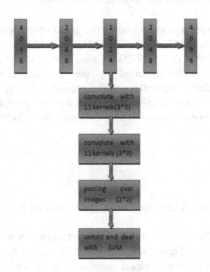

Fig. 4. The process of the experiment

Table 2. The parameters of the architecture

	Learning rate	Activation unction	Poach	Batch size	Alpha
AE	0.01	Sigmoid	10	100	1

The result of this experiment is shown in Table 3, and in the following 4 methods. Hybrid deep learning performs the best among the four methods.

Table 3. The accuracy of the architecture

Method	Accuracy
SVM	88 %
AE + SVM	90 %
CNN + SVM	91 %
Hybrid deep learning + SVM	93 %

5 Conclusion and Future Work

CNN can extract good features through hybrid deep learning, it can help reduce the complexity of the network, it is a good way to extract features in leaves classification. Our experiment show that hybrid deep learning can get better feature, so the result better than pure SVM, pure AE and pure CNN.

Although we have seen the advantage of hybrid deep learning method which the paper experiment is about. But hybrid deep learning can contain many types of deep learning, we may try RBM and so on, so in for the future work, we can combine more deep learning methods, I think it can obtain better result.

Acknowledgments. This work was supported by the grants of the National Science Foundation of China, Nos. 61133010, 61373105, 61303111, 61411140249, 61402334, 61472282, 61472280, 61472173, 61373098 and 61272333, China Postdoctoral Science Foundation Grant, Nos. 2014M561513, and partly supported by the National High-Tech R&D Program (863) (2014AA021502 & 2015AA020101), and the grant from the Ph.D. Programs Foundation of Ministry of Education of China (No. 20120072110040), and the grant from the Outstanding Innovative Talent Program Foundation of Henan Province, No. 134200510025.

References

1. Davis, L.S.: Polarograms—a new tool for image texture analysis. Pattern Recognit. **13**(3), 219–223 (1981)
2. Mavroforakis, M.E., Theodoridis, S.: A geometric approach to support vector machine (SVM) classification. IEEE Trans. Neural Netw. **17**(3), 671–682 (2006)
3. Cai, C.Z., et al.: SVM-Prot: web-based support vector machine software for functional classification of a protein from its primary sequence. Nucleic Acids Res. **31**(13), 3692–3697 (2003)
4. Schuldt, C., Laptev, I., Caputo, B.: Recognizing human actions: a local SVM approach. In: Proceedings of the 17th International Conference on Pattern Recognition, 2004, ICPR 2004, vol. 3 IEEE. (2004)
5. Cires, D.C., Meier, U., Masci, J. et al.: Flexible, high performance convolutional neural networks for image classification. In: Proceedings of the Twenty-Second International Joint Conference on Artificial Intelligence, vol. 2, 1237—1242 (2011)
6. Bengio, Y., Courville, A., Vincent, P.: Representation learning: a review and new perspectives, 1–1, (2013)

7. Krizhevsky, A., Sutskever, I., Hinton, G.E.: ImageNet classification with deep convolutional neural networks. In: NIPS (2012)
8. Larochelle, H. et al.: An empirical evaluation of deep architectures on problems with many factors of variation. In: Proceedings of the 24th International Conference on Machine Learning, ACM. (2007)
9. Lee., H. et al.: Convolutional deep belief networks for scalable unsupervised learning of hierarchical representations. In: Proceedings of the 26th Annual International Conference on Machine Learning. ACM (2009)
10. Goodfellow, I.J., Courville, A.: Deep Learning (2014)
11. Guyon, I.M., Vapnik, V.N.: A training algorithm for optimal margin classifiers. In: Proceedings of the Fifth Annual Workshop on Computational Learning Theory. ACM (1992)
12. http://mlsbl.tongji.edu.cn/chinese/file-database.asp
13. Li, B., Huang, D.S.: Locally linear discriminant embedding: an efficient method for face recognition. Pattern Recognit. 41(12), 3813–3821 (2008)
14. Huang, D.S., Du, J.-X.: A constructive hybrid structure optimization methodology for radial basis probabilistic neural networks. IEEE Trans. Neural Netw. 19(12), 2099–2115 (2008)
15. Huang, D.S., Chi, Z.-R.: A neural root finder of polynomials based on root moments. Neural Comput. 16(8), 1721–1762 (2004)
16. Huang, D.S.: A constructive approach for finding arbitrary roots of polynomials by neural networks. IEEE Trans. Neural Netw. 15(2), 477–491 (2004)
17. Huang, D.S.: Systematic Theory of Neural Networks for Pattern Recognition (in Chinese). Publishing House of Electronic Industry of China, China (1996)
18. Huang, D.S.: Radial basis probabilistic neural networks: model and application. Int. J. Pattern Recognit. Artif. Intell. 13(7), 1083–1101 (1999)

A Container Bin Packing Algorithm
for Multiple Objects

Fei Luo[1(✉)], Chunhua Gu[1], Xu Fang[1], and Yi Tang[2]

[1] School of Information and Engineering,
East China University of Science and Technology, Shanghai, China
{luof, chgu}@ecust.edu.cn, lpeer@163.com
[2] Shanghai ChinaNetCenter Development Co., Ltd., Shanghai, China
tangy@chinanetcenter.com

Abstract. The bin packing problem considered in this paper has the character with items arriving in batches, such as the task dispatching process in the cloud. The problem has two packing objects, one of which is to minimize the number of used bins, and the other is to reduce the packing time. To meet the objects, a container packing algorithm is presented, which consists of rounds of packing. In one round the number of the necessary bins is first calculated, and those bins are opened in batch to pack items in parallel. This round will be iterated until there are no unpacked items. The proposed algorithm is compared with other classical bin packing algorithms through experiments, which have the same property of simplicity for implementation. It is shown that the novel algorithm not only saves the packing time, but also achieves better combined performance than others.

Keywords: Bin packing · Batch packing · Multiple objects · Priority

1 Introduction

The classical bin packing (BP) problem is to pack items into a minimum number of bins so that the sum of the sizes in each bin is no greater than the capacity of the bins [1]. This problem has been greatly applied in diverse applications like logistics, transportation and telecommunications [2]. Direct application like task scheduling in a cloud computing system is also worth mentioning, where items correspond to tasks, bins correspond to virtual machines in the cloud. Then the task dispatching is in correspondence to the bin packing process [3, 4].

As a NP-hard problem [5], the research focus of the BP problem is on the study of approximation algorithms, which do not guarantee an optimal solution for every instance, but attempt to find a near-optimal solution. The problem can be classified as offline problems with all known information regarding the items in advance and online problems without knowing any information about future items. Therein, online algorithms and their performance were reported in a series of papers. One type of the most well-known algorithms are Any-Fit algorithms [1], where a new bin is opened only when the new item does not fit in any open (i.e., non-empty) bin, such as Next Fit (NF) [11], First-Fit (FF) [6], Best-Fit (WF) [7], and Worst-Fit (WF) [8]. The variants of

© Springer International Publishing Switzerland 2015
D.-S. Huang et al. (Eds.): ICIC 2015, Part II, LNCS 9226, pp. 124–134, 2015.
DOI: 10.1007/978-3-319-22186-1_12

these algorithms are also well known as NF Decreasing (NFD), FF Decreasing (FFD), BF Decreasing (BFD) and WF Decreasing (WFD) with items in decreasing order.

Currently in most of the well-known online problems it is the assumption that the items are arriving one by one. In the contrast, in most applications utilizing the BP strategies, the items can arrive in batches without any known information in advance. One of the typical examples is the scheduling process in the cloud, where the tasks are submitted to the cloud in batches. Those tasks should be dispatched to the virtual machines (VMs) in the scheduler center of the cloud. The process of dispatching tasks into VMs corresponds to the bin packing process.

In the cloud's circumstances, the object of the BP solution in correspondence to the dispatching process is not only to minimize the number of the used bins, but also to reduce the time cost of the solution itself, e.g., to minimize the completion time of the packing process. It is known that the current typical Any-Fit algorithms try to achieve a single optimizing object, which cannot be directly applied in those applications with multiple objects. Furthermore, in the classic packing algorithms, the indexes are used to evaluate the performance for its single object, such as the competitive ratio (CR) [9], the absolute worst-case ratio (AR), as well as the asymptotic worst-case ratio (or asymptotic performance ratio, APR) [10], which cannot be used to evaluate comprehensive performance for multiple objects in this paper.

Therefore, to resolve the applications with items arriving in batches for multiple objects, a novel *container bin packing* (CBP) algorithm is proposed in the paper, where each batch is denoted by a container containing many items. The main idea of the algorithm is to pack items in a container into bins in parallel to reduce the packing time and the number of used bins. To further evaluate the comprehensive performance with multiple objects, a new performance index is also presented.

The proposed algorithm will finally be utilized in a practical cloud system in the future, where the algorithm should be simple to be implemented with limited space complexity and time complexity. Although other newly BP algorithms in recent years maybe can achieve better performance with complex data structures, they cannot have the same property of simplicity for implementation as the classic Any-Fit algorithms, such as NFD, FFD, BFD and WFD. Therefore, the performance of the proposed algorithm is just compared with NFD, FFD, BFD and WFD.

The rest of the paper is formulized as follows. Related works are represented in Sect. 2. Illustrated with an application instance, the packing model for the applications with items in batches is abstracted in Sect. 3. The proposed algorithm is detailed to resolve the problem in Sect. 4. To evaluate the performance of the algorithm, the experiments and analysis are carried out in Sect. 5. The final conclusions are drawn in Sect. 6.

2 Related Works

The BP problems are classified as online and offline problems and the main goal of the classic BP problem is to minimize the number of used bins [8]. The problem is known to be NP-hard [4], and then it is unlikely that efficient (i.e., polynomial-time) optimization algorithms can be found for its solution [1]. Researchers have thus turned to the study of approximation algorithms, which do not guarantee an optimal solution for every instance, but attempt to find a near-optimal solution within polynomial time.

Some approximation algorithms have been developed for bin packing problem, and the well-known online algorithms are NF, FF, BF, WF, and so on. In addition to meet the basic goal of minimizing the number of bins, the algorithms try to reduce the time complexity. The time complexity of NF is clearly o(n) and it has a relatively poor APR. The corresponding offline algorithms have the same properties, such as NFD, FFD, BFD and WFD, where the items are assumed to be in descending order with their sizes. By adopting appropriate data structures for representing packing, it is easy to verify that the time complexity of these algorithms is o($n \log n$), such as FFD, BFD and WFD.

A linear time approximation scheme is shown in [12], and a sublinear time o($\sqrt{n}$) with weighted sampling and a sublinear time o($n^{1/3}$) with a combination of weighted and uniform samplings were shown for bin packing problem [13]. Based on the classic online BP algorithms, many refined ones have been proposed. One of them is RFF [14], which uses unbounded space and has time complexity o($n \log n$). In the algorithm of Refined Harmonic-Fit (RHF), the time complexity is o(n), but the algorithm is no longer bounded space [15].

Gambosi et al. [16] gave two algorithms via repacking. Both of them use an unusual step to repack the elements. Therein, the time complexity of the second algorithm is o($n \log n$). Johnson [17] made an interesting attempt to obtain a better APR. His Most-k-Fit (MFk) algorithm takes elements from both ends of the sorted list, packing bins one at a time. The algorithm has time complexity o($n^k \log n$), so it is practical only for small k.

With special limitation conditions, the online algorithms described above can get almost linear time complexity, such as o(n), o($\sqrt{n}$), etc. Furthermore, such performance can be embodied only when n tends to infinity. However, the performance of those algorithms is presented with independent performance indexes, where these algorithms try to be optimized for a single object. In the circumstances in this paper, there are multiple optimizing objects, and the time characteristics for high performance should be embodied for each batch of items, where the number of the items in each batch is limited. Therefore, those algorithms cannot be directly applied in the applications in this paper.

3   Problem Description

As an instance to illustrate the performance requirement of the BP problem for multiple objects, the task dispatching process in the scheduler center in the cloud system is detailed. Then based on the application requirement, the optimizing objects of the problem are formulized.

3.1   Application Requirement

In a cloud computing system, the tasks are submitted to the scheduler center in bathes by the user, and the scheduler center will dispatch the tasks into VMs after the reception of them. It is known that the interest of the cloud providers is to maximize profits, which is modeled as follows. We use the assumption that the income for the reception of a batch of tasks is proportional to the necessary computing resources for the tasks. Specially, the income for a unit computing resource is denoted by a constant income as α, and the computing resources for a batch of tasks are denoted by I_n. Then the income $f(in)$ for the reception of a batch of tasks is represented in formula 1. Therein, the provided VMs are homogenous and a VM's computing capacity is C. Moreover, the average filllevel of the used VMs is represented as β, where the filllevel is defined as the rate of the required computing power of all the dispatched tasks in a VM to its computing capacity. And then the necessary number of VMs n is shown in formula 2. Furthermore, the cost of the VM is metered according to the time of the used VMs. The dispatching time of the tasks in each VM is set as t_l, the computing time for its dispatched tasks is t_r, and the cost for each unit computing time is γ. Accordingly, the cost $f(out)$ for each batch of tasks is depicted in formula 3.

$$f(in) = \alpha \times I_n \tag{1}$$

$$n = \frac{I_n}{\beta} \times C \tag{2}$$

$$f(out) = n \times (t_l + t_r) \times \gamma \tag{3}$$

Based on formula 1–3, the return f_p of each batch of tasks for the cloud provider is represented as formula 4. In the circumstances, it is assumed that α, C, γ, I_n, t_r are constants. Then the profit f_p depends on the values of dispatching time t_l and the filllevel β. While t_l and β affect each other, the relationship between them will be further analyzed in the following sections.

$$f_p = f(in) - f(out) = I_n \times [\alpha - (t_l + t_r) \times \frac{\gamma}{\beta} \times C] \tag{4}$$

To focus on the packing algorithm itself, the VM, the task and the scheduling process in the cloud system are denoted by the bin, the item and the bin packing respectively in the following sections.

3.2   Optimizing Objects

As the items in this paper are submitted in batches, each batch is represented as a container containing many items. Then the application in this paper is abstracted as a number of containers continuously submitting to the scheduler center. Each container R_i (i = 1, 2, ..., n) consists of a list L_i of items (or elements), where $L_i = (a_{ij})$ (j = 1, .., m). Here n and m are the maximum number of containers and that of items in a container respectively. A value $s_{ij} = s(a_{ij})$ gives the size of items a_{ij} and satisfies that $0 < s_{ij} \leq C$, $0 < j \leq m$, where C is the capacity of a bin.

If the lists in the containers are all concatenated together, the problem is the same as the classic BP problem, packing a long list of items into a minimum number of bins. Therefore, there are two main participants in the packing process, e.g., the container and the items. For the items the object is to make use of bins as few as possible, while for the container the object is to pack each container as soon as possible.

Let $k_i(A, L_i)$ be the number of bins used by an algorithm A, and $t_i(A, L_i)$ be the time used to pack L_i into bins. Then the object with minimum number of bins is represented as $\min\{k_i(A, L_i)|1 \leq i \leq n\}$, and the other object with fast packing process is represented as $\min\{t_i(A, L_i)|1 \leq i \leq n\}$. To differentiate the possible conflictions between these two packing objects, each object is assigned with a priority, represented as P_1, P_2 respectively. Then the comprehensive target of the two packing objects is described in formula 5.

It is observed that the BP problem with two objectives in formula 5 is a superset of the classic BP problem, e.g., it is degenerated as a classic BP problem when $P_1 = 0$, $P_2 = 1$. In this paper, the discussed problem is much more complicated. It is assumed that $P_1 > P_2 > 0$, which means that the problem first ensures the realization of first objective by packing the batch of items as soon as possible, and then considers the realization of the problem with minimal number of used bins in the packing process.

$$
\begin{cases}
\min\{z\} = P_1 \times k_i(A, L_i) + P_2 \times t_i(A, L_i) & \text{(a)} \\
\text{s.t.} \\
1 \leq i \leq n & \text{(b)} \\
P_1 + P_2 = 1 & \text{(c)} \\
0 \leq P_1, P_2 \leq 1 & \text{(d)}
\end{cases}
\tag{5}
$$

4 Container Bin Packing Algorithm

As the items are arriving in batches, the main idea in this paper is to pack items based on the container, not just one item. To further describe the proposed algorithm, some definitions are prepared.

Definition 1. The items in the container are processed in batch, while the corresponding necessary bins will also be supplied in batch. During the process, a batch of bins are opened for the rest of the items, and those bins cannot accommodate new items should be closed in batch. Therein, the number of bins to be closed is represented as B_c, that to be opened is recorded as B_o, while the number of these opened bins that can still pack new items is represented as B_r.

Definition 2. The number of the rest items is represented as R_T, while the average size of the rest items is called as A_T, as defined in formula 6. Afterwards, the minimum number of the necessary bins for the rest items is represented as B_{min}, as defined in formula 7.

$$
A_T = \frac{\sum_{i=1}^{R_T} k_i}{R_T}
\tag{6}
$$

$$B_{min} = \left\lceil \frac{\sum_{i=1}^{R_T} k_i}{C} \right\rceil \qquad (7)$$

Definition 3. Some queues are provided to temporarily store the items and the bins during the processing. First, the items needing to be packed in a container are preloaded in a queue named as Q_{item}, while the open bins are prepared in the queue Q_{open}. Then a mapping relationship is established among the elements in Q_{item} and those in Q_{open}, where the item in Q_{item} tries to be packed in the bin in Q_{open} with the same sequence number as the item. If the item in Q_{item} cannot be packed into the corresponding open bin in Q_{open}, it will be put into the recovered queue, Q_{item_v}, while the bin that cannot accommodate the item in Q_{item} is put into the resetting queue, Q_{open_r}.

Based on the prepared definitions above, the CBP algorithm is depicted as follows:

Input: the item set (a_i) in the container R. The size of each item is represented as s (a_i). Here $|R| = m$, and $0 < s_i = s(a_i) <= C$, $0 < i <= m$.

Output: packing the items in the container into bins with the capacity C.

Procedure:

(1) Initialize the parameters, e.g., $B_c = 0$, $B_o = 0$, $B_r = 0$, $R_T = m$, $A_T = \frac{\sum_{i=1}^{m} k_i}{R_T}$, Init_Queue (Q_{item}), Init_Queue (Q_{open}), Init_Queue (Q_{item_v}), and Init_Queue (Q_{open_r}). Here Init_Queue represents the initialization to the queue.

(2) B_{min} is calculated according to formula 7. Then put B_{min} items into Q_{item} and open B_{min} bins which are stored in Q_{open}.

(3) Try to pack the ith element in Q_{item} into the ith bin in Q_{open}, where $0 \leq i < |Q_{item}|$. Here $\|$ represents the operation to get the number of elements in the queue. $r'_i = k_i - s(a_i)$, where k_i is the rest capacity of the ith element in Q_{open}. Then check the packing mapping between Q_{item} and Q_{open} in the following.

    (i) If $r'_i < 0$, it means the ith bin cannot accommodate the corresponding item, the item and the bin will be redrawn back into Q_{item_v} and Q_{open_r} respectively.

    (ii) If $0 < r'_i < A_T$, it means the ith bin packs the item and will be closed, and B_c ++.

    (iii) If $r'_i \geq A_T$, it means the ith bin packs the item and will keep open.

(4) Calculate the number of the failing packing items in step 3-i, represented as f_q. Here $f_q = |Q_{item\_v}|$.

(5) Refresh B_r and R_T. Here $B_r = B_{min} - B_c$ and $R_T = R_T - (B_{min} + f_q)$.

(6) A_T and B_{min} are refreshed according to formula 6 and 7. If $B_{min} = 0$, terminate the algorithm.

(7) A new queue Q'_{item} and a new queue Q'_{open} will be initialized with the size of B_{min}.

(8) $B_o = B_{min} - B_r$. Then B_o bins will be opened and put into Q'_{open} and the remaining element in Q_{open} will be sequentially put into Q'_{open}.

(9) The element in Q_{item_v} will be sequentially put into Q'_{item}, and $(B_{min} - f_q)$ new items from the rest in the container will be added into Q'_{item}.

(10) Reset the parameters, e.g. $Q_{open} = Q'_{open}$, $Q_{item} = Q'_{item}$, $B_c = 0$, $B_o = 0$, $B_r = 0$, Q_{item_v} and Q_{open_r} will be reset and initialized.

(11) Branch to (3) to continue the packing.

5 Analysis and Evaluation

To specify the advantage of the algorithm, its performance is compared with the typical algorithms through experiments.

5.1 Experiment

Since the BP problem is NP-hard, no benchmark exists in literature as far as we know. Therefore, a new experiment is constructed to evaluate the performance of the algorithm. The experiment environment is illustrated in this section, and the simulations are carried out and evaluated in Sect. 5.2.

First the basic performance indexes of the algorithm, such as the number of used bins and the time cost t_l are calculated. A random container with a limited number of items is applied in the algorithm, where items are randomly generated. For the same container, those indexes among the typical BP algorithms, e.g., NFD, FFD, BFD, and WFD, are also acquired to be compared with the CBP algorithm. As the explanation in Sect. 3.1, the fill-level β is also calculated, and the relationship between t_l and β will also be studied to maximize the profit shown in formula 4.

Based on formula 5, the comprehensive performance is defined as z. To compare the combined performance of the container packing algorithm in this paper with that of other typical algorithms, a new comparative performance index (CPI) is defined in formula 8. Therein, for each container, k^B and t^B are the number of bins and time cost used in the container packing algorithm, while k^i and t^i are those used in the such comparative BP algorithms, i.e., FFD, NFD, WFD and BFD. It is seen that the term $\frac{k^B - k^i}{k^i}$ is the saved bins used by the container packing algorithm comparative to other algorithms, while $\frac{t^B - t^i}{t^i}$ is the corresponding saved time. Also the combined performance in formula 8 can be adjusted according to the values of P_1 and P_2.

$$CPI = P_1 \times \frac{k^B - k^i}{k^i} + P_2 \times \frac{t^B - t^i}{t^i} \qquad (8)$$

In the following experiments, the capacity C of each bin is set as an integer, and specially C = 100. The items in a container are submitted to be packed in batches, where the size of each item is a random integer between 1 and C. The container packing algorithm and those typical algorithms are implemented in Java. The number of items in a container is variable, from 500 to 10000.

5.2 Results

Firstly, the number of used bins in different packing algorithms is evaluated, as seen in Fig. 1. It is shown that with the rise of the number of items, the number of used bins increases. However, with the same items, the bins used in the CBP algorithm is less than those used in NFD, while more than those used in FFD, BFD and WFD. At the same time, the number of bins used in FFD, BFD and WFD are almost the same.

Secondly, the average fill-level of the bins in different packing algorithms is evaluated, as shown in Fig. 2. As illustrated in Sect. 3.1, the filllevel of a bin is the rate of the sum of items' size in the bin to the bin's capacity. Then the average filllevel $\overline{FB}$ is the average value of the fill-level of all the bins used to pack a batch of items, as shown in formula 9, where m is the number of used bins for a container.

$$\overline{FB} = \sum_{i=1}^{m} FB_i \qquad (9)$$

It is shown that with the rise of items' number, the filllevel of each BP algorithm is relatively stable. The values of $\overline{FB}$ for FFD, BFD and WFD are almost the same and approach 1. But the value of $\overline{FB}$ for the container algorithm is less than that for FFD, BFD and WFD, while it is much bigger than that for NFD.

The relationship between $\overline{FB}$ and the number of used bins is depicted in formula 10, where the number of items in a container is n, and the number of used bins is m. It is seen that when the average filllevel $\overline{FB}$ increases, the corresponding used bins will be decreased. Both Fig. 1 and 2 verify the idea in formula 10.

Thirdly, the time cost of each algorithm used to pack the items in a container is evaluated, as shown in Fig. 3. From the figure it is known that the time cost for FFD, BFD and WFD increases greatly with the rise of the number of items in the container, and that of WFD is the biggest among those algorithms. However, the time cost of NFD and the CBP algorithm increases slowly. When the number of items is more than 4000, the time cost of the CBP algorithm is almost half that of NFD.

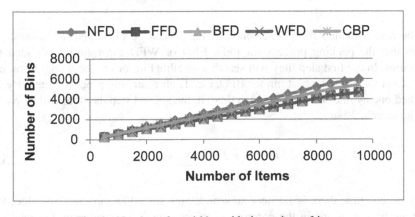

Fig. 1. Number of used bins with the variety of items

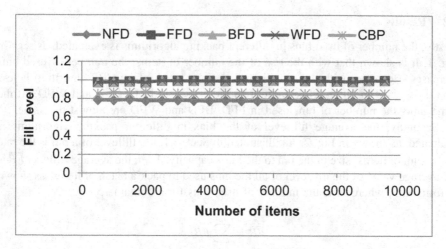

Fig. 2. Average FillLevel of the bins with the variety of items

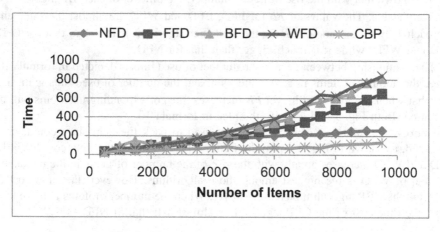

Fig. 3. Time cost of different BP algorithms

The reason behind the phenomenon in Fig. 3 can be explained as follows. It is known that the packing process for FFD, BFD or WFD algorithms is divided two procedures. In the first step they will search a suitable bin for the incoming items, e.g., the first bin for FFD, the best bin for BFD, etc. Their searching process is iterative and launched one by one, which cost much time. In the second step the item will be packed into the searched bin.

$$m \times C \times \overline{FB} = \sum_{j=1}^{n} s(a_j) \tag{10}$$

In the contrast, there is no need to search bins for NFD and the CBP algorithm, while the first open one in the corresponding position is the candidate. Especially, in the CBP algorithm the items in a container are packed in parallel, while in the NFD algorithm the items are packed one by one. The parallel packing process saves much more time than NFD. Anyway, the CBP algorithm has the best time performance compared with other classical online BP algorithms.

6   Conclusions

The application background for the BP problem in this paper is that tasks are submitted to the cloud computing system batches. To meet the multiple objects of such scheduling problem, e.g., minimizing the number of the used bins and the packing time, a novel CBP algorithm is proposed. The multiple objects are given with different priorities, and the main idea of the algorithm is to pack items based on the container. Therein, the number of the necessary bins is calculated, and they are opened in batch to pack the items in parallel. The basic properties of the CBP algorithm are analyzed. To illustrate the performance of the CBP algorithm, it is compared with the typical BP algorithms, such as FFD, BFD and WFD, through experiments. The number of used bins and time cost for the packing procedure to the container with limited items are calculated and compared.

Through the experiments and analysis, it is verified that the CBP algorithm cost less time than other typical online algorithms. Furthermore, the better comprehensive performance can be attained in the CBP algorithm when the priority of minimizing the packing time is greater than that of minimizing the number of used bins during the packing procedure.

Acknowledgement. This paper is supported by the Nature Science Fund of China (NSFC) (No. 61472139) and China Postdoctoral Science Foundation (No. 2013M541487).

References

1. Coffman, E.G., Jr., Csirik, J., Galambos, G., Martello, S., Vigo, D.: Bin Packing Approximation Algorithms Survey and Classification, Handbook of Combinatorial Optimization, Kluwer Academic Publishers, Boston (2011)
2. Bergner, M., Caprara, A., Ceselli, A., Furini, F., Lübbecke, M.E., Malaguti, E., Traversi, E.: Mathematical programming algorithms for bin packing problems with item fragmentation. Comput. Oper. Res. **46**, 1–11 (2014)
3. Burns, A., Davis, R.I., Wang, P., Zhang, F.: Partitioned EDF scheduling for multiprocessors using a C = D task splitting scheme. Real-Time Syst. **48**, 3–33 (2012)
4. Garey, M.R., Johnson, D.S.: Computers and Intractability: A Guide to the Theory of NP-Completeness. W.H Freeman, New York (1979)
5. Dosa, G., Tuza, Z., Ye, D.: Bin packing with Largest In Bottom constraint tighter bounds and generalizations. J Comb. Optim. **26**, 416–436 (2013)
6. Garey, M.R., Johnson, D.S.: Approximation algorithm for bin-packing problems: a survey, Analysis and Design of Algorithm. In: Ausiello, G., Lucertini, M. (eds.) Combinatorial Optimization, pp. 147–172. Springer, New York (1981)

7. Kenyon, C.: Best-fit bin-packing with random order. In: Proceedings of the 7th Annual ACM-SIAM Symposium on Discrete Algorithms Atlanta, GA, pp. 359–364. ACM/SIAM, Philadelphia (1996)
8. Johnson, D.S.: Fast algorithms for bin packing. J. Comput. Syst. Sci. **8**, 272–314 (1974)
9. Demange, M., Grisoni, P., Paschos, V.T.: Differential approximation algorithms for some combinatorial optimization problems. Theor. Comput. Sci. **209**, 107–122 (1998)
10. Boyar, J., Favrholdt, L.M.: The relative worst order ratio for online algorithms. ACM Trans. Algorithms **3**, 1–24 (2007)
11. Boyar, J., Epstein, L., Levin, A.: Tight results for Next Fit and Worst Fit with resource augmentation. Theor. Comput. Sci. **411**, 2572–2580 (2010)
12. de la Vega, W.F., Lueker, G.S.: Bin packing can be solved within 1 + epsilon in linear time. Combinatorica **1**(4), 349–355 (1981)
13. Batu, T., Berenbrink, P., Sohler, C.: A sublinear-time approximation scheme for bin packing. Theor. Comput. Sci. **410**, 5082–5092 (2009)
14. Yao, A.C.-C.: New algorithms for bin packing. J. ACM **27**, 207–227 (1980)
15. Lee, C.C., Lee, D.T.: A new algorithm for on-line bin-packing. In: Technical report 83-03-FC-02, Department of Electrical Engineering and computer Science Northwestern University, Evanston, IL (1983)
16. Gambosi, G., Postiglione, A., Talamo, M.: Algorithms for the relaxed online bin-packing model. SIAM J. Comput. **30**, 1532–1551 (2000)
17. Johnson, D.S.: Near-Optimal Bin Packing Algorithms. Ph.D thesis, MIT, Cambridge, MA (1976)

Distance Sensor Fusion for Obstacle Detection at Night Based on Kinect Sensors

Alexander Filonenko[1], Danilo Cáceres Hernández[1], Andrey Vavilin[2],
Taeho Kim[2], and Kang-Hyun Jo[1(✉)]

[1] Graduate School of Electrical Engineering, University of Ulsan,
Ulsan 680-749, Republic of Korea
{alexander,danilo}@islab.ulsan.ac.kr,
acejo@ulsan.ac.kr
[2] FuturIST, Ulsan, Republic of Korea
andrey.korea@gmail.com, thkim@futist.co.kr

Abstract. This paper introduces a vehicle safety system for moving backwards during night time. The system consists of two RGB-D cameras supported by two computers connected via Gigabit Ethernet. The cameras are mounted on the back side of the vehicle at different altitudes and tilt angles allowing detection of objects behind the vehicle and defects of the road surface. The safety system triggers an alarm if an object appears in dangerous proximity to the car. The dangerous proximity is decided according to the mass of the vehicle and its speed. Obstacles are tracked to predict if ones may become dangerous in the near future. Experiments have shown that the system is able to detect obstacles and holes on the road pavement when there is not enough light for common cameras.

Keywords: Vehicle safety · Kinect · RGB-D camera · Kalman filter

1 Introduction

Industrial complexes are highly automated nowadays. Many operations are performed by robots working at the assembly line. However, some tasks are still done by individuals e.g. transporting tasks. When a driver controls a vehicle, the line of sight may be significantly limited by a cargo or when the vehicle is moving backwards. Furthermore, many production cycles last 24 h a day. That means people have to work with poor lighting conditions if they move outside the buildings. Safety in these cases becomes crucial and it should be guaranteed by assistance systems.

This work presents the driver assistance system for scenarios when a truck is moving backwards at night. This is a usual case when goods are transported to or from a warehouse. One type of danger for the vehicle are pits and ruts on the road surface if they are large enough to damage a car, a driver, or cargo. They are hardly visible at night. Another type of danger is the possible harm a vehicle can deal to people behind it or damage inflicted to a truck by obstacles like lifting hook which may appear at some altitude. The system detects all the types of hindrances discussed above. To cope with bad lighting conditions and to detect obstacles at different altitudes, two RGB-D cameras (Kinect 2.0 model) were utilized. If obstacles should be detected at day time,

© Springer International Publishing Switzerland 2015
D.-S. Huang et al. (Eds.): ICIC 2015, Part II, LNCS 9226, pp. 135–144, 2015.
DOI: 10.1007/978-3-319-22186-1_13

existing methods for CCD cameras discussed in the Related Works section should be utilized for Kinect sensors since they also provide the RGB data. If the relative pose between sensors is known, then they can serve as stereo camera.

1.1 Related Works

Human factors are the leading reason of traffic accidents. Evolutionary necessary features like ignoring static objects may interfere safe driving when vehicles moving towards the driver remain almost immobile related to environment. In this case human brain filters out the real vehicle. The human errors and deficiencies which caused accidents primarily involved recognition and decision errors [1]. In case of specific human direct causes, the most frequent reason of accidents is "improper lookout" which includes changing lanes, passing, or pulling out from an intersecting alley, street, or driveway without looking carefully enough for incoming traffic. It was noted in [1] that half of the individuals cited for "improper lookout" had looked but failed to see oncoming traffic which should have been visible.

In industrial areas where a single mistake may lead to an emergency situation, every introducing system should be fully tested according to ISO 26262. In order to prevent putting drivers at risk in real systems, possible errors should be found in simulation [2]. Before doing a big step to the autonomous system, the driver assistance system which allows human to make decisions is to be developed and tested first.

To make sure that the road surface is safe to move, its condition should be checked. Authors in [3] and [4] use stereoscopic vision to cope with this problem. The weak point of this system is it inability to perform well at night when CCD cameras produce much noise or even fail to show any objects. Cracks on the road surface were detected by [5] using color information. The problems may appear when strong shadows of road sign poles overlay the road pavement.

In order to decide the distance when obstacles, ruts, and pits become dangerous, car safety system should know how fast vehicle can stop. The time spend until the velocity becomes zero depends on vehicle weight and speed. Also, human reaction time is considered for non-autonomous systems [6]. If high precision in calculating the stopping time is required, then vision-based approach introduced in [7] should be used to determine the type of road surface and, accordingly, its friction factor. But again, it will work well at day time only.

2 System Design

The system contains the vehicle of height up to 4 meters. Two Kinect 2 sensors are mounted as shown in Fig. 1(a). The upper sensor (K1) should be attached at the altitude which equals to the height of the vehicle. It is rotated down at specific angle from the horizon which lets K1 to see the road pavement behind the vehicle.

The bottom sensor (K2) is mounted at the bottom part of the car and rotated at a specific angle above the horizon. K2 is intended to detect obstacles behind the vehicle at altitudes from the altitude of K2 and higher. K1 and K2 together should provide wide

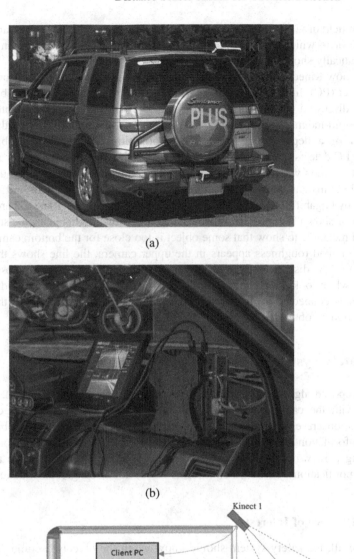

(a)

(b)

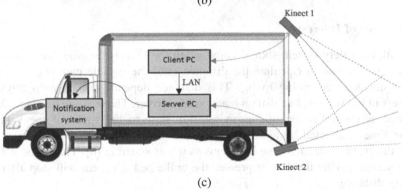

(c)

Fig. 1. Safety system design: (a) mounting position of Kinects; (b) processing and monitoring hardware; (c) system design scheme.

enough field of view to detect any possible obstructions which may be dangerous to the vehicle or to which the vehicle can be dangerous. The relative position of sensors is schematically shown in Fig. 1(c).

By now Kinect SDK 2.0 allows connection of a single Kinect 2 sensor to a personal computer (PC). In order to utilize two sensors, two separate PCs are to be used. In the design discussed in this paper, one PC (called client PC) processes depth image from the bottom-mounted sensor and sends color and infrared images to the server PC. Instead of a depth image itself, the information about detected objects is sent. Server PC detects objects on depth images from the top-mounted sensor and combines this information with data from client PC. The combined information is used to predict possible collisions and notify a driver. Connection between client and server is supported by Gigabit Ethernet. Results of processing data from K1 and K2 are displayed at a monitor shown in Fig. 1(b). When a dangerous situation happens, the software draws the red rectangle to show that some object is too close for the bottom camera. When an object or road roughness appears in the upper camera, the line shows the distance to that entity. As discussed above, if driver is tired, one may not detect obstacles or road cracks while looking at the monitor. To attract attention of the individual, an alarm speaker is connected to the server. It produces a sound in parallel to the visual representation of obstacles.

3 Safety System

The proposed algorithm consists of four main parts as shown in Fig. 2. The process starts with the calibration. During this step, parameters essential for correct object localization are estimated. Data obtained in this step is used in the object detection part. Information about detected objects from both Kinect sensors is combined in the tracking part, which is used to predict collisions. Finally, all data are aggregated in driver notification system which informs driver about dangerous situations.

3.1 Distance of Interest

First of all, the safety system should know when the objects become dangerous. To do that, it is required to calculate the distance until the vehicle can stop. The calculations are done according to [6]. This distance depends not only on physical parameters of the vehicle, but also on reaction of a driver. The reaction distance DR is introduced as multiplication of human reaction time to the speed of the vehicle. Reaction time is affected by age, emotions, and physical condition of a driver. In general, the average reaction time is 750 ms as it was stated in [6]. This value will be taken in account. After the drivers pressed the brake pedal, the car will stop after the following distance:

$$D_S = \frac{W}{2gC_{ae}} \ln\left(\frac{1 + C_{ae}V^2}{\eta\mu W + f_r W \cos\theta + W \sin\theta}\right) \tag{1}$$

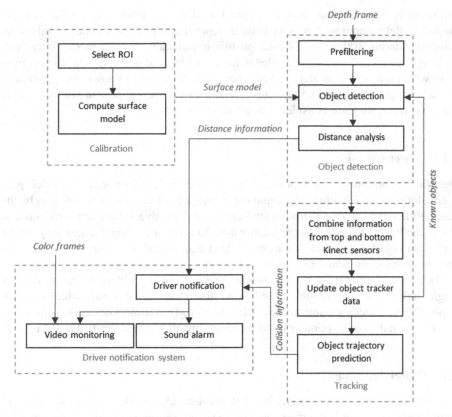

Fig. 2. Safety system flowchart

$$C_{ae} = \frac{\rho A_f C_d}{2} \tag{2}$$

where W: vehicle weight, g: gravity, V: vehicle speed, μ: friction factor, η: brake efficiency, f_r: roll factor, θ: road slope. For example, for Mitsubishi Freeca 2.0 parameters are the following: $W = 1495$ kg, $g = 9.81$ m/s^2, $C_{ae} = 0.64$ kg/m, $A_f = 2.562$ m^2, $C_d = 0.4$, $\eta = 0.6$, $\mu = 0.8$, $f_r = 0.015$ as discussed in [6].

After the final stopping distance D_F is calculated by (3), the safety distance knows when it should consider object as potentially dangerous.

$$D_F = 0.75V + D_S = D_R + D_S \tag{3}$$

3.2 Calibration

In order to do calibration, the vehicle has to be moved to the flat surface with no objects around. The calibration process starts with selecting regions of interest (ROI) for both Kinect sensors. ROIs are used to ignore objects located not on vehicle's trajectory.

Additionally, a surface model is generated for ROI. This process is similar to background model generation. For every pixel in depth image mean and variance values are computed during 100 frames. Based on this information, surface model is created. Although, use of two sensors is needed to have a wider field of view, data streams from them were processed separately. Since a pose of each Kinect is known, the position of obstacles is given in world coordinates. Thus, if two objects appear in both sensors at the same place, they are considered as an identical object.

3.3   Object Detection

Detection algorithm for top-mounted Kinect sensor is based on surface model, generated during the calibration step. Input depth image from the sensor is denoised by the median filter and subtracted from a surface image. Negative values, are considered as object candidates. Objects that are smaller than 50 pixels are ignored unless they were a part of an object in the previous frame. Object candidates, which appear in two consecutive frames, are added to tracker.

Detection algorithm for bottom-mount Kinect is slightly different. Due to the low height, only a small part of the infrared light reflects back to the sensor, which makes it impossible to detect a ground surface stable enough. Thus, object detection is based on differences in distances. In order to do so, clustering algorithm described in [9] is used.

3.4   Object Tracking

If an object appears in the camera, but it is situated outside vehicle dimensions, it still may collide with a car if the object had some speed. Detection of this type of possible collisions is based on trajectories. For all detected objects in image sequence, different paths are constructed. For robust object tracking, the camera frame rate should be sufficient in order to capture movements of the object so that the estimation process may be considered as linear in time. In this case time consuming techniques such as extended Kalman filter are not necessary. In Kalman filter [8] utilized in this work the state vector where position and velocity of an object on the ground plane is shown in (4):

$$\begin{bmatrix} X_k \\ Y_k \\ X_k' \\ Y_k' \end{bmatrix} \tag{4}$$

Since the tracking area is small, the motion of an object can be explained by a constant velocity model. Kalman filter is the recursive procedure which consists of two stages: prediction and correction. The first one is intended to estimate the motion. The second stage corrects the error of prediction. Equation (5) represents the predicted state at current time from the model defined by matrix A (6) which defines a constant velocity model, where T is the time between two frames.

$$\hat{X}_k^* = A\hat{X}_{k-1}^* + W_n \tag{5}$$

$$A = \begin{bmatrix} 1 & 0 & T & 0 \\ 0 & 1 & 0 & T \\ 0 & 0 & 1 & 0 \\ 0 & 0 & 0 & 1 \end{bmatrix} \tag{6}$$

Matrix W_n is the noise in the process. Usually it is considered as a Gaussian noise in form of $N(0, Q)$, where Q is the covariance of the noise of the estimation process. The prediction of the error is calculated by Eq. (7).

$$P_k^* = AP_{k-1}A^T + Q \tag{7}$$

The correction stage is done in the steps explained as follows. The value of Kalman gain (8) is calculated from the a priori error prediction, matrix H representing the relation of measurements and states, and R showing the measurement noise covariance.

$$K_k = P_k^* H^T \left(HP_k^* H^T + R \right)^{-1} \tag{8}$$

The estimate is updated with new measurements obtained from the process. In this work only x and y positions can be directly measured. Considering this information, the matrix H is obtained in Eq. (9).

$$H = \begin{bmatrix} 1 & 0 & 0 & 0 \\ 0 & 1 & 0 & 0 \end{bmatrix} \tag{9}$$

The estimate is updated by the measurement using Eq. (10)

$$\hat{X}_k = \hat{X}_k^* + K_k \left(Z_k - H\hat{X}_k^* \right) \tag{10}$$

where Z_k is the current measurement. The error covariance is updated in (11)

$$P_k = (I - K_k H)P_k^* \tag{11}$$

Since Kinect 2 provides depth information, it is easy to build the vertical projection of object to the ground plane and measure the real world coordinates of the object.

4 Experimental Results

4.1 Experimental Setup

The vehicle safety system assembled for tests consists of a car with two Kinect 2 sensors mounted at different altitudes and tilt angles as shown in Fig. 1(c). The upper camera is placed at the same height which the car has. It is tilted down 30 degrees to make it possible to scan the road surface. The lower camera is mounted at the bottom part of the car and tilted up 15 degrees. Together both cameras can see the whole volume that equals to car dimensions. The vehicle was moving backwards at low speed (5–10 kmph) to imitate the cargo moving scenario. If much faster movement is required, then Kinect sensors cannot be applied due to low detection distance. They should be replaced by much more expensive laser range finders. Tests were done at places with low light condition; however, for taking a screenshot of the working system, the vehicle was moved to the brighter area (Fig. 3). Each Kinect 2 is connected to a separate PC due to Kinect SDK 2.0 limitations (Only a single camera can be connected to a computer). The first PC with Intel Core i3 processor receives data and transmits it to the second PC with Intel Core i5 processor via Gigabit Ethernet.

Fig. 3. Detection results. Horizontal yellow line in the upper image represents the distance to the closest obstacle or pavement corruption. Red rectangle in the bottom image informs the driver that there is an obstacle above the road surface (Color Figure Online).

An alarm speaker is also connected to the second PC. Active IR mode of Kinect 2 devices is used. In the experimental setup, Kinect sensors were mounted using external connectors which leads to re-calibration if the cameras are moved, which is expected to happen in long rides and/or with strong winds. In the real industrial systems it is proposed to mount Kinect sensors into preliminary prepared embedded hardware sockets.

4.2   Test Results

Tests have shown that the minimum height of the object to be detected is 4 cm. However, due to sensor noise, detection is guaranteed if the object height is at least

10 cm. The maximum distance to the object when it can be robustly detected is 4.5 meters. The minimum size of the pit or rut is 10 cm in diameter at 2 meters distance. The level of danger of pits and ruts should be chosen according to the size of tires and clearance of the specific vehicle. If all calculations are performed on a single PC, then another PC should send raw sensor data over network. Translation of data from client to server occupies 564 megabit per second bandwidth if no compression applied. The average processing time of each frame is 32.25 ms on Intel Core i5 processor. An example of the working detection system is shown in Fig. 3. Kinect 2, as many other sensors, provides information with errors which depends on many factors including the type of surface the signals reflects on. For the night urban scene the measurements errors are shown in Fig. 4. The distance in this case is the distance between the vehicle and the object. According to the blue line it is clear that the accuracy is higher if the object appears close to the center of the image. Thanks to using active IR mode, absence or presence of light sources will not affect prediction errors unless they generate strong infrared waves. This approach cannot be used at day time since the direct sunlight prevents getting distance information.

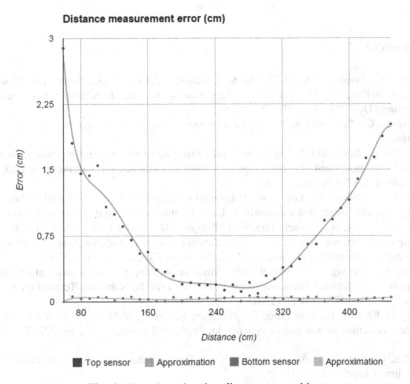

Fig. 4. Error in estimating distance to an object

5 Conclusion

The vehicle safety system for backward movement was developed which is able to work at night. This system detects the obstacles behind the truck and informs a driver if a dangerous situation occurs. The system can not only detect object in the current time, but also predicts the obstacles which will appear in the near future thanks to the tracking algorithm.

In future the following improvements are planned to be introduced to the system:

1. Detect moving objects if they do not satisfy the constant speed model.
2. Adapt region of interest based on the current steering wheel orientation by applying the truck kinematic model.
3. Search obstacles along the newly predicted vehicle trajectory.
4. Address dealing with noise given by unfavorable weather conditions, e.g., rain, which could significantly impact on performance.
5. Use the experimental setup proposed in this work for day time using visual information of Kinect sensors (stereovision) for scene reconstruction.

References

1. Treat, J.R., Tumbas, N.S., McDonald, S.T., Shinar, D., Hume, R.D., Mayer, R.E., Stansifer, R.L., Castellan, N.J.: Tri-level study of the causes of traffic accidents: final report. Executive summary, DOT-HS-034-3-535-79-TAC(S)
2. Edwards, C.: Car safety with a digital dashboard, Engineering & Technology, pp. 60–64 (2014)
3. Mandici, S., Nedevschi, S.: Aggregate road surface based environment representation using digital elevation maps. In: Intelligent Computer Communication and Processing (ICCP), pp. 149–156, Cluj Napoca (2014)
4. Lee, K.-Y., Park, J.-M., Lee, J.-W.: Estimation of longitudinal profile of road surface from stereo disparity using Dijkstra algorithm. Int. J. Control Autom. Syst. 12(4), 895–903 (2014)
5. Premachandra, C.H., Premachandra, W.H., Parape, C.D., Kawanaka, H.: Road crack detection using color variance distribution and discriminant analysis for approaching smooth vehicle movement on non-smooth roads, Int. J. Mach. Learn. Cybern. (2014)
6. Chen, Y-L., Wang, C-A.: Vehicle safety distance warning system: a novel algorithm for vehicle safety distance calculating between moving cars. In: Vehicular Technology Conference, pp. 2570–2574. Dublin (2007)
7. Raj, A., Krishna, D., Hari Priya, R., Shantanu, K., Niranjani Devi, S.: Vision based road surface detection for automotive systems, In: Applied Electronics (AE), pp. 223–228, Pilsen (2012)
8. Welch, G., Bishop, G.: An Introduction to the Kalman Filter. Tech. Rep. University. North Carolina, Chapel Hill, NC (1995)
9. Jo, K.-H., Vavilin, A.: HDR image generation based on intensity clustering and local feature analysis. Comput. Hum. Behav. 27(5), 150–1511 (2011)

Plant Leaf Recognition Based on Contourlet Transform and Support Vector Machine

Ze-Xue Li[1(✉)], Xiao-Ping Zhang[1], Li Shang[2], Zhi-Kai Huang[3],
Hao-Dong Zhu[4], and Yong Gan[4]

[1] College of Electronics and Information Engineering, Tongji University,
4800 Cao'an Highway, Jiading, Shanghai, China
lizexue0@163.com
[2] Department of Communication Technology, College of Electronic Information
Engineering, Suzhou Vocational University, Suzhou 215104, Jiangsu, China
[3] College of Mechanical and Electrical Engineering,
Nanchang Institute of Technology, Nanchang 330099, Jiangxi, China
[4] College of Computer and Communication Engineering,
Zhengzhou University of Light Industry, Zhengzhou, China
ganyong@zzuli.edu.cn

Abstract. Plants are essential to the balance of nature and in people's lives as the fundamental provider for food, oxygen and energy. The study of plants is also essential for environmental protection and helping farmers increase the production of food. As a fundamental task in botanical study, plant leaf recognition has been a hot research topic in these years. In this paper, we propose a new method based on contourlet transform and Support Vector Machine (SVM) for leaf recognition. Contourlet Transform is a promising multi-resolution analysis technique, which provides image with a flexible anisotropy and directional expansion. By basing its constructive principle on a non-subsampled pyramid structure and related directional filter banks, contourlet transform decomposes input images into multi-scale factors which also enjoys additional advantages such as shift invariance and computational efficiency. Compared with one-dimensional transforms, such as the Fourier and wavelet transforms, Contourlet Transform can capture the intrinsic geometrical structure. In order to ameliorate the influence of unwanted artefacts such as illumination and translation variations, in this paper, the contourlet transform was firstly applied to extract feature with high discriminative power. Then the extracted features are classified by SVM. The experimental results show that the proposed method has high sensitivity of directionality and can better capture the rich features of natural images such as edges, curves and contours.

Keywords: Leaf recognition · Contourlet Transform (CT) · Support vector machine (SVM) · Contourlet Transform · Frames · Image denoising · Image enhancement · Multidimensional filter banks · Nonsubsampled filter banks

© Springer International Publishing Switzerland 2015
D.-S. Huang et al. (Eds.): ICIC 2015, Part II, LNCS 9226, pp. 145–154, 2015.
DOI: 10.1007/978-3-319-22186-1_14

1 Introduction

Studying and exploiting the plant recognition has been one of the most important tasks in plant protection. As we know, the crucial stage of plant taxonomy is a genuine scientific and technical challenge, due not only to the huge number of plant species, but also to their highly specialized and diverse taxonomic properties. For this reason, the efficiency of manual plant recognition is too low and we should introduce pattern recognition technology to carry out this work. One key distinguishing feature for the identification of plant species is plant information obtained from leaf images [1–6]. Considering the different extracted features of leaf images, the recognition method can be roughly divided into 3 kinds - based on texture features, subspace projection and statistical features [7–10].

The method based on the structure feature need pre-processing and extracts the texture feature of leaf images. Such methods require complex pretreatment process, and the pretreatment results will influence the accuracy of recognition seriously. Although the texture features can obtain certain recognition accuracy, they are sensitive to the position and orientation changes during the collection process, which lacks stability and robustness [11, 12].

The subspace projection method applies Principal Component Analysis (PCA), Independent Component Analysis (ICA) or linear classification analysis in a certain transform domain of the leaf images. Then choose a proper classifier to take this recognition work by the projection coefficient as the feature. Compared with the method based on the structure, subspace projection method has anti-noise-interference ability without complex pretreatment process. But the changes of location and direction maybe interfered the ability of recognition.

This paper is organized as follows. We throw the concept and principle of Contourlet Transform and propose in Sect. 2 Contourlet Transform to decompose the leaf image, and then introduce the low frequency sub-band feature extraction and high frequency sub-band feature extraction. In Sect. 3, we provide a classifier - Support Vector Machine (SVM) Classifier. The logic of SVM is expatiated in Sect. 3. And then in Sect. 4, the experimental database and results are list. At last, we conclude this paper in Sect. 5.

2 Contourlet Transform

The research results by neural physiologist show that acceptant fields in the visual cortex are characterized as being localized, oriented, and bandpass for the visual system of human. There are experiments suggested that it will be efficient for a computational image representation, if it based on a local, directional, and multiresolution expansion.

As two-dimensional wavelets are constructed from tensor products of one-dimensional wavelets, with the finer resolution, we can clearly find the limitation of the wavelet that it needs to use many special dots to capture the contour, as show in the left of Fig. 1. The new scheme (in the right of Fig. 1.) shows that the support interval of baseband should behave as a long strip shape in order to make full use of the geometrical transformation of the original function and achieve with the least

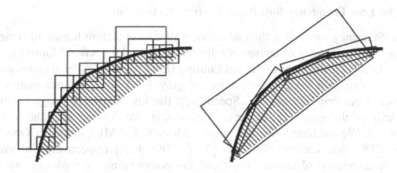

Fig. 1. Wavelet versus new scheme

coefficients to approximate the singular curve. In fact, the elongated support interval of baseband is a reflection of directionality, and this also called multi-scale geometric analysis.

2.1 Feature Extraction

The Contourlet Transform is implemented by a double filter bank named pyramidal directional filter bank (PDFB). PDFB can be seen as a cascade of two steps. Firstly the original image is multi-scaled decomposed into low frequency and high frequency subbands by Laplacian Pyramid (LP) transform. Then the Directional Filter Banks (DFB) decompose bandpass signal of each level in Pyramid into tree structure of L layer, and the band will be divided into two directions in each layer. The singular points distributed in the same direction will be synthesized as one coefficient. It achieved the image sparse representation more effectively that Contourlet Transform combined LP and DFB into the double filter group structure. The Contourlet Transform can be implemented iteratively applying PDFB on the coarse scale of image, as shown in Fig. 2 [20, 21].

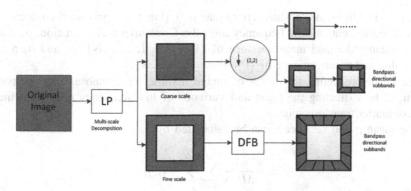

Fig. 2. Contourlet filter bank

2.2   The Low Frequency Sub-Band Feature Extraction

The low-frequency subband is the embodiment of coarse texture feature of image. In this paper, the uniformity of texture is reflected by Angular Second Moment (ASM), Contrast (CON), Correlation (COR) and Entropy (ENT) of the gray level co-occurrence matrix. We extracted these four features of gray level co-occurrence matrix after Contourlet Transform decomposed. Specifically the low frequency image coefficient matrix will be the gray level co-occurrence matrix transformed after the image is decomposed. We calculated Angular Second Moment (ASM), Contrast (CON), Correlation (COR) and Entropy (ENT) in $[0°, 45°, 90°, 135°]$ respectively. In order to reduce the dimension of feature vector and the computational complexity, we get a feature vector in 8 dimensions $f_1 = [a_1, a_2, a_3, a_4, a_5, a_6, a_7, a_8]$ through calculating the mean and variance of each parameter in four directions. The calculating formulae of the specific parameters are as following [22, 23].

Angular Second Moment (ASM):

$$ASM = \sum_i \sum_j P(i, j)^2 \tag{1}$$

Contrast (CON):

$$CON = \sum_i \sum_j (i, j)^2 P(i, j) \tag{2}$$

Correlation (COR):

$$COR = [\sum_i \sum_j i \times j \times P(i, j) - \mu_x \mu_y]/\sigma_x \sigma_y \tag{3}$$

Entropy (ENT):

$$ENT = -\sum_i \sum_j P(i, j) \text{lb}[P(i, j)] \tag{4}$$

where $P(i, j)$ is the elements whose coordinate is (i, j) that is in gray level co-occurrence matrix of coefficients of low frequency after the Contourlet transformation. μ_x and σ_x are the mean value and mean variance of $\{P_x(i)|i = 1, 2, \ldots, N\}$. μ_y and σ_y are the mean value and mean variance of $\{P_y(i)|i = 1, 2, \ldots, N\}$.

For further reflecting the extent of image texture, we composed the eigenvector $f_2 = [\mu, \sigma]$ by extracting the mean and variance of low frequency coefficient matrix after contourlet transformation.

The mean μ and variance σ can be calculated by:

$$\mu = \frac{1}{M \times N} \sum_{i=1}^{M} \sum_{j=1}^{N} P(i, j) \tag{5}$$

$$\sigma = \frac{1}{M \times N} \sum_{i=1}^{M} \sum_{j=1}^{N} (P(i,j) - \mu)^2 \tag{6}$$

In (5) and (6), P(i, j) is the decomposition coefficient whose coordinate is (i, j) in M × N low frequency subband coefficient matrix after the contourlet transformation.

2.3 High Frequency Sub-Band Feature Extraction

The high-frequency directional subband of Contourlet Transform contains the image edges and fine texture feature. In this paper, the image is decomposed into 4 layers. From the first layer to the third layer are intermediate frequency band and the fourth layer is high frequency band.

The intermediate frequency band contains part texture information of image. The mean and variance can reflect not only the unevenness of gray image, but also the depth degree of the texture. Considering these factors, the mean and variance of intermediate frequency coefficient matrix are extracted as texture features of intermediate frequency sub-band. The three intermediate frequency sub-bands contain 3, 4 and 8 direction respectively, and we can compose a 30-dimensional feature vector $f_3 = [\mu_1, \mu_2, \ldots, \mu_{15}, \sigma_1, \sigma_2, \ldots, \sigma_{15}]$ by the mean and variance of sub-band coefficients in these 15 directions.

The energy distribution is sparser at the highest level sub-band. Because energy distribution at different scales and directions can effectively distinguish the texture, the energy of the coefficient matrix is extracted as high frequency characteristics. The high frequency image contains 16 directions after Contourlet Transform and we can extract the energy of sub-band coefficients in these 16 directions to form a 16-dimensional feature vector $f_4 = [b_1, b_2, \ldots, b_{16}]$.

The energy can be calculated as following:

$$E = \sum_{i=1}^{M} \sum_{j=1}^{N} P(i,j)^2 \tag{7}$$

In (7), P(i, j) is the decomposition coefficient whose coordinate is (i, j) in M × N high frequency sub-band coefficient matrix after the contourlet transformation.

In addition, the mean and variance of the high-frequency sub-band coefficient matrix can reflect the depth degree of the texture and it is also an important high-frequency image texture features. We extract the mean and variance of high frequency image sub-band decomposed to form a 32-dimensional feature vector $f_5 = [\mu_1, \mu_2, \ldots, \mu_{16}, \sigma_1, \sigma_2, \ldots, \sigma_{16}]$.

As we have pointed out above, we extracted 5 feature vectors from different frequency band. These feature vectors can fully represent the uniformity, depths of color and energy distribution and other characteristics of image texture. However, if we group the entire feature vector into vector set as the input of identification system, the speed of recognition is bound to be affected due to the dimension of the vector is too large. It is a pretty important part that how to determine the optimal texture feature representation in Contourlet Transform, which means choosing as little feature vectors as possible to characterize the texture on the condition of ensuring the recognition accuracy. In this paper, we put the extracted feature vectors as input of the identification system, and finally determine the optimal feature vector after repeated recognition.

3   Support Vector Machine (SVM) Classifier

As mentioned above, the recognition process is as Fig. 3 showed. We choose Support Vector Machine (SVM) as classifier. Support vector machine is a machine learning method based on development of statistical learning theory [24]. It has very strong generalization ability and less depends on the quantity and quality of samples. It has the best generalization ability for the classification of unknown samples through constructing the optimal hyperplane. It promotes the problem to be processed in linear form that Support Vector Machine (SVM) maps the data from input space to a high-dimensional feature space by support vector (SV) kernel. As SVM usually tries to minimize a bound on the structural risk but not the empirical risk, it always can get a global minimum value.

Empirical Risk Minimization (ERM) is a formal term for a simple concept: find the function $f(x)$ that minimizes the average risk on the training set. Empirical risk is defined as bellow:

$$R_{emp}(f) = \frac{1}{N} \sum_{i=1}^{N} C(f(x_i), y_i) \tag{8}$$

where $C(f, y)$ is a suitable cost function, e.g., $C(f, y) = (f(x) - y)^2$.

Minimizing the empirical risk is not a bad thing to do, provided that sufficient training data is available, since the law of large numbers ensures that the empirical risk will asymptotically converge to the expected risk for $n \to \infty$. However, for small samples, one cannot guarantee that ERM will also minimize the expected risk. This is the all too familiar issue of generalization.

The Vapnik-Chervonenkis dimension (VC dimension) is a measure of the complexity (or capacity) of a class of functions $f(\alpha)$. The VC dimension measures the largest number of examples that can be explained by the family $f(\alpha)$. The basic argument is that high capacity and generalization properties are at odds. If the family $f(\alpha)$ has enough capacity to explain every possible dataset, we should not expect these functions to generalize very well. On the other hand, if functions $f(\alpha)$ have small capacity but they are able to explain our particular dataset, we have stronger reasons to

Fig. 3. Recognition Process

believe that they will also work well on unseen data. The VC dimension is the size of the largest dataset that can be shattered by the set of functions f(α). One may expect that models with a large number of parameters would have high VC dimension, whereas models with few parameters would have low VC dimensions. The VC dimension is a more "sophisticated" measure of model complexity than dimensionality or number of free parameters.

Because the VC dimension provides bounds on the expected risk as a function of the empirical risk and the number of available examples. It can be shown that the following bound holds with probability $1 - \eta$.

$$R(f) \leq R_{emp}(f) + \sqrt{\frac{h(\ln(\frac{2N}{h}) + 1) - \ln(\frac{\eta}{4})}{N}} \tag{9}$$

where h is the VC dimension of $f(\alpha)$, N is the number of training examples, and $N > h$.

Structural Risk Minimization (SRM) is another formal term for an intuitive concept: the optimal model is found by striking a balance between the empirical risk and the VC dimension. SVM achieves SRM by minimizing the following Lagrangian formulation:

$$L_P(\omega, b, \alpha) = \frac{1}{2}||\omega||^2 - \sum_{i=1}^{N} \alpha_i[y_i(\omega^T x_i + b) - 1] \tag{10}$$

where α_i is positive Lagrange multipliers [25, 26].

As the ratio N/h gets larger, the VC confidence becomes smaller and the actual risk becomes closer to the empirical risk. This and other results are part of the field known as Statistical Learning Theory or Vapnik-Chervonenkis Theory, from which Support Vector Machines originated.

4 Experimental Results

To evaluate the effectiveness of the proposed method, we carried out a series experiments on two large and comprehensive texture databases: the Sweden leaves database [27] and the ICL database[1] which is established by the Intelligent Computing Laboratory.

All images in the ICL database are taken by cameras or scanners on a white background paper under vary illumination conditions after the leaves are picked from plants. Two sides of every kind of leaf are respectively taken to images. Images are tended to be in an agreed size and colorful which are taken through this process. To guarantee the background smooth and clear, only one leaf is constructed an image. The ICL database includes 200 species of plants. Each species includes 30 samples of leaf images (15 per side). Hence there are totally 6000 images. Figure 4 shows some

---

[1] http://www.intelengine.cn/dataset/index.html

samples from the ICL database in which the top images are the front side of plant leaves and the bottom images are the opposite side.

The Sweden leaves database contains 15 the monolithic leaves pictures of different Swiss tree and each type has 75 image files. The original image data of Swiss plant leaves contains petiole. This does not have robust feature that the shape of the leaves should have, because the direction and length of petiole deeply depend on the process of the leaf collection when intercept the plant leaf image samples. While leafstalk will provide certain difference information, we can remove some kind of noise to build another data set.

In order to implement the proposed method, we choose 5 sets of data and set the relevant parameters empirically.

4.1   Experimental Results on the ICL Database and the Sweden Leaves Database

The experimental results of the ICL Database are as follows: (Table 1).

Table 1. The classification rates (%) on ICL database

Method	Recognition rate (%)					Equal error rates (EER)
	Set 1	Set 2	Set 3	Set 4	Set 5	
LBP	62.5774	62.4313	58.9223	61.5548	56.7119	0.3845
Curvelet	60.9754	62.7850	65.4688	69.5751	69.6489	0.3323
Contourlet	**69.4189**	**72.2176**	**77.1574**	**75.9221**	**72.7874**	**0.2754**

The experimental results of the Sweden Leaves Database are as follows: (Table 2)

Table 2. The classification rates (%) on Sweden leaves database

Method	Recognition rate (%)					Equal error rates (EER)
	Set 1	Set 2	Set 3	Set 4	Set 5	
LBP	79.0605	72.3183	74.7692	72.4716	72.9713	0.2932
Curvelet	84.2346	82.9483	79.1720	85.5022	76.3445	0.2254
Contourlet	**86.3874**	**85.8156**	**89.6552**	**89.9520**	**83.8635**	**0.1723**

5   Conclusions

In this paper, we studied a hybrid approach based on Contourlet Transform and Support Vector Machine (SVM) Classifiers for plant recognition. By decomposing input images into multi-scale factors which have attractive properties such as shift invariance and computational efficiency, we can extract discriminative features which are not sensitive to the variations of illumination and translation and capture the intrinsic geometry structure of images. By combining the crafted feature with large margin classifiers (more specifically, SVM), the proposed recognition method has higher experimental performance and can better capture the rich features of natural images such as edges, curves and contours. In the future work, we plan to improve the efficiency of the proposed method, and implement it as recognition software which is suitable for real-word applications.

Acknowledgments. This work was supported by the grants of the National Science Foundation of China, Nos. 61133010, 61373105, 61303111, 61411140249, 61402334, 61472282, 61472280, 61472173, 61373098 and 61272333, China Postdoctoral Science Foundation Grant, Nos. 2014M561513, and partly supported by the National High-Tech R&D Program (863) (2014AA021502 & 2015AA020101), and the grant from the Ph.D. Programs Foundation of Ministry of Education of China (No. 20120072110040), and the grant from the Outstanding Innovative Talent Program Foundation of Henan Province, No. 134200510025.

References

1. Guyer, D., Miles, G., Schreiber, M., Mitchell, O., Vanderbilt, V.: Machine vision and image processing for plant identification. Trans. ASAE **29**, 1500–1507 (1986)
2. Wang, X.-F., Huang, D.-S., Xu, H.: An efficient local Chan-Vese model for image segmentation. Pattern Recognit. **43**, 603–618 (2010)
3. Huang, D.-S.: Systematic Theory of Neural Networks for Pattern Recognition, vol. 28, pp. 323–332. Publishing House of Electronic Industry of China, Beijing (1996)
4. Yu, H.-J., Huang, D.-S.: Normalized feature vectors: a novel alignment-free sequence comparison method based on the numbers of adjacent amino acids. IEEE/ACM Trans. Comput. Biol. Bioinform. (TCBB) **10**, 457–467 (2013)
5. Huang, D.-S., Jiang, W.: A general CPL-AdS methodology for fixing dynamic parameters in dual environments. IEEE Trans. Syst. Man Cybern. Part B Cybern. **42**, 1489–1500 (2012)
6. Huang, D.-S., Du, J.-X.: A constructive hybrid structure optimization methodology for radial basis probabilistic neural networks. IEEE Trans. Neural Netw. **19**, 2099–2115 (2008)
7. Wang, X.-F., Huang, D.-S.: A novel density-based clustering framework by using level set method. IEEE Trans. Knowl. Data Eng. **21**, 1515–1531 (2009)
8. Shang, L., Huang, D.-S., Du, J.-X., Zheng, C.-H.: Palmprint recognition using FastICA algorithm and radial basis probabilistic neural network. Neurocomputing **69**, 1782–1786 (2006)
9. Zhao, Z.-Q., Huang, D.-S., Sun, B.-Y.: Human face recognition based on multi-features using neural networks committee. Pattern Recogn. Lett. **25**, 1351–1358 (2004)
10. Huang, D., Ip, H., Chi, Z.: A neural root finder of polynomials based on root moments. Neural Comput. **16**, 1721–1762 (2004)
11. Huang, D.-S.: A constructive approach for finding arbitrary roots of polynomials by neural networks. IEEE Trans. Neural Netw. **15**, 477–491 (2004)
12. Huang, D.-S.: Radial basis probabilistic neural networks: model and application. Int. J. Pattern Recognit Artif Intell. **13**, 1083–1101 (1999)
13. Xiangbin, Z.: Texture classification based on contourlet and support vector machines. In: 2009 ISECS International Colloquium on Computing, Communication, Control, and Management, pp. 521–524 (2009)

14. Liu, Z., Fan, X., Lv, F.: SAR image segmentation using contourlet and support vector machine. In: Fifth International Conference on Natural Computation, 2009. ICNC 2009, pp. 250–254. IEEE (2009)
15. Wang, J., Ge, Y.: Texture feature recognition based on contourlet transform and support vector machine. Jisuanji Yingyong J. Comput. Appl. **33**, 677–679 (2013)
16. Haralick, R.M., Shanmugam, K., Dinstein, I.H.: Textural features for image classification. IEEE Trans. Syst. Man Cybern. **3**(6), 610–621 (1973)
17. Vapnik, V.: The nature of statistical learning theory. Springer Science & Business Media, New York (2000)
18. Burges, C.J.: A tutorial on support vector machines for pattern recognition. Data Min. Knowl. Disc. **2**, 121–167 (1998)
19. Schölkopf, B., Burges, C.J., Smola, A.J.: Advances in kernel methods: support vector learning. MIT press, Massachusetts (1999)
20. Soderkvist, O.J.O.: Computer Vision Classication of Leaves from Swedish Trees (2001)

Implementation of Plant Leaf Recognition System on ARM Tablet Based on Local Ternary Pattern

Gong-Sheng Xu[1(✉)], Jing-Hua Yuan[1], Xiao-Ping Zhang[1], Li Shang[2],
Zhi-Kai Huang[3], Hao-Dong Zhu[4], and Yong Gan[4]

[1] College of Electronics and Information Engineering,
Tongji University, Shanghai, China
haiyi1919@163.com
[2] Department of Communication Technology,
College of Electronic Information Engineering,
Suzhou Vocational University, Suzhou 215104, Jiangsu, China
[3] College of Mechanical and Electrical Engineering,
Nanchang Institute of Technology, Nanchang 330099, Jiangxi, China
[4] College of Computer and Communication Engineering,
Zhengzhou University of Light Industry, Zhengzhou, China

Abstract. The Local Binary Pattern (LBP) and its variants is powerful in capturing image features and computational simplicity, However LBP's sensitivity to noise, particularly in near-uniform image regions has stimulated many transformations of LBP to improve the ability of feature description. The Local Ternary Pattern (LTP) extends the conventional LBP to ternary codes and makes a significant improvement. LTP is more resistant to noise, but no longer strictly invariant to gray-level transformations. In this paper, by adopting the Average Local Gray Level (ALG) to take place of the traditional gray value of the center pixel and taking an auto-adaptive strategy on the selection of the threshold, we propose the Enhanced Local Ternary Pattern (ELTP) to improve the performance of LTP and implement an android application to recognize plant-leaf image and identify the species of the plant.

Keywords: LBP · LTP · ELTP · Texture classification · Android application

1 Introduction

In [1], Ojala et al. proposed the Local Binary Pattern (LBP) to address rotation invariant texture classification. LBP is efficient to represent local texture and invariant to monotonic gray scale transformations. Heikkila et al. [2] presented center-symmetric LBP (CS-LBP) by comparing center-symmetric pairs of pixels. Liao et al. [3] proposed Dominant LBP (DLBP) for texture classification. Tan and Triggs [4] presented Local Ternary Pattern (LTP), extending the conventional LBP to 3-valued codes. However LTP is no longer strictly invariant to gray-level transformations due to the simple strategy on the selection of the threshold. Zhang et al. [5] proposed Local Derivative Pattern (LDP) to capture more detailed information by introducing high order

© Springer International Publishing Switzerland 2015
D.-S. Huang et al. (Eds.): ICIC 2015, Part II, LNCS 9226, pp. 155–164, 2015.
DOI: 10.1007/978-3-319-22186-1_15

derivatives. However, if the order is greater than three, LDP is more sensitive to noise than LBP. Guo et al. [6] proposed Completed LBP (CLBP) by combining the original LBP with the measures of local intensity difference and central pixel gray level. Zhao et al. [7] proposed the Local Binary Count (LBC), to address rotation invariant texture classification by totally discarding the micro-structure which is not absolutely invariant to rotation under the huge illumination changes. However, LBC (or LBP) is sensitive to random noise and quantization noise in the near-uniform regions, as the LBC (or LBP) threshold is the value of the central pixel as mentioned in [4]. Zhao et al. [8] presented Completed Robust Local Binary Pattern (CRLBP) by modifying the center pixel gray level to improve the LBP, but a more parameter has to be tuned.

Motivated by [4, 7, 8], this paper tries to address these potential difficulties by proposing the Enhanced Local Ternary Pattern (ELTP). An auto-adaptive strategy for threshold selection is adopted and a novel coding process is introduced. Further more, the Completed Enhanced Local Ternary Pattern (CELTP) is presented.

The remainder of this paper is organized as follows. Section 2 presents the ELTP and CELTP. Section 3 introduces the different models of the application of plant-leaf recognition system. Section 4 concludes the paper.

2   Brief Review of Enhanced Local Ternary Pattern

In order to address the demerits of LBPs, we propose Enhanced Local Ternary Pattern (ELTP). Gray-levels in a tolerance zone of width $\pm t^e$ around g_c^e are quantized to zero, ones above this are quantized to $+1$ and ones below it to -1, i.e., the indicator $s(x)$ used in LBP is replaced with a 3-valued function:

$$s^e(g^p, g^e, t^e) = \begin{cases} 1, & g^p - g_c^e \geq t^e \\ 0, & |g^p - g_c^e| < t^e \\ -1, & g^p - g_c^e \leq -t^e \end{cases} \quad p = 0, 1, 2 \ldots P-1$$

$$g_c^e = mean(G), t^e = mad(G), G = \{g^i | i = 0, 1, \ldots, 8\}$$

where P is the size of the neighbor set of pixels, g_p $(p = 0, 1, \ldots, P\text{-}1)$ denotes the gray value of the neighbor, G is the set of the gray-level values in a 3×3 local region, $mean$ (G) is the mean of the set G, $mad(G)$ is the median absolute deviation of the set G.

g_c is replaced by g_c^e, which make the ELTP more robust to noise. On the other hand, instead of using the simple user-specified threshold, we adopt new strategy to set the threshold t^e. To make the selection of threshold auto-adaptive, the threshold is set as the median absolute deviation (MAD), which not only reflect derivation of the local region but make the ELTP code invariant to gray-level transformations and insensitive to noise. In addition, the use of MAD does not affect the complexity of the model.

For simplicity, the experiments use a coding scheme that splits each ternary pattern into its positive half (ELTP_P) and negative half (ELTP_N) as illustrated in Fig. 1, subsequently combining these two separate channels of LBC descriptors to form the final ELTP descriptor and finally computing the histogram and similarity metric.

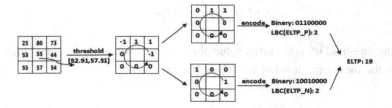

Fig. 1. The process of forming the ELTP code.

Obviously, ELTP is also rotation invariant. Using the same notations as in (1), the ELTP descriptor used in the experiment is defined as follows:

$$ELTP_{P,R} = ELTP\_P_{P,R} * (P+2) - (ELTP\_P_{P,R} * (ELTP\_P_{P,R} + 1))/2$$
$$+ ELTP\_N_{P,R}$$

$$ELTP\_P_{P,R} = \sum_{p=0}^{P-1} e(s^e(g_p, g_c^e, t^e), 1)$$

$$ELTP\_N_{P,R} = \sum_{p=0}^{P-1} e(s^e(g_p, g_c^e, t^e), -1), e(x,y) = \begin{cases} 1, & x = y \\ 0, & x \neq y \end{cases}$$

Aiming to improve the discriminative capability of the local structure, the enhanced image local differences d_p^e are decomposed into two complementary components [6], i.e., the signs (s_p^e) and the magnitudes (m_p^e) respectively

$$d_p^e = m_p^e, s_p^e = s^e(g_p, g_c^e, t^e)$$

$$m_p^e = ((g_p - g_c^e) - s_p^e * t^e) * s_p^e, p = 0, 1, \ldots, P.$$

Similar to the CLBP, we also proposed Completed Enhanced Local Ternary Pattern (CELTP) which contain three operators: CELTP_S, CELTP_M and CELTP_C. Generally, the CELTP_S equals to the original ELTP described above. And the CELTP_M and CELTP_C can be defined as:

$$CELTP\_M_{P,R} = CELTP\_MP_{P,R} * (P+2) - (CELTP\_MP_{P,R} * (CELTP\_MP_{P,R}$$
$$+ 1))/2 + CELTP\_MN_{P,R}$$

$$CELTP\_MN_{P,R} = \sum_{p=0}^{P-1} s(m_p^e - m_I^e) * e(s_p^e, -1) \ CELTP\_MP_{P,R}$$

$$= \sum_{p=0}^{P-1} s(m_p^e - m_I^e) * e(s_p^e, 1)$$

$$CELTP\_C_{P,R} = s(g_c^e - c_t)$$

where the threshold m_l^e is the mean value of m_p^e of the whole image and the threshold c_l is set as the mean gray level of the whole image.

3 Experimental Results

To evaluate the effectiveness of the proposed method, we carried out a series experiments on a large and comprehensive texture databases: the Outex database [10], which includes 24 classes of textures collected under three illuminations and at nine angles.

The Outex database includes two test suites, that is, Outex_TC_00010 (TC10) and Outex_TC_00012 (TC12). The two test suites contain the same 24 classes of textures, which were collected under three different illuminants ("horizon," "inca," and "t184") and nine different rotation angles (0°,5°,10°,15°,30°,45°,60°,75°,90°). There are 20 non-overlapping texture samples for each class under a given illumination and rotation angle condition. For TC10, the classifier was trained with samples of illumination "inca" and angle 0° in each class and the testing dataset was constructed by the other eight rotation angles with the same illuminant. Therefore, there are 480 (24 * 20) and 3840 (24 * 20 * 8) images used for models and validation samples respectively. For TC12, same dataset as TC10 were used for classifier training and all samples captured under illumination "t184" or "horizon" were used as testing samples. Hence, there are 480 (24 * 20) models and 4320 (24 * 20 * 9) validation samples.

Table 1 lists the experimental results by different schemes, from which we could make the following findings. Firstly, the proposed CELTP method is inferior to CLBP_S/M/C, CLBC_S/M/C and CRLBP_S/M/C on the condition (R = 1, P = 8). This mainly because we use the mean of the gray-level values in a 3×3 region to replace the value of the center pixel. Secondly, except for the experiment (R = 1, P = 8) on TC10 dataset, ELTP performs much better than LBP; except for the experiment (R = 3,

Table 1. Classification Rate (%) on TC10 and TC12 Using Different Methods

	TC10	R = 1. P = 8 TC12		Average	TC10	R = 2. P = 16 TC12		Average	TC10	R = 3. P = 24 TC12		Average
		t184	horizon			t184	horizon			t184	horizon	
LBP[15]	84.82	65.46	63.68	71.32	89.40	82.27	75.21	82.29	95.08	85.05	80.79	86.97
LTP[15]	76.06	62.56	63.42	67.34	96.11	85.20	85.87	89.06	98.64	92.59	91.52	94.25
CLBP_M[15]	81.74	59.31	62.78	67.94	93.67	73.80	72.41	79.96	95.52	81.18	78.66	85.12
CLBP_S/M/C[15]	96.56	90.30	92.29	93.05	98.72	93.54	93.91	95.39	98.93	95.32	94.54	96.26
CLBC_S/M/C[16]	97.16	89.79	92.92	93.29	98.54	93.26	94.07	95.29	98.78	94.00	93.24	95.67
CRLBP_S/M/C(α=1)[17]	96.54	91.16	92.06	93.25	98.85	96.67	96.97	97.50	99.48	97.57	97.34	98.13
CRLBP_S/M/C(α=8)[17]	97.55	91.94	92.45	93.98	98.59	95.88	96.41	96.96	99.35	96.83	96.16	97.45
ELTP	79.06	67.82	76.34	74.41	97.14	90.95	91.00	93.03	96.54	93.96	93.22	94.57
CELTP_M	80.83	70.81	74.63	75.42	97.06	90.88	91.50	93.15	98.59	96.46	96.25	97.10
CELTP_M/C	87.99	78.31	83.17	83.16	92.27	86.92	89.58	89.59	94.51	92.08	93.54	93.38
CELTP_S_M/C	87.81	76.67	82.94	82.47	93.67	90.56	91.34	91.86	95.55	94.65	95.76	95.32
CELTP_S/C	87.58	76.64	86.32	83.51	98.18	94.93	95.76	96.29	98.46	93.63	95.72	95.94
CELTP_M_S/C	87.89	77.50	86.20	83.86	98.83	96.71	97.29	97.61	99.40	98.56	98.61	98.86
CELTP_S_M	82.68	71.25	77.94	77.29	98.39	96.00	95.97	96.79	99.40	98.45	98.08	98.64
CELTP_S_M_C	87.50	76.39	84.56	82.82	98.54	95.58	95.70	96.64	99.43	98.17	97.92	98.51

P = 24) on TC10 dataset, ELTP outperform LTP, too. It should be noticed that ELTP is much more robust to illumination variations. Thirdly, CELTP_M achieves impressive performance as compared with CLBP_M. This means that CELTP_M could capture more discriminative gray-level information than CLBP_M. Finally, except for the experiment (R = 1, P = 8), CELTP_M_S/C and CELTP_S_M both perform much better than other mentioned methods on TC12 dataset and achieve impressive performance. In one word, CELTP, with low feature dimensionality, achieves higher classification rates than other methods and is less sensitive to illumination variations.

4 Plant Leaf Recognition System

All of the models of the application can be inferred from the main interface of the application, which is shown in Fig. 2. The main function of the application is to identify the species of the plant according to its leaf image with the method of ELTP, which is implemented in the model of identification. Apart from this, some other models to facilitate the use of the application and extend its function are available to users. These different models will be presented as follows.

Fig. 2. Main interface of plant-leaf recognition application

4.1 Identification Model

Identification model implements various functions, including leaf image gray processing, image noise reduction, image segmentation, morphological processing and pre-contour detection, etc., and after extracting target leaf images' feature with LBP, the server call the embedded classifier, based on the Nearest Neighbor classifier to identify, and finally retrieves matches in the database. The detail procedure of identification is listed as follows:

I. Image capture: We have two ways to obtain the plant-leaf image. One is to take a picture of the plant-leaf by tablet camera and the other one is from local storage, which is shown in Fig. 3.

Fig. 3. Two ways of plant-leaf image capture

II. Image upload: In local identification, we don't need this step. Otherwise, in cloud identification model, the application uploads the image of the plant-leaf to cloud server by network connection for further processing.

III. Plant-leaf identification: The plants description is stored in the MySQL database in the server. After receiving the image from the application, server needs several procedures including preprocessing, feature extraction and classification to implement the identification function. Then it retrieves detail information of the plant identified from plants description database (ICL shown in Fig. 4 and Table 1) to send the message to the application.

IV. Information display: After receiving the detail information of the plant from the server, the application analyzes the data with specific format and display the information with android components.

4.2　Favorite Model

The favorites model is shown in Fig. 5. The model is used to put away plants information that users have an interest in and need to check the plants information in the future. After users identify the species of the plant and the result is shown to users, there is a star that you can choose to press in the right of the interface. The model is implemented by maintaining a database that can save and delete the item of the favorite list.

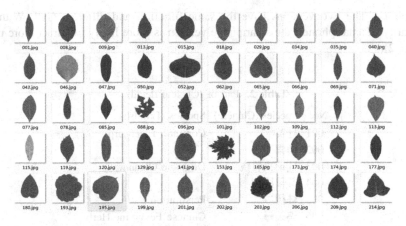

Fig. 4. Part of ICL leaf images

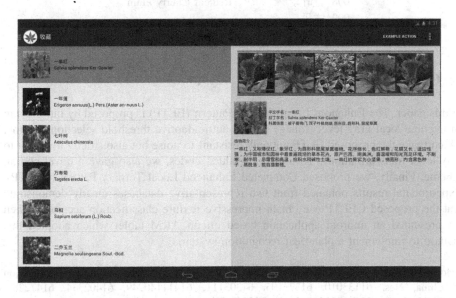

Fig. 5. The plants species shown in the favorite model

4.3 Neighbor Model

The application can access the geography location information of the user. The neighbor model display plants information near the location by the help of a local plants information database.

4.4 Smartphone Version

We developed the application in the android monitor which has a lot of versions of android devices. Apart from the ARM version, we also adapted it to the android smartphone. The two versions have the same function and different GUI. With the popularity of smartphones, the smartphone version is more likely to attract more users (Table. 2).

Table 2. Part of ICL leaf details

No.	Species (Chinese)	Species (English)
·001	杜英	Common Elaeocarp
·002	海棠	
·003	红花夹竹桃	Dogbane Oleander
·004	桑科拓树	
·005	柿树	Persimmon
·006	鸡矢藤	Chinese Fevervine Herb
·007	服藤子	
·008	红叶李	Redleaf Cherry Plum
·009	木瓜	
·010	蛇莓	India Mockstrawberry

5 Conclusions

In this paper, The Enhanced Local Ternary Pattern (ELTP) is proposed by introducing an average local gray-level strategy and an auto-adaptive threshold selection scheme, which make the proposed ELTP not only resistant to noise but also strictly invariant to gray-level transformations. To use the 3-valued codes, we also gave a novel coding scheme. Finally, we proposed Completed Enhanced Local Ternary Pattern (CELTP). Experimental results obtained from two representative databases clearly demonstrate that the proposed CELTP can obtain impressive texture classification accuracy. Then we presented an android application based on an ARM tablet which adopted this method to implement a plant-leaf recognition system.

Acknowledgments. This work was supported by the grants of the National Science Foundation of China, Nos. 61133010, 61373105, 61303111, 61411140249, 61402334, 61472282, 61472280, 61472173, 61373098 and 61272333, China Postdoctoral Science Foundation Grant, Nos. 2014M561513, and partly supported by the National High-Tech R&D Program (863) (2014AA021502 & 2015AA020101), and the grant from the Ph.D. Programs Foundation of Ministry of Education of China (No. 20120072110040), and the grant from the Outstanding Innovative Talent Program Foundation of Henan Province, No. 134200510025.

References

1. Ojala, T., Pietikainen, M., Maenpaa, T.: Multiresolution gray-scale and rotation invariant texture classification with local binary patterns. IEEE Trans. Pattern Anal. Mach. Intell. **24** (7), 971–987 (2002)

2. Heikkilä, M., Pietikäinen, M., Schmid, C.: Description of interest regions with center-symmetric local binary patterns. In: Kalra, P.K., Peleg, S. (eds.) ICVGIP 2006. LNCS, vol. 4338, pp. 58–69. Springer, Heidelberg (2006)
3. Liao, S., Law, M.W.K., Chung, S.: Dominant local binary patterns fortexture classification. IEEE Trans. Image Process. 18(5), 1107–1118 (2009)
4. Tan, X.Y., Triggs, B.: Enhanced local texture feature sets for face recognition under difficult lighting conditions. IEEE Trans. Image Process. 19(6), 1635–1650 (2010)
5. Zhang, B., Gao, Y., Zhao, S., Liu, J.: Local derivative pattern versus local binary pattern: Face recognition with high-order local pattern descriptor. IEEE Trans. Image Process. 19(2), 533–544 (2010)
6. Zhao, Y., Huang, D.-S., Jia, W.: Completed local binary count for rotation invariant texture. IEEE Trans. on Image Process. 21(10), 4492–4497 (2012)
7. Zhao, Y., Jia, W., Hu, R.-X., Min, H.: Completed robust local binary pattern for texture classification. Neurocomputing 106, 68–76 (2013)
8. Ojala, T., Maenpaa, T., Pietikainen, M., Viertola, J., Kyllönen, J., Huovinen, S.: Outex - new framework for empirical evaluation of texture analysis algorithms. In: Proceedings of 16th International Conference on Pattern Recognition, vol. 1, pp. 701–706 (2002)
9. Dana, K.J., van Ginneken, B., Nayar, S.K., Koenderink, J.J.: Reflectance and texture of real world surfaces. ACM Trans. Graph. 18(1), 1–34 (1999)
10. Varma, M., Zisserman, A.: A statistical approach to texture classification from single images. Int. J. Comput. Vis. 62(1–2), 61–81 (2005)
11. Li, B., Huang, D.S.: Locally linear discriminant embedding: an efficient method for face recognition. Pattern Recognit. 41(12), 3813–3821 (2008)
12. Huang, D.S.: Radial basis probabilistic neural networks: model and application. Int. J. Pattern Recognit. Artif. Intell. 13(7), 1083–1101 (1999)
13. Huang, D.S., Ma, S.D.: Linear and nonlinear feed forward neural network classifiers: a comprehensive understanding. J. Intell. Syst. 9(1), 1–38 (1999)
14. Huang, D.S.: A constructive approach for finding arbitrary roots of polynomials by neural networks. IEEE Trans. Neural Netw. 15(2), 477–491 (2004)
15. Huang, D.S., Horace, HS.Ip, Chi, Z.-R.: A neural root finder of polynomials based on root moments. Neural Comput. 16(8), 1721–1762 (2004)
16. Zhou, H., Wang, R., Wang, C.: A novel extended local binary pattern operator for texture analysis. Inf. Sci. 178(22), 4314–4325 (2008)
17. Tan, X., Triggs, B.: Enhanced local texture feature sets for face recognition under difficult lighting conditions. In: Zhou, S., Zhao, W., Tang, X., Gong, S. (eds.) AMFG 2007. LNCS, vol. 4778, pp. 168–182. Springer, Heidelberg (2007)
18. Zhao, Y., Huang, D.-S., Jia, W.: Completed local binary count for rotation invariant texture. IEEE Trans. Image Process. 21(10), 4492–4497 (2012)
19. Heikkilä, M., Pietikäinen, M., Schmid, C.: Description of interest regions with local binary patterns. Pattern Recognit. 42(3), 425–436 (2009)
20. Tan, X., Triggs, B.: Fusing Gabor and LBP feature Sets for Kernel-based Face Recognition. In: Proceedings of the IEEE International Workshop on Analysis and Modeling of Face and Gesture, pp. 235–249 (2007)
21. Huang, D.S.: Systematic Theory of Neural Networks for Pattern Recognition (in Chinese). Publishing House of Electronic Industry of China, Beijing (1996)
22. Huang, D.S., Yu, H.-J.: Normalized feature vectors: a novel alignment-free sequence comparison method based on the numbers of adjacent amino acids. IEEE/ACM Trans. Comput. Biol. Bioinf. 10(2), 457–467 (2013)
23. Huang, D.S., Jiang, W.: A general CPL-AdS methodology for fixing dynamic parameters in dual environments. IEEE Trans. Syst. Man Cybern. Part B 42(5), 1489–1500 (2012)

24. Wang, X.-F., Huang, D.S., Xu, H.: An efficient local Chan-Vese model for image segmentation. Pattern Recognit. **43**(3), 603–618 (2010)
25. Huang, D.S., Du, J.-X.: A constructive hybrid structure optimization methodology for radial basis probabilistic neural networks. IEEE Trans. Neural Netw. **19**(12), 2099–2115 (2008)
26. Huang, D.S.: Radial basis probabilistic neural networks: model and application. Int. J. Pattern Recognit. Artif. Intell. **13**(7), 1083–1101 (1999)
27. Wang, X.-F., Huang, D.S.: A novel density-based clustering framework by using level set method. IEEE Trans. Knowl. Data Eng. **21**(11), 1515–1531 (2009)
28. Shang, L., Huang, D.S., Du, J.-X., Zheng, C.-H.: Palmprint recognition using FastICA algorithm and radial basis probabilistic neural network. Neurocomputing **69**(13–15), 1782–1786 (2006)
29. Huang, D.S., Horace, H.S.Ip, Chi, Z.: A neural root finder of polynomials based on root moments. Neural Comput. **16**(8), 1721–1762 (2004)
30. Huang, D.S.: A constructive approach for finding arbitrary roots of polynomials by neural networks. IEEE Trans. Neural Netw. **15**(2), 477–491 (2004)

A BRMF-Based Model for Missing-Data Estimation of Image Sequence

Yang Liu, Zhan-Li Sun$^{(\boxtimes)}$, Yun Jing, Yang Qian, and De-Xiang Zhang

School of Electrical Engineering and Automation, Anhui University, Hefei, China
zhlsun2006@126.com

Abstract. How to effectively deal with occlusion is an important step of structure from motion (SFM). In this paper, an accurate missing data estimation method is proposed by combining Bayesian robust matrix factorization (BRMF) and particle swarm optimization (PSO). Experimental results on several widely used image sequences demonstrate the effectiveness and feasibility of the proposed algorithm.

Keywords: Missing-data estimation · Structure from motion · Matrix factorization · Weaker estimator

1 Introduction

Recovering 3D shape from a sequence of 2D point tracks, known as structure from motion (SFM), is one focus in the areas of computer vision [1, 2]. As the occlusion is frequently encountered in SFM [3–5], the low-rank property is often destructed by the missing feature points. In order to get an accurate reconstruct result, it is essential to recovery the missing data before the non-rigid SFM algorithm is applied.

By now, some methods have been proposed to deal with the occlusion encountered in SFM. In [6], an iterative multi-resolution scheme is proposed for SFM with missing data. In [7], a modified constrained model is presented to solving the low-rank matrix factorization. In [8], the evolutionary agents and an incremental bundle adjustment strategy are used to improve both the efficiency of data recovery and the robustness to outliers. A new affine factorization algorithm is proposed in [9], which employs a robust factorization scheme to handle outlying and missing data. In [10], an adaptive online learning algorithm for non-rigid SFM is proposed, and it is capable to handle missing data. A method that exploits temporal stability and low-rank property of

The work was supported by grants from National Natural Science Foundation of China (Nos. 61370109, 61272025), a grant from Natural Science Foundation of Anhui Province (No. 1308085MF85), and 2013 Zhan-Li Sun's Technology Foundation for Selected Overseas Chinese Scholars from department of human resources and social security of Anhui Province (Project name: Research on structure from motion and its application on 3D face reconstruction).

© Springer International Publishing Switzerland 2015
D.-S. Huang et al. (Eds.): ICIC 2015, Part II, LNCS 9226, pp. 165–169, 2015.
DOI: 10.1007/978-3-319-22186-1_16

motion data is proposed in [11], and it has been proved to be effective to deal with missing data.

In this paper, a Bayesian robust matrix factorization (BRMF) [12] based model is proposed for missing-data estimation of image sequence. For the proposed method, the missing entries of the observation matrix are firstly replaced with the values that are significantly larger than the non-missing entries. A BRMF-based weaker estimator is then constructed by inputting the observations to the BRMF with a specific rank value. Further, a particle swarm optimization (PSO)-based weighting strategy is designed to obtain the weighting coefficients for the weaker estimators. Finally, the weighted outputs of the weaker estimators are computed and used as the final estimations. Experimental results on several widely used image sequences demonstrate the effectiveness and feasibility of the proposed algorithm.

The remainder of the paper is organized as follows. In Sect. 2, we present the proposed approach. Experimental results are given in Sect. 3, and our concluding remarks are presented in Sect. 4.

2 Methodology

2.1 BRMF-Based Weaker Estimator

Denote the missing entries as *, the observation matrix with invisible feature points can be written as follows:

$$\mathbf{W} = \begin{pmatrix} x_{1,1} & * & \cdots & x_{1,n} \\ y_{1,1} & * & \cdots & y_{1,n} \\ \vdots & \vdots & \ddots & \vdots \\ * & x_{F,2} & \cdots & x_{F,n} \\ * & y_{F,2} & \cdots & y_{F,n} \end{pmatrix}, \tag{1}$$

In order to estimate the missing entries, $\mathbf{W}$ is represented as a product of two low-rank matrices $\mathbf{U} \in \mathbb{R}^{m \times r}$ and $\mathbf{V} \in \mathbb{R}^{m \times r}$, i.e.,

$$\mathbf{W} = \mathbf{U}\mathbf{V}^T. \tag{2}$$

Assume that the rank r is set as r_k, the corresponding $\mathbf{U}^k$ and $\mathbf{V}^k$ can be obtained via the following optimization model [12, 13]:

$$\min_{\mathbf{U}^k, \mathbf{V}^k} \left\| \mathbf{I}. * \left(\mathbf{W} - \mathbf{U}^k \mathbf{V}^{kT} \right) \right\|_1 + \frac{\lambda_u}{2} \left\| \mathbf{U}^k \right\|_2^2 + \frac{\lambda_v}{2} \left\| \mathbf{V}^k \right\|_2^2, \tag{3}$$

where λ_u and $\lambda_v > 0$ are regularization parameters. $\mathbf{I}$ is an $m \times n$ binary matrix, in which $\mathbf{I}_{i,j}$ is 0 if $\mathbf{W}_{i,j}$ is missing. The optimization problem can be formulated as a *maximum a posteriori* (MAP) estimation problem [12, 13],

$$\log p\left(\mathbf{U}^k, \mathbf{V}^k | \mathbf{W}, \lambda, \lambda_u, \lambda_v\right) = - \lambda \left\| \mathbf{W} - \mathbf{U}^k \mathbf{V}^{kT} \right\|_1$$
$$- \frac{\lambda_u}{2} \left\| \mathbf{U}^k \right\|_2^2 - \frac{\lambda_v}{2} \left\| \mathbf{V}^k \right\|_2^2 + C, \tag{4}$$

The observation matrix including the estimated values of missing entries can be given by:

$$\mathbf{W}^k = \mathbf{U}^k \mathbf{V}^{kT}. \tag{5}$$

A weaker estimator is constructed by inputting the observation matrix of (1) to the BRMF algorithm [12] to estimate the missing entries for a fixed rank value r. When the rank r is set as $r_k(k = 1, \cdots, K)$, we can get a set of estimations $\mathbf{W}^k(k = 1, \cdots, K)$ of weaker estimators.

2.2 Weighting Coefficients Optimization with PSO

For the kth weaker estimator, the estimation error of non-missing entries (ε_n) and missing entries (ε_m) can be computed as

$$\varepsilon_n = \left\| \mathbf{I} . * \left(\mathbf{W} - \mathbf{W}^k \right) \right\| \tag{6}$$

and

$$\varepsilon_n = \left\| (1 - \mathbf{I}) . * \left(\mathbf{W} - \mathbf{W}^k \right) \right\|, \tag{7}$$

respectively. Generally speaking, a small ε_n means a relatively low ε_m. After obtaining the outputs of weaker estimators, we first select K_s estimations according to the estimation errors of non-missing entries. The weighted sum of the selected outputs are then computed, and used as the final estimation $\mathbf{W}_f$, i.e.,

$$\mathbf{W}_f = \frac{\sum_{k=1}^{K_s} \omega_k \mathbf{W}^k}{\sum_{k=1}^{K_s} \omega_k}, \tag{8}$$

where ω_k denotes the weighting coefficient of the kth weaker estimator.

In our proposed method, the weighting coefficients ω_k are obtained by PSO. For a practical application, the non-missing entries of the observations are known to us. Thus, the estimation errors of the non-missing entries between $\mathbf{W}$ and $\mathbf{W}_f$ are designed as the object function of PSO, i.e.,

$$f(\mathbf{x}) = \left\| \mathbf{I} . * \left(\mathbf{W} - \mathbf{W}_f \right) \right\|, \tag{9}$$

where $\mathbf{x} = [\omega_1, \ldots, \omega_k]^T$ denotes the unknown variables. In order to search for the optimized $\mathbf{x}$, a population of n individuals, named particle swarm, are first randomly initialized. Then, we use PSO to get an optimized $\mathbf{x}$. Finally, $\mathbf{W}_f$ is computed via (8).

3   Experimental Results

3.1   Experimental Data

In order to assess the performance of the proposed method (denoted as BRMF-PSO), two image sequences (*dance, shark*) are used in the experiments. Table 1 shows the number of image frames (T) and the number of points tracked (n) of these sequences.

Table 1. The number of frames (T) and the number of points tracked (n) for two image sequence.

Sequences	T	n
Dance	264	75
Shark	240	91

3.2   Experimental Results

Table 2 shows the estimation errors of four methods for two sequences. Note that, for BRMF, we cannot know which rank value can provide a high estimation accuracy. Therefore, we perform a series of experiments by varying the rank values from 1 to 20. Then, the 5 trials are selected according to the estimation errors of non-missing entries. As shown in Table 2, the results of BRMF are the mean (μ) and standard deviation (σ) of the 5 trials with the minimum estimation errors of non-missing entries. We can see from Tables 2 that the estimation errors of BRMF-PSO are generally lower than that of other methods. Therefore, the proposed method is competitive to these methods.

Table 2. The estimation errors of four methods for two image sequences when the percentage of missing entries is 30 %.

	BALM	CSF	BRMF($\mu \pm \sigma$)	BRMF-PSO($\mu \pm \sigma$)
Dance	3.15	3.09	2.41 ± 0.029	2.30 ± 0.0001
Shark	0.41	0.25	0.13 ± 0.0092	0.10 ± 1.09 × 10^{-5}

Moreover, we can see from Table 3 that, the estimation errors of BRMF-PSO are lower than the minimum errors of BRMF (BRMF$_m$). As the true optimal rank values are unknown, this means that BRMF-PSO can achieve relatively good results. Thus, the strategy of PSO-based weaker estimator integration is very effective to improve the performance of BRMF.

Table 3. Comparisons of the estimation errors of BRMF-PSO and the minimum estimation errors of BRMF (BRMF$_m$).

	BRMF$_m$	BRMF-PSO
Dance	2.38	2.30 ± 0.0001
Shark	0.123	0.10 ± 1.09 × 10^{-5}

4 Conclusions

In this paper, an accurate missing data estimation method is proposed by combining BRMF and PSO. The experimental results of several widely used image sequences demonstrated the effectiveness and feasibility of the proposed algorithm.

References

1. Costeria, J.P., Kanade, T.: A multibody factorization method for independently moving objects. Int. J. Comput. Vis. **29**, 159–179 (1998)
2. Hartley, R., Zisserman, A.: Multiple View Geometry in Computer Vision, 2nd edn. Cambridge University Press, Cambridge (2003)
3. Torresani, L., Hertzmann, A., Bregler, C.: Nonrigid structure-from-motion: estimating shape and motion with hierarchical prior. IEEE Trans. Pattern Anal. Mach. Intell. **30**, 878–892 (2008)
4. Gotardo, P.F.U., Martinez, A.M.: Computing smooth time-trajectories for camera and deformable shape in structure from motion with occlusion. IEEE Trans. Pattern Anal. Mach. Intell. **33**, 2051–2065 (2011)
5. Hamsici, O.C., Gotardo, P.F., Martinez, A.M.: Learning spatially-smooth mappings in non-rigid structure from motion. In: Fitzgibbon, A., Lazebnik, S., Perona, P., Sato, Y., Schmid, C. (eds.) ECCV 2012, Part IV, LNCS, vol. 7575, pp. 260–273. Springer, Heidelberg (2012)
6. Julia, C., Sappa, A.D., Lumbreras, F., Serrat, J., Lopez, A.: An iterative multiresolution scheme for SFM with missing data: single and multiple object scenes. Image Vis. Comput. **28**, 164–176 (2010)
7. Zhao, K., Zhang, Z.: Successively alternate least square for low-rank matrix factorization with bounded missing data. Comput. Vis. Image Underst. **114**, 1084–1096 (2010)
8. Hu, M.X., Penney, G., Figl, M., Edwards, P., Bello, F., Casula, R., Rueckert, D., Hawkes, D.: Reconstruction of a 3D surface from video that is robust to missing data and outliers: application to minimally invasive surgery using stereo and mono endoscopes. Med. Image Anal. **16**, 597–611 (2012)
9. Wang, G.H., Zelek, J.S., Wu, Q.M.J.: Robust structure from motion of nonrigid objects in the presence of outlying and missing data. In: International Conference on Computer and Robot Vision, vol. 39, pp. 159–166 (2013)
10. Tao, L., Mein, S.J., Quan, W., Matuszewski, B.J.: Recursive non-rigid structure from motion with online learned shape prior. Comput. Vis. Image Underst. **117**, 1287–1298 (2013)
11. Feng, Y., Xiao, J., Zhuang, Y.T., Yang, X.S., Zhang, J.J., Song, R.: Exploiting temporal stability and low-rank structure for motion capture data refinement. Inf. Sci. **277**, 777–793 (2014)
12. Wang, N.Y., Yeung, D.Y.: Bayesian robust matrix factorization for image and video processing. In: IEEE International Conference on Computer Vision, 1785–1792 (2013)
13. Wang, N.Y., Yao, T.S., Wang, J.D., Yeung, D.-Y.: A probabilistic approach to robust matrix factorization. In: Fitzgibbon, A., Lazebnik, S., Perona, P., Sato, Y., Schmid, C. (eds.) ECCV 2012, Part VII. LNCS, vol. 7578, pp. 126–139. Springer, Heidelberg (2012)

Image Splicing Detection Based on Markov Features in QDCT Domain

Ce Li[1,2(✉)], Qiang Ma[1], Limei Xiao[1], Ming Li[1], and Aihua Zhang[1]

[1] College of Electrical and Information Engineering,
Lanzhou University of Technology, Lanzhou 730050, China
Xjtulice@gmail.com
[2] School of Electronic and Information Engineering, Xi'an Jiaotong University,
Xi'an 710049, China

Abstract. Image splicing is very common and fundamental in image tampering. Therefore, image splicing detection has attracted more and more attention recently in digital forensics. Gray images are used directly, or color images are converted to gray images before processing in previous image splicing detection algorithms. However, most natural images are color images. In order to make use of the color information in images, a classification algorithm is put forward which can use color images directly. In this paper, an algorithm based on Markov in Quaternion discrete cosine transform (QDCT) domain is proposed for image splicing detection. The support vector machine (SVM) is exploited to classify the authentic and spliced images. The experiment results demonstrate that the proposed algorithm not only make use of color information of images, but also can achieve high classification accuracy.

Keywords: Markov model · QDCT · Image-splicing detection · Color image forgery detection

1 Introduction

In recent years, image splicing forgery detection is a hot topic in image processing. Splicing forgery detection need to depend on the concept of hypothesis, although any trace on the vision may not be left in tampering images, the underlying statistics are likely to be changed. It is these inconsistencies that detection techniques are used to detect the tampering. Various passive image splicing detection approaches have been proposed. Shi et al. [1] employed wavelet moment characteristics and Markov features to identify splicing images on DVMM dataset [2]. Wang et al. used gray level co-occurrence matrix (GLCM) of threshold edge image to classify splicing image [3] and employed Markov chain for tampering detection [4] on CASIA V2.0 Dataset [5]. Patchara et al. [6] had pointed out two problems of the data sets [5] and had shown how to rectify the dataset and the experimental results showed the detection rate of the causal Markov model-based features was reduced to 78 % on the rectified dataset. Recently, Xudong Zha et al. used non-causal Markov model domain on DCT and DWT domains to obtain a high accuracy [7]. He et al. [8] proposed a method based on Markov features in DCT and DWT domain. The detection accuracy was up to 93.42 %

© Springer International Publishing Switzerland 2015
D.-S. Huang et al. (Eds.): ICIC 2015, Part II, LNCS 9226, pp. 170–176, 2015.
DOI: 10.1007/978-3-319-22186-1_17

on DVMM dataset [2], 89.76 % on CASIA V2.0 dataset [5]. Amerini I *et al.* [9] used the DCT coefficients first digit features to distinguish and then localize a single and a double JPEG compression in portions of an image.

These algorithms mainly use the gray information of images or certain color component information to detect whether the images have been spliced, whole color information is not used effectively. In order to make use of the whole color information, QDCT is introduced into image splicing detection algorithm in this paper. The superiority of QDCT in color image processing is fully reflected in these applications [10, 11]. Therefore, an image tamper detection algorithm under QDCT transform domain is proposed in this paper. A new idea for image tamper detection is proposed, which has a certain theoretical and practical significance.

2 The Proposed Approach

In this section, the whole framework of the proposed algorithm is presented with detailed description of each part. The framework of the proposed algorithm based on Markov features is shown in Fig. 1. Compared with literature [8], we extract Markov features in QDCT domain. Expanded Markov features in QDCT domain are obtained in this paper. Besides, Ref. [1], we introduce main diagonal difference matrices, minor diagonal difference matrices, main diagonal transition probability matrices and minor diagonal transition probability matrices in QDCT. Different the literature [1, 8], We also consider QDCT coefficients correlation between the intra-block correlation and the inter-block in main diagonal and minor diagonal direction. Finally, the feature vector obtained is used to distinguish authentic and spliced images with Primal SVM [12] as the classifier.

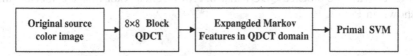

| Original source color image | → | 8×8 Block QDCT | → | Expangded Markov Features in QDCT domain | → | Primal SVM |

Fig. 1. Illustration of the stages of our algorithm

2.1 Quaternion Discrete Cosine Transform

The research of QDCT is prompted by the precedents of the successful application of real number and complex number domain, the basic principle of QDCT was proposed by Feng and Hu [10], and the actual algorithm was given. $h_q(m, n)$ is a two-dimensional $M \times N$ quaternion matrix, m and n is row and column of the matrix respectively, here, $m \in [0, M - 1]$, $n \in [0, N - 1]$, the definition of L-QDCT as follows:

$$\text{L - QDCT}: J_q^L(p, s) = \alpha(p)\alpha(s) \sum_{m=0}^{M-1} \sum_{n=0}^{N-1} u_q \cdot h_q(m, n) \cdot T(p, s, m, n) \quad (1)$$

In Eq. (1), u_q is a unit pure quaternion, it represents the direction of axis of transformation, and it satisfies $u_q^2 = -1$, p and s are row and column of the transform matrix, respectively. It is similar to DCT in real number and complex number domain, the definition of $\alpha(p)$, $\alpha(s)$ and $T(p, s, m, n)$ are shown as follows:

$$\alpha(p) = \begin{cases} \sqrt{1/M} & p = 0 \\ \sqrt{2/M} & p \neq 0 \end{cases} \quad \alpha(s) = \begin{cases} \sqrt{1/N} & s = 0 \\ \sqrt{2/N} & s \neq 0 \end{cases}$$

$$T(p, s, m, n) = \cos\left[\frac{\pi(2m+1)p}{2M}\right] \cos\left[\frac{\pi(2n+1)s}{2N}\right]$$

The spectral coefficient of $J(p, s)$ through transformation is still a quaternion matrix of $M \times N$, and its representation is by Eq. (2).

$$J(p, s) = J_0(p, s) + J_1(p, s)i + J_2(p, s)j + J_3(p, s)k \tag{2}$$

2.2   Block QDCT

The Expanded Markov features in DCT domain proposed in [8] are very remarkable in capturing the differences between authentic and spliced images. They can be calculated by seven steps. Unlike the first step, the original color images are blocked into 8×8 non-repeatedly, and each block is still color image. Secondly, three color components R, G and B of blocked images are used to construct quaternion matrix, and the quaternion matrix is processed by QDCT transform to obtain QDCT coefficient matrix of each block, then the square root of the real part (r) and three imaginary parts (i, j, k) are calculated. Then all calculated matrices need to reassemble according to the site of blocking, thus a 8×8 blocked QDCT matrix F of original color image can be acquired. It is shown in Fig. 2.

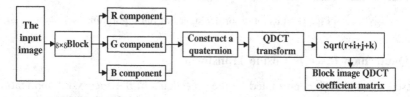

Fig. 2. Illustration of the stages of 8×8 block QDCT.

2.3   Expanded Markov Features in QDCT Domain

Compared with the literature [8], Expanded Markov features in QDCT domains are obtained in this paper. Besides, Ref. [1], we introduce main diagonal difference matrices, minor diagonal difference matrices, main diagonal transition probability matrices and minor diagonal transition probability matrices in QDCT. Different the literature [1, 8], we also consider QDCT coefficients correlation between intra-block correlation and inter-block in main diagonal and minor diagonal direction. In addition to these different other steps are roughly the same. The following steps are shown.

Firstly, 8×8 block QDCT is applied on the original image pixel array following Part 2.2, and the corresponding QDCT coefficient array G is obtained. Then, the round QDCT coefficients G to integer and take absolute value (denote as arrays F).

Secondly, calculate the horizontal, vertical, main diagonal and minor diagonal intra-block difference 2-D arrays F_h, F_v by applying Eqs. (1) and (2), respectively in [8], F_d and F_{-d} by applying Eqs. (3) to (4) as follow:

$$F_d(u, v) = F(u, v) - F(u + 1, v + 1) \tag{3}$$

$$F_{-d}(u, v) = F(u + 1, v) - F(u, v + 1) \tag{4}$$

and calculate the horizontal, vertical, main diagonal and minor diagonal inter-block difference 2-D arrays G_h, G_v, by applying Eqs. (7) and (8), respectively in [8], G_d and G_{-d} by applying Eqs. (5) and (6) as follow:

$$G_d(u, v) = F(u, v) - F(u + 8, v + 8) \tag{5}$$

$$G_{-d}(u, v) = F(u + 8, v) - F(u, v + 8) \tag{6}$$

Thirdly, introduce a threshold T ($T \in N_+$), if the value of an element in F_h (or F_v, G_h and G_v) is either greater than T or smaller than $-T$, replace it with T or $-T$.

Fourthly, calculate the horizontal, vertical, main diagonal and minor diagonal transition probability matrices of F_h, F_v, G_h, G_v, F_d, F_{-d}, G_d and G_{-d}. That $P1_h(i,j)$, $P1_v(i,j)$, $P2_h(i,j)$, $P2_v(i,j)$, $P3_h(i,j)$, $P3_v(i,j)$, $P4_h(i,j)$, $P4_v(i,j)$ by applying Eqs. (3), (4), (5), (6), (9), (10), (11) and (12), respectively in [8]. $P1_d(i,j)$, $P1_{-d}(i,j)$, $P3_d(i,j)$, $P3_{-d}(i,j)$ by applying Eqs. (7) and (10) as follow:

$$P1_d(i,j) = \frac{\sum_{u=1}^{S_u-2} \sum_{v=1}^{S_v-2} \delta(F_d(u, v) = i, F_d(u + 1, v + 1) = j)}{\sum_{u=1}^{S_u-2} \sum_{v=1}^{S_v-2} \delta(F_d(u, v) = i)} \tag{7}$$

$$P1_{-d}(i,j) = \frac{\sum_{u=1}^{S_u-2} \sum_{v=1}^{S_v-2} \delta(F_{-d}(u + 1, v) = i, F_{-d}(u, v + 1) = j)}{\sum_{u=1}^{S_u-2} \sum_{v=1}^{S_v-2} \delta(F_{-d}(u, v) = i)} \tag{8}$$

$$P3_d(i,j) = \frac{\sum_{u=1}^{S_u-16} \sum_{v=1}^{S_v-16} \delta(G_d(u, v) = i, G_d(u + 8, v) = j)}{\sum_{u=1}^{S_u-16} \sum_{v=1}^{S_v-16} \delta(G_d(u, v) = i)} \tag{9}$$

$$P3_{-d}(i,j) = \frac{\sum_{u=1}^{S_u-16} \sum_{v=1}^{S_v-16} \delta(G_{-d}(u, v) = i, G_{-d}(u, v + 8) = j)}{\sum_{u=1}^{S_u-16} \sum_{v=1}^{S_v-16} \delta(G_{-d}(u, v) = i)} \tag{10}$$

where $i, j \in \{- T, - T + 1,..., 0,...T - 1, T\}$, thus $(2T + 1) \times (2T + 1) \times 12$ dimensionality Markov features in QDCT domain are obtained.

3   Experiments and Results

In this section, we introduce image datasets for experiment, and present a set of experiments to demonstrate high performance and effectiveness of proposed algorithm.

3.1   Image Dataset

Two public image datasets for tampering and splicing detection are provided by DVMM, Columbia University [2]. However, the color information is not provided by their data of gray images. The number of color images is too small. In addition, it doesn't closer to reality tamper image. In order to provide more challenging evaluation database for evaluation, two color image data sets CASIA V1.0 and CASIA V2.0 are chosen, which are shown in Fig. 3.

Fig. 3. Some example images of CASIA dataset (the left from V1.0, and the right from V2.0, authentic images in the top row, their forgery counter parts in the bottom row).

Two problems of the dataset CASAI V1.0 and CASAI V2.0 have been pointed out by Patchara *et al.* [6]. First one is the JPEG compression applied to authentic images is one-time less than that applied to tampered images; the second one is for JPEG images, the size of chrominance components of 7140 authentic images is only one quarter of that of 2061 tampered images. For fairness purpose, we accord to the processing method to modify the database in the article [6]. Because we can't use YCbCr color space, we just only need solve the first problem of CASAI TIDE V2.0. We used Matlab for standard JPEG compression to alleviate the influence of the difference in the number of JPEG compression by the following procedure: (1) Re-compressing 7437 JPEG authentic images with quality factor of 84; (2) Compressing 3059 TIFF tampered images by Matlab with quality factor of 84; (3) Leaving 2064 JPEG tampered images untouched. The data set CASAI V1.0 is also processed by same method.

3.2   Classification

In our experiment, Primal SVM [12] is used for classification. The Primal SVM has more kernel functions, such as RBF, linear, histogram intersection and so on. After the comparison with several experiments, the histogram intersection kernel has the highest classification accuracy. Therefore, histogram intersection is chosen as kernel function. The threshold T is fixed in all of the experiments. For CASIA V1.0, CASIA V2.0

dataset, we choose $T = 4$, which will produce 972-dimensional feature vector. To be fair, all the experiments and comparisons are tested on the dataset mentioned above with the same classifier. The testing platform is Matlab R2012b, and the hardware platform is a PC with a 2G, 32 bit operating system, and Intel i3 processor. In each experiment, the average rate of 30 repeating independent tests is recorded. In each of the 30 runs, 5/6 of authentic images and 5/6 of spliced images in the dataset are randomly selected to train SVM classifier. Then the remaining is used to test.

3.3 The Detection Performance of the Proposed Approach

Some common experiments are conducted first to assess the detection ability of the proposed algorithm. The detailed results are shown in Tables 1 and 2. TP (true positive) rate is the ratio of correct classification of authentic images. TN (true negative) rate is the ratio of correct classification of spliced images. Accuracy of detection is the weighted average value of TP rate and TN rate.

Table 1. Experiment results obtained on CASIA V2.0 dataset

Fv	Dn	Ac (%)	FT (s)	FS (s)	TT (s)
NIM[1]	266	84.86	4.479	0	4.479
He[8]	100	89.76	2.218	2.158	4.376
Our method	972	92.38	3.61	0	3.61

Table 2. Results of the proposed algorithm with different threshold T on CASIA dataset

Threshold T	C1-3	C1-4	C1-5	C2-3	C2-4	C2-5
Dimensionality n	588	972	1452	588	972	1452
TP (%)	95.881	95.440	95.735	88.371	89.185	89.426
TN (%)	95.073	96.470	97.131	95.637	95.567	95.914
Accuracy (%)	95.478	95.958	96.435	92.003	92.377	92.668

3.3.1 Comparison with Other Algorithms

To evaluate the proposed algorithm comprehensively, a comparison between the proposed algorithm and some state-of-the-art image splicing detection methods were performed. The gray image algorithms of NIM [1] and He [8] were chosen to compare. Our color image algorithm obtained a higher accuracy than their algorithms. Fv, Dn, Ac, FT, FS and TT represent features vector, dimensionality, accuracy, feature extraction time, feature selection time and total time respectively.

3.3.2 Choice of Threshold T

Another issue is the choice of the threshold T, which is used to reduce the dimension of Markov features. Based on past experience of Markov threshold selection [1, 8], we choose $T = 3, 4, 5$. In Table 2, we provide the performance of Markov features with three different T. Ci-T (i = 1, 2 represent CASIA V1.0 dataset, CASIA V2.0 dataset respectively, $T = 1, 2, 3$).

4 Conclusion

In the most of current image tamper detection algorithms, color image is converted to gray image before detection. Therefore, the whole color information is not taken into account. Images can be processed by quaternion in a whole manner to improve the accuracy of the image tamper detection algorithm. The experiment results demonstrate that the proposed algorithm not only make use of color information of images, but also can achieve high classification accuracy. Because the tamper images are mostly color in real life, this new idea for image tamper detection research has a certain theoretical and more practical significance.

Acknowledgments. The paper was supported in part by the National Natural Science Foundation (NSFC) of China under Grant Nos. (61365003, 61302116), National High Technology Research and Development Program of China (863 Program) No. 2013AA014601, Natural Science Foundation of China in Gansu Province Grant No. 1308RJZA274.

References

1. Shi, Y.Q., Chen, C., Chen, W.: A natural image model approach to splicing detection. In: ACM Proceedings of the 9th Workshop on Multimedia and Security, pp. 51–62 (2007)
2. Ng, T.T., Chang, S.F., Sun, Q.: A data set of authentic and spliced image blocks. Columbia Univ. 203 (2004)
3. Wang, W., Dong, J., Tan, T.: Effective image splicing detection based on image chroma. In: IEEE International Conference on Image Processing, pp. 1257–1260 (2009)
4. Wang, W., Dong, J., Tan, T.: Image tampering detection based on stationary distribution of Markov chain. In: IEEE International Conference on Image Processing, pp. 2101–2104 (2010)
5. http://forensics.idealtest.org/
6. Sutthiwan, P., Shi, Y.Q., Zhao, H., Ng, T.-T., Su, W.: Markovian rake transform for digital image tampering detection. In: Shi, Y.Q., Emmanuel, S., Kankanhalli, M.S., Chang, S.-F., Radhakrishnan, R., Ma, F., Zhao, L. (eds.) Transactions on Data Hiding and Multimedia Security VI. LNCS, vol. 6730, pp. 1–17. Springer, Heidelberg (2011)
7. Zhao, X., et al.: Passive image-splicing detection by a 2-D noncausal Markov model. IEEE Trans. Circ. Syst. Video Technol. 2(25), 185–199 (2014)
8. He, Z., Wei, L., Sun, W., et al.: Digital image splicing detection based on Markov features in DCT and DWT domain. Pattern Recogn. 45(12), 4292–4299 (2012)
9. Amerini, I., Becarelli, R., et al.: Splicing forgeries localization through the use of first digit features. In: IEEE Conference on Parallel Computing Technologies, pp. 143–148 (2015)
10. Feng, W., Hu, B.: Quaternion discrete cosine transform and its application in color template matching. IEEE Congr. Image Sig. Process. 5(2), 252–256 (2008)
11. Schauerte, B., Stiefelhagen, R.: Quaternion-based spectral saliency detection for eye fixation prediction. In: Fitzgibbon, A., Lazebnik, S., Perona, P., Sato, Y., Schmid, C. (eds.) ECCV 2012, Part II. LNCS, vol. 7573, pp. 116–129. Springer, Heidelberg (2012)
12. Chapelle, O.: Training a support vetor machine in the primal. Neural Comput. 19(5), 1155–1178 (2007)

Implementation of 3-D RDPAD Algorithm on Follicle Images Based CUDA

Hongwei Jiang[1,2], Jun Liu[1,2(✉)], Keyang Luo[1], Wei Hu[1,2], and Xiaoming Liu[1,2]

[1] College of Computer Science and Technology, Wuhan University of Science and Technology, Wuhan, China
ljwhcn@gmail.com
[2] Hubei Province Key Laboratory of Intelligent Information Processing and Real-time Industrial System, Wuhan, China

Abstract. The 3-D Robust Detail Preserving Anisotropic Diffusion (3-D RDPAD) is an anisotropic diffusion which has many advantages on 3-D ultrasound image processing, especially in 3-D cattle follicle ultrasound Images. The 3-D RDPAD algorithm has good performance both in noise suppression and edge preserving. However, the computation of 3-D RDPAD algorithm requires a huge amount of time to complete because it directly operates on the 3-D image and it has logical judgment in every iteration. In this paper, we proposed a 3-D Parallel RDPAD algorithm using CUDA to solve this problem. By optimizing the logical judgment in diffusion stop condition, the speed of 3-D RDPAD algorithm has a notable improvement. We performed our algorithm on 3-D cattle follicle ultrasound Images and the result of the experiments showed that it revealed a great speedup over the original 3-D RDPAD.

Keywords: Follicle · Ultrasound images · Anisotropic diffusion · CUDA

1 Introduction

3-D ultrasound image has been used in medical diagnosis for many purposes. The speckle reduction is a very important part of ultrasound image preprocessing. In [1], Sun proposed the 3-D anisotropic diffusion filter called 3-D SRAD algorithm, it is the 3-D version of SRAD algorithm, and this algorithm directly operated on the 3-D image and achieved good results of speckle reduction, then Zhu H Implements on CUDA with this method [2]. We investigated an anisotropic diffusion filtering algorithm called 3-D Robust Detail Preserving Anisotropic Diffusion (3-D RDPAD) for Speckle Reduction. 3-D RDPAD is shown to have better speckle reduction performance to 3-D SRAD and is less sensitive to the diffusion iteration times.

D.-S. Huang et al. (Eds.): ICIC 2015, Part II, LNCS 9226, pp. 177–184, 2015.
DOI: 10.1007/978-3-319-22186-1_18

2 Background

2.1 CUDA Program Model

CUDA (Compute Unified Device Architecture) is a parallel computing platform and programming model created by NVIDIA and implemented by the graphics processing units (GPUs) that they produce [3]. NVIDIA provides a set of C language interface for developers to code on GPUs (CUDA3.0 supports C++ and FORTRAN). CUDA shielded most hardware details for the developers so that it is easy to program on GPUs using CUDA.

CUDA program structure is divided into two parts: the host and the device. Generally, the host is CPU and memory and the device refers GPU. In CUDA, the main program executed by a CPU, and when the portion of data across the parallel processing, CUDA will be compiled into the program that the GPU can perform, and sent to the GPU [4]. This function is called CUDA kernel function. GPU threads create in thread blocks according to the most effective data sharing, they can be more than one dimension. The thread in the same block use the same shared memory. Each thread is identified by the thread ID. Thread blocks can also be specified as a one-dimensional, two-dimensional or higher dimensional arrays, and use multiple index component to identify each thread.

2.2 3-D RDPAD Algorithm

We proposed 3-D RDPAD by modifying the diffusion coefficient and the instantaneous coefficient of variation (ICOV) in 3-D SRAD. The PDE of 3-D anisotropic diffusion filtering algorithms as follows:

$$\frac{\partial I(x,y,z;t)}{\partial t} = div[c(x,y,z;t)\nabla I(x,y,z;t)] \tag{1}$$

where I is three-dimensional image, ∇ is the gradient operator, div is the divergence operator, $c(\cdot)$ is the diffusion coefficient.

In 3-D RDPAD algorithm, $c(\cdot)$ is a sectional function:

$$c(q) = \begin{cases} \frac{1}{2}\left[1 - \frac{q(x,y,z;t)^2 - q_0(t)^2}{q_0(t)^2\left[1 + q(x,y,z;t)^2\right]}\right]^2 & if \quad q(x,y,z;t)^2 \leq \frac{2q_0(t)^2}{\left|1 - q_0(t)^2\right|} \\ 0 & otherwise \end{cases} \tag{2}$$

where $q_0(t)^2$ is the mean value of $q(x,y,z;t)^2$ over a homogeneous speckle area, the $q(x,y,z;t)^2$ is defined as:

$$q(x,y,z;t)^2 = \frac{\frac{1}{\left|\eta_{x,y,z}^U\right| - 1}\sum_{p \in \eta_{x,y,z}^U}\left(I_p - \bar{I}_{x,y,z}\right)^2}{\bar{I}_{x,y,z}^2} \tag{3}$$

Where $\eta_{x,y,z}^{U}$ is the neighbors of image I at (x,y,z), $\bar{I}_{x,y,z}$ is the mean value over pixels in $\eta_{x,y,z}^{U}$.

The 3-D RDPAD algorithm can reduce the speckle effectively while image edges are still get a good preserving, and the modified algorithm is less sensitive to the number of iterations.

3   Implementation of 3-D RDPAD Algorithm on CUDA

3.1   Overview

The 3-D Parallel RDPAD algorithm are is mainly divided into three parts: preprocessing, calculate diffusion coefficient, image diffusion (Fig. 1).

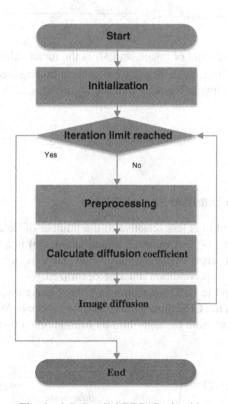

Fig. 1. 3-D Parallel RDPAD algorithm

3.2   Preprocessing Step of 3-D Image

Before the iteration, many variables need to be calculated. In 3-D RDPAD algorithm, there are gradients in six directions:

$$\begin{cases} M_{dFront}(x,y,z) = M_I(x+1,y,z) - M_I(x,y,z) \\ M_{dBack}(x,y,z) = M_I(x-1,y,z) - M_I(x,y,z) \\ M_{dLeft}(x,y,z) = M_I(x,y+1,z) - M_I(x,y,z) \\ M_{dRight}(x,y,z) = M_I(x,y-1,z) - M_I(x,y,z) \\ M_{dUp}(x,y,z) = M_I(x,y,z+1) - M_I(x,y,z) \\ M_{dDown}(x,y,z) = M_I(x,y,z-1) - M_I(x,y,z) \end{cases} \tag{4}$$

where M_I is the matrix of 3-D image, M_{dg} are gradients in specified directions.

Then we calculate the instantaneous coefficient of variation (ICOV). From (3), we can get the following expressions:

$$M_{q\_squared}(x,y,z) = \frac{\frac{1}{|\eta^U_{x,y,z}|-1} \sum\limits_{p\in\eta^U_{x,y,z}} \left(M_{I_p} - M_{\bar I}\right)^2}{M_{\bar I}^2} \tag{5}$$

$\eta^U_{x,y,z}$ is the is the neighbors of image, and $M_{\bar I}$ is the mean value over pixels in $\eta^U_{x,y,z}$.

In (2), $q_0(t)^2$ is the mean value of $q(x,y,z;t)^2$ over a homogeneous speckle area, we consider it is the whole image block so that the algorithm can be more parallelism:

$$M_{q0}(t)^2 = \bar M_{q\_squared}(x,y,z) \tag{6}$$

3.3   Optimization on Diffusion Stop Condition

The 3-D RDPAD algorithm is less sensitive to the number of iterations by introducing diffusion stop condition. As we can see in equation (2), $c(\cdot)$ is a sectional function. It needs logical judgment when computing. However, GPU is not good at logic calculation, so the diffusion stop condition must be optimized.

In order to speed up the GPU calculation, we need make ensure that the GPU has no communication with the CPU during the calculation process. We separated the logic calculation algorithm part from the algorithm by introducing a 0–1 matrix, it defines as follows:

$$M_b(x,y,z) = \begin{cases} 0 & \text{if } M_{q\_squared}(x,y,z) \leq \frac{2M_{q0}(t)^2}{|1-M_{q0}(t)^2|} \\ 1 & \text{otherwise} \end{cases} \tag{7}$$

This matrix is generated in the CPU computing and it will be transferred into GPU in the next step.

Using equation (5) and (7), the diffusion coefficient is as follows:

$$M_c(x, y, z) = \frac{1}{2} \left[1 - \frac{M_{q\_squared}(x, y, z) - M_{q0}(t)^2}{M_{q0}(t)^2 \left[1 + M_{q\_squared}(x, y, z) \right]} \right] gM_b(x, y, z) \tag{8}$$

3.4 The Iteration of Update Function for 3D Image

The update function for 3D Anisotropic Diffusion is as follows:

$$I_{i,j,k}^{n+1} = I_{i,j,k}^n + \frac{\Delta t}{6} d_{i,j,k}^n \tag{9}$$

where

$$
\begin{aligned}
d_{i,j,k}^n = \frac{1}{h^2} & \left[c_{i+1,j,k}^n \left(I_{i+1,j,k}^n - I_{i,j,k}^n \right) + c_{i,j,k}^n \left(I_{i-1,j,k}^n - I_{i,j,k}^n \right) \right. \\
& + c_{i,j+1,k}^n \left(I_{i,j+1,k}^n - I_{i,j,k}^n \right) \bigg] + \frac{1}{h^2} \left[c_{i,j,k}^n \left(I_{i,j-1,k}^n - I_{i,j,k}^n \right) \right. \\
& + c_{i,j,k+1}^n \left(I_{i,j,k+1}^n - I_{i,j,k}^n \right) + c_{i,j,k}^n \left(I_{i,j,k-1}^n - I_{i,j,k}^n \right) \bigg]
\end{aligned}
\tag{10}
$$

Using equation (4), (8), (9), (10), the update function on CUDA is as follows:

$$M_{(x,y,z)}^{n+1} = M_{(x,y,z)}^n + \frac{\Delta t}{6} M_d^n \tag{11}$$

where

$$
\begin{aligned}
M_d^n = \frac{1}{h^2} & \left[M_c^n(x+1, y, z) g M_{dFont}^n(x, y, z) \right. \\
& + M_c^n(x, y, z) g M_{dBack}^n(x, y, z) \\
& + M_c^n(x, y+1, z) g M_{dLeft}^n(x, y, z) \bigg] \\
& + \frac{1}{h^2} M_c^n(x, y, z) g M_{dRight}^n(x, y, z) \\
& + M_c^n(x, y, z+1) g M_{dUp}^n(x, y, z) \\
& + M_c^n(x, y, z) g M_{dDown}^n(x, y, z) \bigg]
\end{aligned}
\tag{12}
$$

4 Result

4.1 Experiment Platform

See Table 1.

Table 1. Platform

CPU	Intel core i5 650
RAM	4 GB
GPU	Tesla C1060
OS	Windows 7 (x64)
Programming tools	Matlab 2012a
	CUDA toolkit 5.0
	Microsoft visual studio 2010

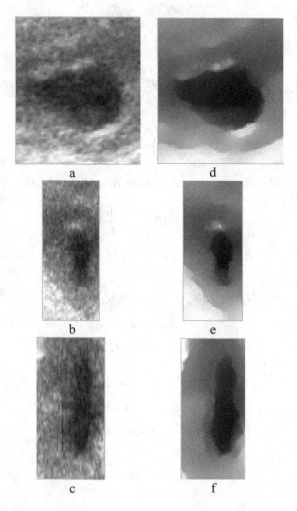

Fig. 2. a-c are the original ultrasound images sliced from X-Y, X-Z, Y-Z planes, Figure d-e are the results using 3DSRAD algorithm.

4.2   Images

The size of the 3-D ultrasound images is 800, 600, and 40 (in units of pixels) in x, y and z directions. We performed our algorithm on several 3-D ultrasound images and statistics running time of program on each image.

4.3   Result

Figure 2 shows one result of the 3-D RDPAD algorithm using CUDA, the number of the iteration is 300.

4.4   Performance

The comparison of performance between 3-D RDPAD using CPU with 3-D RDPAD using GPU is as follows:

We can see that the result of 3-D RDPAD algorithm using CUDA is excellent on GPU, it is more efficient than 3-D RDPAD on CPU. From the Table 2 we find that the 3-D RDPAD algorithm is accelerated by more than 70 times when using GPU.

Table 2. Performance

Image no.	CPU	GPU	Speed up
1	49988 ms	579 ms	86.33
2	44600 ms	511 ms	87.28
3	31930 ms	413 ms	77.32
4	40853 ms	494 ms	82.70
5	46991 ms	603 ms	77.93

5   Conclusion

In this paper, we implemented the 3-D RDPAD algorithm on GPU by using CUDA. By introducing the 0-1 matrix, this avoid the logic calculation on GPU and reduce the data exchange between the Host and GPU memory, which can significantly improve the speed of the algorithm.

Acknowledgement. This work was supported by the National Natural Science Foundation of China (Grant No. 31201121, NO. 61373109 and NO. 61403287), the Natural Science Foundation of Hubei Province (Grant No 2014CFB288) and Open foundation of Hubei Province Key Laboratory of Intelligent Information Processing and Real-time Industrial System (Grant No. ZNSS2013A001 and ZNSS2013A004).

References

1. Sun, Q., et al.: Speckle reducing anisotropic diffusion for 3-D ultrasound images. Comput Med Imaging Graph **28**(8), 461–470 (2004)
2. Zhu, H., Chen, Y., Wu, J., et al.: Implementation of 3D SRAD algorithm on CUDA. In: 2011 4th International Conference on Intelligent Networks and Intelligent Systems (ICINIS), pp. 97–100. IEEE (2011)
3. Che, S., Boyer, M., Meng, J., et al.: A performance study of general-purpose applications on graphics processors using CUDA. J. Parallel Distrib. Comput. **68**(10), 1370–1380 (2008)
4. NVIDIA Coporation: Nividia cuda programming guide 5.1, NVIDIA (2014)

Localization of Pedestrian with Respect to Car Speed

Joko Hariyono, Danilo Cáceres Hernández, and Kang-Hyun Jo[✉]

Graduate School of Electrical Engineering,
University of Ulsan, Ulsan 680–749, Korea
{joko,danilo}@islab.ulsan.ac.kr, acejo@ulsan.ac.kr

Abstract. Pedestrian location with respect to the road and its distance from vehicle are important information for driver assistance system. This work proposes a method to classify location of pedestrians using estimated road region and vehicle movement region. Single camera is used to detect pedestrians, thus classifying their locations to assist the driver in avoiding accidents. The system consists of three stages. First, moving pedestrians are detected using optical flows method. They are extracted from the relative motion by segmenting the region representing the same optical flows after compensating the ego-motion of the camera. Thus, histogram of oriented gradients (HOG) features descriptor and linear support vector machine (SVM) are used to recognize the pedestrian. Second, road lane is detected using combination of color based and Hough based detection method. It is used to define road boundary region by connecting left and right lanes on a vanishing point. Third, a heuristic method according to a vehicle movement region is defined by the typical stopping distance. It is depended on the speed of the car that will be impact to the thinking distance and braking distance. Combination of estimated road region and vehicle movement region are used to classify the location of the detected pedestrians. They are classified into three zones; car movement zone, road zone surrounding car path and the pedestrian track zone. The proposed method is evaluated using ETH and Caltech datasets, and the performance results shown the best pedestrian detection rate is 99.50 % at 0.09 false positive per image. The location classification is applied in our real world driving data, the evaluation result shows correct classification rate is 98.10 %.

Keywords: Pedestrian detection · Driver assistance system · Location classification

1 Introduction

Future intelligent vehicle domain is not only focus on accuracy of pedestrian detection. Location of pedestrian in traffic scene, car movement speed and some context information from environment are also important information. It is more challenging to detect pedestrian crossing road in order to avoid an accident and control locomotion of the vehicle. Thus it can be implementing for autonomous driving and vision based driving assistance system. This work addressed a problem that can have a significant impact on people safety, to detect pedestrian and assist the driver there are any people cross the street in order to avoid the accident.

© Springer International Publishing Switzerland 2015
D.-S. Huang et al. (Eds.): ICIC 2015, Part II, LNCS 9226, pp. 185–194, 2015.
DOI: 10.1007/978-3-319-22186-1_19

Most of works on driver assistance based pedestrian detection [1, 13], follow these kinds of ideas. The first is detection rate improvement by proposing descriptor methods [4, 5]. Second, the complexity reduction with analysis function methods [2, 6, 14]. Third, the combination with people tracking or estimation methods [3] and the other one is to solve the problem on occluded situation.

In the past few years, pedestrian detection methods for moving vehicle have been actively developed. For a practical real-time pedestrian detection system, Gavrila et al. [1] employed hierarchical shape matching to find pedestrian candidates from moving vehicle. Multi-cue vision system is used for the real-time detection and tracking of pedestrians. Nishida et al. [2] applied SVM with automated the selection process of the components by using AdaBoost. It shows that the selection of the components and the combination of them are important to get a good pedestrian detector.

Several local descriptors are proposed for object recognition and image retrieval. Mikolajczyk et al. [3] compared the performance of the several local descriptors and showed that the best matching result is obtained by the Scale Invariant Feature Transform (SIFT) descriptor [4]. Dalal et al. [5] proposed a human detection algorithm using histograms of oriented gradients (HOG). HOG features are calculated by taking orientation histograms of edge intensity in a local region. Dalal et al. extracted the HOG features from all locations of a dense grid on an image and the combined features are classified by using linear SVM. They showed that the grids of HOG descriptors significantly out-performed existing feature sets for human detection [6].

Road lane detection is useful information which has received considerable attention since last three decades [15]. Techniques used varied from using monocular to stereo vision [16], using low level morphological operations [17] to using probabilistic grouping and B-snakes [18]. Our technique is focused on detection of lane markers on highway roads. Challenges are included parked and moving vehicles, shadows cast from trees, buildings and other vehicles, curves, irregular lane shapes, emerging and merging lanes, different pavement materials, and different slopes. Accurate estimation of vehicle ego-motion relative to the road and environment is a key component for autonomous driving [7]. Talukder et al. [8] proposed a qualitative obstacle detection method using the directional divergence of the motion field. The camera ego-motion is measured by [20, 11] using KLT optical flow tracker [9].

The typical stopping distance [19] is proposed according to the speed of vehicle. Given a car speed while it moves in the road, the safety distance for car braking could be defined. The speed of the car will be considered as the important impact to the thinking distance and braking distance. Our challenge is to propose a method for autonomous driving system which has ability to classify location of pedestrians. Combination lane detection and car movement region are used to control car movement and avoid the accident.

In summary, the contribution of this work is to identify and localize the pedestrian from moving vehicle with respect to the road area and the distance from the vehicle. In the next section, the proposed algorithm for moving object detection is showed. The proposed location classification of pedestrians is presented in Sect. 3. The experimental results are described in Sect. 4. The conclusion and future works are given in Sect. 5.

2 Moving Object Detection

In this section, the proposed optical flow analysis is described. The optical flow analysis is needed to prove that the motion of the independent moving object will have difference pattern inside the motion flow caused by the camera ego-motion. A moving object is extracted from the relative motion by segmenting the region representing the same optical flows after compensating the ego-motion of the camera. The regions of moving object are detected as transformed objects which are different from the previously registered background.

2.1 Ego-Motion Compensation

To analyze a motion on the image, features are important things which are needed. They are extracted from the image as a point of interest. Our motion feature is motivated by the fact that strong cues exist in the movements of different body parts when the pedestrian is walking. They include the motion of two legs, those between two parts of an arm, or those between a leg and an arm. They provide useful cues to identify the walking motion. This proposed optical flow method is defined to segment out moving objects from static environment. Landmarks, such as building, road, statue, etc. are used as our reference in the case of outdoor application. Because our camera system is moving, so in order to calculate the camera ego-motion, interest points from landmarks are needed. Sparse optical flow is chosen and corner features are used as interest points.

Corner features are used in this work as the interest point to perform the task for feature tracking. Using Harris corner method [10], corner features are extracted in each image. Then, all the features are defined within patches (cells) which generates by divided an image using n x n cell windows.

Two consecutive images are compared and tracked corresponding features from previous frame to the current frame. The feature which located in a group cell is used to find the motion distance of each pixel in a group of cells. The motion distance d in x and y - $axis$ by of features cell in the previous frame $g_{t-1}(i, j)$ is obtained by finding most similar cell $g_t(i, j)$ in the current frame $g_{t-1}(i, j) = g_t(i + d_x(i, j), j + d_y(i, j))$. It represented as affine transformation of each group as $g_t(i, j) = Ag_{t-1}(i, j) + d(i, j)$. Where A is 2×2 transformation matrix and $d(i, j)$ is 2×1 translation vector. The camera ego-motion compensated frame difference I_d is calculated based on the tracked corresponding pixel groups using $I_d(i, j) = |g_{t-1}(i, j) - g_t(i, j)|$, where $I_d(i, j)$ is a pixel group located at (i, j) in the grid cell.

Forward-backward error is performed for detecting tracking failure [17]. First, a tracker produces a trajectory by tracking the point forward in time. Second, the point location in the last frame initializes a validation trajectory. The validation trajectory is obtained by backward tracking from the last frame to the first one. Third, the two trajectories are compared and if they differ significantly, the forward trajectory is considered as incorrect.

Features which have corresponding are used to calculate the affine parameter. It is calculated by the least square method using three corresponding features in those two consecutive frames. Then, the affine parameter is applied on the all of window cells.

Thus, each pixel in previous frame is corresponded, one to one, with the pixel in current frame. The regions which are difference from previously registered are obtained by subtracting frame difference on those corresponding pixel cells. Those regions are segmented from the motion of camera ego-motion.

2.2 Moving Object Localization

The region segmentation method is devised for accurately locates bounding boxes of the motion in the different image. Each pixel output from the previous step cannot show clearly as silhouette. It just gives information about motion region of moving objects. Due to those regions are obtained from subtracting pixel to pixel according to its corresponding, a hole appears caused by the intersection of two regions. So, those regions should be applied morphological process, opening and closing, to obtain smoothness region of moving objects.

The fact that humans usually appear in upright positions, and conclude that segmenting the scene into vertical strips is sufficient most of the time. Then, detected moving objects are represented by the position in width in x axis. Using projection histogram h_x by pixel voting vertically project image intensities into $x -$ coordinate. Moving object area is detected based on the constraint of moving object existence that the bins of histogram in moving object area must be higher than a threshold and the width of these bins should be higher than a threshold $h_x(i \pm 10) > A \ max(h_x)$. Where A is a control constant and the threshold of bin value is dependent on the maximum bin's value. Adopting the region segmentation technique proposed in [15], the region is defined using boundary salience. It measures the horizontal difference of data density in the local neighborhood. The local maxima correspond to where maximal change in data density occurs. They are candidates for region boundaries of human in moving object detection.

3 Location Classification

This section presents our proposed method for classify location of pedestrians. Road lane detection is used to define road boundary area. A heuristic method according to the image formation from its geometrical coordinates is used to define vehicle movement region. Combination of road boundary area and vehicle movement region is used to evaluate the accuracy of correct location of detected pedestrian.

3.1 Road Lane Detection

Road lane boundary is extracted using combination Hough-based detection and Color-based detection methods. Hough-based detection method employs global thresholding to solve the problem of illumination. It also makes this method less sensitive to scene conditions. Thus, edge detection is performed to detect lane boundary in the image. Finally, Hough line transform is performed in order to find lines in an image. The output of edge detection step as the input of Hough line detection, and this transformation finds lines in an image based on that every point in Hough space is a line in Euclidean space and vice versa.

Color-based detection method uses color segmentation in order to extract more data from lines and their boundaries. HSI color space is used which saturation and Intensity are important for processing to detect the lanes. Because of both values are considerable in variations. High contrast between gray color of road and white color of lines make higher values of saturation and intensity components rather than hue component.

A heuristic method is proposed to combine both methods. Consecutive frames are used to keep only the brightest regions of the image by applied a threshold value. Some of the detected regions are then eliminated having as basis orientation properties. A clustering methodology is used to group the detected points. Lines that are too close are merged. If more than two lines are detected, the two most similar to the ones detected in the previous frame are kept. If there are big jumps in the lines from one frame to the next ones, the corresponding line is eliminated. If less than two lines are detected, lines from the previous frame are retrieved. Finally, if the lines are not being updated for 12 frames they are either not displayed (in the case of the outside lanes) or the line is replaced with a line between the two adjacent lines (in the case of the inside lanes). The projected intersection of these two lines is determined and is referred to as the horizon. The road region is defined by the polygon which is constructed from two lines and bottom part of image.

The method is tested in our real driving data, the performance of correct detection rate is 96.21 %. False positive rate is 10.58 %. Some of them are lanes in curve road and due to color degradation and occluded part.

3.2 Estimating Region of Car Movement

Our vision system is located fix at the top of the vehicle and it look in front align to the vehicle direction. The fact that the center of an image is always align with the camera direction. Then, sequential images which are collected by the camera will give information that the direction objectives of vehicle movement are located on the center of image. The walking pedestrian should be located in right and left sides of the image.

According to the typical stopping distance [19], the vehicle movement region is defined. It is depend on the speed of the car that will be impact to the thinking distance and braking distance. For example, when the car is moving with the speed 48 km/h, the stopping distance minimum [19] is defined at 23 m. It is consist of 9 m for thinking distance and 14 m braking distance. Figure 2 (left) shows typical stopping distance. The car speed is measured by an inertial measurement unit sensor while it is used in our real world driving dataset.

The location on the image is classified into three regions. First is the vehicle movement zone. It is defined as the area in front of the vehicle that it should move forward safely. It depends on the car speed and the typical stopping distance [19]. Second is the estimated road zone which is obtained from the road lane boundaries. Third is pedestrian zone which they should be located. They are located in the left and right location in the image. Figure 1 (right) illustrates our proposed location classification using lane detection and geometric properties of image.

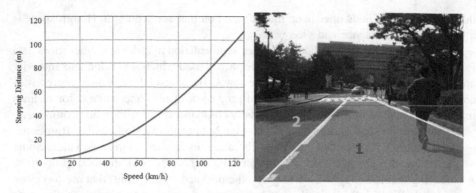

Fig. 1. Typical stopping distance (left) and location classification on the image (right).

3.3   Distance Estimation

Our camera system is looking forward and mounted at the top of vehicle. It is located 170 cm height from the ground. A point in the real world coordinate is converted to the image coordinate by calibration process. The ground plane is assumed a planar, then some distance points in the world coordinate are used to compute those in the image coordinate. The people who have tall approximately the height position of the camera, by our measurement that the top of head will show in the horizon line on the image. The horizon line is a horizontal line through the center coordinate of an image.

Suppose the normal human tall is less than 200 cm. Thus, the maximum height of pedestrian on the image is estimated by geometrical analysis model which is showed in Fig. 3 (top-right).

$$\frac{f}{h} = \frac{d}{H} \tag{1}$$

Where f is the focal length of camera, h is the pixel height of the object on the image, H is actual human height and d is the estimated distance. Suppose the focal length of camera is fixed, and also average of human height is assumed, around 170 cm. So, relation of the pixel height of human on the image and their distances from the vehicle can be calculated. Figure 2 (bottom-right) shows the relationship.

The sky region is eliminated for the top quarter part of image. It should be done by our assumption when our car is moving, the normal distance of pedestrians is five meters far or more. Three regions of classification are used to confirm that our system will recognize whether the pedestrian is walking in the well track or crossing the road. Given the detected pedestrian in the zone 1 → U, zone 2 → V and zone 3 → W, So, function of $f(U, V, W)$ follows,

$$f(U, V, W) = \begin{cases} 1, & if\ U = 1\ or\ U + V = 1 \\ 0, & if\ V = 1\ or\ V + W = 1 \\ -1, & otherwise \end{cases} \tag{2}$$

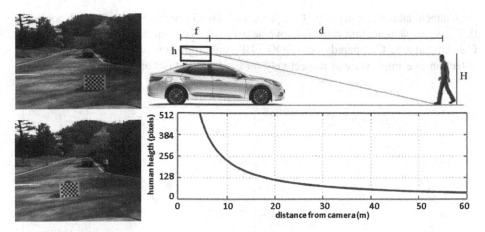

Fig. 2. Several poses of a checkerboard pattern is used for our calibration (left), geometrical analysis (top-right), Pixel height h as a function of distance d (bottom-right).

The driving assistance system will give alert for stop, if the pedestrian is located in the zone 1 or overlapping zone 1 and 2. When the pedestrian is located in the zone 2 or overlapping zone 2 and 3, it will give alert for be careful.

4   Experimental Results

4.1   Experimental Result on Pedestrian Detection

The proposed pedestrian detection algorithm is evaluated on two publicly available datasets: Caltech [13] and ETH [12]. For the experiment on the datasets Caltech and ETH, the INRIA training dataset in [5] is used for training.

In the experiments, we mainly compare with the original HOG [5]. It uses the same feature and the same training dataset. The HOG is used sliding windows in the whole of frame. The HOG selection in [6] and the HOG-LBP in [14] are also included for comparison. The per-image evaluation methodology is used. The labels and evaluation log-average detection rate is used as the evaluation criterion.

The proposed algorithm is programmed in MATLAB and executed on the Intel Pentium 3.40 GHz, 64-bit operating system with 8 GB RAM. The original HOG, which proposed by [5] is implemented. The HOG features are extracted from 16×16 local regions. The first, Sobel filter is performed to obtain gradient orientations from each pixel in this local region. The local region is divided into small cells with cell size is 4×4 pixels. Histograms of gradients orientation with eight orientations are calculated from each of the local cells. Then, the total number of HOG features becomes $128 = 8 \times (4 \times 4)$ and they constitute a HOG feature vector.

Figure 3 shows detection result comparison the original HOG, the HOG selection, HOG-LBP and our proposed method at 0.02 FPPI on the ETH dataset and the Caltech dataset. It can be seen that our proposed pedestrian detection approach, has 2 % and 2.5 % detection rates improvement compared with original HOG on ETH dataset and the Caltech dataset respectively. Compared with HOG selection, our approach achieves 9 % and 10 % detection rates improvement respectively on the ETH dataset and the Caltech dataset. Compared with HOG-LBP, our approach achieves 1 % and 2 % detection rate improvement respectively on the ETH dataset and the Caltech dataset.

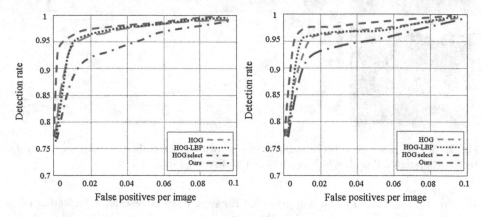

Fig. 3. Experimental results on the ETH dataset (left) and Caltech dataset (right) for HOG [7], HOG-LBP [18], HOG Selection [8], and our ego-motion compensated + HOG.

Figure 4 (left) shows the computational cost comparison. It shows our proposed method has eight times lower comparing with HOG when the small ratio of positive to evaluated data is used. Compared with HOG selection and HOG + LBP, our approach achieves two times and nine times improvement respectively. However, if the number of ratio is increased it also reduces time consuming significantly.

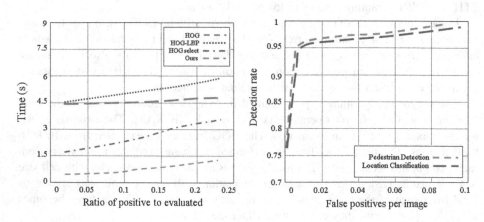

Fig. 4. Time consuming comparison (left), and experimental results on real world driving data for detecting pedestrian and location classification (right).

4.2 Experimental Result on Location Classification with "Real World" Driving Data

Sequential images are collected using our autonomous vehicle which are contains around 2,000 images1 of pedestrians in outdoor scenes. Our vehicle moved with varies speed 10–40 km/h. The number of true detected human crossing the region of street is computed based on the ground truth. The ground truth is obtained by the number of the human crossing the street. Table 1 shows summary of the real world driving data for performance evaluation of location classification. Figure 4 (right) shows, the best recognition rate at 0.09 FPPI, the detection rate and correct location classification of pedestrians are 99.50 % and 98.10 % respectively.

Table 1. Real world driving data

Properties	Value
Number of images	2,581
Number of pedestrian	2,173
Number of pedestrian crossing road	572

5 Conclusion

A method to classify location of pedestrians is presented. Combination lane detection and typical stopping distance are used to control car movement objective and avoid the accident. The proposed method is evaluated using public datasets and our real world driving data, performance results shown the best pedestrian detection rate is 99.50 % at 0.09 false positive per image. The location classification evaluation shown correct detection rate is 98.10 %. In some cases, e.g. a curve road and a road with degradation lane color; it will be impact to the false location classification. For the next challenge, a robust lane detection method is considered to perform. Localization of pedestrian with incorporate additional scenes (e.g. traffic situation) is also considered to implement.

References

1. Gavrila, D.M., Munder, S.: Multi-cue pedestrian detection and tracking from a moving vehicle. Int. J. Comput. Vis. **73**(1), 41–59 (2007)
2. Nishida, K., Kurita, T.: Boosting soft-margin SVM with feature selection for pedestrian detection. In: International Workshop on Multiple Classifier Systems, vol. 13, pp. 22–31, 2005
3. Mikolajczyk, K., Schmid, C.: A performance evaluation of local descriptors. IEEE Trans. Pattern Anal. Mach. Intell. (PAMI) **27**(10), 1615–1630 (2005)
4. Lowe, D.G.: Distinctive image features from scale-invariant keypoints. Int. J. Comput. Vis. **60**(2), 91–110 (2004)
5. Dalal, N., Triggs, B.: Histograms of oriented gradients for human detection. In: IEEE Conference on Computer Vis. and Pattern Recognition, pp. 886–893 (2005)

6. Kobayashi, T., Hidaka, A., Kurita, T.: Selection of histograms of oriented gradients features for pedestrian detection. In: Ishikawa, M., Doya, K., Miyamoto, H., Yamakawa, T. (eds.) ICONIP 2007, Part II. LNCS, vol. 4985, pp. 598–607. Springer, Heidelberg (2008)
7. Stein, G.P., Mano, O., Shashua, A.: A robust method for computing vehicle ego-motion. In: Intelligent Vehicles Symposium (2000)
8. Talukder, A., Goldberg, S., Matthies, L., Ansar, A.: Real-time detection of moving objects in a dynamic scene from moving robotic vehicles. In: Proceedings of the International Conference on Intelligent Robotics and systems, pp. 1308–1313 (2003)
9. Tomasi, C., Kanade, T.: Detection and tracking of point features. Int. J. Comput. Vis. **9**, 137–154 (1991)
10. Harris, C., Stephens, M.: A combined corner and edge detector. In: Proceedings of the 4th Alvey Vis. Conference, pp. 147–151 (1988)
11. Hariyono, J., Hoang, V.D., Jo, K.H.: Moving object localization using optical flow for pedestrian detection from a moving vehicle. Sci. World J. **2014**, 1–8 (2014). doi:10.1155/2014/196415
12. Ess, B. Leibel, L.V. Gool.: Depth and appearance for mobile scene analysis. In: IEEE International Conference on Computer Vision (2007)
13. Dollar, P., Wojek, C., Schiele, B., Perona, P.: Pedestrian detection: an evaluation of the state of the art. IEEE Trans. Pattern Anal. Mach. Intell. **34**(4), 743–761 (2012)
14. Wang, X., Han, X., Yan, S.: An HOG-LBP human detector with partial occlusion handling. In: CVPR (2009)
15. McCall, J.C., Trivedi, M.M.: Video-based lane estimation and tracking for driver assistance: survey, system, and evaluation. IEEE Trans. Int. Transp. Syst. **7**(1), 20–37 (2006)
16. Franke, U., Gavrila, D., Gorzig, S., Lindner, F., Puetzold, F., Wohler, C.: Autonomous driving goes downtown. IEEE Intell. Syst. Appl. **13**(6), 40–48 (1998)
17. Bertozzi, M., Broggi, A., Conte, G., Fascioli, A.: Obstacle and lane detection on argo. In: IEEE Conference on Intelligent Transportation System, pp. 1010–1015, 9–12 November 1997
18. Kim, Z.: Realtime lane tracking of curved local road.: In: Proceedings of the IEEE Intelligent Transportation Systems, pp. 1149–1155, 17–20 September 2006
19. https://www.gov.uk/government/uploads/system/uploads/attachment_data/file/312249/the-highway-code-typical-stopping-distances.pdf
20. Liu, H., Dong, N., Zha, H.: Omni-directional vision based human motion detection for autonomous mobile robots. Syst. Man Cybern. **3**, 2236–2241 (2005)

Plant Leaves Extraction Method Under Complex Background Based on Closed-Form Matting Algorithm

Hao-Dong Zhu$^{(\boxtimes)}$, Di Wu, and Zhen Sun

School of Computer and Communication Engineering,
ZhengZhou University of Light Industry, Henan Zhengzhou 450002, China
zhuhaodong80@163.com

Abstract. The extraction of plant leaves is a technique to separate the captured leaves from the background. In natural scene, the captured picture of plant leaves always contain of complex background and it's hard to extract the leaves. Considering that the drawback of extraction technique based on image segmentation algorithm, this paper introduces Closed-form algorithm into this technique and proposes an improved extraction method which can extract plant leaves from complex background. Firstly, this algorithm requires users to add certain constraints for plant leaves image with complex background. Secondly, the algorithm identifies foreground and background via estimating diaphaneity α of unknown sample. Through these two steps, the plant leaves within complex background could be extracted efficiently. Experimental comparison results show that the proposed method has a better effect.

Keywords: Leaves extraction · Image segmentation · Matting algorithm

1 Introduction

As the largest number of life species on Earth, plant is one of the most basic material conditions of people's survival, which affects the quality of people's life from all sides, But with the development of economic, the environmental resources have been severely damaged and the plant is affected firstly. Recent years, the frequent occur of natural disasters lead people to realize the importance of protecting plants and the necessity of studying plant classification. Therefore, it is more and more urgent to create the database about plant leaves. The features of plant leaves like texture, color and shape characteristic can represent a plant species and the extraction of plant leaves is the prerequisite to study these characteristics [1]. The work of plant leaves extraction lay a good foundation for the identification of plant leaves which based on the Neural Network and Pattern Recognition [2–5]. The extraction of plant leaves' contour will make it easier for the extraction of plant leaf shape feature, then we can use the plant leaf shape feature sort the plants [6–12], and it is the main point that we'll be focusing on. So far, lots of researches have been done on the plant leaves extraction from the perspective of image segmentation. Wang H.J. [13] used the

© Springer International Publishing Switzerland 2015
D.-S. Huang et al. (Eds.): ICIC 2015, Part II, LNCS 9226, pp. 195–202, 2015.
DOI: 10.1007/978-3-319-22186-1_20

Perona and Malik algorithm to deal with the plant leaves' pre-processing stage, and then combined the watershed algorithm to perform the segmentation of plant leaves image. This algorithm has some effects, but Guo X.C. [14] proposed a level set method and graph-cut algorithm, which has had a certain effect and Man Q.K. [15] used a marker-based watershed algorithm and it works well in some complex background. Although the algorithms which have been putted forward both have some good effects, there are still have problems when they function in the plant leaves which have a complex background, so these proposed algorithms need to be further modified.

The rest of the paper is organized as follows. Section 2 discusses the difference between image segmentation algorithm and natural image matting, Sect. 3 gives an overview of the closed-form matting method and Sect. 4 sets up for the experimental results and discussions. Some suggestions regarding future work are given in Sect. 5.

2 The Difference Between Image Segmentation and Image Matting

Image segmentation generally can be described as partitioning an image into multiple segments [16], and the purpose of image matting is to extract the target prospects. Image segmentation and image matting is two similar domain because their essence is the redistribution of the pixel tags. In fact, image matting is a special case of the image segmentation, but their mathematical model is unequal. The difference is as follows:

a. The image matting technology has highly targeted on image extraction and it's designed to extract a foreground object from an image bases on limited input of users. For plant leaves image, it has different color and texture feature when they are in different season, So it's difficult for image segmentation to deal with the situation. But image matting doesn't be affected by these factors [17], based on these characteristics and advantages, image matting is more suitable than image segmentation for the extraction of plant leaves image which are under the complex background.
b. In terms of algorithm complexity, image matting technology is more complex than image segmentation.
c. Image matting requires user interaction [18], but image segmentation needn't.

Against the above two different angles of plant leaves extraction methods, we find image segmentation algorithms could quickly deal with plant leaves image under simple background, but when the plant leaves image are under complex background, most image segmentation algorithms are quite helpless. As this point, the image matting algorithms work. And for this, to add emphasis, we utilize the Closed-form algorithm in the extraction of the plant leaves image with complex background, and our previous studies have shown that the algorithm acquires good result.

3 The Application of Digital Image Matting Technology in Plant Leaf Images Extraction

Natural image matting imply that a technology of extracting the region of interest such as object from an image based on user's interaction like trimap or scribbles [19], and the image matting could be divided into two categories: One is the blue screen matting, in which the image background is a single color. The other is the natural image matting, where the image background is complex and unknown [20]. In this paper, we mainly discuss the latter – natural image matting. So for, the leading methods of image matting algorithms have Knockout algorithm, Ruzon-Tomasi algorithm, Bayesian algorithm and Closed-form algorithm. Our paper mainly discusses Closed-form algorithm [21].

3.1 Gray-Scale Images

For a better deduction, we start our work by deriving a closed-form solution for gray-scale images. The basic model of closed-form algorithm is color synthesis model proposed by Porter, the model is:

$$I_i = \alpha_i F_i + (1 - \alpha_i)B_i \tag{1}$$

We express the matting problem by treating an image I_i as a collection of convex, with foreground image F_i and background image B_i as shown above, where $\alpha_i \in [0, 1]$, which we call "alpha matte". If $\alpha_i = 0$, then I_i belongs to the absolute foreground pixel and if $\alpha_i = 1$, I_i belongs to the absolute background pixel. For gray-scale images, we assume that both F_i and B_i are approximately constant in a small window (a 3×3 square) around every pixel. Based on that assumption we can express α_i as a linear function of I_i:

$$\alpha_i \approx a^c I_i^c + b, \forall i \in \omega \tag{2}$$

In this formula, I_i, α_i, a, b represent the secondary color, alpha matte, scalar and a 3×1 vector respectively. ω is a small window in which a and b is constant. In order to solve alpha α, we minimize the cost function about a and b:

$$J(\alpha, a, b) = \sum_{j \in I} \left[\sum_{i \in \omega_j} (\alpha_i - a_j I_i - b_j)^2 + \varepsilon a_j^2 \right] \tag{3}$$

Where ε is a regularization parameter and ω_j is the small window around pixel j. One of the reasons for adding ε is to keep numerical stability. We can eliminate a_j and b_j because a and b is constant in the small window ω_j, and then we get a quadratic energy function which only has an unknown quantity α:

$$J(\alpha) = \alpha^T L \alpha \tag{4}$$

Where α is a $N \times 1$ vector, L is an $N \times N$ matrix, in which the entry of position (i,j) is:

$$\sum_{k|(i,j)\in\omega_k} [\delta_{ij} - \frac{1}{|\omega_k|}[1 + \frac{1}{\frac{\varepsilon}{|\omega_k|} + \sigma_k^2}(I_i - \mu_k)(I_j - \mu_k)]] \tag{5}$$

In this formula, δ_{ij} is the Kronecker delta, μ_k and σ_k^2 represent the mean and variance in the small window ω_k around k, $|\omega_k|$ represents the number of pixels in ω_k The optimal solution of α is the minimum solution which can minimize the energy function. Therefore, we can obtain α via solving the minimization of the energy function.

3.2   Color Images

For the color image, we apply the gray-level cost to each channel separately, and use a 4D linear model to replace the linear model (2):

$$\alpha_i \approx \sum_c a^c I_i^c + b, \forall i \in \omega \tag{6}$$

Where c is the sum of the color channels. The linear model (6) relaxes the assumption that the foreground color F and the background color B are constant in the small window ω_k, instead we can assume that F and B is a linear combination of two colors:

$$F_i = \beta_i^F F_1 + (1 - \beta_i^F)F_2 \tag{7}$$

Similarly, the background color B can be represented as:

$$B_i = \beta_i^B B_1 + (1 - \beta_i^B)B_2 \tag{8}$$

We substitute model (7) and (8) into the color synthesis model (1):

$$I_i^c = \alpha_i(\beta_i^F F_1 + (1 - \beta_i^F)F_2) + (1 - \alpha_i)(\beta_i^B B_1 + (1 - \beta_i^B)B_2) \tag{9}$$

Let H be expressed as a matrix whose row is $[F_2^c + B_2^c, F_1^c - F_2^c, B_1^c - B_2^c]$ and model (9) can be rewritten as:

$$H \begin{bmatrix} \alpha_i \\ \alpha_i\beta_i^F \\ (1 - \alpha_i)\beta_i^B \end{bmatrix} = I_i - B_2 \tag{10}$$

Then we can obtain model (11):

$$\alpha_i = \sum a^c I_i^c + b, \forall i \in \omega \tag{11}$$

We define the cost function for the color images matting by utilizing the 4D linear model (6):

$$J(\alpha, a, b) = \sum_{j \in \omega_j} [\sum_{i \in \omega_j} [\alpha_i - \sum_c a^c I_i^c - b]^2 + \varepsilon \sum_c a_j^{c^2}] \tag{12}$$

In a similar way, we can eliminate a^c and b from the cost function, and then we get a quadratic cost with the α unknowns alone:

$$J(\alpha) = \alpha^T L \alpha \tag{13}$$

Where L is an $N \times N$ matting Laplacian, where the entry of (i, j) is:

$$\sum_{k|(i,j) \in \omega_k} [\delta_{ij} - \frac{1}{|\omega_k|}[1 + (I_i - \mu_k)^T [\sum_k + \frac{\varepsilon}{|\omega_k|} I_3]^{-1} (I_j - \mu_k)]] \tag{14}$$

Here $\sum_k$ is an 3×3 matrix, which represents the color-covariance of the window ω_k, I_3 is a 3×3 identity matrix, μ_k is a 3×1 vector, which is the mean of the colors in a window ω_k. $|\omega_k|$ represents the number of pixels in ω_k. The important quality of the matrix L we previously mentioned in each row of L sums to zero, so the null space of L contains constant vector, we can guide users to interact via analyzing the minimum feature vector of the Laplacian matrix.

4   Algorithm Implementation and Contrast Experiment

4.1   Add User Constrains

Natural image matting is a severely constrained problem and requires user interaction to extract a good matte. According to the characteristics of the closed-form algorithm, we divide original images by the line drawing. The original plant leaf image is shown in Fig. 1, and the plant leaf image after regional division as the Fig. 2 shown. We add our constrains through the drawing tools, and the region which is marked by black or white lines is the object background or foreground. Respectively, after we marked the original plant leaf image, the closed-form algorithm estimates the diaphaneity α via the acquired information.

4.2   Experiment Results and Analysis

Here, the purpose of extract the blade is to position the target area of the plant leaf, which makes it easier to calculate the shape features of the leaves, so the extraction result

Fig. 1. Original image   **Fig. 2.** Marked image

should be a binary image, and the foreground color appear white, while the background color is dark. The image matting algorithm aims to separate the foreground and the background via user interaction, and the process of algorithm is to solve the foreground and background value of the pixels and the algorithm isn't prone to be influenced by additional factors of leaves. The experimental result is shown in Fig. 3. From the Fig. 3, we can see the plant leaf is entirely extracted. Figure 4 is the canny edge detection experimental results, and Fig. 5 is the results of label watershed experimental, and Fig. 6 is the otsu experimental results and they all belong to the image segmentation algorithms. From the experimental results we can see these image segmentation algorithms performance does not stand out, they don't extract the plant leaves from the complex background very well, especially in the adhesion areas where the leave color is close to its surrounding, and compared with the image matting algorithm Closed-form matting shown in Fig. 3, the closed-form algorithm has a better effect.

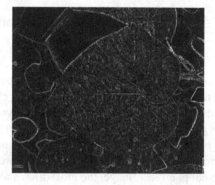

Fig. 3. Experimental results   **Fig. 4.** Canny edge detection experimental

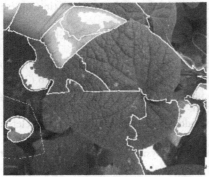

Fig. 5. Label watershed experimental results   **Fig. 6.** Otsu experimental results

5 Conclusions

This paper proposes that Closed-form algorithm in image matting technology can be utilized to extract plant leaf image which is under the complicated background, and compared with conventional image segmentation technology, this method solves the problem which exists in splitting plant leaf under the complicated background. The method which we propose gets better results via simulation experiment. This paper compares the advantages and the disadvantages of image matting algorithm and image segmentation algorithm comprehensively, and provides a novel solution to identify plant leaves. However, since the image matting algorithm requires users' interaction, the computation complexity of it is high, which will affect the efficiency of extracting plant leaves. In the future, it is the key of research how to overcome those limited conditions of image matting algorithm and present a more effective extraction algorithm.

Acknowledgment. The authors would like to thank the editors and anonymous reviewers for their valuable comments. This work was supported in part by the Technology Innovation Outstanding talents Plan Project of Henan Province of China under grant No.134200510025, the National Natural Science Foundation of China under grant No. 61201447, the Youth Backbone Teachers Funding Planning Project of Colleges and Universities in Henan Province of China under grant No. 2014GGJS-084, the Youth Backbone Teachers Training Targets Funded Project of Zhengzhou University of Light Industry of Henan Province of China under grant No. XGGJS02, the Ph.D. Research Funded Project of Zhengzhou University of Light Industry of Henan Province of China under grant No. 2010BSJJ038 and the Science and Technology Innovation Fund project of postgraduate of Zhengzhou University of Light Industry.

References

1. Wang, Z.F., Huang, D.S.: Feature extraction and recognition for leaf images. Comput. Eng. Appl. **42**(03), 190–193 (2006)

2. Huang, D.S.: Systematic Theory of Neural Networks for Pattern Recogintion (in Chinese). Publishing House of Electronic industry of China, Beijing (1996)
3. Li, B., Huang, D.S.: Locally linear discriminant embedding: an efficient method for face recognition. Pattern Recogn. **41**(12), 3813–3821 (2008)
4. Shang, L., Huang, D.S., Zheng, C.-H.: Palmprint recognition using FastICA algorithm and radial basis probabilistic neural network. Neurocomputing **69**(13–15), 1782–1786 (2006)
5. Zhao, Z.-Q., Huang, D.S., Sun, B.-Y.: Human face recognition based on multiple features using neural networks committee. Pattern Recogn. Lett. **25**(12), 1351–1358 (2004)
6. Huang, D.S., J, Wen.: A general CPL-AdS methodology for fixing dynamic parameters in dual environments. IEEE Trans. Syst. Man Cybern. Part B **42**(5), 1489–1500 (2012)
7. Huang, D.S., Zheng, C.-H.: Independent component analysis based penalized discriminant method for tumor classification using gene expression data. Bioinformatics **22**(15), 1855–1862 (2006)
8. Huang, D.S.: Radial basis probabilistic neural networks model and application. Int. J. Pattern Recogn. Artif. Intell. **13**(7), 1083–1101 (1999)
9. Wang, X.-F., Huang, D.S.: A novel density-based clustering framework by using level set method. IEEE Trans. Knowl. Data Eng. **21**(11), 1515–1531 (2009)
10. Huang, D.S., Horace, H.S.Ip, Chi, Z.: A neural root finder of polynomials based on root moments. Neural Comput. **16**(8), 1721–1762 (2004)
11. Huang, D.S.: A constructive approach for finding arbitrary roots of polynomials by neural networks. IEEE Transactions on Neural Networks **15**(2), 477–491 (2004)
12. Huang, D.S., Du, J.-X.: A constructive hybrid structure optimization methodology for radial basis probabilistic neural networks. IEEE Trans. Neural Netw. **19**(12), 2099–2115 (2008)
13. Wang, H.-J., Chen, W., Zhao, H.: Color image segmentation of plant leaves under complex background. J. Chin. Agr. Mech. **34**(2), 207–211 (2013)
14. Guo, X.: Study of Plant Image Segmentation Based on Level Set Method and Graph-Cut Algorithm. Huaqiao University, Fujian (2013)
15. Man, Q.-K.: Research the Segmentation Algorithm on Plant Leaves Images Which Under Complex Background. Qufu Normal University, Qufu (2009)
16. Wang, X.-F., Huang, D.S., Xu, H.: An efficient local Chan-Vese model for image segmentation. Pattern Recogn. **43**(3), 603–618 (2010)
17. Wang, X.-S., Huang, X.-Y.: Study surveys on tree image extraction in a complex background. J. Beijing Forest. Univ. **32**(3), 197–203 (2010)
18. Chuang, Y.Y., Agarwala, A., Curless, B.: Video Matting of Complex Scenes. ACM Trans Graph. **21**(3), 243–248 (2002)
19. Yang, H., Au, O.C.: A comprehensive study on digital image matting. J. Signal Inf. Process. **4**(3), 286–290 (2014)
20. Zhu, Q.-S.: Targeting accurate object extraction from an image: a comprehensive study of natural image matting. IEEE Trans. Neural Netw. Learn. Syst. **26**(2), 185–207 (2015)
21. Levin, A., Lischinski, D., Weiss, Y.: A closed form solution to natural image matting. In: Proceedings of IEEE, CVPR, pp. 228–242. IEEE (2008)

Adaptive Piecewise Elastic Motion Estimation

Huijun Di[1(✉)], Linmi Tao[2], and Guangyou Xu[2]

[1] Beijing Key Laboratory of Intelligent Information Technology,
School of Computer Science, Beijing Institute of Technology,
Beijing 100081, People's Republic of China
ajon@bit.edu.cn
[2] Key Laboratory of Pervasive Computing, Ministry of Education,
Department of Computer Science and Technology,
Tsinghua University, Beijing 100084, People's Republic of China
{linmi,xgy-dcs}@tsinghua.edu.cn

Abstract. This paper proposes a novel probabilistic graphical model, called MTHMM-P, for partitioning general nonrigid motion into piecewise elastic motion, so as to achieve nonrigid motion estimation without the need of any *a priori* shape model. To this end, two interrelated sub-problems have to be addressed: partitioning whole motion sequence into several coherent pieces and estimating elastic motion inside each one. The proposed MTHMM-P jointly formulates these two sub-problems. By means of a competition-cooperation partition mechanism and joint inference algorithm, the MTHMM-P can automatically determine the number of pieces as well as adjust the span of each piece by adapting to the input sequence, and at the same time achieve the estimation of piecewise elastic motion. Experiments on the motion of face and body show the capability of the MTHMM-P.

Keywords: Elastic motion estimation · Probabilistic graphical model

1 Introduction

Nonrigid motion analysis has been attracting more and more research efforts. According to the degree of deformation of the object, nonrigid motion studied in computer vision can be divided roughly into three major categories: articulated, elastic, and fluid motion [1, 2]. Elastic motion is a general type of nonrigid motion, where there are no constraints other than coherence (i.e., some degree of smoothness and continuity [2]). Due to such a generality, we argue that most of nonrigid motion can be treated as elastic, as far as considered in a sufficiently short interval of time. Under this argument, piecewise elastic motion can be a useful and unified representation for general nonrigid motion.

For instance, an articulated human side walking can be viewed as a repeat of three elastic phases: the front leg is moving ahead (along with the synchronized hands motion, the same below), the movement of the back leg is followed but without the

---

This research was supported in part by the National Natural Science Foundation of China under Grant No. 61003098.

D.-S. Huang et al. (Eds.): ICIC 2015, Part II, LNCS 9226, pp. 203–215, 2015.
DOI: 10.1007/978-3-319-22186-1_21

overlap of two legs, and the back leg is moving under the overlap with the front one. To achieve an automatic extraction of such unified representations from an input video sequence without assumption about particular motion type and motion content, a problem of adaptive piecewise elastic motion estimation is addressed in this paper.

The addressed problem actually involves two interrelated sub-problems: partitioning whole motion sequence into several coherent pieces and estimating elastic motion inside each one. In the case of relying on prior knowledge of particular motion structure, e.g., by using *a priori* shape models such as multi-pose 2D face model or articulated model of human body, the addressed problem can be solved with less challenging. However, our intention is not to rely on such prior knowledge, so as to enable the capabilities of automation and of handling general nonrigid motion.

The methods related to elastic motion estimation without a priori shape models can be divided into two groups: explicit and implicit matching methods [2]. Feature extraction is required in the explicit matching [3–5], while intensity information is directly used in the implicit matching (e.g., patch matching or optical flow estimation [6–10]). Although the impressive progresses have been made in the implicit matching methods (e.g. the long term tracking driven by optical flow [11–14]), the lack of explicit geometric information makes implicit matching methods not convenient to be integrated with motion partition.

Our previous work [5] has proposed a probabilistic graphical model MTHMM (a Mixture of Transformed Hidden Markov Models) for elastic motion estimation. However, MTHMM cannot be directly applied to solve the problem in this paper. MTHMM assumes that the shape of the object should be coherent in the whole motion sequence, which cannot be ensured when considering general nonrigid motion. An extension of the MTHMM is in demand.

This paper proposes a valuable extension of the MTHMM, called MTHMM-P, which makes a deep integration of motion partition and elastic motion estimation. The MTHMM-P can automatically determine the number of coherent pieces and adjust the span of each piece by adapting to the input motion sequence, without assumption about particular motion type and motion content. Together with the abilities inherited from the MTHMM, the MTHMM-P can achieve motion partition and piecewise elastic motion estimation simultaneously. The rest of the paper is organized as follows: The basic idea and formal formulation of the MTHMM-P are given in Sects. 2 and 3 respectively. The inference under the MTHMM-P is then described in Sect. 4. Experiments are presented in Sect. 5. Finally, conclusions are elaborated.

2   Basic Idea of the MTHMM-P

It is necessary to briefly review the idea of the MTHMM [5]. In the MTHMM, the overall motion of an elastic object is interpreted by a fiber bundle, where a fiber is a metaphor for the motion trajectory of one particle on the object. In spatial domain, the fiber bundle is defined according to a common Reference Cross Section (RCS) and related nonrigid transformation at each frame. Each cross section of the bundle is defined as a smoothly transformed version of the RCS by the corresponding transformation. In temporal domain, each fiber is modeled as a continuous-state HMM. The constraints of smoothness

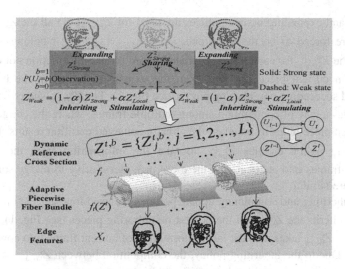

Fig. 1. The basic idea of the MTHMM-P.

and continuity of elastic motion are fully exploited in the MTHMM by a tight integration of motion tracking and shape registration, where motion tracking is formulated along the fibers to account for temporal continuity, while shape registration is formulated across fibers to account for spatial smoothness. The edge features are used as the observations for the MTHMM, where each edge point is viewed as a random sample generated from a Gaussian mixture whose centers are the cross section of the bundle. The matching of edge features is achieved by inferring the MTHMM's parameters which best explain the observation sequence, yielding the estimation of elastic motion.

To be able to handle general nonrigid motion, the proposed MTHMM-P uses adaptive piecewise fiber bundle which is designed based on a competition-cooperation partition mechanism. As shown in Fig. 1, in the adaptive piecewise fiber bundle, each frame is configured with two dynamic $RCSs$ $Z^{t,b}$ ($b = 0$, 1) which compete with each other. A high-level Markov chain is subsequently introduced in the MHTMM-P to deal with the partition, whose binary hidden state U_t at each frame determines the association with current frame of the two dynamic $RCSs$. The partition is then posed as an inference problem of U_t, where the optimal sequence of U_t will give a representation of the achieved partition results. The frames which are continuously in a same optimal partition state form one piece. For instance, a sequence "11100011" of the optimal U_t has three pieces. The inference of U_t is a problem of choosing a best $Z^{t,b}$ at each frame to fit the input motion. Therefore, partition is actually related to $Z^{t,b}$, and the principles of the partition mechanism will be reflected into the update policies of $Z^{t,b}$.

To best explain our idea, we use a notion of strong state b^t_{Strong} and weak state b^t_{Weak} which are defined as $b^t_{Strong} = \arg\max_b P(U_t = b|Observation)$ and $b^t_{Weak} = 1 - b^t_{Strong}$, i.e. the state value where the posteriori of U_t takes maximum and minimum respectively. The sequence of strong state is actually the output of partition. With these definitions, two other notions of strong RCS Z^t_{Strong} and weak RCS Z^t_{Weak} can be used to denote the corresponding $Z^{t,b}$.

Sharing and expanding update policies are used for Z^t_{Strong}: (1) all the Z^t_{Strong} inside one piece are forced to be the same one, i.e. $\{Z^t_{Strong} \equiv Z^k_{Strong}:$ t belongs to the kth piece}, (2) the Z^k_{Strong} for each piece is updated based on the information not only from the frames in current piece but also from the expanded frames in the half of the previous and succeeding neighboring pieces (see Fig. 1). The Z^k_{Strong} shared inside each piece provides a way of cooperation among multiple frames, where the common mode Z^k_{Strong} of each piece helps to maintain the already achieved partition results. In further, the expanding policy enables Z^k_{Strong} to take the information from the expanded neighboring frames into account, and therefore provides a drive to merge some pieces when they are similar.

While inheriting and stimulating update policies are used for Z^t_{Weak}: (1) the Z^t_{Weak} at each frame inherits the $Z^{k'}_{Strong}$ of the nearest neighboring pieces (see Fig. 1), (2) at the same time the Z^t_{Weak} takes also the local information of current frame into consideration (see Fig. 1). Due to the inheriting policy, the competition between Z^t_{Strong} and Z^t_{Weak} at each frame implies a competition between neighboring pieces, and thus provides a way to adjust the span of each piece as well as to merge similar pieces. By local stimulating, the competition between Z^t_{Strong} and Z^t_{Weak} also implies a competition between the common mode of current piece and the local mode at each frame. When one particular piece is not sufficient to represent the actual motion, the local stimulating will cause Z^t_{Weak} to rise up as a new stronger one, and therefore new piece is created.

3 Problem Formulation

In the adaptive piecewise fiber bundle, suppose the two dynamic RCSs at frame t ($t = 1$, $2,..., N$) are denoted as $\{Z^{t,b}: Z^{t,b}_j; b = 0, 1; j = 1, 2, ..., L\}$, where N is the number of frames in the input sequence, L is the number of the fibers contained in the bundle, and $Z^{t,b}_j$ is a coordinate vector of the jth point in the bth RCS. Subsequently, at frame t, when partition state $U_t = b$, the cross section of the bundle is defined as a smoothly transformed version of the associated dynamic RCS, i.e. $f_t(Z^{t,b})$, where f_t is a nonrigid transformation. The smoothness of f_t is measured similarly to [5].

Denote the edge points at frame t as $\{X_t : X^i_t; i = 1, 2,..., N_t\}$, where N_t is the number of the edge points in the current frame and X^i_t is the pixel coordinate of the ith edge point. X^i_t is represented by a column vector with a size of $d \times 1$ (d always equals 2). Analogous to [5], the edge observation X^i_t is viewed here as a random sample generated from a Gaussian mixture whose centers are the cross section of the adaptive piecewise fiber bundle. The density function of X^i_t is given by

$$P(X^i_t|U_t = b, m^i_t = j, f_t, Z^t) = N(X^i_t; f_t(Z^{t,b}_j), \Sigma^{j,b}_t), \tag{1}$$

where the hidden correspondence variable m^i_t denotes an index of the Gaussian which generates the ith edge point X^i_t and $N(X; \mu, \Sigma)$ denotes the normal density function. Similar to [5], for the sake of simplicity and ease of implementation,

$P(m_t^i = j | U_t = b) = 1/L$ and $\Sigma_t^{j,b} = (\sigma_t^{j,b})^2 I_d$ are assumed. The prior state transition of U_t is defined as

$$P(U_t = b | U_{t-1} = r) = a_{b,r} = \begin{cases} \eta & , b = r \\ (1-\eta), & b \neq r \end{cases}, 0.5 < \eta < 1, \qquad (2)$$

where η controls the consistency of partition. Inherited from [5], the temporal continuity of elastic motion is enforced by modeling the state transition in the tracking of each fiber. The motion of each fiber is defined under kinematics, i.e.

$$\left[f_t(Z_j^{t,b}) \quad \xi_t^j \right] = \left[f_{t-1}(Z_j^{t-1,b}) \quad \xi_{t-1}^j \right] A + \left[a_{t,p}^j \quad a_{t,\xi}^j \right] G, \qquad (3)$$

where ξ_t^j is fiber's velocity, $a_{t,p}^j$ and $a_{t,\xi}^j$ are fiber's normally distributed accelerations (with zero mean and standard deviation τ_t^b) in the equations of displacement and velocity, respectively. The definition of the matrixes A and G are the same as [5]. Equation (3) can be rewritten as

$$\Phi(Z_j^{t,b}, f_t, \xi_t^j, Z_j^{t-1,b}, f_{t-1}, \xi_{t-1}^j) = \prod_{l=1}^d N(S_t^{j,b}(l,:) - S_{t-1}^{j,b}(l,:)A; 0, (\tau_t^b)^2 \Psi), \qquad (4)$$

where the definition of Ψ is the same as [5], and an intermediate symbol $S_t^{j,b}$ is introduced for the sake of brevity, which is defined as $S_t^{j,b} \equiv \left[f_t(Z_j^{t,b}) \quad \xi_t^j \right]$. The symbol $S_t^{j,b}(l,:)$ used in (4) refers to the lth row of $S_t^{j,b}$.

The MTHMM-P is finally represented by a factor graph shown in Fig. 2, which involves all above considerations. Our objective is the joint distribution of U, Z, ξ, f, m, and X, given by the product of all local functions in Fig. 2. The joint distribution is

$$P(U, Z, \xi, f, m, X) \propto \prod_{t=1}^N \left(g_u(U_t, U_{t-1}) g_f(f_t, U_t) \right) \prod_{j=1}^L g_z \left(Z_j^t, Z_j^{t-1}, \ldots \right)$$

$$\prod_{t=1}^N \prod_{i=1}^{N_t} \left(g_m(m_t^i, U_t) g_x(X_t^i, U_t, m_t^i, f_t, Z^t) \right) \prod_{j=1}^L g_\xi(U_t, U_{t-1}, Z^t, f_t, \xi_t^j, Z^{t-1}, f_{t-1}, \xi_{t-1}^j). \qquad (5)$$

$$g_u(U_t | U_{t-1}) = P(U_t | U_{t-1}), \; g_m(m_t^i, U_t) = P(m_t^i | U_t), \qquad (6)$$

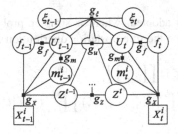

Fig. 2. MTHMM-P represented by a factor graph.

$$g_\xi(U_t, U_{t-1}, Z^t, f_t, \xi_t^j, Z^{t-1}, f_{t-1}, \xi_{t-1}^j) = \begin{cases} \Phi(Z_j^{t,b}, f_t, \xi_t^j, Z_j^{t-1,b}, f_{t-1}, \xi_{t-1}^j) , U_t = U_{t-1} = b \\ c_\xi , U_t \neq U_{t-1} \end{cases},$$

$$(7)$$

$$g_x(X_t^i, U_t, m_t^i, f_t, Z^t) = P(X_t^i | U_t, m_t^i, f_t, Z^t), \tag{8}$$

$$g_f(f_t, U_t) = \exp\left(-\lambda \|Lf_t\|^2 / 2\right), \tag{9}$$

where g_u and g_m are defined by the corresponding priors, g_ξ explains the piecewise temporal continuity, g_x is the likelihood measurement, and g_f is a smooth prior of the nonrigid transformation f_t. The $g_z\left(Z_j^t, Z_j^{t-1}, \dots\right)$ in (5) is the prior of the dynamic *RCS*, which is related to the mechanism discussed in Sect. 2. How to deal with this prior formally will be given in Sect. 4.2.

The problem of simultaneously achieving adaptive motion partition and piecewise elastic motion estimation is finally expressed as an inference under the MTHMM-P: given the observation X, jointly inferring the values of hidden random variables (RVs) U, Z, ξ, f, and m, which maximize the joint distribution given in (5), i.e.,

$$(\hat{U}, \hat{Z}, \hat{\xi}, \hat{f}, \hat{m}) = \arg\max_{U, Z, \xi, f, m} P(U, Z, \xi, f, m, X). \tag{10}$$

The joint inference procedure will be derived from structural EM algorithm and given in the next section.

4   Inference and Optimization

4.1   Inference Under Structural EM Algorithm

Using the notions in [15], denote the hidden RVs by h and the visible RVs by v, where they are defined here as

$$h = \left\{ \left(U_t, Z_j^{t,b}, \xi_t^j, f_t, m_t^i\right), j = 1, 2, \dots, L; b = 0, 1; t = 1, 2, \dots, N; i = 1, 2, \dots, N_t \right\},$$

$$(11)$$

$$v = \left\{ (X_t^i), t = 1, 2, \dots, N; i = 1, 2, \dots, N_t \right\}. \tag{12}$$

Structural EM algorithm is adopted, where a simpler probability distribution $Q(h)$, which is used to approximate the true posterior distribution $P(h|v)$, is defined as

$$Q(h) = \prod_{t=1}^N Q(U_t | U_{t-1}) \prod_{b=0}^1 \prod_{j=1}^L \delta(Z_j^{t,b} - \hat{Z}_j^{t,b})$$

$$\prod_{t=1}^N \left(\delta(f_t - \hat{f}_t | U_t) \prod_{j=1}^L \delta(\xi_t^j - \hat{\xi}_t^j | U_t) \prod_{i=1}^{N_t} Q(m_t^i | U_t) \right) = Q_u \cdot Q_{h/u}.$$

$$(13)$$

True posteriori distributions are maintained for the discrete RVs U and m, while point estimations are used for the continuous RVs Z, ξ, and f. The dependence on U considered in the structural Q-distributions in (13) does not increase too much computational cost but can lead to more exact inference algorithms than the ones without considering the dependence. For the sake of brevity, define

$$\delta(f_t - \hat{f}_t^b) \equiv \delta(f_t - \hat{f}_t | U_t = b), \qquad (14)$$

$$\delta(\xi_t^j - \hat{\xi}_t^{j,b}) \equiv \delta(\xi_t^j - \hat{\xi}_t^j | U_t = b). \qquad (15)$$

The $Q(h)$ in (13) is divided into two terms: Q_u and $Q_{h/u}$, where $Q_u = \prod_{t=1}^{N} Q(U_t | U_{t-1})$, and $Q_{h/u}$ is the remaining parts after removing Q_u from $Q(h)$. Following the way in [15], a free energy to measure the accuracy of $Q(h)$ is written as

$$F(Q, P, \beta_a, \beta_b) = \int_h Q(h) \ln\left(Q_{h/s}{}^{\beta_a} Q_s{}^{\beta_b} / P(h, v)\right), \qquad (16)$$

where β_a and β_b are two temperature parameters which will control the fuzziness of $Q(m_t^i | U_t)$ and $Q(U_t | U_{t-1})$ respectively. Minimizing $F(Q, P, \beta_a, \beta_b)$ w.r.t. $Q(m_t^i | U_t)$, we obtain its update equation

$$Q(m_t^i = j | U_t = b) \leftarrow q_t^{ijb} / \sum_{j=1}^{L} q_t^{ijb}, \qquad (17)$$

$$q_t^{ijb} = \exp(-\left|X_t^i - \hat{f}_t^b(\hat{Z}_j^{t,b})\right|^2 / \beta_t). \qquad (18)$$

In (18), β_a and $\sigma_t^{j,b}$ are merged into one compact temperature β_t to regularize the procedure of annealing.

The remaining optimization problems can be split into three sub-problems

$$SP_1 : \min_{\hat{U}} \sum_{t,b,j} Q(U_t = b) \sum_{i=1}^{N_t} Q(m_t^i = j | U_t = b) \left|X_t^i - \hat{S}_t^{j,b} B\right|^2 / (\sigma_t^{j,b})^2$$
$$+ Q(U_t = b, U_{t-1} = b) tr[(\hat{S}_t^{j,b} - \hat{S}_{t-1}^{j,b} A) \Psi^{-1} (\hat{S}_t^{j,b} - \hat{S}_{t-1}^{j,b} A)^T] / (\tau_t^b)^2, \qquad (19)$$

$$SP_2 : \min_{\hat{f}, \hat{Z}} \sum_{t,b} Q(U_t = b) (\sum_{j=1}^{L} \left|\hat{f}_t^b(\hat{Z}_j^{t,b}) - \hat{S}_t^{j,b} B\right|^2 + \lambda \left\|L\hat{f}_t^b\right\|^2) + E_z, \qquad (20)$$

$$SP_3 : \min_{Q(U_t | U_{t-1})} F(Q, P, \beta_a, \beta_b), \qquad (21)$$

where $\hat{S}_t^{j,b}$ is defined as $\hat{S}_t^{j,b} \equiv \left[\hat{f}_t^b(\hat{Z}_j^{t,b}) \quad \hat{\xi}_t^{j,b}\right]$, the B is the same as the one in [5], and E_z is related to g_z. SP_1 is a piecewise motion tracking problem (can be solved by the analytic Viterbi algorithm proposed in [5]), SP_2 is a competitive shape registration problem, and SP_3 is a motion partition problem. Solutions to SP_2 and SP_3 are presented in Sects. 4.2 and 4.3 respectively.

4.2 Competitive Shape Registration, SP$_2$

In (20), since $Q(U_t = b)$ and E_z are constant w.r.t. $\hat{f}_t^b$, the inference of $\hat{f}_t^b$ is same as the one in [5], and $\hat{f}_t^b$ will be a Thin-Plate Spline (TPS). The inference of $\hat{Z}^{t,b}$ is focused here, and the mechanism discussed in Sect. 2 will be exploited to break down the E_z to give related formal formulations.

Given $Q(U_t = b)$ at each frame, the strong state sequence can be obtained as $\left\{ b_{Strong}^t = \arg \max_b Q(U_t = b); t = 1, 2, \ldots, N \right\}$, which is actually the output of partition at current iteration. Suppose the number of pieces determined by the strong state sequence is K (e.g., K equals 4 for a sequence of "0001110011"). According to the sharing policy, all the Z_{Strong}^t inside one piece are forced to be the same one, i.e. $\{Z_{Strong}^t \equiv Z_{Strong}^k : t$ belongs to the kth piece, $k = 1, 2, \ldots, K\}$. Subsequently, (20) can be rewritten as the one that Z_{Strong}^k is the target variable to be solved:

$$SP_Z^{Strong} : \min_{Z_{Strong}^K} \sum_{t,k,j} I_t^k \bullet Q(U_t = b_k) \left| \hat{f}_t^{b_k} \left(Z_{Strong}^{k,j} \right) - \hat{S}_t^{j,b_k} B \right|^2, \tag{22}$$

where b_k is the strong state value of the kth piece, I_t^k is an indicator function that the tth frame will be used to update the kth Z_{Strong}^k when $I_t^k = 1$. I_t^k is defined based on the expanding policy. Similar to [5], under a linear approximation of $\hat{f}_t^b$ w.r.t. Z_{Strong}^k, the update equation of Z_{Strong}^k can be obtained by solving a least-squares problem.

For the Z_{Weak}^t, (20) can be rewritten as

$$SP_Z^{Weak} : \min_{\hat{z}_j^{t,b'_{Weak}}} \sum_{j=1}^L \left(\begin{array}{l} \Gamma_{Local}^t \bullet \left| \hat{f}_t^{b'_{Weak}} \left(\hat{Z}_j^{t,b'_{Weak}} \right) - \hat{S}_t^{j,b'_{Weak}} B \right|^2 \\ + \left(1 - \Gamma_{Local}^t \right) \left| \hat{Z}_j^{t,b'_{Weak}} - Z_{Strong}^{k_{Inherit}^t j} \right|^2 \end{array} \right), \tag{23}$$

where $k_{Inherit}^t$ is an index of the inherited strong RCS from the nearest neighboring pieces, and Γ_{Local}^t is a weight of local stimulating. Γ_{Local}^t is defined as

$$\Gamma_{Local}^t = \varsigma \cdot \varphi(|Q(U_t = 1) - Q(U_t = 0)|), \tag{24}$$

where ζ is a control parameter, and $\{\varphi(x); 0 \leq x \leq 1\}$ is a monotone decreasing function (e.g., $\varphi(x) = \sqrt{1 - x^2}$) to account for the fuzziness of $Q(U_t)$. The local stimulating will be more significant when $Q(U_t)$ is amphibolous, and thus more new pieces are allowed to be created at the earlier stage of the iteration while the creation is disabled near the convergence. The update equation of Z_{Weak}^t can also be obtained by solving a least-squares problem.

4.3 Motion Partition, SP_3

The problem given by (21) is a motion partition problem: inferring the optimal posteriori transition $Q(U_t|U_{t-1})$ to best fit the input motion sequence. Minimizing $F(Q, P, \beta_a, \beta_b)$ w.r.t. $Q(U_t|U_{t-1})$ and ignoring minor factors, we obtain

$$Q(U_t = b|U_{t-1} = r) \leftarrow \theta_t^{br} / \sum_{b=0}^{1} \theta_t^{br}, \tag{25}$$

$$\theta_t^{br} = \exp \left(\begin{array}{c} -\sum_{i,j} Q(m_t^i = j|U_t = b) \left| X_t^i - \hat{f}_t^b(\hat{Z}_j^{t,b}) \right|^2 / \left(2(\sigma_t^{j,b})^2 \right) \\ -\frac{\lambda}{2} \| L\hat{f}_t^b \|^2 + I_{t > 1} \ln a_{b,r} - I_{t < N} F_{t+1}'(U_t = b) \end{array} \right) / \beta_b. \tag{26}$$

The $F_t'(U_{t-1} = r)$ in (26) is defined recursively as

$$F_t'(U_{t-1} = r) = \sum_{b=0}^{1} Q(U_t = b|U_{t-1} = r)$$

$$\left[\begin{array}{c} \beta_b \ln Q(U_t = b|U_{t-1} = r) - \ln a_{b,r} + \frac{\lambda}{2} \| L\hat{f}_t^b \|^2 + F_{t+1}'(U_t = b) \\ + \sum_{i,j} Q(m_t^i = j|U_t = b) \left| X_t^i - \hat{f}_t^b(\hat{Z}_j^{t,b}) \right|^2 / \left(2(\sigma_t^{j,b})^2 \right) \end{array} \right]. \tag{27}$$

From (26), the criteria of motion partition are concluded as: (1) try to select the partition state b where the corresponding cross section $\hat{f}_t^b(\hat{Z}_j^{t,b})$ matches the edges as better as possible, (2) the related nonrigid transformation $\hat{f}_t^b$ should be as smooth as possible, (3) at the same time, the temporal consistency of partition should be satisfied. The form $Q(U_t|U_{t-1})$ indicates that the partition at current frame depends on the one at previous frame, and the $F_{t+1}'(U_t = b)$ in (26) implies that the partition at current frame is achieved by considering the dependence on it at succeeding frames. In fact, the procedure of calculating $Q(U_t|U_{t-1})$ and $Q(U_t)$ makes up a Viterbi algorithm [16] where

---

Initialize $Q(U_t|U_{t-1})$, $Q(U_{t-1}|U_t)$, $Q(U_t)$, $\hat{f}_t^b$, $\hat{Z}_j^{t,b}$ and $Q(m_t^i|U_t)$

Initialize β_a, β_b, λ and α (missing data handling [[5]])

Begin: Deterministic Annealing

    Determine the pieces according to $Q(U_t)$

    For each state value b (b=0, 1)

        Local Softassign: Update $Q(m_t^i|U_t)$

        Piecewise Motion Tracking: Solve SP_1

    Competitive Shape Registration: Solve SP_2

  Motion Partition: Solve SP_3

  Calculate $Q(U_t)$ and $Q(U_{t-1}|U_t)$ based on $Q(U_t|U_{t-1})$

  Decrease β_a, β_b, λ and α

End

Fig. 3. The inference algorithm of the MTHMM-P.

globally optimal partition results $Q(U_t)$ can be achieved. The handling of outlier and missing data is same as [5]. The inference algorithm is shown in Fig. 3.

5   Experiments

5.1   Results on Facial Motion

In data set PF, the subject is talking under a right-to-left head yaw movement. The results achieved by the MTHMM-P are presented in Fig. 4. As shown in Fig. 4a, the motion is partitioned into three pieces. If check the strong *RCS* of each piece (can be understood as the mean shape of each piece) shown in Fig. 4b, the three pieces actually correspond to the phases where head is moving near the right, middle, and left head pose respectively. Thus the obtained partition results are consistent with our intuition. Inside each piece, the estimated elastic motion can be represented either by the obtained nonrigid transformations, or by the trajectories of the fibers. Based on the achieved motion estimation results, a mean face image of each piece can be calculated by warping the related input images into the common coordinate of the strong *RCS*. The mean face will be blurred if the motion is misestimated. From Fig. 4c, we can observe that the mean faces are pretty clear. To directly validate the results of motion estimation, some facial landmarks in the mean faces are manually labeled. Subsequently

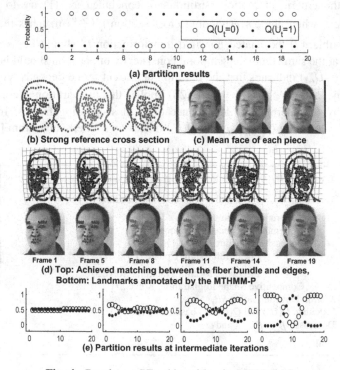

(a) Partition results

(b) Strong reference cross section   (c) Mean face of each piece

Frame 1   Frame 5   Frame 8   Frame 11   Frame 14   Frame 19
(d) Top: Achieved matching between the fiber bundle and edges,
Bottom: Landmarks annotated by the MTHMM-P

(e) Partition results at intermediate iterations

Fig. 4. Results on PF achieved by the MTHMM-P.

the landmarks in the related input images can be automatically annotated according to the estimated transformations. The bottom row of Fig. 4d shows the annotated landmarks where the motion estimation achieved by the MTHMM-P can be validated by checking the consistency of each landmark inside same piece. The top row of Fig. 4d shows the matching between the fiber bundle and edges at the same frames with the bottom row. In Fig. 4e, the partition results at intermediate iterations are shown, where the creation and mergence of pieces can be seen.

5.2   Results on Body Motion

Data set PB is from a ballet performance, where the performer's arms and legs are moving under a rotation of the body. From Fig. 5a – c, the motion is partitioned into four pieces: (1) the motion phase where the body is rotating and the right arm is visible, (2) the motion phase where the body keeps rotating but the right arm is occluded, (3) the motion phase where the body keeps rotating and the right arm is visible again, at the same time the right leg is moving down, (4) the motion phase where both arms are moving down and the right leg keeps moving toward the ground. It is interesting that the MTHMM-P is capable of estimating the nonrigid motion even when the performer wears skirt. The validations of the achieved piecewise elastic interpretation of the motion are given in Fig. 5d.

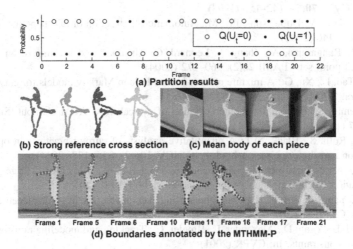

(a) Partition results

(b) Strong reference cross section   (c) Mean body of each piece

Frame 1   Frame 5   Frame 6   Frame 10   Frame 11   Frame 16   Frame 17   Frame 21
(d) Boundaries annotated by the MTHMM-P

Fig. 5. Results on PB achieved by the MTHMM-P.

6   Conclusion

Piecewise elastic motion is used as a unified representation for general nonrigid motion, so as to estimate the nonrigid motion without assumption about particular motion type and motion content. This paper proposes a novel probabilistic graphical model, called

MTHMM-P, for partitioning the general nonrigid motion into piecewise elastic motion. Motion partition and elastic motion estimation are jointly formulated under the MTHMM-P. By means of a competition-cooperation partition mechanism and joint inference algorithm, the MTHMM-P can automatically determine the number of pieces as well as adjust the span of each piece by adapting to the input motion sequence, and at the same time achieve the estimation of piecewise elastic motion. Experiments on the motion of face and body show the capability of the MTHMM-P.

Rich applications of the MTHMM-P can be considered in the future owing to its advantages: no requirement of *a priori* shape models, no heavy dependence on feature extraction, and the capability of handling general nonrigid motion. For instance, one can consider the potential applications such as automatically constructing multi-pose/multi-class shape/appearance model for face and human body (even with cloth), making use of the MTHMM-P as a capable bottom-up estimation approach to guide the top down approaches, or using the estimation results directly for recognition (e.g., recognition of expression and gait).

References

1. Huang, T.: Modeling, analysis, and visualization of nonrigid object motion. In: ICPR. (1990)
2. Aggarwal, J., Cai, Q., Liao, W., Sabata, B.: Nonrigid motion analysis: articulated and elastic motion. CVIU **70**(2), 142–156 (1998)
3. Duta, N., Jain, A., Dubuisson-Jolly, M.: Automatic construction of 2D shape models. PAMI **23**(5), 433–446 (2001)
4. Chui, H., Rangarajan, A., Zhang, J., Leonard, C.: Unsupervised learning of an atlas from unlabeled point-sets. PAMI **26**(2), 160–172 (2004)
5. Di, H., Tao, L., Xu, G.: A mixture of transformed hidden Markov models for elastic motion estimation. PAMI **31**(10), 1817–1830 (2009)
6. Beauchemin, S., Barron, J.: The computation of optical flow. ACM Comput. Surv. **27**(3), 433–467 (1995)
7. Sun, D., Roth, S., Black, M.J.: A quantitative analysis of current practices in optical flow estimation and the principles behind them. IJCV **106**(2), 115–137 (2014)
8. Revaud, J., Weinzaepfel, P., Harchaoui, Z., Schmid, C.: EpicFlow: Edge-preserving interpolation of correspondences for optical flow. In: CVPR (2015)
9. Bao, L., Yang, Q., Jin, H.: Fast edge-preserving patchmatch for large displacement optical flow. In: CVPR (2014)
10. Torresani, L., Yang, D., Alexander, E., Bregler, C.: Tracking and modeling non-rigid objects with rank constraints. In: CVPR (2001)
11. Wang, H., Schmid, C.: Action recognition with improved trajectories. In: ICCV (2013)
12. Wang, H., Klaser, A., Schmid, C., Liu, C.L.: Dense trajectories and motion boundary descriptors for action recognition. IJCV **103**(1), 60–79 (2013)
13. Sundaram, N., Brox, T., Keutzer, K.: Dense point trajectories by GPU-accelerated large displacement optical flow. In: Daniilidis, K., Maragos, P., Paragios, N. (eds.) ECCV 2010, Part I. LNCS, vol. 6311, pp. 438–451. Springer, Heidelberg (2010)
14. Rubinstein, M., Liu, C., Freeman, W.T.: Towards longer long-range motion trajectories. In: BMVC (2012)

15. Frey, B., Jojic, N.: A comparison of algorithms for inference and learning in probabilistic graphical models. PAMI **27**(9), 1392–1416 (2005)
16. Rabiner, L.: A tutorial on hidden Markov models and selected applications in speech recognition. Proc. IEEE **77**(2), 257–286 (1989)

A Steganography Algorithm Using Multidimensional Coefficient Matching for H.264/AVC

Liyun Qian[1], Zhitang Li[1,2(✉)], and Bing Feng[1]

[1] Department of Computer Science and Technology,
Huazhong University of Science and Technology, Wuhan, China
[2] Network and Computing Center,
Huazhong University of Science and Technology, Wuhan, China
lzt@hust.edu.cn

Abstract. Diamond encoding (DE) and adaptive pixel pair matching (APPM) are two data-hiding methods recently proposed. The APPM offers lower distortion than DE by providing more compact neighborhood sets and allowing embedded digits in any notational system. This paper proposes a new data-hiding method based on multidimensional coefficient matching (MCM). The proposed algorithm increases the dimension of coefficient, but reduces the magnitude of the coefficient modification. So it can provide better performance and Security than APPM. The proposed method also embeds more messages per modification and thus increases the embedding efficiency and the visual quality of the video.

Keywords: Diamond encoding (DE) · Adaptive pixel pair matching (APPM) · Multidimensional coefficient matching (MCM) · DCT coefficients

1 Introduction

In the Digital Age, the security requirements of information transmission have become increasingly higher. Traditional encryption technology directs to encrypt the information to make it incomprehensible. But incomprehensible cipher exposes encryption behavior itself. Information Hiding Technology conceals data into ordinary carrier such as public digital media for conveying secret messages confidentially, which is not only to protect the message, but also to protect the transmission behavior. H.264/AVC video coding standard rapidly gained popularity because of its excellent compression efficiency and outstanding network affinity. So the design and implementation of information hiding algorithm based on H.264/AVC standard video is very meaningful.

After embedding the secret information into the video, the video will be modified and distortion occurs [3, 4, 9]. The specific prediction mode of H.264/AVC video

Foundation Item: The National Natural Science Foundation of China (61272407).

coding standard will diffuse distortion to adjacent blocks and adjacent frames. So the key of information hiding algorithm is to modify the carrier as small as possible to evade visual and statistical detection on the basis of secret information being able to be extracted exactly [10]. This is just the advantage of the algorithm proposed in this paper in comparison to APPM [1].

So far, many scholars have done a lot of work in information hiding algorithm for H.264/AVC video [7, 8]. According to existing research, Modifying quantized DCT coefficients have better experimental results [17, 18]. On the one hand, because the quantization is a lossy process, choosing quantized coefficients can avoid the loss of embedded secret information due to quantify, on the other hand, we can effectively control distortion drift by choosing embed position reasonably [3, 4]. What is more, there are many existing steganographic method for Images by changing pixels can directly be used for video by changing quantized coefficients [11–14]. My team had already made many outstanding achievements in this area as early as 2009. In this paper, the following experiments will refer to the method of limited frame distortion drift of my team [5, 6].

Section 2 is a brief review of DE and APPM. The proposed method is given in Sect. 3, and the Performance analysis and comparison is presented in Sect. 4. The last part Sect. 5 is the conclusions.

2   Related Works

2.1   Diamond Encoding (DE)

In 2009, Chao et al. [2] proposed a diamond encoding (DE) method. DE Conceals a secret digit in a L -ary notational system into two pixels and it uses a parameter k to control the payload, where $L = 2 \times k^2 + 2k + 1$. To conceal a message digit s_l in a L-ary notational system into pixel pair(x, y). The neighborhood set $N(\tilde{x}, \tilde{y})$ is defined as following formula: $N(\tilde{x}, \tilde{y}) = \{(\tilde{x}, \tilde{y}) | |\tilde{x} - x| + |\tilde{y} - y| \leq k\}$. A diamond function is to calculate the DCV of (x, y), where $f(x, y) = ((2k + 1)x + y) \bmod L$. Then calculate the distance between the secret digit s_l and $f(x, y)$, where $d = |s_l - f(x, y)| \bmod L$. According to the value of d and template D_K which is also determined by parameter k, we can find a coordinate $(\tilde{x}, \tilde{y})$ which belongs to the set $N(\tilde{x}, \tilde{y})$ satisfying $f(\tilde{x}, \tilde{y}) = s_l$ and then (x, y) is replaced by $(\tilde{x}, \tilde{y})$. When extracts information, we only need to calculate the value of $f(\tilde{x}, \tilde{y})$.

Here is a simple example. Let $k = 2$ and two pixels $(x, y) = (1, 2)$, then $L = 2 * 2^2 + 2 * 2 + 1 = 13$, digits in a 13-ary notational system can be concealed. If we suppose that secret digit s_l to be embedded is 10_{13}, then $f(1, 2) = ((2 * 2 + 1) * 1 + 2) \bmod 13 = 7$, $d = (|10_{13} - 7|) \bmod 13 = 3$. According to the following template (diamond encoding template when k = 2), we can know $(\tilde{x}, \tilde{y}) = (x + x_d, y + y_d) = (x + x_3, y + y_3) = (x - 2, y + 0) = (-1, 2)$, so we replace $(1, 2)$ with $(-1, 2)$. It is easy to test and verify that $f(\tilde{x}, \tilde{y}) = f(-1, 2) = ((2 * 2 + 1) * (-1) + 2) \bmod 13 = 10 = s_l$ (Fig. 1).

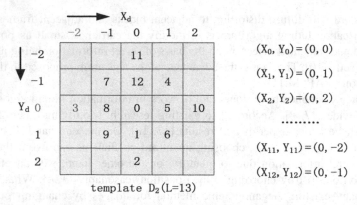

Fig. 1. Template of diamond encoding ($K = 2$)

2.2 Adaptive Pixel Pair Matching (APPM)

Literature [1] proposed a new embedding method APPM. APPM redefined $N(\tilde{x}, \tilde{y})$ as well as $f(x, y)$ to allow concealing digits in any notational system and provide smaller embedding distortion than DE, where $f(x, y) = (c_l * x + y) \bmod L$. Parameter c_l is determined by notational system of embedded secret digit. The templates of hexadecimal and duotricemary notation are as follows. The principles of embedding and extracting are same to DE (Figs. 2 and 3).

3   Multidimensional Coefficient Matching

AAPM greatly enhances the payload compared to DE. However, there is a big problem. When embedding a message digit in a 16-ary notational system or above, single coefficient modifier reach 3 at most. This is not conducive to safety. The proposed method in this paper considers security as a guiding principle for developing a less detectable embedding scheme by increasing the dimension of coefficient, but reducing the magnitude of the coefficient modification. When embedding a message digit in any notational system, we control every coefficient in coefficient set to modify at most one unit up or down.

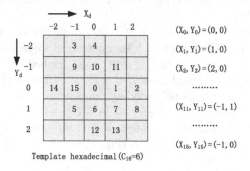

Fig. 2. The templates of hexadecimal

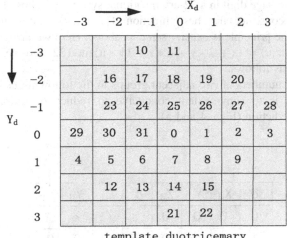

template duotricemary
notation (C_{32}=7)

Fig. 3. The templates of duotricemary

3.1 Extraction Function and Neighborhood Set

We chose two coefficients (x, y) to conceal a message digit in a 8-ary notational system by controlling every coefficient to modify at most one unit up or down. It is exactly the same with the APPM.

- To conceal a message digit in a 16-ary notational system, we chose three coefficients (x, y, z) to construct the best function $f(x, y, z)$ and neighborhood set $N(\tilde{x}, \tilde{y}, \tilde{z})$. Suppose a digit c_{16} is to be concealed. The range of c_{16} is between 0 and 15, and a coordinate $(\tilde{x}, \tilde{y}, \tilde{z})$ in $N(\tilde{x}, \tilde{y}, \tilde{z})$ has to be found such that $f(\tilde{x}, \tilde{y}, \tilde{z}) = c_{16}$. Therefore, the range of $f(x, y, z)$ must be integers between 0 and 15, because $d = |c_{16} - f(x, y, z)| \bmod 16$, so the range of d is also between 0 and 15, and each integer d must occur once and only once. In addition, $f(x_d, y_d, z_d) = d$. This is because, during embedding, (x, y, z) is replaced by one of the coordinates $(\tilde{x}, \tilde{y}, \tilde{z})$ in $N(\tilde{x}, \tilde{y}, \tilde{z})$ and it meet the condition $f(\tilde{x}, \tilde{y}, \tilde{z}) = f(x + x_d, y + y_d, z + z_d) = f(x,y,z) + f(x_d, y_d, z_d) = c_{16}$, so we can know $f(x_d, y_d, z_d) = c_{16} - f(x,y,z) - d$. We control every coefficient to modify at most one unit up or down, so $|x_d| \le 1$ and $|y_d| \le 1$ and $|z_d| \le 1$. Let $f(x, y, z) = (k_1 * x + k_2 * y + k_3 * z) \bmod 16$. So the best function $f(x, y, z)$ and neighborhood set $N(\tilde{x}, \tilde{y}, \tilde{z})$ must satisfy the following conditions:

1. $\min \sum_{d=0}^{15} |x_d| + |y_d| + |z_d|$
2. $1 \le k_1, k_2, k_3 \le 15; -1 \le x_d, y_d, z_d \le 1; k_1, k_2, k_3, x_d, y_d, z_d \in Z$
3. $f(x_d, y_d, z_d) = d; d \in \{0, 1..15\}$; Each d appears once and only once.
4. $f\left(x_{d_i}, y_{d_i}, z_{d_i}\right) \ne f\left(x_{d_j}, y_{d_j}, z_{d_j}\right)$ if $d_i \ne d_j$, for $0 \le d_i, d_j \le 15$

Enumerate and search by programming. Optimal solution: $f(x, y, z) = (x + 2 * y + 6 * z) \bmod 16$ and neighborhood set $N(\tilde{x}, \tilde{y}, \tilde{z})$ is listed in the following table.

- To conceal a message digit in a 32-ary notational system, we chose four coefficients (x, y, z, u) to construct the best function $f(x, y, z, u)$ and neighborhood set $N(\tilde{x}, \tilde{y}, \tilde{z}, \tilde{u})$. The principle is exactly same as above. Here we just give the optimal solution: $f(x,y,z,u) = (x + 2*y + 6*z + 11*u) \bmod 32$ and neighborhood set $N(\tilde{x}, \tilde{y}, \tilde{z}, \tilde{u})$ is as follows.
- As we can see, numbers of two adjacent groups in the following table are mutually opposite. This is to some extent conducive to reduce distortion drift due to Inter-Frame prediction (Figs. 4 and 5).

d	X_d	Y_d	Z_d	d	X_d	Y_d	Z_d
0	0	0	0				
1	1	0	0	15	−1	0	0
2	0	1	0	14	0	−1	0
3	1	1	0	13	−1	−1	0
4	0	−1	1	12	0	1	−1
5	−1	0	1	11	1	0	−1
6	0	0	1	10	0	0	−1
7	1	0	1	9	−1	0	−1
8	0	1	1	或	0	−1	−1

MCM **hexadecimal template**

Fig. 4. MCM hexadecimal template

d	x_d	y_d	z_d	u_d	d	x_d	y_d	z_d	u_d	d	x_d	y_d	z_d	u_d	d	x_d	y_d	z_d	u_d
0	0	0	0	0															
1	1	0	0	0	31	−1	0	0	0	9	0	−1	0	1	23	0	1	0	−1
2	0	1	0	0	30	0	−1	0	0	10	−1	0	0	1	22	1	0	0	−1
3	1	1	0	0	29	−1	−1	0	0	11	0	0	0	1	21	0	0	0	−1
4	0	−1	1	0	28	0	1	−1	0	12	1	0	0	1	20	−1	0	0	−1
5	−1	0	1	0	27	1	0	−1	0	13	0	1	0	1	19	0	−1	0	−1
6	0	0	1	0	26	0	0	−1	0	14	−1	0	−1	−1	18	1	0	1	1
7	1	0	1	0	25	−1	0	−1	0	15	0	0	−1	−1	17	0	0	1	1
8	0	1	1	0	24	0	−1	−1	0	16	1	0	−1	−1	或	−1	0	1	1

MCM **duotricemary notation template**

Fig. 5. MCM duotricemary notation template

3.2 Embedding Procedure

Here we take MCM with three coefficients to conceal a hexadecimal digit for example.

- Hierarchical B frame prediction structure with two viewpoints is shown as below. We choose p macroblock in p frame in viewpoint s_l [5] to embed information, because it will not be as a reference frame for inter-view prediction (Fig. 6).
- The pixel values of the 4×4 luma blocks may be used as the adjacent blocks' reference. It is only the last row and the last column as shown below. So we only embed information in Non-reference sub-MBs to prevent the inter distortion drift [5] (Fig. 7).
- Construct a random embedding sequence Q.
- Three quantized DCT coefficients (x,y,z) in non-reference sub-MBs which have shown above are selected to embed a digit c_{16}, and calculate the distance $d = (c_{16} - f(x,y,z)) \bmod 16$, then (x,y,z) is replaced by $(\tilde{x}, \tilde{y}, \tilde{z})$, where $(\tilde{x}, \tilde{y}, \tilde{z}) = (x + x_d, y + y_d, z + z_d)$.
- Repeat step d until all the digits are embedded.

3.3 Extraction Procedure

- Get the values of three quantized DCT coefficients (x,y,z).
- Calculate $f(x, y, z)$, the result is the embedded digit.
- Repeat Steps a and b until all the digits are extracted.

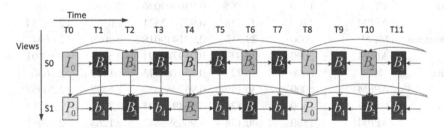

Fig. 6. Hierarchical B frame prediction structure with two viewpoints

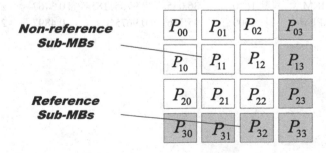

Fig. 7. Reference sub-MBs and non-reference sub-MBs

4 Performance Analysis and Comparison

In experiments, for the same video, we compared the quality of the video by embedding the same number of bits. The quality of video is measured by a variety of factors such as PSNR and so on. The algorithm Choose MCM with three coefficients to conceal a hexadecimal digit and APPM with $c_l = 6$.

4.1 Experiment Results Comparison of MCM with Three Coefficients to Conceal a Hexadecimal Digit and APPM with $C_l = 6$

See Table 1 and Figs. 8 and 9.

Table 1. Performance analysis and comparison:

Video sequence	Algorithm	Capacity (bit/P frame)	PSNR	SSIM	Bit rate increase	Embedding efficiency
Akko & Kayo	MCM	1500	39.459	0.968140006	0.5238	2.64768
	APPM	1500	39.167	0.966420293	0.7461	2.19437
Ballroom	MCM	4000	35.874	0.934869885	0.702	2.66347
	APPM	4000	35.195	0.928102076	1.0044	2.19676
Crowd	MCM	10000	34.34	0.959443331	0.8097	2.66738
	APPM	10000	33.415	0.951664448	1.1681	2.20431
Exit	MCM	1400	37.765	0.930469036	0.7195	2.65369
	APPM	1400	37.336	0.927463531	1.033	2.18784
Flamenco	MCM	2000	39.318	0.971449018	0.4009	2.6694
	APPM	2000	38.977	0.969740391	0.5761	2.19901
Race	MCM	6000	36.72	0.956786573	1.1419	2.6656
	APPM	6000	36.179	0.95296365	1.6447	2.20057
Rena	MCM	1500	40.529	0.968925595	0.794	2.65189
	APPM	1500	40.178	0.966596842	1.1264	2.19085
Vassar	MCM	2000	35.782	0.90674293	0.8781	2.65793
	APPM	2000	35.211	0.899612486	1.2564	2.19146
Objects	MCM	1600	36.045	0.970511854	0.3467	2.65281
	APPM	1600	35.136	0.967510462	0.4892	2.1978

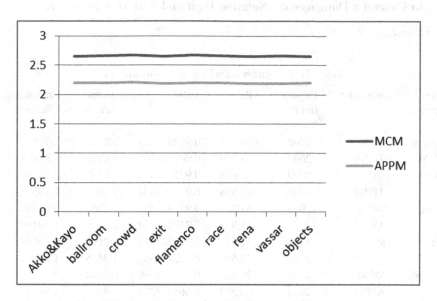

Fig. 8. The line chart of embedding efficiency

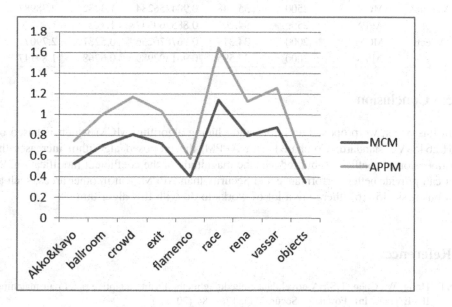

Fig. 9. The line chart of bit rate increase

4.2 Experiment Results Comparison of MCM with Four Coefficients to Conceal a Duotricenary Notation Digit and AAPM with $c_l = 7$.

See Table 2.

Table 2. Performance analysis and comparison:

Video sequence	Algorithm	Capacity (bit/P frame)	PSNR	SSIM	Bit rate increase	Embedding efficiency
Akko & Kayo	MCM	2000	39.355	0.967574418	0.8382	2.79512
	APPM	2000	38.509	0.96275878	1.0788	1.89801
Ballroom	MCM	5000	35.608	0.932761073	1.078	2.80767
	APPM	5000	33.984	0.914568424	1.3866	1.90525
Crowd	MCM	12000	33.932	0.957386434	1.205	2.81358
	APPM	12000	31.935	0.937911332	1.5679	1.90259
Exit	MCM	1700	37.57	0.929373622	1.0649	2.80158
	APPM	1700	36.586	0.922239661	1.3655	1.89987
Flamenco	MCM	2600	39.187	0.970811069	0.6464	2.79851
	APPM	2600	38.303	0.965865374	0.8268	1.90798
Race	MCM	7500	36.449	0.95560509	1.7629	2.80647
	APPM	7500	35.212	0.945377469	2.2607	1.90359
Rena	MCM	1800	40.442	0.968250036	1.1263	2.79315
	APPM	1800	39.425	0.961310625	1.456	1.90215
Vassar	MCM	2500	35.548	0.904256284	1.3482	2.78987
	APPM	2500	34.202	0.885269642	1.722	1.90124
Objects	MCM	2000	34.31	0.967876256	0.5287	2.7907
	APPM	2000	33.595	0.961692989	0.6768	1.89747

5 Conclusion

In this paper, we proposed an information hiding algorithm MCM which is based on H.264/AVC standard. Compared with APPM, the proposed algorithm increases the dimension of coefficient, but reduces the magnitude of the coefficient modification. So it can provide better performance and Security than APPM. But in other areas, such as robustness [15, 16], there are a lot of works to do with this algorithm.

References

1. Hong, W., Chen, T.S.: A novel data embedding method using adaptive pixel pair matching. IEEE Trans. Inf. Forensics Secur. 7(1), 176–184 (2012)
2. Chao, R.M., Wu, H.C., Lee, C.C., Chu, Y.P: A novel image data hiding scheme with diamond encoding. EURASIP J. Inf. Security 2009, doi: 10.1155/2009/658047, Article ID 658047 (2009)

3. MA, X.J., Li, Z.T., Tu, H., Zhang, B.C.: A data hiding algorithm for H.264/AVC video streams without intra-frame distortion drift. IEEE Trans. Circ. Syst. Video Technol. **20**(4), 1320–1330 (2010). (A SCI-indexed journals, signed by: Huazhong University of Science and Technology)

4. Ma, X.J., Li, Z.T., Lv, J., Wang, W.D.: Data hiding in H.264/AVC streams with limited intra-frame distortion drift. (EI included,signed by: Huazhong University of Science and Technology)

5. Song, G.H., Li, Z.T., Zhao, J., Tu, H., Cheng, J.X.: A video steganography algorithm for MVC without distortion drift. In: 4th International Conference on Audio, Language and Image Processing (ICALIP). (EI included,signed by: Huazhong University of Science and Technology)

6. Song, G.H., Li, Z.T., Zhao, J., Zou, M.G.: A data hiding algorithm for 3D videos based on inter-MBs. In: 2014 Tenth International Conference on Intelligent Computing (ICIC2014). (EI included,signed by: Huazhong University of Science and Technology)

7. Tew, Y., Wong, K.: An overview of information hiding in H.264/AVC compressed video. IEEE Trans. Circ. Syst. Video Technol. **24**, 305–319 (2014)

8. Kim, S.-M., Kim, S.B., Hong, Y., Won, C.S.: Data hiding on H.264/AVC compressed video. In: Kamel, M.S., Campilho, A. (eds.) ICIAR 2007. LNCS, vol. 4633, pp. 698–707. Springer, Heidelberg (2007)

9. Cheddad, A., Condell, J., Curran, K., McKevitt, P.: Digital image steganography: Survey and analysis of current methods. Signal Process. **90**, 727–752 (2010)

10. Filler, T., Judas, J., Fridrich, J.: Minimizing embedding impact in steganography using trellis-coded quantization. In: Proceedings SPIE Media Forensics and Security, vol. 7541, doi: 10.1117/12.838002 (2010)

11. Lee, C.F., Chang, C.C., Wang, K.H.: An improvement of EMD embedding method for large payloads by pixel segmentation strategy. Image Vis. Comput. **26**(12), 1670–1676 (2008)

12. Wu, D.C., Tsai, W.H.: A steganographic method for images by pixel value differencing. Pattern Recogn. Lett. **24**(9), 1613–1626 (2003)

13. Wang, C.M.: A high quality steganographic method with pixel value differencing and modulus function. J. Syst. Softw. **81**(1), 150–158 (2008)

14. Hong, W., Chen, T.S., Luo, C.W.: Data embedding using pixel value differencing and diamond encoding with multiple base notational system. J. Syst. Softw. **85**(5), 1166–1175 (2012)

15. Miller, M.L., Doërr, G.J., Cox, I.J.: Applying informed coding and embedding to design a robust high-capacity watermark. IEEE Trans. Image Process. **13**(6), 792–807 (2004)

16. Chen, C., Ni, J., Huang, J.: Temporal statistic based video watermarking scheme robust against geometric attacks and frame dropping. In: Ho, A.T., Shi, Y.Q., Kim, H.J., Barni, M. (eds.) IWDW 2009. LNCS, vol. 5703, pp. 81–95. Springer, Heidelberg (2009)

17. Lin, C.C., Shiu, P.F.: DCT-based reversible data hiding scheme. J. Softw. **5**(2), 214–224 (2010)

18. Lin, Y.K.: A data hiding scheme based upon DCT coefficient modification. Comput. Stand. Interfaces **36**, 855–858 (2014)

A P300 Clustering of Mild Cognitive Impairment Patients Stimulated in an Immersive Virtual Reality Scenario

Vitoantonio Bevilacqua[1(✉)], Antonio Brunetti[1], Davide de Biase[1],
Giacomo Tattoli[1], Rosario Santoro[2], Gianpaolo Francesco Trotta[1],
Fabio Cassano[1], Michele Pantaleo[2], Giuseppe Mastronardi[1],
Fabio Ivona[3], Marianna Delussi[4], Anna Montemurno[4],
Katia Ricci[4], and Marina de Tommaso[4]

[1] Dipartimento di Ingegneria Elettrica e dell'Informazione, Politecnico di Bari,
Bari, Italy
vitoantonio.bevilacqua@poliba.it
[2] AMT Services Srl, Bari, Italy
[3] Trait d'Union Srl, Bari, Italy
[4] Dipartimento di Scienze Mediche di base, Neuroscienze ed organi di senso,
Università di Bari, Bari, Italy

Abstract. In this paper, we present an innovative framework useful for clustering patients affected by a mild cognitive impairment and designed to improve the living environment and the lifestyle of patients in order to delay their cognitive state decline. The cognitive state changes are evaluated by means the event - related potentials elicited by environmental stimuli administered in several Virtual Reality scenarios. In particular, we formerly describe our innovative Virtual Reality environment, the protocol of stimuli administration, the procedure to measure the P300 latency in the response signal and finally the Self Organizing Map used to cluster the data. This research finds application in the fields of re-qualification of the environments for patients and healthcare introducing a new method for evaluation of best living conditions through VR.

Keywords: Virtual reality · Human computer/machine interaction · Cognitive decline · Event related potentials · P300 · Clustering · Self-organizing maps neural networks

1 Introduction

The cognitive decline problem, which increases with age, needs a constant monitoring especially in old patients. Overall, the symptoms can be associated for example, to forgetfulness and to a progressive loss of problem solving skills.

The progressive loss of cognitive status may be an indicator of diseases, like Alzheimer. Cognitive decline does not affect all individuals equally, but it depends on lifestyle, cognitive training, nutritional style and genetic inheritance.

The requalification of domestic and healthcare environments is an important methodology to allow a fast and economic solution to improve the habitability of the

© Springer International Publishing Switzerland 2015
D.-S. Huang et al. (Eds.): ICIC 2015, Part II, LNCS 9226, pp. 226–236, 2015.
DOI: 10.1007/978-3-319-22186-1_23

environment according to personal needs. Before structural changes are made, it is necessary to analyze the user cognitive status, in base of some environmental parameters that can affect user status, e.g. lights color and intensity.

With electroencephalography (EEG) it is possible to evaluate and estimate cognitive state, as described in [1], in relationship to the brainwaves. As in [2] a P300 and Event-Related Potentials (ERP) protocol are used to evaluate the cognitive status of users. In the first case, four tasks are submitted to the user divided between motory activity and mental tasks, while in the second case the external stimuli are proposed in the form of creative mental process.

During the presentation of a stimulus, there is a significant increase in synaptic activity in a synchronized way. The electrical responses of neurons can be recorded by an EEG amplifier. The potentials are called Evoked Potential (EP) [3] or ERP that consist of a series of positive and negative waves that can be named numerically or according to their latency. The third positive wave of the ERP is named P3 or P300. Potentials evoked by external visual stimuli (exogenous) are called Visual Evoked Potential (VEP).

The main analysis are focused on P300 VEP elicited by a Virtual Environment (VE). The P300 is a positive deflection in the human EEG [4], appearing approximately 300–500 ms after the presentation of rare task-relevant stimuli. These stimuli can be visual, auditory or tactile. An ERP can be correctly evoked using a VE, in fact a user can watch a virtual scenario instead of an image sequence, according to experiment stimulation protocol, to elicit a P300 with frequent and rare stimuli.

An example of using a Virtual Reality as tool for assessing the cognitive status of a patient is presented in [5]; in this experiment, Okahashi et al. use a VE to reproduce a Virtual Shopping Test (VST) to evaluate the cognitive status of persons with dysfunctions caused by brain injury. Moreover, in [6] a quantitative analysis for assessing driver's cognitive responses by investigating the neurobiological information provided by EEG in traffic - light experiments in a VR dynamic driving environment is presented. In [7], it is used a scenario designed in a Virtual Reality to evaluate the correlation between the age of a person with the ability of way finding, while in [8] Visual Stimuli in a VE are used to develop a Brain Computer Interface to navigate in a Virtual Environment.

The aim of this work is to reach new methodologies for realizing living environments dedicated to users with residual capabilities in the initial stage of autonomy loss. The analysis provides an objective evaluation of the cognitive deficit, which is used to configure a living environment in base of the previous cognitive evolution - state. The target patients of the study are evaluated firstly by Mini-Mental State Evaluation (MMSE) [9] to have a relative score ranging between 18 and 30.

The virtual prototyping of life environments is greatly simplified through VR that is relatively easy to change and modify without structural changes and costs. With a domotic system, installed in the environment, it is possible to change the lights color and their intensity simply modifying these parameters on a computer interface. In this way, the best environment parameters can be selected to improve the users' cognitive state.

In this preliminary study, a new methodology is proposed, combining VE, EEG, cognitive analysis and re-qualification of living environment, to improve life of patients with cognitive deficit. In order to evaluate the main P300 parameters, this system is based on Oddball paradigm.

2   Workflow Description

The project workflow is constituted by four stages, shown in Fig. 1. In the first stage, a VE is created from real environment adding the stimulation system to perform the cognitive experiment. The next phase is to analyze the patient data and evaluate his cognitive state. Afterward a clustering stage is performed, in order to identify relationships between cognitive status and VE changes. The final stage is to adapt environment light parameters, evaluated from the VE experiment, using domotic technologies, to improve the life experience in living environment.

2.1   Making Virtual Environment

The re-qualification starts with acquisition of the real parameters of the environment such as dimension of the rooms and their shape. The starting real environment is a laboratory of Politecnico of Bari. The rooms, after they have been molded in VR, are furnished as a living environment.

The steps needed to create a virtual model are:

- Through a high precision laser (Leica 3D Disto) were acquired measurements of the real environment;
- The points acquired by the laser meter have been exported to a CAD file which has been processed using Autodesk Inventor from which was created a rough 3D model of the environment;
- The rough model was enhanced with textures and models of furniture, so the model was made look like a home environment containing two corridors, three rooms (living room, bedroom and a kitchen) and a bathroom (Fig. 2).

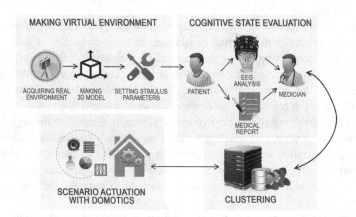

Fig. 1. Project workflow

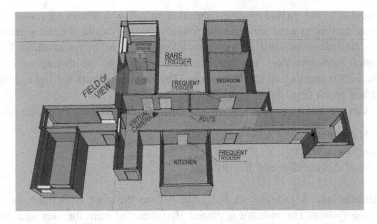

Fig. 2. Plan layout

From the basic model (Fig. 3A), other two different models were created, where each one contains the bathroom in a different position. In Fig. 3B (also shown in Fig. 2) the bathroom position is in the living room, while in Fig. 3C the bathroom is in the bedroom.

Starting from each object, three different scenes were created, each containing one of the three models mentioned above.

2.2   Cognitive State Evaluation

The proposed system aims to understand which are the best living conditions in home environments for subjects with residual capabilities; this is done through the navigation in a VE and the response evaluation of a subject visually stimulated, according to the oddball paradigm.

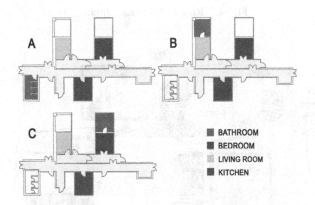

Fig. 3. The three scenarios developed (Color figure online).

In particular, in the virtual environment, that simulates a home, all rooms that are different from bathroom were identified as frequent stimuli, while the rare stimulus was assigned to the bathroom itself. In order to differentiate the target stimulus from the frequent ones, bathroom door has been lit with a spotlight, so this room is clearly identifiable with respect to all the other ones. The default colours for light stimuli are white, red and green (Fig. 4).

As we can see in Fig. 5, stimulation and data acquisition system is composed by:

- Visual Stimulation;
- Operator Monitoring;
- EEG Signal Acquisition and Recording.

Visual Stimulation. A monitor showing the VE to the user constitute the stimuli administration system. The monitor is placed 50–60 cm far from the user and the room where the experiment is executed hasn't a controlled brightness. Medical technicians indicate to the patient to reduce to the bare minimum their movements, in order to limit noise effects on the EEG system during recording.

During the navigation, user sees VE with a first person view, without see body parts (Fig. 5).

Fig. 4.   Rare and frequent stimuli

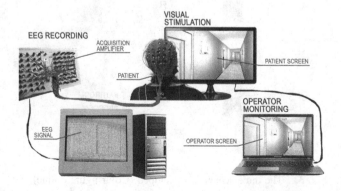

Fig. 5.   Stimulation, data acquisition and operator display

The reference protocol is the Oddball Paradigm that is used within event related potential research, where presentations of sequences of repetitive audio/visual stimuli are infrequently interrupted by a deviant (rare) stimulus. To improve attention during the experiment, a motion task is required from user. In detail, a button has to be pressed when a rare stimulus appears in the field of view of the subject. In this way, ERP response is better in terms of amplitude and latency and is also possible to evaluate the reaction time.

In the virtual environment, frequent stimuli are all the opened door encountered during the path. It is assumed that the patient ignores the closed doors present in virtual scene, but rooms whose doors are open could stimulate attention because they could contain the bathroom inside (Fig. 4).

The choice to associate the rare stimulus to the bathroom is justified by its presence in all houses and health facilities, where a subject with a Mild Cognitive Impairment (MCI) could have difficulties in finding that room.

To make the experiment easier, navigation inside the environment is automatically controlled and with a fixed duration; in this way rare and frequent stimuli were proposed in succession and in a controlled number. In particular, stimuli were presented in blocks containing stimuli sequences constituted by one rare stimulus and up to 4 frequent ones in an 80–20 frequent-rare ratio respectively. Each block corresponds to a random path between bathroom position in VE and another place of the house; all these positions are different in the three environments (Fig. 3) to prevent the effects linked to the expectation potential that user could have in case he remember the bathroom position [10].

At the beginning of the experiment, the first scene is selected; thanks to the anchors inserted in VE the system compute random virtual paths that, alternatively, start from bathroom, end up to another place of the house and vice versa, in order to have a rare stimulus followed by a certain number of frequent ones, as explained before. Virtual paths were generated in order to link 42 destinations: 21 are located in front of bathroom door and the others 21 are located in 7 different other places of the house, equally divided.

During the experiment, a virtual agent, equipped with a virtual camera that reproduces its visualization on user display, follows the route. During the automated navigation in VE, whenever an open door enters the field of view of the virtual camera (Fig. 2) a trigger was launched to the EEG recording system according to the door type; in particular, bathroom door throws rare trigger while doors of the remaining three rooms (living room, bedroom, kitchen) launch frequent triggers. During each stimuli sequence block, bathroom spotlight may randomly assume one of the three colours (white, red and green) for 7 times respectively.

Operator Monitoring. In order to allow the health technicians to supervise the experiment evolution, an operator mode of the application was developed. In this modality, operator sees the same virtual environment visualization of patient and other information related to acquisition protocol.

A second monitor, used exclusively by the operator (Fig. 5), is necessary. In operator display, trigger information shows which type of stimulus is visible in the field of view and accordingly which trigger code is sent to EEG recording system. Other

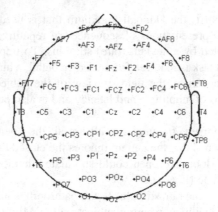

Fig. 6. EEG Scalp map

information provided to the operator are the number of stimuli sequence block, from 1 to 21, and the next destination, e.g. "bathroom" and "other place".

EEG Channels Setup and Processing. EEG was recorded by 63 scalp electrodes, according to enlarged 10–20 system as shown in Fig. 6, impedance below 5000 ohms, referring to the nasion with the ground at Fpz. Two electrodes were placed above the right and left eye to record the EOG. Signals were amplified, filtered (0.5 to 80 Hz, at a sampling rate of 256 Hz), and stored on a biopotential analyzer [11].

Data acquired during the EEG recording phase were processed with EEGLAB. The first step consists of data filtering; in particular it were applied a band pass filter from 0.5 Hz to 30 Hz; in order to remove frequencies produced by power supply, is applied a Notch filter at 50 Hz. In addition, it were applied some techniques based on ICA (Independent Component Analysis) in order to reduce the influence of ocular artefact on the EEG signal. In the next phase, starting from the all channels of EEG record, it were selected only the most important channels for data elaboration, that are Fz, Cz and Pz. The third phase is the epoch creation: from the entire signal, it were extracted windows of samples from 300 ms before trigger to 800 ms after trigger; this operation was repeated for the four types of triggers (three triggers for rare stimuli and one trigger for the frequent one). Moreover, in each window, an operation of baseline removal was made. Next, for each type of trigger was made an operation of averaging between the windows extracted before.

2.3 Methods and Data of Clustering

At the end of experiment, after the EEG processing phase, from each subject a value of P300 latency is extracted for each colour. In particular, it has been created a table where each record is formed by the following attributes:

- Sex: {Male, Female}
- Age: {≥18}
- Schooling Age: {[0–4], [5–7], [8–12], [13–17] years}

- Mini Mental Test: {[18–30]}
- Color: {White, Red, Green}
- P300 Latency: {[250–700] ms}

The first four attributes are obtained by interviewing the patient, while the remaining two attributes are obtained after processing the EEG signals. All these records are input for a Self-Organizing Maps (SOM) [12] Neural Networks that search for the number of different classes by means the positions of their centroids.

A self-organizing map (SOM) is a type of artificial neural network that is trained using unsupervised learning to produce a low-dimensional, discretized representation of the input space of the training samples, called a map. Self-organizing maps are different from other artificial neural networks in the sense that they use a neighborhood function to preserve the topological properties of the input space. This makes SOMs useful for visualizing low-dimensional views of high-dimensional data, akin to multidimensional scaling. Like most artificial neural networks, SOMs operate in two modes: training and mapping. "Training" builds the map using input examples (a competitive process, also called vector quantization), while "mapping" automatically classifies a new input vector. A self-organizing map consists of components called nodes or neurons; a weight vector with the same dimension of the input data vectors, and a position in the map space are associated to each node. The usual arrangement of nodes is a two-dimensional regular spacing in a hexagonal or rectangular grid. The self-organizing map describes a mapping from a higher-dimensional input space to a lower-dimensional map space. The procedure for placing a vector from data space onto the map is to find the node with the closest (smallest distance metric) weight vector to the data space vector.

3   Experimental Results

For this study 48 persons, tested after giving their consensus, were recruited: 18 subjects aged between 20 and 39 years, and 30 subjects older than 50 years.

For each subject, after an interview with the doctor, was registered the EEG signal related to the 21 scenarios previously presented.

At the end of the experiment, the acquired signal was processed with the methods described in the previous section and the P300 signal was extracted. Figures 7 and 8

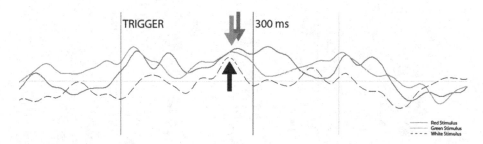

Fig. 7. Latency of P300 for each colour in a subject aged between 20 and 39 years (Color figure online).

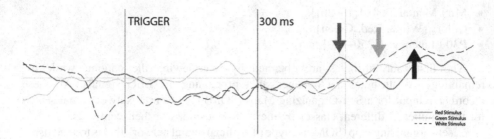

Fig. 8. Latency of P300 for each colour in a subject aged from 50 years.

show the P300 signal related to each color administered in the experiment for a subject in the first age range and for a subject in the second range respectively.

The P300 latencies, that were extracted from signals, and all the records in the table described in Sect. 2.3, were provided as input to the self - organizing neural network. The most important result was the relationship between the age of the subject and the latency of P300 linked to a specific color. As shown in the figures below, for those subjects aged between 20 and 39 years there is no difference between the different colours of the stimulation; in fact, in all three cases the clusters are positioned in the lower left. For the second group of persons, the ones older than 50 years, things are different: as we can see from the figures below, in cases where the color of stimulation is white (Fig. 9A) or red (Fig. 9B), these subjects respond in the same way, that is a

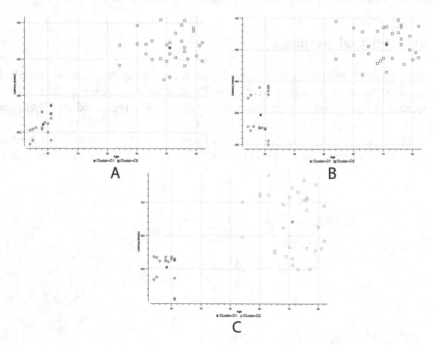

Fig. 9. Cluster with red stimulation (A), white stimulation (B) and green stimulation (C) (Color figure online)

latency of P300 very far from the average latency of P300 relative to the first group. Moreover, when the stimulation color is green, some subjects of these group have a response time comparable with that of the subjects of the first group (Fig. 9C).

4 Conclusions

Final results aim to demonstrate that the latency of P300 signal elicited by oddball stimuli is strictly dependent from their colours and correlates only with the age.

In particular, the classes obtained clustering the data are clearly distinct among young and older subjects, in fact, the resulting classes have a clear separation between them with the respective centroids mutually far.

Moreover, the results presented before show that the best color, among white, red and green, to be chosen for painting the bathroom door in order to retrain a living environment is the latter able to elicit fastly a P300 signal. The virtual reality environment and the relative coloured stimuli provided by our project seem to be feasible, sustainable and then useful to implement an innovative framework to learn how it is possible to reach a personalized design of living environment in general. In particular, we suppose that patients with different grading of cognitive impairment could take advantage to be hosted in houses or hospital rooms where the colour of doors or other furniture of vital interest, could be changed personally by using commercial equipment provided by a domotic controller. Furthermore, the visual perception and the movements of impaired patients can be modeled [13, 14] to implement in virtual reality more realistic interactions with domestic objects and personalized environments.

Acknowledgments. This work is partially funded by the Italian POR 2007-2013 Apulian ICT Living Labs Project called RESCAP.

References

1. Lan, T., Adami, A., Erdogmus, D., Pavel, M.: Estimating cognitive state using EEG signals. J. Mach. Learn. **4**, 1261–1269 (2003)
2. Srinivasan, N.: Cognitive neuroscience of creativity: EEG based approaches. Methods **42**(1), 109–116 (2007)
3. Veiga, H., Deslandes, A., Cagy, M., McDowell, K., Pompeu, F., Piedade, R., Ribeiro, P.: Visual event-related potencial (P300): a normative study. Arq. Neuropsiquiatr. **62**(3A), 575–581 (2004)
4. Hoffmann, U., Vesin, J.M., Ebrahimi, T., Diserens, K.: An efficient P300-based brain–computer interface for disabled subjects. J. Neurosci. Methods **167**(1), 115–125 (2008)
5. Okahashi, S., Seki, K., Nagano, A., Luo, Z., Kojima, M., Futaki, T.: A virtual shopping test for realistic assessment of cognitive function. J. Neuroeng. Rehabil. **10**(1), 59 (2013)
6. Lin, C.T., Chung, I.F., Ko, L.W., Chen, Y.C., Liang, S.F., Duann, J.R.: EEG-based assessment of driver cognitive responses in a dynamic virtual-reality driving environment. Trans. Biomed. Eng. IEEE **54**(7), 1349–1352 (2007)

7. Morganti, F., Riva, G.: Virtual reality as allocentric/egocentric technology for the assessment of cognitive decline in the elderly. Stud. Health Technol. Inform. **196**, 278–284 (2014)
8. Bevilacqua, V., Tattoli, G., Buongiorno, D., Loconsole, C., Leonardis, D., Barsotti, M., Frisoli, A., Bergamasco, M.: A novel BCI-SSVEP based approach for control of walking in virtual environment using a convolutional neural network. In: International Joint Conference on Neural Networks (IJCNN), pp. 4121–4128. IEEE (2014)
9. Folstein, M.F., Folstein, S.E., McHugh, P.R.: "Mini-mental state": a practical method for grading the cognitive state of patients for the clinician. J. Psychiatr. Res. **12**(3), 189–198 (1975)
10. Wang, G., Jin, J., Wang, X.: Target to efficient stimulus presentation sequence for P300 BCI. J. Cent. S. Univ. Sci. Technol. **42**, 746 (2011)
11. Micromed SystemPlus; Micromed, Mogliano Veneto, Italy. http://www.micromedit.com/
12. Bevilacqua, V., Mastronardi, G., Marinelli, M.: A neural network approach to medical image segmentation and three-dimensional reconstruction. In: Huang, D.-S., Li, K., Irwin, G.W. (eds.) ICIC 2006. LNCS, vol. 4113, pp. 22–31. Springer, Heidelberg (2006)
13. Debernardis, S., Fiorentino, M., Gattullo, M., Monno, G., Uva, A.E.: Text readability in head-worn displays: Color and style optimization in video versus optical see-through devices. IEEE Trans. Visual Comput. Graph. **20**(1), 125–139 (2014)
14. Loconsole, C., Stroppa, F., Bevilacqua, V., Frisoli, A.: A robust real-time 3D tracking approach for assisted object grasping. In: Auvray, M., Duriez, C. (eds.) EuroHaptics 2014, Part I. LNCS, vol. 8618, pp. 400–408. Springer, Heidelberg (2014)

Correcting and Standardizing Crude Drug Names in Traditional Medicine Formulae by Ensemble of String Matching Techniques

Duangkamol Pakdeesattayapong and Verayuth Lertnattee[✉]

Faculty of Pharmacy, Silpakorn University,
Muang 73000, Nakhon Pathom, Thailand
{g56363202,lertnattee_v}@su.ac.th
verayuths@hotmail.com

Abstract. Common problems of representing crude drug names in traditional herbal formulae are spelling errors, grammatical variants, synonyms and various formats. In order to make these names more obvious and useful, correcting and standardizing of these names should be applied. In this work, crude drug names in various forms were corrected and standardized by string matching techniques. A set of experiments were done using crude drug names from a database of registered traditional medicines in Thai Food and Drug Administration as the test set. Two well-known algorithms, i.e., similar text and Levenshtein were investigated. However, the results from each algorithm indicated that crude drug names in the test set were moderately matched with those of the standard set. To increase performance of these single algorithms, the ensemble algorithm was proposed. From the results, the ensemble algorithm outperforms single algorithms to match crude drug names, especially crude drug names with the modifier that have no significant meaning.

Keywords: String matching technique · Levenshtein · Similar text · Crude drug name · Ensemble algorithm

1 Introduction

Nowadays, herbs are widely used in cooking and processed into various products. Most of herbs have been blended into health products, for example health supplements, cosmetics and pharmaceutical products. Both modern medicine and traditional medicine use herbs to treat and prevent diseases. Mostly they are common components in traditional medicine formulae. According to Thailand Drug Act of B.E. 2510 [1], products of traditional medicine have to be registered and owners also have to declare composition(s) of those products (crude drug names) to the Thai Food and Drug Administration (Thai FDA). The names of the crude drugs are frequently represented by the common names or scientific names with or without parts used of their sources. However, criteria for the registration of products have not been specified the format of crude drugs. It causes a variety of data formats. Nevertheless, several names may be referred to the same source of natural product when they are written in different formats. From this reason, huge records of the crude drug names (in different formats)

© Springer International Publishing Switzerland 2015
D.-S. Huang et al. (Eds.): ICIC 2015, Part II, LNCS 9226, pp. 237–247, 2015.
DOI: 10.1007/978-3-319-22186-1_24

have been kept in a database. In addition, spelling or typing errors are causes of data inconsistency. Therefore, it is hard and time consuming for human to correct and/or standardize patterns of these data. To alleviate these problems, string matching, a technique for correcting, checking and spelling words [2], plays a role in helping to correct and standardize patterns of crude drug names. Two well-known algorithms are usually applied for string matching, i.e., similar text and Levenshtein. To the best of our knowledge, there is no any research investigating the usefulness of these string matching algorithms on crude drug names. Therefore, effectiveness of these two algorithms is investigated in this work. Moreover, the ensembles of these algorithms with various parameter settings, is introduced. Experiments are done using a set of crude drug names, sampling from the list of components for registration medicines. In the rest of this paper, the background about crude drugs and string matching technique are given in Sect. 2. The method including our proposed method is made in Sect. 3. The experimental results are reported in Sect. 4. The Sect. 5 provides conclusion and future work.

2 Background

2.1 Crude Drugs

In herbal pharmacopoeias such as Thai Herbal Pharmacopoeia, Pharmacopoeia of the People's Republic of China and WHO monograph, they consist of information about crude drugs which are used in traditional medicine in the pattern of crude drug monograph. The topics in crude drug monographs generally include crude drug names in various languages including scientific name, description, identification and so on [3–5]. Crude drugs are natural products derived from plants, animals and minerals that can be used as medicine. They may be fresh, dried or traditionally processed material. Almost all of crude drugs in this research obtain from plants and only few crude drugs obtained from animals. Crude drugs from minerals are excluded. Normally, a crude drug name is represented by source names with/without their parts used. For example, the ginseng rhizome is a pharmaceutical crude drug name. The ginseng is a plant which is considered as a source of crude drugs, and the rhizome is a part used of this plant. Several problems of crude drug names may be occurred, e.g., the real part used and its recognized name, the problems of synonyms and homonyms, as well as typing errors. Normally, the real part used is a part of crude drug name. However, some crude drugs represent their part used names which are not the real part used, such as long pepper fruit. Long pepper fruit in Thai is Dok-Di-Pli. In Thai language, "dok" means flower but the real part used of Dok-Di-Pli is fruit. For problems of homonyms, the one common name may be referred to several herbs such as Khun. With this name, it may be either a plant whose scientific name is "*Colocasia gigantean* Hook.f." or the other plant whose scientific name is "*Cassia fistula* L." As the opposite of homonyms, various herbs have synonyms from the same language and different languages. For example, Betel leaf in English name is Bai-Pluu in Thai name and Piperis folium in Latin name. If a pharmacopoeia does not have scientific names to identify the exact source of crude drugs, it may cause confusion about crude drug names. In Thailand, various crude drugs are used in traditional Thai medicine. Moreover, traditional Chinese medicine is also popular. Therefore, several compositions are represented with Chinese names. Two dialects of Chinese are commonly used in Thailand, i.e., Chaozhou and Mandarin. Information about crude drugs which are components of pharmaceutical products has been recorded in both paper form and electronic form (database). Various problems can be found such as typing errors, synonyms, homonyms and several formats are causes of data redundancy. The registered crude drug names have been recorded with three patterns of characters, i.e., Thai, English and combination of Thai and English. Thai characters are used to present Thai crude drug names (both Thai common name and Thai local names) as well as crude drug names which are transliterated to Thai from dialects of Chaozhou and Mandarin. While the English characters are used to present English crude drug names, pharmaceutical Latin names, scientific names and crude drug names which are romanized from the Mandarin dialect (Hanyu Pinyin). The formats of crude drug names are also described in the topic of standardized crude drug names.

2.2   String Matching Techniques

String matching technique is used for various applications such as text correction, information retrieval, and bioinformatics [6–8]. Spelling errors, grammatical variants and various formats are the problem of text correction and information retrieval. In the past, research using string matching technique was conducted to solve scientific name problems such as biomedical text retrieval by Smith-Waterman algorithm [7] and verification of scientific names using Taxamatch algorithm [9, 10]. String matching techniques were conducted to solve clinical drug name problems such as Look Alike Sound Alike Drug (LASA Drug) by phonetic string matching technique and orthographic string matching techniques [11]. In addition, drug centric algorithm was developed for solving the problem of drug name variants found in local drug formularies. This algorithm is based on approximate matching which is a set-based similarity measure. Therefore, the problem of this algorithm is that misspelled drug names cannot be matched. However, there is no any research investigating the usefulness of these string matching algorithms on crude drug names. In this work, two popular string matching algorithms, i.e., similar text algorithm and Levenshtein algorithm, are applied.

Similar Text Algorithm. This is a basic function for comparison between the string and returns value of similarity as a number, The function counts character in one string match character in the other [12, 13]. Similarity value between strings will be calculated in percentage. For example, string A = "*Rheum palmatum* Linn.", string B = "*Rheum Palmatum* L.", and string C = "*Rhum palmatum* L." String A, B and C are scientific names of an herb, a difference between string A and B is the author name while differences between string A and C are both the author name as well as a spelling error of genus name in string C. When similar text algorithm is used, percentage of similarity between A and B is 91.8919 but percentage of similarity between A and C is 86.4865. From this example, the proximity between A and B is more than A and C. The similar text algorithm was used to evaluate translation from scratch and editing [14].

Levenshtein Algorithm. This is an algorithm for calculating similarity between two strings by edit distance. Minimum value to transform string S to string T by insertion, deletion or replacement characters, is indicated that S is similar to T. This function can be represented by Levenshtein (S, T, insertion, replacement, deletion), for short Levenshtein (i, r, d) [15, 16]. We can weight cost of insertion, replacement or deletion. In this work, two sets of values i, r, d were set to 1,1,1 and 1,1,0, respectively. For example, string S = "*Cuninum cyminum*" and string T = "*Cuminum cyminum*", distance between S and T is equal to 1 by a transition from "n" to "m". To take the value of edit distance from this algorithm to compare with other algorithms which the proximate value is presented by similarity, the function to change the distance to similarity in percentage (%Simlev) [17] is shown as follow.

$$\%Simlev\ (S,\ T) = [1.0 - (distance\ (S,\ T)/\max(|S|, |T|))] * 100 \qquad (1)$$

The distance (S, T) is minimum value to transform S to T and max ($|\ S\ |, |\ T\ |$) is the maximum length of string between S and T.

Levenshtein algorithms is used to solve several problems of string matching such as expansion of abbreviations in clinical texts and extraction of relations between medicals concepts in clinical texts [18, 19]. The Levenshtein (1,1,1) is often used. However, Levenshtein (1,1,0) may take advantages in some situations. For example, Chinese angelica root is often processed with honey stir-fried. The names of this crude drug are either Chinese angelica root (processed with honey stir-fried) or Chinese angelica root (processed with honey). The term "stir-fried" can be omitted. When these two strings are matched by the Levenshtein (1,1,1), the percentage of similarity is less than 100. In spite of the fact that these two strings refer to the same thing, the percentage of similarity should be 100. The Levenshentein (1,1,0) can handle this problem.

2.3   Ensemble Algorithm

In some situations, single algorithm may not be the best way to resolve crude drug names problems. Using the ensemble algorithm may be more efficient than using one algorithm. Ensemble algorithm is the combination of more than one algorithm for improved efficiency of basement algorithm. Some algorithms were used to combine with Levenshtein algorithm such as Stoilos distance, longest common substring and SmithWaterman algorithm [20–22]. The OATS algorithm and SAPLE fuzzy are ensemble algorithms used for matching schema and they are suitable for integrating services information, web service data aggregation and algorithm for easily and quickly search the personnel [22, 23].

3   Method

3.1   Data Set

Data were randomly selected from composites of traditional medicines in Thai FDA database from 1983 A.D. until September 2013. The number of sample was calculated by Taro Yamane equation [24], i.e., $n = N/(1 + Ne2)$. Here, N = number of ingredients in traditional medicines from Thai FDA database = 15402, e = the error is acceptable in this case, the value = 0.05. With this formula, number of samples from the calculation is 390. Therefore, the number of samples = 400 was used as a test set. A list of crude drug names from Thai FDA is called the test set. These names in the test set exclude English and Hanyu Pinyin crude drug names. Inactive excipients are also excluded. The 400 crude drug names come from 234 crude drugs.

3.2 Standardized Crude Drug Names

The list of crude drug database which was used for matching crude drug names in Thai FDA database, was created and called the standard set. Crude drug names in the standard set were prepared from Herbal Pharmacopoeia and Thai herbal text books. Crude drug names in the standard set were verified by two pharmacists. These names are usually composed of source names (plants, animals or minerals) and their parts used. However, various crude drugs have particular names which traditional practitioners recognize. In this work, the crude drug names are both single attributes and composite attributes. For single attribute, crude drug names or sources of crude drugs are represented in one language. Eight single attributes were used in the work, i.e., name of source, scientific name with author name (SCA name), scientific name without author name (SC name), correct crude drug name in Thai, common crude drug name in Thai, pharmaceutical Latin name, crude drug name in Mandarin and Chaozhou. The Mandarin and Chaozhou crude drug names are transliterated to Thai alphabets. For the composite attributes, two single attributes are combined. Thirty six composite attributes of crude drug names were created, eighteen composite attributes and sample of crude drug names are shown in Table 1. Moreover, values of swapping from a pair of attributes were also used. The number of crude drug names in standard set is 8,538 from 278 crude drugs.

3.3 Ensemble Algorithm for Crude Drug Names

Three algorithms, i.e., similar text, Levenshtein algorithms with weight cost of (1,1,1) and (1,1,0), were investigated. Moreover, the ensemble of these three algorithms so-called the MLLS (Mixed Levenshtein (1,1,1), Levenshtein (1,1,0) and similar text), was evaluated. The concept of the ensemble algorithm is presented as follow Fig. 1.

3.4 Criteria for Decision Crude Drug Names

When a pair of strings (crude drug names in the test set and the standard set), is compared. If the two strings are identical, the percentage of similarity is 100 %. In this case, a crude drug name in test set is fully correct. However, several crude drug names in the test set are not exactly match to the crude drug names in the standard set but it can be identified as correct. For these cases, a set of criteria is used to consider which

Table 1. Examples of composite attributes of crude drug names

Composite attributes	Sample of crude drug names
Latin name + Thai correct name	Granati Pericarpium เปลือกผลทับทิม
Latin name + Thai common name	Granati Pericarpium เปลือกลูกทับทิม
Latin name + Chaozhou name	Granati Pericarpium เจียะหลิ่วฮ้วย
Latin name + Mandarin name	Granati Pericarpium ชีหลิ่วผี
Thai correct name + Chaozhou name	เปลือกผลทับทิม เจียะหลิ่วฮ้วย
Thai correct name + Mandarin name	เปลือกผลทับทิม ชีหลิ่วผี
Thai common name + Chaozhou name	เปลือกลูกทับทิม เจียะหลิ่วฮ้วย
Thai common name + Mandarin name	เปลือกลูกทับทิม ชีหลิ่วผี
name of source + SCA name	ทับทิม *Punica granatum* L.
Chaozhou name + SCA name	เจียะหลิ่วฮ้วย *Punica granatum* L.
Mandarin name + SCA name	ชีหลิ่วผี *Punica granatum* L.
Thai correct name + SCA name	เปลือกผลทับทิม *Punica granatum* L.
Thai common name + SCA name	เปลือกลูกทับทิม *Punica granatum* L.
name of source + SC name	ทับทิม *Punica granatum*
Chaozhou name + SC name	เจียะหลิ่วฮ้วย *Punica granatum*
Mandarin name + SC name	ชีหลิ่วผี *Punica granatum*
Thai correct name + SC name	เปลือกผลทับทิม *Punica granatum*
Thai common name + SC name	เปลือกลูกทับทิม *Punica granatum*

Run basic algorithms
similar text algorithm, Levensthein (1,1,1) algorithm, Levensthein (1,1,0) algorithm

Select the algorithm to use is based algorithm from the best algorithm

Analyze weak points of the best algorithm

Using strong points of other algorithms to enhance weak point of the best algorithm

Ensemble of the three algorithms to improve performance of single algorithm

Fig. 1. Ensemble algorithm from the three algorithms

crude drug names in the test set are identified as correct. These criteria are listed as follow.

- *Fully correct name:* correct common name, scientific name, part used, dosage form, process preparation and may be few typing errors. For example, senna leaf extract, senna leaves extract and semna leaves extract are accepted as a fully correct name.
- *Correct meaning:* If crude drug names have process preparation, it was written in different terms but these different terms are synonyms. Moreover, a crude drug name may have modifiers. The modifier is used to emphasize the condition of the crude drug explicitly. These crude drugs are not different in meaning between crude drug with and without modifiers, e.g., fresh, dry, stir fire and carbonization. Few typing errors may be occurred. For example, turmeric rhizome and fresh turmeric rhizome, in Thai language, the "fresh" is written as an adjective. Meaning of turmeric rhizome and fresh turmeric rhizome are identical.

Crude drug names in various forms in the test set were matched with crude drugs names in the standard set. From above criteria, 243 of 400 crude drug names in the test set should be matched to 243 of 8538 crude drug names in the standard set. Due to the fact that a crude drug may have several crude drug names, numbers of crude drugs in the test set and the standard set are also presented. Numbers of crude drug names and crude drugs in both the test set and the standard set are shown in Fig. 2.

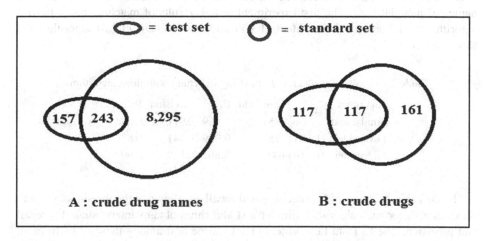

Fig. 2. (A) numbers of crude drug names in the test set and the standard set (B) numbers of crude drugs in the test set and the standard set

3.5 Evaluation Measures

Evaluation measurement of information retrieval [25] is recall (R), precision (P) and F1. The R is the fraction of relevant names that are retrieved. The P is the fraction of retrieved names that are relevant. The formulas are $P = a/(a + b)$ and $R = a/(a + c)$.

Here, a is the number of crude drugs which are expected correct names and they are correctly assigned, b is the number of crude drugs which are expected correct names and they are incorrectly assigned, c is the number of crude drugs which are expected incorrect names and they are incorrectly assigned. The performance measures above may be misleading when examined alone. Usually there is a measurement which represents a trade-off between recall and precision, called F1 measure. By combining recall and precision with the equal weight of these two factors, the F1 measure is defined as $2PR/(P + R)$. Crude drug names in the test set are compared to those in the standard set. The similarity between two strings is calculated based on the algorithm used to compare. A level at which we set to consider the result is called threshold. For example, when the threshold of Levenshtein (1,1,1) is set to 75 %. This means the pairs of strings which obtain the percentage of similarity greater than or equal to 75 %, is considered as correct or true and will be considered in the next step to be true positive or true negative. On the contrary, the pairs of strings which obtain the percentage of similarity less than 75 % is considered as incorrect or false and will be evaluated again as false positive or false negative.

4   Experimental Results

4.1   Performance of the Standard Algorithms

In summary, total of 243 correct names in Thai FDA database were matched crude drug names of new database. The first experiment tested, results of matching name from 3 algorithms: similar text, Levenshtein (1,1,1) and Levenshtein (1,1,0) algorithm are shown in Table 2.

Table 2.   Results of evaluation from string matching with three algorithms

Algorithm	Threshold	F_1	Precision	Recall
Similar text	85	0.889	0.911	0.868
Levenshtein (1,1,1)	75	0.909	0.941	0.878
Levenshtein (1,1,0)	100	0.861	0.819	0.907

From Table 1, values of F_1, precision and recall are shown at the threshold given to maximum F_1 for each algorithm. Similar text algorithm obtains intermediate F_1, recall and precision. The F_1 from Levenshtein (1,1,1) is the best among those of F_1 from all algorithms. However, the recall is not as good as Levenshtein (1,1,0). Some observations are made when we look at the details in error analysis. A few pairs of matching are false positives and some are false negatives such as *Angelica sinensis* extract and *Angelica sinensis* radix are the same herb. Although they are recognized as different crude drugs, Levenshtein (1,1,1) predicted them as the same crude drug. The result of prediction is false positive. The dry galangal rhizome and galangal rhizome are the same due to the fact that galangal rhizome is usually used in dry form. At threshold 75, the results from similar text and Levenshtein (1,1,0) algorithms are true positives while the results from Levenshtein (1,1,1) are false negative. Therefore, these two algorithms should be improve a weak point of Levenshtein (1,1,1) in case of a modifier with no significant meaning is added to the basic term. In our test set with 400 crude drug names, we found out that 31 crude drug names had some modifiers with no significant meaning (7.75 %).

4.2   Performance of the Ensemble String Matching Algorithm

In this experiment, the ensemble of the three algorithms so-called the MLLS, was evaluated. As the result from previous experiment, the Levenshtein (1,1,1) was set as the main algorithm. The Levenshtein (1,1,0) and the similar text were taken priority to make a decision when similarity from the Levenshtein (1,1,1) is less than 90 %. The results of the MLLS are shown in Table 3. Performance of the Levenshtein (1,1,1) (Lev111) was compared to the MLLS between the threshold of 85 % and 65 % with the gap of 5 %.

Table 3. Results of evaluation from MLLS compare with Levenshtein (1,1,1) algorithm

Threshold (%)	F_1		Precision		Recall	
	Lev_{111}	MLLS	Lev_{111}	MLLS	Lev_{111}	MLLS
85	0.845	0.891	0.994	0.995	0.735	0.807
80	0.896	0.931	0.971	0.973	0.832	0.893
75	0.909	0.940	0.941	0.950	0.878	0.930
70	0.901	0.934	0.900	0.913	0.903	0.955
65	0.897	0.919	0.867	0.880	0.929	0.963

From the result, the MLLS can solve the problem of crude drug names with no significant meaning extended modifiers. The similarities of the MLLS are higher than those of Levensteine (1,1,1) and crude drug name match with correct name. Although only a small portion of crude drug names with no significant meaning modifiers was included in the test set, the ensemble algorithm outperformed the Levenshtein (1,1,1) especially the recall with the gap of 5 %. The F1 of the MLLS is higher than F1 of the other algorithms and higher than the Levenshtein (1,1,1) with the gap of 3 %. In summary, our proposed method is useful for correction and standardization the crude drug names and outperforms the single algorithms.

5   Conclusion and Future Work

In this work, problems of completing for crude drug name are addressed. These problems may be occurred from spelling errors, various forms of writing, synonyms and names with or without modifiers in Thai FDA database. String matching algorithms were used to alleviate these problems. The similar text and Levenshtein algorithms were investigated. However, the results from single algorithm showed that crude drug names in the test set did not well match with those of the standard set.

The ensemble technique so-called MLLS, was proposed. The MLLS outperformed single algorithms to complete crude drug names, especially crude drug names with modifiers that have no significant meaning or implicit meaning. The MLLS should be used for standardizing crude drug names. However, crude drug names in some formats cannot be standardized with the MLLS algorithm, such as crude drug names which their parts used may be appeared in various forms. For example, Radix Ginseng, Ginseng Radix and Ginseng (Radix), these three names refer to the same crude drug. However, positions or formats of parts used in crude drug names are different. Therefore, ensemble algorithm should be adapted for solving these problems.

Acknowledgments. This work was supported by the Higher Education Research Promotion and National Research University Project of Thailand, Office of the Higher Education Commission and also supported by the Graduate of Silpakorn University as well as the Research and Creative Funding Scheme, Faculty of Pharmacy, Silpakorn University (2014).

References

1. Food and Drug Administration Thailand. http://www.fda.moph.go.th/fda_eng/frontend/theme_1/info_data_main.php?ID_Info_Main=4
2. Navarro, G.: A guide tour to approximate string matching. ACM Comput. Surv. **33**(1), 31–88 (2001)
3. Bureau of Drug and Narcotic, Department of Medical Sciences: Thai Herbal Pharmacopoeia, vol. 3. Office of National Buddishm Press, Bangkok (2009)
4. Ministry of Health of the People's Republic of China: Pharmacopoeia of the People's Republic of China. China Medical Science Press, Beijing (2010)
5. World Health Organization: WHO Monograph on Selected Medicinal Plants, vol. 4. WHO Press, Geneva (2005)
6. Klaus, U.S., Stoyan, M.: Fast string correction with Levenshtein automata. Int. J. Doc. Anal. Recogn. **5**, 67–85 (2002)
7. Wang, J.F., Li, Z.R., Cai, C.Z., Chen, Y.Z.: Assessment of approximate string matching in a biomedical text retrieval problem. Comput. Biol. Med. **29**, 717–724 (2005)
8. Tilo, B., Leonid, V.B.: Levenshtein error-correcting barcodes for multiplexed DNA sequencing. BMC Bioinform. **14**, 272–281 (2013)
9. Rees, T.: Fuzzy matching of taxon names for biodiversity informatics applications. Poster session presented at the meeting of e-Biosphere Conference, UK (2009)
10. Brad, B., et al.: The taxonomic name resolution service: an online tool for automated standardization of plant names. Bioinformatics **14**(16), 1–14 (2013)
11. Grzegorz, K., Bonnie, D.: Automatic identification of confusable drug names. Artif. Intell. Med. **36**, 29–42 (2006)
12. PHP. http://php.net/manual/en/function.similar-text.php
13. Oliver, I.: Programming Classics: Implementing the World's Best Algorithms. Prentice Hall Inc., Englewood Cliffs (1993)
14. Ilse, D., Nathalie, D.S., Arda, T.: Post-editing of machine translation: a case study. In: Laura, W.B., Michael, C. (eds.) Processes and Applications, pp. 78–108. Cambridge Scholar publishing, Newcastle (2014)
15. Levenshtein, V.I.: Binary code capable of correcting deletions, insertions, and reverals. Sov. Phys. Dokl. **10**(8), 707–710 (1966)

16. Andres, M., Enrique, V.: Computation of normalized edit distance and applications. IEEE Trans. Pattern Anal. Mach. Intell. **15**(9), 1091–1095 (1993)
17. Peter, C.: A comparison of personal name matching. In: Sixth IEEE International Conference on Data Mining Workshop, pp. 290–294. The Printing House Publication, USA (2006)
18. Lisa, T., Beata, M., Aron, H., Martin, D., Maria, K.: EACL - expansion of abbreviations in CLinical text. In: The 3rd Workshop on Predicting and Improving Text Readability for Target Reader Populations, pp. 2085–2090. Association for Computational Linguistics (ACL), Pennsylvania (2014)
19. Bryan, R., Sanda, H., Kirk, R.: Automatic extraction of relations between medicals concepts in clinical texts. J. Am. Med. Inform. Assoc. **18**, 594–600 (2011)
20. Zied, M., Lina, F.S., Elise, P.-G., Thierry, L., Stefan, J.D.: Spell-checking queries by combining Levenshtein and Stoilos distances. In: Oral presentation session presented at Network Tools and Applications in Biology Clinical Bioinformatics (NETTAB) Workshop, Italy (2011)
21. Shaun, J.G., Overhage, J.M., Clement, M.: Real world performance of approximate string comparators for use in patient matching. In: Medinfo 2004 Proceedings of the 11th World Congress on Medical Informatics, pp. 43–47. IOS Press (2004)
22. Johnston, E., Kushmerick, N.: Aggregating web services with active invocation and ensembles of string distance metrics. In: Motta, E., Shadbolt, N.R., Stutt, A., Gibbins, N. (eds.) EKAW 2004. LNCS (LNAI), vol. 3257, pp. 386–402. Springer, Heidelberg (2004)
23. Michael, J.P.: SAPLE: Sandia advanced personnel locator engine. In: U.S. Department of Energy (ed.) Sandia Report. U.S. Department of Energy, Springfield (2010)
24. Taro, Y.: Elementary Sampling Theory. Prentice Hall Inc., Englewood Cliffs (1967)
25. Christopher, D.M., Prabhakar, R., Hinrich, S.: Introduction to Information Retrieval. Cambridge University Press, New York (2008)

Clustering High-Dimensional Data via Spectral Clustering Using Collaborative Representation Coefficients

Shulin Wang$^{(\boxtimes)}$, Jinchao Gu, and Fang Chen

College of Computer Science and Electronics Engineering, Hunan University,
Changsha 410082, Hunan, China
smartforesting@gmail.com

Abstract. Clustering high-dimensional data is challenging for traditional clustering methods. Spectral clustering is one of the most popular methods to cluster high-dimensional data, in which the similarity matrix plays an important role. Recently, sparse representation coefficients have been proposed to construct the similarity matrix via the cosine similarity between each pair of coefficient vectors for spectral clustering and showed promising results. However, the sparse representation emphasizes too much on the role of ℓ_1-norm sparsity and ignores the role of collaborative representation, which makes its computational cost very high. In this paper, we propose a spectral clustering method based on the similarity matrix which is constructed based on the collaborative representation coefficient vectors. Extensive experiments show that the proposed method has a strong competitiveness both in terms of computational cost and clustering performance.

Keywords: High-dimensional data · Spectral clustering · Sparse representation · Collaborative representation · Affinity matrix

1 Introduction

With the improvement of technology, it is increasingly easier to collect data in various fields, thus resulting in increasingly large and complex datasets. Many real-world datasets have very high-dimensional features, such as document data, user ratings data, multimedia data, financial time series data, gene expression data, and so on. However, clustering high-dimensional data is extraordinarily difficult for traditional clustering methods due to the "curse of dimensionality" [1]. K-means [2] is the most popular traditional clustering methods which can achieve good results on the low-dimensional data but is prone to deal with non-spherical distribution of data and cannot identify clusters which are not linearly separable in feature space [3].

In recent years, spectral clustering (SC) [4] has attracted more and more interest due to the high performance in data clustering and simplicity in implementation. Compared with other clustering algorithms, spectral clustering has the advantage of finding clusters of arbitrary shape and getting the global optimal solution. Its basic idea is to find a cluster membership of the data points by using the spectrum of the affinity matrix

© Springer International Publishing Switzerland 2015
D.-S. Huang et al. (Eds.): ICIC 2015, Part II, LNCS 9226, pp. 248–258, 2015.
DOI: 10.1007/978-3-319-22186-1_25

that depicts the similarity among data points. The success of spectral clustering depends heavily on the choice of the similarity measure, thus the affinity matrix plays an important role in spectral clustering. There are two ways to measure the similarity between samples: pairwise distance and representation coefficients [5]. Pairwise distance is very sensitive to noises and outliers, for pairwise distance value only depends on the distance between two data points; alternatively, representation coefficients-based methods assume that each data point can be denoted as a linear combination of other data points, which not only consider the relationship between the two points, but also consider other points.

Recently, many researchers have proposed to construct affinity matrix in spectral clustering via representation coefficient. Wright et al. [6] proposed a spectral clustering based on individual sparse representation coefficient, however, it cannot exploit more contextual information from the whole coefficient vectors. Wu et al. [7] proposed a novel spectral clustering based on the sparse representation coefficients (SRSC), which constructs the similarity matrix via the cosine similarity between each pair of coefficient vectors and showed promising results in clustering high-dimensional data. Although SRSC algorithm achieved very good clustering results, the sparse representation emphasizes too much on the role of ℓ_1-norm sparsity and ignores the role of collaborative representation, which makes its computational cost very high [8].

Inspired by the collaborative representation (CR) [8], we propose a novel spectral clustering algorithm which constructs the affinity matrix via collaborative representation coefficient and call it collaboratively representation based spectral clustering (CRSC). Compared with the SRSC algorithm, CRSC has lower computational cost and higher clustering performance. Extensive experiments conducted on real-world datasets validate that our algorithm is very efficient and effective in clustering high-dimensional data.

2   Methods

Our proposed clustering algorithm include three steps: it firstly obtains the collaborative representation vectors; then constructs the affinity matrix via the cosine measure between the collaborative representation coefficients vectors; lastly runs the spectral clustering algorithm on the affinity matrix.

2.1   Collaborative Representation

Assuming that each data point $y = \mathbb{R}^m$ can be represented as a linear combination of other points. In math empathically, $y = X\alpha$, where X is a dictionary whose columns consist of some data points, and α is the representation coefficients of y over X It can be obtained by solving:

$$(\hat{\alpha}) = \arg \min \|\alpha\|_0 \quad s.t. \quad y = X\alpha \tag{1}$$

where $\|\cdot\|_0$ denotes the ℓ_0-norm of a vector, which counts the number of nonzero entries in a vector, However, the problem of finding the sparse solution of Eq. (1) is difficult to solve since it is a NP-hard problem. The theory of compressive sensing [9] reveals that if the solution α is a few non-zero coefficients, this problem can be approximately characterized as:

$$(\hat{\alpha}) = \arg \min \|\alpha\|_1 \quad s.t. \quad y = X\alpha \tag{2}$$

where $\|\cdot\|_1$ denotes the ℓ_1-norm of a vector.

Since real-world data contain noise, the sparse solution α can be approximately obtained by solving:

$$(\hat{\alpha}) = \arg \min \|\alpha\|_1 \quad s.t. \quad \|y = X\alpha\|_2 \leq \varepsilon \tag{3}$$

where $\|\cdot\|_2$ denotes the ℓ_2-norm of a vector, and $\varepsilon > 0$ represents the maximum residual error.

However, in many applications, the noise level ε is unknown in advance. In such cases, the Lasso optimization algorithm [10] can be used to recover the sparse solution from:

$$(\hat{\alpha}) = \arg \min\{\|y - X\alpha\|_2 + \lambda\|\alpha\|_1\} \tag{4}$$

where λ can be viewed as an inverse of the Lagrange multiplier.

While the role of collaborative representation is ignored and its computational cost is very high. This difficulty is resolved by choosing the minimum ℓ_2-norm solution [8]:

$$(\hat{\alpha}) = \arg \min\{\|y - X\alpha\|_2^2 + \lambda\|\alpha\|_2^2\} \tag{5}$$

where λ is the regularization parameter.

The solution of collaborative representation with regularized least square in Eq. (5) can be easily and analytically derived as:

$$\hat{\alpha} = (X^T X + \lambda \cdot I)^{-1} X^T y \tag{6}$$

Let $P = (X^T X + \lambda \cdot I)^{-1} X^T$. Obviously, P can be calculated beforehand, which is a projection matrix. The speed of collaborative representation is very fast since P and y are irrelevant.

2.2   Constructing Affinity Matrix Based on Cosine Similarity

The distance function can be used to represent the similarity between samples in the cluster analysis, such as Euclidean distance, Minkowski distance, and Cosine distance.

Cosine is used to measure the difference between two vectors, which is adopted in our algorithm to construct the affinity matrix based on cosine similarity between the collaborative representation coefficients vectors. The cosine similarity is defined as:

$$COS_{ij} = \max \left\{ 0, \frac{\alpha_i \cdot \alpha_j}{\|\alpha_i\|_2 \times \|\alpha_j\|_2} \right\} \tag{7}$$

where $\alpha_i = \left(\alpha_{i,1}, \alpha_{i,2}, \cdots, \alpha_{i,n}\right)^T$ and $\alpha_j = \left(\alpha_{j,1}, \alpha_{j,2}, \cdots, \alpha_{j,n}\right)^T$.

2.3   Spectral Clustering

Spectral clustering algorithm is a clustering method based on graph theory. Its essence is to find the clustering structure by using the graph optimal partition problems. We represent the dataset as an undirected graph $G = (V, E)$, where the vertex V represents the data points, the edge E represents the connections between points. So the clustering problem could be solved by using the undirected graph partitioning problem. The main tool for spectral clustering algorithm is Laplacian matrix, one of which is the unnormalized graph Laplacian L:

$$L = D - W \tag{8}$$

where W is the affinity matrix, which represents the weight between points, D is the degree matrix, which is defined as the diagonal matrix with $d_{ii} = \sum_{j=1}^{n} W_{ij}$. The normalized graph Laplacian L_{sym} is defined as:

$$L_{sym} = D^{-\frac{1}{2}} L D^{-\frac{1}{2}} \tag{9}$$

The basic steps of the spectral clustering are summarized as the following three steps: the first is constructing the affinity matrix based on data; the second is building eigenvectors space, which consists of some eigenvectors of graph Laplacian matrix; the third is clustering the eigenvectors of eigenvectors space with k-means algorithm or other methods.

2.4   Algorithm Description

Our algorithm is summarized as follows. The collaborative representation coefficients of each object are obtained as the Algorithm 1.

Algorithm 1. Extracting collaborative representation coefficients

---

Input: A high-dimensional dataset $X = (x_1, x_2, \cdots, x_n) \in R^{m \times n}$,
where $x_i = (x_{i1}, x_{i2}, \cdots, x_{im})^T \in R^m$ **represents the** i **-th data object**

Param: penalty coefficient λ **for Lasso optimization**

   for dataset X **do**

$$P = \left(X^T X + \lambda \cdot I \right)^{-1} X^T$$

   end

   for each data point y **do**

$$\hat{\alpha} = Py$$

   end

Output: Collaborative representation coefficients α .

After extracting collaborative representation coefficients, we build the collaborative representation coefficients vectors β using the Algorithm 2.

Algorithm 2. Building collaborative representation coefficients vectors β

Input: Collaborative representation coefficients α

   for $i \leftarrow 1$ **to** n **do**

      for $j \leftarrow 1$ **to** n **do**

         if $j = i$ **then** $\beta_{i,j} \leftarrow 0$

         else if $j > i$ **then** $\beta_{i,j} \leftarrow \alpha_{i,j-1}$

         else $\beta_{i,j} \leftarrow \alpha_{i,j}$

         end

      end

   end

Output: Collaborative representation coefficients vectors β .

We construct the affinity matrix according to Eq. (7) using the cosine distance, the pseudo-code is described in Algorithm 3.

Algorithm 3. Constructing affinity matrix based on cosine similarity of coefficient vectors

---

Input: Collaborative representation coefficients vectors β

for $i \leftarrow 1$ to n do

    for $j \leftarrow 1$ to n do

        if $j = i$ then $w_{i,j} \leftarrow 0$ else

        $\cos \leftarrow \dfrac{\beta_i' \cdot \beta_i'}{\|\beta_i'\|_2 \times \|\beta_i'\|_2}$;

        if $\cos > 0$ then $w_{i,j} \leftarrow \cos$ else $w_{i,j} \leftarrow 0$

        end

      end

    end

end

Output: Affinity matrix W .

After constructing the affinity matrix W we select the classical spectral clustering algorithm [11] to discover the cluster structure of high-dimensional data.

3 Experimental Results

3.1 Datasets

Table 1 provides an overview of these datasets, which are typical high-dimensional datasets. It includes seven biological data datasets, of which the THCACancer (Thyroid carcinoma cancer) is produced by Next Generation Sequencing technology, and are downloaded from The Cancer Genome Atlas (http://cancergenome.nih.gov/).

Table 1. The statistics of the datasets used our experiments

Datasets	#Instance	#Attributes	#Classes
ALL [12]	248	12625	6
Colon [13]	62	2000	2
DLBCL [14]	77	5469	2
Gliomas [15]	50	102625	2
MLL [16]	72	12582	3
SRBCT [17]	83	2308	4
THCACancer	112	20531	2

3.2   Experiment Setting

All the experiments were carried out using MATLAB on a 3.10 GHz machine with 4 GB RAM. We compare our method with 5 clustering algorithms, including k-means, SRK (k-means runs on sparse representation coefficients), CRK (k-means runs on collaborative representation coefficients), SC, and SRSC. We set different Lasso optimization in SRK, CRK, SRSC and our method. Three commonly used external cluster validation metrics are used to evaluate the clustering methods, including clustering accuracy (CA), F-measure (F) and normalized mutual information (NMI) [10], the larger value represents the better clustering performance. Lastly, we compare the computational time of the SRSC and CRSC.

3.3   Results and Discussion

To illustrate the affinity matrices from sparse representation and collaborative representation, we demonstrate the visual property of the affinity matrices in Fig. 1, taking the Colon dataset and MLL dataset as examples. In Fig. 1, each subfigure is a matrix with n-by-n entries.

From Fig. 1, we can see that the affinity matrices of collaborative representation is more similar to the ground truth compared with that of sparse representation, which means that collaborative representation can encode more discriminative information and will be more effective in spectral clustering.

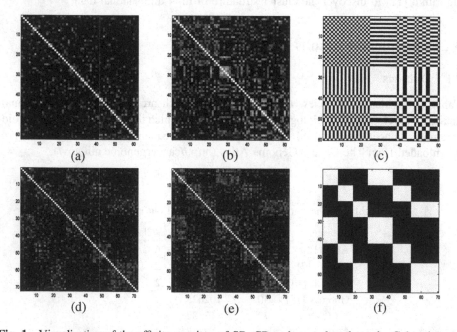

Fig. 1. Visualization of the affinity matrices of SR, CR and ground truth on the Colon dataset and MLL dataset; Colon dataset: (a) SR, (b) CR, (c) the ground truth; MLL dataset: (d) SR, (e) CR, (f) the ground truth.

For each dataset, we repeat spectral clustering and k-means 100 times and record the average result. The best clustering results obtained from the six clustering algorithms with different Lasso optimization parameters are reported in Tables 2, 3 and 4, each of which corresponds to one evaluation metric. For each dataset, the best results are in bold. The last row in each table presents the average performance of each algorithm over all seven datasets.

From Tables 2, 3 and 4, we can clearly see that, generally, the CRK algorithm gets better mean clustering performance than the SRK algorithm; and the CRSC algorithm

Table 2. Evaluation of all algorithms with CA as metric

Datasets	k-means	SRK	CRK	SC	SRSC	CRSC
ALL	0.3846	0.5278	0.6114	0.5753	**0.7214**	0.6914
colon	0.7158	0.6625	0.7383	0.5000	0.8621	**0.9032**
DLBCL	0.6883	0.7111	0.6883	0.6364	0.7273	**0.7922**
Gliomas	0.6100	0.6380	0.6182	0.6002	**0.7410**	0.7400
MLL	0.7446	0.7130	0.7582	0.7083	0.8750	**0.9306**
SRBCT	0.4819	0.4474	0.5435	0.3747	0.6068	**0.6571**
THCACancer	0.8304	0.6172	0.8182	0.8393	**0.8929**	**0.8929**
Average	0.6365	0.6167	0.6823	0.6049	0.7752	**0.8011**

Table 3. Evaluation of all algorithms with F as metric.

Datasets	k-means	SRK	CRK	SC	SRSC	CRSC
ALL	0.3982	0.4453	0.6301	0.6364	0.7286	**0.7344**
colon	0.7000	0.5976	0.7338	0.5102	0.8622	**0.9032**
DLBCL	0.7096	0.6565	0.7096	0.6611	0.6343	**0.8064**
Gliomas	0.5994	0.5901	0.6102	0.5806	0.7372	**0.7393**
MLL	0.7389	0.6819	0.7512	0.7144	0.8753	**0.9311**
SRBCT	0.4933	0.3991	0.5447	0.3870	0.6006	**0.6618**
THCACancer	0.8264	0.5455	0.8037	0.8380	0.8823	**0.8920**
Average	0.6380	0.5594	0.6833	0.6182	0.7601	**0.8097**

Table 4. Evaluation of all algorithms with NMI as metric.

Datasets	k-means	SRK	CRK	SC	SRSC	CRSC
ALL	0.2409	0.3619	0.5384	0.5005	**0.6712**	0.6228
colon	0.1923	0.0972	0.2411	0.0012	0.3982	**0.5184**
DLBCL	0.1238	0.0493	0.1238	0.0567	0.2880	**0.3954**
Gliomas	0.0590	0.1120	0.0620	0.0213	**0.2164**	0.2050
MLL	0.5478	0.5204	0.5773	0.4543	0.6131	**0.7551**
SRBCT	0.2262	0.2267	0.3341	0.1111	0.3881	**0.4353**
THCACancer	0.4296	0.1740	0.4230	0.3902	0.5402	**0.5614**
Average	0.2600	0.2202	0.3285	0.2193	0.4450	**0.4991**

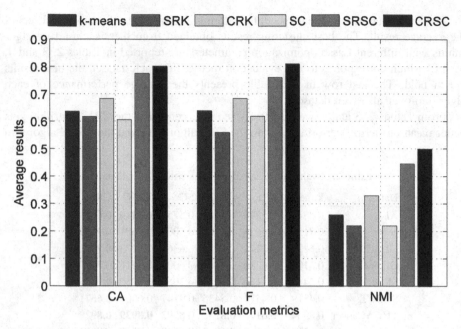

Fig. 2. A comparison of average clustering results from six algorithms. (Color figure online)

gets better mean clustering performance with other five algorithms in terms of CA, F and NMI. However, there are also some particular cases where CRSC does not get the best result.

We plot the average of the mean value (from the last rows of the three tables) for comparing clustering algorithms, as shown in Fig. 2, which clearly demonstrates that the CRSC algorithm outperforms other algorithms; CRK has better performance than SRK.

Finally, In order to further illustrate the advantages of our algorithm, the computational time are listed in the Table 5, in which the second and third columns represent the runtime of SRSC and CRSC, respectively, and the last column represents the ratio of the runtime of SRSC to CRSC.

Table 5. Speed on the nine datasets of SRSC and CRSC.

Datasets	SRSC_time(sec.)	CRSC_time(sec.)	Times
ALL	1572	53.86	29.19
Colon	15.46	0.477	32.41
DLBCL	64.81	2.006	32.31
Gliomas	34.68	1.660	20.89
MLL	70.73	3.722	19.00
SRBCT	30.74	1.164	26.41
THCACancer	1136	17.11	66.39

From Table 2 and 5, we can see that the clustering accuracy rate of CRSC is not better than SRSC on the ALL, Gliomas and THCACancer, but the speed of CRSC is about 38 times faster than SRSC. On the other datasets, CRSC is about 28 times fastest than SRSC and has 5 % improvement in clustering accuracy rate.

4 Conclusion

Clustering high-dimensional data is challenging in data mining research. In this paper, we propose a novel spectral clustering algorithm based on collaborative representation coefficient vectors (CRSC) for clustering high-dimensional data, in which the affinity matrix is constructed based on the collaborative representation coefficient vectors. In the collaborative representation coding, each feature can be well represented by its associated dictionary, and P can be calculated beforehand, which makes its computational cost very low, so it is very efficient to handle large datasets. The experimental results clearly demonstrate that this new approach can reduce the time cost of clustering and also achieves improvement in clustering accuracy, which is reasonable and efficient for high-dimensional data clustering problem. However, CRSC is sensitive to the parameter λ, how to determine the optimal value of the parameter λ needs further research.

Acknowledgement. This work is supported by the National Natural Science Foundation of China under (Grant no. 61474267, 60973153 and 61471169) and Collaboration and Innovation Center for Digital Chinese Medicine of 2011 Project of Colleges and Universities in Hunan Province.

References

1. Steinbach, M., Ertöz, L., Kumar, V.: The challenges of clustering high dimensional data. In: Wille, L.T (ed.), New Directions in Statistical Physics, pp. 273–309. Springer, Heidelberg (2004)
2. Cai, D.: Litekmeans: the fastest matlab implementation of kmeans (2011). http://www.zjucadcg.cn/dengcai/Data/Clustering.html
3. Phoungphol, P., Zhang, Y.: Multi-source kernel k-means for clustering heterogeneous biomedical data. In: IEEE International Conference on Bioinformatics and Biomedicine Workshops (BIBMW) 2011, pp. 223–228 (2011)
4. Qin, G., Gao, L.: Spectral clustering for detecting protein complexes in protein–protein interaction (PPI) networks. Math. Comput. Model. **52**, 2066–2074 (2010)
5. Peng, X., Zhang, L., Yi, Z.: An Out-of-sample Extension of Sparse Subspace Clustering and Low Rank Representation for Clustering Large Scale Data Sets. arXiv preprint arXiv:1309.6487 (2013)
6. Wright, J., Ma, Y., Mairal, J., Sapiro, G., Huang, T.S., Yan, S.: Sparse representation for computer vision and pattern recognition. Proc. IEEE **98**, 1031–1044 (2010)
7. Wu, S., Feng, X., Zhou, W.: Spectral clustering of high-dimensional data exploiting sparse representation vectors. Neurocomputing **135**, 229–239 (2014)

8. Zhang, D., Yang, M., Feng, X.: Sparse representation or collaborative representation: Which helps face recognition? In: IEEE International Conference on Computer Vision (ICCV) 2011, pp. 471–478 (2011)
9. Donoho, D.L.: Compressed sensing. IEEE Trans. Inf. Theor. **52**, 1289–1306 (2006)
10. Tibshirani, R.: Regression shrinkage and selection via the lasso. J. Roy. Stat. Soc. B (Methodol.) **58**(1), 267–288 (1996)
11. Ng, A.Y., Jordan, M.I., Weiss, Y.: On spectral clustering: analysis and an algorithm. Adv. Neural Inf. Process. Syst. **2**, 849–856 (2002)
12. Yeoh, E.-J., Ross, M.E., Shurtleff, S.A., Williams, W.K., Patel, D., Mahfouz, R., Behm, F. G., Raimondi, S.C., Relling, M.V., Patel, A.: Classification, subtype discovery, and prediction of outcome in pediatric acute lymphoblastic leukemia by gene expression profiling. Cancer Cell **1**(2), 133–143 (2002)
13. Alon, U., Barkai, N., Notterman, D.A., Gish, K., Ybarra, S., Mack, D., et al.: Broad patterns of gene expression revealed by clustering analysis of tumor and normal colon tissues probed by oligonucleotide arrays. Proc. Nat. Acad. Sci. **96**, 6745–6750 (1999)
14. Shipp, M.A., Ross, K.N., Tamayo, P., Weng, A.P., Kutok, J.L., Aguiar, R.C., et al.: Diffuse large B-cell lymphoma outcome prediction by gene-expression profiling and supervised machine learning. Nat. Med. **8**, 68–74 (2002)
15. Golub, T.R., Slonim, D.K., Tamayo, P., Huard, C., Gaasenbeek, M., Mesirov, J.P., et al.: Molecular classification of cancer: class discovery and class prediction by gene expression monitoring. Science **286**, 531–537 (1999)
16. Armstrong, S.A., Staunton, J.E., Silverman, L.B., Pieters, R., den Boer, M.L., Minden, M. D., et al.: MLL translocations specify a distinct gene expression profile that distinguishes a unique leukemia. Nat. Genet. **30**, 41–47 (2001)
17. Khan, J., Wei, J.S., Ringner, M., Saal, L.H., Ladanyi, M., Westermann, F., et al.: Classification and diagnostic prediction of cancers using gene expression profiling and artificial neural networks. Nat. Med. **7**, 673–679 (2001)

Classifying Stances of Interaction Posts in Social Media Debate Sites

Xiaosong Rong[✉], Zhuo Wang, Ziqi Wang, and Ming Zhang

Institute of Network Computing and Information Systems,
School of Electronics Engineering and Computer Science,
Peking University, Beijing 100871, China
{rxs,wangzhuo,zikkiwang,mzhang_cs}@pku.edu.cn

Abstract. Social media debate sites provide a rich collection of public opinions on various controversial issues. In an online debate, participants express their stances by replying directly to the main topic or indirectly to other participants. We observe that the majority of the posts in online debates are interaction posts. In this paper, we propose a new method for the task of stance classification of interaction posts in online debates. We mine the historical activities and textual content of posts to learn the relationships between participants in online debates. Then we build the interaction graph and develop a greedy algorithm to classify participants by stance. Empirical evaluation shows that our method performs better than the baseline methods.

Keywords: Online debates · Stance classification · Local attitudes · Interaction graph · Greedy algorithm

1 Introduction

Online discussion forums provide individuals a social platform to participate in popular ideological debates and express their opinions. Many corporations and governmental organizations use these kinds of communication tools to aggregate public opinions, collect ideas and understand the trends of opinions about popular topics. Classifying a user's stance, i.e., recognizing which debate-side a user is supporting in an online debate, is an important task for getting a quick overview of users' opinions.

Online public debates are different from congressional debates [1]. Users in online debate forums or social media debate sites frequently use informal language, which makes it more difficult to classify users' stances from linguistic features. Besides, social media debate sites contain more social behaviors, such as interaction expressions exchanged between participants. In online debates, users write posts to express their opinions, where each post is either a reply to another post or directly to the main topic. Interactions are quite rich in social media debate sites. One example of such websites is *createdebate*[1], in which users create debates and participate in discussions on different kinds of topics, such as politics and entertainment. Nearly 80 % of the posts in *createdebate* are interaction posts, i.e., posts that are responses to other posts [2].

---

[1] http://www.createdebate.com

© Springer International Publishing Switzerland 2015
D.-S. Huang et al. (Eds.): ICIC 2015, Part II, LNCS 9226, pp. 259–269, 2015.
DOI: 10.1007/978-3-319-22186-1_26

These interaction posts indicate if the authors agree or disagree with each other. In this paper, we focus on the task of classifying users' stances of interaction posts in social media debate sites.

It's difficult to identify stances of interaction posts because the authors may go off-topic when they reply to other participants and cause the discussion to drift off to an unrelated subject [3]. The content of interaction posts doesn't have to be related to the main topic, since it is prepared for supporting or opposing the previous authors. Luckily, we can extract the local attitudes, i.e., whether the authors agree or disagree with the adjacent posts or users, which can help induce the global positions of participants. Table 1 shows sample posts of an online debate from *createdebate*.

Table 1. Sample posts from *createdebate*

Title: Does God Exist?
Sides: *Yes* vs. *No*
User A:
I believe in God so I guess my answer …I know I feel it all around me and see it in my children and try to reflect in my actions. I do not go to church on Sunday …
Side: Yes
User B:
You do have proof my friend. Let your faith be reinforced in the proof of …
Side: Yes
User C:
What evidence do you have that the universe was created by anything…
Side: No
User D:
That's not true!
Side: No
User E:
God merely exists in people beliefs. Without people, or more accurately believers, he would not exist nor be known of.
Side: No

In this example, there are two root posts directly replying to the main topic and three interaction posts replying to other posts. The post of user A supports the side "Yes", so the position for this post is "Yes". However, the post of user B is not a response to the main topic. It supports the opinion of user A, therefore indirectly implies the stance of use B is also "Yes". Analogously, we can infer the positions of the other posts or users.

To classify user stances in online debates, we need to extract the local attitudes between participants. Previous work has covered this task of agree/disagree classification [2, 4, 5]. The work in [4] extracts local agreements and disagreements by creating a simple pattern dictionary that contains agreement/disagreement expressions. It's impossible to list all the patterns manually, so their method cannot cover all the cases. We propose a bootstrapping method to collect agreement/disagreement expressions semi-automatically. Moreover, we observe that users participate in multiple topics and form stable allies or enemies if they frequently support or oppose each other. Based on this observation, we mine the historical activities in social media debate sites and learn the opinion similarity between any two participants. We combine this knowledge with the local attitudes, and compute the consistency score representing the degree of agreement between two users.

Participants may create one or more posts in the same topic. One important assumption in this paper is user consistency: a user takes a unique stance across different posts in the same debate. Therefore, we can infer the stances of posts by identifying the user-level positions. For every debate, we build an interaction graph in which the consistency scores are used as the link weight between nodes corresponding to participants in the debate. At last, we propose an efficient method to classify participants into two opposing opinion groups.

We conduct experiments on real-world datasets for evaluation. We collect debate posts from createdebate. When users participate in a debate on this platform, they also choose the stance label according to which side they support, i.e., self-labeling the stance of the post. We use the stance labels to evaluate the performance of our proposed method for the task of stance classification. The experimental results show that our approach achieves higher accuracy than the competing baseline methods.

The rest of this paper is organized as follows. We describe related work in Sect. 2 and propose our method in Sect. 3. In Sect. 4 we explain the datasets and present the experimental results. We conclude this paper and describe the future work in Sect. 5.

2 Related Work

Several researchers have worked on the task of modeling stances. Previous work covers different debate settings, such as congressional debates [6–8] and social media debate sites [1, 2, 9, 10]. Debates in social media platforms are different from congressional debates because users use informal language to express their opinions. Social media debate sites also contain more social behaviors such as interactions between participants.

The work in [11] proposed a supervised approach for classifying stances in ideological debates based on the discourse structure. Another work in [3] observed that people more frequently respond to messages when they disagree than they agree, and developed a graph-theoretic algorithm for stance classification that completely ignores the text of postings and only uses the link structure of the network of interactions. The study in [4] utilized the textual content of the remarks into the link-based method, and formulated the task of identifying general positions as a max cut problem. Classifying agree/disagree opinions in conversational debates using Bayesian networks was presented in [12]. The work in [10] studied two tasks of rebuttal classification and debate-side classification. They extracted different types of linguistic features such as unigrams, bigrams, cue words, and syntactic dependency. Differently, we investigate both structural and linguistic features to model the participants' stances.

Other researchers have worked on a similar task called subgroup detection, i.e., clustering users by opinions. An unsupervised approach based on unsupervised clustering techniques was proposed in [13] to detect the split of discussions into subgroups of opposing opinions. The work in [14] identified subgroups in discussions by modeling positive and negative relations among participants. The work in [5] proposed a method to automatically discover opposing opinion networks of users from forum discussions using both textual content and social interactions. Our model can classify users into two opposing groups and determine the stance labels of the two sets, which is not the goal of subgroup detection.

3   Method

In order to recognize stances of interaction posts, we first calculate the consistency scores between participants by exploiting the historical activities and the textual content of interaction posts. The consistency scores are used as weight of links between nodes representing participants in the interaction graph. We then design a greedy algorithm applied to the interaction graph for the task of stance classification.

3.1   Consistency Score

Consistency score measures the degree of agreement between two participants when they interact with each other. It has two components: the historical consistency score and the timely consistency score, which are described as follows.

Historical Consistency Score: The historical consistency score estimates the overall opinion coherence or similarity between two users independently of any single debate. The score is estimated from all the opinions expressed in the historical debates. This is inspired by the idea of *collaborative filtering* in recommendation systems [15]. Users express their preferences by creating posts to support particular topics. The associated stances can be viewed as an approximate representation of the user's overall opinions. When comparing one user's opinions with other users', we can find opinion allies with most similar opinions and opinion enemies with opposing opinions. We assume that opinion allies are very likely to support each other in a new debate, while opinion enemies will oppose each other in the future.

To compute the historical consistency score, we extract the stance information associated with the posts in previous debates and generate an $m \times n$ matrix R, where m is the number of users and n is the number of topics. The element of $R(i,j)$ indicates the stance of *user i* towards *topic j*. Let $R(i,j)$ be 1, -1 or 0 when *user i* supports, opposes, or doesn't participate in *topic j*, respectively. We then apply Singular Value Decomposition (SVD) [16] to factor the $m \times n$ matrix R into three matrices as the following:

$$R = U \cdot S \cdot V^T \tag{1}$$

The matrix U represents the feature vectors corresponding to the users in the hidden feature space and the matrix V represents the feature vectors corresponding to the topics in the hidden feature space. For any two users i and j, we get the opinions vectors u_i and u_j from the user matrix U. The historical consistency score between i and j is defined as the cosine similarity of their opinion vectors:

$$HC(i,j) = cosine(u_i, u_j) \tag{2}$$

Timely Consistency Score: The timely consistency score estimates the degree of agreement between participants by analyzing the posts in the present debate. For any pair of participants, we extract the associated posts and label them as agreement or disagreement expressions. We use a bootstrapping method to do the labeling work.

We first create a small pattern dictionary that contains only a handful of agreement and disagreement patterns, e.g., "I agree with you" in the agreement set and "I disagree with you" in the disagreement set. Then, we collect other meaningful patterns based on the assumption of *user relation consistency*, i.e., the local attitude between the same two participants is identical in different posts in the same debate. For example, if there is a post between *user i* and *user j* that contains the agreement expressions, we assume that all the other posts between the same two users also express the agreement attitude in the first sentence. Iterating over the posts, we can extract rich attitudinal expressions to fill the agreement set E_{agr} or the disagreement set E_{dis}. For a new post p, we can determine its attitude by calculating $sim(p, E_{agr})$ and $sim(p, E_{dis})$, where $Sim(p, E)$ is defined as following:

$$sim(p, E) = max_{e \in E}\{cosine(p_s, e)\} \tag{3}$$

Here, p_s is the first sentence of post p. The sentences and expressions are represented as tf-idf vectors. Following is the definition of post-level consistency score, which represents the degree of agreement of a single post:

$$C_p = \frac{sim(p, E_{agr}) - sim(p, E_{dis})}{2} \tag{4}$$

The user-level consistency score or the timely consistency score is the mean value of post-level scores of all the posts between two users:

$$TC(i,j) = \frac{\sum_{p \in P(i,j)} C_p}{|P(i,j)|} \tag{5}$$

where $P(i,j)$ is the set of interaction posts between *user i* and *user j*, and $|P(i,j)|$ is the size of $P(i,j)$.

Overall Consistency Score: The overall consistency score between two participants is a linear combination of the historical consistency score and the timely consistency score:

$$C(i,j) = \alpha HC(i,j) + (1 - \alpha)TC(i,j) \tag{6}$$

where parameter $\alpha \in [0, 1]$ determines the weights of the two scores.

3.2   Stance Classification

We build an interaction graph according to the reply network in a debate. Let the graph be $G(V, E)$ where V represents the set of vertices corresponding to the participants within the debate. There is an edge connecting vertex v_i and v_j if and only if there are interaction posts between user i and j. For an edge $e \in E$, $e = (i,j)$, the weight $w(i,j)$ is determined by the consistency score $C(i,j)$. We design a greedy algorithm on the interaction graph to classify the nodes into two opposing groups: S_{sup} and S_{opp}, which represent the sets of participants supporting or opposing the main topic of the debate. We first initialize S_{sup} and S_{opp} with the users extracted from the root posts whose stances labels are assumed to be given. Then we greedily choose a new user with the most evident attitude towards the users already inside the two sets, and add him/her to the corresponding set. Table 2 shows the algorithm.

Table 2. The greedy algorithm on interaction graph for stance classification

The Greedy Algorithm on Interaction Graph
1: `Input:An interaction graph` $G(V,E)$`, stance labels of the root posts.`
2: `Initialize:Filling` S_{sup} `and` S_{opp} `with the users extracted from the root posts.`
3: `Repeatuntil` $S_{sup} \cup S_{opp} = V$`:`
4:     `Choose an edge (u, v) with maximum absolute weight` $\|w(u,v)\|$ `suchthat` $u \in S_{sup} \cup S_{opp}$ `and` $v \in V - (S_{sup} \cup S_{opp})$`.`
5:     `If(` $u \in S_{sup}$ `&&` $w(u,v) > 0$`) || (` $u \in S_{opp}$ `&&` $w(u,v) \leq 0$`):`
6:         `Add` v `to` S_{sup}`.`
7:     `Else:`
8:         `Add` v `to` S_{opp}
9: `Output:` S_{sup} `and` S_{opp} `describe the users foror against the main topic.`

4 Experiments

4.1 Dataset

We use the two-sided debates in createdebate as our datasets. Each two-sided debate (also called for/against debate) contains a title as the main topic and two side labels that can be interpreted as supporting or opposing the main topic according to the semantics. There may be several threads in a debate. Each thread is a tree with one or more nodes corresponding to the posts: one root post replying directly to the main topic and several interaction posts replying to other posts. A post p_i is the parent of post p_j if p_j is a reply to p_i. Note that the interaction graph $G(V, E)$ of a debate can be a disconnected graph, since a discussion thread may generate an isolated subgraph. Each post in the debate receives one stance label indicating which side it supports. We use these labels of interaction posts as the ground truth to evaluate the performance of our approach for the task of stance classification. Labels of root post are assumed to be given.

We crawled over 20,000 two-sided debates from createdeabte. We used the top-50 popular debates as the test data. Table 3 shows some statistics for the test data. All the other debates were used as the development data to learn the opinion vectors and build the agreement/disagreement expression sets. We discarded the posts whose authors have no clear opinions, e.g., some authors do not choose the specified labels to declare their stances.

Table 3. Some statistics of the debates in the test data

Avg# of posts	Avg# of root-posts	Avg# of interaction-posts	Avg# of participants	Avg# of posts per participant
367.9	79.1	288.8	82.3	4.5

4.2 Baselines

4.2.1 Text-Based Classification

Stance classification in online debates can be viewed as normal text classification problem. We apply one of the state-of-the-art text classification methods Support Vector Machines (SVM) [17] and use the software libsvm [18] for the experiments.

There are two methods to build the training and testing datasets for text-based classification: post-level or user-level. We treat each post as one document for the post-level method. For the user-level method, the integration of the textual data of all the posts created by the same user in a debate is a document. Both methods have advantages and disadvantages. There are more documents in the post-level method, but each document doesn't contain as much linguistic features as a document in the user-level method. We adopt both methods as the baselines and use 10-fold cross-validation to evaluate the performance.

4.2.2 Link-Based Approach

We employ a pure link-based method that ignores the text of posts and only uses the link structure of interactions [3]. This approach formulated the stance classification task as a max cut problem on the interaction graph. The relationships between participants were all treated as disagreements in their work. We also implement another approach described in [4], which classifies the local attitudes into "agree", "disagree" and "neutral", and uses BiqMac [19] to solve the max cut problem. The link weight represents the degree of disagreement in their work, which is contrary to our work. So we need to reverse the weight in our interaction graph before applying the max cut algorithm.

4.3 Results

4.3.1 Local Attitude Classification

To evaluate effectiveness of the consistency score, we use it to directly classify the relationship between two participants. We apply a simple threshold-based method: classifying the attitude as "agree" if the consistency score is above the threshold and "disagree" otherwise. We also conduct experiments on local attitude classification by only using the historical consistency score ($\alpha = 1$) or the timely consistency score ($\alpha = 0$). We summarize the results in Table 4. Another baseline is to treat every reply as disagreement [3].

Table 4. Accuracy of local attitude classification

All-disagreement	Historical-only	Timely-only	Historical + Timely
0.711	0.744	0.796	0.837

The results indicate that the historical and timely features are insightful for local attitude classification. Although most interactions are antagonistic (over 70 %), utilizing features extracted from historical activities and textual content of the posts can further improve the accuracy of agree/disagree classification.

4.3.2 Debate-Side Classification

We classify the participants into "support" and "oppose" groups, and then label the interaction posts with appropriate labels according to their authors' stances. In addition to accuracy, we also adopt another evaluation metric called estimation index accuracy described in [4], since the general accuracy doesn't work well when the posts are extremely biased. The estimation index accuracy is defined as follows:

$$EI\_Accuracy = \frac{1}{2}\left(\frac{|T_{sup} \cap S_{sup}|}{|T_{sup}|} + \frac{|T_{opp} \cap S_{opp}|}{|T_{opp}|}\right) \tag{7}$$

where T_{sup} and T_{opp} are the sets of "support" and "oppose" participants in the ground truth and S_{sup} and S_{opp} are the predicted sets of "support" and "oppose" participants. The experimental results are shown in Table 5.

Table 5. Accuracy of stance classification of interaction posts

	Accuracy	EI_Accurary
SVM (post-level)	0.613	0.607
SVM (user-level)	0.646	0.638
MaxCut1	0.754	0.751
MaxCut2	0.793	0.797
Our method	0.858	0.859

Our method achieves better accuracy than all other baseline methods. The text-based methods show low accuracy and the main reasons may be (1) posts in both debate sides talk about the same topic and use similar vocabulary, so it is hard to distinguish between the two classes using "bag-of-words" models; (2) interaction posts reply to other posts, not to the main topic directly, which may cause "topic drift". MaxCut1 and MaxCut2 are two link-based methods described in [3, 4]. The interaction graph corresponding to the online debate is not always a connected graph, since the participants in a single discussion thread may form an isolated subgraph that is not connected to the participants in other discussion threads. Therefore, we have to apply the max cut algorithm to different subgraphs separately. This may be the reason why the max cut methods do not perform as good as our greedy algorithm.

5   Conclusion and Future Work

We have shown a novel method for stance classification of interaction posts in online debates. Our approach is based on the observation that social media debate sites contain rich social behaviors and links carry less noisy information than text. We mine the historical activities to learn the opinion representations for participants. We combine this knowledge with textual contents of replies, and calculate the consistency scores between participants to determine the weight of the corresponding links in the inter-action network. A greedy algorithm is applied to the interaction graph for the task of stance classification of interaction posts. Empirical evaluation indicates that our method performs better than the competing baseline methods.

In social media debate sites, users participate in different debates on different topics. We observe that users may form stable relationships, i.e., they become opinion allies or enemies if they often support or oppose each other. We assume that opinion allies/enemies are more likely to support/oppose each other in future debates, which help build a better interaction graph for every debate.

There are several directions to extend our method. For example, we can deal with more sophisticated linguistic features to classify the agree/disagree attitudes. When replying to a post, participants may also acknowledge the viewpoint of the previous post even if they don't really support it. For example, from the sentence "Although you are right in a way, I support the other side", we can see that the author is express-ing disagreement though the sentence also contains expressions in the agreement set. Therefore, it's meaningful to distinguish between concessionary and non-concessionary opinions when determining the overall attitude of a post.

The social media debate sites contain other information that is not considered in this paper. For example, users may fill their personal information (such as age, religion) in the personal pages. Some debate websites also allow users to declare their friends, enemies, favorite topics, or other interesting features. All the information may boost the accuracy of classifying users into different parties in the debates. We leave the incorporation of the additional information for stance classification as future work.

Acknowledgement. This paper is partially supported by the National Natural Science Foundation of China (NSFC Grant Number 61472006), the Doctoral Program of Higher Education of China (Grant No. 20130001110032) and the National Basic Research Program (973 Program No. 2014CB340405).

References

1. Sridhar, D., Getoor, L., Walker, M.: Collective stance classification of posts in online debate forums. In: ACL 2014, p. 109 (2014)

2. Qiu, M., Yang, L., Jiang, J.: Modeling interaction features for debate side clustering. In: Proceedings of the 22nd ACM International Conference on Conference on Information & Knowledge Management, pp. 873–878. ACM (2013)
3. Agrawal, R., Rajagopalan, S., Srikant, R., Xu, Y.: Mining newsgroups using networks arising from social behavior. In: Proceedings of the 12th International Conference on World Wide Web, pp. 529–535. ACM (2003)
4. Murakami, A., Raymond, R.: Support or oppose?: classifying positions in online debates from reply activities and opinion expressions. In: Proceedings of the 23rd International Conference on Computational Linguistics: Posters, pp. 869–875. Association for Computational Linguistics (2010)
5. Lu, Y., Wang, H., Zhai, C., Roth, D.: Unsupervised discovery of opposing opinion networks from forum discussions. In: Proceedings of the 21st ACM International Conference on Information and Knowledge Management, pp. 1642–1646. ACM (2012)
6. Yessenalina, A., Yue, Y., Cardie, C.: Multi-level structured models for document-level sentiment classification. In: Proceedings of the 2010 Conference on Empirical Methods in Natural Language Processing, pp. 1046–1056. Association for Computational Linguistics (2010)
7. Bansal, M., Cardie, C., Lee, L.: The power of negative thinking: exploiting label disagreement in the min-cut classification framework. In: COLING (Posters), pp. 15–18 (2008)
8. Thomas, M., Pang, B., Lee, L.: Get out the vote: determining support or opposition from Congressional floor-debate transcripts. In: Proceedings of the 2006 Conference on Empirical Methods in Natural Language Processing, pp. 327–335. Association for Computational Linguistics (2006)
9. Somasundaran, S., Wiebe, J.: Recognizing stances in online debates. In: Proceedings of the Joint Conference of the 47th Annual Meeting of the ACL and the 4th International Joint Conference on Natural Language Processing of the AFNLP: Volume 1, vol. 1, pp. 226–234. Association for Computational Linguistics (2009)
10. Anand, P., Walker, M., Abbott, R., Tree, J.E.F., Bowmani, R., Minor, M.: Cats rule and dogs drool!: classifying stance in online debate. In: Proceedings of the 2nd Workshop on Computational Approaches to Subjectivity and Sentiment Analysis, pp. 1–9. Association for Computational Linguistics (2011)
11. Somasundaran, S., Wiebe, J.: Recognizing stances in ideological on-line debates. In: Proceedings of the NAACL HLT 2010 Workshop on Computational Approaches to Analysis and Generation of Emotion in Text, pp. 116–124. Association for Computational Linguistics (2010)
12. Galley, M., McKeown, K., Hirschberg, J., Shriberg, E.: Identifying agreement and disagreement in conversational speech: use of bayesian networks to model pragmatic dependencies. In: Proceedings of the 42nd Annual Meeting on Association for Computational Linguistics, p. 669. Association for Computational Linguistics (2004)
13. Abu-Jbara, A., Radev, D.: Subgroup detector: a system for detecting subgroups in online discussions. In: Proceedings of the ACL 2012 System Demonstrations, pp. 133–138. Association for Computational Linguistics (2012)
14. Hassan, A., Abu-Jbara, A., Radev, D.: Detecting subgroups in online discussions by modeling positive and negative relations among participants. In: Proceedings of the 2012 Joint Conference on Empirical Methods in Natural Language Processing and Computational Natural Language Learning, pp. 59–70. Association for Computational Linguistics (2012)
15. Sarwar, B., Karypis, G., Konstan, J., Riedl, J.: Application of dimensionality reduction in recommender system-a case study. Technical report (No. TR-00-043), Department of Computer Science and Engineering, University of Minnesota, Minneapolis (2000)

16. Golub, G.H., Reinsch, C.: Singular value decomposition and least squares solutions. Numer. Math. **14**(5), 403–420 (1970)
17. Joachims, T.: Text categorization with support vector machines: learning with many relevant features. In: Nédellec, C., Rouveirol, C. (eds.) ECML 1998. LNCS, vol. 1398, pp. 137–142. Springer, Heidelberg (1998)
18. Chang, C.C., Lin, C.J.: LIBSVM: a library for support vector machines. ACM Trans. Intell. Syst. Technol. (TIST) **2**(3), 27 (2011)
19. Rendl, F., Rinaldi, G., Wiegele, A.: Solving max-cut to optimality by intersecting semidefinite and polyhedral relaxations. Math. Program. **121**(2), 307–335 (2010)

Mining Long Patterns of Least-Support Items in Stream

Qinhua Huang[✉] and Weimin Ouyang

Modern Education Technique Center, Shanghai University of Political
Science and Law, Shanghai 201701, China
{hqh, oywm}@shupl.edu.cn

Abstract. The mining task of finding long sequential pattern has been well
studied for years. Typical algorithms often apply vary cascading support
counting methods, including the basic apriori algorithm, FP-growth, and other
derived algorithms. It is commonly known that during the mining process the
items with very high support may lead to poor time performance and very huge
useless branch search space, especially when the items in fact are not the
member of the end long pattern. On the other hand the items with least user
specified support, but will be the member of long pattern, might be discarded
easily. This problem could be more challenging in scenarios where the data
source is stream data, for data stream being unbounded, time-varied and
un-revisited. We carefully considered the role of hidden Markov chain structure
and then checked the item frequency evolution in stream mining context. In this
paper we presented a method of mining long patterns for data stream application
scenarios. Our algorithm can well overcome the negative effects generated in
stream scenarios.

Keywords: Stream · Long pattern · Least-support item · HMM

1 Introduction

Knowledge discovering in the stream data scenario has now become a basic research
problem in data mining. During the earlier past decade data mining and knowledge
discovery has well developed. The technical successes can be found in many fields
including science, industries, especially business application. And it is one of the basic
techniques of the prevalent current big data application. Among plenty of branches of
mining techniques and algorithms, frequent sequential pattern mining is a very interesting application. Many applications based on sequential pattern have come to the
world, such as DNA analysis, financial data analysis, varied kinds of human behavior
analysis, etc. With the big data application becoming widespread, the online and instant
analysis mining on data is urgently demanded. Big data are generated continuously in
vast volumes by diverse sources. The cost of revisiting this sort of big data is rather
expensive from aspects of both finance and time. Thus the research scenario of data
stream is been purposed and getting wide attention. Data in stream may have the same
attributes with in traditional data set. But notice data stream also may evolve, which
would probably change the aggregation features of the current data. This is called

© Springer International Publishing Switzerland 2015
D.-S. Huang et al. (Eds.): ICIC 2015, Part II, LNCS 9226, pp. 270–282, 2015.
DOI: 10.1007/978-3-319-22186-1_27

concept drift and it seriously challenged the techniques developed in data mining. In this paper we are going to discuss problem mining sequential patterns while reduce the bad impact bought by non-result but with high occurrence items in the scenario of data stream application.

Frequent pattern mining has been the basic subject due to substantial research, countless extensions of the problem scope, and broad application studies. Start from basic frequent pattern concept, there are different diverse patterns. We list some useful types here, association rules mining, high-dimension data mining, top-k patterns, and so on. Pattern mining applications are abundance, such as multimedia data, time-series, temporal data, spatial data, DNA and biological, etc. In a simple word, a frequent pattern is a pattern that satisfies a minimum support threshold. We are given a data base $\mathcal{D}$ of customer transactions. Each transaction may have contain some items, which are subset of item dictionary, $I = \{i_1, i_2, \ldots, i_n\}$. Denote the transaction with TId. These transactions are appeared in time order. And within a transaction TId, the items are arranged in lexicographic order. If a item occurs in many transactions, and it's occurrences is accumulated to be great than a user-specified frequency $\theta * |\mathcal{D}|$, we call this item frequent item and name the user-specified frequency, θ, as support. For simplicity, sometimes we define support to be $\theta * |\mathcal{D}|$. If a set of items occurred frequently in transactions, we call the itemset frequent pattern. A vertical format of transaction often used in application, described in Table 1.

Table 1. A demo transaction database $\mathcal{D}$

TID	Itemset
1	2, 3, 4, 5
2	1, 3, 4, 5
3	1, 2, 4, 5
4	1, 2, 3, 5
5	1, 2, 3, 4
6	6, 7, 8, 9
7	6, 7, 8, 9
8	6, 7, 8, 9
9	6, 7, 8, 9

From Table 1, the item dictionary is {1,2,3,4,5,6,7,8,9}. The supports of all of the single items are same, 4. Let's specify the threshold support to be 2. Only one 4-item set, {6,7,8,9}, is frequent. This is the longest frequent pattern in the database.

For the simplicity of calculation, the transactions database can be presented in binary format. As Table 2 shows.

The above mentioned term itemset can be taken as sequence of items. In the sequence, items occurred must follow the lexicographic order. For example, sequence $\langle 6, 7, 8, 9 \rangle$ and sequence $\langle 9, 8, 7, 6 \rangle$ are the same. This simplification from itemset to sequence will benefit our mining method in this context.

Table 2. Binary format of transaction database

TID	1	2	3	4	5	6	7	8	9
1	0	1	1	1	1	0	0	0	0
2	1	0	1	1	1	0	0	0	0
3	1	1	0	1	1	0	0	0	0
4	1	1	1	0	1	0	0	0	0
5	1	1	1	1	0	0	0	0	0
6	0	0	0	0	0	1	1	1	1
7	0	0	0	0	0	1	1	1	1
8	0	0	0	0	0	1	1	1	1
9	0	0	0	0	0	1	1	1	1

When coming to describe relations between items in a sequence, the term association is often used. An association rule takes the form of $\alpha \Rightarrow \beta$, where $\alpha \subset I$, $\beta \subset I$, and $\alpha \cap \beta = \phi$. Association can be expressed in probability representation: $\Pr\{\beta|\alpha\}$. And support and confidence are two measures of rule interestingness. An association rule is considered interesting if it satisfies both a user-specified support threshold and a minimum conditional probability threshold.

With associations between item-pairs, we can get a whole chain connect all of the items. HMM is a widely used model to deal with connections within sequence. Now we briefly introduce the Hidden Markov Model.

A discrete HMM, denoted by a triplet $\lambda = (A, B, \pi)$, is defined by the following elements: [10, 11].

A including set $S = \{S_1, S_2, \ldots, S_N\}$ of hidden states; transition matrix $A = \{a_{ij}\}$, representing the probability to go from state S_i, to state S_j.

$$a_{ij} = P[q_{t+1} = S_j | q_t = S_i], \ a_{ij} \geq 0, \ 1 \leq i, j \leq N, \Sigma_{j=1}^N a_{ij} = 1;$$

set $V = \{v_1, v_2, \ldots, v_T\}$ of observation symbols

B is a emission matrix, $B = \{b_j(k)\}$, indicating the probability of emission of symbol v_k when system state is S_j.

$$b_{j(k)} = P[v_k \text{ at time } t | q_t = S_j], 1 \leq j \leq N, 1 \leq k \leq T,$$

with $b_i(k) \geq 0$ and $\Sigma_{j=1}^T b_j(k) = 1$.

π is the initial state probability distribution, $\pi = \{\pi_i\}$, representing probabilities of initial states. $\pi_i = P[q_1 = S_i], 1 \leq i \leq N, \pi_i \geq 0, \Sigma_{i=1}^N \pi_i = 1$.

Our context of using HMM is that given a sequential pattern, say $\{1,1,1,1\}$, which items in Table 2 emitted the bit 1. Note that under the representation of binary itemset, we do not need to care about bit 0, because there is no 0 symbol in our pattern.

Data mining employ HMM is far from scarce. But in the following problem, it will meet big challenge.

Consider a 40×40 square table with each row being the integers from 1 to 40 in increasing order. Remove the integers on the diagonal, and this gives a 40×39 table, which we call $Diag_{40}$. Add to $Diag_{40}$ 20 identical rows, each being the integers 41 to 79 in increasing order, to get a 60×39 table. Take each row as a transaction and set the support threshold at 20. Obviously, it has an exponential number (i.e., $\binom{40}{20}$) of midsized closed/maximal frequent patterns of size 20, but only one that is the longest: $\alpha = (41, 42, \ldots, 79)$ of size 39 [9].

Literature [9] has analyzed the challenge when mining using apriori-based method. The exponential increasing in searching space makes mining process task unaffordable in calculation. But if we consider HMM, another problem appeared. The items that with high support, but are not the member of the longest result pattern, will take very bad effect on deciding pattern member items.

In stream mining, data flows infinitely, unbounded, and time-varied. Stream data are often with high dimensions. Researches on data streaming focused on mining frequent and sequential patterns, multidimensional analysis, classification, clustering, outlier analysis, and the online detection of rare events. To address the infinity problem of data stream, many single-scan or a few-scan algorithms using sliding windows, coarser granularity, micro-clustering, limited aggregation has been developed. Typical stream data mining applications include real-time detection of anomalies in computer network traffic, botnets, text streams, video streams, power-grid flow, web searches, sensor networks, cyber-physical systems and background protein domain ranking in structural biology [14].

2 Related Work

The problem of sequential patterns mining was originally proposed by Agrawal and Srikan [1]. A lot of algorithms have been proposed during the past decade, such as Apriori [1], GSP [2], PSP [3], PrefixSPAN [4], FPtree [5, 6], Spade [7], Spam [8]. Apriori and GSP both need to generate candidate set, while PrefixS-pan and Spam do not. In comparison with PrefixSpan, Spam prevails in running speed, but with lower efficiency in storage context, relatively. In mining progress, the problem of search space increases in a exponential speed greatly worse performance of algorithms. Pattern Fusion [9] try to overcome this difficulty by fusing the core patterns in the database, thus find some long frequent patterns with a certain probability. If mining using a random method, the result pattern could be constructed in a probability. That also means if an item have very high frequency, but not the member of the result long pattern, it might be kept with high probability to construct searching space.

HMM is a very popular framework in processing sequence. In our sequential pattern mining context, HMM could be possibly employed. But, our experiment showed, using HMM directly could not overcome the bad effect of high frequency as well. We studied the Viterbi algorithm and find that, if we want to employ HMM, the emission probability calculation should be made some modification. Furthermore in data stream scenario some data aggregation features may change as time flows, which will pose great challenge for the HMM calculation.

3 HMM Calculation and Streaming Strategy

In this section we describe our method in details. First, we present the framework of our algorithm.

3.1 Overview

The aim of our algorithm is to discover the longest pattern, while overcome the bad effect from high frequency of items other than the result items. Here we refer result item to items in the discovered long frequent patterns. For simplicity, we define the term non-result item to be item that is not in the destination frequent pattern.

Our proposed algorithm can be divided into 3 steps. First, the raw data should be preprocessed so as to simplify the calculation of support counting and connection building. After preprocessing, the single items are kept and the connections between any two items are constructed for future processing. Second, to evaluate the connections between more items, a modified HMM calculation mode is presented. The basic HMM model have some problems in our mining context. Third, we backtrack the chain to get the most probably item of long frequent pattern.

3.2 Preprocessing

According to apriori principle, if a itemset is frequent, every sub itemset of the itemset must be frequent. This is the most basic way to discover frequent pattern. But we can't rely on apriori principle all the way in discovering process, or else it will bring unaffordable calculation cost. To make the best of the principle, we employ it to get all of the frequent 1-item set, thus remove items do not reach the specified threshold of support.

After weeding out the unqualified items, next we will try to build connections between items. The connection between 2 items is within the simplest case. Naturally the conditional probability can well describe this relation. Let $Pr\{1|j\}$ be the probability of item i in existence of item j.

Now we are considering the simplification of counting. In transactions databases, transactions or sequences often have unequal length. This adds difficulty to our counting job. We proposed the following method to get this done easier.

Algorithm 1: Building binary transaction data matrix.

```
1 For each sequence in dataset
2   Generate an empty binary sequence with size of item
    dictionary.
3   For each item in sequence
4     Set the relative bit to be 1
5   End for
6 End for
```

Algorithm 2: Counting the support of each item and the conditional probability of between each two-items

```
1   For each column in binary transaction data matrix
2     Sum all of the bits to get support of each item.
3     If sum < specified threshold support
4       Delete the item from matrix
5     End if
6   End for
//we get a binary matrix and a support list, both have
// same column size, each column represents an element.
7   let L = the columns of the binary matrix
8   build a empty matrix (Trans) of size L×L to store the
      2-item connections, name this matrix CMAT
9   for each 2 column pair (column_i, column_j) in the bi-
      nary matrix, where i< j
10        Set product to be 0
11        For each row
12          product = product + bit(i) × bit(j)
13        End for
14      set CMAT(I,j) to be procuct/support(i)
15  End for
```

When finishing the processing, we get a upper-triangle matrix CMAT, which stores the transition probability between i to j, (i < j). Beware that the CMAT is a conditional probability matrix and items in pattern are in lexicographic order.

3.3   Constructing the Hidden Connection

The purpose of this step is to get a sequential pattern of length n as long as possible. When dealing with sequence processing problem, a very popular way is to consider HMM. To calculate in HMM, the conditional probability of any two items is needed as well as the emission probability and the initial states. From the definition of support, support in fact is the probability emission. We take the initial states to be equal distributed.

First, initialization step:

$$\begin{cases} \delta_1(i) = \pi_i b_i(o_1) & 1 \leq i \leq N \\ \psi_1(i) = 0 & (no\ previous\ states) \end{cases}, \tag{1}$$

where N is the is the size of item dictionary after preprocessing, π equal distributed.

Second, Recursion step:

$$\left.\begin{array}{l} \delta_t(j) = \max_{1 \le i \le N} [\delta_{t-1}(i)a_{ij}] b_j(o_t) \\ \psi_t(j) = \operatorname*{argmax}_{1 \le i \le N} [\delta_{t-1}(i)a_{ij}] \end{array}\right\} 2 \le t \le T; 1 \le j \le N, \qquad (2)$$

where a_{ij} is the transition probability in matrix CMAT.

The aim of our mining algorithm is to find the sequential pattern with max length, while this pattern should be existed in a curtain number (threshold) of transactions. But when checking the definition of emission rate b_j, we find that it can't be the taken as the frequency of item j. The reason can be seen intuitively from Eq. 2. As we know, as j increasing, which means sequential pattern increasing in size, item with higher support will get more weight for staying in the hidden chain. If the item is a non-result item, it will most probably have higher P^* by multiply a higher support. To minimize the negative effects of b_j and simplify the calculation, we set b_j to be constant during recursion. The experiment result proved it benefits us a lot.

$$\text{Third, Termination step:} \left\{ \begin{array}{l} P^* = \max_{1 \le i \le N} [\delta_T(i)] \\ q_T^* = \operatorname*{argmax}_{1 \le i \le N} [\delta_T(i)] \end{array} \right. .$$

In termination step, a problem will arise. When to terminate? Actually we do have strategy options when mining. Option 1, we can choose a size T, which is enough big, by user experience. When T is set too big, it is easy to find the P^* get values of 0 in last several stages. Let the length of mined longest pattern be nFP, N be size of item dictionary. Obviously we have nFP $\le$ n. Table 3 shows the calculation map in a matrix. This condition can easily be loosed to nFP $\ll$ n, because rarely we have most of the items in a frequent pattern. The new calculation map is described in Table 4. Clearly a great of calculation cost can be reduced. Option 2, keep calculating along the route on Table 3, terminate till we get all zeros in one column.

Table 3. Calculation map matrix when itemset size equal pattern length

Item	pattern	1	2	3	···	nFP
1		→				
2		→	→			
3		→	→	→		
⋮		···				
N		→	→	→	→	→

3.4 Back Tracking

And the optimal state sequence can be retrieved by backtracking.

$$q_t^* = \psi_{t+1}(q_{t+1}^*) \quad t = T - 1, T - 2 \ldots 1$$

Table 4. Calculation map matrix when itemset size great greater than pattern length

Item	Pattern	1	2	3	....	nFP	
1		➔					
2		➔	➔				
3		➔	➔	➔			
:		....			➔		
n-nFP		➔	➔	➔		➔	
			➔	➔		➔	
				➔		➔	
					➔	....	
N						➔	

By checking q_t^* we can decide which item is the most probable item in each bit of a pattern like $\{1,1,1,1\}$.

3.5 Stream Using Time-Tilted Window

In this section we will describe how to stream our method. In stream scenario, data flows infinitely. To let the aggregation calculation run, we can build data chunks and do calculation on the data chunks. By human experience, especially for tasks of real time application, current data are generally of most importance. We employed a logarithmic tilted-time window [13] to stream our method.

The basic tilted-time window is called natural tilted-time window, where the frequency of the most recent item is kept, take a example, with a precision of a quarter of an hour, in another level of granularity in the last 24 h and then again at another level in the last 31 days. As new data items arrive over time, the history of items will be shifted back in the tilted-time to reflect the changes. The logarithmic time scale is shown in Fig. 1. The current window holds data for the latest quarter. And the remaining windows contain last quarter, past 2 quarters, 4 quarters, 8 quarters,.... As a result, to represent a year, there are $\log_2(365 \times 24 \times 4) + 1 \approx 17$ units of time needed, instead of $366 * 24 * 4 = 35136$ units. Thus the space demanded is dramatically reduced.

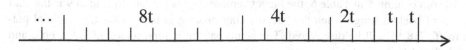

Fig. 1. Time tilted windows model

3.6 Windows Updating

As data stream flows continuously, the windows move forward. The size of for the windows is fixed. To avoid the overflow, some transaction sequences will be dropped. We adopted the same pruning strategy as proposed in FP-Stream [12]:

$$\exists l, \forall i, l \leq i \leq n, f_I(t_i) < \sigma \omega_i \ and \ \forall l', l \leq m \leq l' \leq n, \sum_{i=l}^{l'} f_I(t_i) < \in \sum_{i=l}^{l'} \omega_i$$

In equation, $f_I(t_i)$ is the frequency of a given itemset I in windows t_i, $\sigma \omega_i$ is the specified minimum support threshold in windows t_i.

4 Experiments

In this section, we tested our algorithms on synthetic and real datasets. First we run our algorithm on a small demo dataset to illustrate the accuracy of our algorithm. Second, we adjust the support of items with high frequency in a larger synthetic data set, which are not member items of result long patterns, to demonstrate the performance of our strategy. All experiments were done on a PC with a 2.6G dual-core CPU and 6 GB main memory running Linux of Ubuntu 64-bit 14.04 LTS. All algorithms were programmed using Matlab.

4.1 Experimental Datasets

We use synthetic datasets to test our proposed algorithm. They are described in the following subsections.

4.2 Small Synthetic Demo

In order to demonstrate the algorithm process in an intuitive way, we first run our algorithm on a synthetic small demo data set. This data set has 8 transaction sequence, contains 9 items. All the items in the first 5 transactions have same support of 4, while all the items in the last 3 transactions have same support of 3. Let the support threshold to be 3. It can be easily find that the longest frequent pattern is $\langle 6, 7, 8, 9 \rangle$, that is same with the last 3 transaction sequence (Table 5).

Table 6 shows the result probability matrix of algorithm before backtracking. The cells in the table head row label the member order of discovered pattern, the cells in left column are the items of dataset.

From column 5 in Table 6, the biggest probability is 0.3750, thus item 9 is the start point of backtracking. When the backtracking process is being called, sequential pattern $\langle 6, 7, 8, 9 \rangle$ will be discovered. The result pattern is represented in italic, bold and underlined font.

4.3 Attenuating the Importance of Non-result Items with High Support

If an item is not a member of the discovered result frequent pattern, we denote is non-result item.

Table 5. Small data set with 7 transactions

TID	Sequence
1	2, 3, 4, 5
2	1, 3, 4, 5
3	1, 2, 4, 5
4	1, 2, 3, 5
5	1, 2, 3, 4
6	6, 7, 8, 9
7	6, 7, 8, 9
8	6, 7, 8, 9

Table 6. The probability matrix of pattern members

item Pattern	1	2	3	4
1	0.5000	0	0	0
2	0.5000	0.3750	0	0
3	0.5000	0.3750	0.2813	0
4	0.5000	0.3750	0.2813	0.2109
5	0.5000	0.3750	0.2813	0.2109
6	*0.3750*	0	0	0
7	0	*0.3750*	0	0
8	0	0	*0.3750*	0
9	0	0	0	*0.3750*

To analysis the effect of non-result items but with high support, we generate a series of datasets based on dataset described in Introduction. Let us call the base dataset dataset_1. We will run our proposed algorithm on these datasets to check the impacts of high supports of non-result items on our proposed algorithm. Remember the last sequence, $\langle 41, \ldots, 79 \rangle$, has a support of 20. And it is easy to find the 39th sequence is $\langle 1, \ldots, 38, 40 \rangle$. The support of item 40 is 39. We repeat the 39th sequence one more time in order get item 40's support of 40. Name this new dataset to be dataset_2. Be aware that the support of item 40 is twice of support of sequential pattern $\langle 41, \ldots, 79 \rangle$.

Now let's continue to generate the rest synthetic datasets. To generate synthetic dataset_3, dataset_4, dataset_5, dataset_6, we repeat the 39th sequence in dataset_1 20, 40, 60 and 80 times respectively. Thus the frequencies of item 40 in the 39th sequence in dataset_2, dataset_3, dataset_4, dataset_5 and dataset_6 are 2, 3, 4, 5, 6 fold of sequential pattern $\langle 41, \ldots, 79 \rangle$'s frequency, respectively. Table 7 shows the relative multiples in value.

Table 8 presents difference of probability being the 39th result pattern element between item 79 and item 40. When the frequency of item 40 increases from 2-fold to 6-fold, item 79 remains greater probability of being the 39th element in the result pattern.

Table 7. The multiples of frequency of item 40 compared to frequency of sequential pattern $\langle 41, \ldots, 79 \rangle$ in each dataset

Dataset	Item 40' frequency represented in multiples of $\langle 41, \ldots, 79 \rangle$'s frequency
Dataset_2	2X
Dataset_3	3X
Dataset_4	4X
Dataset_5	5X
Dataset_6	6X

Table 8. The Probability comparison of item 79 and item 40, where item 79 is the last element of result long pattern, while 40 is not but with several-fold higher support.

Times	Probability of item 79	Probability of item 40
2	0.3279	0.0957
3	0.2469	0.1090
4	0.1980	0.1171
5	0.1653	0.1224
6	0.1418	0.1263

4.4   Stream Data Generation

The stream data was generated by the IBM synthetic market-basket data generator [13]. There were 10 M transactions generated using 1 K distinct items. The batch size of time-tilted window was set to 100 k. The $\in$ was set to 0.01σ. Other parameters were set using default values.

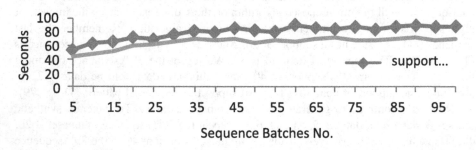

Fig. 2. The time performance of the method in stream scenario.

Figure 2 shows the performance of our stream strategy. As can be seen, with support varied from 0.005 to 0.0075, the time performance keeps nearly stable after batch 50.

5 Conclusion

In this paper we introduced a data mining problem that takes into account of the negative-effect of frequent items which might not be the members of resulting long discovered patterns. The problem of non-pattern member items with high frequency might make most algorithms lost in useless huge branch search space. To address this problem we proposed a method, focusing on the evolution of items instead of merely caring about high frequency support of items. Further we streamed our method to suit for stream scenario using time-tilted window. Our algorithm was tested on both synthetic fixed-size datasets and generated stream. The experimental results showed our proposed algorithm was able to be well suited for mining task to discover long patterns and had stable time performance in data stream scenario.

References

1. Agrawal, R., Srikant, R.: Mining sequential patterns. In: Proceedings of the Eleventh International Conference on Data Engineering, pp. 3–14. IEEE Computer Society Press (1995)
2. Srikant, R., Agrawal, R.: Mining sequential patterns: generalizations and performance improvements. In: Apers, P.M., Bouzeghoub, M., Gardarin, G. (eds.) EDBT 1996. LNCS, vol. 1057, pp. 3–17. Springer, Heidelberg (1996)
3. Masseglia, F., Cathala, F., Poncelet, P.: The PSP approach for mining sequential patterns. In: Żytkow, J.M. (ed.) PKDD 1998. LNCS, vol. 1510, pp. 176–184. Springer, Heidelberg (1998)
4. Pei, J., Han, J., Mortazavi-Asl, B., Pinto, H.: Prefixspan: mining sequential patterns efficiently by prefix-projected pattern growth. In: Proceedings of the 17th International Conference on Data Engineering (ICDE 2001), pp. 215–226 (2001)
5. Han, J., Pei, J., Yin, Y., Mao, R.: Mining frequent patterns without candidate generation. In: Proceedings of the 2000 ACM-SiGMOD International Conference Management of Data (SIGMOD 2000), pp. 1–12 (2000)
6. Han, J., Pei, J., Mortazavi-Asl, B., Chen, Q.: FreeSpan: frequent pattern-projected sequential pattern mining. In: Proceedings of the 2000 International Conference Knowledge Discovery and Data Mining (KDD'00), pp. 355–359. Boston, MA (2000)
7. Zaki, M.J.: Spade: an efficient algorithm for mining frequents sequences. Mach. Learn. **42**, 31–60 (2001)
8. Ayres, J., Gehrke, J., Yiu, T., Flannick, J.: Sequential pattern mining using a bitmap representation. In: SIGKDD 2001, Edmonton, Alberta, Canada (2001)
9. Zhu, F., Yan, X., Han, J., Yu, P.S., Cheng, H.: Mining colossal frequent patterns by core pattern fusion. In Proceedings of the International Conference Data Engineering (ICDE) (2007)
10. Rabiner, L.R.: Proc. IEEE **77**(2), 257–286 (1989)

11. Panuccio, A., Bicego, M., Murino, V.: A hidden Markov model-based approach to sequential data clustering. In: Caelli, T.M., Amin, A., Duin, R.P., Kamel, M.S., de Ridder, D. (eds.) SPR 2002 and SSPR 2002. LNCS, vol. 2396, pp. 734–743. Springer, Heidelberg (2002)
12. Giannella, C., Han, J., Pei, J., Yan, X., Yu, P.S.: Mining frequent patterns in data streams at multiple time granularities. In: Kargupta, H., et al. (eds.) Data Mining: Next Generation Challenges and Future Directions. MIT Press, Cambridge (2003). Ch. 3
13. www.almaden.ibm.com/cs/quest/syndata.html/#assocSynData

Network Guide Robot System Proactively Initiating Interaction with Humans Based on Their Local and Global Behaviors

Md. Golam Rashed[1(✉)], Royta Suzuki[1], Toshiki Kikugawa[1],
Antony Lam[1], Yoshinori Kobayashi[1,2], and Yoshinori Kuno[1]

[1] Graduate School of Science and Engineering, Saitama University,
Saitama, Japan
{golamrashed, suzuryo, kikugawa-tk, antonylam, kobayashi,
kuno}@cv.ics.saitama-u.ac.jp
[2] Japan Science and Technology Agency (JST), PRESTO, Kawaguchi, Japan

Abstract. In this paper, we present a Human Robot Interaction (HRI) system which can determine people's interests and intentions concerning exhibits in a museum, then proactively approach people that may want guidance or commentary about the exhibits. To do that, we first conducted observational experiments in a museum with participants. From these experiments, we have found, mainly three kinds of walking trajectory patterns that characterize *global behavior,* and visual attentional information that indicates the *local behavior* of the people. These behaviors ultimately indicate whether certain people are interested in the exhibits and could benefit from the robot system providing additional details about the exhibits. Based on our findings, we then designed and implemented a network enabled guide robot system for the museum. Finally, we demonstrated the viability of our proposed system by experimenting with a set of *Desktop Robots* as guide robots. Our experiments revealed that the proposed HRI system is effective for the network enabled *Desktop Robots* to proactively provide guidance.

Keywords: HRI · Local and global behavior · Network robot system

1 Introduction and Motivation

With the development of HRI systems over the last decades, social robots have started to move from laboratories to real-world environments [1], where a robot interacts with ordinary people. But, in most of these typical HRI systems, the interaction partners (human and robot) are restricted to controlled conditions. There has been a great deal of research to extend to more interaction modalities such as voice and gesture [2]. In addition, Yamazaki et al. proposed that robots should show their availability and recipiency by nonverbal behaviors so that people could more easily ask robots for help [3]. But in real-world environments, the identification of people in need is also an important task for service robots. In daily life, for example, if we find a person who is looking around with a map in his/her hand at a train station, we may offer assistance to that person. Dealing with such types of situations for a robot is quite difficult. However,

© Springer International Publishing Switzerland 2015
D.-S. Huang et al. (Eds.): ICIC 2015, Part II, LNCS 9226, pp. 283–294, 2015.
DOI: 10.1007/978-3-319-22186-1_28

very few research studies (for example, [4–6]) have been conducted in the fields of HRI that consider cases where robots are expected to offer their services to people who seem to need or want their potential services.

We note, there are several systems that address related issues in HRI. For example, robots have been deployed in public spaces, including day-care service centers [7], hospitals [8], train stations [9], office buildings [10, 11], museums [12–14], shopping centers [6] that address navigational and perceptual problems. Service robots deployed in child care centers [1], autism therapy center [15, 16] and in schools [17, 18] have addressed the quality of interaction between humans and robots. In addition, estimation of the intentions of the surrounding humans towards the robots is also highlighted in these works. Moreover, many robots have also been deployed in public spaces with the capability to encourage people to initiate interaction with them [19, 20]. What differentiates our paper is that we develop an HRI system for public spaces in which service robots can proactively serve people as opposed to the conventional reactive approach where robots wait until people explicitly request them for their service. The main research topic here is to find people who may want robots to serve them.

We can often tell what other people would like to do from their behaviors. There are various such behaviors: *global behaviors* such as walking trajectories, *and local behaviors* such as gaze patterns. It is difficult to devise a single sensor system to detect and recognize all these behaviors. Thus, we propose a network sensor system and a network robot system that uses all sensor data. The sensor system consists of a global sensor subsystem observing a large area and a local sensor subsystem with a number of distributed sensors each observing a small area. The robot system estimates people's intentions by combining the bits of information from the global and local sensor subsystems and proactively offers help to people in need. In our present study, we address this issue by taking museum guide robots as an example and develop a network-enabled HRI system which can find people who seem to be interested in a particular exhibit and offer them accordingly, more detailed explanations about that exhibit by any one of the guide robots in the network robot system.

In addition, there has been much research on technologies that are employed to track people's *local* and *global behavior* in the fields of robotics and computer vision [21, 22]. In ubiquitous computing, positioning devices are often used. These include the use of GPS, or the signal strength of radios (GSM, WiFi, Bluetooth, RFID) [5]. These technologies all used wearable or mobile personal devices, but these approaches have a number of weaknesses for applications in large public spaces. Thus, in our present work, we are interested in a wearable-free solution where people do not need to attach markers to themselves or carry special devices, as we wish to observe the unrestricted interests and intentions of all the people in public spaces. Hence, we use a global sensor subsystem based on Laser Range Finder (LRF) poles [23] to track people freely and to obtain their *global behavior* information (e.g. walking trajectories). We use a set of low cost USB video cameras as a local sensor subsystem to obtain people's *local behavior* information (e.g. visual attention). That is, our HRI system tracks the interests and intentions of people in public spaces using technologies that incorporate global and local sensor subsystems that are independent of traditional wearable sensor systems.

In the preliminary stage of our present work, we conducted observational experiments in an art museum and observed the behaviors of people with various levels of interest in the exhibits by using the technologies that we have discussed above. By analyzing the observed results, we have found that we can detect people who may desire the robot's service from their walking trajectory patterns and visual attention information in the museum. We then developed a network enabled multi-robot based HRI system that can find such people to offer them guidance using the guide robot for their exhibit of interest. Finally, we implemented our HRI system by incorporating a set of four *Naoko Desktop Robots* [24] as museum guide robots and tested the system with an art museum designed in our laboratory to confirm its effectiveness in proactively approaching selected people to provide guidance.

2 Observational Experiments

To conduct observational experiments in a museum scenario, we set up an art museum room sized 8 m × 10 m in our laboratory where we hung six paintings. The overall setup of our designed art museum is illustrated in Fig. 1(a).

The people at a museum may be there for various purposes with different intentions. Their interests in the exhibits may also vary. We considered the following three cases as the most typical cases.

- *Case 1.* They are interested in the collection in the museum and would like to look at all the paintings carefully.
- *Case 2.* They know that a famous painting by some painter is in this museum and would like to specifically appreciate it.
- *Case 3.* They are not particularly interested in any of the exhibits. For example, they could have just been in town and decided to visit the museum to pass the time.

We used four laser range sensor poles in the four corners of our designed museum room to track people and to obtain *global behavior* information. The details of the

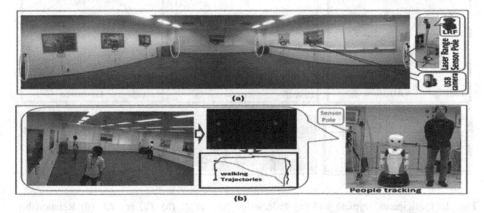

Fig. 1. (a) Setup of our designed art museum including the paintings, LRF poles and USB cameras, (b) People's walking trajectories tracked using laser sensor poles.

tracking method can be found in our previous work [23]. This method tracks the locations and orientations of people by using a particle filter framework, which is an important requirement for maintaining the continuity of walking trajectories in public spaces. Our tracking of people's walking trajectories via laser range sensor poles is illustrated in Fig. 1(b). We also placed low cost USB cameras just beneath each of the six paintings. The purpose of using the cameras is to obtain the *local behavior* information of people. In order to observe the above three types of people, we divided 48 participants (42 males, 6 females, average age 25.2 years) from Saitama University into three groups and instructed each group as follows.

- *Group 1* (14 participants). Look at all paintings carefully and choose one that you most like. We will ask about the painting later.
- *Group 2* (18 participants). We show them one of the paintings and tell them to look at that painting. We also ask about that painting later.
- *Group 3* (16 participants). No specific instruction is given.

In this study, we treat people's walking trajectory patterns as *global behavior* and visual attention (for example, indicated via face detection) as *local behavior*. In our observational experiments, we have found that there are three major trajectories *T1*, *T2*, and *T3* where *T1*: "*trajectory where all exhibits were viewed sequentially followed by a return back to any one of the previously viewed exhibits*", *T2*: "*trajectory where the person went straight to view only one exhibit before leaving the museum*", and *T3*: "*trajectory where all exhibits were viewed sequentially*". Examples of found walking trajectory patterns of type *T1*, *T2*, and *T3* are illustrated in Fig. 2(a), (b) and (c), respectively.

Figure 2(d) illustrates the relationships between the three groups and the walking trajectory patterns. We observed in videos that the faces of people viewing a given painting would always be frontal views in the painting specific USB video camera.

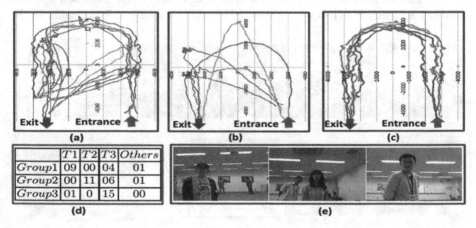

	T1	T2	T3	Others
Group1	09	00	04	01
Group2	00	11	06	01
Group3	01	0	15	00

(d)   **(e)**

Fig. 2. Participants' typical walking trajectories for (a) *T*, (b) *T2*, (c) *T3*. (d) Relationship between different groups of participants and their walking trajectory patterns. (e) Examples of visual attention observation of participants in three different cameras.

Figure 2(e) shows the examples of visual attention observation of different participants in three different painting specific USB video cameras.

People with trajectory pattern *T2* were found to be interested in a specific painting. Such people are the first candidates for the robot to offer its guidance. It is also natural for people to see paintings that they like multiple times. Thus people with trajectory pattern *T1* are assumed to be interested in the painting's that they returned to and are also candidates for receiving extra commentary from the robot about the painting.

In addition, after detecting the trajectory patterns of the people in order to observe their *global behavior*, we can be more sure about their interest towards any specific paintings if we can detect their visual attention as *local behavior* near the painting's specific USB video camera. It can be seen from our stored USB video camera footage that all the video cameras successfully captured the frontal faces of the participants, if and only if they really viewed the paintings for some extended period of time.

3 Our Proposed HRI System

Based on the findings from the observational experiments, we proposed a network enabled HRI system which is illustrated in Fig. 3. Basically, our system consists of two main types of sub-systems: the *Server Sub-System (SSS)* and painting specific individual client system-called *Client Sub-Systems (CSSs)*. We incorporated a total of three software units in these two types of sub-systems: **(a)** The human positions and walking trajectory pattern tracking unit-called the *Global Behavior Tracking Unit (GBTU)* in the *SSS*, **(b)** Visual attention detection and tracking units-called *Local Behavior Tracking Units (LBTU)*, and **(c)** the *Robot Control Unit (RCU)*. The latter two units are in each individual *CSS*. We employed one *CSS* for each of the paintings. For example, if we have *n* -exhibits in a museum then we will have one *SSS* and *n-CSSs* to implement our proposed system. Our HRI system detects people's interests and intentions toward the exhibits and guides them accordingly, if and only if the three units in two sub-systems work in order in real time. The communication between the *SSS* and *CSSs*

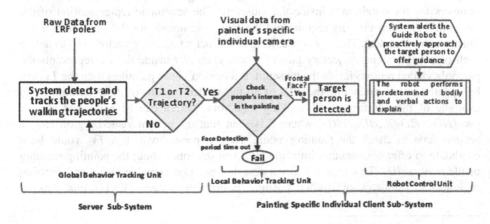

Fig. 3. States of our proposed HRI system

units are done by a TCP network connection. The functionalities of different units of our proposed system are described in the following subsections.

We assume that people are free to move inside the museum to view exhibits based on their own interests and intentions. If any person with *local* and *global behaviors* indicating strong interest in an exhibit is found, our HRI system immediately sends a command to the assigned guide robot to proactively approach him/her to offer its commentary on the exhibit. The potential steps of our proposed HRI system are stated below.

- *Step-1*: The *SSS* tracks the *global behavior* of people using the *GBTU* (described in Sect. 3.1).
- *Step-2*: If the *GBTU* detects interest from a person's *global behavior* (e.g. either walking trajectory pattern *T1* or *T2*) in front of any painting viewing region then **goto** *Step-3*, otherwise **goto** *Step-1*.
- *Step-3*: The *SSS* sends commands to the *LBTU* (described in Sect. 3.2) of a specific *CSS* to track the *local behavior* of that person.
- *Step-4*: If the *LBTU* detects interest from the person's *local behavior*, then **goto** *Step 5*, otherwise **STOP** and wait to receive future commands from the *SSS*.
- *Step-5*: The *CSS* sends commands to its *RCU* (described in Sect. 3.2) to trigger the assigned guide robot to proactively offer extra commentary about the painting to the target person.
- Step-6: After finishing its commentary about the paintings, the *CSS* makes the *LBTU* and *RCU* go idle and wait for future commands from the *SSS* if it identifies interest from the *global behavior* of other people (i.e. **goto** *Step 2*).

3.1 Server Sub-System (SSS)

Global Behavior Tracking Unit (GBTU). Our basic aim in introducing this unit is to detect walking trajectories patterns of type *T1* and *T2* as the *global behavior* of the people in the museum. We defined painting viewing regions[1] with region-ID (*rID*) numbers for the *n*-paintings inside the museum. The schematic representation of the museum's painting viewing regions and the *rID* assignments for the six-paintings are illustrated in Fig. 4(a). The *rIDs* are used to construct a trajectory vector (*TV*) to define each person's walking trajectory pattern where each *rID* inside the *TV* represents the person's visited regions inside the museum. If a person views paintings then the *TV* will be updated by concatenating the *rIDs* of all the visited regions of those paintings. An example of the TV for the *T1* walking trajectory pattern can be expressed as $TV_{T1} = [rID_1, rID_2, ..., rID_n, rID_2]$ where it is seen that the person visited region rID_2 a second time to check the painting again. Thus a person with this *TV* would be a candidate to offer commentary immediately from the robot about the painting residing inside region rID_2. To check whether a person's position is stable inside any viewing region, the variances of his/her position coordinate values ($Var(x)$ and $Var(y)$)

---

[1] The regions from where a person typically views exhibits in a museum.

for every past 10 consecutive frames are calculated and combined using $D = \sqrt{(Var(x))^2 + (Var(y))^2}$. For any person, if his/her position is inside any viewing region with body orientation toward the painting, and $D < D_{Thres}$ for the next 100 consecutive frames, then the system will assume that the person is stable at that region. Here D_{Thres} defines some threshold integer values. Again, if any person goes directly to any region rID to view the painting without viewing other paintings at other regions in the museum, we treat the person as having a $T2$ walking trajectory. However, our conditions of stability in this case require that his/her position to be inside any painting viewing region with body orientation toward the painting, and $D < D_{Thres}$ for next 200 consecutive frames. This is because it was observed from experiments that people with $T2$ trajectories spend more time viewing only one particular painting than the average time people take to view a painting.

Performance Evaluation. We examined the performance of our method to recognize the $T1$ and $T2$ walking trajectory patterns (*global behavior*) by applying it to the stored data recorded in the observational experiments. Our method correctly recognized 80 % (8 out 10), 81 % (9 of 11) of the $T1$, and $T2$ walking trajectory patterns, respectively, thereby recognizing participants that were initially interested in any particular paintings.

3.2 Client Sub-System (CSS)

Local Behavior Tracking Unit (LBTU). To estimate the visual attention (*local behavior*) of the person (whether s/he is looking at the painting or not), in this unit, we employ the Viola-Jones AdaBoost Haar-like frontal face detector [25] to the continuously capture the image plane from each USB video camera placed beneath each painting. This unit can detect the person's frontal face, which is when s/he really looks at the painting (as seen in our observational experiment, described in Sect. 2), thereby estimating that the visual attention of the person (*local behavior*) is high toward any particular painting. In this study, we experimentally set this face detection time to 5 frames. We applied the face detector to the stored video footage from our observational experiments and found fruitful detection of frontal faces of the participants. Examples of the participants' visual attention tracking are illustrated in Fig. 4(b). From the *local* and *global behavior* information of the people, our HRI system can accurately estimate the interest level of a person towards any given painting. For example, when a person's specific walking trajectory pattern is detected in the *GBTU* but his/her frontal face is not detected in the *LBTU* by the painting specific USB video camera, the person can be considered to be less interested in the painting. In such a case, the *CSS* will not alert the *RCU* to offer commentary about the painting to him/her.

Robot Control Unit (RCU). After successfully detecting a target person using the *GBTU* and *LBTU*, our HRI system immediately sends signals to the *RCU* of the corresponding *CSS*. Then the *RCU* triggers the guide robot to proactively perform predetermined verbal and gestural actions to attract the attention of the target person. After finishing its explanation on its assigned painting, the guide robot returns back to

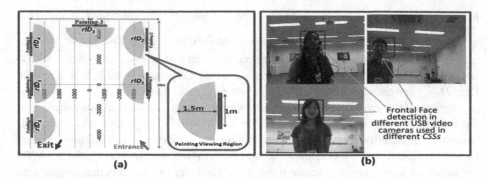

Fig. 4. (a) Schematic representation of the paintings, painting viewing regions and their *rID* assignment, (b) Examples of Visual Attention tracking

its idle state to receive future alert signals from the *RCU* to offer explanations about the painting for future interested people.

Naoko Desktop Robot. We used the *Naoko Desktop Robot* as the type of guide robot for every *CSS*. It is 40 cm in height and 3.0 kg in weight. It has 7 degrees of freedom (2 for each arm, 2 for the head in the form of pitch and yaw, and 1 for its body base's movement). The robot itself does not have any built-in artificial intelligence, but in the implementation of our proposed system, we programmed it to perform various kinds of functions for holding a conversation in a fixed location. We then placed each robot on a desk beside each of the paintings so it could proactively approach target people that are interested in particular paintings. If the attention level of any given person is detected as high in the *GBTU* and *LBTU* towards any specific painting, then the guide robot exhibits its verbal and gestural (head shaking, hand waving and body base movement in between the target person and the painting) actions to draw attention and offer commentary about the paintings.

4   Experiments

In order to confirm the effectiveness and accuracy of our proposed HRI system, we conducted experiments by implementing it using four painting specific Naoko desktop robots as guide robots for four paintings at the self-designed art museum in our laboratory where we previously conducted our observational experiments. A schematic representation of the experiment setup of our proposed HRI system is illustrated in Fig. 5(a). A panoramic view of the considered museum scenario under our proposed HRI system is illustrated in Fig. 5(b).

The experiments were conducted where people moved inside the museum to view the paintings, while our system estimated their *local* and *global behavior* to determine their intention and interest in the paintings. If any desired *local* and *global behaviors* of the people are found then one of the network enabled guide robots will proactively offer guidance to them.

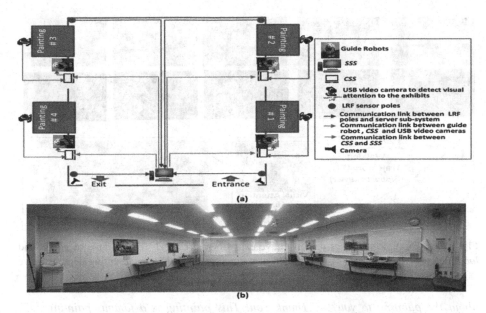

Fig. 5. (a) Experiment setup of our proposed HRI system, (b) Panoramic view of our considered museum scenario under our proposed HRI system.

Demonstration Using Guide Robots. We conducted demonstrative experiments with attendees from our university. In the demonstration, the attendees were asked to visit that museum under our proposed HRI system according to the instructions from our observational experiments (stated in Sect. 2). Two video cameras were placed at appropriate positions inside the museum to capture all the activities of the attendee and the guide robots during the demonstration. Example scenes are shown in Figs. 6(a) and 7(a). In both figures, the lower rows show the tracking results corresponding to the scenes in the upper row where each scene defines the potential steps of a typical demonstration with an attendee. Figures 6(b) and 7(b) show the *local* and *global behavior* tracking results of two typical cases under our proposed HRI system.

Case-1. Consider Fig. 6(a) where an attendee came inside the museum and briefly viewed painting #1 (scene #1) then moved to the next painting #2 (scene #2). In this case, he briefly viewed all the other remaining paintings (i.e. painting #3, and #4, as shown in scene #3 and #4, respectively). It is seen from the recorded video that he viewed every painting for a very short time. After that, it is also seen in scene #5 that the attendee moved again to view painting #2, and viewed that painting carefully for much more time than earlier. In such a case, our system detected the type of walking trajectory pattern (global behavior) of that attendee as *T1*, which is shown in the left side of Fig. 6(b). Accordingly, his visual attention (frontal face) toward that painting was detected as his *local behavior* which is shown in the middle of Fig. 6(b). Finally, the guide robot assigned to painting #2 approached him proactively offering more explanation about that painting by saying *"Excuse me, may I explain more details*

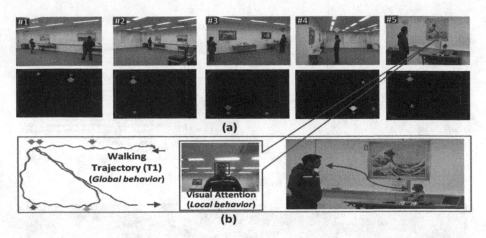

Fig. 6. (a) Example scenes of demonstrative experiments: *Case-1*, (b) Example of *global* and *local behavior* tracking results and session of a guide robot and attendee during demonstrative experiments: *Case-1*.

about the painting to you?…….Thank You! This painting is a famous painting…..". A snapshot of the session with the guide robot and the attendee is depicted in the right side of Fig. 6(b).

Case-2. In Fig. 7(a), it is seen that an attendee just came inside the museum and did not go to the painting viewing regions of all the paintings (scene #1, and scene #2) but went to the painting viewing region of painting #3 (scene #3) and stayed there for while to view that painting carefully. After a few moments, the walking trajectory pattern (*global behavior*) of that attendee was detected as *T2* and this is shown in the left side of Fig. 7(b). With a very short delay, his visual attention (frontal face) toward that

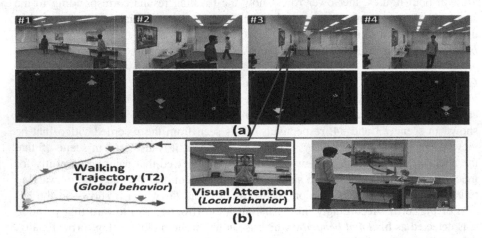

Fig. 7. (a) Example scenes of demonstrative experiments: *Case-2*, (b) Illustration of *global* and *local behavior* tracking result and a commentary session of between the guide robot and the attendee during demonstrative experiments: *Case-2*.

painting was detected as his *local behavior* which is shown in the middle of Fig. 7(b). Finally, the guide robot assigned to painting #2 approached him proactively offering extra commentary about that painting by saying *"Hello, may I explain to you more details about the painting?..........".* A snapshot of the commentary session between the guide robot and the attendee is shown in the right side of Fig. 7(b). After listening to the guide robot's commentary, the attendee left the museum room without making additional stops at other paintings (scene #4).

Conclusively, we can say that our proposed HRI system is effective for detecting the interests and intentions of people to paintings in museums from their *local* and *global behaviors* using our local and global sensor systems and is able to offer proactive guidance to them accordingly using the guide robots in our network robot system.

5 Conclusions and Future Work

In this study, we proposed a network enabled multi-robot based HRI system. Before doing that, we conducted observational experiments on the behaviors of people with varied interests in the exhibits inside a museum. Furthermore, we analyzed people's walking trajectories as *global behavior* information and visual attention as *local behavior* information, and finally proposed and implemented a new HRI system for museum environments to find people who are potentially interested in a particular exhibit and to offer them extra commentary using the *Naoko desktop robot* platform as guide robots. We found from the experiment's demonstration that our proposed HRI system effectively identifies people's interests and intentions to the paintings and proactively approaches them accordingly using the *Naoko desktop robot* to offer explanations about the paintings. We are now planning to deploy our proposed network guide robot system using the multi-robot setup in a real museum to detect potential people who actually need guidance. We will also evaluate our proposed network based HRI system by comparing it with existing conventional HRI systems. These are left for our future work.

Acknowledgments. This work was supported by JST, CREST.

References

1. Tanaka, F., et al.: Socialization between toddlers and robots at an early childhood education center. Proc. Natl. Acad. Sci. **104**(46), 17954–17958 (2007)
2. Papaefstathiou, E., et al.: Providing remote gestural and voice input to a mobile robot, US Patent App. 13/154,468 (2012). http://www.google.com/patents/US20120316679
3. Yamazaki, K., et al.: Prior-to-request and request behaviors within elderly day care: implications for developing service robots for use in multiparty settings. In: Bannon, L.J., Wagner, I., Gutwin, C., Harper, R.H.R., Schmidt, K. (eds.) ECSCW 2007, pp. 61–78. Springer, London (2007)

4. Glas, D.F., et al.: The network robot system: enabling social human-robot interaction in public spaces. J. Hum.-Robot Interact. **1**(2), 5–32 (2012)
5. Kanda, T., et al.: Abstracting people's trajectories for social robots to proactively approach customers. IEEE Trans. Robot. **25**(6), 1382–1396 (2009)
6. Kato, Y., et al.: May I help you?: design of human-like polite approaching behavior. In: Proceedings of the 10th Annual ACM/IEEE International Conference on Human-Robot Interaction, pp. 35–42. ACM, New York, NY, USA (2015)
7. Gyoda, M., et al.: Mobile care robot accepting requests through nonverbal interaction. In: 17th Korea-Japan Joint Workshop on FCV, pp. 1–5 (2011)
8. Engelberger, J.F.: Health-care robotics goes commercial: the helpmate experience. Robotica **11**, 517523 (1993)
9. Shiomi, M., et al.: Field trial of a networked robot at a train station. Int. J. Soc. Robot. **3**(1), 27–40 (2011)
10. Arras, K., et al.: Hybrid, high-precision localisation for the mail distributing mobile robot system mops. In: Proceedings of the 1998 IEEE International Conference on Robotics and Automation, vol. 4, pp. 3129–3134 (1998)
11. Asoh, H., et al.: Socially embedded learning of the office-conversant mobile robot jijo-2. In: Proceedings of 15th IJCAI 1997, pp. 880–885 (1997)
12. Burgard, W., et al.: Experiences with an interactive museum tour-guide robot. Artif. Intell. **114**(1–2), 3–55 (2000)
13. Nourbakhsh, I.R., et al.: The mobot museum robot installations: a five year experiment. In: 2003 IEEE/RSJ International Conference on Intelligent Robots and Systems, Las Vegas, Nevada, USA, pp. 3636–3641 (2003)
14. Thrun, S., et al.: Probabilistic algorithms and the interactive museum tour-guide robot minerva. Int. J. Robot. Res. **19**(11), 972–999 (2000)
15. Kozima, H., et al.: Interactive robots for communication-care: a case-study in autism therapy. In: IEEE International Workshop on ROMAN 2005, pp. 341–346 (2005)
16. Begum, M., et al.: Measuring the efficacy of robots in autism therapy: how informative are standard HRI metrics. In: Proceedings of the 10th Annual ACM/IEEE International Conference on Human-Robot Interaction, pp. 335–342. ACM, New York, NY, USA (2015)
17. Kanda, T., et al.: Interactive robots as social partners and peer tutors for children: a field trial. Hum.-Comput. Interact. **19**(1), 61–84 (2004)
18. Leite, I., et al.: Emotional storytelling in the classroom: individual versus group interaction between children and robots. In: Proceedings of HRI 2015, pp. 75–82 (2015)
19. Dautenhahn, K., et al.: How may I serve you?: a robot companion approaching a seated person in a helping context. In: Proceedings of the 1st ACM SIGCHI/SIGART Conference on Human Robot Interaction, pp. 172–179. ACM Press (2006)
20. Bergstrom, N., et al.: Modeling of natural human-robot encounters. In: IEEE/RSJ International Conference on Intelligent Robots and Systems, pp. 2623–2629 (2008)
21. Brscic, D., et al.: Person tracking in large public spaces using 3-D range sensors. IEEE Trans. Hum. Mach. Syst. **43**(6), 522–534 (2013)
22. Moeslund, T.B., et al.: A survey of advances in vision-based human motion capture and analysis. Comput. Vision Image Underst. **104**(2), 90–126 (2006)
23. Oyama, T., et al.: Tracking visitors with sensor poles for robots museum guide tour. In: The 6th International Conference on HSI, pp. 645–650 (2013)
24. Naoko x Robot. http://www.naoko-robot.net
25. Viola, P., et al.: Robust real-time face detection. Int. J. Comput. Vision **57**(2), 137–154 (2004)

Hierarchical Features Fusion for Salient Object Detection in Low Contrast Images

Nan Mu[1], Xin Xu[1,2(✉)], and Ziheng Li[1]

[1] School of Computer Science and Technology, Wuhan University of Science and Technology, Wuhan 430081, China
xuxin0336@163.com
[2] Hubei Province Key Laboratory of Intelligent Information Processing and Real-Time Industrial System, Wuhan University of Science and Technology, Wuhan 430081, China

Abstract. Salient object detection has drawn much attention recently. The extracted salient objects can be used for tasks including recognition, segmentation and retrieval. Most existing models for saliency detection can only be applied to the high contrast and normal contrast scenes. They are not robust enough to handle objects in low contrast images. Due to low lightness, low Signal to Noise Ratio (SNR) and low appearance information, saliency detection in low contrast scenes remains a challenge. To solve this problem, this paper proposes a saliency detection model based on the feature extraction, in which optimal features have been learned in different contrast environment to improve the effectiveness of salient object detection. Compared with other existing saliency models, the proposed method has strong adaptability for different scenes, and gives superior performance in detecting the salient object especially in low contrast images.

Keywords: Salient object detection · Feature extraction · Image contrast · Adaptive weight selection

1 Introduction

Visual saliency refers to the perceptual ability to select the most important visual information in the scene for further processing. The visual attention mechanism of human is capable of allocating the limited information resources effectively. Once introduced to the field of machine vision, the computing resources can be preferentially allocated to the regions which are apt to draw the observers' attention; this mechanism will greatly improve the efficiency of image analysis and processing methods. On this basis, salient object detection is proposed and developed. Using visual attention model to extract the salient object, can lay a foundation for further optimizing the image retrieval technology based on interest regions.

Most existing saliency models are based on the bottom-up approach because the visual attention is generally driven by the low-level stimuli such as edge, color, orientation, symmetry, etc. The approaches to detect the salient objects typically contain two steps. The first step is to extract the low-level features from the input image. The second step is to compute the saliency map by fusing the optimal features. After years

© Springer International Publishing Switzerland 2015
D.-S. Huang et al. (Eds.): ICIC 2015, Part II, LNCS 9226, pp. 295–306, 2015.
DOI: 10.1007/978-3-319-22186-1_29

of research, people have extensively studied the low-level features of the image, but the saliency features are still complex and difficult to determine. And how to choose the optimal features for salient object detection in low contrast scene is still the key point in current research. Up to now, a number of saliency models compute the image saliency primarily by measuring the contrast feature of the image. However, all of them have their weakness and perform poorly on low contrast scenes.

To gain insight into various scenes of different contrast, we have analyzed the influence of feature contrast in natural scenes. Here, we measure the contrast of the image by computing the root-mean-square (RMS) contrast [1]. This work first extracts various low-level features (lightness, color, edge, orientation, gradient, sharpness, shape, and coarseness) of the image in a global contrast method, which can fully represent the saliency information of the object. Then, we learn the properties of these features in high contrast, normal contrast and low contrast scenes. At last, the adaptive weight selection method is utilized to fuse these various features. To analyze the role of contrast in salient object detection, we conduct the experiment on the public MSRA dataset provided by [2], and divide these images into three categories: (i) the high contrast images, (ii) the normal contrast images, and (iii) the low contrast images. In comparison with other existing saliency models, the proposed method has better robustness in the low contrast images as well as the high contrast and normal contrast images.

The rest of this paper is organized as follows. Section 2 reviews the relevant salient object methods. Section 3 describes the feature extraction methods and analyzes the performance of these various features. Section 4 puts forward the adaptive feature fusion method and analyzes the properties of various features in different contrast scene. Section 5 presents the comparison of the proposed model and other saliency models. Finally, the conclusion draws in Sect. 6.

2 Related Works

Some representative salient object detection models are introduced in this section. For saliency is the attribute that an object stands out from its surrounding, most of the previous methods were devoted to measuring the difference of image features; they mainly extract the certain low-level features to detect the salient object.

Itti et al. [3] put forward the visual attention model, which is inspired by the feature fusion theory proposed by Treisman et al. [4] and the neurobiological framework proposed by Koch et al. [5], this model is implemented by extracting three low-level features (luminance, color and direction) at different scales, and the center-surround operator is utilized for the inner fusion with multi-scale features. Achanta et al. [6, 7] used the luminance and color features to detect the salient region of the image, the main method is calculating the contrast between the local image region and its surrounding region, and the saliency value of the input image is obtained by utilizing the average color vector difference. Cheng et al. [8] proposed the contrast model based on the region segmentation, this model utilizes the image segmentation algorithm to divide the image into different regions according to the homogeneity of different features such as luminance, texture, color, etc.

For various salient object detection models, they only extract some certain low-level features. Although they can successfully detect the salient objects in high contrast and normal contrast images, for the low contrast images, the detection effect is not ideal. Most current saliency models introduce the feature selection strategy for object detection. Klein et al. [9] detected the salient object by reconstituting the cognitive visual attention model. They computed the saliency of all feature channels in an information-theoretic way, and the optimal features are determined by using Kullback-Leibler Divergence (KLD). Qian et al. [10] constructed the saliency model by using a new concept of optimal contrast; this model is based on the sparse coding principle. Lu et al. [11] put forward the diffusion-based saliency model; they learned the optimal saliency seeds by combining two kinds of features: the bottom-up saliency maps and mid-level vision cues.

In order to improve the overall saliency detection performance and especially enhance the applicability on low contrast images. This research compares the effectiveness of saliency features in different contrast scenes and makes a further study about the optimal feature selection. The main contribution of this work is analyzing the impact of contrast change on various saliency features, and learning the optimal features which can better detect the salient object in low contrast images. The proposed model can retain the important information of each feature to achieve the best detection effect. Furthermore, the multi-scale, center priority and filtering method are also utilized to optimize the final salient map. As the experimental results shown in this work, the proposed method fused the optimal low-level features achieves better performance than other state-of-the-art salient region detection algorithms. Figure 1 shows the overview of the proposed model.

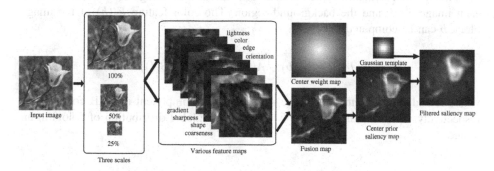

Fig. 1. Overview of the proposed saliency computational model.

3 Visual Features Extraction

By simulating the process of human visual system (HVS), it can be found that the stimuli which capture human visual attention are based on the low-level features. For this reason, to fully and accurately describe an image object, this paper extracts eight critical features: lightness, color, edge, orientation, gradient, sharpness, shape, and coarseness. In this section, we describe the specific method to extract these features, and analyze the contribution of each feature for salient object detection.

This research utilizes the block-based method to extract the global saliency features of the input image (denoted as I), which is divided into 8×8 blocks (denoted as b) with 50 % overlap. Let $F_n(b)$, $n = 1, \cdots, 8$ denote the saliency value of the block b, and $F_n(I)$ denote the saliency map of each feature.

3.1   The Extraction of Lightness Feature

Lightness is the perception attributes of HVS for visible objects radiation or glow amount, and the lightness feature measures how the average brightness of each block is different from the whole image, it can affect the performance of other low-level features. Let $F_1(b)$ denote the Euclidean distance of the average lightness value between image block b and the image I, in which the larger distance means the greater lightness value difference. The lightness feature is obtained by:

$$F_1(b) = |\bar{L}(b) - \bar{L}(I)|, \tag{1}$$

where $\bar{L}$ represents the average lightness component L in LAB color space.

3.2   The Extraction of Color Feature

The salient object is the image region which has a strong contrast to its surrounding; the reaction in the color space is that the color value of this image region is relatively far from the average color value of the whole image. Therefore, the color salient region can be extracted by calculating the Euclidean distance of the average color value between each image block and the background region. The color feature $F_2(b)$ of the image block b can be computed via:

$$F_2(b) = \sqrt{(\bar{A}(b) - \bar{A}(I))^2 + (\bar{B}(b) - \bar{B}(I))^2}, \tag{2}$$

where $\bar{A}$ and $\bar{B}$ represent the average A and B color component in LAB color space, respectively. Color feature can simply describe the global distribution of colors in an image.

3.3   The Extraction of Edge Feature

The edge refers to a collection of the pixel in which the grayscale intensity has a strong contrast change. Thus, we use the edge feature to capture the image region with dramatic brightness variations. Let $E(b)$ denote the binary map obtained through the Roberts edge detection. The edge feature $F_3(b)$ is then achieved by computing the average value of $E(b)$ as:

$$F_3(b) = \mu(E(b)), \tag{3}$$

The edge feature of the image is often associated with the discontinuity of image gray scale.

3.4   The Extraction of Orientation Feature

Based on the feature extraction method of Itti's model, the proposed method calculates the orientation feature by running the Gabor filter on the grayscale image of I. The two-dimensional Gabor function $G(x, y)$ is shown as follows (refer to [12]):

$$G_{\lambda,\theta,\varphi,\gamma}(x, y) = \exp(-\frac{x'^2 + \gamma^2 y'^2}{2\sigma^2}) \cos(2\pi\frac{x'}{\lambda} + \varphi), \tag{4}$$

where $x' = x\cos\theta + y\sin\theta$, $y' = -x\sin\theta + y\cos\theta$, $\gamma = 0.5$ represents the spatial aspect ratio and $\sigma = 0.56\lambda$ represents the standard deviation of the Gaussian factor. The parameter λ, φ, and θ denote the wavelength, the phase offset, and the angle, respectively.

The image processing based on the 2-D Gabor operator to calculate orientation image (denoted as $O_\theta(I)$, $\theta \in \{0°, 45°, 90°, 135°\}$) can be expressed as follows:

$$O_\theta(I) = I(x, y) * G_\theta(x, y), \tag{5}$$

Let $F_4(b)$ denote the orientation feature of image block b, which is obtained by computing the Euclidean distance between the orientation image block $O_\theta(b)$ and the orientation image $O_\theta(I)$ via:

$$F_4(b) = \sum_{\theta=\{0°,45°,90°,135°\}} |\bar{O}_\theta(b) - \bar{O}_\theta(I)|, \tag{6}$$

where $\bar{O}(b)$ and $\bar{O}(I)$ represent the average orientation value of $O(b)$ and $O(I)$, respectively. The orientation feature has global properties, so it makes saliency information have very good feasibility and stability even in the low contrast scenes.

3.5   The Extraction of Gradient Feature

The gradient feature is sensitive to the gradient variation; however, it is not sensitive to the grayscale of the image. Let $g(x, y)$ denote the grayscale at pixel (x, y) in image I, the size of which is $M \times N$, the gradient feature (denoted as $F_5(b)$) of image block b is acquired by averaging the abscissa squared gradient (denoted as $g_x(b)$) and the ordinate squared gradient (denoted as $g_y(b)$) through:

$$g_x(b) = \sum_{x=0}^{M-2} \sum_{y=0}^{N-1} g(x+1, y) - g(x, y)^2, \tag{7}$$

$$g_y(b) = \sum_{x=0}^{M-1} \sum_{y=0}^{N-2} g(\mathrm{x}, \mathrm{y}+1) - g(\mathrm{x}, \mathrm{y})^2, \tag{8}$$

$$F_5(b) = \mu\left(\frac{g_x(b) + g_y(b)}{2}\right), \tag{9}$$

Gradient value can describe the magnitude of the dramatic changes of the pixel values, for that the gradient map constituted by the pixel gradient values can reflect the local gray scale changes of the image.

3.6   The Extraction of Sharpness Feature

The sharpness feature measures how the acutance of each region is different from its surrounding, which can indicate the contrast of the adjacent region. The proposed method extracts the sharpness value $\phi(p)$ at position p by computing the convolution between the input image I and the first-order derivatives of the Gaussian via:

$$\phi(p) = \sum_x \sum_y \left[g(x,y) * G_x^\sigma(x,y)\right]^2 + \left[g(x,y) * G_y^\sigma(x,y)\right]^2, \tag{10}$$

where $g(x,y)$ is the grayscale at pixel (x,y) in the image region. $G_x^\sigma(x,y)$ and $G_y^\sigma(x,y)$ represent the first-order derivatives of the Gaussian in the vertical and the horizontal direction, respectively. σ is the scale of the Gaussian filter.

Let $\overline{\phi}(b)$ and $\overline{\phi}(I)$ denote the average sharpness value of the image block b and the input image I, respectively. Thus, the sharpness feature $F_6(b)$ can be expressed as the Euclidean distance between $\overline{\phi}(b)$ and $\overline{\phi}(I)$.

$$F_6(b) = |\overline{\phi}(b) - \overline{\phi}(I)|, \tag{11}$$

Sharpness feature can represent the image definition and the edge acuteness. It is less susceptible to local variations.

3.7   The Extraction of Shape Feature

The main method to extract the shape feature is using Hu's moment invariants [13], which has the invariant properties of rotation, translation, and scale. The proposed method calculates the 7 moments (denoted as $Mi, i = 1, \cdots, 7$) by Hu's method, which are obtained by normalizing the central moments through order two and three.

Let $M_i(b)$ and $M_i(I)$, $i = 1, \cdots, 7$ denote the 7 moments of the image block b and the input image I, respectively. Thus, the shape feature (denoted as $F_7(b)$) of b can be expressed as the Euclidean distance between $M_i(b)$ and $M_i(I)$.

$$F_7(b) = \sum_{i=\{1,2,\cdots,7\}} |M_i(b) - M_i(I)|, \tag{12}$$

The shape feature is not sensitive to the lightness and the contrast changes.

3.8   The Extraction of Coarseness Feature

Coarseness was seen by Tamura et al. [14] as the most fundamental perceptual texture feature, it measures the particle size of the texture pattern, of which the larger the particle size means the coarser the texture image, and in a certain sense, the coarseness implies the texture. Method of coarseness calculation is shown as follows:

First, calculate the average grey value (denoted as $A_k(x,y)$, $i = 1, 2, \cdots, 5$) of the neighborhood of size $2^k \times 2^k$ in I.

$$A_k(x,y) = \sum_{i=x-2^{k-1}}^{x+2^{k-1}-1} \sum_{j=y-2^{k-1}}^{y-2^{k-1}-1} g(x,y)/2^{2k}, \tag{13}$$

where $g(x,y)$ is the gray value at pixel (x,y) in the active window.

Then, for each pixel, calculate the average intensity difference (denoted as $E_{k,h}(x,y)$ and $E_{k,v}(x,y)$) between the non-overlapping neighborhoods in image I at the horizontal and vertical direction, respectively.

$$E_{k,h}(x,y) = |A_k(x+2^{k-1},y) - A_k(x-2^{k-1},y)|, \tag{14}$$

$$E_{k,v}(x,y) = |A_k(x,y+2^{k-1}) - A_k(x,y-2^{k-1})|, \tag{15}$$

At last, set the best size by k which gives the highest output value of E, and the coarseness (denoted as C) is the average of $S_{best}(x,y) = 2^k$.

$$C = \frac{1}{m \times n} \sum_{x=1}^{m} \sum_{y=1}^{n} S_{best}(x,y), \tag{16}$$

Let $C(b)$ and $C(I)$ denote the coarseness of the image block b and image I, respectively. Thus, the coarseness feature (denoted as $F_8(I)$) of b can be expressed as the Euclidean distance between $C(b)$ and $C(I)$.

$$F_8(b) = |C(b) - C(I)|, \tag{17}$$

Coarseness feature represents the surface properties of the whole image, and can well describe the integrity of the salient object. Meanwhile, the coarseness feature has good rotation invariance; it can effectively resist the interference of noise.

4   Adaptive Multi-feature Fusion

According to the discrete degree and clarity of the each feature maps, different weights are assigned to different features to merge into the final saliency map, thus can make better complementary between different feature information, and improve the robustness. Let v_n denote the statistical validity and ω_n denote the weights of different feature maps $F_n(I)$, $n = 1, 2, \cdots, 8$. The v_n is defined as (refer to [15]):

$$v_n = \sigma_n^2 + \kappa_n, \tag{18}$$

where σ_n^2 and κ_n represent the variance and the kurtosis of $F_n(I)$, respectively. The weights ω_n of different feature maps $F_n(I)$ are determined by the numerical magnitudes of the statistical validity vn via:

$$\omega_n = \begin{cases} 1, & \text{if } v_n = v_1^* \\ 3/4, & \text{if } v_n = v_2^* \\ 2/4, & \text{if } v_n = v_3^* \\ 1/4, & \text{if } v_n = v_4^* \\ 0, & \text{otherwise} \end{cases} \tag{19}$$

where $v_i^* = \text{sort}\{v_i\}$ use the descending order. The proposed method assigns different weights according to numerical sort of v_n. The final fusion map F_I is calculated by the weighted sum of the eight feature maps $F_n(I)$, $n = 1, 2, \cdots, 8$.

$$F_I = \frac{\sum_n w_n F_n(I)}{\sum_n w_n}, \tag{20}$$

To enhance the robustness of detection and achieve a preferable visual effect, the proposed method is performed in three scales: $\{100\%, 50\%, 25\%\}$, which can better suppress the interference of background information.

This research also uses the center prior principle to enhance the visual effect of salient objects. When humans watch a picture, they will naturally focus on the objects next to the center of the picture. Thus, in order to obtain the saliency objects closer to the human visual gaze, more weight is needed to add in the center of the image region. To realize the center prior, a feature (denoted as $f_c(b)$) is included to indicate the distance between each image block and the center of the image. For each image block feature $F_n(b)$, $n = 1, \cdots, 8$, they can be recalculated via:

$$F_n^*(b) = F_n(b)f_c(b), \tag{21}$$

Finally, the feature map $F_n(I)$ is generated by combining the image feature $F_n^*(b)$ of all the blocks. The generated saliency map is smoothed by a Gaussian filter (the template size is 10×10, and σ is 2.5).

5 Experimental Results

The experiment is implemented on the MSRA dataset created by Liu et al. [2] which includes two parts: (i) Image set A, contains 20,000 images and the principle salient objects are labeled by three users, and (ii) Image set B, contains 5,000 images and the principle salient objects are labeled by nine users. The proposed method which fuses the eight features is compared with other seven state-of-the-art methods: Itti's (IT) method [3], spectral residual (SR) method [16], saliency using natural statistics (SUN) method [17], frequency-tuned (FT) method [18], non-parametric (NP) method [19], context-aware (CA) method [20], and patch distinction (PD) method [21].

We choose the low contrast images for testing, of which the RMS contrast is less than 0.5. Figure 2 displays the subjective performance comparison of these various salient region detection methods. From Fig. 2, the salient objects extracted by the proposed method are more similar to the real salient objects, and the saliency maps have a uniform salient region same with the ground-truth binary masks. The approaches [16, 17] can not detect the real salient object under the condition of complicated background. The saliency maps of method [3, 19] look rather blurry, and are difficult to clearly distinguish the salient region. The saliency maps of method [18] retain a lot of background information, and do not have a high accuracy for detecting the texture images. The CA and PD method [20, 21] have a good detection effect, but the salient objects they detect are not uniform.

The performance assessment is implemented by computing the True Positive Rate (TPR) and the False Positive Rate (FPR). Given the ground-truth binary mask and the

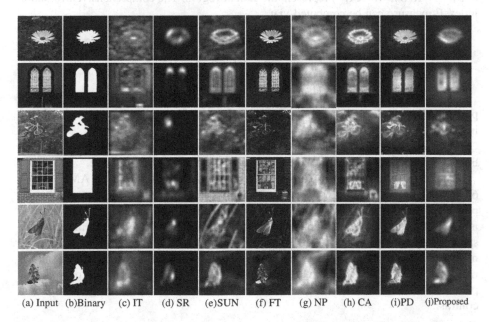

(a) Input   (b)Binary   (c) IT   (d) SR   (e)SUN   (f) FT   (g) NP   (h) CA   (i)PD   (j)Proposed

Fig. 2. The saliency maps comparison of the proposed method and other salient region detection methods in low contrast images (RMS contrast is less than 0.5) of MSRA dataset.

obtained saliency map $F_I(x, y)$ $(0 \leq F_I(x, y) \leq 1)$ of this method. The performance comparison of TPR and FPR results of the seven methods and the proposed method can be intuitively seen in Fig. 3, which shows that the proposed method has a better performance.

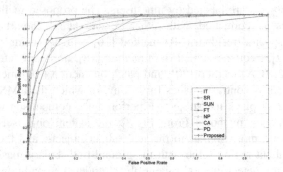

Fig. 3. ROC curves performance comparison of the proposed models with other seven saliency models.

Given the generated saliency map $F_I(x, y)$, we set a threshold T (computed by Otsu's method) to segment the saliency objects. The binary marks is denoted as $B_I(x, y)$. Let G and B_I denote the ground-truth binary mask and binary map of the proposed approach, respectively. The $Precision = R(B_I \cap G)/R(B_I)$ and $Recall = R(B_I \cap G)/R(G)$, where $R(\cdot)$ represent the salient region. The comprehensive evaluation index F-measure can be computed via:

$$F_\beta = \frac{(1 + \beta^2) \, Precision \times Recall}{\beta^2 \times Precision + Recall},$$ (22)

The proposed method uses $\beta^2 = 0.5$ to weigh the precision and recall. The comparison of precision, recall, and F-measure of these various saliency models are shown in Fig. 4.

As seen in Fig. 4, the F-measure value of the proposed method is relatively higher than the other seven methods, which indicates an excellent performance.

The proposed algorithm is carried on in MATLAB environment on AMD Athlon 64 X2 Dual Core CPU PC with 2 GB RAM. Table 1 shows the computational complexity of the proposed method and other seven methods. In relative terms, this method is more efficient than CA which also uses the multi-scale spatial domain method. The spectrum domain methods [16, 18] are time-saving, but their accuracy is not high. The method [19] combines the spectrum domain and the spatial domain method, and its time consumption is also high. Other methods under a single scale, which only choose parts of the image features, are difficult to adapt to the complex image of different contrast.

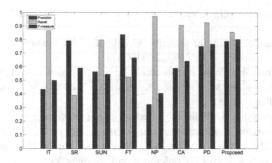

Fig. 4. The precision, recall, and F-measure performance comparison of various saliency models.

Table 1. The run-time performance (in seconds) comparison of various saliency models.

IT [3]	SR [16]	SUN [17]	FT [18]	NP [19]	CA [20]	PD [21]	Proposed
2.4384	0.4697	3.2196	0.5777	4.5013	118.2615	19.8938	20.4390

6 Conclusions

In this paper, we explore the connections between the influence of different contrast scenes and the effectiveness of different features in salient object detection. The proposed model is based on adaptive method to fuse the eight features (lightness, color, edge, orientation, gradient, sharpness, shape, and coarseness) of the input image; the generated saliency map can highlight the salient region uniformly. From the comparison with other seven state-of-the-art salient region detection algorithms on MSRA dataset, the proposed method has high robustness and achieves the optimal performance in different contrast scenes; it can further accurately extract the saliency objects from a complex low contrast background.

Acknowledgement. This work was supported by the Natural Science Foundation of Hubei Provincial of China (2014CFB247), and the National Natural Science Foundation of China (No. 61440016, 61273303).

References

1. Moulden, B., Kingdom, F.A., Gatley, L.F.: The standard deviation of luminance as a metric for contrast in random-dot images. Perception **19**(1), 79–101 (1990)
2. Liu, T., Yuan, Z., Sun, J., Wang, J., Zheng, N., Tang, X., Shum, H.Y.: Learning to detect a salient object. In: IEEE Conference on Computer Vision and Pattern Recognition, pp. 1–8 (2007)
3. Itti, L., Koch, C., Niebur, E.: A model of saliency-based visual attention for rapid scene analysis. IEEE Trans. Pattern Anal. Mach. Intell. **20**(11), 1254–1259 (1998)

4. Treisman, A.M., Gelade, G.: A feature-integration theory of attention. Cogn. Psychol. **12**(1), 97–136 (1980)
5. Koch, C., Ullman, S.: Shifts in selective visual attention: towards the underlying neural circuitry. Hum. Neurobiol. **4**(4), 219–227 (1985)
6. Achanta, R., Estrada, F.J., Wils, P., Süsstrunk, S.: Salient region detection and segmentation. In: Gasteratos, A., Vincze, M., Tsotsos, J.K. (eds.) ICVS 2008. LNCS, vol. 5008, pp. 66–75. Springer, Heidelberg (2008)
7. Achanta, R., Susstrunk, S.: Saliency detection using maximum symmetric surround. In: IEEE 17th International Conference on Image Processing, pp. 2653–2656 (2010)
8. Cheng, M.-M., Zhang, G.-Xi., Mitra, N.J., Huang, X., Hu, S.-M.: Global contrast based salient region detection. in: IEEE Conference on Computer Vision and Pattern Recognition, pp. 409–416 (2011)
9. Klein, D.A., Frintrop, S.: Center-surround divergence of feature statistics for salient object detection. In: IEEE International Conference on Computer Vision, pp. 2214–2219 (2011)
10. Qian, X., Han, J., Cheng, G., Guo, L.: Optimal contrast based saliency detection. Pattern Recogn. Lett. **34**(11), 1270–1278 (2013)
11. Lu, S., Mahadevan, V., Vasconcelos, N.: Learning optimal seeds for diffusion-based salient object detection. In: IEEE Conference on Computer Vision and Pattern Recognition, pp. 2790–2797 (2014)
12. Daugman, J.G.: Uncertainty relation for resolution in space, spatial frequency, and orientation optimized by two-dimensional visual cortical filters. J. Opt. Soc. Am. A Opt. Image Sci. Vision **2**(7), 1160–1169 (1985)
13. Hu, M.-K.: Visual pattern recognition by moment invariants. IRE Trans. Inf. Theor. **8**(2), 179–187 (1962)
14. Tamura, H., Mori, S., Yamawaki, T.: Texture features corresponding to visual perception. IEEE Trans. Syst. Man Cybern. **8**(6), 460–473 (1978)
15. Vu, C.T., Chandler, D.M.: Main subject detection via adaptive feature selection. In: IEEE 16th International Conference on Image Processing, pp. 3101–3104 (2009)
16. Hou, X., Zhang, L.: Saliency detection: a Spectral residual approach. In: IEEE Conference on Computer Vision and Pattern Recognition, pp. 1–8 (2007)
17. Zhang, L., Tong, M.H., Marks, T.K., Shan, H., Cottrell, G.W.: SUN: a Bayesian framework for saliency using natural statistics. J. Vision **8**(7), 32:1–32:20 (2008)
18. Achanta, R., Hemami, S., Estrada, F., Susstrunk, S.: Frequency-tuned salient region detection. In: IEEE Conference on Computer Vision and Pattern Recognition, pp. 1597–1604 (2009)
19. Murray, N., Vanrell, M., Otazu, X., Parraga, C.A.: Saliency estimation using a non-parametric low-level vision model. In: IEEE Conference on Computer Vision and Pattern Recognition, pp. 433–440 (2011)
20. Goferman, S., Zelnik-Manor, L., Ayellet, T.: Context-aware saliency detection. IEEE Trans. Pattern Anal. Mach. Intell. **34**(10), 1915–1926 (2012)
21. Margolin, R., Tal, A., Zelnik-Manor, L.: What makes a patch distinct? In: IEEE Conference on Computer Vision and Pattern Recognition, pp. 1139–1146 (2013)

Big Data Load Balancing Based on Network Architecture

Yu Bin[✉]

Faculty of Information Engineering, City College Wuhan University of Science
and Technology, Wuhan 430083, China
yub2k@163.com

Abstract. The Big Data load balancing technology involves the establishment of cross the cluster network channel, cluster deployment, task scheduling optimization. These should be assessed in a longitudinal study based on network architecture. This paper is to summarize the current situation of load balancing research about Big Data, and provide a basis for management and research of Big Data in the future.

Keywords: Big data · Network architecture · Load balancing

1 Network Architecture and Load Balancing at the Age of Big Data

BDaaS (BigData-as-a-Service) is a kind of data usage patterns, it encapsulates all kinds of data, it is the unified data modeling based on various types of data, and it will provide ubiquitous, standardized, on-demand retrieval, analysis and visualization service delivery. BDaaS is not only a new technology, but also a new data resource usage patterns and a new economic model [1, 2].

The main features of Big Data in the network include the off-line calculation, high bandwidth cluster communication, task scheduling and fault-tolerant storage ability [3]. The challenge to Big Data in the network includes: low cost non-blocking network structure, SDN cluster deployment optimization, TCP Incast solution, the optimization of task scheduling.

Load balancing is a major technique to improve network performance and network service quality. Load balancing is based on network architecture. It provides a cheap and effective method to expand the bandwidth and increase throughput capacity of server, strengthen data processing ability of network, increase network flexibility and availability. It is mainly to complete the following tasks: to solve the problem of network congestion and to provide service without having to go far to implement independent geographical location, to provide better access quality for the user, to improve the server response speed, to improve the utilization efficiency of the server and other resources, to avoid that key part of the network is a single point of failure.

© Springer International Publishing Switzerland 2015
D.-S. Huang et al. (Eds.): ICIC 2015, Part II, LNCS 9226, pp. 307–317, 2015.
DOI: 10.1007/978-3-319-22186-1_30

2 Non-blocking Network Architecture

With the arrival of Big Data, there are more and more abundant business that produced a huge shock to data flow model of the data center. Big Data business, such as searching, parallel computing, need a cluster of large servers to complete the work in coordination. Then the data flow becomes very large between the servers.

The complicated and changeable demands also brought traffic uncertainty. The server traffic cannot be accurately predicted, the network bandwidth cannot be planned through the design. At the same time, virtualization brought dynamic live migration between virtual machines, further leading to the network traffic model become more complex, and more across flow increased [4].

Because of the changing of data center traffic model, traditional convergence network will no longer meet the data center business needs. We need a non-blocking network deployed in the data center, and data flow can be interacted at line speed between any servers inside the data center.

At present, Fat tree architecture (Fat-Tree, proposed by Charles E. Leiserson in the last century eighty's) is the non-blocking network technology which the industry generally recognized to achieve. The basic idea is that a lot of low performance switches construct non-blocking network scale.

In order to achieve the no convergence of network bandwidth, each node of the fat tree network (except for the root node) need to ensure the uplink bandwidth and downlink bandwidth equal, and each node must have the ability to provide access bandwidth wire-speed forwarding. In the fat tree architecture, in order to achieve a flexible extension, the root node sets aside the same uplink bandwidth as access downlink.

In the theoretical model of the fat tree architecture, the whole network uses the same performance switches. But in the data center network, Access Switch is only responsible for accessing a small amount of servers, its forwarding capacity requirement is much lower than the Aggregation Switches and the Core Switches.

Therefore, we usually use cassette switch in the TOR (Top of Rack, at the top of the switch cabinet, uplink accessing bandwidth, each port connected to a single server.) position, but use high performance frame type switch in the Convergence Zone and the Core Zone to meet the demand of the network. At the same time, it can enhance the network performance, simplified deployment.

In the fat tree network as shown below, we can also use the frame type switch with high performance on the root, to reduce the number of switches in the network, and then reduce the complexity of network deployment and maintenance, simplify wiring (Fig. 1).

From the topology of fat tree architecture, it can be seen, there is a loop in the fat tree network. In order to achieve non-blocking switching, STP this "blocking link" way cannot be used, to make full use of all link resources. In current relatively mature technology, in addition to the traditional routing protocols (running in 3rd layer of IP network), there is the TRILL protocol (running in 2nd layer of ETH network). TRILL (Transparent Interconnection of Lots of Links) is an IETF standard implemented by devices called RBridges (Routing Bridges) or TRILL Switches. TRILL combines the

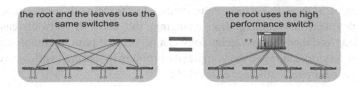

Fig. 1. The use of high performance switch

advantages of bridges and routers and is the application of link state routing to the VLAN-aware customer-bridging problem.

In order to reduce the network delay, data center network deployment is becoming flatter. The data center network is directly using the Core switch and TOR switch, Fat tree can be deployed between Core and TOR, two layer ETH network is composed of non-blocking. Flatter tree network can fully meet the deployment of large scale, super large scale data center network [5].

3   SDN Cluster Deployment Optimization

3.1   Software Defined Network

Networking changing, network devices N: 1 cluster virtualization, on a certain scale optimized the network architecture within a certain range, and then network design simplified. The Large 2 Layer network technology, by eliminating the loop factors, supports that virtual machines carry out two layers diffusion computing in a wide range under the virtualization conditions.

New technologies for commercial will cause equipment upgrading, and along with the great changes in flow, deployment and change on the network become more and more complex. It is difficult to have good expectation in response to variation in network traffic. In this way, once the service deployment completed, servers connected to the network through cable, it is difficult to control the influence of application traffic throughput on the network. Adjusting on the network becomes quite lag.

Software Defined Network, the emergence and evolution of the concept SDN, began to change the network passive status. SDN makes the network has "definition" ability of greater degree of flexibility. This definability, is the network active "processing" flow rather than passive "bearing" flow, and the relationship between the network and computing is not just "butt to butt", but "interaction".

In SDN, the Control Layer and the Physical Data Forwarding Layer are separated, this has a profound impact on the network switch work mode. The high-end users are not satisfied with using the original preset function on the network. They hope to adjust according to own demand rapidly while their business continue to enrich. After the Control Layer separated, Control Layer can open out. It can realize virtualization flexibility easier. User can program, and the quick response based on application and flow change does not need completely dependent on the long period hardware and software upgrades by equipment suppliers.

The idea of SDN is to provide more control to the network users. In addition to design deployment and configuration changing, the network software can also be reconstructed. And then the new technology can be verified prior to commercialization. This kind of network can solve the complexity problem by abstract way, and can study and meet business needs at a higher level [6].

3.2   SDN Classic Architecture and OpenFlow

The most classic SDN architecture description is from ONF (Open Network Foundation). Figure 2 is SDN architecture diagram.

A standardized interaction protocol used to decouple the Data Forwarding Layer and Control Layer, this is currently OpenFlow. Figure 3 is a graph from Stanford showed OpenFlow decoupling model.

SDN introduces the concept of Centralized Control. In SDN open architecture, the network within a certain range (or SDN domain), is managed by the Centralized Control Logic Unit. Thus the issue operating and managing independently dispersed large number devices will be solved, and then the network design, deployment, operation and maintenance, management will be completed at a control point, and also the differences in the underlying network will be eliminated by architecture decoupling. Centralized Control introduces SDN Controller, different from traditional network

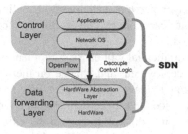

Fig. 2. SDN architecture

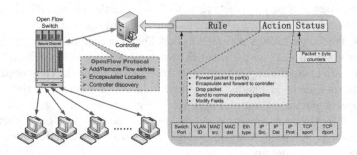

Fig. 3. OpenFlow work mode

architecture role, running SDN network operating system and controlling the control unit of all the network nodes. SDN can provide the interface of network applications, on basis of it can carry out software design and programming in accordance with business needs. Being loaded on SDN Controller, new network features of the whole network will quickly upgrade, without having to operate independent on each network element node.

Network device is composed of standard network hardware and OpenFlow Agent software. A variety of network applications running on the Controller, are converted into OpenFlow "Instruction Set" to be issued to the OpenFlow agent software on the device. So it is easy to achieve a standardized model, OpenFlow becomes an important technology in SDN architecture.

3.3 H3C SDN Framework and Strategy

H3C, Hangzhou H3C Technologies Co., Ltd. H3C SDN is based on the end to end overall network architecture. H3C SDN provides three schemes: Controller/Agent, Open API, OAA based on custom network platform [6].

H3C provides standardized series Controller components, OpenFlow equipments can be centralized controlled using OpenFlow protocol, and provides flexible open interface to upper layer, to meet the needs of a variety of network applications call demand.

IRF (Intelligent Resilient Framework) is H3C equipment proprietary virtualization technology, the actual physical devices will be turned into logical virtual device for users. IRF has a strong role of horizontal integration, and does not change the physical topology of the network connection (Fig. 4).

However, large scale data center needs for large capacity Large 2 Layer network. Based on IRF, the VCF technology is realized by using private RIPC protocol. VCF is a vertical stack management technology, can make the original server access equipments in two layer IRF network integrated into a large stack system, to form a logic

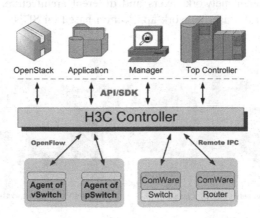

Fig. 4. Controller/Agent SDN network

greater single management system. VCF using SDN architecture N: 1 network virtu-alization, not only multiple devices on the same network layer integrated, but also devices on another layer, the entire network runs as a large box-type equipment, various operations of operating and management are virtualized within a large device.

H3C realized multilevel Open API scheme. Basic equipment layer can provide the depth of the SDK standard VCC (Virtual Computing Container) network application, and provide advanced XML access NETCONF as standard interface system, Open-Flow also provided by equipment layer.

Equipment control layer (SDN Controller) can provide VCC, REST/SOAP, NETCONF, OpenFlow etc. to the outside. In the Open API interface, REST/SOAP is a routine high-level protocol programming interface, NETCONF is an emerging XML language programming interface on the network device, OpenFlow is a protocol of SDN, above all is implemented of general technology, VCC is the standard realization in bottom formed in the process of H3C long-term accumulating of network software technology.

In the VCC environment, users can loaded the package to run on the device independently, the software can be upgraded uninterrupted service.

H3C put forward to OAA (Open Application Architecture) network model, which provides card computing in H3C network equipment, users can develop their own special application thereon, get on data interaction through the H3C OAA association protocol in the network.

OAA architecture has two open interface models, shown in Fig. 5. OAA loosely coupled architecture, using the SDN model, the user data processing device is operated under SDN Controller mode, the particular traffic directed to user's computing device custom processing. OAA tightly coupled architecture, two modes: Model 1, Users design their own daughter card providing high performance computing unit, H3C provides OAA line card Mother board; Model 2, H3C provides the overall OAA line cards, users develop their own software based H3C hardware. OpenFlow protocol is still used for the specific data processing.

SDN is a broad network architecture, through the flexible and open structure to achieve user demands. H3C technology model can provide interface and business environment in different network layers and different architecture, at the same time H3C itself also provides user network application based on SDN.

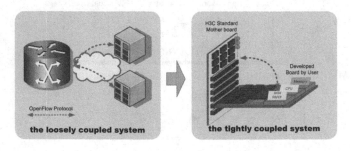

Fig. 5. Custom network platform based on OAA

4 TCP Incast Countermeasures

Incast more specific is known as TCP Incast, this communication mode is heavily dependent on TCP. The Incast traffic can be very short lived flows (micro burts), depending on the application. And Incast may also be used in a context referring to the detrimental effect on TCP throughput caused by network congestion in the Incast communication patterns.

TCP Incast, appears in many-to-one communication model, the link packet loss leads to the network performance suffer sharp collapse. After Incast happened, the throughput is far below the link bandwidth. TCP Incast occurs in some specific application scenarios in data center, such as distributed storage, MapReduce and web-search (Fig. 6).

The Incast caused by congestion has the effect of increasing the delay observed by the application and its users. There's a "give" and "take" with application performance as node cluster sizes grow. Larger clusters provide more distributed processing capacity allowing more jobs to complete faster. However larger clusters also mean a wider fan-in source for Incast traffic, increasing network congestion and lowering TCP throughput per node.

The "give" of a larger cluster is a lot more obvious than the "take", as Big Data centers continue to build larger cluster sizes to increase application performance. The question is: How much of the "take" (Incast congestion) is affecting the true performance potential of the application? And what can be done to measure and reduce the impact of Incast congestion?

DCTCP (Data Center TCP), a new variation of TCP, has been proposed specifically for the data center computers. DCTCP optimizes the way to detect congestion and can quickly minimize the amount of in flight data at risk during periods of congestion, minimizing packet loss and time outs. DCTCP optimizations have the effect of reducing buffer utilization on switches. While DCTCP need to have longer live flows before it can accurately detect and react to congestion.

Another method is taking a closer look at the switch architecture. For example, a switch with sufficient egress buffer resources will be able to absorb the Incast micro

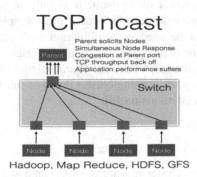

Fig. 6. TCP incast occurs

burts better, preventing the performance detrimental packet loss and timeouts. Then the fabric chip is needed inside the switch to detect and manage micro burst congestion.

With the right switch architecture to handle the micro bursts, in combination with optimizations like DCTCP to provide better performance for longer lived flows, the "take" can be minimized and the "give" accentuated.

In addition, research is carrying out on centralized control mechanism for TCP Incast based on OpenFlow protocol, and implementation designed on OpenFlow [7]. And the mechanism is a flexible and effective solution to TCP Incast in SDN network.

The core idea is: the network centralized controller monitors real-time information on the exchange network, when network conflict occurs caused by resource contention, the network controller configuration information of flow control will be issued to the network switching nodes over OpenFlow protocol, the flow control mechanism is triggered. And the switching nodes send flow control information to the source nodes, indicating the source servers contained all of the requested data will be grouped based on the current state of the nodes. Then each group of servers interacts data in sequence. This is a dynamic process in real time, ad infinitum. The use of centralized control over Openflow protocol to solve the TCP Incast problem is a highly innovative solution.

The above solutions are proposed for particular configuration (such as the number of servers, data block size, link capacity, microsecond timer, Switches support), exist various restrictions of its own, to solve the problem in depth still take time to study and explore.

5 Task Scheduling Optimization

5.1 Network Load Balancing Algorithm

In traditional distributed systems, the main goal of load balancing is balanced on the server and the application requests, the current latest user requests will be processed by the lightest load node in order to improve the quality of services across distributed systems.

Balancing algorithm designed quality directly determines the performance of the cluster load balancing. The main task of the general balancing algorithm is to decide how to choose the next cluster node, and then the new service requests to him. A good load balancing algorithm is not omnipotent, it is only in some special application environment to achieve maximum effectiveness. Some simple algorithm can be used independently, some must be used in combination with other simple or advanced algorithm. At the same time of studying load balancing algorithm, the algorithm itself applicability must be pay attention to, and the characteristics of the cluster must be also considered comprehensively while using cluster deployment. Different algorithms and techniques can be used in combination [8].

5.2 Dynamic Feedback Load Balancing

When customers access the cluster resources, time required and computing resources consumed are vastly different of the submitted tasks. It is dependent on many factors.

For example: service type of requested task, the current network bandwidth, resources utilization of the current server etc. Request processing time for different tasks may lead to processing node utilization tilt (Skew), namely the processing node load imbalance. Therefore, it is necessary to use a mechanism, the load balancer can learn about the load status of each node in real time, and can be adjusted according to changes in load.

The specific approach is using the dynamic load balancing algorithm based on negative feedback mechanism, the algorithm takes into account the real-time load and response capacity of each node, and constantly adjusts the ratio of task distribution, to avoid some overload nodes still receive a lot of requests, thereby improving the overall throughput of a single cluster.

Within the cluster, server monitor process running on the load balancer, monitor process is responsible for monitoring and collecting load information from each node within cluster; while the client processes running on each node is responsible for regularly reporting their situation to the load balancer. Monitor process will synchronize according to the load information received from all nodes, the tasks to be allocated will be re-distributed in proportion to the weight. It is primarily based on CPU utilization, available memory and disk I/O status of each node, to calculate a new weight value, if the difference between the new weight and current weight is greater than the set threshold, the monitor uses the new weight to re-distribute tasks within a cluster range, until the next load information synchronization arrivals. Load balancer can be used with dynamic weights, using Weighted Polling Algorithm to dispatch network service requests accepted.

Practice has proved that the use of dynamic load balance has improved the overall throughput of the cluster system, in particular in the case of the performance of each node in the cluster varies and resources accessed by network services program diversifies, the effect of the negative feedback mechanism is obvious. In other types of clusters, the negative feedback mechanism of dynamic load balancing is also able to get a good application, while the load balancer operating unit is unlike the network connection, and the specific load algorithms will be different.

5.3   Load Balancing Strategy Research

In various existing load balancing strategy, the different assumptions have been put forward in terms of task granularity, topology, localized and optimized performance indicators, a comparative study of these strategies is not a simple issue. Direct comparison of its performance pros and cons of various strategies needed to achieve a algorithm well, however, it is a very challenging task to develop effective software and get computing performance from parallel machines [9].

Load balancing technology has made great progress the past 30 years, it is not only driven by emerging applications, but also driven by emerging parallel architecture. Software oriented gigabit computing and Many-Cores is a challenge, to take full advantage of such hardware parallelism, applications, libraries and algorithms, and even new programming languages need to be redesigned. Therefore, the load balancing technology, at least in the following aspects should be further studied.

The traditional load balancing strategy is often overlooked machine or network topology. For emerging massively parallel machines, large network diameter, message passing hops between tasks, transmission delay and communication spending increasing, the network topology can no longer be ignored.

With the Multi-Core brought a surge in hardware parallelism, also brings the load balancing strategy new challenge. In the hierarchy of large-scale systems, the study of how to use mixed strategy to take advantage of distributed strategy and centralized strategy, digging the system parallel performance is beneficial [10].

6 Summary and Outlook

Load balancing can be implemented by the use of specific hardware gateway, also can be implemented by the use of special software and protocol. In general, load balancing technologies include global and local two categories. Global load balancing technology is to solve problems of load balancing in different areas, multiple clusters between different architectures. These are mostly mirrored server cluster nodes with the same content, can avoid data congestion brought by the site high concurrent access. Partial load balancing technique is mainly to solve the load balancing between nodes within a cluster, it is primarily responsible for scheduling and allocation of reasonable network connection requests, real-time transferring the work on the failed server to the properly server, so that no sewing takeover come true.

The development of information technology is rapid, a variety of new technologies are emerging. Study of load balancing also needs the latest technology to seek new solutions to problems. Moment, Big Data and Cloud technology are the most popular technologies, considering using Cloud computing technology into the load balancing cluster system, merging, learn, to explore better load balancing mode.

Acknowledgements. I would like to express my gratitude to all those who helped me during the writing of this thesis. I gratefully acknowledge the help of my supervisor, Professor Chaoqing Yin, who has offered me valuable suggestions in the academic studies. In the preparation of the thesis, he has provided me with inspiring advice. And I gratefully acknowledge the help of our president of City College, Professor Jianxun Chen, without his patient instruction, expert guidance and strong support, the finish of this thesis would not have been possible.

I should also thank all of my colleagues, from whose devoted teaching and enlightening lectures I have benefited a lot and academically prepared for the thesis.

I should finally like to express my gratitude to my family who have always supported me and stood by me.

References

1. Jing, H.: Research on some key technologies of BIG DATA-AS-A-Service. Beijing University of Posts and Telecommunications (2013)
2. Watt, S.: Certified IT architect: deriving new business insights with big data. Developer Works, IBM, 28 June 2011 (First published 29 June 2010). http://www.ibm.com/developerworks/opensource/library/os-bigdata/index.html

3. Zhihua, Y.: The age of big data network architecture. In: 2013 SACC, July 2013
4. Zhenqian, F.: Research on network bandwidth isolation of data center for cloud computing. Graduate School of National University of Defense Technology, June 2012
5. Zhongfeng, H.: Cloud computing era, non-blocking switching. ZDNET Network Channel, July 2012. http://net.zdnet.com.cn/network_security_zone/2012/0710/2101257.shtml
6. Xinmin, L.: Depth of openness and integration - H3C SDN analytical framework. IP Navigator, H3C, March 2013. http://www.h3c.com.cn/About_H3C/Company_Publication/IP_Lh/2013/01/Home/Catalog/201303/777797_30008_0.htm
7. Jianhua, X., Dongxu, Z., Hongxiang, G.: Research to flow control mechanism for TCP incast based on openflow. Chinese Scientific Papers Online, Beijing, January 2014. http://www.paper.edu.cn/releasepaper/content/201401-158
8. Dengwei, C., Zhiyong, L.: Arithmetic analysis for network dynamic load balance. Mod. Electron. Technol. **164** (2003)
9. Rongsheng, W., Jixiang, Y., Fan, W.: Survey of load balancing strategies. J. Chin. Comput. Syst. **31**, 1681–1686 (2010)
10. Dawei, S., Guangyan, Z., Weimin, Z.: Large data flow computing: key technologies and system instance. J. Softw. (2014)

Fuzzy Adaptive Controller Design for Control of Air in Conditioned Room

Štefan Koprda and Martin Magdin[✉]

Department of Computer Science, Faculty of Natural Sciences,
Constantine the Philosopher University in Nitra, Tr. A. Hlinku 1,
949 74 Nitra, Slovakia
{skoprda, mmagdin}@ukf.sk

Abstract. In the paper we describe an adaptive controller for control of air in conditioned room. We know, that exist more procedures for management of change of climate air in room (for setup and maintain of optimal temperature) - classical control by using the button on/of, management by bimetallic sensors, temperature management through control elements and the like. The human body perceives the minimum temperature difference (+3/−3) in your area. Where it is necessary to maintain a constant temperature for a very long time (e.g. in offices, server rooms), we need to consider the controlled system. Control system is part of the control circuit and may be considered as transfer element with certain characteristics. In the choice of the controller and its configuration is important to know the dynamic properties of the system. The simplest is to determine the dynamic characteristics of the system obtained from the its step response. We may also use the frequency response, which is much more difficult to obtain. Dynamic properties of the control system can we mathematically to express as the relationship between the change in input (Action) and a change the output value of controlled variable. On the basis of the current research, as the most appropriate way to set up and maintain a constant temperature in room at minimum operating costs may be solution by using fuzzy adaptive controller.

Keywords: Fuzzy controller · Controlled system · Adaptive control, thermo-dynamic system · MATLAB

1 Introduction

Given restrictions that comfort conditions in the interior of a building are satisfied, it becomes obvious that the problem of energy conservation is a multidimensional one. Important research was conducted on optimum and predictive control strategies during the 1980s and 1990s. However, no industrial development has followed these scientific studies, especially due to implementation issues [11].

Indoor temperature regulation and energy resources management in buildings require the design and the implementation of efficient and readily adaptable control schemes. One can use standard schemes, such as "on/off" [25].

Classic controllers represent feedback (or closed-loop) controllers with constant parameters and no direct knowledge of the considered process. When used alone, they

© Springer International Publishing Switzerland 2015
D.-S. Huang et al. (Eds.): ICIC 2015, Part II, LNCS 9226, pp. 318–329, 2015.
DOI: 10.1007/978-3-319-22186-1_31

usually give poor control performance for large time-delay process, in case of process noise or in the presence of non-linearities [15, 27].

The primary goals of control strategies for the Heating, ventilating and Air Conditioning (HVAC) systems are to maintain occupants' thermal comfort and energy efficiency [7].

Thermal comfort is a vague and subjective concept and varies from one person to another [7]. Thermal comfort is difficult to define as a range of environmental and personal factors. That is why a common approach deals with the "Predicted Mean Vote" (PMV) (on the following thermal sensation scale: +3 is "very hot", +2 is "hot", +1 is "relatively hot", 0 is "neither hot nor cold", −1 is "relatively cold", −2 is "cold" and −3 is "very cold") of a large population of people exposed to a certain environment [25]. Research done by American Society of Heating, Refrigerating and Air-conditioning Engineers (ASHARE) during last year's identified the most important parameters that influence thermal comfort as Temperature, Relative Humidity, Air Velocity and Radiant Temperature. However, activity level and clothing insulation of occupants are effective in thermal comfort, but they are variable and usually not measurable [7].

All this implies that the systems to be controlled are not only difficult to model but also time variant. In the light of these general observations we can assign the primary objective to this work which is to propose a general algorithm which will be able to control such systems. It is easily understandable that a satisfactory behavior will be obtained only with the use of an adaptive control algorithm [8]). In this sense, fuzzy logic allows controlling temperature and managing energy sources, taking advantage of the flexibility offered by linguistic reasoning. With this kind of approaches, both the specific use of a building and the specificities of a proposed energy management strategy can be easily taken into account when designing or adjusting the control scheme, without having to model the process to be controlled [25].

In this sense, fuzzy adaptive controllers have been successfully applied to heating [5, 21], with the aim of maximizing both energy efficiency and thermal comfort, visual comfort [1, 3] or natural ventilation [20], one of the most interesting ways for improving buildings' energy performance [2, 10, 12, 22]. Predictive control [6, 16, 17, 24] is very important because it includes a model for future disturbances (e.g. solar gains, presence of humans, etc.) [11].

In order to use optimal control [14, 26], or adaptive control, [9] a model of the building is necessary [11].

2 Materials and Methods

The design of controller for air climate systems is a big challenge for practical engineers [6]. Many control methods have been proposed in air climate systems from traditional controllers to advanced and recently intelligent controllers during the last years [7, 13, 18, 19, 23]. It is easily understandable that a satisfactory behavior will be obtained only with the use of an adaptive control algorithm, that is realized in adaptation model (Fig. 1).

Fig. 1. Generic adaptation model. a/i – action or control input variables of object, o – output variables of controlled object, c – controllable parameters of adaptation object, i – input parameters for providing optimization, ch – desired change

The required behaviour of individual objects is characterized by purpose function, which is also the criteria of optimization:

$$J = J(a/i, o, ch, i, c) \qquad (1)$$

The disadvantage of current adaptive system (AS) is the necessity of obtaining input data from the users of the system and in the determination of the relevant adaptive rules.

The solution is AS which uses fuzzy logic and eliminates the disadvantages of the necessity of obtaining the input data - knowledge. If we want to use this model in the structure of our regulator, we must count with a large number of functions and the inferential rules (rules for dealing with problems), which is usually determined on the basis of trial and error. The decision-making process can take place according to the following scheme (Fig. 2):

Fig. 2. Decision-making process of adaptive system

As an example of decision-making process we can mention generic rule of adaptation:

$$\text{IF } <\text{PREREQUISITE}> \text{ THEN } <\text{ACTION}> \qquad (2)$$

which determines the action <ACTION> if the condition <PREREQUISITE> is true. Moreover we can define sufficient conditions which do not necessarily meet all the conditions, but sufficient set of them:

$$\text{IF ENOUGH } (<\text{PREREQUISITES}>) \text{ THEN } <\text{ACTION}> \qquad (3)$$

The effectivity of control algorithm obviously depends on how accurately identifier predicts the dynamics of the process. As our proposed regulator has the ability to learn, the fuzzy rules and membership functions will be adjusted automatically by learning algorithms. Learning will be made on the basis of the output errors, so it is necessary to know the error, which is determined by comparing the attained output and the desired outcome. We can thus claim that our regulator will be trained by using data with the desired properties. In general, such regulation may not be available. In this case, it is possible to train the our regulator by applying methods of individual learning. In the approach of individual learning two systems are used: one of them will have the function of a regulator - an optimizer and the second one will have the function of predictor.

Control system is part of the control circuit and may be considered as transfer element with certain characteristics. In the choice of the controller and its configuration is important to know the dynamic properties of the system. The simplest is to determine the dynamic characteristics of the system obtained from the its step response. We may also use the frequency response, which is much more difficult to obtain. Dynamic properties of the control system can we mathematically to express as the relationship between the change in input (Action) and a change the output value of controlled variable.

This dependence can be described by using the differential equation, which is compiled on the basis of mass or energy balance of the regulated system with infinitely small changes in the input variable. On the basis of solution of the differential equation we can obtained the dynamic characteristics of the controlled system. Each control system is characterized by the ability to accumulate energy. We say that the system has capacity. Accumulation of energy is realized by a certain time dependency and in this time occurs to the delay the impact of input variables to the output variable. Degree of delayed is the time constant T. Depending from the size effect on the time course of the input signal is controlled systems are divided into prime, first, second and n-th order. Reinforcement of control system usually indicates as k.

3 The Design of the Thermodynamic System of First Order

Thermodynamic system, which is a model of the building (Fig. 3), was built using wood frame with dimensions x = 1.0 m, y = 0.5 m and z = 0.5 m. This frame we have subsequently topped from the inside by using with drywall (has the same dimensions as a frame). The thickness of the drywall is 1 cm. From the outside was frame lined with polystyrene the same dimensions as a frame. The thickness of the polystyrene is 5 cm. Volume of this system is 0.25 m^3. The figure represents realistic model of the thermodynamic system.

Fig. 3. The design of building used for testing the proposed hybrid control scheme and real model of thermodynamic system of first order

4 The Analytical Solution of the Differential Equation of System of the First Order

The basic form of the differential equation for heat conduction:

$$\frac{\partial}{\partial x}\left(k_X \frac{\partial T}{\partial x}\right) + \frac{\partial}{\partial y}\left(k_Y \frac{\partial T}{\partial y}\right) + \frac{\partial}{\partial z}\left(k_Z \frac{\partial T}{\partial z}\right) + q = \rho c \frac{\partial T}{\partial t}, \ W.m^{-3} \tag{4}$$

q – Internal source of heat,
k_x, k_y, k_z – thermal conduction in the direction of axis x, y, and z.

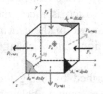

Fig. 4. Model of thermodynamic system with thermal balance of elemental volume

The solution, if we do not consider the internal sources of heat:

$$\frac{\partial t}{\partial \tau} = \frac{\lambda}{\delta c}\left(\frac{\partial^2 t}{\partial x^2} + \frac{\partial^2 t}{\partial y^2} + \frac{\partial^2 t}{\partial z^2}\right), \quad t(x, y, z, \tau) \tag{5}$$

t – temperature
τ – time
t_1 – ambient temperature
t_2 – temperature in system
l1, l2, l3 – dimensions of the cuboid
boundary conditions:

$$t(0, y, z, \tau) = t_1 - \text{ambient temperature} - (0 \leq y \leq l_2, 0 \leq z \leq l_3) \tag{6}$$

$$t(l_1, y, z, \tau) = t_1 \tag{7}$$

$$t(x, 0, z, \tau) = t(x, l_2, z, \tau) = t_1 \tag{8}$$

$$t(x, y, 0, \tau) = t(x, y, l_3, \tau) = t_1 \tag{9}$$

Initial conditions:

$$t(x, y, z, 0) = t_2 \quad (\text{x, y, zin cuboid}) \tag{10}$$

(while $t_2 > t_1$)

$$t(x,y,z,\tau) = t_1 + \sum_{k,l,k=1}^{\infty} A_{k,l.m} \sin\left(\frac{k\pi}{l_1}x\right) \sin\left(\frac{l\pi}{l_2}y\right) \sin\left(\frac{m\pi}{l_3}z\right)$$
$$* e^{-\frac{\lambda}{\varsigma c}\left[\left(\frac{k\pi}{l_1}\right)^2 + \left(\frac{l\pi}{l_2}\right)^2 + \left(\frac{m\pi}{l_3}\right)^2\right]\tau} \tag{11}$$

$$t_1 + \sum_{k,l,k=1}^{\infty} A_{k,l.m} \sin\left(\frac{k\pi}{l_1}x\right) \sin\left(\frac{l\pi}{l_2}y\right) \sin\left(\frac{m\pi}{l_3}z\right) = t_2 \tag{12}$$

$$\sum_{k,l,k=1}^{\infty} A_{k,l.m} \sin\left(\frac{k\pi}{l_1}x\right) \sin\left(\frac{l\pi}{l_2}y\right) \sin\left(\frac{m\pi}{l_3}z\right) = t_2 - t_1 \tag{13}$$

Where

$$A_{k,l,m} = \frac{8}{l_1 l_2 l_3} \int_0^{l_1} \int_0^{l_2} \int_0^{l_3} f(x,y,z) * \sin\left(\frac{k\pi}{l_1}x\right) \sin\left(\frac{l\pi}{l_2}y\right) \sin\left(\frac{m\pi}{l_3}z\right) * d_x d_y d_z \tag{14}$$

$$f(x,y,z) = t_2 - t_1 \tag{15}$$

Process solutions A_{klm}

$$\int_0^{l_1} \sin\left(\frac{k\pi}{l_1}x\right) \sin\left(\frac{k'\pi}{l_1}x\right) dx \tag{16}$$

Substitution:

$$\frac{\pi}{l_1}x = \tilde{\tilde{x}} \tag{17}$$

$\frac{\pi}{l_1}dx = d\tilde{\tilde{x}}$ after substituting

$$= \frac{l_1}{\pi} \int_0^{\pi} \sin\left(k\tilde{\tilde{x}}\right) \sin\left(k'\tilde{\tilde{x}}\right) d\tilde{\tilde{x}} = 0 \text{ if } k \neq k' \tag{18}$$

$$= \frac{\pi}{2} \frac{l_1}{\pi} = \frac{l_1}{2} \text{ if } k = k' \tag{19}$$

$$\sum_{k=1}^{\infty} a_k \int_0^{l_1} \sin\left(\frac{k\pi}{l_1}x\right) \sin\left(\frac{k'\pi}{l_1}x\right) dx = c \int_0^{l_1} \sin\left(\frac{k'\pi}{l_1}x\right) dx \tag{20}$$

$$a_{k'} \frac{l_1}{2} = c \frac{l_1}{\pi} \int_0^\pi \sin\left(k' \overset{\approx}{x}\right) d\overset{\approx}{x} \tag{21}$$

$$a_{k'} = \frac{2}{l_1} c \int_0^{l_1} \sin\left(\frac{k\pi}{l_1} x\right) dx = 0 \text{ if } ``k'' \text{ is even} \tag{22}$$

$$= \frac{4c}{k\pi} = \frac{2l_1}{k\pi} \text{ if } ``k'' \text{ is odd} \tag{23}$$

After substituting into the formula (6) we obtain:

$$A_{klm} = \frac{8(t_2 - t_1)}{l_1 l_2 l_3} \left(\int_0^{l_1} \sin\left(\frac{k\pi}{l_1} x\right) dx \right) \left(\int_0^{l_2} \sin\left(\frac{l\pi}{l_2} y\right) dy \right) \left(\int_0^{l_3} \sin\left(\frac{m\pi}{l_3} z\right) dz \right) \tag{24}$$

$$= \frac{8(t_2 - t_1)}{l_1 l_2 l_3} \left(\frac{2l_1}{k\pi}\right) \left(\frac{2l_2}{l\pi}\right) \left(\frac{2l_3}{m\pi}\right) = \frac{64(t_2 - t_1)}{klm\pi^3} \tag{25}$$

This is result of analytical solutions.

5 Determination of Transient Characteristics of the Proposed Thermodynamic System

To determine of transient characteristic can be used MATLAB simulation environment. Before the recording of temperature in MATLAB, we used for sensing temperature and humidity device with a name DN 20. It is a device that has been designed at the Department of Electrical Engineering and Automation in cooperation with Power - One Ltd. [4]. This device consists of a microprocessor DN 20, of connectors for connecting temperature and humidity sensors and outputs to control the heater and fan. Connectivity to the PC is provided by serial communication via serial interface RS 232 interface or modern way connection - USB. The outputs are programmed and configured by using environment MATLAB and provides a smooth voltage regulation by PWM modulation at the output connector. The design of this solution can be seen in Fig. 5.

Device properties DN 20:

4 × analog 10-bit inputs, digital inputs/outputs
1 × inputs of external references
2 × independent 10-bit analog power outputs
4 × 16-bit PWM independent power outputs
1 × electronic potentiometer
4 × relay
1 × RS232
1 × USB2.0
1 × SD Card

Fig. 5. The design of solution of the device DN 20

1 × I2C/SMBus
1 × Ethernet 10 Mbps
1 × two-line display
1 × graphic display
2 × comparator

The device is connected via a serial bus with a computer (in the PC is running a simulation program. This simulation program performed records, regulates temperature and moisture and is able to control the heat source and the fan in the thermodynamic system.

In determining of transient characteristic we proceed as follows. Thermodynamic system we basked by the lamp (bulb) with the power consumption 55 W. Heating time was about 23.55 h. Temperature sensors were placed at the bottom and top of the thermodynamic system. For sensing temperature, we used four sensors labeled Dallas DS18B20. Sensors that we placed in the thermodynamic system are connected to device that managed microprocessor DN 20. The device was connected to the computer on which was running environment MATLAB with the control program for the measurement (Fig. 6).

The equation of thermodynamic system has the form:

$$G(p) = \frac{K_P}{Tp + 1} \tag{26}$$

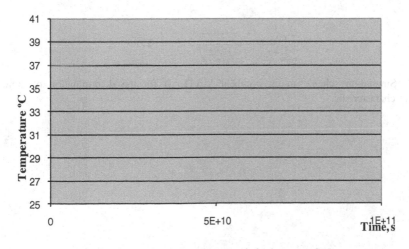

Fig. 6. Real transient characteristic (measured)

where T - is calculated as $\left(1 - \frac{1}{e}\right)$ multiple of steady value,
  K_P - is steady value,
  p - Laplace operator.
  e - Euler number

To calculate the T we chose the mean of characteristics.
  T = 14290 s
  K_P = 1

$$\text{Simulation of transient characteristic of } G(p) = \frac{1}{14290s + 1} \tag{27}$$

To determine the transient characteristic in environment MATLAB we use the following commands (Fig. 7):

```
sys = tf(1,[14290 1])
step(sys)
```

The similarity is 99.77 % (Fig. 8). Modelling of thermodynamic system is need to the obtain basic algorithm of behaviour of system. On this basis, the model will be modified to the specifically conditions. They will be examined selected algorithms to control and optimize the system. We expect these algorithms suitable PID regulation, and the fuzzy control. From these algorithms will be created a control system that may be a prerequisite to the specific technical implementation. To achieve the greatest efficiency is need the optimization of the system. In general, it is in this way possible to design a universal control system that is capable of implementation in different types of environments (offices, hotels, houses), in various systems of ventilation and by difference construction design (Fig. 9).

Fuzzy logic of control system works on the basis on implications of two fuzzy assertion: If ascendant so consequent (by implication Mandy alternative If <fuzzy assertion> Then <sharp output value>). We using fuzzy linguistic model with the following hierarchy:

Fig. 7. Simulation scheme in environment MATLAB for the determination of simulated transient characteristic

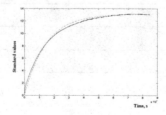

Fig. 8. The final course of real and simulated transient characteristic

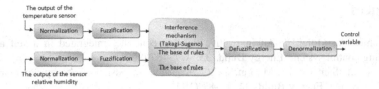

Fig. 9. The block structure of the proposed fuzzy control system

If (OT is ZOT) and (OV is ZOV) then (u is V)
If (OT is ZOT) and (OV is 0OV) then (u is 0)
If (OT is ZOT) and (OV is KOV) then (u is 0)
If (OT is OOT) and (OV is ZOV) then (u is V)
If (OT is OOT) and (OV is 0OV) then (u is 0)
If (OT is OOT) and (OV is KOV) then (u is 0)
If (OT is KOT) and (OV is ZOV) then (u is V)
If (OT is KOT) and (OV is 0OV) then (u is Z)
If (OT is KOT) and (OV is KOV) then (u is Z)

If we using rules of quantitative expert knowledge's, this rules must be based on theoretical knowledge's and on the basis of vague input information. It is necessary to ensure the conditions of completeness, consistency, continuity and interaction of a set of rules (Jura, 2003). Local changes can be achieved by changing of the relevant rules and the size of the area where there is a change, which can be to influence the width of fuzzy sets.

6 Conclusion

In this paper we describe with the development of a control of air in conditioned room, allowing managing energy resources in buildings.

The variety of control methodologies applied for air control in conditioned room were reviewed and investigated in the present study. Conventional control methods are still the first choice for air control systems today. Their traditional use is due to their simplicity of implementations and low initiation cost. However, these methods suffer from a high cost of maintenance and energy consumption. An alternative approach would be to revert to the more modern and intelligent methodologies. It is noted that thermal comfort is inherently a vague issue in humans since they have different definitions for comfort [23].

Because the hybrid control scheme proposed has been developed in simulation, it needs now to be tested using more complex models and/or situations, with the aim of being finally implemented in real buildings.

Acknowledgment. This publication is supported thanks to the project KEGA 015UKF-4/2013 Modern computer science – New methods and forms for effective education.

References

1. Guillemin, A., Morel, n: An innovative lighting controller integrated in a self adaptive building control system. Energy Build. **33**, 477–487 (2001)
2. Dounis, A.I., Santamouris, M., Lefas, C.C., Argiriou, A.: Design of a fuzzy set environment comfort system. Energy Build. **22**, 81–87 (1994)
3. Dounis, A.I., Santamouris, M.J., Lefas, C.C.: Building Visual comfort control with fuzzy reasoning. Energy Convers. Manag. **34**, 17–28 (1993)
4. Amrich, M., Hrubý, D.: Možnosti simulácie a návrhu Fuzzy Riadiacich Algoritmov. In: International Professional Seminar Topical Questions of Instruction of Electricaal Engineering Subjects at Faculties with Non-electrical Orientation "SEKEL2006. 12–14 Sept 2006, Vrátna, Slovakia. pp. 29–33. ISBN 80-8070-584-4 (2006)
5. Altrock, C.V., Arend, H.O., Krause, B., Steffens, C., Behrens-Rommler, E.: Adaptive fuzzy control applied to home heating system. Fuzzy Sets Syst. **61**, 29–35 (1994)
6. Chen, T.: Real-time predictive supervisory operation of building thermal systems with thermal mass. Energy Build. **33**(2), 141–150 (2001)
7. Chiang, M.L., Fu, L.C.: Hybrid Systém based adaptive control for the nonlinear hvac system. In: Proceeding of the Conference on American Control Minneapolis, 14–16 June 2006 Minnesota, USA pp. 5324–5329 (2006)
8. Collewet, C., Rault, G., Quellec, S., Marchai, P.: fuzzy adaptive controller design for the joint space control of an agricultural robot. Fuzzy Sets Syst. **99**(1), 1–25 (1998)
9. Curtis, P.S., Shavit, G., Kreider, K.: neural networks applied to buildings—a tutorial and case studies in prediction and adaptive control. ASHRAE Trans. **102**(1), 24 (1996)
10. Kolokotsa, D., Tsiavos, D., Stavrakakis, G.S., Kalaitzakis, K., Antonidakis, E.: Advanced fuzzy logic controllers design and evaluation for buildings' occupants thermal visual comfort and indoor air quality satisfaction. Energy Build. **33**, 531–543 (2001)
11. Dounis, A.I., Caraiscos, C.: Advanced control systems engineering for energy and comfort management in a building environment-a review. Renew. Sustain. Energy Rev. **13**(6–7), 1246–1261 (2009)
12. Calvino, F., Gennusca, M.L., Rizzo, G., Scaccianoce, G.: The control of indoor thermal comfort conditions: introducing a fuzzy adaptive controller. Energy Build. **36**, 97–102 (2004)
13. Huaguang, Z., Cai, L.: Decentralized nonlinear adaptive control of an HVAC system. Trans. Syst. Part C. **32**, 493–498 (2002)
14. Inoue, T., Kawase, T., Ibamoto, T., Takakusa, S., Matsuo, Y.: The development of an optimal control system for window shading devices based on investigations in office buildings. ASHRAE Trans. **104**, 1034–1049 (1998)
15. Ang, K.H., Chong, G.C.Y., Li, Y.: PID Control system analysis, design, and technology. IEEE Trans. Control Syst. Technol. **13**(4), 559–576 (2005)
16. Kummert, M., Andre, P., Nicolas, J.: Optimal heating control in a passive solar commercial building. Solar Energy **69**(1–6), 103–116 (2001)
17. Lam, H.N.: Stochastic modeling and genetic algorithm based optimal control of air conditioning systems. In: Proceedings of the 3rd International Conference of the International Building Performance Simulation Association, Adelaide, Australia, pp. 435–441 (1993)
18. Li, S., Zhang, X., Xu, J., Cai, W.: An improved fuzzy RBF based on cluster and its application in HVAC system. In: Proceedings of the 6th World Congress on Intelligent Control and Automation, 21–23 June 2006 Dalian, China, pp. 6455–6459 (2006)

19. Liang, J., Du, R.: Thermal comfort control based on neural network for HVAC application. In: Proceeding of the IEEE Conference on Control Applications, 28–31 Aug 2005, Toronto, Canada, pp. 819–824 (2005)
20. Eftekhari, M., Marjanovic, L., Angelov, P.P.: Design and performance of a rulebased controller in a naturally ventilated room. Comput. Ind. **51**, 299–326 (2003)
21. Gouda, M., Danaher, S., Underwood, C.: Thermal comfort based fuzzy logic controller. Build. Serv. Eng. Res. Technol. **22**(4), 237–253 (2001)
22. Lah, M.T., Borut, Z., Krainer, A.: Fuzzy control for the illumination and temperature comfort in a test chamber. Build. Environ. **40**, 1626–1637 (2005)
23. Mirinejad, H., Sadati, S.H., Ghasemian, M., Torab, H.: Control techniques in heating, ventilating and air conditioning systems. J. Comput. Sci. **4**, 777–783 (2008)
24. Morel, N., Bauer, M., Khoury, M., El-Krauss, J.: Neurobat a predictive and adaptive heating control system using artificial neural networks. Int. J. Solar Energy **21**, 161–201 (2000)
25. Paris, B., Eynard, J., Grieu, S., Polit, M.: Hybrid PID-fuzzy control Scheme for managing energy resources in buildings. Appl. Soft Comput. J. **11**(8), 5068–5080 (2001)
26. Winn, C.B.: Controls in solar energy systems. Adv. Solar Energy (Am. Solar Energy Soc.) **1**, 209–220 (1982)
27. Li, Y., Ang, K.H., Chong, G.C.Y.: PID control system analysis and design – problems, remedies, and future directions. IEEE Control Syst. Mag. **26**(1), 32–41 (2006)

DBH-CLUS: A Hierarchal Clustering Method to Identify Pick-up/Drop-off Hotspots

XueJin Wan[✉], Jiong Wang, Yuan Zhong, and Yong Du

Beijing Transportation Information Center, Beijing, China
{wanxuejin,wangjiong,zhongyuan,duyong}@bjjtw.gov.cn

Abstract. Travelling by taxi is more convenient and effective. With an over-crowding population and a much terrible traffic, the traditional way of hailing a taxi encounters many challenges like where to pick-up/drop-off passengers reasonably and where to find potential passengers quickly. More cities have established taxi stands to advocate and to guide passengers to hail a taxi. However, most of them have low rate of usage. The reason lies in that to determine where to establish reasonably is a big problem. In this paper, we are the first to propose a DFA to identify data signifying pick-up/drop-off events. We propose a DBH-CLUS method to identify pick-up/drop-off hotspots. The method applies hierarchal clustering based on agglomerative clustering analysis method. We have conducted three experiments to verify the DFA, to analyze the region agglomeration and to analyze the accuracy. The experimental results manifest that our method can precisely identify hotspots from the original GPS data and provide an excellent tool to facilitate taxi stand planning.

Keywords: Identify hotspots · DFA · Hail a taxi · Pick-up · Drop-off

1 Introduction

For a long time, travelling by taxi is convenient and effective to actively promote a healthy and low-carbon travel. However, the traditional way of hailing a taxi encounters many challenges with a growing population and a heavy traffic. Traditionally, people are accustomed to hail a taxi along the roadside or at an intersection. Taxi drivers have no way to find potential passengers but to drive a vacant taxi along the road. In addition, taxis pick up or drop off passengers casually during restricted hours which will cause a traffic jam. Challenges present much acute in some central cities with an overcrowding population and a terrible traffic like Beijing and Singapore. Taxi stands are established to advocate and to guide passengers to hail a taxi. For instance, hundreds of taxi stands are established in key areas in Beijing last year. However, a majority of taxi stands are of low-usage or abandoned totally due to their unreasonable locations. Hence, we are motivated to identify pick-up/drop-off hotspots to help establish reasonable taxi stands.

To identify pick-up/drop-off hotspots, the following challenges need to be addressed: (1) Precise data related to pick-up/drop-off location. The original GPS data may contain a plenty of irrelevant taxi status data or even error data. Therefore, to extract the precise data related to the pick-up/drop-off status from the original data is a

© Springer International Publishing Switzerland 2015
D.-S. Huang et al. (Eds.): ICIC 2015, Part II, LNCS 9226, pp. 330–341, 2015.
DOI: 10.1007/978-3-319-22186-1_32

big challenge. (2) Suitable original data set. The larger the data set is, the more noise data there will be. Selecting a suitable data set is another challenge to be addressed. (3) An algorithm to balance the spatial-temporal data and the periodic data. Pick-up/drop-off hotspots are the result of processing the data in spatial-temporal correlation and the periodic data. However, periodic data neither can be transformed into spatial-temporal data equally nor can be neglected because it reflects familiar but not common traffic condition. Hence, it is a great challenge to process the GPS data with an appropriate algorithm.

Some notable clustering methods have been proposed to extract cluster data from original datasets. The partitioning clustering algorithms perform well on small data sets while have defects to find medoids for large ones. The locality-based clustering algorithm like DB-SCAN performs without fixed centroids and is robust to outliers. Grid-based algorithms such as BIRCH, STING and DENCLUE focus on the problem of an increasing dimension. The idea of the algorithms has an impact on the curse of dimensionality and performs poorly on noisy data. To the best of our knowledge, establishing taxi stands is a requirement for an overcrowded city in the last five years. However, all the above work aiming at mining the data of taxi distribution can be summed to analyze the trajectory of taxi. They have not centered on identifying pick-up/drop-off locations with taxi GPS data. Therefore, we have to create a new method to identify the hotspots and to meet the strong demands especially for the taxi stand planning.

In this paper, we are the first to propose the DBH-CLUS(Density Based Hierarchical CLUStering) method to conjecture pick-up/drop-off locations and to identify pick-up/drop-off hotspots. We construct a DFA(deterministic finite automaton) to extract precise data related to pick-up/drop-off locations. The DFA can recognize error GPS data and extract the data representing pick-up/drop-off status from the trajectory data. We adopt the proposed algorithms to mine pick-up/drop-off status data, to construct clusters, and to identify hotspots from the potential hotspots.

We have conducted three experiments to verify our method. The first experiment analyzes the trajectory data of one taxi and verifies the error data correcting algorithm and the DFA. The second experiment verifies the regional agglomeration by adopting a large amount of data to adjust the parameters involved in our method. The third experiment implements the whole method. According to the result of identifying pick-up/drop-off hotspots, we analyze the accuracy of the proposed method.

Comparing the formers' work, we have made three main contributions as below:

We construct a DFA to conjecture the pick-up/drop-off position. The DFA can precisely recognize and correct error GPS data in the dataset. It can also accurately extract the data related to pick-up/drop-off location and prepare data for the method to identify hotspots.

We are the first to propose an innovative method to identify pick-up/drop off hotspots. To the best of our knowledge, we innovatively propose the DBH-CLUS method to identify hotspots with inferring pick-up/drop-off GPS data. The method is proved to be accurate according to experiment results.

We propose a reference standard to evaluate the reasonability of establishing taxi stands. It is a critic criterion to evaluate a high-quality service of hailing a taxi whether the location of one taxi stand is reasonable or not. The pick-up/drop-off hotspots can precisely reflect the reasonability of taxi stands. Our work will be helpful to evaluate the reasonability of taxi stands and to transportation management and planning.

The rest of this paper is organized as follows: In Sect. 2, we summarize some previous work. In Sect. 3, we make an overview including preliminary concepts and the framework. Section 4 introduces the methodology. Section 5 reports experimental results and related discussions. Finally, we make a conclusion in Sect. 6.

2 Related Work

Clustering of data in different industries is an important problem. There are plenty of clustering techniques proposed in literature. Here, we will only review some notable work due to space limitation. Existing clustering algorithms can be divided into partitioning, hierarchical, locality-based and grid-based algorithms. We describe some of those algorithms stated in brief as follows.

The idea of partitioning algorithms is to construct a partition and to divide the whole dataset into k clusters. These clusters are represented by one representative object named k-medoid. PAM proposed by Kaufman and Rousseeuw takes k-clustering on medoids to identify clusters. It performs well on small data sets, but is extremely costly for large ones. CLARANS [2] is a k-medoid algorithm which sources from PAM and CLARA. The algorithm identifies medoids by a randomized search of a graph. Due to their defects to specify the medoids, all the above algorithms perform badly on convex clusters.

DBSCAN [3] proposed by Ester et al. is a locality-based clustering algorithm which assumes that points inside clusters distribute randomly but do not require one to specify the number of clusters. It has a notion of noise data and is robust to outliers. DBCLASD [4] is also a locality-based clustering algorithm, while dissimilar to DBSCAN which performs without any input parameters.

Grid-based clustering algorithms such as BIRCH, STING and DENCLUE [5–7], are proposed to resolve the problem that existing algorithms perform poorly with an increasing dimension. BIRCH is based on Cluster-Feature-Tree. STING relies on a quadtree-like structure. DENCLUE depends on a regular grid. However, all the three algorithms have a serious impact on the curse of dimensionality and will be impacted by noisy data.

The cluster algorithms above help to think out an idea to process the data points presenting regional agglomeration.

Yamamoto et al. [8] propose a heuristic routing recommending algorithm using fuzzy clustering to improve taxi dispatching by assigning vacant taxis to pathways with many potential passengers expecting. Santi et al. [9] propose a model to identify and to predict vacant taxis relying on the temporal characteristics and weather condition in a given area using the taxi data. It constructs the prior probability distributions based on the naïve Bayesian classifier. This method emphasizes on identifying dynamic vacant distribution but not static pick-up/drop-off hotspots. The work of Chang et al. [10] is similar to the former one. They focus on constructing a model that predicts taxi demand distribution based on time, weather condition, and location. Yang et al. [11] aim to construct an urban taxi service model in a network context. The model describes vacant and occupied taxi's movement in a road network and the relationship between passengers and waiting time.

Markus Loecher et al. [12] center on scanning for both elliptical and rectangular clusters and adjusting for covariates in different temporal dimensions. It adopts an empirical Bayes procedure to account for parameter uncertainties in the framework. Jerry H. Ratcliffe et al. [13] outline a hotspots matrix to identify spatial and temporal broad categories of crime hotspots. Ke Hu et al. [14] present a service – oriented online system which can inform the real-time location and its trajectory with a color-like icon. It shows passengers where can hail a taxi with high probabilities. The system implements the visualizing of taxi GPS data but with little data analysis.

In a word, all the work aiming at mining the data of taxi distributing can be summarized to analyze the trajectory of taxi. Nevertheless, they have not centered on identifying pick-up/drop-off locations with the taxi GPS data. Therefore, we have to create a new method to identify the hotspots and to meet the strong demands especially the taxi stand planning.

3 Overview

In this section, we primarily make several definitions involved in this paper. Afterwards, we introduce the method to preprocess the original taxi GPS data. Finally, we describe the DBH-CLUS method in detail.

3.1 Pick-up/Drop-off DFA

The overall traffic map is partitioned into regions which consist of road-links in different levels.

One taxi equipped with GPS will upload a piece of record at a certain frequency or when it comes to an event like pick-up, drop-off or stop. Each taxi GPS record is a six-tuples which contains following fields i.e. TIMESTAMP, VEHICLE_ID, LONGITUDE, LATITUDE, SPEED, and STATE.

To extract GPS data of pick-up/drop-off, we propose a DFA to define a state transition rule. The DFA is described as $M = (Q, \Sigma, \delta, q_0, F)$. Q is a finite set of Σ, where $Q = \{S, U, V, F\}$. Σ is a finite set of input symbols where $\Sigma = \{0, 1, 2\}$. 0, 1 and 2 represent states i.e. vacant, occupied and stop. δ is a transition function where $\delta : Q \times \Sigma \rightarrow Q$. S is a start state where $S \in Q$. F is a set of accept states where $F \in Q$. U and V are the intermediate set of accept states.

Figure 1 illustrates the DFA. Taking time segments $(t_0, t_1, \cdots, t_i, t_{i+1}, t_{i+2}, \cdots, t_k)(0 \leq i \leq k - 2)$ as an instance, the data point at t_i is ε_i. The whole input symbols can be written by $Z = \{\varepsilon_0, \varepsilon_1, \cdots, \varepsilon_i, \varepsilon_{i+1}, \varepsilon_{i+2}, \cdots, \varepsilon_k\}$ $(0 \leq i \leq k - 2)\}$. With the DFA of pick-up/drop-off, we regard ε_i as a pick-up data point where $(\varepsilon_i, \varepsilon_{i+1}, \varepsilon_{i+2}) = (0, 1, 1)$ and ε_{i+2} as a drop-off data point where $(\varepsilon_i, \varepsilon_{i+1}, \varepsilon_{i+2}) = (1, 1, 0)$.

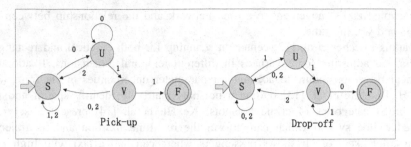

Fig. 1. Pick-up/drop-off DFA

3.2   Framework

The main structure of our work is shown in Fig. 2. The method firstly preprocesses original GPS data and road network data by classifying GPS data to trajectory data and then extracting regional edge data. Secondly, with the pick-up/drop-off DFA, the method extracts the pick-up/drop-off location data. Afterwards, with the algorithms of DBH-CLUS, the hotspots are obtained from the potential hotspots by adjusting parameters. Details are signified in the following section.

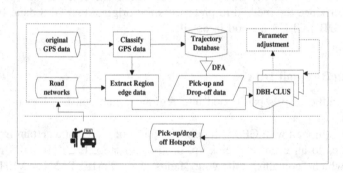

Fig. 2. Framework of our method

4   Methodology

In this section, the preprocessing of data and the clustering algorithm are described in detail. Specifically, we center on the application of DFA to extract the data of pick-up/drop-off. Afterwards, we focus on the clustering algorithm to identify hotspots.

4.1   Extract Trajectory Data

In order to conjecture pick-up/drop-off locations, we primarily extract pick-up/drop-off data of one region from the original data. The original GPS data can be transferred to trajectory data if sorted by timestamps ascending.

A trajectory of one taxi contains all the events created by that taxi. Hence, we can extract pick-up/drop-of data from its trajectory. We adopt the regional edge data consisting of GPS data points, each of which consists of a set of longitude and latitude.

We regard the location of p_i represented by GPS record, as the pick-up position where $p_i, p_{i+1}, p_{i+2} = (0, 1, 1)$.

We extract and sort to construct trajectory data from original data by the following steps.

Classify records by ID: Due to that collected GPS records are out of order both in time and in ID, we primarily classify the original GPS records by VEHICLE_ID. The whole data of one unique taxi ID in one day are inserted into one queue. Each piece of record contains six fields i.e. TIMESTAMP, VEHICLE_ID, LONGITUDE, LATITUDE, SPEED and STATE.

Extracting trajectory data: Suppose that there are n pieces of records in queue marked as taxi ID, each of which is represented by $p_i = (lng_i, ltd_i, s_i, t_i), 1 \leq i \leq n$. We sort all the records by timestamp t_i. After sorting, the trajectory data of one taxi can be written by $p_1 \rightarrow p_2 \rightarrow \cdots p_i \rightarrow \cdots \rightarrow p_n, 1 \leq i \leq n$.

4.2 Detect Outliers and Extract Pick-up/Drop-off Data

We take the following steps to obtain pick-up/drop-off data.

Correcting error trajectory data: Since the collected GPS data is not accurate absolutely, we have to clean the original data and correct the error.

Extracting pick-up/drop-off records: These data exist in trajectory data. As proposed in pick-up DFA, all the pick-up data points have to satisfy $(\varepsilon_i, \varepsilon_{i+1}, \varepsilon_{i+2}) = (0, 1, 1)$.

Partitioning regions: The whole road network of one city is partitioned into regions by administrative districts and main roads. The regional edge data consists of GPS data points.

Extracting records in regions: We extract all the records in the region by the regional edge data. Then we utilize the proposed algorithm [15] to filter the records within the region.

4.3 Identify Pick-up/Drop-off Hotspots

To reduce the impact of noisy data on our method, we adopt the dataset in evening peak. To identify the hotspots, we adopt DBH-CLUS method to process the pick-up/drop-off records. The DBH-CLUS method represented by Algorithm 2 contains first-step clustering and second-step clustering. Details will be conducted as follows:

Extracting data in specific time frame

In consideration of noisy data in the original dataset, we select data in certain time frame and reduce its impact on the result of clustering radically. Here we partition time T of the whole dataset into time frame sequence written by $TF_i = [tf_i, tf_{i+1}), 0 \leq i \leq floor(T/TF)$, where TF is a time interval, tf_i is a timestamp. $T = \bigcup\limits_{i=1}^{floor(T/TF)} TF_i, 0 \leq i \leq floor(T/TF)$. We extract data in tf_i which belongs to peak hours.

Identify pick-up/drop-off clusters

To exclude the impact of noise data, we adopt DBH-CLUS described in Algorithm 2 to cluster core data points and extract data points to construct clusters. We apply DB-HCLUS algorithm to the original pick-up/drop-off dataset.

The first step is clustering data into several clusters. All the data points included in one cluster are mutually density-connected [3]. The algorithm called uncertain-center-based agglomerative clustering analysis is described in Algorithm 1.

In Algorithm 1, eps is the maximum radius of its neighbors. $minPts$ is the minimum number of data points required to construct a dense region. To get core data points, we define a core distance $\omega_{i,j} = \sqrt{(lng_i - lng_j)^2 + (ltd_i - ltd_j)^2}$ between two points $P_i(lng_i, ltd_i)$ and $P_j(lng_j, ltd_j)$. For each data point $P_i(lng_i, ltd_i)$, compute its core distance set $S_\omega = \{w_{i,j} | w_{i,j} = \sqrt{(lng_{p_i} - lng_{p_i})^2 + (ltd_{p_j} - ltd_{p_j})^2}, 1 \leq j \leq n, i \neq j\}$, $Count(S_\omega) = \sum\limits_{j=1}^{n} \lambda_j (1 \leq j \leq n, i \neq j)\lambda_j = \begin{cases} 1 & w_j \leq eps \\ 0 & w_j > eps \end{cases} . P_i(lng_i, ltd_i)$ is a core data point if and only if $Count(S_\omega) \geq minPts$. We define a core point list LC to merge data points. Each cluster is depicted by $LC_i(1 \leq i \leq Count(clusterlist))$. So LC_i satisfies the condition: $\forall i, j, \exists w_{i,j} \leq eps, w_{i,j} = \sqrt{(lng_{p_i'} - lng_{p_j})^2 + (ltd_{p_i'} - ltd_{p_j})^2}$ where $P_i'(lng_i, ltd_i), 0 \leq i \leq n_0$.

The process of identifying hotspots from the core data points in certain dataset is shown in Algorithm 2. Given a pick-up dataset $D_{pick-up}$, a drop-off dataset $D_{drop-off}$ and a time set T, we firstly initialize the dataset DL with data (Line 1 of Algorithm 2). Secondly, we apply UACA (Uncertain-center-based agglomerative clustering analysis) (Algorithm 1) to the objective data in certain date and to obtain the core data points (Line 2 to 5). Then, we adopt to merge core data points to clusters (Line 6). Finally, we take UACA to the data of clusters again to obtain the clusters which represent the potential hotspots (Line 8).

Algorithm 1. UACA(TF_j , $D_{pick-up}$, eps , $minPts$)

Input: pick-up and drop-off dataset, time set

Output: Hotspots Clusterlist TC

---

1. Intialize clusterlist $C \leftarrow null$, datapoints $\leftarrow TF_j, D_{pick-up}$

2. for each p_i in datapoints do

3. if $IsKey(p_i, datapoints, eps, minPts)$ then

4. add $p_i \rightarrow C$

5. end for

6. for each c_i in C do

7. for each c_i in C do

8. $C' \leftarrow$ mergelist(c_i , c_j)

9. end for

10. end for

11. return clusterlist C'

Algorithm 2. DBH-CLUS	
Input: pick-up dataset, drop-off dataset,time set	
Output: Hotspots Clusterlist TC	
1: Intialize dateList DL ;	6: readString ← $MergeAreaData(DL_i, TF_j)$
2: for each DL_i in DL do	7: end for
3: for each TF_j in time frame T do	8: TC ← UACA (readString, $D_{pick-up}$,eps,minPts)
4: UACA (TF_j , $D_{pick-up}$,eps,minPts)	9: Return TC
5: end for	

To identify hotspots from the results of DBH-CLUS, we utilize δ to select the clusters. δ is defined as $\delta_i = \frac{N_i}{S_{cluster_i}}$, N_i is the number of data points in cluster i, $S_{cluster_i}$ is the minimum area of the convex polygon which consists of the whole data points in cluster i. The algorithm extracts the data points consisting of the convex polygon which has the minimum area. We regard all the polygons as hotspots where $\delta_i \geq \alpha, (\alpha = 1.5 in default)$.

5 Experimental Results

In this section, we firstly make a description of testing data. Then we conduct experiment I to verify DFA. Finally, we make an accuracy analysis in experiment II.

5.1 Dataset Description

We implement the algorithm with a real dataset sourcing from 12,000 taxis during a period of 3 months (01/01/2014-31/03/2014). It can be downloaded by the link: http://pan.baidu.com/s/1bnlcCeN. The number of taxi GPS records is around 2.5 billion.

Haidian District within 4th ring is the testing region. The performances of all the evaluations are based on the region.

We set 2 h as a time frame. We define $eps = 20m$ and $minPts = 4$ in Algorithm 2.

5.2 Validity of DFA

In experiment I, to analyze the accuracy of the DFA proposed in 3.1, we apply the DFA to the trajectory data of experimental taxis to detect and correct the error data. Then we conjecture the pick-up/drop-off positions with the corrected data.

Taking one taxi (VEHICLE_ID: 13321180414) on Jul 10th, 2014 for instance, the result of the DFA recognition is described as follows. In Fig. 4, the lower red line shows the original trajectory data and the upper blue line shows the trajectory data after correcting. The original trajectory data is constructed with 7977 records totally. We

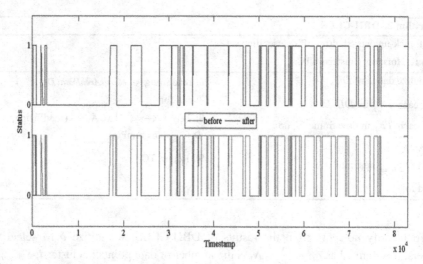

Fig. 3. Status data of one taxi

have detected 38 error records among them. Figure 5 shows the pick-up and drop/off times of the taxi on that day. The blue bars represent 63 pick-up events and 62 drop-off events identified from the original data set. The red bars represent 30 pick-up events and 29 drop-off events identified from the corrected data set. The result of 32.5 pick-up/drop-off events daily verifies the reliability of our experimental result [15] (Fig. 3).

Though error records occupy a low proportion in total, they will lead to a serious result in identifying pick-up/drop-off events. The error proportion can reach up to 50 %.

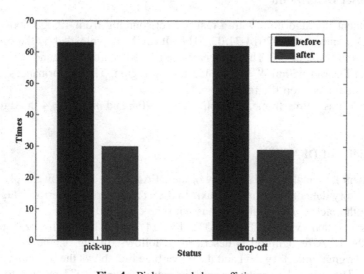

Fig. 4. Pick-up and drop-off times

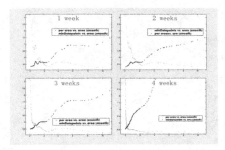

Fig. 5. Tendency in four data sets

The result shows that our method can validly correct error records. Besides, the DFA can identify pick-up/drop-off events validly.

5.3   Accuracy Analysis

In this experiment, we apply data from 01/01/2014 to 01/31/2014 to analyze two factors' effects on identifying hotspots. The factors are data volume and characteristic date.

Taking identifying pick-up hotspots as an instance, we take the follow parameters to identify hotspots from clusters. We set testing data points in Algorithm 2 shown in Table 1. Each of the parameters means the minimum data points of one cluster. We adopt four kinds of dataset to conduct the experiment i.e. one week, two weeks three weeks and four weeks.

In Fig. 5, four sub-figures show the results of 4 datasets. We can find that when the dataset are 3 weeks and 4 weeks, the results are stable and familiar. As we all know, the larger the dataset is, the more noise data there will be. The experimental result has shown that a data set beyond one method will lead to more impacts. Given the conclusion, we adopt four weeks as a scale of the original data set.

We apply our method to the original dataset in four weeks to identify pick-up hotspots. To describe the results clearly, we select the result in the section beyond 200

Table 1. Parameters of minimum data points

Data set	Minimum data points				
1 week	10	20	30	40	50
	60	70	80	90	100
2 weeks	20	40	60	80	100
	120	140	160	180	200
3 weeks	40	80	120	160	200
	240	280	320	360	400
4 weeks	100	150	200	350	300
	350	400	450	500	550

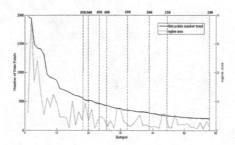

Fig. 6. Hotspots changing with minimum data points

minimum data points. As is shown in Fig. 6, the blue line represents the changing trend of data point number in one region. The red dot line represents the region area changing trend. Each green dash-dot line shows the minimum data point to identify hotspots.

The experimental result illustrates that our method can accurately identify the hotspots distributed in different agglomeration region for pick-up/drop-off. Besides, we can determine the number of hotspot depending on the demand of traffic planning.

6 Conclusions and Future Work

In this paper, we revisit the problem of hailing a taxi in downtown and analyze the GPS data to conjecture the pick-up/drop-off positions. We have presented a DFA to automatically identify the data of pick-up/drop-off events. With the DFA, we have proposed a method to identify pick-up/drop-off hotspots. The method applies hierarchal clustering based on agglomerative clustering analysis method to extract pick-up/drop-off position data and to identify hotspots. We have experimented on a real GPS data set to verify the DFA, to adjust the parameters and to analyze the accuracy. Our study shows that the DFA is valid to recognize the pick-up/drop-off events automatically, and our results represent that the proposed method can identify the hotspots accurately. We believe that our method can precisely identify hotspots from the original GPS data and provide an excellent tool to facilitate taxi stand planning. As future research, we plan to study self-adapting parameter adjustment.

References

1. Ester, M., Kriegel, H.P., Sander, J., Xu, X.: A density-based algorithm for discovering clusters in large spatial databases with noise. In: Kdd, vol. 96, pp. 226–231 (1996)
2. Ng, R.T., Han, J.: CLARANS: a method for clustering objects for spatial data mining. Knowl. Data Eng. IEEE Trans. **14**(5), 1003–1016 (2002)
3. Chuang, K.S., Tzeng, H.L., Chen, S., Wu, J., Chen, T.J.: Fuzzy C-means clustering with spatial information for image segmentation. Comput. Med. Imaging Graph. **30**(1), 9–15 (2006)

4. Xu, X., Ester, M., Kriegel, H.P., Sander, J.: A distribution-based clustering algorithm for mining in large spatial databases. In: Proceedings, 14th International Conference on Data Engineering, 1998. pp. 324–331. IEEE (1998)
5. Yu, X.G., Jian, Y.: A new clustering algorithm based On KNN and DENCLUE. In: Machine Learning and Cybernetics, 2005. Proceedings of 2005 International Conference on, vol. 4, pp. 2033–2038. IEEE (2005)
6. Wang, W., Yang, J., Muntz, R.: STING: a statistical information grid approach to spatial data mining. In: VLDB, vol. 97, pp. 186–195 (1997)
7. Zhang, T., Ramakrishnan, R., Livny, M.: BIRCH: an efficient data clustering method for very large databases. ACM SIGMOD Rec. 25(2), 103–114 (1996)
8. Yamamoto, K., Uesugi, K., Watanabe, T.: Adaptive routing of cruising taxis by mutual exchange of pathways. In: Lovrek, I., Howlett, R.J., Jain, L.C. (eds.) KES 2008, Part II. LNCS (LNAI), vol. 5178, pp. 559–566. Springer, Heidelberg (2008)
9. Phithakkitnukoon, S., Veloso, M., Bento, C., Biderman, A., Ratti, C.: Taxi-aware map: identifying and predicting vacant taxis in the city. In: de Ruyter, B., Wichert, R., Keyson, D. V., Markopoulos, P., Streitz, N., Divitini, M., Georgantas, N., Mana Gomez, A. (eds.) AmI 2010. LNCS, vol. 6439, pp. 86–95. Springer, Heidelberg (2010)
10. Chang, H.W., Tai, Y.C., Hsu, J.Y.J.: Context-aware taxi demand hotspots prediction. Int. J. Bus. Intell. Data Min 5(1), 3–18 (2010)
11. Wong, K.I., Wong, S.C., Bell, M.G.H., Yang, H.: Modeling the bilateral micro-searching behavior for urban taxi services using the absorbing markov chain approach. J. Adv. Transp. 39(1), 81–104 (2005)
12. Loecher, M., Jebara, T.: CitySense: multiscale space time clustering of gps points and trajectories. In: Proceedings of the Joint Statistical Meeting (2009)
13. Ratcliffe, J.H.: The hotspot matrix: a framework for the spatio-temporal targeting of crime reduction. Police Pract. Res. 5(1), 5–23 (2004)
14. Hu, K., He, Z., Yue, Y.: Taxi-Viewer: Around The Corner Taxis are!. In: Ubiquitous Intelligence and Computing and 7th International Conference on Autonomic & Trusted Computing (UIC/ATC), 2010 7th International Conference on, pp. 498–500. IEEE (2010)
15. Wan, X., Kang, J., Gao, M., Zhao, J.: Taxi Origin-destination areas of interest discovering based on functional region division. In: 2013 Third International Conference on Innovative Computing Technology (INTECH), pp. 365–370. IEEE (2013)

A New Representation Method of H1N1 Influenza Virus and Its Application

Wei-Wei Li, Yang Li, and Xu-Qing Tang[✉]

School of Science, Jiangnan University,
Wuxi 214122, China
txq5139@jiangnan.edu.cn

Abstract. Based on the 38,899 pieces of H1N1 virus protein sequences from 1902 to 2013 in the world, the 1805 H1N1 virus sequences with HA and NA protein are selected according to viruses occurred at the same time and place. A new representation of feature vector for protein sequences is proposed by the physicochemical properties of amino acids and coarse graining theories. The 20 kinds of amino acids are divided into 4 classes and connected with each other to construct 16-dimensional feature vectors to represent HA and NA protein sequence, respectively. The whole protein sequence is represented by a 32-dimensional feature vector, which combines the feature vectors of HA and NA protein sequences, and the optimal cluster of the H1N1 influenza virus is obtained by the structural clustering. The relationship between HA and NA protein structures and the outbreak of H1N1 virus protein sequences is analyzed by selecting the representative elements and constructing evolutionary tree. The results show that the new representation of feature vector for protein sequences is reasonable, and large amount of data confirms that HA and NA protein sequences play a direct and important role in the outbreak of H1N1 influenza virus.

Keyword: HA and NA protein sequences · Coarse grain · Structural clustering · Evolutionary tree

1 Introduction

Influenza virus is RNA virus [1] which can make human and animal sick with flu. It can spread widely in the world, especially the outbreak of H1N1 influenza virus in Mexico in March, 2009, which had already been found infected cases in many countries [2]. Influenza virus become seasonal pandemic even widely in the world because its proteins have the constant mutation of antigenicity, specially the variation of surface HA and NA [3]. HA is a major surface protein of influenza virus. It is more prone to convert the antigenic drift and antigen. The main protective immunity antigen protein to influenza virus induced by influenza virus; NA is one of the important fiber inlaid with virus particles with enzyme activity. In order to get an effective way for preventing and treating H1N1 influenza virus, many scholars have done a lot of researches on HA and NA. Rudneva et al. (1993) studied the influence on

© Springer International Publishing Switzerland 2015
D.-S. Huang et al. (Eds.): ICIC 2015, Part II, LNCS 9226, pp. 342–350, 2015.
DOI: 10.1007/978-3-319-22186-1_33

overall properties of virus by combined gene of HA and NA [4]. Bhoumik et al. (2010) explored the relations between recombination of ancient HA sequence and influenza virus HA sequence burst in 2009 [5], and found that HA protein in H1N1 virus was derived from NA protein or from more diverse proteins. Christopher et al. (2013) discovered that HA protein of influenza virus bursting in 2008 had a striking similarity in construction with HA protein bursting in 1918 [6]. So it is important to study and analyze the evolution processes of HA and NA proteins in H1N1 influenza virus sequence.

A large amount of data generated by systems biology can be processed by different methods [7, 8]. Constructing the feature vector is the main method for studying the property of a protein sequence. According to the physicochemical property of amino acids, Xia Wu [9] constructed the 6-dimensional feature vector to represent the protein sequence, which is studied from the angle that base translated into amino acids. Panpan Qian [10] constructed the 40-dimensional feature vector of the protein sequence. The protein sequences' similarity was compared with the property of acids encoding the entire sequence, and the calculation was complex. However, the studies mentioned above studied the protein's structure by its amino acid's properties. The new representation of virus studies the protein's structure and property by the amino acid's properties in other proteins, which make connections between different protein's structure and function.

Granular Computing (GrC) is an emerging conceptual and computing paradigm of information processing. It examples to all aspects of theories, methodology, technology and tools that make use of granules, i.e. groups, classes, or clusters of a universe [11]. Yao et al. conducted a series of research work [12, 13], and solved the classification problems by using the lattice consisted of all classifications.

Based on the branch level and topology graph, the protein evolutionary tree can directly reflect the evolutionary process of virus proteins, and show the difference which is caused by gene replication [14, 15]. Cai et al. constructed virus evolutionary tree through cluster analysis method to study the relationship between viral proteins and variation characteristics of the outbreak of flu [16]. Qiu et al. studied the virus gene of Hepatitis C virus gene evolutionary tree and analyzed the virus infection strain of different genotypes [17]. These show that the evolution tree can reflect the relationship between the virus protein sequences and the nature of virus occurrence processes.

The rest of the paper is organized as follows. In Sect. 2, we give the basic methods, especially a new representation of feature vector for protein sequences and the structural clustering. In Sect. 3, the above mentioned methods are applied into analyzing the properties of H1N1 influenza virus and the result is gotten. In Sect. 4, the result obtained above is reported and discussed, and in Sect. 5 the paper is concluded.

2 Data Processing Method

In this paper, to study the relationship between the physicochemical properties of amino acids of HA and NA protein sequences and the outbreak of H1N1 virus protein sequences, the 22455 HA protein sequences and 16444 NA protein sequences occurred from 1902 to 2013 in the world are downloaded from Protein Sequence in Molecular Databases of NCBI website.

2.1 A New Feature Vector for Protein Sequences

It is well known that protein sequence consists of twenty kinds of amino acids, which can be divided into four types according to the physicochemical properties of amino acids [9], namely the polarity and hydrophilicity (pq), polar and hydrophobic(pr), nonpolar and hydrophilic(sq) and nonpolar and hydrophobic (sr), and $pq = \{G\}$, $pr = \{A, V, L, I, F, P\}$, $sq = \{S, C, N, E, T, D, K, R, H\}$, $sr = \{W, Y, M\}$. The four types of amino acid are connected with each other, and the 16-dimensional feature vector is constructed by calculating the two kinds of amino acids adjacent occurrence frequency. It can be used to represent the HA (NA) protein sequences. It not only reflects the frequency of different amino acids, but also explains the proportion of different physicochemical properties of amino acids in the protein sequence. In this paper, feature vectors representing the HA and NA characteristics are combined to describing entire protein sequence, in other words, combining 16-dimensional feature vector gets a 32-dimensional feature vector that is used to study the properties of H1N1 influenza virus protein sequence. For example the H1N1 influenza virus burst at the South Carolina in 1918. Its feature vector of the HA protein sequence can be expressed as (0.317, 0.279, 0.221, 0.183, 0.333, 0.313, 0.181, 0.171, 0.181, 0.311, 0.351, 0.156, 0.472, 0.264, 0.170, 0.094) and the feature vector of the NA protein sequence can be expressed as (0.317, 0.279, 0.221, 0.183, 0.333, 0.313, 0.181, 0.171, 0.181, 0.311, 0.351, 0.156, 0.472, 0.264, 0.170, 0.094). Therefore, the feature vector of the whole protein sequence can be expressed as (0.317, 0.279, 0.221, 0.183, 0.333, 0.313, 0.181, 0.171, 0.181, 0.311, 0.351, 0.156, 0.472, 0.264, 0.170, 0.0940.317, 0.279, 0.221, 0.183, 0.333, 0.313, 0.181, 0.171, 0.181, 0.311, 0.351, 0.156, 0.472, 0.264, 0.170, 0.094).

2.2 Clustering Method for Obtaining the Optimal Cluster

Based on the granular computing getting a reasonable algorithm, the clustering optimization principle is established to obtain the optimal cluster [18]. Let d be a normalized (pseudo-)metric on finite set $X = \{x_1, x_2, \ldots x_n\}$, $D = d(x, y)|x, y \in X = d_0, d_1, \ldots d_m$, where $d_0 = 0 < d_1 < \ldots < d_m$. Let k_0 be the finite value between two classes. The algorithm is following:

Algorithm A

Step1: $i \Leftarrow 0, x(d_i) = \{a_1, a_2, \ldots a_N\}, (N \leq n)$. For $i = 1$ to N;

Step2: $c(d_i) = k_0$;

Step3: $A \Leftarrow C, i \Leftarrow i+1, C \Leftarrow \varnothing$;

Step4: $B \Leftarrow \varnothing$;

Step5: Taken $a_j \in A, B = B \cup a_j, A \Leftarrow A \backslash a_j$;

Step6: $\forall a_k \in A$ if there exists $x_j \in a_j, y_k \in a_k$, such that $d(x_j, x_k) \leq d_i, B \Leftarrow B \cup a_k, A \Leftarrow A \backslash a_k$;

Step7: $C \Leftarrow \{B\} \cup C$;

Step8: If $A \Leftarrow \varnothing$, then goto Step4, otherwise if $x(d_i) \neq x(d_{i-1})$, output $x(d_i) = c$;

Step9: $\bar{a} = \sum_{i=1}^{n} \dfrac{d_i}{n}$, $a_k = a_{k_1}, a_{k_2}, \ldots, a_{k_{J_k}}$, $\overline{a_k} = \sum_{i=1}^{J_k} \dfrac{d_{kj}}{J_k}$;

Step10: $S_{between} = \sum_{i=1}^{n} J_i \dfrac{\left\| \overline{a_i} - \overline{a} \right\|_2^2}{n}$, $S_{in} = \sum_{i=1}^{n} \sum_{j=1}^{J_i} \dfrac{\left\| \overline{d_{ij}} - \overline{a_j} \right\|_2^2}{J_i}$;

Step11: $c_{d_i} = \max\{\alpha \times s_{between} + (1-\alpha)s_{in}\}$;

Step12: If $c(d_{i-1}) > c(d_i)$, then goto Step14;

Step13: If $i \neq m$ and $c \neq \{x\}$, then goto Step 3;

Step14: Output $X(d_i)$;

Step15: End

Where the $\bar{a}$ denotes the center of the n-dimensional distance vectors, $\overline{a_k}$ stands for the center of the classification a_k, and α is 1/2 in our article. The system can be simplified by the structural clustering and coarse granularity, which can reduce the computational complexity and provide the simpler data for constructing evolutionary tree.

2.3 Choose Representative Sequence in Every Cluster

Based on the structural clustering method, the influenza virus sequences are clustered. Every cluster contains lots of elements, which have a similar distance. Therefore, they are seen to have the same properties. Considering the complexity of the evolutionary tree of large data, the representative sequence is picked out to represent all the sequences in this cluster, which is the nearest to the class center. The specific method is:

$$\begin{cases} \overline{C_{iJ_i}} = \dfrac{1}{J_i} \sum_{k=1}^{j_i} C_{iJ_k} \\ d(C_{iJ_{i0}}) = \min d(C_{iJ_i}, \overline{C_{iJ_i}}) \quad J_i = 1, \ldots, j_i \end{cases} \tag{1}$$

Where $\overline{C_{iJ_i}}$ is the center of the class C_{iJ_i}, and $C_{iJ_{i0}}$ stands for the representative sequence in class C_{iJ_i}.

3   The Results of Data Processing

3.1   Data Processing and Preliminary Results

The 22455 HA protein sequences and 16444 NA protein sequences from 1902 to 2013 in the world are downloaded from Protein Sequence in Molecular Databases of NCBI website. However, not every H1N1 virus sequence contains the HA and NA protein sequences because of the protein variation, the 1805 H1N1 virus sequences, occurred at the same time and place, with HA and NA protein sequences are picked out by MATLAB, For example, there are 15 viruses containing HA protein and 2 viruses containing NA protein, while only 2 viruses contain HA protein and NA protein at the same time, so they are picked out from all the influenza viruses occurred in 1918.

3.2   Constructing Evolutionary Tree of Representative Sequence

The outbreak of influenza virus has a close relation with the variation of HA and NA protein antigenicities. In fact, the pathogenicity of influenza virus and activity enhancement are due to the amino acid substitutions in the HA and NA proteins sequences [19]. In this paper, by using 16-dimensional feature vectors of HA and NA protein sequences are combined into a 32-dimensional feature vector influenza viruses are divided into different clusters. The classifications are analyzed, then the relationship between the HA and NA protein sequences and the outbreak of H1N1 influenza virus is further studied.

Firstly, according to the clustering optimization principle in algorithm A, the 1805 influenza virus sequences can be divided into 22 categories where c_{d_i} takes the maximum value, as shown in Fig. 1 (the abscissa represents the number of clusters, and the ordinate is the c_{d_i}).

Secondly, in order to reduce the computational complexity, the 22 representative sequences are selected by the method of choosing representative elements. The results are shown in Table 1:

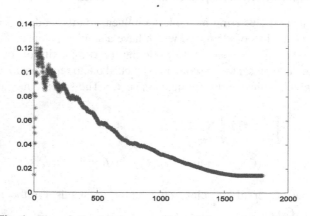

Fig. 1. The relationship between C_{d_i} and the numbers of clusters

Table 1. Representative sequences of 22 clusters of H1N1 influenza virus sequences

Serial number	Virus
1	[Influenza A virus (A/swine/Minnesota/2005)]
2	[Influenza A virus (A/Niigata/2007)]
3	[Influenza A virus (A/California/2009)]
4	[Influenza A virus (A/swine/Italy/2006)]
5	[Influenza A virus (A/swine/Lidkoeping/2002)]
6	[Influenza A virus (A/swine/Malaysia/1984)]
7	[Influenza A virus (A/Navarra/2009)]
8	[Influenza A virus (A/duck/Victoria/1980)]
9	[Influenza A virus (A/Liverpool/2008
10	[Influenza A virus (A/Tunisia/2011)]
11	[Influenza A virus (A/Mallard/Norway/2007)]
12	[Influenza A virus (A/Saudi Arabia/2009)]
13	[Influenza A virus (A/swine/Austria/2003)]
14	[Influenza A virus (A/northern pintail/Minnesota/2006)]
15	[Influenza A virus (A/Chelyabinsk/2009)]
16	[Influenza A virus (A/Mexico/2011)]
17	[Influenza A virus (A/bh/1935)]
18	[Influenza A virus (A/camel/Mongolia/1982)]
19	[Influenza A virus (A/swine/Dongen/1996)]
20	[Influenza A virus (A/turkey/France/2010)]
21	[Influenza A virus (A/Patna-India/2009)]
22	[Influenza A virus (A/South Carolina/1918)]

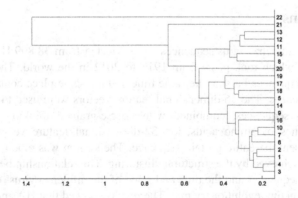

Fig. 2. Evolutionary tree of 22 representative sequences of H1N1 influenza virus

Finally, to further study relationships among different influenza virus portrayed by HA and NA proteins, we construct evolutionary tree of the 22 influenza virus sequences by Algorithm A in combination with coarse graining, as shown in Fig. 2 (Note: The abscissa represents the distance, and the ordinate is the number of the protein sequences; The numbers in Finger 2 are consistent with the viruses' numbers in Finger 1).

4   Result Analysis and Discussion

It can be seen from Fig. 2, viruses occurred at the same place, such as 11 ([Influenza A virus (A/Mallard/Norway/2007(H1N1))]) and 13 (Influenza A virus (A/swine/Austria/2003 (H1N1))), are clustered early, which shows that they have high similarity. Therefore, the virus' similarity is related to the place. What's more, (1, 2, 3, 4, 6, 7, 10, 19, 9, 14), (5, 18, 17, 19), (8, 15, 11, 12, 13) separately form the branches, and the virus in the same branch occurs at the close time. For example, viruses occurred between 2005 and 2011 in (1, 2, 3, 4, 6, 7, 10, 19, 9, 14) except 6 ([Influenza A virus (A/swine/Malaysia/1984(H1N1))]) shows that the similarity among viruses is bounded with the time. However, the 17([Influenza A virus (A/bh/1935(H1N1))]) is far from the rest of viruses' time of occurrence in (5, 18, 17, 19), but it has high similarity with the others. That means the other viruses evolve from it. While the time and place of occurrence are quite different and the virus host is same, the viruses have a high similarity, such as 8 ([Influenza A virus (A/duck/Victoria/1980(H1N1))]) and 11 ([Influenza A virus (A/Mallard/Norway/2007(H1N1))]) in (8, 15, 11, 12, 13). They have different occurrence times and places, but their host are both the duck. So the virus' similarity is closed to the virus' host, which is consistent with the results in [20, 21]. The Fig. 2 also shows the host, the times and places of virus' occurrence are all different, but the viruses are in the same branch, such as (18, 19). This may be caused by migratory birds, artificial breeding and so on [21]. The study mentioned above indicates that the new feature vector of virus is quite reasonable.

5   Conclusions

The 1805 H1N1 influenza virus sequences were selected from 38,899 H1N1 influenza virus sequences, which took place from 1918 to 2012 in the world. The 1805 H1N1 influenza virus sequences burst at the same time and the same place contained HA and NA protein sequences. The 16-dimensional feature vectors were used to represent HA and NA protein sequences, combined with coarse-grained thinking and physico-chemical properties of amino acids, the 32-dimensional feature vectors were constructed to represent the entire protein sequence. The system was modularized and the simple data was achieved by the structural clustering. The relationship between the HA and NA protein sequences and the time and scale of the influenza virus' outbreak were studied by constructing evolutionary tree. The results showed that HA and NA proteins play a direct and important role in the outbreak and proliferation of the H1N1 influenza virus. A theoretical support is provided to predict the time and scale of outbreak of the influenza virus by studying the structure of the HA and NA proteins.

Acknowledgements. The work was supported by National Natural Science Foundation of China (Grant No. 11371174), Fundamental Research Funds for the Central Universities of China (Grant No. JUSRP51317B), International Technology Collaboration Research Program of China (Grant No. 2011DFA70500), and Colleges and Universities in Jiangsu Province Plans to Graduates Research and Innovation (Grant No. 1145210232141170).

References

1. Anhlan, D., Grundmann, N., Makalowski, W.: Origin of the 1918 pandemic H1N1 Influenza A virus as studied by codon usage patterns and phylogenetic analysis. RNA **17**(1), 64–73 (2001)
2. Fraser, C., Donnelly, C.A., Cauchemez, S.: Pandemic potential of A strain of Influenza A (H1N1): early findings. Science **324**(5934), 1557–1561 (2009)
3. Gao, J., Zhang, L., Jin, P.-X.: Influenza pandemic early warning research on HA/NA protein sequences. Curr. Bioinform. **9**(3), 228–233 (2014)
4. Rudneva, I.A., Kovaleva, V.P.: Influenza A virus reassortants with surface glycoprotein genes of the avian parent virus: effects of HA and Na gene Combinations on Virus aggregation. Arch. Virol. **133**(3–4), 437–450 (1993)
5. Bhoumik, P., Hughes, A.L.: Ressortment of ancient neuraminidase and recent hemagglutinin in pandemic (H1N1) 2009 virus. Emerg. Infect. Dis. **16**(11), 1748–1750 (2010)
6. Christopher, J.V., Liu, Y.: Special features of the 2009 pandemic swine-origin Influenza A H1N1 hemagglutintin and neuraminidase. Chin. Sci. Bull. **56**(17), 1747–1752 (2013)
7. Wang, J.J.-Y., Wang, X., Gao, X.: Non-negative matrix factorization by maximizing correntropy for cancer clustering. BMC Bioinform. **14**(1), 107 (2013)
8. Wang, J.J.-Y., Bensmail, H., Gao, X.: Multiple graph regularized protein domain ranking. BMC Bioinform. **13**(1), 307 (2012)
9. Wu, X., Liao, B.: 6-D representation of protein sequences and the analysis of similarity/dissimilarity based on it. Lett. Biotechnol. **15**(4), 366–368 (2004)
10. Qian, P.-P.: A novel representation of protein sequences. Shandong Univ., China (2011)
11. Tang, X.-Q., Zhu, P., Cheng, J.-X.: The structural clustering and analysis of metric based on granular space. Pattern Recogn. **43**(11), 3768–3780 (2010)
12. Yao, Y.-Y.: Granular computing: basic issues and possible solutions. In: Proceedings of the 5th Joint Conference. Science, vol. 5, no. 1, pp. 186–189 (2000)
13. Yao, Y.-Y., Yao, J.-T.: Granular computing as a basis for consistent classification problems. In: Proceedings of PAKDD Workshop Found, Data Mining, vol. 5, pp. 101–106 (2002)
14. Chanderbali, A.S., Wong, G.K., Soltis, D.E.: Phylogeny and evolutionary history of glycogen synthase kinase 3/SHAGGY-like kinase genes in land plants. BMC Evol. Biol. **13**(1), 143–145 (2013)
15. Luo, J.-W., Yin, Z.H.-Q., Liu, S.H.-Y.: Novel method for evolutionary tree construction based on correlation feature and fuzzy clustering. Appl. Res. Comput. **28**, 2844–2847 (2011)
16. Cai, B., Peng, J., Jiang, H.: Tracking the spread of avian influenza in china: a model based on evolutionary genetics analysis and geographic visualization. Chin. J. Emerge. Med. **21**, 887–891 (2012)
17. Qiu, G.-H., Zhang, R., Du, Sh-C: The evolutionary trees analysis of HCV genotypes in China. Chin. J. Clin. Lab. Sci. **23**(3), 165–167 (2005)
18. Tang, X.-Q., Zhu, P.: Hierarchical clustering problems and analysis of fuzzy proximity relation on granular space. IEEE Trans. Fuzzy Syst. **21**(5), 814–824 (2013)

19. Guo, Y.-J., Wen, L.-Y.: Origin of hemagglutinin and neuraminidase gene of swine Influenza A H1N1 viruses. Chin. J. Exp. Clin. Virol. **17**(4), 315–318 (2013)
20. Li, J., Shao, T.-J., Yu, X.-F., et al.: Molecular evolution of HA gene of the Influenza A H1N1 Pdm09 strain during the consecutive seasons 2009–2011 in Hangzhou, China: several immune-escape variants without positively selected sites. J. Clin. Virol. **55**(4), 363–366 (2012)
21. Mullick, J., Cherian, S.S., Potdar, V.A., et al.: Evolutionary dynamics of the Influenza A Pandemic (H1N1) 2009 virus with emphasis on Indian isolates: evidence for adaptive evolution in the HA gene. Infect Genet Evol. **11**(5), 997–1005 (2011)

Using Restricted Additive Tree Model for Identifying the Large-Scale Gene Regulatory Networks

Bin Yang[✉] and Wei Zhang

School of Information Science and Engineering, Zaozhuang University,
Zaozhuang, China
batsi@126.com

Abstract. S-system model has been proposed to identify gene regulatory network in the past years. However, due to the computation complexity, this model is only used for reconstruction of small-scale networks. In this paper, the restricted additive tree model (RAT) is proposed to infer the large-scale gene regulatory networks. In the method, restricted additive tree model is used to encode the S-system model, and the hybrid evolution approach is used to optimize the structure and parameters of restricted additive tree. The large-scale gene regulatory network containing 30 genes is used to test the performance of our method. The results reveal that our method could identify the large-scale gene regulatory network correctly.

Keywords: Gene regulatory network · Restricted additive tree · S-system model · Hybrid evolution approach

1 Introduction

Many scientific and engineering problems can be finally formulated as a complex nonlinear map problem, in which local linear models and local nonlinear models play an important role in analyzing the characteristics of the system. S-system model could capture various dynamics and has a good compromise between accuracy and mathematical flexibility, so it is applied widely to infer the complex regulatory relationship among genes [1–3].

The existed methods focused on the optimization of parameters of S-system model, containing Genetic Algorithms (GA), Simulated Annealing (SA), Differential Evolution (DE), Particle Swarm Optimization (PSO), Artificial bee colony algorithm (ABC), and so on [4–8]. Kikuchi et al. [9] proposed a method to predict not only the network structure but also its dynamics using a unified extension of a Genetic Algorithm and an S-system formalism. Gonzalez et al. [10] described how the heuristic optimization technique simulated annealing could be effectively used for estimating the parameters of S-system from time-course biochemical data, and used three artificial networks designed to simulate different network topologies and behavior to demonstrate the method. Palafox et al. [11] implemented a variation of particle swarm optimization, called dissipative PSO (DPSO) to optimize the parameters of the popular non-linear

© Springer International Publishing Switzerland 2015
D.-S. Huang et al. (Eds.): ICIC 2015, Part II, LNCS 9226, pp. 351–359, 2015.
DOI: 10.1007/978-3-319-22186-1_34

differential equation model named S-system in order to infer small-scale gene regulatory networks.

The past study work could lead to the overwhelming computational complexities, and has only been applied to infer the small-scale networks. Because the number of S-system parameters is proportional to the square of the number of variable, a large number of parameters need to be simultaneously estimated, when the number of variable is very large. However gene regulatory network is spare, there is no need to estimate all the parameters simultaneously. To solve this problem we proposed the restricted additive tree model for inferring the S-system models. This model has two advantages: (1) the form of model is very similar to S-system; (2) the structure and parameters could be optimized simultaneously. But we only applied this method for inferring small and medium gene regulatory network [12].

In this paper, we investigate the performance of restricted additive tree model for inferring the large-scale gene regulatory networks. In the method, restricted additive tree model is used to encode the S-system model, and the hybrid evolution approach combining evolution algorithm based on tree-structure with dissipative particle swarm optimization is used to optimize the structure and parameters of restricted additive tree. The large-scale gene regulatory network containing 30 genes is used to test the performance of our method.

The paper is organized as follows. In Sect. 2, we describe the details of S-system as well as our method, containing restricted additive tree model and DPSO. In Sect. 3, the example is performed to validate the effectiveness and precision of the proposed method. Conclusions are reported in Sect. 4.

2   Method

2.1   Structure Optimization of Models

S-System Model. The S-system model is a popular model of biochemical system theory, which consists of n nonlinear ordinary differential equations (ODEs) and the generic form of Eq. 1 is given as follows:

$$X_i^{'}(t) = \alpha_i \prod_{j=1}^{N} X_j^{g_{ij}}(t) - \beta_i \prod_{j=1}^{N} X_j^{h_{ij}}(t) \tag{1}$$

Where X_i is a vector element of dependent variable, N is the number of variables, α_i and β_i are vector elements of non-negative rate constants, and g_{ij} and h_{ij} are matrix elements of kinetic orders.

The Restricted Additive Tree Model. In our past study, the additive tree models were proposed for the system identification especially the reconstruction of polynomials and the identification of linear/nonlinear systems. To be more near to the representation of the S-system model to be reconstructed, restricted additive tree model was proposed. The example of restricted additive tree model is described as in Fig. 1. To create the tree model, the operator set I is determined firstly.

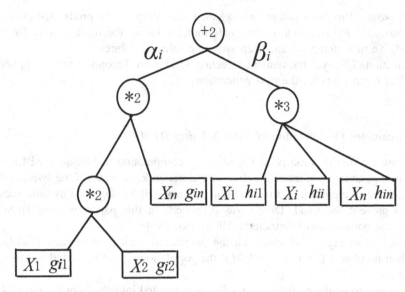

Fig. 1. An example of the restricted additive tree model, its symbolic expression: $f_i = \alpha_i X_1^{g_{i1}} X_2^{g_{i2}} X_n^{g_{in}} + \beta_i X_1^{h_{i1}} X_i^{h_{ii}} X_n^{h_{in}}$

$$I = \{*2, *3, \ldots, *n, x_1, x_2, \ldots, x_n, R\} \tag{2}$$

Where $*n$ presents the product of n variables, n is the number of variables [12]. The root node is set as +2, the other nodes could be selected from the operator set I. When $*i$ ($i = 1,\ldots, n$) is selected, i branches are created.

Evolving Structure of Equation. Finding an optimal or near-optimal restricted additive tree is an evolution process. In this study, the additive tree operators are as following.

(1) Mutation. In this paper, we choose four mutation operators to generate off-springs from the parents, which are as following:

   (1) Change one terminal node: according to the predefined mutation probability P_m, select one terminal node in the tree and replace it with another terminal node which is generated randomly.
   (2) Grow: according to P_m select a leaf in the hidden layer of the tree and replace it with a newly generated subtree.
   (3) Prune: according to P_m select a function node in a tree and replace it with a newly generated terminal node.
   (4) Change all terminal nodes: replace all terminal modes in the tree with other terminal nodes which are generated randomly.

(2) Crossover. First two parents are selected according to the predefined crossover probability P_c. And select one nonterminal node in the hidden layer for each additive tree randomly, and then swap the selected subtree.

(3) Selection. EP-style tournament selection [13] with 12 opponents is applied to select the parents for the next generation.

2.2   Parameter Optimization of Models Using DPSO

Particle swarm optimization is an evolutionary computation technique used for optimization of parameters of model, such as neural network. However if the swarm which is going to be in equilibrium, the evolution process will be stagnated as time goes on [14]. To prevent the trend, DPSO was proposed. In this paper, we use DPSO to optimize the parameters of restricted additive tree model.

According to Fig. 1, we check all the parameters contained in each model, and count their number n_i ($i = 1, 2,..., M$, M is the population size of restricted additive tree model).

According to n_i, the particles are randomly generated initially from the range $[l, u]$. Each particle x_i represents a potential solution. A swarm of particles moves through space, with the moving velocity of each particle represented by a velocity vector v_i. At each step, each particle is evaluated and keeps track of its own best position, which is associated with the best fitness it has achieved so far in a vector $Pbest_i$. The best position among all the particles is kept as $Gbest$. A new velocity for particle i is updated by

$$v_i(t+1) = w * v_i(t) + c_1 r_1 (Pbest_i - x_i(t)) + c_2 r_2 (Gbest(t) - x_i(t)). \tag{3}$$

Where w is the inertia weight and impacts on the convergence rate of DPSO, which is computed adaptively as $w = (max\_generation - t)/(2*max\_generation) + 0.4$ ($max\_generation$ is maximum number of iterations, and t is current iteration), c_1 and c_2 are positive constant, and r_1 and r_2 are uniformly distributed random number in [0,1]. Based on the updated velocities, each particle changes its position according to the following equation:

$$x_i(t+1) = x_i(t) + v_i(t+1) \tag{4}$$

In this step, DPSO adds negative entropy through additional chaos for each particle using the following equations.

$$IF(rand() < c_v) \text{ THEN } v_i(t+1) = rand() * v_{max} \tag{5}$$

$$IF(rand() < c_i) \text{ THEN } x_i(t+1) = random(l, u) \tag{6}$$

Where c_v and c_l are chaotic factors randomly created in the range [0, 1], v_{max} is the maximum velocity, which specified by the user [14].

2.3   Fitness Function

The sum of squared relative error (SSRE) is usually used to search the optimization gene regulatory network according to the actual data and predicted data.

$$SSRE = \sum_{i=1}^{N} \sum_{j=1}^{T} \frac{(x'_{ij} - x_{ij})^2}{x'_{ij}} \tag{7}$$

Where N is the number of genet, M is the number of sample point, x_{ij} is the raw experimental data of gene j in the i-th time point, and x'_{ij} is an output data of gene j in the i-th time point.

2.4   Summary of Algorithm

The overall method is illustrated in Fig. 2. The optimal design of each ordinary differential equation could process in parallel.

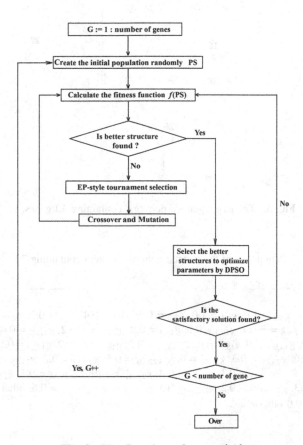

Fig. 2. The flowchart of our method.

3 Experiment

Our proposed method is tested on the simulated data from the S-system in [15]. The large-scale S-system model containing 30 genes is used as the target model, and the network structure is described in Fig. 3. The Table 1 gives the corresponding parameters of the S-system model. The experiment is executed on a PC (Intel Pentium Dual 1.80 GHz processor and 2 GB memory). We use the sum of squared error (SSE) as the fitness function. The parameters used in the method are listed in Table 2. According to inference rule of gene regulatory network using S-system model, we make some

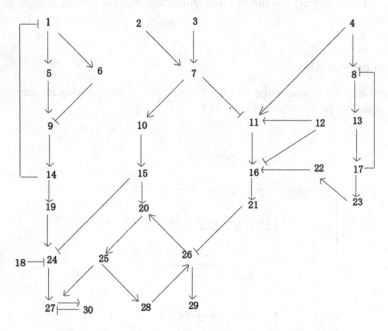

Fig. 3. The target genetic network containing 30 genes.

Table 1. Parameters of the genetic network system containing 30 genes.

α_i	1.0
β_i	1.0
$g_{i,j}$	$g_{1,14} = -0.1$, $g_{5,1} = 1.0$, $g_{6,1} = 1.0$, $g_{7,2} = 0.5$, $g_{7,3} = 0.4$, $g_{8,4} = 0.2$, $g_{8,17} = -0.2$, $g_{9,5} = 1.0$, $g_{9,6} = -0.1$, $g_{10,7} = 0.3$, $g_{11,4} = 0.4$, $g_{11,7} = -0.2$, $g_{11,22} = 0.4$, $g_{12,23} = 0.1$, $g_{13,8} = 0.6$, $g_{14,9} = 1.0$, $g_{15,10} = 0.2$, $g_{16,11} = 0.5$, $g_{16,12} = -0.2$, $g_{17,13} = 0.5$, $g_{19,14} = 0.1$, $g_{20,15} = 0.6$, $g_{20,26} = 0.3$, $g_{21,16} = 0.6$, $g_{22,16} = 0.5$, $g_{23,17} = 0.2$, $g_{24,15} = -0.2$, $g_{24,18} = -0.1$, $g_{24,19} = 0.3$, $g_{25,20} = 0.4$, $g_{26,21} = -0.2$, $g_{26,28} = 0.1$, $g_{27,24} = 0.6$, $g_{27,25} = 0.3$, $g_{27,30} = -0.2$, $g_{28,25} = 0.5$, $g_{29,26} = 0.4$, $g_{30,27} = 0.6$, other $g_{i,j} = 0.0$
$h_{i,j}$	1.0 if $i = j$, 0.0 otherwise

Table 2. Parameters for experiment.

Parameters	Values
Population size	100
Generation	100
Crossover rate	0.7
Mutation rate	0.3
PSO Population size	200
PSO Generation	100
PSO V_{max}	2
PSO $[l, u]$	[0, 5]
step	0.05
Time points	11

restrictions in the process of creating tree models. For example variable X_i contributes to its own degradation, so X_i is usually a part of the beta-term and thus h_{ii} is usually not zero (h_{ii} = not 0 for $i = 1, ..., n$).

The initial conditions are created randomly and the synthetic time series are generated by solving the differential equations. In this experiment, the time series is from 0 to 10, including 11 time point. The used instruction sets to create an optimal restricted additive expression tree model is $I = \{*2, *3, *4, x_1, x_2, ..., x_{29}, R\}$. The used execution time for each experiment was ~3.25 h. We get the best parameters of the corresponding S-system models (Table 3). Compared Table 1 with Table 3, we can see that our method could identify all the regulatory relationships among genes (68 true-positive interactions), and the predicted parameters of model are very close to the targeted one. The sum of squared relative error obtained from our method averaged about 1.23×10^{-3}. Figure 4 shows the SSRE convergence respect to generations. From Fig. 4, we could see that our method could gain the best model at about 20-th generation.

Table 3. Parameters of the genetic network system containing 30 genes by our method.

α_i	1.0
β_i	1.0
$g_{i,j}$	$g_{1,14} = -0.1$, $g_{5,1} = 1.0$, $g_{6,1} = 1.0$, $g_{7,2} = 0.5$, $g_{7,3} = 0.4$, $g_{8,4} = 0.2$, $g_{8,17} = -0.2$, $g_{9,5} = 1.0$, $g_{9,6} = -0.1$, $g_{10,7} = 0.3$, $g_{11,4} = 0.4$, $g_{11,7} = -0.2$, $g_{11,22} = 0.4$, $g_{12,23} = 0.1$, $g_{13,8} = 0.6$, $g_{14,9} = 1.0$, $g_{15,10} = 0.2$, $g_{16,11} = 0.5$, $g_{16,12} = -0.2$, $g_{17,13} = 0.5$, $g_{19,14} = 0.1$, $g_{20,15} = 0.6$, $g_{20,26} = 0.3$, $g_{21,16} = 0.6$, $g_{22,16} = 0.5$, $g_{23,17} = 0.2$, $g_{24,15} = -0.2$, $g_{24,18} = -0.1$, $g_{24,19} = 0.3$, $g_{25,20} = 0.4$, $g_{26,21} = -0.2$, $g_{26,28} = 0.1$, $g_{27,24} = 0.6$, $g_{27,25} = 0.3$, $g_{27,30} = -0.2$, $g_{28,25} = 0.5$, $g_{29,26} = 0.4$, $g_{30,27} = 0.6$, other $g_{i,j} = 0.0$
$h_{i,j}$	1.0 if $i=j$, 0.0 otherwise

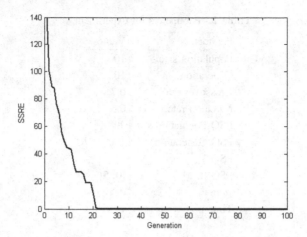

Fig. 4. SSRE convergence respect to generations.

Table 4. Performance comparison with different noise.

Noise	0.05	0.1	0.15	0.2	0.25	0.3	0.4
True-positive	60.1	55	34.5	20	18.6	15.4	15

In [15], Kimura proposed the method based on the problem decomposition strategy and a cooperative coevolutionary algorithm for inferring S-system models of large-scale genetic networks. But in this method 2460 parameters should be estimated. This method could identify 58.4 true-positive interactions on average, but the computational time required ~57.5 h on a single-CPU personal computer (Pentium III 933 MHz). The sum of squared relative error obtained from Kimura's method averaged about 27.72. Compared with our computational time and results, our method is superior to Kimura's method in term of runtime and accuracy of network.

To test the noise-tolerance of the proposed algorithm, normally distributed noise with zero mean and 0.05, 0.1, 0.15, 0.2, 0.25, 0.3, 0.4 standard deviations are added to the simulated time series, respectively. The parameters are listed in Table 2. Through several runs, the results are listed in Table 4. When the standard deviation of noise is increased beyond 0.15, our method performs badly.

4   Conclusion

In this paper, the restricted additive tree model is proposed to infer the large-scale gene regulatory networks. The S-system model is encoded by restricted additive tree model. And the hybrid evolution approach based on tree-structure and dissipative particle swarm optimization is used to optimize the structure and parameters of restricted additive tree. The large-scale gene regulatory network containing 30 genes is used to test the performance of our method. The results reveal that our method could identify the large-scale gene regulatory network correctly.

In the future research, we plan to apply our method to the inference of large-scale real biological networks.

Acknowledgments. This work was supported the PhD research startup foundation of Zaozhuang University (No. 1020702).

References

1. Srinath, S., Gunawan, R.: Parameter identifiability of power-law biochemical system models. J. Biotechnol. **149**(3), 132–140 (2010)
2. Wang, H., Qian, L., Dougherty, E.: Inference of gene regulatory networks using system: a unified approach. IET Syst. Biol. **4**(2), 145–156 (2010)
3. Akutsu, T., Miyano, S., Kuhara, S.: Inferring qualitative relations in genetic networks and metabolic pathways. Bioinformatics **16**, 727–734 (2000)
4. Liu, L.Z., Wu, F.X., Zhang, W.J.: Inference of biological S-system using the separable estimation method and the genetic algorithm. IEEE/ACM Trans. Comput. Biol. Bioinform. (TCBB). **9**(4), 955–965 (2012)
5. Yeh, W.C., Lin, W.B., Hsieh, T.J., Liu, S.L.: Feasible prediction in S-system models of genetic networks. Expert Syst. Appl. **38**(1), 193–197 (2011)
6. Yeh, W.C., Hsieh, T.J.: Artificial bee colony algorithm-neural networks for S-system models of biochemical networks approximation. Neural Comput. Appl. **21**(2), 365–375 (2012)
7. Meskin, N., Nounou, H., Nounou, M., Datta, A., Dougherty, E.R.: Parameter estimation of Biological phenomena modeled by S-systems: an extended Kalman filter approach. In: 2011 50th IEEE Conference on Decision and Control and European Control Conference (CDC-ECC), pp. 4424–4429. IEEE Press, New York (2011)
8. Tian, L.P., Mu, L., Wu, F.X.: Alternating constraint least squares parameter estimation for S-system models of biological networks. In: 4th International Conference on Bioinformatics and Biomedical Engineering (iCBBE), pp. 1–5 (2011)
9. Kikuchi, S., Tominaga, D., Arita, M., Takahashi, K.: Dynamic modeling of genetic networks using genetic algorithm and S-system. Bioinformatics **19**, 643–650 (2003)
10. Gonzalez, O.R., Kuper, C., Jung, K., Naval, P.C., Mendoza, E.: Parameter estimation using simulated annealing for S-system models of biochemical networks. Bioinformatics **23**, 480–486 (2007)
11. Palafox, L., Noman, N., Iba, H.: Reverse engineering of gene regulatory networks using dissipative particle swarm optimization evolutionary computation. IEEE Trans. **17**(4), 577–587 (2013)
12. Yang, B., Jiang, M.Y., Chen, Y.H.: A fast and efficient method for inferring structure and parameters of S-system models. HIS **2011**, 235–240 (2010)
13. Chellapilla, K.: Evolving computer programs without subtree crossover. IEEE Trans. Evol. Comput. **1**, 209–216 (1997)
14. Xie, F.X., Zhang, W.J., Yang, Z.L.: A dissipative particle swarm optimization. In: Proceedings of Congress on Evolutionary Computation (CEC), pp. 1456–1461 (2002)
15. Kimura, S., Ide, K., Kashihara, A.: Inference of S-system models of genetic networks using a cooperative coevolutionary algorithm. Bioinformatics **21**, 1154–1163 (2005)

Identification of Osteosarcoma-Related Genes Based on Bioinformatics Methods

Ze-Bing Si[1,2] and Ji-Gong Wu[2(✉)]

[1] Clinical College of the 306th Hospital of PLA, Anhui Medical University,
Hefei 230032, Anhui, China
[2] Department of Orthopaedics, Spinal Center of PLA, The 306th Hospital of
PLA, No 9, Anxiangbeili, ChaoYang District, Beijing 100101, China
{moonzebing, Jgwu306}@163.com

Abstract. Osteosarcoma is one of the malignant cancers. Improving diagnosis and treatment of this disease is urgent. Microarray data have revealed abnormal expression of several genes implicated in osteosarcoma. From the mRNA data, differentially expressed (DE) genes were selected by Wilcox test. Using Endeavour online service, 200 DE genes were selected as osteosarcoma related genes. Combing the module extraction algorithm, we identified 90 osteosarcoma candidate genes. Nine (DCC, FGFR, Fzd1, NFKBIA, CCNE1, CASP3, APC, and LRP6) out of the 90 candidate genes have great potential roles in osteosarcoma.

Keywords: Osteosarcoma · Network · Protein

1 Introduction

Most osteosarcoma is a malignant bone tumor that occurs most frequently in adolescents [1]. Although the multi-agent chemotherapy conjunction with surgical resection has improved the prognosis, the treatment of osteosarcoma is still a long way due to the risk of recurrence [2]. In recent years, some valuable prognostic biomarkers have been identified in osteosarcoma [3]. It is known that early detection of osteosarcoma is crucial to prolong the survival time of the patient and improve the effect of treatment. Therefore, there is urgent for discovering valuable novel biomarkers for early diagnosis.

Biomarkers can contribute to investigations into tumorgenesis. However, discovery of novel candidate genes related to osteosarcoma only by biological experiments is challenging, which is time-consuming and high cost. Computational methods make up the shortage and effectively identify novel cancer-related genes.

Recently, several computational methods were applied to uncover new candidate genes. We adopted a computational method to discover new osteosarcoma candidate genes. Additionally, several newly identified genes have a great potential to be candidate genes related to osteosarcoma. Our results provide new directions for uncovering new osteosarcoma related genes.

© Springer International Publishing Switzerland 2015
D.-S. Huang et al. (Eds.): ICIC 2015, Part II, LNCS 9226, pp. 360–366, 2015.
DOI: 10.1007/978-3-319-22186-1_35

2   Method

2.1   Dataset

The gene expression profiling of 7 osteosarcoma and 5 healthy patients were down-loaded from NCBI Gene Expression Omnibus (GEO) (http://www.ncbi.nlm.nih.gov/geo/) with the series of GSE42572.

2.2   Selection of Differentially Expressed Genes

Wilcox test was applied to select all possible differentially expressed genes from the gene expression data. Genes with p-value less than 0.05 were maintained. At last, 999 differentially expressed (DE) genes were obtained after filtering.

2.3   Gene Prioritization

First, We extracted 12 osteosarcoma seed genes (CDKN2A,CDKN2B, CYP3A4, CYP3A5, BIRC5, MDM2, MYC, ACTA2, RB1, TSPAN31, TP53, and EZR) from KEGG database [4]. These genes were supported by mounts of literature evidences that involved in osteosarcoma. Then, we prioritized the 999 DE genes using the Endeavour web server [5]. Endeavor [5] needs user to input training genes and candidate genes. 12 osteosarcoma seed genes were selected as training genes, while the 999 DE genes were selected as candidate genes. Endeavour combines the 12 training genes and the relevant significance from multiple genomic data source to rank the 999 DE genes. We selected the top 200 ranking genes as osteosarcoma related genes.

2.4   Driver Core Module Identification

We employed the NetBox [6] software to identify core modules. We input the top 200 ranking osteosarcoma related genes into NetBox. A shortest path threshold was assigned to 2, which means the candidate linker gene can connect at least two altered genes. To evaluate the statistical significance of the linker gene in the network, the p-value cut-off is 0.05. To ensure the osteosarcoma network is more modular than a null model of random network, 1000 random iterations were performed.

2.5   KEGG Pathway Enrichment Analysis

KEGG pathway enrichment analysis was performed in the DAVID [7] website, p values were adjusted for multiple testing using the method of Benjamini and Hochberg method [8]. P-value less than 0.05 was considered statistically significant.

3 Results

3.1 Selection of Differentially Expressed Genes and Osteosarcoma Altered Genes

The expression data was retrieved from GEO with the series of GSE42572, which contained 12 samples and 48701 probes. 7 osteosarcoma samples belonged to experimental group while the 5 health samples belonged to control group. We first identified 999 genes that were significant differentially expressed ($p < 0.05$) in tumors. Then, we extracted 12 osteosarcoma seed genes from KEGG database. We prioritized the 999 DE genes using the Endeavour web server. Endeavour adopts many algorithms to integrate 7 genomic data sources to rank the candidate genes. The top 200 ranking genes were selected as osteosarcoma related genes.

3.2 Core Module Related to Development

To obtain a global view of gene interaction network in osteosarcoma, we selected the top 200 ranking osteosarcoma related genes in Endeavour. The detailed information was found in the method. Using the module detection algorithm in NetBox, we finally identified 16 modules, with a modularity of 0.592. Based on 1000 random trials, the average modularity of random networks is 0.458, with a standard deviation of 0.02. This indicated that osteosarcoma network is more modular than random (the scaled network modularity is 6.535). 90 osteosarcoma candidate genes were included in the 16 modules (Table 1).

3.3 KEGG Enrichment Analysis of the Significant Candidate Genes

DAVID is a widely used functional annotation tool, which provides a convenient to perform Gene Ontology (GO) and KEGG pathway enrichment analysis. Here, we analyzed the 90 osteosarcoma candidate genes by carrying out the KEGG pathway enrichment analysis. The KEGG enrichment results indicated that these genes were significantly enriched in cancer pathway, T cell receptor signaling pathway and Wnt signaling pathway (Table 2). It is instructive to note that T cell receptor signaling pathway and Wnt signaling pathway have been proved to be implicated in osteosarcoma.

3.4 Analysis of the Potential Candidate Genes Associated with Osteosarcoma

In this study, we obtained 90 genes that possibly associated with osteosarcoma, among them, 23 genes were KEGG enriched genes. These 23 genes are listed in Table 2. Among the 23 genes, 9 genes (DCC, FGFR, Fzd1, NFKBIA, CCNE1, CASP3, APC, and LRP6) have direct evidence for their roles in osteosarcoma. In the following paragraphs, based on existing experimental evidences in literature, we will discuss the

Table 1. Sixteen modules were detected by NetBox.

Module ID	Number of Genes	Genes
0	24	IL2RA PRKCD NFKB1 PRKCZ IL6R PTPN11 IL2 CAV1 PML ACTB TNFRSF4 REL FOXP3 NFKBIA NFKBIB CD33 STAT3 NFAT5 GZMB CD3E IL2RB BDKRB2 UBE2D3 NGFR
1	6	RPL10 RPL3L RPS26 RPS29 EIF4E ETF1
2	3	ADRA1A DLG4 NOS1
3	6	AKAP1 PPP1R11 PPP1R1B PPP1CA KIF18A C1orf114
4	2	THRB BRD8
5	5	AIFM1 DCC HCLS1 CASP3 CDC2L1
6	12	AZI1 CSNK1E PPP2R5C AXIN1 PPP2R5D ANP32A APC GSK3B FZD1 CHEK2 LRP6 MACF1
7	10	NCK2 PARVG KCNRG FXR2 ITGB1 CCNK LIMS1 ILK FGFR1 LGALS1
8	3	MITF PAX3 PAX6
9	2	GTF2F1 CPSF1
10	2	F2 FGA
11	2	GNG2 GNAS
12	5	MAP2K6 GADD45G DST DCTN1 MAP3K4
13	2	AP1G2 AP1S2
14	3	ARFIP1 ARF1 ARF3
15	3	CCNE1 PPP2R3A RBL1

Note: Count represents the number of genes in each GO category; *P* value was adjusted by using the Benjamini method.

Table 2. KEGG pathways analysis for 90 identified osteosarcoma candidate genes.

Term	Genes
hsa05200:Pathways in cancer	DCC, FGFR1, MITF, FZD1, PML, NFKBIA, NFKB1, ITGB1, STAT3, CCNE1, CASP3, GSK3B, APC, AXIN1
hsa04660:T cell receptor signaling pathway	NCK2, CD3E, GSK3B, NFKBIB, NFAT5, NFKBIA, NFKB1, IL2
hsa04310:Wnt signaling pathway	CSNK1E, GSK3B, PPP2R5D, PPP2R5C, NFAT5, LRP6, FZD1, AXIN1, APC

relationship between these 9 potential candidate genes and the biological process of osteosarcoma.

DCC (deleted in colorectal carcinoma) gene implicated as a tumour suppressor gene that encodes a netrin 1 receptor. This protein is a member of the immunoglobulin superfamily of cell adhesion molecules. Reduction or loss of DCC gene expression could be a specific event in many osteosarcomas development and progression [9].

FGFR is essential to normal cell growth. Abnormal FGF signaling is usually observed in multiple cancers. Overexpression of FGFR activate MAPK signaling and engagement of PI3 K-AKT signaling, which was observed in amplified cell lines [10]. FGFR1 gene amplification was reported in osteosarcoma and associated with poor clinical outcome [11].

Fzd1 (frizzled class receptor 1) is a receptor for Wnt protein. In a previous study,it showed that the expression of Fzd1 was up-regulated in human osteosarcoma tissues [12], which is consistent with our results.

NFKBIA inhibits NF-k-B signaling and involved in cell adhesion, apoptosis, differentiation and growth. NFKBIA expression was up-regulated after treating osteosarcoma cells with decitabine [13]. NFKBIA is one of the most important genes associated with cell proliferation and apoptosis implicated in osteosarcoma [14].

The Stat3 signaling plays an important role in the proliferation of cancer cells. Previous studies have showed that Stat3 expression is involved in cell-cycle regulation and cellular transformation and anti-apoptosis [15]. A negative correlation between Stat3 expression and survival rate has been found in osteosarcoma [16].

Cyclin E1 (CCNE1) is cell cycle genes and has been showed to be implicated in various cancer, such as ovarian cancer [17], breast cancer [18] and osteosarcoma [19]. The amplification of CCNE1 gene is related to clinical outcome in ovarian cancer and has a potential to become a therapy target [17].

CASP3 (Caspase-3) is involved in the signal transduction pathways of apoptosis. One study showed that genetic variant in CASP3 was significantly associated with the risk of esophageal squamous cell carcinoma [20].

APC is a tumor suppressor gene and inhibit the Wnt signaling pathways. It also participated in other biological process, such as cell migration, apoptosis and adhesion. It is significant signaling factors in normal bone development and loss of the gene expression in human osteosarcomas with a poor prognosis [21]. APC may serve as a possible target for osteosarcoma therapy [22].

The Wnt signaling pathway is crucial to embryogenesis and tumorigenesis. LRP6 (low density lipoprotein receptor-related protein 6), acts as a cell-surface coreceptor for the Wnt signaling pathway, which plays an important role in bone biology [23]. LRP6 has been extensively investigated for its function in the bone. Expression of LRP6 gene promotes tumorigenesis and accelerates tumor proliferation by changing the β-catenin subcellular distribution [24].

4 Conclusions

This study utilized the module detection algorithm to discover 90 osteosarcoma candidate genes that are associated with osteosarcoma. Furthermore, a randomization test was used to prune core modules from the random network. Among the final 90 identified genes, nine genes have direct evidence for their relationship between the gene and osteosarcoma, whereas several others may play crucial roles in bone development. We believe that our method provides new insights into predicting novel osteosarcoma related genes, and that it might useful for osteosarcoma research.

References

1. Miao, J., Wu, S., Peng, Z., Tania, M., Zhang, C.: MicroRNAs in osteosarcoma: diagnostic and therapeutic aspects. Tumor Biol. **34**, 2093–2098 (2013)
2. Hameedand, M., Dorfman, H.: Primary malignant bone tumors—recent developments. Semin. Diagn. Pathol. **28**, 86–101 (2011)
3. Zou, C., Shen, J., Tang, Q., Yang, Z., Yin, J., Li, Z., Xie, X., Huang, G., Lev, D., Wang, J.: Cancer-testis antigens expressed in osteosarcoma identified by gene microarray correlate with a poor patient prognosis. Cancer **118**, 1845–1855 (2012)
4. Kanehisaand, M., Goto, S.: KEGG: kyoto encyclopedia of genes and genomes. Nucleic Acids Res. **28**, 27–30 (2000)
5. Tranchevent, L.-C., Barriot, R., Yu, S., Van Vooren, S., Van Loo, P., Coessens, B., De Moor, B., Aerts, S., Moreau, Y.: ENDEAVOUR update: a web resource for gene prioritization in multiple species. Nucleic Acids Res. **36**, W377–W384 (2008)
6. Cerami, E., Demir, E., Schultz, N., Taylor, B.S., Sander, C.: Automated network analysis identifies core pathways in glioblastoma. PLoS ONE **5**, e8918 (2010)
7. Alvord, G., Roayaei, J., Stephens, R., Baseler, M.W., Lane, H.C., Lempicki, R.A.: The DAVID gene functional classification tool: a novel biological module-centric algorithm to functionally analyze large gene lists. Genome Biol. **8**, 183 (2007)
8. Benjamini, Y., Hochberg, Y.: Controlling the false discovery rate: a practical and powerful approach to multiple testing. J. Roy. Stat. Soc. Ser. B (Methodological) **57**, 289–300 (1995)
9. Horstmannand, M., PÃ, M.: Frequent reduction or loss of DCC gene expression in human osteosarcoma. Br. J. Cancer **75**, 1309 (1997)
10. Turner, N., Pearson, A., Sharpe, R., Lambros, M., Geyer, F., Lopez-Garcia, M.A., Natrajan, R., Marchio, C., Iorns, E., Mackay, A.: FGFR1 amplification drives endocrine therapy resistance and is a therapeutic target in breast cancer. Cancer Res. **70**, 2085–2094 (2010)
11. Fernanda Amary, M., Ye, H., Berisha, F., Khatri, B., Forbes, G., Lehovsky, K., Frezza, A. M., Behjati, S., Tarpey, P., Pillay, N.: Fibroblastic growth factor receptor 1 amplification in osteosarcoma is associated with poor response to neo-adjuvant chemotherapy. Cancer Med. **3**, 980–987 (2014)
12. Liu, H., Chen, Y., Zhou, F., Jie, L., Pu, L., Ju, J., Li, F., Dai, Z., Wang, X., Zhou, S.: Sox9 regulates hyperexpression of Wnt1 and Fzd1 in human osteosarcoma tissues and cells. Int. J. Clin. Exp. Pathol. **7**, 4795 (2014)
13. Al-Romaih, K., Somers, G.R., Bayani, J., Hughes, S., Prasad, M., Cutz, J.-C., Xue, H., Zielenska, M., Wang, Y., Squire, J.A.: Modulation by decitabine of gene expression and growth of osteosarcoma U2OS cells in vitro and in xenografts: identification of apoptotic genes as targets for demethylation. Cancer Cell Int. **7**, 14 (2007)
14. Sollazzo, V., Galasso, M., Volinia, S., Carinci, F.: Prion proteins (PRNP and PRND) are over-expressed in osteosarcoma. J. Orthop. Res. **30**, 1004–1012 (2012)
15. Calò, V., Migliavacca, M., Bazan, V., Macaluso, M., Buscemi, M., Gebbia, N., Russo, A.: STAT proteins: from normal control of cellular events to tumorigenesis. J. Cell. Physiol. **197**, 157–168 (2003)
16. Ryu, K., Choy, E., Yang, C., Susa, M., Hornicek, F.J., Mankin, H., Duan, Z.: Activation of signal transducer and activator of transcription 3 (Stat3) pathway in osteosarcoma cells and overexpression of phosphorylated-Stat3 correlates with poor prognosis. J. Orthop. Res. **28**, 971–978 (2010)
17. Nakayama, N., Nakayama, K., Shamima, Y., Ishikawa, M., Katagiri, A., Iida, K., Miyazaki, K.: Gene amplification CCNE1 is related to poor survival and potential therapeutic target in ovarian cancer. Cancer **116**, 2621–2634 (2010)

18. Luo, Q., Li, X., Li, J., Kong, X., Zhang, J., Chen, L., Huang, Y., Fang, L.: MiR-15a is underexpressed and inhibits the cell cycle by targeting CCNE1 in breast cancer. Int. J. Oncol. **43**, 1212–1218 (2013)
19. Lockwood, W.W., Stack, D., Morris, T., Grehan, D., O'Keane, C., Stewart, G.L., Cumiskey, J., Lam, W.L., Squire, J.A., Thomas, D.M.: Cyclin E1 is amplified and overexpressed in osteosarcoma. J. Mol. Diagn. **13**, 289–296 (2011)
20. Zhang, Z., Yu, X., Guo, Y., Song, W., Yu, D., Zhang, X.: Genetic variant in CASP3 affects promoter activity and risk of esophageal squamous cell carcinoma. Cancer Sci. **103**, 555–560 (2012)
21. Entz-Werle, N., Lavaux, T., Metzger, N., Stoetzel, C., Lasthaus, C., Marec, P., Kalita, C., Brugieres, L., Pacquement, H., Schmitt, C.: Involvement of MET/TWIST/APC combination or the potential role of ossification factors in pediatric high-grade osteosarcoma oncogenesis. Neoplasia **9**, 678–688 (2007)
22. Guimarães, A.P.G., Rocha, R.M., da Cunha, I.W., Guimarães, G.C., Carvalho, A.L., de Camargo, B., Lopes, A., Squire, J.A., Soares, F.A.: Prognostic impact of adenomatous polyposis coli gene expression in osteosarcoma of the extremities. Eur. J. Cancer **46**, 3307–3315 (2010)
23. Williamsand, B.O.: K.L. Insogna.: Where Wnts went: the exploding field of Lrp5 and Lrp6 signaling in bone. J. Bone Miner. Res. **24**, 171–178 (2009)
24. Li, Y., Lu, W., He, X., Schwartz, A.L., Bu, G.: LRP6 expression promotes cancer cell proliferation and tumorigenesis by altering β-catenin subcellular distribution. Oncogene **23**, 9129–9135 (2004)

A New Protein-Protein Interaction Prediction Algorithm Based on Conditional Random Field

Wei Liu[1,2(✉)], Ling Chen[1,3], and Bin Li[1,3]

[1] Institute of Information Engineering, Yangzhou University, Yangzhou, China
yzliuwei@126.com, yzulchen@163.com, bli@yzu.edu.cn
[2] Jiangsu Co-Innovation Center for Prevention and Control of Important Animal Infectious Diseases and Zoonoses, Yangzhou, China
[3] National Key Laboratory of Novel Software Technology, Nanjing University, Nanjing, China

Abstract. It is very important to study on protein-protein interaction (PPI) because proteins carry out their cellular functions through their concerted interactions with other proteins. In order to overcome the shortcomings such as the strong independence assumptions and the label-bias problem in the existing methods for PPI prediction, we propose a novel method for PPI prediction in PPI network based on conditional random fields (CRF) model. The method transforms PPI prediction into the problem of sequential data labeling. Based on the properties of protein sequences, we design the methods of model constructing, training and decoding in CRF accordingly. Our method can determine the labeled sequence with maximum probability and thereby detect PPI. Experimental results on benchmark data sets show that our method is more accurate and efficient than other traditional methods.

Keywords: Protein-protein interaction · PPI network · Conditional random fields model

1 Introduction

Protein-Protein interaction (PPI) is the hot and difficulty area in the research of molecular biology [1]. Many important physiological or pathological cell activities are all finished through PPI and its network. Therefore, studies of protein interactions and its network have become more necessary and urgent [2].

Recently a variety of methods have been proposed for predicting protein-protein interactions. Those methods can be classified into two categories: experimental methods and computational methods. However, the experimental results on PPI data suffer from both false negatives and false positives. In order to solve the defects of biological experiments, recently a large amount of algorithms [3, 4] have been designed for PPI prediction. Unfortunately, until now, there was no prediction method which can provide convincing results because the prediction results still have higher rate of false positive and false negative. Therefore, it becomes the most important issue in PPI prediction to develop fast and accurate computational method.

© Springer International Publishing Switzerland 2015
D.-S. Huang et al. (Eds.): ICIC 2015, Part II, LNCS 9226, pp. 367–378, 2015.
DOI: 10.1007/978-3-319-22186-1_36

A PPI network can be denoted as a graph $G = (V, E)$,where each vertex in V corresponds to a protein and each edge (v_i, v_j) in E indicates that there exist physically interactions between protein v_i and v_j. Supposed that there are n vertexes in the network, G can be denoted as $A = [aij]$ which is an n*n adjacency matrix. The value of element aij is defined as follows:

$$a_{ij} = \begin{cases} 1 & (v_i, v_j) \in E \\ 0 & (v_i, v_j) \notin E \end{cases}$$

In addition, there is also biological similarity between proteins. Namely, the two proteins with higher similarity will be more possible to exist physically interactions. We represent protein similarity using an $n*n$ order similarity matrix $S = [s_{ij}]$. The element s_{ij} of the matrix indicates the similarity between proteins v_i and v_j. The value of element s_{ij} is classified into four levels l_1, l_2, l_3 and l_4 shown as follows:

Level	l_1	l_2	l_3	l_4
Similarity	[0, 25 %)	[25 %, 50 %)	[50 %, 75 %)	[75 %, 100 %]

Therefore, given adjacency matrix A and similarity matrix S, the PPI prediction problem is to predict potential links. The predicting result is represented as an n order label matrix $Y = [y_{ij}]$, where $y_{ij} = $ '+' indicates that proteins v_i and v_j linked by edge (v_i, v_j) actually exist interactions while $y_{ij} = $ '-' indicates that v_i and v_j have no interaction.

Conditional random fields (CRFs) model is a new model for sequence labeling problem in recent years. Compared with some existing models, the CRFs model can overcome the shortcomings of strong independence assumptions and the label-bias problem. In this paper, we present a new method for identifying interactions in PPI network based on CRFs model. The method converts PPI prediction into a sequence labeling problem. We design the corresponding feature functions based on attributive characteristics of proteins. By training the sample data, the algorithm obtains the weights of the feature functions based on the joint distribution over the label sequence. Then according to the distribution model obtained, we can identify the interactions between proteins. Experimental results on benchmark data sets show that our algorithm is more efficient and accurate.

2 Conditional Random Fields

2.1 Definitions

Definition 1. Let $G = (V, E)$ be a graph, here V and E are the sets of vertexes and edges in G respectively. If a sequence X observed and its labeling sequence $Y = (Y_v)_{v \in V}$ obeys the Markov property with respect to the graph G, namely $P(Y_v|X, Y_w, v \neq w) = P(Y_v|X, Y_w, v \sim w)$ where $v \sim w$ means that v and w are neighbors in G and conditionally depend on sequence X, then we call (X, Y) a conditional random field [5] on G.

Let $X = (X1, X2, \ldots, Xn)$ and $Y = (Y1, Y2, \ldots, Yn)$ be the observation and labeling sequences respectively. Suppose the graph $G = (V, E)$ of Y is a tree or, in the simplest case, a chain, where every edge is a cut of the graph. By the fundamental theorem of random fields, the conditional probability over the sequences X and Y has the form:

$$p_\theta(y|x) \propto \exp\left(\sum_k \left(\sum_{e \in E} \lambda_k f_k(e, y|e, x) + \sum_{v \in V} u_k g_k(v, y|v, x) \right) \right) \qquad (1)$$

where V and E are the sets of vertexes and edges in G respectively, k is the feature parameter, f_k is a transitional feature function and g_k is a state feature function. The parameter $\theta = (\lambda_1, \lambda_2, \ldots; \mu_1, \mu_2, \ldots)$ is estimated from training data $D = \{(x^{(i)}, y^{(i)})\}_{i=1}^{N}$.

2.2   The Chain-Structured CRFs

When modeling sequences, the simplest and most common graph structure encountered is that in which the nodes corresponding to elements of Y form a simple first-order chain.

In the CRFs model, the parameters f_k, g_k in (1) can be unified into the form of $f_k(y_i, s_i, x_i, x_i')$, and the conditional probability of chain-structured CRFs is written as:

$$p\lambda(y|x) = \frac{1}{Z(x)} \exp\left(\sum_i \sum_j \lambda_j f_j(y_i, s_i, x_i, x_i') \right) \qquad (2)$$

Here, $Z(x)$ is an instance-specific normalization function as follows

$$Z(x) = \sum_y \exp\left(\sum_i \sum_j \lambda_j f_j(y_i, s_i, x_i, x_i') \right) \qquad (3)$$

In (2), $f_k(y_i, s_i, x_i, x_i')$ is a feature function with value 1 or 0. In our algorithm, the feature function is represented as Fig. 1 which is for vertex x_i and its connected vertex x_i', that is x_i and x_i' contain a common protein. For instance, suppose that x_i represents a pair of vertices (v_1, v_2) and x_i' represents a pair of vertices (v_1, v_3), then y_i and s_i are respectively the label and similarity for the predicted vertex pair (v_2, v_3). For the vertices (v_1, v_2) represented by x_i, we denote $x_i = 1$ if $(v_1, v_2) \in E$, $x_i = 0$ otherwise.

Fig. 1. Predicting y_i from vertices x_i and x_i'

Definition 2. The feature function can be defined as:

$$f\left(y_i, s_i, x_i, x_i^{'}\right) = \mathrm{I}_{(cond(y_i))}\mathrm{I}_{(cond(s_i))}\mathrm{I}_{(cond(x_i))}\mathrm{I}_{\left(cond(x_i^{'})\right)}$$

where $cond(x_i)$ is a logical function of x_i, the binary function I is defined as

$$I_{(condition)} = \begin{cases} 1 & condition = true \\ 0 & otherwise \end{cases} \qquad (4)$$

For instance, in the feature function $f\left(y_i, s_i, x_i, x_i^{'}\right) = \mathrm{I}_{(y_i='+')}\mathrm{I}_{(s_i=l_1)}\mathrm{I}_{(x_i=1)}\mathrm{I}_{\left(x_i^{'}=1\right)}$, assuming that x_i stands for a pair of vertices (v_1, v_2), $x_i^{'}$ represents vertices (v_1, v_3), if $(v_1, v_2) \in E$, $(v_1, v_3) \in E$, $S(2,3) = 1$, and $Y(2,3) = 1$, then the value of the feature function is 1, otherwise it is 0.

The parameter λj in (2) is used to indicate the weight corresponding to each feature function. Its value reflects the possibility of the occurrence of the event involving the feature function. A CRF model is specified by a group of feature functions fi and the corresponding weights λi, which can be obtained by the training procedure.

3   Identifying Interactions in PPI Network Using CRFs

We adopt CRFs model to identify interactions in PPI network. Three main problems must be solved in using the CRFs model: feature selection, parameter training and decoding.

3.1   Data Preprocessing

In our algorithm, firstly we can obtain the set of protein pairs $X = (x_1, x_2, \ldots x_m)$ and the records of interactions between protein pairs from the database. We use the following method to record the interactions between two proteins: for the vertices (v_1, v_2) represented by xi, we denote $x_i = 1$ if $(v_1, v_2) \in E$, $x_i = 0$ otherwise. In the prediction structure shown in Fig. 1, we can see that the solution of yi in the labeled sequence set $Y = (y_1, y_2, \ldots, y_n)$ depending on $\left(x_i, x_i^{'}\right)$, $x_i, x_i^{'} \in \{0, 1\}$ and the similarity si for (v_2, v_3), $s_i \in \{l_1, l_2, l_3, l_4\}$. Now we consider the value of $\left(x_i, x_i^{'}\right)$. Since $x_i, x_i^{'} \in \{0, 1\}$, their values have four combinations: (0,0), (0,1), (1,0) and (1,1). To convert the problem into the label problem of CRFs model, we treat those combinations of similarity values between protein pairs as different states and assign a character to each state. Table 1 shows those combinations and their character representations.

Considering the symmetry of x_i and $x_i^{'}$, we use character "A" to denote the states of (1,0) and (0,1), thus actually we only consider three states. Based on such character representation of the states, the input sequence set can be converted into the state sequence set $T=(t_1, t_2, \ldots, t_n)$, where $t_i \in \{A, B, C\}$.

Table 1. The transform table of similarity values about protein pairs

The value of (x_i, x_i')	Character representation
(0,1)	A
(1,0)	A
{1,1}	B
(0,0)	C

3.2 Feature Selection

3.2.1 Defining the States of Each Position

We first transform the PPI prediction problem into the label problem of character sequences. We define two states in each position of the sequence, namely, the states of "interaction" or "non-interaction". Therefore, there are six states in the CRFs model as illustrated in Table 2, which are $A^+, B^+, C^+, A^-, B^-, C^-$. We can use the CRFs model mentioned above to give the most suitable labels for an observation sequence.

Table 2. All kinds of states for the observation sequence

x_i	x_i'	(v_2, v_3)	state
1	1	Interaction	B^+
0	1	Interaction	A^+
0	0	Interaction	C^+
1	1	Non-interaction	B^-
0	1	Non-interaction	A^-
0	0	Non-interaction	C^-

In the prediction structure shown as Fig. 1, the solution of yi is also depending on the similarity si for (v_2, v_3), $s_i \in \{l_1, l_2, l_3, l_4\}$. Therefore, there are 24 feature functions based on six states of the CRFs model.

3.2.2 Feature Functions

Based on the above defined six states and the nature of the problem, we set 24 feature functions as follows:

$$f_1\left(y_i, s_i, x_i, x_i'\right) = I_{(y_i='-')}I_{(s_i=l_1)}I_{(x_i=1)}I_{(x_i'=1)} \quad \text{(State B}-\text{)}$$

$$f_2\left(y_i, s_i, x_i, x_i'\right) = I_{(y_i='-')}I_{(s_i=l_1)}I_{(x_i=0)}I_{(x_i'=1)} \quad \text{(State A}-\text{)}$$

$$f_3\left(y_i, s_i, x_i, x_i'\right) = I_{(y_i='-')}I_{(s_i=l_1)}I_{(x_i=0)}I_{(x_i'=0)} \quad \text{(State C}-\text{)}$$

$$f_4\left(y_i, s_i, x_i, x_i'\right) = I_{(y_i='+')}I_{(s_i=l_1)}I_{(x_i=1)}I_{(x_i'=1)} \quad \text{(State B}+\text{)}$$

$$f_5\left(y_i, s_i, x_i, x_i'\right) = \mathbf{I}_{(y_i='+')}\mathbf{I}_{(s_i=l_1)}\mathbf{I}_{(x_i=0)}\mathbf{I}_{\left(x_i'=1\right)} \quad \text{(State A+)}$$

$$f_6\left(y_i, s_i, x_i, x_i'\right) = \mathbf{I}_{(y_i='+')}\mathbf{I}_{(s_i=l_1)}\mathbf{I}_{(x_i=0)}\mathbf{I}_{\left(x_i'=0\right)} \quad \text{(State C+)}$$

$$f_7\left(y_i, s_i, x_i, x_i'\right) = \mathbf{I}_{(y_i='-')}\mathbf{I}_{(s_i=l_2)}\mathbf{I}_{(x_i=1)}\mathbf{I}_{\left(x_i'=1\right)} \quad \text{(State B−)}$$

$$f_8\left(y_i, s_i, x_i, x_i'\right) = \mathbf{I}_{(y_i='-')}\mathbf{I}_{(s_i=l_2)}\mathbf{I}_{(x_i=0)}\mathbf{I}_{\left(x_i'=1\right)} \quad \text{(State A−)}$$

$$f_9\left(y_i, s_i, x_i, x_i'\right) = \mathbf{I}_{(y_i='-')}\mathbf{I}_{(s_i=l_2)}\mathbf{I}_{(x_i=0)}\mathbf{I}_{\left(x_i'=0\right)} \quad \text{(State C−)}$$

$$f_{10}\left(y_i, s_i, x_i, x_i'\right) = \mathbf{I}_{(y_i='+')}\mathbf{I}_{(s_i=l_2)}\mathbf{I}_{(x_i=1)}\mathbf{I}_{\left(x_i'=1\right)} \quad \text{(State B+)}$$

$$f_{11}\left(y_i, s_i, x_i, x_i'\right) = \mathbf{I}_{(y_i='+')}\mathbf{I}_{(s_i=l_2)}\mathbf{I}_{(x_i=0)}\mathbf{I}_{\left(x_i'=1\right)} \quad \text{(State A+)}$$

$$f_{12}\left(y_i, s_i, x_i, x_i'\right) = \mathbf{I}_{(y_i='+')}\mathbf{I}_{(s_i=l_2)}\mathbf{I}_{(x_i=0)}\mathbf{I}_{\left(x_i'=0\right)} \quad \text{(State C +)}$$

$$f_{13}\left(y_i, s_i, x_i, x_i'\right) = \mathbf{I}_{(y_i='-')}\mathbf{I}_{(s_i=l_3)}\mathbf{I}_{(x_i=1)}\mathbf{I}_{\left(x_i'=1\right)} \quad \text{(State B−)}$$

$$f_{14}\left(y_i, s_i, x_i, x_i'\right) = \mathbf{I}_{(y_i='-')}\mathbf{I}_{(s_i=l_3)}\mathbf{I}_{(x_i=0)}\mathbf{I}_{\left(x_i'=1\right)} \quad \text{(State A−)}$$

$$f_{15}\left(y_i, s_i, x_i, x_i'\right) = \mathbf{I}_{(y_i='-')}\mathbf{I}_{(s_i=l_3)}\mathbf{I}_{(x_i=0)}\mathbf{I}_{\left(x_i'=0\right)} \quad \text{(State C−)}$$

$$f_{16}\left(y_i, s_i, x_i, x_i'\right) = \mathbf{I}_{(y_i='+')}\mathbf{I}_{(s_i=l_3)}\mathbf{I}_{(x_i=1)}\mathbf{I}_{\left(x_i'=1\right)} \quad \text{(State B+)}$$

$$f_{17}\left(y_i, s_i, x_i, x_i'\right) = \mathbf{I}_{(y_i='+')}\mathbf{I}_{(s_i=l_3)}\mathbf{I}_{(x_i=0)}\mathbf{I}_{\left(x_i'=1\right)} \quad \text{(State A+)}$$

$$f_{18}\left(y_i, s_i, x_i, x_i'\right) = \mathbf{I}_{(y_i='+')}\mathbf{I}_{(s_i=l_3)}\mathbf{I}_{(x_i=0)}\mathbf{I}_{\left(x_i'=0\right)} \quad \text{(State C+)}$$

$$f_{19}\left(y_i, s_i, x_i, x_i'\right) = \mathbf{I}_{(y_i='-')}\mathbf{I}_{(s_i=l_4)}\mathbf{I}_{(x_i=1)}\mathbf{I}_{\left(x_i'=1\right)} \quad \text{(State B−)}$$

$$f_{20}\left(y_i, s_i, x_i, x_i'\right) = \mathbf{I}_{(y_i='-')}\mathbf{I}_{(s_i=l_4)}\mathbf{I}_{(x_i=0)}\mathbf{I}_{\left(x_i'=1\right)} \quad \text{(State A)}$$

$$f_{21}\left(y_i, s_i, x_i, x_i'\right) = \mathbf{I}_{(y_i='-')}\mathbf{I}_{(s_i=l_4)}\mathbf{I}_{(x_i=0)}\mathbf{I}_{\left(x_i'=0\right)} \quad \text{(State C−)}$$

$$f_{22}\left(y_i, s_i, x_i, x_i'\right) = \mathbf{I}_{(y_i='+')}\mathbf{I}_{(s_i=l_4)}\mathbf{I}_{(x_i=1)}\mathbf{I}_{\left(x_i'=1\right)} \quad \text{(State B+)}$$

$$f_{23}\left(y_i, s_i, x_i, x_i'\right) = I_{(y_i='+')}I_{(s_i=l_4)}I_{(x_i=0)}I_{(x_i'=1)} \quad \text{(State A+)}$$

$$f_{24}\left(y_i, s_i, x_i, x_i'\right) = I_{(y_i='+')}I_{(s_i=l_4)}I_{(x_i=0)}I_{(x_i'=0)} \quad \text{(State C+)}$$

3.2.3 Parameter Training

In the CRFs model, the parameter λ_j in (2) indexes the weight corresponding to each feature function, and its value can be obtained by training from the known samples. In the CRFs model, we use maximum likelihood to train the samples for parameter estimation.

Given a similarity sequence set $\{x^{(i)}\}$, the corresponding label sequence set $\{y^{(i)}\}$ $(i = 1, \cdots, N)$ and twenty four feature functions defined above. In parameter training, we should select some proper value of weights so as to maximize the conditional log likelihood

$$L(\lambda) = \sum_{i=1}^{N} \log P\lambda\left(y^{(i)}|x^{(i)}\right) \propto \sum_{x,y} \widetilde{P}(x,y) \log P\lambda(y|x) \tag{5}$$

Here, $\widetilde{P}(x,y) = \frac{1}{N}\sum_{i=1}^{N} I_{\{y=y^{(i)}\}} I_{\{x=x^{(i)}\}}$

Therefore our goal is to search for the appropriate parameter λ_{ML} satisfying

$$\lambda_{ML} = \arg\max L(\lambda) \tag{6}$$

It is not possible to analytically determine the parameter values that maximize the log-likelihood, since setting the gradient to zero and solving for λ_{ML} does not always yield a closed form solution. Therefore, we use the Quasi-Newton method to optimize the parameters of CRFs model. In the Quasi-Newton method, we use the second-order deviation technique, and avoid computing the inverse of Hessian matrix by using the information included in gradients of object functions to train parameters.

3.2.4 Decoding Problem

We solve the decoding problem of CRFs model using Viterbi algorithm which is originally designed for model decoding in HMM. In the Viterbi algorithm, a Viterbi variable $\Phi(i, y)$ is defined to denote the possibility of optimal label sequence y' on the processed subsequence $x = x_1, x_2, \ldots, x_i$ in the observation sequence $x = x_1, x_2, \ldots, x_l (i \leq l)$. Accordingly, $\Psi(i, y)$ is defined to denote the label sequence with the maximal possibility. The formal definitions of $\Phi(i, y)$ and $\Psi(i, y)$ are described as follows:

$$\Phi(i, y) = \max_{y_1, y_2, \ldots, y_i} P_\lambda(y_1, y_2, \ldots, y_i | x_1, x_2, \ldots, x_i)$$

and

$$\Psi(i, y) = \underset{y_1, y_2, \ldots, i}{\arg \max} P_\lambda(y_1, y_2, \ldots, y_i | x_1, x_2, \ldots, x_i)$$

The algorithm gradually gets the optimal label in each position i based on $\Phi(i, y)$ until the whole label sequence is obtained. The framework of the algorithm is depicted as follows.

Algorithm 1 PPI-Viterbi

> **Input:** $f_k\left(y_i, s_i, x_i, x_i'\right)$: feature functions;
>
> λ_k, $k=1,..,24$: the corresponding weights;
>
> $X="x_1, x_2, ..., x_l"$: Sequence to be labeled;
>
> **output:** $Y="y_1, y_2, ..., y_l"$: The label sequence over x;

> **Begin**
> 1、Initialization
>   For all labels y, set $\Phi(1, y) = 1$;
> 2、**for** i=2 **to** l **do**
>   For all labels y, calculate

$$\Phi(i, y) = \max_{y'} \left\{ \Phi(i-1, y') * \exp\left(\sum_k \lambda_k f_k\left(y_i, s_i, x_i, x_i'\right) \right) \right\}$$

$$\Psi(i, y) = \arg\max_{y'} \left\{ \Phi(i-1, y') * \exp\left(\sum_k \lambda_k f_k\left(y_i, s_i, x_i, x_i'\right) \right) \right\}$$

> **endfor**
> 3、Termination:
> $y_l = \Psi(l, y)$
> 4、Computing label sequence $y_1, \ldots, y_l$:
>   **for** i=l-1 **downto** 1 **do**
> $y_i = \Psi(i+1, y_{i+1})$
>   **endfor**
> **End**

4   Experimental Results and Analysis

4.1   Test Data and Experimental Environment

The test data we used in our experiments are as follows: (1) H.Pylori and Human created by Martin et al. [6]; (2) three popular used yeast databases include DIP Core [7], MIPS Core [8] and BIND [9] which are PPI standard databases recognized by international.

We use CRF ++0.54 (http://sourceforge.net/projects/crfpp/files/) as the tool to train CRFs model. All the experiments were conducted on a 1.8 GHz Intel Core i5 with

6 GB memory, Windows7 operating system. All codes were compiled using Microsoft Visual C ++ 6.0.

4.2 Measurements for Evaluating the Experimental Results

To evaluate the quality of our prediction results, we use the 4 measurements: sensitivity, precision, accuracy and Matthew's correlation coefficient. The calculation of those evaluation measurements are defined as follows:

(1) Sensitivity

$$\text{Sensitivity} \ = \ \frac{TP}{TP+FN}$$

(2) Precision

$$\text{Precision} \ = \ \frac{TP}{TP+FP}$$

(3) Accuracy

$$\text{Accuracy} \ = \ \frac{TP+TN}{TP+TN+FP+FN}$$

(4) Matthew's Correlation Coefficient (MCC)

$$MCC = \frac{TP \times TN - FN \times FP}{\sqrt{(TP+FN) \times (TP+FP) \times (TN+FN) \times (TN+FP)}}$$

4.3 Experimental Results

We randomly select data from the datasets mentioned above and construct five datasets to test. The prediction results are shown in Table 3. From Table 3, we can obviously find that our algorithm can obtain high quality results in terms of sensitivity, precision,

Table 3. The experimental results obtained in different datasets

Test set	Accuracy (%)	Sensitivity (%)	Precision (%)	MCC
1	87.97	91.05	85.79	0.761
2	87.95	91.18	85.69	0.763
3	87.85	90.47	85.95	0.759
4	88.19	89.69	87.17	0.765
5	88.17	90.48	86.53	0.764
Average	88.03 ± 0.16	90.57 ± 0.59	86.23 ± 0.61	0.762 ± 0.003

accuracy and MCC. Especially the average accuracy of our algorithm has reached 88.03 ± 0.16 % and the average MCC has achieved 0.762 ± 0.003.

To illustrate the effectiveness of our algorithm for PPI prediction, we compare the performance of our algorithm with some other traditional algorithms on H.Pylori and Human datasets. The traditional algorithms compared are SVM_Sequence [10], SVM_Signature [6] based on SVM, Fusion_Classifie [11] based on classifiers fusion, and HKNN_Add [12] based on HKNN. We use 10-fold cross validation to test the algorithms' performance in the two datasets, namely, we first randomly divide the datasets into 10 partitions in average, and then choose one partition as tested set and use the other nine partitions as training sets in each test. We test all data once and take the statistics of all test results as the dataset's results. The prediction results of each method is listed in Table 4.

From Table 4, we can see that the prediction performance of our algorithm method is higher than other algorithms. On the H.Pylori and Human dataset, our algorithm can get respectively 5.1–16.5 % and 8.3–26.3 % higher accuracy than the other algorithms. The superior performance in two datasets shows that our algorithm has strong generalization ability and is effective in PPI prediction.

4.4   The Predicting Results of Independent Test Data Sets

We verified whether our algorithm is effective by considering non-interacting pairs of proteins with identical cellular localization [13]. The location information of proteins is extracted from 20110531 release version of UniProtKB/Swiss-Prot which contains 529056 records. Non-interacting pairs of proteins are generated among Cytoplasm, Nucleus, Mitochondrion, Endoplasmic reticulum, Golgi apparatus, Peroxisome and Vacuole. Table 5 illustrates the number of proteins and the accuracy of predicting result on each dataset. From the table we can see that the predicting accuracies of our algorithm on Mitochondrion, Peroxisome, Endoplasmic reticulum and Vacuole are more than 80 %, and the average accuracy of the seven different cellular localizations achieves 77.66 %. The results demonstrate that our algorithm can efficiently predict the interaction of the pair of proteins in the same cellular localization.

Table 4. Performance comparison of PPI prediction among different algorithms

Methods	H.pylori			Human		
	Precision (%)	Sensitivity (%)	Accuracy (%)	Precision (%)	Sensitivity (%)	Accuracy (%)
SVM_Sequence	80.2	69.8	75.8	-	-	-
SVM_Signature	85.7	79.9	83.4	72	66	70
Fusion_Classifier	85.1	80.6	83	60	59	60
HKNN_Add	84	86	84	65	68	66
Our method	86.4	89.95	88.3	80.5	83.1	80.8

Table 5. The predicting accuracy obtained on different cellular localizations

Cellular localization	The number of interacting pairs of proteins	Accuracy of predicting result
CytoPlasm	8000	5988(74.85 %)
Nucleus	8000	4661(58.26 %)
Mitochondrion	8284	7216(87.11 %)
Endoplasmic reticulum	1953	1695(86.79 %)
Golgi apparatus	300	225(75 %)
Peroxisome	171	138(80.70 %)
Vacuole	496	434(87.5 %)
Total	27204	21126(77.66 %)

5 Conclusions

In this paper, we present a novel method for PPI prediction based on the model of conditional random fields (CRF). Based on the properties of protein sequences, we define the corresponding feature functions and design the methods of model constructing, training and decoding in CRF accordingly. Then according to the distribution model obtained, we can determine the labeled sequence with maximum probability and thereby detect whether a protein interacts with another. Experimental results on standard datasets show that our algorithm is more practicable and efficient than other methods. Meanwhile, our algorithm also can obtain higher quality results for independent test datasets, and illustrates its strong generalization ability.

Acknowledgements. This research was supported in part by the Chinese National Natural Science Foundation under grant Nos. 61379066, 61379064, 61472344, Natural Science Foundation of Jiangsu Province under contracts BK20130452, BK2012672, BK2012128, and Natural Science Foundation of Education Department of Jiangsu Province under contract 12KJB520019, 13KJB520026. Science and Technology Innovation Foundation of Yangzhou University under grant No. 2014CXJ022.

References

1. Cannataro, M., Guzzi, P.H., Veltri, P.: Protein-to-protein interactions: technologies, databases, and algorithms. ACM Comput. Surv. **43**(1), 1–36 (2010)
2. Lage, K.: Protein-protein interactions and genetic diseases: the interactome. Biochim. Biophys. Acta **1842**, 1971–1980 (2014)
3. Shen, J., Zhang, J., Luo, X., Zhu, W., Yu, K., Chen, K., Li, Y., Jiang, H.: Predicting protein-protein interactions based only on sequences information. PNAS **104**(11), 4337–4341 (2007)
4. Guo, Y., Yu, L., Wen, Z., Li, M.: Using support vector machine combined with auto covariance to predict protein-protein interactions from protein sequences. Nucleic Acids Res. **36**(9), 3025–3030 (2008)

5. Jiang, X., Fei, W., et al.: The classification of multi-modal data with hidden conditional random field. Pattern Recogn. Lett. **51**(1), 63–69 (2015)
6. Martin, S., Roe, D., Faulon, J.L.: Predicting protein-protein interactions using signature products. Bioinformatics **21**(2), 218–226 (2005)
7. Xenarios, I., Salwinski, L., Duan, X.J., et al.: DIP: The database of interacting proteins. Nucleic Acids Res. **30**, 303–305 (2002)
8. Mewes, H.W., Frishman, D., et al.: MIPS: a database for genomes and protein sequences. Nucleic Acids Res. **30**(1), 31–34 (2002)
9. Bader, G.D., Donaldson, I., Wolting, C., et al.: BIND-the biomolecular interaction network database. Nucleic Acids Res. **29**(l), 242–245 (2001)
10. Bock, J.R., Gough, D.A.: Whole-proteome interaction mining. Bioinformatics **19**(l), 125–134 (2003)
11. Nanni, L.: Fusion of classifiers for predicting protein-protein interactions. Neurocomputing **68**, 289–296 (2005)
12. Nanni, L.: Hyperplanes for predicting protein-protein interactions. Neurocomputing **69**, 257–263 (2005)
13. Ben-Hur, A., Noble, W.S.: Choosing negative examples for the prediction of protein-protein interactions. BMC Bioinform. **7**(Suppl l), S2 (2006)

Sequence-Based Random Projection Ensemble Approach to Identify Hotspot Residues from Whole Protein Sequence

Peng Chen[1](✉), ShanShan Hu[1], Bing Wang[2], and Jun Zhang[3]

[1] Institute of Health Sciences, Anhui University, Hefei 230601, Anhui, China
bigeagle@mail.ustc.edu.cn
[2] School of Electronics and Information Engineering, Tongji University,
Shanghai 804201, China
[3] College of Electrical Engineering and Automation, Anhui University,
Hefei 230601, Anhui, China

Abstract. Hot spot residues of proteins are key to performing specific functions in many biological processes. However the identification of hot spots by experimental methods is costly and time-consuming. Computational method is an alternative to identify hot spots by using sequential and structural information. However, structural information of protein is not always available. In this paper, the issue of identifying hot spots is addressed by using statistically physicochemical properties of amino acids only. Firstly, 34 relatively independent physicochemical properties are extracted from the 544 properties in AAindex1. Since the hot spots data set is extremely imbalanced, the ratio of the number of hot spots to that of non-hot spots is about 1.4 %, the hot spot set and a set of non-hot spot subset with roughly the number of that hot spots forms an initial input matrix. Random projection on the matrix achieves an input to a REPTree classifier. Several random projections and different sets of non-hot spots build an ensemble REPTree system. Experimental results showed that although our method performed worse it is a complement to the experiments on hot spot determination, on the commonly used hot spot benchmark sets.

Keywords: Random projection · Hot spots · Reptree · Ensemble system

1 Introduction

Hot spots, or hot spot residues, are the residues that contribute a large portion of the binding energy in complexes of proteins [1, 2], since the binding energy of proteins is not uniformly distributed over their interaction surfaces [2]. In the binding interface, hot spots are surrounded by energetically less important residues and packed more tightly than others [1]. Hot spots are key to understanding binding mechanisms and the stability of protein-protein interactions [3, 4]. There are a few of hot spots in a

This work was supported by the National Natural Science Foundation of China (Nos. 61300058, 61271098 and 61472282).

complexes of proteins, in a typical 1,200 to 2,000 interface, less than 5 % of the interface residues contribute more than 2.0 kcal/mol energy to binding. Therefore, there is about 1.4 % hot spots in a protein.

Experimentally, a hot spot can be identified by alanine scanning mutagenesis experiments that says a residue as hot spot if a change in its binding free energy is larger than a certain threshold when the residue is mutated to alanine. There are several public databases that used the method: Alanine Scanning Energetics Database (ASEdb) [5], Binding Interface Database (BID) [6], PINT database [7] and SKEMPI dataset that has collected 3,047 binding free energy changes from 85 protein-protein complexes from the literature [8].

As the identification of hot spots by experimental methods is costly and time-consuming. Computational method is an alternative to identify hot spots by using sequential and structural information of proteins [9]. Many computational prediction methods of hot spots were proposed, including energy function-based physical models [10–13], molecular dynamics simulation-based approaches [14–16], evolutionary conservation-based methods [4, 17, 18], graph-based approaches [19], docking-based methods [20, 21], and machine learning methods that combine features such as solvent accessibility, conservation, sequence profiles and pairing potential [22–32]. Among them, different machine learning methods have been applied or developed to the prediction, such as neural network [22], decision tree [10, 23], SVM [27], random forest [31] and the consensus of different machine learning methods [25].

Despite significant progress on hot spot prediction, the state-of-the-art methods depend on structural information [10–12, 27, 31]. However, not all of the protein structures are publicly available. Therefore, accurately identifying hot spots from "pure" physicochemical characteristics of a protein sequence remains challenging, to the best of our knowledge. Addressing the issue would have a broader range of applications than do structure information-based methods and it would establish simpler principles about binding mechanisms [22].

Here, we propose a method that predicts hot spots using only physicochemical characteristics extracted from amino acid sequences. Firstly, 34 relatively independent physicochemical properties are extracted from the 544 properties in AAindex1 [33]. Since the hot spots data set is extremely imbalanced, the ratio of the number of hot spots to that of non-hot spots is about 1.4 %, the hot spot set and a set of non-hot spot subset with roughly the number of that hot spots forms an initial input matrix. Random projection on the matrix achieves an input to a REPtree classifier. Several random projections and different sets of non-hot spots build an ensemble REPtree system. Experimental results showed that our method significantly outperformed other state-of-the-art hot spot predictors, on the commonly used hot spot benchmark sets.

2   Methods

2.1   Feature Vector Representing a Residue

In AAindex1 data base [33], there are 544 amino acid properties. Most of them are relevant, so like as our previous work [32], irrelevant ones with a correlation coefficient (CC) of 0.5 are extracted, as did in AAindex1 itself. The CC of each two properties is computed and the number of relevant properties is counted. Ranking the relevant number in descend, a list of properties is obtained. For the top one property, we remove all of the next properties related to the top one. Step by step, each property related to the previous one is removed from the list. Finally, 34 properties are retained, where each two properties have a CC less than 0.5.

For a given residue in a protein chain, the association among the neighboring residues is considered in this work. Experimentally, a sliding window involving totally 7 residues centered at the residue i is used to encode our classifiers. An encoding schema integrating amino acid properties and sequence profile is used to represent the residue. The sequence profile for one residue created by PSI-Blast with default parameters [34] is then multiplied by each amino acid property, where the property for one amino acid is multiplied by the score of the sequence profile for the same amino acid. Therefore, the profile SP^k, $k = 1,...,7$ for residue k in the 7 residues window and one amino acid property scale, Aap, are both vectors with 1×20 dimensions. Thereafter, $MSK_i^k = SP^k \times Aap$ for residue k represents the multiplication of the corresponding sequence profile by the scale, whose jth element $MSK^{k,ji} = SP^{k,j} \times Aap^j$, $j = 1,...,20$. A similar vector representation can be found in our previous work [32, 35, 36].

For the residue, therefore, it is represented by a 1×4760 (20 sequence profile $\times$ 7 residues window $\times$ 34 AAindex1 properties) vector V_i; the corresponding target value T_i is 1 or 0, denoting whether the residue is hot spot or not. Actually, our method expects to learn the relationship between input vectors V and the corresponding target array T and try to make its output as close to the target T as possible.

2.2 Random Projection on REPTree

Random projection is a data reduction technique that projects a high dimensional data onto an low-dimensional subspace [37, 38]. Given the original data vector, $X \in \mathbb{R}^{N \times L_1}$, the linear random projection is to multiply the original vector by a random matrix $R = \mathbb{R}^{L_1 \times L_2}$. The projection

$$X^R = XR = \sum_i x_i r_i \tag{1}$$

yields a dimensionality reduced vector $X^R \in \mathbb{R}^{N \times L_2}$, where x_i is the ith sample of the original data, r_i is the ith column of the random matrix, and $L_2 \ll L_1$. The matrix R consists of random values and each column has been normalized to unity. In the Eq. 1, each original data sample with dimension L_1 has been replaced by a random, non-orthogonal direction L_2 in the reduced-dimensional space [38]. Therefore, the dimensionality of the original data is reduced from 4760 to a rather small value.

REPTree is a fast tree learner that uses reduced-error pruning [39], based on information gain/variation reduction as the splitting principle, and sorts values for numeric attributes once that is optimized for speed. The REPTree has a "numFolds" parameter (default 3 in WEKA software) that determines the size of the pruning set: the data is divided equally into that number of parts and the last one used as an independent test set to estimate the error at each node.

Previous results showed that the generalization error caused by one classifier can be compensated by other classifiers, therefore using tree ensembles can lead to significant improvement in prediction accuracy [40]. For our hot spot prediction problem, the ensemble of simple trees votes for the most popular hot spot class. Given the set of training data $D_{tr}^k = \{(X_i^{R^k}, Y_i)\}, i = 1, \ldots, N$, after multiplied by kth random projection R^k, let the number of training instances be N, the number of features in the classifier be L_2, and the number of random projection be K. For kth random projection, the data D^k is generated as an input to a REPTree and thus it forms a classifier $CF_x(x)$, where x is a training instance.

After all of REPTree classifiers with random projection are generated, they vote for the most popular class and thus the prediction of the ensemble is,

$$R(X) = majority\,vote\,\{CF_k(x)\}_{k=1}^K, \tag{2}$$

where x is a query instance.

Since the hot spot data set is extremely imbalanced, only 1.4 % of instances belonging to hot spot class, balancing the positive (hot spot class) and the negative (non-hot spot class) data is necessary to avoid the overfitting of classifier. Finally, the original data matrix $X \in \mathbb{R}^{N_1 \times L_1}$ consists of the positive data and the different negative data with roughly the same size. Thus data matrix D^k after the kth random projection has N_1 instances.

Results showed that the majority vote with independent classifiers can often make a dramatic improvement [41]. Here a residue is labeled as being hot spot if all of the classifiers identified it as positive class 1, otherwise it is identified as a non-hot spot residue.

3 Materials

3.1 Definition

In definition, a hot spot is an interface residue whose change in the binding free energy is larger than a predefined threshold, if mutated to alanine. However, there are many definitions for this threshold in the literature. Some works used 2.0 kcal/mol as the threshold to differentiate hot spots and non-hot spots [23, 27, 31], whereas [22] chose anything above 2.5 kcal/mol to define hot spots and 0 kcal/mol (i.e., no change in binding energy) to define non-hot spots. In [25], Tuncbag and colleagues defined hot spots to be interface residues with higher than 2.0 kcal/mol and non-hot spots to be the ones with lower than 0.4 kcal/mol.

In our previous work, we investigated several definitions on hot spots [32]. Actually, different definitions for hot spots and non-hot spots change the performance of the prediction methods. Results showed that most of lower thresholds on hot spots yields relatively higher performance of the methods since it contains more hot spots than those with upper threshold [26, 32]. In this paper, we will mainly use the same definition of hot spots as the one used in [25, 32] but consider all other residues as non-hot spots. We expect to identify hot spots from the whole sequence of one protein.

3.2 Data Sets

Two commonly used benchmark hot spot data sets are used in this work. Following [25], we used the ASEdb data set [5] as the training set. Protein sequences in ASEdb were filtered so that the sequence identity between any pair of sequences was below 35 %. The binding free energy of each interface residue was extracted. The interface residues with binding energies higher than 2.0 kcal/mol were considered as hot spots and all other residues in sequence were considered as non-hot spots. This resulted in 4015 residues in the training set, including 58 hot spots and 3957 non-hot spots.

Similar to [25], we used the BID data set [6] as one of our test sets. This data set was filtered by the same sequence identity threshold and hot spots were extracted using the same binding energies while all other residues are non-hot spots as in the training set. As a result, the BID test set consisted of 2947 residues with 52 hot spots and 2895 non-hot spots. The data in the training and test sets came from different complexes and were mutually exclusive.

3.3 Hot Spot Prediction Evaluation

In this work we adopted four evaluation measures to show the ability of our model objectively, criteria of sensitivity (Sen), precision (Prec), F-measure (F1), and Matthews correlation coefficient (MCC) [35, 42]. They are defined as follows:

$$
\begin{aligned}
Sen &= TP/(TP + FN) \\
Prec &= TP/(TP + FP) \\
F1 &= 2 \times \frac{Prec \times Sen}{Prec + Sen} \\
MCC &= \frac{TP \times TN - FP \times FN}{\sqrt{(TP + FN)(TP + FP)(TN + FP)(TN + FN)}},
\end{aligned}
\tag{3}
$$

where TP (True Positive) is the number of correctly predicted hot spot residues; FP (False Positive) is the number of false positives (incorrectly over predicted non-hot spot residues); TN (True Negative) is the number of correctly predicted non-hot spot residues; and FN (False Negative) is false negative, i.e., incorrectly under predicted hot spot residues.

4 Results

The amino acid preferences for hot spots and non-not spots are investigated first. Figure 1 illustrates the preference comparison. It can be seen that amino acids Tyr and Phe frequently occur in the hot spot group, while amino acids Ala, Gly, Val, and Pro

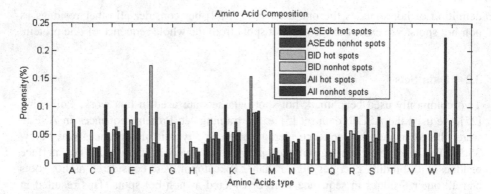

Fig. 1. Amino acid preferences for hot spots and non-hot spots. Each bar denotes the ratio of the number of each amino acid to that of the corresponding group (such as "ASEdb hot spots", "ASEdb non-hot spots", etc.). Moreover, the latter two groups stand for the total statistics from the ASEdb and BID data sets.

are often regarded as non-hot spots. However it is not always the case because our used data set is relatively small. Despite of that, Tyr is always deemed as hydrophilic amino acids while Ala and Val as hydrophobic ones in literature [43], which is partially in agreement with our statistics. Actually, hydrophilic amino acids seem to be hot spots more likely.

4.1 Performance of the Hot Spot Prediction

In this work, the ASEdb is regarded as training data set and the BID as the test one. There are 7000 random projections on the original data matrix, therefore 7000 REP-Tree classifiers are obtained. The ensemble of the 7000 REPtree classifiers yields the final predictions on hot spots. The majority vote is applied to the ensemble, i.e., for one residue it will be identified as hot spot if all of the 7000 classifiers identify it as hot spot. Since the number of positive instances, i.e., the hot spot instances, are much less than that of the negative instances, i.e., the non-hot spot instances, results of some classifiers don't change drastically. The variance of the results on BID data set from one classifier will be a useful factor such that results with little change of variance is not good for the classifier. Some of the 7000 REPTree classifiers are removed from the ensemble if their variances of the results are less than 0.1. Table 1 shows the performance comparison of the ensemble ones with different number of random projections. Here the dimensionality of the original data is reduced from 4760 to 63. From the Table 1, it can be seen that the ensemble with 7000 random projections with majority vote yields a good performance compared with other ensembles. It yields an MCC of 0.131 as well as a precision of 0.178 at a recall of 0.264. It seems that more number of random projections does not perform better. Moreover, the ensemble with 11000 random projections yields a second best performance with an F1 of 0.131 and a comparative lower "Prec" of 0.146 compared with that with 7000 random projections.

Table 1. Prediction performance of the REPTree classifier ensemble with majority vote technique, i.e., the ensemble predicts a residue to be hots spot if all of REPTree classifiers in the ensemble predict it to be a hot spot residue. The performance is averaged across 25 proteins in BID test set.

No.	Sen	MCC	Prec	F1
1000	0.581	0.085	0.079	0.111
3000	0.322	0.089	0.104	0.122
5000	0.382	0.082	0.102	0.112
7000	0.264	0.131	**0.178**	0.13
9000	0.28	0.106	0.104	0.114
11000	0.282	0.124	0.146	**0.131**
13000	0.322	0.089	0.124	0.11
15000	0.228	0.11	0.104	0.129

Table 2. Prediction performance of the REPTree classifier ensemble with different number of reduced dimensions on BID test data set.

No. dimension	Sen	MCC	Prec	F1
9	0.278	0.075	0.096	0.117
18	0.244	0.081	0.113	0.107
27	0.365	0.086	0.102	0.11
36	0.227	0.073	0.101	0.106
45	0.294	0.118	0.131	0.127
54	0.459	0.066	0.079	0.109
63	0.264	0.131	**0.178**	**0.13**
72	0.219	0.08	0.116	0.103
81	0.256	0.093	0.085	0.106
90	0.366	0.071	0.096	0.111
99	0.29	0.085	0.107	0.114

For the ensemble with 7000 random projections, performance on different number of reduced dimensions by random projection technique is investigated. The ensembles on 45 and 63 reduced dimensions yield good performance and that on 63 reduced dimension performs best. Table 2 shows the performance comparison of the REPTree classifier ensemble with different number of reduced dimensions.

4.2 Comparison with Other Methods

As our best knowledge, there is no method for the prediction of hot spots in the whole protein sequence used only the sequence information. Most of hot spot prediction methods applied structural information of homologous proteins to the identification. So the random predictor is implemented here and ran 100 times. The average performance is compared with our method. Table 3 lists the comparison for details. Our method

Table 3. Performance comparison of the two methods on BID data set by training on ASEdb data set.

Method	Type	Sen	MCC	Prec	F1
Our method	REPTree	0.264	0.131	**0.178**	0.13
Random predictor		0.983	0.000	0.018	0.035

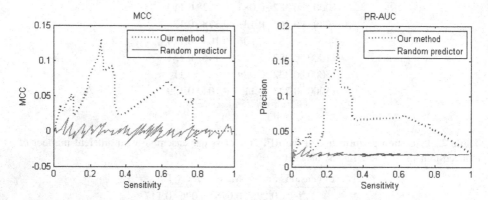

Fig. 2. Comparison of our method with random predictor on the BID data set. The left graph illustrates the MCC curves and the right one shows the PR curves.

yields a precision of 0.178 at 0.264 recall by only the use of sequence information. Results showed that our method outperforms the random predictor by 10 times of Prec score and 37 times of F1 score.

The performance of ensemble classifier with majority vote is illustrated in Fig. 2. Also, the performance of random predictor is shown in Fig. 2. Although it is much difficult to identify hot spots, our method yields a good result based on sequence information only. Our method obtains an area under the PR curve (Precision-Recall) of 0.064 while the random predictor only has a PR-AUC of 0.017.

5   Conclusions

This paper proposes an ensemble of REPTree classifiers by random projection with only sequence information to identify hot spots. In order to balance the hot spot data set, a data set consisting of roughly the same number of hot spots (positive set) and non-hot spots (negative subset) is constructed. Different random projections on the balanced data set are input into a REPTree classifier. The ensemble of these REPTree classifiers can yield good prediction on hot spots. Moreover, our method is simple for only statistical amino acid properties are applied and the dimensionality reduction of random projection is adopted here. Actually, the random projection technique provides a useful mechanism such that it reduces the high dimensional original data and makes the data more diverse and thus, the method yields a good prediction on hot spots.

Although the sequence based method performs worse than experimental methods and even those structure-based methods, it is a good complement to experiment identification when there is no homologous structure available in current data banks.

References

1. Bogan, A.A., Thorn, K.S.: Anatomy of hot spots in protein interfaces. J. Mol. Biol. 280((1), 1–9 (1998)
2. Clackson, T., Wells, J.A.: A hot spot of binding energy in a hormone-receptor interface. Science 267(5196)), 383–386 (1995)
3. Kortemme, T., Baker, D.: A simple physical model for binding energy hot spot in protein-protein complex. Proc. Natl. Acad. Sci. USA 99(22), 14116–141121 (2002)
4. Keskin, O., Ma, B., Nussinov, R.: Hot regions in protein-protein interactions: the organization and contribution of structurally conserved hot spot residues. J. Mol. Biol. 345(5), 1281–1294 (2005)
5. Thorn, K.S., Bogan, A.A.: Asedb: a database of alanine mutations and their Effects on the free energy of binding in protein interactions. Bioinformatics 17(3), 284–285 (2001)
6. Fischer, T.B., Arunachalam, K.V., Bailey, D., Mangual, V., Bakhru, S., Russo, R., Huang, D., Paczkowski, M., Lalchandani, V., Ramachandra, C., Ellison, B., Galer, S., Shapley, J., Fuentes, E., Tsai, J.: The binding interface database (bid): a compilation of amino acid hot spots in protein interfaces. Bioinformatics 19(11), 1453–1454 (2003)
7. Kumar, M.D.S., Gromiha, M.M.: Pint: protein-protein interactions thermodynam-Ic database. Nucleic Acids Res. 34, D195–D198 (2006)
8. Moal, I.H., Fernández-Recio, J.: Skempi: A structural kinetic and energetic database of mutant protein interactions and its use in empirical models. Bioinformatics 28(20), 2600–2607 (2012)
9. DeLano, W.L.: unraveling hot spots in binding interfaces: progress and challenges. Curr. Opin. Struct. Biol. 12(1), 14–20 (2002)
10. Kortemme, T., Baker, D.: A simple physical model for binding energy hot spots in protein–protein complexes. Proc. Natl. Acad. Sci. 99(22), 14116–14121 (2002)
11. Guerois, R., Nielsen, J.E., Serrano, L.: Predicting changes in the stability of proteins and protein complexes: a study of more than 1000 mutations. J. Mol. Biol. 320(2), 369–387 (2002)
12. Gao, Y., Wang, R., Lai, L.: Structure-based method for analyzing protein-protein interfaces. J. Mol. Model. 10(1), 44–54 (2004)
13. Schymkowitz, J., Borg, J., Stricher, F., Nys, R., Rousseau, F., Serrano, L.: The foldx web server: an online Force field. Nucleic Acids Res. 33(Web Server issue), W382–W388 (2005)
14. Huo, S., Massova, I., Kollman, P.A.: Computational alanine scanning of the 1:1 human growth hormone-receptor complex. J. Comput. Chem. 23(1), 15–27 (2002)
15. Rajamani, D., Thiel, S., Vajda, S., Camacho, C.J.: Anchor residues in protein-Protein interactions. Proc. Natl. Acad. Sci. USA 101(31), 11287–11292 (2004)
16. Gonzlez-Ruiz, D., Gohlke, H.: Targeting protein-protein interactions with small molecules: challenges and perspectives for computational binding epitope detection and ligand finding. Curr. Med. Chem. 13(22), 2607–2625 (2006)
17. Ma, B., Elkayam, T., Wolfson, H., Nussinov, R.: Protein-protein interactions: structurally conserved residues distinguish between binding sites and exposed protein surfaces. Proc. Natl. Acad. Sci. USA 100(10), 5772–5777 (2003)

18. del Sol, A., O'Meara, P.: Small-world network approach to identify key residues in protein-protein interaction. Proteins **58**(3), 672–682 (2005)
19. Brinda, K.V., Kannan, N., Vishveshwara, S.: Analysis of homodimeric protein interfaces by graph-spectral methods. Protein Eng. **15**(4), 265–277 (2002)
20. Guharoy, M., Chakrabarti, P.: Conservation and relative importance of residues across protein-protein interfaces. Proc. Natl. Acad. Sci. USA **102**(43), 15447–15452 (2005)
21. Grosdidier, S., Fernndez-Recio, J.: identification of hot-spot residues in protein-protein interactions by computational docking. BMC Bioinform. **9**, 447 (2008)
22. Ofran, Y., Rost, B.: Protein-protein interaction hotspots carved into sequences. PLoS Comput. Biol. **3**(7), e119 (2007)
23. Darnell, S.J., Page, D., Mitchell, J.C.: An automated decision-tree approach to predicting protein interaction hot spots. Proteins **68**(4), 813–823 (2007)
24. Guney, E., Tuncbag, N., Keskin, O., Gursoy, A.: Hotsprint: database of computational hot spots in protein interfaces. Nucleic Acids Res. **36**(Database issue), D662–D666 (2008)
25. Tuncbag, N., Gursoy, A., Keskin, O.: Identification of computational hot spots in protein interfaces: combining solvent accessibility and inter-residue potentials improves the accuracy. Bioinformatics **25**(12), 1513–1520 (2009)
26. Cho, K.I., Kim, D., Lee, D.: A feature-based approach to modeling protein-protein interaction hot spots. Nucleic Acids Res. **37**(8), 2672–2687 (2009)
27. Lise, S., Archambeau, C., Pontil, M., Jones, D.T.: Prediction of hot spot residues at protein-protein interfaces by combining machine learning and energy-based methods. BMC Bioinform. **10**, 365 (2009)
28. Xia, J.F., Zhao, X.M., Song, J., Huang, D.S.: Apis: accurate prediction of hot spots in protein interfaces by combining protrusion index with solvent accessibility. BMC Bioinform. **11**, 174 (2010)
29. Tuncbag, N., Keskin, O., Gursoy, A.: Hotpoint: hot spot prediction server for protein interfaces. Nucleic Acids Res. **38**(Web Server issue), W402–W406 (2010)
30. Lise, S., Buchan, D., Pontil, M., Jones, D.T.: Predictions of hot spot residues at protein-protein interfaces using support vector machines. PLoS ONE **6**(2), e16774 (2011)
31. Wang, L., Liu, Z.P., Zhang, X.S., Chen, L.: Prediction of hot spots in protein interfaces using a random forest model with hybrid features. Protein Eng. Des. Sel. **25**(3), 119–126 (2012)
32. Chen, P., Li, J., Wong, L., Kuwahara, H., Huang, J.Z., Gao, X.: Accurate prediction of hot Spot residues through physicochemical characteristics of amino acid sequences. Proteins **81**(8), 1351–1362 (2013)
33. Kawashima, S., Pokarowski, P., Pokarowska, M., Kolinski, A., Katayama, T., Kanehisa, M.: Aaindex: amino acid index database, progress report 2008. Nucleic Acids Res. **36**(Database issue), D202–D205 (2008)
34. Altschul, S.F., Madden, T.L., Schäffer, A.A., Zhang, J., Zhang, Z., Miller, W., Miller, D.J.: Gapped blast and psi-blast: a new generation of protein database search programs. Nucleic Acids Res. **25**(17), 3389–3402 (1997)
35. Chen, P., Li, J.: Sequence-based identification of interface residues by an integrative profile combining hydrophobic and evolutionary information. BMC Bioinform. **11**, 402 (2010)
36. Chen, P., Wong, L., Li, J.: Detection of outlier residues for improving interface prediction in protein heterocomplexes. IEEE/ACM Trans. Comput. Biol. Bioinform. **9**(4), 1155–1165 (2012)
37. Papadimitriou, C.H., Raghavan, P., Tamaki, H., Vempala, S.: Latent semantic indexing: a probabilistic analysis. In: Proceedings of the 17th ACM Symposium on the Principles of Database Systems, pp. 159–168 (1998)

38. Kaski, S.: dimensionality reduction by random mapping: fast similarity computation for clustering. In: Neural Networks Proceedings, 1998. IEEE World Congress on Computational Intelligence. The 1998 IEEE International Joint Conference, vol. 1, pp. 413–418 (1998)
39. Esposito, F., Malerba, D., Semeraro, G., Tamma, V.: The Effects of pruning methods on the predictive accuracy of induced decision trees (1999)
40. Chen, P., Huang, J.Z., Gao, X.: Ligandrfs: random forest ensemble to identify ligand-binding residues from sequence information alone. BMC Bioinform. 15(Suppl 15), S4 (2014)
41. Kuncheva, L.I., Whitaker, C.J., Duin, R.P.W.: Limits on the majority vote accuracy in classifier fusion. Pattern Anal. Appl. 6(1), 22–31 (2003)
42. Wang, B., Chen, P., Huang, D.S., Li, J.J., Lok, T.M., Lyu, M.R.: Predicting protein interaction sites from residue spatial sequence profile and evolution rate. FEBS Lett. 580(2), 380–384 (2006)
43. Kyte, J., Doolittle, R.F.: A simple method for displaying the hydropathic character of a protein. J. Mol. Bio. 157(1), 105–132 (1982)

Identification of Hot Regions in Protein-Protein Interactions Based on SVM and DBSCAN

Xiaoli Lin[1,2(✉)], Huayong Yang[1], and Jing Ye[2]

[1] City College, Wuhan University of Science and Technology,
Wuhan 430083, Hubei, China
[2] School of Computer Science and Technology,
Wuhan University of Science and Technology, Wuhan 430065, Hubei, China
{Aneya,Levi_wanlin,Yuxin_wan}@163.com

Abstract. Hot regions are the key factor to maintain stability and coordination of protein-protein interactions. In this paper, combining evolutionary information and support vector machine (SVM), we have developed an improved method for predicting binding sites in a protein sequence. The prediction models developed in this study have been trained and tested on binding protein chains and evaluated using fold cross validation technique. The performance of this SVM model further improved. Based on the predicted hot spots, DBSCAN method is used to predict the hot regions in protein-protein interactions. The experimental results demonstrate that the proposed method improves the predictive accuracy of hot regions and is more reliable compared with previous method.

Keywords: Hot regions · Protein–protein interactions · SVM · DBSCAN

1 Introduction

Prediction of functionally important regions is one of the most challenging and intriguing problems in structural biology and in computational molecular biology [1–3]. As more and more protein sequences are stored without available structural information, it is strongly desirable to predict protein hot regions by protein sequences [4]. Protein–protein interactions (PPI) play an important role in many biological processes such as gene expression control, enzyme inhibition, antibody–antigen recognition and so on. The principles that govern the interaction of two proteins and the general properties of their interacting interfaces remain uncovered [5], resulting in the difficulties of predicting interface regions directly from protein sequences.

A number of experiment results indicate that only a small subset of contact residues contribute significantly to the binding free energy [5]. These residues have been termed "hot spots". The prediction of hot spot residues is a very difficult but important problem. Hot spot residues are closely related to protein function, and the identification of protein binding sites is very essential to understand their interactions. Hot regions help proteins to exert their biological function and contribute to understand the molecular mechanism, which is the foundation of drug designs. Many research

© Springer International Publishing Switzerland 2015
D.-S. Huang et al. (Eds.): ICIC 2015, Part II, LNCS 9226, pp. 390–398, 2015.
DOI: 10.1007/978-3-319-22186-1_38

methods for the analysis [6] and prediction [7] of hot spots and hot regions [4] have been reported.

Hot spots are typically defined as those residues for which $\Delta\Delta G >= 2.0$ kcal/mol ($\Delta\Delta G$ is the change in free energy of binding). In the recent years, several computational approaches have been developed to identify hot spot residues at protein-protein interfaces [8, 9]. Hus [4], Tuncbag [10], Cukuroglu [11] and Carles Pons [12] do many research works to identify protein-protein binding regions and analyze the organization structure of hot regions conformation with physical and chemical characteristics.

In this paper, a novel strategy is proposed to predict hot spot residues at protein-protein interfaces. Similarly to other previous methods, we consider the basic terms that contribute to hot spot interactions. Then, DBSCAN algorithm is used to predict the hot regions based on the predicted hot spots.

2 Methods

2.1 Data Sets

In our study, we use the same data set as in [13]. The data set consists of protein complexes for which alanine mutational data are available. Alanine mutation data are collected from the Alanine Scanning Energetics Database (ASEdb) http://www.asedb. org, which is a searchable database of single alanine mutations in protein–protein, protein–nucleic acid, and protein–small molecule interactions for which binding affinities have been experimentally determined [14]. In cases where structures are available, it contains surface areas of the mutated side chain and links to the PDB entries.

Apart from these data, an independent test set, composed of 127 alanine-mutated interface residues in complexes, is also constructed from the Binding Interface Database (BID) http://tsailab.org/BID/, of which 39 residues are hot spots. In this database, the alanine mutation data are listed as either "strong", "intermediate", "weak" or "insignificant". Only "strong" mutations are regarded as hot spots, and the other mutations are regarded as unimportant residues. Table 1 lists the 17 protein-protein complexes analyzed [15].

2.2 Support Vector Machine Models

Machine learning methods are widely used for classification tasks [16]. This paper uses support vector machine classifier which is playing an increasingly important role in the field of computational biology and is a well-known classifier to demonstrate the success of classifying interfaces using hot region characteristics [15]. SVM has been used to learn from a training set to classify residues as hot spots ($\Delta\Delta G >= 2.0$ kcal/mol) or non-hot spots ($\Delta\Delta G < 2.0$ kcal/mol).

We have used the program package SVM, which is available at the website http://www.csie.ntu.edu.tw/~cjlin/libsvm. The program provides several standard kernel functions and we have experimented with linear, polynomial and Gaussian kernels, and with different combinations of input features.

Table 1. The 17 protein-protein complexes analyzed

PDB ID	First molecule	Second molecule
1a4y	RNase inhibitor	Angiogenin
1a22	Human growth hormone	Human growth hormone binding protein
1ahw	Immunoglobulin Fab5G9	Tissue factor
1brs	Barnase	Barstar
1bxi	Colicin E9 Immunity Im9	Colicin E9 DNase
1cbw	BPTI Trypsin inhibitor	Chymotrypsin
1dan	Blood coagulation factor VIIA	Tissue factor
1dvf	Idiotopic antibody FV D1.3	Anti-idiotopic antibody FV E5.2
1f47	Cell division protein ZIPA	Cell division protein FTSZ
1fc2	Fc fragment	Fragment B of protein A
1fcc	Fc (IGG1)	Protein G
1gc1	Envelope protein GP120	CD4
1jrh	Antibody A6	Interferon-gamma receptor
1nmb	N9 Neuramidase	Fab NC10
1vfb	Mouse monoclonal antibody D1.3	Hen egg lysozyme
2ptc	BPTI	Trypsin
3hfm	Hen Egg Lysozyme	Ig FAB fragment HyHEL-10

2.3   Measures of Prediction Performance

To assess the performances of the classification, let TP, FP, FN refer to the number of true positives, false positives and false negative respectively. Precision and recall are defined as

$$Precision = \frac{TP}{TP + FP} \tag{1}$$

$$Recall = \frac{TP}{TP + FN} \tag{2}$$

The F1 score is the harmonic mean of precision and recall

$$F1 = \frac{2 \times Recall \times Precision}{Recall + Precision} \tag{3}$$

To evaluate the performances of hot regions prediction, two evaluation criterions C1 and C2 are used in the hot regions prediction. The C1 is the accuracy in the hot region prediction

$$C1 = \frac{TP\_PHR}{RNUM\_PHR} \tag{4}$$

The true positive of the predicted hot region (TP_PHR) is the number of TP in the hot regions of prediction results, and the RNUM_PHR is the number of residues in the hot regions of prediction results. The C2 is the coverage of hot regions prediction on real hot regions

$$C2 = \frac{TP\_PHR}{RNUM\_RHR} \tag{5}$$

The RNUM_RHR is the number of hotspots in the real hot regions. The C represents the balance performance between C1 and C2

$$C = \frac{2 * C1 * C2}{C1 + C2} \tag{6}$$

2.4 Feature Selection

The feature selection techniques preserve the original characteristic and enhance the precision of experimental results. With the help of feature selection, the over fitting can be avoided. The feature selection can also help to improve model performance and provide faster and more effective models. To obtain the best subset of features for predicting hot spots, the feature selection is performed using the Non-dominated Sorting Genetic Algorithm (NSGA).

The NSGA is a popular non-domination based genetic algorithm for multi-objective optimization [17]. It has been widely applied to solve the problem of parameters selection for SVM classification due to its ability to discover good solutions quickly for complex searching and optimization problems. The general idea of our NSGA is as follows.

Step 1:Generate the population randomly.

Step 2:Evaluate the objective function value of each individual, and sort the population.

Step 3:Repeat the following steps till the maximum.

(a) Select elitist individual.
(b) Perform crossover strategy and evaluate objective for offspring individual.
(c) Perform mutation strategy and evaluate objective for the mutated chromosomes.

The feature selection based on NSGA is shown in Fig. 1.

2.5 Evaluation of the Hot Spot Prediction

In our study, the hot region is predicted based on the predicted hot spots, which combines some important physicochemical properties and structure properties. The hot spot residues measured experimentally are directly compared with other residues.

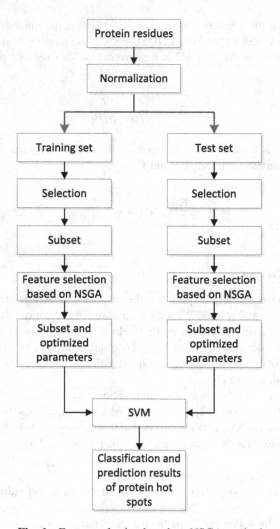

Fig. 1. Feature selection based on NSGA method

It can be seen from Table 2 that our method is better than those of other four methods for Recall value. For the precision, our result is lower than KFC method and Robetta method, but is identical to that of Nurcan and is better than FOLDEF method.

Table 2. Comparison of the hot spots results with the independent test set

Method	Recall	Precision	F1-Score
KFC	0.31	0.85	0.37
FOLDEF	0.22	0.4	0.28
Robetta	0.33	0.87	0.40
Nurcan	0.59	0.73	0.65
Our method	0.82	0.73	0.77

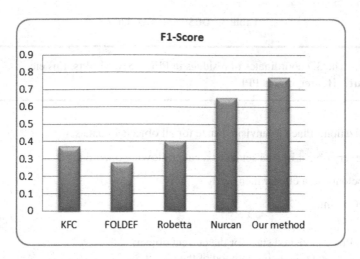

Fig. 2. Comparison of the F1-score

Especially, as shown in the Fig. 2, the F1-score of our method are much better than those of KFC, FOLDEF and Robetta. These indicate that our method give better predictive performance compared with other methods.

3 Identification of Hot Region

3.1 DBSCAN Algorithm

The Density-Based Spatial Clustering of Applications with Noise (DBSCAN) is a data clustering algorithm proposed by Martin Ester [18]. It is a density-based clustering algorithm: given a set of points in some space, it groups together points that are closely packed together [19].

DBSCAN does not need to define the number of clusters, and it can handle discretional shaped clusters and can even search a cluster absolutely surrounded by other different cluster. The determination of two parameters *Eps* and *MinPts* is very key and even sometimes sensitive to the effectiveness of clustering for the reduced set. According to the definition of protein hot region, a hot region contains at least three hot spot residues, and each hot spot has at least two hot spot neighbors in this hot region [5], some of hot spots form hot region [20], while the isolated hot spots disperse in PPI. Here, the DBSCAN algorithm is used to predict the hot regions in PPI. The steps of our algorithm are shown in the following Table 3.

3.2 Evaluation of the Hot Region Prediction

Based on previous studies on hot spots by using NSGA-SVM classifier, we predict the hot regions with the density-related features and solvent accessibility-related features.

Table 3. DBSCAN Algorithm

Input: The 3D coordinates of residues in PPI, ASA, MinPts, Eps etc.
Output: Hot regions in PPI

Begin

Initialization: Place to unvisited state for all objects in data set.

Visit every object in data set and execute following operations:

If it belongs to a cluster then

      Continue;

 Else
      1. Place to visited state for the current object;
      2. Generate the neighborhood of the object

      If the residues in the neighborhood less than the threshold then

          Remove the object from data set to noise set;

      Else

          1) Generate a new cluster with current object as it's center point;
          2) Add all residues in the neighborhood to the cluster;
          3) Visit the residues in the cluster, generate neighborhood of each residues and create new cluster depending on residues in the neighborhood

End

Table 4 lists the accuracy and the coverage in the hot region prediction. Our results are better than those of other two methods. Tuncbag [21] uses some physicochemical attributes to predict hot spots and form hot regions. Nan [22] used complex network and community method to predict hot regions. It can be seen from the Fig. 3 that our C-Score is much better than that of Tuncbag and is slightly better than that of Nan. This indicates that our method has the good performance for predicting hot regions in PPI.

Table 4. Comparison of the coverage of hot region

Method	C1	C2	C
Nurcan	0.154	0.36	0.216
Nan	0.443	0.413	0.427
Our method	0.581	0.614	0.596

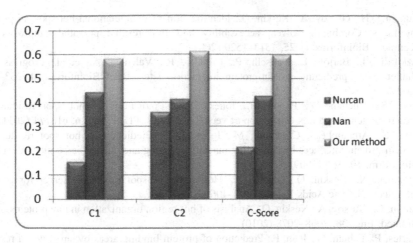

Fig. 3. Performance comparison of Nurcan, Nan and our method

4 Conclusion

In this paper, the feature selection is performed using the NSGA method to identify the best subset of features for prediction hot spots. Based on the selected features, a prediction model for hot spots is created using support vector machine and tested on an independent test set. Then, DBSCAN algorithm is used to locate hot regions based on the predicted hot spots. The experimental results demonstrate that our method shows better overall prediction accuracy than previous methods. As one of the further work, the new features that describe the different energetic contributions to binding interactions are still greatly needed, and we will make available method to improve the prediction accuracy of hot regions.

References

1. Hsu, C.M., Chen, C.Y., Liu, B.J.: MAGIIC-PRO: detecting functional signatures by efficient discovery of long patterns in protein sequences. Nucleic Acids Res. **34**, W356–W361 (2006)
2. Casari, G., Sander, C., Valencia, A.: A method to predict functional residues in proteins. Nat. Struct. Biol. **2**, 171–178 (1995)
3. Armon, A., Graur, D., Ben-Tal, N.: ConSurf: an algorithmic tool for the identification of functional regions in proteins by surface mapping of phylogenetic information. J. Mol. Biol. **307**, 447–463 (2001)
4. Hsu, C.M., Chen, C.Y., Liu, B.J., Huang, C.C.: Identification of hot regions in protein-protein interactions by sequential pattern mining. BMC Bioinform. **8**(Suppl 5), S8 (2007)
5. Keskin, O., Ma, B.Y., Mol, R.J.: Hot regions in protein-protein interactions: the organization and contribution of structurally conserved hot spot residues. J. Mol. Biol. **345**, 1281–1294 (2005)

6. Tuncbag, N., Gursoy, A., Keskin, O.: Identification of computational hot spots in protein interfaces: Combining solvent accessibility and inter-residue potentials improves the accuracy. Bioinformatics **25**, 1513–1520 (2009)
7. Ezkurdia, I., Bartoli, L., Fariselli, P., Casadio, R., Valencia, A., et al.: Progress and challenges in predicting protein-protein interaction sites. Brief Bioinform. **10**, 233–246 (2009)
8. Lise, S., Buchan, D., Pontil, M., Jones, D.T.: Predictions of hot spot residues at protein-protein interfaces using support vector machines. PLoS ONE **6**, e16774 (2011)
9. Lise, S., Archambeau, C., Pontil, M., Jones, D.T.: Prediction of hot spot residues at protein-protein interfaces by combining machine learning and energy-based methods. BMC Bioinform. **10**, 365 (2009)
10. Tuncbag, N., Keskin, O., Gursoy, A.: HotPoint: hot spot prediction server for protein interfaces. Nucleic Acids Res. **38**, 402–406 (2010)
11. Engin, C., Gursoy, A., Keskin, O.: Analysis of hot region organization in hub proteins. Ann. Biomed. Eng. **38**, 2068–2078 (2010)
12. Carles, P., Fabian, G., Juan, F.: Prediction of protein-binding areas by small world residue networks and application to docking. BMC Bioinform. **12**, 378–388 (2011)
13. Dongfang, N., Xiaolong, Z.: Prediction of hot regions in protein-protein interactions based on complex network and community. In: The 4thWorkshop on Interative Data Analysis in System Biology (IDASB), in conjunction with BIBM2013 (2013)
14. Thorn, K.S., Bogan, A.A.: ASEdb: a database of ananine mutations and their effects on the free energy of binding in protein interactions. Bioinformatics **17**(3), 284–285 (2001)
15. Kyu-il, C., Dongsup, K., Doheon, L.: A feature-based approach to modeling protein-protein interaction hot spots. Nucleic Acids Res. **37**(8), 2672–2678 (2009)
16. Noble, W.S.: What is a support vector machine? Nat. Biotechnol. **24**, 1565–1567 (2006)
17. Susmita, B., Ranjan, B.: NSGA-II based multi-objective evolutionary algorithm for a multi-objective supply chain problem. In: IEEE International Conference on Advances In Engineering, Science and Management (ICAESM-2012), pp. 126–130 (2012)
18. Campello, R.J.G.B., Moulavi, D., Zimek, A., Sander, J.: A framework for semi-supervised and unsupervised optimal extraction of clusters from hierarchies. Data Min. Knowl. Disc. **27**(3), 344 (2013)
19. Campello, R.J., Moulavi, D., Sander, J.: Density-based clustering based on hierarchical density estimates. In: Pei, J., Tseng, V.S., Cao, L., Motoda, H., Xu, G. (eds.) PAKDD 2013, Part II. LNCS, vol. 7819, pp. 160–172. Springer, Heidelberg (2013)
20. Ahmad, S., Keskin, O., Sarai, A., Nussinov, R.: Protein–DNA interactions: structural, thermodynamic and clustering patterns of conserved residues in DNA-binding proteins. Nucleic Acids Res. **36**, 5922–5932 (2008)
21. Tuncbag, N., Keskin, O., Gursoy, A.: HotPoint: hot spot prediction server for protein interfaces. Nucleic Acids Res. **38**(2), 402–406 (2010)
22. Nan, D.F., Zhang, X.L.: Prediction of hot regions in protein-protein interactions based on complex network and community detection. In: IEEE International conference on Bioinformatics and Biomedicine, pp. 17–23 (2013)

Identification of Hot Regions in Protein Interfaces: Combining Density Clustering and Neighbor Residues Improves the Accuracy

Jing Hu[1,2] and Xiaolong Zhang[1,2(✉)]

[1] School of Computer Science and Technology, Wuhan University of Science and Technology, Wuhan 430065, Hubei, China
[2] Hubei Province Key Laboratory of Intelligent Information Processing and Real-Time Industrial System, Wuhan 430065, Hubei, China
{hujing,xiaolong.zhang}@wust.edu.cn

Abstract. Discovering hot regions in protein–protein interaction is important for understanding the interactions between proteins, while because of the complexity and time-consuming of experimental methods, the computational prediction method can be very helpful to improve the efficiency to predict hot regions. In hot region prediction research, some models are based on structure information, and others are based on a protein interaction network. However, the prediction accuracy of these methods can still be improved. In this paper, a new method that uses density-based incremental clustering to predict hot regions and optimizes the predicted hot regions using neighbor residues is proposed. Experimental results show that the proposed method significantly improves the prediction performance of hot regions.

Keywords: Hot region · Density · Clustering · Neighbor residue · Protein–protein interaction

1 Introduction

Hot regions [1, 2] of protein–protein interactions play important roles in the functions and stability of protein complexes. In the past, many attempts have been made to predict hot regions. The research group [1, 3, 4] in Koc University of Turkey made many contributions to the prediction of hot regions. In 2007, Hsu [5] presented a pattern-mining approach for the identification of hot regions in protein–protein interactions. In [6], Pons studied a network-based method and used small-world residue networks to predict protein-binding areas. Although the proposed method has potential applications for protein docking as a complement to energy-based approaches, it shows limitations in many cases with certain topological features, like spherical or very large proteins.

In this paper, we propose a method to identify hot regions by combining density clustering and neighbor residues (DCNR), which can predict hot regions in protein–protein interactions by combining density-based incremental clustering and neighbor

© Springer International Publishing Switzerland 2015
D.-S. Huang et al. (Eds.): ICIC 2015, Part II, LNCS 9226, pp. 399–407, 2015.
DOI: 10.1007/978-3-319-22186-1_39

residues. Experimental results show that the proposed method significantly improves the prediction performance for hot regions.

2 Method

2.1 Density-Based Incremental Clustering

Clustering is a process to group data into multiple sub-groups or clusters so that objects within a cluster may have strong similarities [7]. The proposed algorithm connects core residues and their neighborhoods to form dense regions as clusters. In order to use the clustering method to cluster the hot spot residues, we need to adopt several concepts from [7] (all distances in this paper are Euclidean distance):

- **Neighborhood:** a user-specified parameter $\varepsilon > 0$ is used to specify the radius of a neighborhood for every residue.
- **Dense Region:** to determine whether a neighborhood is dense or not, another user-specified parameter "Min" is used to specify the density threshold of dense regions.
- **Core Residue:** a residue is a core residue if the ε-neighborhood of that residue contains at least "Min" residues.

The process of density-based incremental clustering is described as follows: Initially, all residues in D are marked as "unvisited". Then an unvisited residue p is selected randomly and marked as "visited". After that, the ε-neighborhood of p is checked to see whether it contains at least "Min" residues. If not, p is marked as a noise point. Otherwise, a new cluster C is created for p, and all the residues in the ε-neighborhood of p are added to a candidate set, N. The algorithm iteratively adds to C those residues in N that do not belong to any cluster. In this process, for a residue p in set N that is labeled "unvisited", we mark it as "visited" and check its ε-neighborhood. If the ε-neighborhood of p' has at least "Min" residues, those residues in the ε-neighborhood of p' are added to N. Clustering continues adding residues to C until C can no longer be expanded, that is, until N is empty. At this time, cluster C is completed, and then we randomly select an unvisited residue from the remaining ones. The clustering process continues until all residues in the set D are visited.

2.2 Predicted Result Optimization Using Neighbor Residues

A residue is defined as a "neighbor residue" if the distance between its C^α atom and a C^α atom of a residue is under 6.5 Å [1]. Packing around hot spots is significantly tighter than non-hot spots. Residues whose C^α atoms are closer than the threshold are defined to be neighbors. Close neighbors along the chain (i.e. for the ith, the $i - 1$ and $i + 1$ residues) are not summed in the calculation. Thus a contact between ith and jth residues is defined using:

$$Contact_{i,j} = 0 \quad if\,|i - j| \leq 1;$$
$$Contact_{i,j} = 1 \quad if\,|i - j| > 1 \; and \; d_{i,j} \leq 6.5A; \tag{1}$$

Where $d_{i,j}$ is the distance between two C^α atoms of the ith and jth residues. The neighbor number of a residue i (number of residues around it) is calculated by:

$$NN_i = \sum_{j=1}^{res} Contact_{i,j} \tag{2}$$

Where NN is the neighbor number and res is the number of residues in the protein chains. Since interface residues have neighbors both from their own chain and the complementary chain, we count the neighbors from both chains.

The number of neighbor is different between hot spots and non-hot spots. The average neighbor residue number of the hot spot is about 5.0, 6.0 and 7.0, while the non-hot spot usually below than 5.0 [3]. We adopt a threshold of 5.0 to distinguish hot spot and non-hot spot. If the neighbor number of residue i is more or equal than five, the residue i is more likely to be hot spot and we retain residue i in predicted hot regions. While if the neighbor number of residue i is less than five, then we check all neighbor residues of residue i, if there is at least one hot spot, we retain residue i else remove it from predicted hot regions. The optimized rule as follows:

$$NN_i \geq 5 \; retain \; i;$$
$$NN_i < 5 \begin{cases} retain \; i \; if \; hotspot \in Set\{N_i\}; \\ remove \; i \; if \; hotspot \notin Set\{N_i\}; \end{cases} \tag{3}$$

Where NNi is the neighbor number of residue i, and Set $\{N_i\}$ are set of all neighbor residues of residue i.

2.3   The DCNR Algorithm

Input:

- D: a data set containing 450 residues and their three-dimensional coordinates.
- ε: the radius parameter.
- Min: the neighborhood density threshold.

Method:

- Calculate structure feature values of residues in set D using PSAIA;
- Classify all residues in set D using LIBSVM to remove non-hot spots;
- Create the core residues set C;
- Visit residues in the ε-neighborhood of set C, and add all residues that match the condition into set C;
- The clustering process continues until all residues in set D are visited, then mark C as original hot region and mark all residues outside set C as outliers;

- Calculate neighbor number of each residue in set C and use set $\{N_i\}$ to store all neighbor residues of residue i;
- Remove residues that don't match the optimized condition from set C.

Output: predicted hot regions from DCNR.

3 Experiment Results and Evaluation

3.1 Datasets

The dataset we use in this paper are from the latest database SKEMPI (Structural database of Kinetics and Energetics of Mutant Protein Interactions) [8]. We select alanine mutation residues from SKEMPI, and then replace the repetitive items with the average value of their different binding free energy values. After that we had 31 complexes, each complex is composed of a bunch of interface residues. An interface residue will be defined as a hot spot if its corresponding binding free energy is higher than or equal to 2.0 kcal/mol, while an interface residue with binding free energy less than or equal to 0.4 kcal/mol is considered as a non-hot spot, according to which there are 112 hot spots and 338 non-hot spots in the dataset. The other 254 unlabeled residues with binding free energy between 0.4 kcal/mol and 2.0 kcal/mol are excluded from the training set because their hot spot or non-hot spot features are not very clear [9]. Finally, we got 450 interface residues in 31 complexes as our dataset, whose structure features are obtained by PSAIA [10]. Table 1 summarizes the 31 complexes and the interface residue for each complex.

Table 1. Number of interface residues, hot spots, non-hot spots and other unlabeled residues

Complex	Interface residues	Labeled residues		Unlabeled residues
		Hot spots	Non-hot spots	
1A22	129	4	85	40
1A4Y	28	3	16	9
1AHW	8	0	4	4
1AK4	7	6	0	1
1BRS	14	9	2	3
1CBW	15	1	8	6
1CHO	9	5	4	0
1DAN	81	2	59	20
1DVF	23	8	1	14
1EAW	21	1	12	8
1EMV	46	9	16	21
1F47	10	3	3	4
1FCC	7	3	3	1
1FFW	8	1	4	3

(Continued)

Table 1. (*Continued*)

Complex	Interface residues	Labeled residues		Unlabeled residues
		Hot spots	Non-hot spots	
1GC1	53	0	40	13
1IAR	17	1	10	6
1JCK	7	3	0	4
1JRH	32	9	9	14
1JTG	30	11	10	9
1KTZ	21	4	0	17
1LFD	7	0	4	3
1PPF	9	2	4	3
1VFB	27	3	8	16
1Z7X	15	4	1	10
2G2U	22	2	13	7
2J0T	4	2	1	1
2JEL	6	1	4	1
2PCC	5	2	2	1
2VLJ	9	0	3	6
3HFM	25	11	8	6
3SGB	9	2	4	3
TOTAL	704	112	338	254

3.2 Experimental Results

In this paper, an extended version of SVM called LIBSVM [11] is adopted to remove non-hot spots from the dataset. In our work, we select five best features RctASA, RcsASA, BsASA, UsASA and RctmPI to distinguish hot spots and non-hot spots. The principal of selecting comes from previous study of our lab [12, 13].

After using feature-based classification to remove non-hot spot residues from the dataset, we get 110 predicted hot spots. Then we conducted the density-based incremental clustering on the 110 predicted hot spots. We used the coordinates of the C^α atom of a protein residue to represent the residue's coordinates. All the values of coordinates of the residues were from the protein data bank (PDB) [14]. In the clustering experiments, the density threshold "Min" was set to three when performing the density-based incremental clustering because every hot region contains at least three hot spots [1]. In order to determine the value of the neighborhood radius ε, we studied the relationship between neighborhood radius ε and the number of clusters. The relationship is shown in Fig. 1, from which we can find that different ε values will result in different cluster numbers. When $\varepsilon = 13$, the cluster number reaches the maximum value, which is 11.

After clustering with $\varepsilon = 13$ and Min = 3, we get 11 predicted hot regions are compounded from 70 residues. Then we using neighbor residues to further optimize predicted result, Table 2 provides the prediction results after optimizing. For a clear

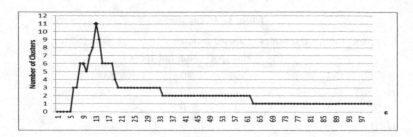

Fig. 1. Relationship between neighborhood radius ε and the number of clusters. The horizontal axis represents neighborhood radius and the vertical axis represents number of clusters. The dots show the value of the number of clusters, and the diamond shows the maximum value of the number of clusters.

Table 2. Prediction result of DCNR in this paper

Complex	Hot region	Residues in hot region
1BRS	1	(D 35) (A 59) (A 102) (D 39)
1CHO	2	(I 17) (I 18) (I 20) (I 21)
1DAN	3	(T 17) (T 18) (T 20) (U 106) (U 133)
1DVF	4	(B 52) (B 54) (B 98) (D 98) (A 32)
1EMV	5	(A 33) (A 50) (A 54) (B 74)
1JRH	6	(I 47) (I 49) (I 52) (I 53) (L 92) (L 93) (L 94)
1JTG	7	(A 105) (A 107) (A 110) (B 53) (B 112) (B 148) (B 36)
1VFB	8	(C 121) (B 99) (C 19)
3HFM	9	(H 33) (H 50) (H 53) (H 98) (Y 96) (Y 97) (Y 100) (L 96) (L 31) (L 32) (L 50) (Y 20)

*(A 168), 'A' is chain ID and 168 is residue ID

discrimination, we call the method with optimization as DCNR (Density Clustering and Neighbor Residues) and the method without optimization as DC (Density Clustering).

3.3   Performance Comparison with Other Methods

The prediction result will be compared to the standard hot regions. In this study, we adopted the standard definition of hot regions from Ozlem Keskin [1]. We compare the proposed method DCNR with DC and Tuncbag's methods [4]. In order to evaluate the performance of the proposed method, three criteria (4)–(6) are used for predicting both hot spots and hot regions:

$$Recall = \frac{TP}{TP + FN} \tag{4}$$

Table 3. Comparison results with different methods to predict hot regions

Method	Hot spot			Hot region		
	Recall	Precision	F-measure	Recall	Precision	F-measure
Tuncbag	0.13	0.6	0.21	0.31	0.8	0.45
DC	**0.51**	0.51	0.51	**0.62**	0.73	0.67
DCNR (this work)	0.46	**0.65**	**0.54**	**0.62**	**0.89**	**0.73**

The highest value in each column is shown in bold

$$Precision = \frac{TP}{TP + FP} \tag{5}$$

$$F - measure = \frac{2 \times Recall \times Precision}{Recall + Precision} \tag{6}$$

Where TP is true positive, FN is false negative and FP is false positive. Table 3 summarizes the performance of these methods on the same data set. Regarding the number of hot regions, DCNR and DC are both able to correctly predict 8 (recall = 0.62) hot regions from 13 standard hot regions while Tuncbag only predict 4 (recall = 0.31). After optimizing by neighbor residue, recall, precision and F-measure of hot region all improved.

On the hot spot coverage of predicted hot regions in standard ones, the two proposed method in this paper predicts larger proportion of hot regions, which reaches 46 % and 57.1 %, while Tuncbag's methods achieve no more than 20 %. Here the recall of DCNR decreased after neighbor residue optimization, because both the false positive residues and the true positive residues were removed. From the above comparison, we can conclude that the proposed method DC and DCNR provide better prediction performance than Tuncbag's methods, and DCNR with neighbor residue optimization further improved prediction performance compare to DC without neighbor residue optimization. Figure 2 describes the coverage status of the predicted hot regions in standard hot regions.

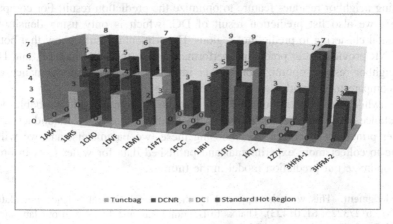

Fig. 2. Number of standard hot regions correctly predicted by different methods. The number above each bar is the number of residues correctly predicted by the different methods.

1JTG

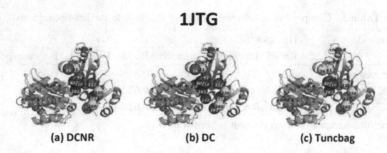

(a) DCNR   (b) DC   (c) Tuncbag

Fig. 3. Visual prediction results of 1JTG from different methods. The green belt represents chain A, the cyan belt represents chain B. The related residues are shown as small spheres: red ones are the real hot regions and are correctly predicted; yellow ones are the real hot regions but missed out by the prediction; blue ones are the fake hot regions but predicted as hot regions. The number on each sphere is its residue ID.

To further illustrate the effectiveness of the proposed approach (DCNR) for predicting hot regions, we randomly selected one example for visualizing using Pymol [15]. The example is the interaction between chain A and chain B of complex 1JTG, which is shown in Fig. 3. There are nine hot spots in standard hot regions, DC is able to correctly predict five of these nine hot spots, and DCNR is able to correctly predict four of these nine hot spots. While Tuncbag's method failed to predict any residues in this hot region. Hot spots (B 41), (B 74), (B 150) and (B 160) are not be predicted by all three methods, suggesting that the mutations of these residues might contribute to protein destabilization.

4 Conclusions

In this paper, we proposed a method called DCNR, which predicts hot regions by applying the density-based incremental clustering from the predicted hot spots, and then using neighbor residues feature to optimize the prediction result. For comparison purposes, we also list prediction result of DC, which is only using density-based incremental clustering to predict hot regions. Experimental results show that both DC and DCNR provide better prediction performance than the other method, and DCNR with neighbor residue optimization further improved prediction performance of hot region compare to DC without neighbor residue optimization.

Meanwhile, the limitation of DCNR is that the clustering results probably depend on the choice of the parameters "ε" and "Min". In the future, we will conduct research on better principles or criteria for choosing these two parameters, and we will also continue to collect more experimental and published data for wider tests in order to further optimize our prediction model in the future.

Acknowledgment. This work is supported by the National Natural Science Foundation of China (No. 61273225, 61201423). Thanks to Bingqing Tan and Jing Ye in our lab, and Shen Peng and Qi Mo for their meaningful discussion.

References

1. Keskin, O., Ma, B., Nussinov, R.: Hot regions in protein-protein interactions: the organization and contribution of structurally conserved hot spot residues. J. Mol. Biol. **345**, 1281–1294 (2005)
2. Ma, B., Nussinov, R.: Druggable orthosteric and allosteric hot spots to target protein-protein interactions. Curr. Pharm. Des. **20**, 1293–1301 (2014)
3. Cukuroglu, E., Gursoy, A., Keskin, O.: HotRegion: A database of predicted hot spot clusters. Nucleic Acids Res. **40**, 829–833 (2012)
4. Tuncbag, N., Gursoy, A., Keskin, O.: Identification of computational hot spots in protein interfaces: combining solvent accessibility and inter-residue potentials improves the accuracy. Bioinformatics **25**, 1513–1520 (2009)
5. Hsu, C.M., Chen, C.Y., Liu, B.J., Huang, C.C., Laio, M.H., Lin, C.C., et al.: Identification of hot regions in protein-protein interactions by sequential pattern mining. BMC Bioinformatics **8**(5), S8 (2007)
6. Pons, C., Glaser, F., Fernandez-Recio, J.: Prediction of protein-binding areas by small-world residue networks and application to docking. BMC Bioinform. **12**, 378 (2011)
7. Han, J.W., Kamber, M., Pei, J.: Data Mining Concepts and Techniques. China Machine Press, Beijing (2012)
8. Moal, I.H., Fernandez-Recio, J.: SKEMPI: A structural kinetic and energetic database of mutant protein interactions and its use in empirical models. Bioinformatics **28**, 2600–2607 (2012)
9. Xia, J.F., Zhao, X.M., Song, J., Huang, D.S.: APIS: accurate prediction of hot spots in protein interfaces by combining protrusion index with solvent accessibility. BMC Bioinform. **11**, 174 (2010)
10. Mihel, J., Sikic, M., Tomic, S., Jeren, B., Vlahovicek, K.: PSAIA - protein structure and interaction analyzer. BMC Struct. Biol. **8**, 21 (2008)
11. A Library for Support Vector Machines. http://www.csie.ntu.edu.tw
12. Hu, J., Zhang, X.L., Liu, X.M., Tang, J.S.: Prediction of hot regions in protein-protein interaction by combining density-based incremental clustering with feature-based classification. Comput. Biol. Med. **61**, 127–137 (2015)
13. Liu, X.M., Tang, J.S.: Mass classification in mammograms using selected geometry and texture features, and a new SVM-based feature selecting method. IEEE Syst. J. **99**, 1932–8184 (2013)
14. Berman, H.M., Westbrook, J., Feng, Z., Gilliland, G., Bhat, T.N., Weissig, H., et al.: The protein data bank. Nucleic Acids Res. **28**, 235–242 (2000)
15. Python Molecule, http://www.pymol.org

Prediction of Protein Structural Classes Based on Predicted Secondary Structure

Fanliang Kong, Dong Wang, Wenzheng Bao, and Yuehui Chen[✉]

School of Information Science and Engineering, University of Jinan,
Jinan, People's Republic of China
yhchen@ujn.edu.cn

Abstract. Prediction of protein structural classes is an important area in bioinformatics, it is beneficial to research protein function, regulation and interactions. In this paper, a 20-dimensional feature vector is extracted based on the predicted secondary structure sequence and the corresponding E-H sequence. Hierarchical classification model based on flexible neural tree (FNT) which is a special kind of artificial neural network with flexible tree structures is used to complete the experiment. 640 dataset and 25 pdb dataset with low homology are chosen as the test dataset. The 10-fold cross validation test is used to test and compare this method with other existing methods. The overall accuracies of our method are 2.7 % and 3.2 % higher for the two datasets respectively.

Keywords: Protein structural classes · Predicted secondary structure · Hierarchical classification model · Flexible neural tree

1 Introduction

In the post-genomic era prediction of protein structural classes plays an important role in bioinformatics. Protein function, regulation and interactions are associated with the structure of the protein [1, 2].

Now, the researchers are generally believed that Protein spatial structure information can be obtained by protein sequences. Nearly half a century, with the development of the human genome project, biology experimental data accumulated mass of protein structure data which of them most are protein primary structure. Faced with massive data of protein primary structure, how to predict protein tertiary structure from these sequences has become one of the key contents of proteomics research. The protein tertiary structure refers to a specific spatial structure that bonded by the covalent and polypeptide chain coiled and folded on the basis of secondary structure. According to the definition by Levitt and Chothia [3], there are four major structural classes: all-α, all-β, α/β, and α + β. The all-α and all-β classes mean structures that involve mainly α-helices and β-strands, respectively. The α/β and α + β classes include both α-helices and β-strands which are mainly interspersed and segregated, respectively [4].

These authors contributed equally to this work and should be considered co-first authors.

© Springer International Publishing Switzerland 2015
D.-S. Huang et al. (Eds.): ICIC 2015, Part II, LNCS 9226, pp. 408–416, 2015.
DOI: 10.1007/978-3-319-22186-1_40

The prediction accuracy has much to do with protein sequence homology. For protein sequences with high-homology, many methods can achieve highly accuracy, but for protein sequences with low-homology (20 %–40 %), these methods perform relatively poorly [5]. In recent years, in order to improve the accuracy of low-homology protein sequences, many researchers complete the extracted features based on predicted secondary structure sequence [5–10]. Using this method can improve the prediction accuracy effectively.

After experiments, two methods that can be used to improve the accuracy of prediction were summed up: (1) Expressing the protein sequence based on a more accurate feature extraction method; (2) Selecting a more reasonable classification model.

In this paper, a 20-dimensional feature vector is extracted which based on the predicted secondary structure sequence and the corresponding E-H sequence. Hierarchical classification model based on flexible neural tree (FNT) is used to complete the experiment. The experimental results on two low-homology datasets show that this method is can obtain ideal effect.

2   Materials and Methods

2.1   Materials

Two widely used datasets with low homology were used to design and test the proposed method. The 25PDB dataset with 25 % sequence homology that contains 1673 sequences where 443 of them belong to the all-α class, 443 to all-β class, 346 to α/β class and 441 to $\alpha + \beta$ class [11]. The 640 dataset, which contains 640 sequences with 25 % sequence homology and it includes 138 all-α, 154 all-β, 177 α/β and 171 $\alpha + \beta$ sequences [9].

2.2   Feature Extraction

When researchers tried to predict the structure directly from the protein sequence, they found that this is difficult and prediction accuracy is not satisfactory. In this case, the protein predicted secondary structure is particularly important. It is not only become bond that linked to the tertiary structure and protein primary structure, but also a key step in predicting the three dimensional space structure from the primary structure [12]. In practical work, protein secondary structure prediction also has wide range of uses. Such as: can be used for the design of new proteins or protein mutations, help determine the relationship between protein spatial structure and its function, contribute to identify the multidimensional nuclear magnetic resonance (NMR) in the secondary structure and crystal structure analysis, etc. [17].

Features are extracted can through the secondary structure which is predicted from protein sequence. The secondary structure has three elements: H(Helix), E(Strand), and C(Coil). There are two methods can obtain the secondary structure: PSIPRED [13] and PSI-BLAST [14]. In this paper, the first method was selected to predict the secondary structure.

Through the previous experimental results, a phenomenon can be found that the prediction accuracy of α/β class has little relationship with α + β class. Then, remove C from the predicted secondary structure sequence to reflect the distributions of α-helices and β-strands effectively. This sequence is denoted by E-H sequence [10]. The predicted secondary structure sequence length is denoted by N_1, and the corresponding E-H sequence length is denoted by N_2. The 20-dimensional feature vector can be expressed as $P = (p_1, p_2, \cdots, p_{20})^T$.

(1) $P_1 = N(H)/N_1$, $P_2 = N(E)/N_1$ and $P_3 = N(H)/N_2$, where $N(E)$ and $N(H)$ denote the counts of E and H respectively.

(2) $P_4 = MaxSeg_H/N_1$, $P_5 = MaxSeg_E/N_1$, $P_6 = MaxSeg_H/N_2$ and $P_7 = MaxSeg_E/N_1$, where $MaxSeg_H$ and $MaxSeg_E$ denote the length of the longest α-helix and β-strand respectively.

(3) $P_8 = AvgSeg_H/N_1$, $P_9 = AvgSeg_E/N_1$, $P_{10} = AvgSeg_H/N_2$, and $P_{11} = AvgSeg_E/N_2$, where $AvgSeg_H$ and $AvgSeg_E$ denote the average length of α-helix and β-strand respectively.

(4) $P_{12} = CountSeg_H/N_1$, $P_{13} = CountSeg_E/N_1$, $P_{14} = CountSeg_H/N_2$ and $P_{15} = CountSeg_H/N_1$, where $CountSeg_H$ and $CountSeg_E$ denote the count of residues H and E respectively.

(5) $P_{16} = N(EH)/N_2$ and $P_{17} = N(HE)/N_2$, where $N(EH)$ and $N(HE)$ denote the counts of EH and HE in E-H sequence respectively.

(6) $P_{18} = Altn/N_1$, where P_{18} represents the normalized alternating frequency of α-helices and β-strands. This feature was designed because α-helices and β-strands are usually separated in α + β proteins, but are usually interspersed in α/β proteins, α-helices and β-strands alternate more frequently in α/β proteins than in α + β proteins as shown in Fig. 1 [8]. Take the Fig. 2 as an example, the α-helices and β-strands alternate 5 times, so Altn = 5.

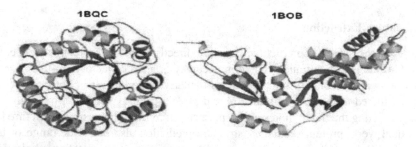

Fig. 1. Structures of proteins from α/β class (1BQC) and α + β class (1BOB)

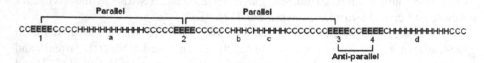

Fig. 2. The protein secondary structural sequence. The α-helices and β-strands are labeled from a to b and 1 to 4, respectively.

(7) P_{19} = pne/(pne + Apne) and P_{20} = Apne/(pne + Apne), where P_{19} and P_{20} represent the proportion of parallel β-sheets and anti-parallel β-sheets, respectively. These features were designed because in α + β proteins, β-strands are usually composed of parallel β-sheets, and in α/β proteins β-strands are usually composed of anti-parallel β-sheets as shown in Fig. 1. For example, if two β-strands (segments of E) are separated by α-helix (segments of H), these two β-strands would form parallel β-sheets. Otherwise, they would form anti-parallel β-sheets [8]. Take the Fig. 2 as an example, β-strand 1 and β-strand 2, as well as β-strand 2 and β-strand 3 constitutes parallel β-sheets. β-strand 3 and β-strand 4 constitutes anti-parallel β-sheets. So, pne = 3 and Apne = 2.

2.3 Flexible Neural Tree

In the early 1980s, the large-scale biological data calculation started using computers [18]. Experimental biologists began to model complex biological problems using computational methods, and began to cooperate with other scientists in other field. Researchers realized the importance of computer technology and the potential value increasingly, especially in the aspect of simulation and analysis of biological data. Biology data is typically a large number, the traditional data processing method can not deal with these data effectively. Machine learning can overcome the above difficulties effectively, and be applied in the field of bioinformatics gradually. Machine learning has become the more commonly used methods in the field of bioinformatics currently. The machine learning methods used in protein tertiary structure prediction include the following methods: artificial neural networks, Bayesian networks, hidden Markov model, genetic algorithm and support vector machines. But in this test, the chosen machine learning method is Flexible neural tree. Because of FNT possesses the advantages of flexible-structure and feature selection.

Flexible neural tree (FNT) is put forward by Chen and it is a special kind of artificial neural network with flexible tree structures [15, 16]. A FNT model is easy to obtain near-optimal structure using tree structure optimization algorithms relatively. Here, a FNT model is employed as the predictor. A tree-structural based encoding method with specific instruction set is selected to represent a FNT model.

A typical FNT model is showed in Fig. 3. Its overall output can be computed from left to right by a depth-first method recursively.

General learning algorithm of FNT

- Step 1. Initialize the values of parameters used in the particle swarm optimization (PSO) algorithms. Set the elitist program as NULL and its fitness value as the biggest positive real number of the computer at hand. Create the initial population.
- Step 2. Construct optimization using PSO algorithm, in which the fitness function is calculated by root mean square error (RMSE).
- Step 3. If the better structure is found, then go to step 4, otherwise go to step 2.
- Step 4. Optimize parameters using PSO algorithm.
- Step 5. If the maximum number of local search is reached, or no better parameter vector is found for a significantly long time (100 steps), then go to step 6; otherwise go to step 4.

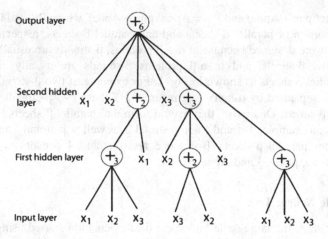

Fig. 3. Typical representation of FNT with function instruction set {+2, +3, +4, +5, +6} and terminal set {X1,X2,X3}, which has four layers.

– Step 6. If the satisfied solution is found, then stop; otherwise go to step 2.

2.4 Hierarchical Classification Model

Hierarchical classification model is a multi-level tree classification model actually, and is also a kind of method to solve the multi-classification problem. This model divides all classes into two categories, and then each sub-class will be divided into two sub-classes successor, Each sub-class then will be divided into two sub-classes successor, after several division, the sub-class contains only a single class [19–23].

This paper built a hierarchical classification model based on FNT as shown in Fig. 4.

Protein structure prediction is a four classification problem, on the first layer, the four classes are divided into two categories (α, β and α/β, α + β) by the first classifier.

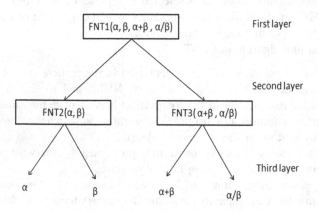

Fig. 4. Hierarchical classification model

Then on the second layer, the two categories are classified by the second classifier and the third classifier respectively.

3   Results and Discussion

In this paper, the experiments are tested on 640 and 25PDB datasets using 10-fold cross validation. The overall accuracy and each classifier accuracy can be seen in the following. The overall accuracy is shown in Table 1, from which we can see that, the overall accuracies for the two datasets are all improved. Just see the 25PDB dataset, the accuracy of α/β class is only 3 % lower than the best performing method RKS-PPSC, but higher than other methods, the accuracy of α + β class is the highest among all methods, the overall accuracy is higher 2.73 % than RKS-PPSC. For 640 dataset, only β class accuracy is lower than RKS-PPSC, other classes accuracy are all higher than other methods, the overall accuracy is higher 3.23 % than the best performing method RKS-PPSC.

As shown in Table 2, we can see that, the accuracy of three classifier which named FNT1, FNT2, FNT3 respectively. FNT1 is the classifier in the first layer, FNT2 is the classifier for α class and β class in the second layer and FNT3 is another classifier for α + β class and α/β class in the second layer.

The FNT2 has the best accuracy. It shows that this method for the classification of α and β is the best. But the accuracy of FNT1 is not very high, if this accuracy can be improved, the overall accuracy will be greatly improved. Although the accuracy of FNT3 is the lowest in these three classifiers, it has improved a lot than other methods. The improvement of FNT3 accuracy is the most important reason to improve the overall accuracy.

Table 1. Performance comparison of different methods on 2 test datasets

Dataset	Method	α(%)	β(%)	α/β(%)	α + β(%)	Overall(%)
25PDB	SCPRED	92.60	80.10	74.00	71.00	79.70
	MODAS	92.30	83.70	81.20	68.30	81.40
	RKS-PPSC	92.80	83.30	85.80	70.10	82.90
	This paper	92.18	86.36	82.75	79.81	85.63
640	SCPRED	90.60	81.80	85.90	66.70	80.80
	RKS-PPSC	89.10	85.10	88.10	71.40	83.10
	This paper	92.86	81.04	91.33	80.88	86.33

Table 2. The accuracy of three classifier on 2 test datasets

Dataset	FNT1	FNT2	FNT3
25PDB	92.06	98.14	87.60
640	92.185	99.00	89.37

Table 3. The accuracies of four classes in FNT1

Dataset	α(%)	β(%)	α/β(%)	α + β(%)
25PDB	93.30	88.07	89.67	97.10
640	92.86	82.60	98.24	92.65

The accuracy of four classes in FNT1 is shown in Table 3. We can see that the accuracy of β class is the lowest in FNT1, this led to the overall accuracy of β class is not ideal.

From the Tables 1 and 3, can we come to the conclusion that the improvement of the overall accuracy is mainly attributed to the increase of the accuracy of α + β class. In FNT1, the accuracy of α + β is relatively high, so, in FNT3, only distinguish between α/β and α + β can get the higher accuracy of α + β than other methods which without hierarchical classification model. But the accuracy of β class in FNT1 is relatively low and this is the key point for future experiments to improve the overall accuracy.

The result shows that this method improved the overall accuracy significantly. In the next step, experiment should focus on improving the accuracy of β class in FNT1. According to this idea, different feature vectors can be constructed for every classifier and a more reasonable feature extraction which can better describe the different of α class, β class and α/β class, α + β class.

4   Conclusions

In this paper, a 20-dimension feature vector and hierarchical classification model based on flexible neural tree are used to predict structural classes for low-homology protein sequences. The test results are the highest accuracy compared with other methods. The main reasons that achieve such results are FNT was used, feature extraction is reasonable and hierarchical classification model was designed in this test. FNT possesses the advantages of flexible structure and feature selection. Hierarchical classification model can further improve the accuracy.

Next, in order to improve the accuracy, we can start from the following two aspects: (1) Find a more reasonable feature extraction method for protein sequences. (2) Build a more reasonable classification model.

Acknowledgements. Fanliang Kong and Dong Wang contributed equally to this work and should be considered co-first authors. This research was partially supported by Program for Scientific research innovation team in Colleges and universities of Shandong Province 2012–2015, the Key Project of Natural Science Foundation of Shandong Province (ZR2011FZ001), the Natural Science Foundation of Shandong Province (ZR2011FL022, ZR2013FL002), the Key Subject Research Foundation of Shandong Province and the Shandong Provincial Key Laboratory of Network Based Intelligent Computing. This work was also supported by the National Natural Science Foundation of China (Grant No. 61302128, 61201428, 61203105). The scientific research foundation of University of Jinan (xky1109).

References

1. Chou, K.C.: Structural bioinformatics and its impact to biomedical science. Curr. Med. Chem. **11**, 2105–2134 (2004)
2. Chou, K.C., Wei, D.Q., Du, Q.S., Sirois, S., Zhong, W.Z.: Progress in computational approach to drug development against SARS. Curr. Med. Chem. **13**, 3263–3270 (2006)
3. Levitt, M., Chothia, C.: Structural patterns in globular proteins. Nature **261**, 552–558 (1976)
4. Murzin, A.G., Brenner, S.E., Hubbard, T., Chothia, C.: SCOP: A structural classification of protein database for the investigation of sequence and structures. J. Mol. Biol. **247**, 536–540 (1995)
5. Yang, J., Peng, Z., Chen, X.: Prediction of protein structural classes for low-homology sequences based on predicted secondary structure. BMC Bioinform. **11**, S9 (2010)
6. Kurgan, L., Cios, K., Chen, K.: SCPRED: accurate prediction of protein structural class for sequences of twilight-zone similarity with predicting sequences. BMC Bioinform. **9**, 226 (2008)
7. Mizianty, M.J., Kurgan, L.: Modular prediction of protein structural classes from sequences of twilight-zone identity with predicting sequences. BMC Bioinform. **10**, 414 (2009)
8. Liu, T., Jia, C.: A high-accuracy protein structural class prediction algorithm using predicted secondary structural information. J. Theor. Biol. **267**, 272–275 (2010)
9. Yang, J.Y., Peng, Z.L., Yu, Z.G., Zhang, R.J., Anh, V., Wang, D.S.: Prediction of protein structural classes by recurrence quantification analysis based on chaos game representation. J. Theor. Biol. **257**, 618–626 (2009)
10. Ding, S., Zhang, S., Li, Y., Wang, T.: A novel protein structural classes prediction method based on predicted secondary structure. Biochimie **94**, 1166–1171 (2012)
11. Kurgan, L.A., Homaeian, L.: Prediction of structural classes for protein sequences and domains-Impact of prediction algorithms, sequence representation and homology, and test procedures on accuracy. Pattern Recogn. **39**, 2323–2343 (2006)
12. Wang, Z.: The current situation and prospect of protein structure prediction. Chem. Life **18** (6), 19–22 (1998)
13. Jones, D.T.: Protein secondary structure prediction based on position specific scoring matrices. J. Mol. Biol. **292**, 195–202 (1999)
14. Altschul, S.F., Madden, T.L., Schäffer, A.A., Zhang, J., Zhang, Z., Miller, W., Lipman, D.J.: Gapped BLAST and PSI-BLAST: a new generation of protein database search programs. Nucleic Acids Res. **25**, 3389–3402 (1997)
15. Chen, Y., Yang, B., Dong, J., Abraham, A.: Time-series forecasting using flexible neural tree model. Inf. Sci. **174**, 219–235 (2005)
16. Yang, B., Chen, Y., Jiang, M.: Reverse engineering of gene regulatory networks using flexible neural tree models. Neurocomputing **99**, 458–466 (2013)
17. Zhao, G., et al.: Bioinformatics, pp. 158–160. Science Press, Beijing (2002)
18. Tan, A.C., Gilbert, D.: Machine learning learning and its application to bioinformatics: an overview [M]. 31 Aug. 2001
19. He, Y., Chen, Z., Yang, Y.: Research on a hierarchical text categorization method based on centroid. J. Inf. Technol. **12**, 116–118 (2007)
20. Hsu, C.W., Lin, C.J.: A comparison of methods for multi-class support vector machines. J. IEEE Trans. Neural Netw. **13**(2), 415–425 (2002)
21. Ma, X., Huang, X., Chai, Y.: 2PTMC classification algorithm based on support vector machines and its application to fault diagnosis. J. Control Decis. **18**(3), 272–276 (2003)

22. An, J., Wang, Z., Ma, Z.: A new SVM multiclass classification method. J. Inf. Control **33**(3), 262–267 (2004)
23. Lu, Y., Lu, J., Yang, J.: A survey of hierarchical classification methods. J. Patt. Recog. Artif. Intell. **26**(12), 1131–1139 (2013)

Interplay Between Methylglyoxal
and Polyamine in *Candida Albicans*

Ji-Hyun Lee[1], Hyun-Young Cho[1], Yi-Jin Ahn[1],
and Taeseon Yoon[2(✉)]

[1] Department of International Studies, Hankuk Academy of Foreign Studies,
Yongin, Republic of Korea
{allyl999,mnl634,yijini98}@naver.com
[2] Hankuk Academy of Foreign Studies, Yongin, Republic of Korea
tsyoon@hafs.hs.kr

Abstract. It has been documented that Methylglyoxal (MG) concentrations in cells from diabetic patients are higher than cells from non-diabetics. Despite ongoing research efforts, a clear understanding of the mechanisms involving MG as well as methods to reduce MG concentrations in diabetic patients have yet to be obtained. Consistent growth in the patient demographic affected by adult disease has led to increased interest in the causes of the disease. Implications of MG with widespread adult diseases such as diabetes, obesity, etc. have necessitated the investigation of the role MG in these diseases, methods to regulate MG concentration levels, possible and cures that involve MG. In this paper, we propose a cure through manipulating MG generation rates by either using the SPE1 gene or handling the Polyamine externally.

Keywords: Candida albicans · Methylglyoxal · Polyamine · FBA1 · SPE1

1 Introduction

Methylglyoxal, a highly reactive dicarbonyl produced by cells during glucose and fructose metabolism, is a major ingredient in the formation of advanced glycation end products (AGEs) and free radicals that are known to be virulent and cause premature aging in cells. Ubiquitous in cells are Polyamines which are organic compounds having two or more primary amino groups and they include putrescine, spermidine, and spermine which are known to be important factors in the growth of a cell. When the gene encoding putrescine generating ornithine decarboxylase, SPE1 was isolated from Candida albicans, the development rate of the altered specimens were reduced compared to the wild type while overexpression resulted in the opposite. Such phenomena is strong evidence that putrescine plays a vital role in the life cycle of the Candida albicans. Furthermore, the FBA1 gene, which was recently implicated in the generation of methylglyoxal, was over-expressed. It was observed that FBA1 differentiated in rich medium YPD and didn't differentiate in the spider differentiation medium. Considering its association with the development of methylglyoxal and putrescine within in the cell, we compared development curve of its mutations. Compared to wild specimens, SPE1-free mutants showed significantly slower growth while FBA1 overexpressed

© Springer International Publishing Switzerland 2015
D.-S. Huang et al. (Eds.): ICIC 2015, Part II, LNCS 9226, pp. 417–425, 2015.
DOI: 10.1007/978-3-319-22186-1_41

mutants exhibited the slowest growth but demonstrated a full recovery with the additional 1 mM of putrescine. Comparing mutant cell methylglyoxal concentrations via fast liquid chromatography, FBA1 overexpressed mutants exhibited higher concentrations of methylglyoxal compared to the wild specimens while concentrations plummeted to almost half with the additional 1 mM of putrescine. The described results provide sound evidence that methylglyoxal plays a significant role in the life cycle of the Candida albicans and ultimately show that the possibility that methylglyoxal can be manipulated via putrescine.

2   Materials and Methods

2.1   pGEM-T-EASY Vector (Promega, Commercial) Ligation

T-vector ligation test method: 1 μL ligase enzyme, 2 μL 10X ligase buffer, 8 μL DNA fragment (gene cleaned), 0.5 μL T-vector, was mixed thoroughly with 8.5 μL of triple distilled water in a test tube and stored at 4 °C for at least 16 h.

2.2   E.Coli Transformation Method

Ligated specimens were inserted to E.coli and stabilized in (E.coli DH5α) ice. Heat shock at 42 °C for 80–100 s was applied to soften the cell wall and then ice incubated for 10 min. Using a clean bench, 400 μl of E.coli was applied to the LB medium. One end of an inoculation loop was dipped in alcohol and then sterilized by heating over an alcohol lamp. The inoculation loop was heated in solid medium, cooled, then E.coli was applied. The E.coli was transferred to the LB plate by scraping the tip of the inoculation loop with the E.coli against the plate. Shaking incubation was performed for 50 min at 37 °C (optimal temperature for E.coli).

2.3   Colony Selection

20 μL of ampicillin and X-gal was smeared onto an LB plate. The transformed E.coli was smeared and stored in the incubator at 37 °C for 16 h. Only white colonies were selected.

2.4   Lithium Acetate Transformation Definition

Approximately 50–100 of CAF4-2 and linear DNA of transformation is applied for every μg. The transformation has effects similar to plasmid containing CARS1. It is typically used to transform C.tropcalis, however, in this work, it was used to transform the DNA of the Candida albicans. It is a widely used and relatively trivial transformation method where alkaline cautions affect the cell walls of the yeast to allow the introduction of DNA.

< Reagent Used >

10x LiAC(Lithium Acetate)=1M sterilized LiAc solution and set to 7.5 pH to get the desired reaction to watered acetic acid. 1. PEG solution = 100 μg 10x TE, 100 μL 10x LiAc, 800 μL, solution is made with triple distilled water to maintain 1:1:8 ratio. 10x Lithium Acetate= to serve as the buffer during DNA or RNA electrophoresis or to allow for the penetration of cell walls during DNA transformation. LiAc is used to obstruct hydrogen bonding or to denature DNA, RNA, and protein.

2.5   Lithium Acetate Transformation

Colony was inoculated in YPD and TE(10x TE) at 28 °C is inserted, then centrifuged. 100 μg 10x TE, 100 μL 10x LiAc, and 800 μL TDW were used maintaining 1:1:8 ration to make LiAc/TE solution. 10x TE 30 μL, 10x LiAc 30 μL, PEG 240 μL that promotes transformation is inserted to result in a viscous PEG solution. 6 μLof carrier DNA was inserted into the solution to assist with DNA penetration into cell wall. The cleaned cells were congregated using a centrifuge then mixed with supernatant-free LiAc/TE solution to make 100 μL. 50 μL of the 100 μL were inserted into a solution containing PEG solution and carrier DNA. Gene cleaned DNA fragments to transform were inserted together and gently tapped, then incubated in a shaking incubator for an hour at 28 °C. Supernatants were eliminated using a centrifuge and 1xTE buffer solution is mixed, then smeared to a selection plate.

2.6   UV-Visible Spectrophotometer

Device used to measure RNA, DNA, protein, cell levels. 1. A260/A280 ratio is necessary to evaluate purity since the absorption peak of several important photometers including nucleic acid concentration is 280 nm. Ideal sample ratio is between 1.8 ~ 2.0 and anything less or more means protein or phenol is present. Samples equal to A230 means the presence pollutants including phenol, carbohydrate, peptide, etc. Cells were measured using optical density using 600 nm visible light. Method to measure the specimen using UV-visible spectrophotometer; 1 mL of the specimen were inserted into a quartz cuvette to be measured using 1 mL of water as the control group.

2.7   RNA Extraction

RNA extraction uses electrophoresis to extract RNA of desired size for RNA analysis. RNA size and manifestation level information obtained through the extraction can be used to determine mRNA manifestation levels. During RNA extraction, RNA must be protected from RNase. Since RNase is present in the hand, poly gloves must be worn when handling the RNA. Also, it must be protected from endogenous RNase produced during cell destruction. DNA and RNA are selectively categorized after proteins are eliminated using Phenol and Chloroform.

Colonies were inoculated on the medium and stored for at least 16 h, then congregated using a centrifuge. Cells were thoroughly mixed with 1.2 mL commercial phenol (Takara) and inserted into a micro tube. A bead is inserted and a bead beater is used to penetrate the cell wall of the candida. Supernatants were moved to a new micro tube and vortexed with chloroform to remove proteins. Supernatants were moved to a new micro tube and isopropyl alcohol was inserted and RNA were cooled at −20 °C for 1 h. Then, a centrifuge was used to wash in 70 % ethanol after the RNA had been settled and supernatants discarded. It was then dried at room temperature and RNA was melted using formamide.

2.8   Cell Methylglyoxal Level Measurement

1,2-diaminobenzene(=o-Phenylenediamine) transforms methylglyoxal into quinoxaline derivate thereby making it a stable compound. It is also an important precursor to the heterocyclic compound. HClO4 helps penetration of the cell wall as well as prevent the reaction of 1, 2-diaminobenzene.

Cells are cultivated and congregated into a tube for measurement. 0.5 M Perchloric acid five times the mass of the cell is inserted and placed on ice for 20 min. A centrifuge was used three times to congregate the supernatants in their entirety. Cell methylglyoxal in the solution were made into derivatives by reacting with 1 mM of 1, 2-diamino benzene for 3–4 h. A syringe was used to react C-18 solid phase extraction (SPE column) in dark conditions. SPE columns were activated, cleaned, and filled with specimen. Quinoxaline derivate stuck on the SPE column are washed off with 70 % ethanol. Specimens eluted with methanol are evaporated and concentrated using a rotary vacuum evaporator. Concentrated specimen was dissolved by the solution of mobile phase. The quantity of dissolved specimen was measured with HPLC (High Performance Liquid Chromatography).

3   Materials and Methods

3.1   *SPE1* Gene Construct Schematic Production Process and Electrophoresis Results

The primer to the intron portion of the SPE1 gene was made and inserted before and after the vector containing hph-ura3-hph to produce a SPE1 gene-deficient mutant by making the primer portion of the intron gene SPE1 to make knockout mutants SPE1 hph-ura3-hph was inserted before and after a vector that contains the cassette (Fig. 1).

Primer sequences of the primers are:

5' → 3' (*SacI*):
GAGCTCGTTGTTTAAACAATGAACAATTA
3' → 5' (*KpnI*):
GGTACCCTTATATTCTCTTTCTCTGATCTT
5' → 3' (*SalI*):
GTCGACGTGTTTACTTATTTATGTATTCAA
3' → 5' (*HindIII*):
AAGCTTTATTAACTATTTCGTCAACCAACC

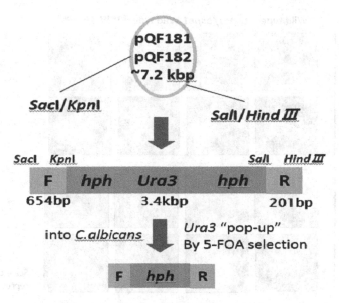

Fig. 1. SPE1 gene construct schematic production process

The results were obtained with plasmid that was created using the electrophoresis. Given that the size of 3.4 kb for the hph-ura3-hph cassette is correctly registered 654 bp-s01 bp restricted enzymes processed with SacI and HindIII resulted in a total of 4.255 kb (Fig. 2).

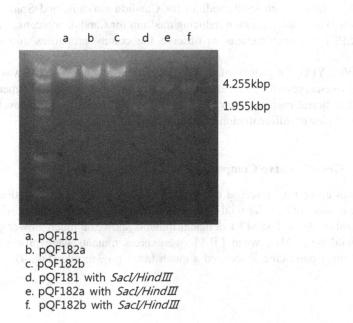

a. pQF181
b. pQF182a
c. pQF182b
d. pQF181 with *SacI/HindIII*
e. pQF182a with *SacI/HindIII*
f. pQF182b with *SacI/HindIII*

Fig. 2. Electrophoresis **results**

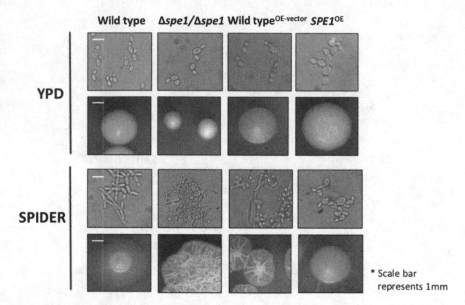

* Scale bar represents 1mm

Fig. 3. Observation results regarding colony morphology and cell phenotype of the *SPE1* mutants

3.2 Observations Regarding Colony Morphology and Cell Phenotype of the *SPE1* Mutants

YPD, a nutrient rich solid medium for Candida albicans, and Spider solid medium, known to be a differentiation inducing medium for Candida albicans, was smeared onto the SPE1 deficient mutants to observe the colony properties and cell phenotypes (Fig. 3).

With YPD, the growth of the SPE1 deficient mutants in YPD was inhibited while SPE1 overexpressed exhibited faster growth compared to wild specimens. With Spider, SPE1 deficient mutants showed faster differentiation while SPE1 overexpressed cells showed slower differentiation compared to wild specimens.

3.3 Growth Curve Comparison

Growth curve was observed every two hours by measuring the optical density of the same number of cells for wild-type and mutants that were inoculated with SD minimum nutrient medium. The SPE1 deficient mutants showed a much slower growth rate than the wild type. Also, when FBA1 overexpress mutants were processed with 1 mM polyamine putrescine, it showed a much faster growth rate (Fig. 4).

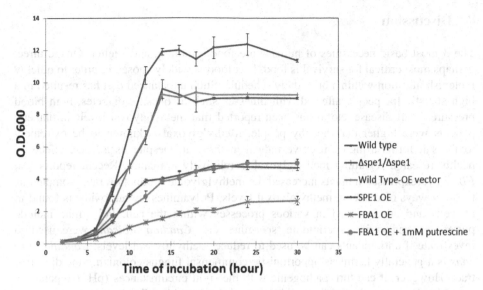

Fig. 4. Growth curve comparison results

3.4   Methylglyoxal Level Measurement Results

Measuring the amount of methylglyoxal, SPE1 deficient mutants showed approximately 31.43 more methylglyoxal compared to the wild type. FBA1 overexpressed mutants processed with 1 mM putrescine showed 50 % less methylglyoxal (Fig. 5).

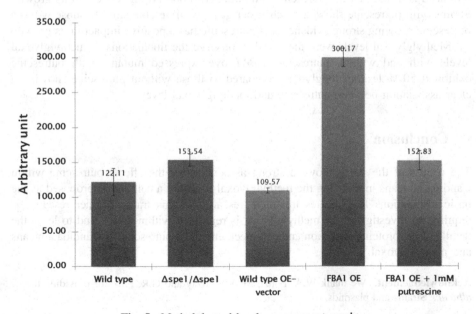

Fig. 5. Methylglyoxal level measurement graph

4 Discussion

The 3 most basic necessities of humans are food, clothing, and shelter. Of the three, perhaps most critical for survival is food. Fast food is widely chosen in order to quickly replenish nutrition within a busy daily schedule. Such unbalanced diet has resulted in a high statistic for people affected with illnesses such as obesity, diabetes, high blood pressure, heart disease, etc. It has been reported that methylglyoxal levels in diabetic patients were higher than healthy people. Methylglyoxal is known to be extremely toxic as it forms a highly reactive radical in the cell. Despite its adverse effects to health, research regarding methylglyoxal is relatively immature. Recent reports that *FBA1* overexpressed mutants increased the methylglyoxal production rate prompted us to study ways of reducing methylglyoxal levels. Polyamines are antioxidants found in all cells and are involved in various processes within the cell. Polyamine include putrescine, cadaverine, spermidine, spermine, etc. *Candida albicans* were used to investigate if antioxidants can be used to reduce methylglyoxal levels. *Candida albicans* is a generally harmless opportunistic commensal fungus colonizing the digestive tract. However, it can turn pathogenic with the right circumstances (pH, temperature, etc.).The *SPE1* gene in *Candida albicans* produces ornithine dicarboxylase, an enzyme that enables ornithine to putrescine conversion. Observing the colony properties and growth curve, *SPE1* gene deficient mutants showed slower growth rates compared to wild types. This implies *SPE1* is critical in the growth of *Candida albicans*. *SPE1* overexpressed mutants were observed to have a slightly larger colony morphology and slow differentiation rate compared to wild types. This implies that even when dicarboxylate is overexpressed, it remains pathogenic.

FBA1 overexpressed mutants, reported to increase methylglyoxal levels, were produced and placed with putrescine antioxidants to observe the effects in its growth. Strains with putrescine showed much faster growth curve compared to those without putrescine showing strong evidence that putrescine has a positive impact in its growth.

Methylglyoxal levels were measured to observe the fluctuations in methylglyoxal levels with and without putrescine. *FBA1* overexpressed mutants with putrescine exhibited 50 % less methylglyoxal compared to those without putrescine showing a clear association between putrescine and methylglyoxal levels.

5 Conclusion

The results of the study shows a strong association of the effects putrescine within Candida albicans in reducing the methylglyoxal levels in a cell and is proposed as the basic foundations for diabetes treatment research in this track. Further research is required to investigate how methylglyoxal is regulated within a cell, and to learn the details of the protein interaction and cell mechanisms of putrescine in Candida albicans and methylglyoxal.

Acknowledgement. We thank W.A. Fonzi, M. Whiteway and G.R. Fink for providing the *C. albicans* Strains and plasmids.

References

1. Asanuma, N., Yoshii, T., Kikuchi, M., Hino, T.: Effects of the overexpression of Fructose-1, 6-bhisphosphate aldolase on fermentation pattern and transcription of the genes encoding lactate dehydrogenase and pyruvate formate-lyase in a ruminal bacterium, Streptococcus bovis. J. Gen. Appl. Microbiol. **50**, 71–78 (2004)
2. Berman, J., Sudbery, P.E.: *Candida albicans*: a molecular revolution built on lessons from budding yeast. Nat. Rev. **3**, 918–930 (2002)
3. Carlisle, P.L., Kadosh, D.: A genome-wide transcriptional analysis of morphology determination in *Candida albicans*. Mol. Biol. Cell **24**, 246–260 (2013)
4. Chan, W.H., Wu, H.J., Shiao, N.H.: Apoptotic signaling in methylglyoxal-treated human osteoblasts involves oxidative stress, c-Jun N-terminal kinase, caspase-3, and p21-activated kinase 2. J. Cell. Biochem. **100**, 1056–1069 (2007)
5. Hanahan, D.: Studies on transformation of Escherichia coli with plasmids. J. Mol. Biol. **166**, 557–580 (1983)
6. Kalapos, M.P.: Methylglyoxal in living organisms: chemistry, biochemistry, toxicology and biological implications. Toxicol. Lett. **110**, 145–175 (1999)
7. Kwak, M.K., Ku, M., Kang, S.O.: NAD+-linked alcohol dehydrogenase regulates methylglyoxal concentration in *Candida albicans*. FEBS Lett. **588**, 1144–1153 (2014)
8. Matafome, P., Sena, C., Seica, R.: Methylglyoxal, obesity, and diabetes. Endocrine **43**, 472–484 (2013)
9. Ramasamy, R., Yan, S.F., Schmidt, A.M.: Methylglyoxal comes of AGE. Cell **124**, 258–260 (2006)
10. Rodaki, A., Young, T., Brown, A.J.: Effects of depleting the essential central metabolic enzyme fructose-1,6-bisphosphate aldolase on the growth and viability of *Candida albicans*: implications for antifungal drug target discovery. Eukaryotic cell. **5**, 1371–1377 (2006)
11. Sambrook, J., Fritsch, E.F., Maniatis, T.: Molecular Cloning: A Laboratory Manual, 2nd edn. Cold Spring Harbor, New York (1989)

Analysis of Human Herpes Viruses with the Application of Data Mining

Yusin Kim[1], Sung Min Kim[1(✉)], Jiwoo Lee[1], Ann Jeong[2], Jaeuiy Lim[2], and Taeseon Yoon[1]

[1] Natural Science Department, Hankuk Academy of Foreign Studies, Yongin, Korea
{ushin612,danielkim98}@naver.com, tsyoon@hafs.kr
[2] International Department, Hankuk Academy of Foreign Studies, Yongin, Korea

Abstract. The human herpes Virus, categorized into 8 types from HHV-1 to HHV-8 is in a DNA virus family, Herpesviridae. They are highly infectious and responsible for many human related diseases. Through this research, we applied data mining, which is the process of finding and analyzing the relationship among the raw data, for the first time to analyze the human herpes virus family. We discovered the patterns regarding the sequence data and analyzed them in associated with the clinical aspects and phylogenetic study results. In addition, we applied decision tree algorithm to support our data. As a result, we were able to stabilize the proof for the bioinformatical categorization. The rules we found were consists of 11 amino acids, Alanine, Leucine, Proline, Valine, Asparagine, Threonine, Phenylalanine, Arginine, Isoleucine, Serine and Lysine. Alanine and Leucine were the two major amino acids from the rules.

Keywords: Human herpes virus · Data mining · Apriori algorithm · Decision tree algorithm

1 Introduction

1.1 The Human Herpes Viruses

The Herpes Virus (also known as Herpesviridae) is a family of DNA viruses and one of the most infectious pathogens in existence. Distinguished by 8 categories, the Human Herpes Virus (HHV) can be analyzed according to its protein amino acid sequence. We believed that in order to acquire the appropriate means of eradicating the symptoms, it is necessary to comprehend the similarities and differences of the protein sequence and relate these comparisons to the characteristics of each virus based on the type. Thus, we analyzed the sequences to find patterns for categorizing the Herpes viruses with the application of data mining, using Apriori algorithm and decision tree algorithm to extract rules.

The Human Herpes Viruses represent a group of DNA viruses that share common biological features that account for oral pathology. The eight categories are named: Herpes Simplex Virus (HSV, HHV-1, 2), Varicella-zoster Virus (VZV, HHV-3),

© Springer International Publishing Switzerland 2015
D.-S. Huang et al. (Eds.): ICIC 2015, Part II, LNCS 9226, pp. 426–435, 2015.
DOI: 10.1007/978-3-319-22186-1_42

Epstein-Barr Virus (EBV, HHV-4), Cytomegalovirus (CMV, HHV-5), and the remaining HHV-6, HHV-7, and HHV-8.

The most common oral HHV infections, HHV-1, 2, as known as Herpes Simplex Virus (alpha herpes viruses), are the causes of Primary Herpetic Gingivostomatitis. This diagnosis is what brings about skin, lip, oral mucosal ulcerations, fever, and lymphadenopathy. It is related to Gingivostomatitis in that it is related to gingival erythma, edema, and erosive lesions. Usually sexually transmitted, HSV causes ulcerations and blistering of the external genitalia and cervix. [4] The practical thera-peutic control method for this infection is natural antibodies produced from the patient's body. The recurring episodes of HSV get milder as the presence of antibodies to previous viruses are made. The control process extends over a few weeks; during the first week the illness exacerbates, starting from the second week the condition begins to improve, and on the fourth week, symptoms disappear completely. Another symptom that occurs as a consequent of latent ganglionic HHV-1 infection is the Recurrent Herpes Labialis. This typically affects the lips, bringing cold sores to the infected areas. In this condition, stress, fever, and trauma commonly takes place.

The Varicella-zoster virus, HHV-3 (alpha herpes virus), leads to chicken pox. It is related to oral manifestations of vesicles and ulcerations. The infection usually appears at childhood near the respiratory tract. During the process of secondary multiplication, the vesicles are filled with infectious viruses which may lead to Central Nervous System infection.

The Epstein-Barr virus, HHV-4 (gamma herpes virus), is associated with infectious mononucleosis, lymphomas, nasopharyngeal carcinoma, and oral manifestations. However, in most cases, patients do not realize that they are infected. This virus imposes symptoms that result usually in slight fatigueness or a minor flu. HHV-4 is highly contaminous, spreading through the contact of saliva. HHV-4 has an affinity for the cells of the mouth and throat. Once it enters the B lymphocyte, it controls the DNA. When these cells slough off, the virus is gets dispersed inside the saliva, thus the cause for its communicable aspects.

The Cytomegalovirus, HHV-5 (beta herpes virus) is widespread but usually imposes no harm on adults. Because there are rarely any symptoms or signs of the infection, most people who are infected do not know that they are infected. However, the one detrimental effect that the Cytomegalovirus may have upon humans is during the state of pregnancy. Its infection during pregnancy may cause birth defects such as hearing loss, blindness, epilepsy, mental retardation, or in extreme cases, death.

Human Herpesvirus 6, HHV-6 (beta herpes virus), is a group of two viruses: HHV-6 alpha and HHV-6 beta. These are widespread and infect humans in infantry. Acquiring these viruses may lead to various symptoms such as fever, diarrheas, rashes, and seizures. Like the former herpes viruses depicted above, the HHV-6 is potentially long-lasting and may reactivate after some time.

Human Herpesvirus 7 (beta herpes virus), is similar to the HHV-6. Both viruses affect humans in their childhood; HHV-7 tending to affect children generally 2 years older than the patients of HHV-6. Both viruses show symptoms of fever and rash [3]. It is unknown whether the HHV-7 infection is primarily caused by the HHV-7 virus, or if it is affected by the reactivating HHV-6 virus. Thus, the main inquiry is whether HHV-7 is wholly independent with the HHV-6 virus.

The Human Herpes Virus 8, HHV-8 (gamma herpes virus), also known as the Kaposi's sarcoma-associated herpesvirus, causes Kaposi's sarcoma, a cancer of the blood vessel, and Lymphoma, a cancer of the lymphocyte. It also causes Castleman's disease, which is an infection that enlarges the lymph node. HHV-8 can be transmitted through sexual interaction or oral contact. It can also be transmitted through organ transplantation. Similar to the other Herpes viruses, this virus is widely infected by many people; however, most do not have symptoms.

1.2   The Control Group

We selected control groups for the experiment. Herpes viruses associated with animals from the main three subfamilies of Herpes viruses; alpha, beta, gamma, were selected in consideration of symptoms and significances. These were the Alcelaphine herpes-virus, Bovine herpesvirus 2, and Elephant Endotheliotropic Herpes Virus. These viruses were used to compare the characteristics and to check the uniqueness of the extracted rules from human herpes virus.

Alcelaphine herpesvirus is responsible for malignant Catarrhal fever. Bovine her-pesvirus 2 causes bovine mammillitis and pseudo-lumpy skin disease. BoHV-2 is similar in structure to the Human Herpes simplex virus.

Symptoms of pseudo-lumpy skin disease include fever and skin nodules on the face, back, and perineum. Bovine mammillitis is characterized by lesions restricted to the teats and udder. This virus may spread through an arthropod vector, but can also be spread through milkers and milking machines.

The Elephant Endotheliotropic Herpes Virus (EEHV) has affected many elephants, usually infecting young elephants. Early symptoms include bloodshot eyes, head and neck swelling, cyanosis of the tongue and ulcers in the mouth. Later on, EEHV spreads to the heart, liver, and tongue, infecting the microvascular endothelial cells and bringing cardiac failure that may lead to death. The transmission method of EEHV is unknown, and more research is necessary.

Alcelaphine Herpesvirus 1 is a fatal lymphoproliferative disease caused by a group of ruminant gamma herpes viruses. The most common areas of infection are the head and the eyes. Symptoms appear in forms of fever, depression, discharge from the eyes and nose, lesions of the buccal cavity and muzzle, swelling of the lymph nodes, and opacity of the corneas leading to blindness, inappetance and diarrhea. Death usually occurs within ten days. The mortality rate in symptomatic animals is 90 to 100 percent.

1.3   Subfamily of Herpes Virus

There are three known subfamily, alpha, beta, gamma in herpes virus family.

Alphaherpesvirinae (HHV-1, 2, 3) and Bovine herpesvirus 2 (BoHV-2) have many similarities. Both groups of viruses show symptoms of blisters and occur on the face and perineum. However, there is a difference that we can see between HHV and BoHV-2- whether the disease accompanies fever or not. HHV-1, 2, 3 do not cause fever while BoHV does.

Betaherpesvirinae (HHV-5, 6, 7) and Elephantid herpesvirus 1 (EEHV) are similar in that they both target infants or children. (The main target of HHV-5 is the embryo, the main targets of HHV-6 and 7 are infants and children, and the main target of EEHV is a young elephant.) On the other hand, there is a difference that HHV-5, 6, and 7 have similar symptoms of red spots like roseola or "owl eye" while EEHV does not.

Gammaherpesvirinae, (HHV-4, 8) and Alcelaphine herpesvirus 1 (AIHV-1) are common; both groups are associated with symptoms of the lymphoma. Otherwise, the two groups are fairly distinguishable. HHV-4 involves lymphoma and causes diseases such as gastric cancer or colorectal cancer. HHV-8 causes skin cancer which is called kaposi's sarcoma. Lastly, AIHV-1 causes Bovine malignant catarrhal fever which includes fever, blindness, and diarrhea. The features and similarities.

2 Materials and Methods

2.1 Sequence Data

We obtained the protein sequence data from the NCBI (National Center for Biotechnological Information). We chose the Major Capsid Protein as common in all types of the Herpes Virus, which consist about 1000 amino acids.

The structure of the major capsid protein plays an important role in virus categorization. It forms the capsomere structures of the herpes virus, the pentons and hexons, which contain five and six VP5 monomers, respectively. Since major capsid protein is common in all types of the herpes virus, and is known as they take part in crucial functions of virus assembly, we made a conclusion that it can represent the properties of each herpes virus partly. As the data contained the structure information, we could expect that the rules we extracted using apriori algorithm from the sequence can explain the characteristics and similarities of the herpes virus.

2.2 Apriori Algorithm

The Apriori algorithm is a data mining algorithm for finding associated rules. It is well known for simple methods and effective performance. It is related to a set theory and uses breadth-first search and a hash tree structure to count the frequency of the item sets. The basic principle of this algorithm is that the subsets of highly frequent item sets are also highly frequent.

We divided each sequence set into 7, 9 and 11 windows. Using the apriori algorithm, we could extract rules from each MCP sequence data. The rules we found consist of a number of amino acids found in the digit of each divided window. For example, "amino 5L 17" means 17 Leucine were found on the fifth section. We obtained the protein sequence data from the NCBI(National Center for Biotechnological Information). We chose the Major Capsid Protein as common in all types of the Herpes Virus, which consist about 1000 amino acids.

2.3   Decision Tree Algorithm

The Decision Tree Algorithm is one of the most common and effective algorithms for data mining analysis. This algorithm is usually used for data classification. This algorithm is a tree shaped graph (model) of decisions and shows the possible consequences. The application of the algorithm in bioinformatics is mainly for classifying the sequences. The Decision Tree Analysis can extract rules, classify each experimental class and print out the accuracy of the classification. If the classification error rate is high, it means that the classification is difficult, and that the experimental classes are similar. In this research, using the See 5.0 program, we applied the decision tree to support our apriori algorithm analysis and to analyze the provided classification table.

3   Result

3.1   Apriori Algorithm Analysis

Apriori algorithm analysis can extracts rules regarding the index of the divided window. The divided window number is related with the amino acid binding and protein formation. One rule is consists of three information, index, amino acid and the size. For example, let's slice a randomly created sequence ...LAAVNGLLEARFPNKII... in 7 windows, ...LAAVNGL/LEARFPN/KIIP..., or in 9 windows, ... LAAVNGLLE/ARFPNKIIP. There are two Leucine at the first place in the 7window sliced sequence and two Asparagine at the forth place in the 9window sliced sequence. The rules extracted from apriori algorithm are same as the example, amino acids repeated constantly.

The results of the diagrams that display the rules were drawn. The x axis is the amino acids abbreviation (Table 1) and the y axis is the number of rules found. To make a clear description of our results, we combined the rules made of same amino acids.

The similarities of rules are quite clear between HHV1 and 2, the simplex viruses. But HHV3 and bovine2 are not that similar consists of other rules (Figs. 1, 2 and 3).

We analyzed the distribution of the rules using diagram and detailed rules, The rules were abbreviated to index-amino acid. For example, 1-A indicates the alanine repeated at the first place of the sliced window. The list of the extracted rules are omitted in this paper because the amount of data. Approximately 500 rules were extracted.

To begin with, Alanine, Leucine, and Valine each have their own rules that have effects on symptoms, stages of infection, and the viruses themselves. 1-A was found in the 7th window experiment, 1-A, 2-A, 4-A, and 6-A from the 9th window experiment, and 4-A, 5-A, 6-A, and 11-A from the 11th window experiment from HHAv-1, 2 and Bovine2. These patterns of alanine seem to appear widely for Herpes Simplex Virus-HHV-1, 2 and Bovine2; however, this is not true for viruses HHV-6, 7 and Elephantid1, which are the betaherpesvirinae. Leucine usually appears at location 10 and 11 in the 11th window. 5-V, 6-V, 7-V, 8-V and 9-V show obvious stakes, with high frequency in several experiments.

Table 1. Amino acid abbreviation

Amino acid	One-letter abbreviation	Three-letter abbreviation	Amino acid	One-letter abbreviation	Three-letter abbreviation
Alanine	A	Ala	Methionine	M	Met
Cysteine	C	Cys	Asparagine	N	Asn
Aspartic acid	D	Asp	Proline	P	Pro
Glutamic acid	E	Glu	Glutamine	Q	Gln
Phenylalanine	F	Phe	Arginine	R	Arg
Glycine	G	Gly	Serine	S	Ser
Histidine	H	His	Threonine	T	Thr
Isoleucine	I	Ile	Selenocysteine	U	Sec
Lysine	K	Lys	Valine	V	Val
Leucine	L	Leu	Tryptophan	W	Trp
			Tyrosine	Y	Tyr

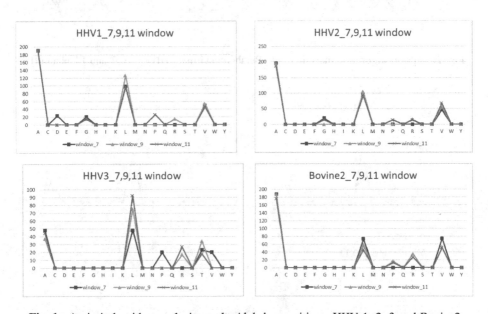

Fig. 1. Apriori algorithm analysis result. Alphaherpesvirinae, HHV 1, 2, 3 and Bovine2

Secondly, HHV-6 and HHV-7 seem to have similarities. The common rules of these two viruses show similar distributions of Asparagine located in the 4th, 7th place and Threonine located in the 6th and 7th place. They also show common rules on 5-L, 7-L, 8-L and 9-N in the 9 window experiment. In the 11 window experiment, HHV-6 and HHV-7 show similar distributions of 1-L, 6-L, 9-L, 10-L and 5-I. Thus, they partly shared their rules with Elphantid1. This relationship between HHV-6 and HHV-7, plus Elephantid1, is not the common features of the betaherpesvirinae and other herpesviruses. They didn't have a large common alanine rules and other minor rules were

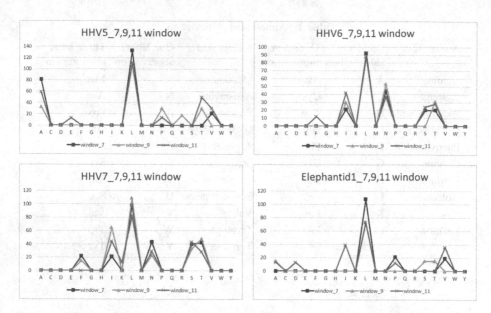

Fig. 2. Apriori algorithm analysis result. Betaherpesvirinae, HHV5, 6, 7 and Elephantid1

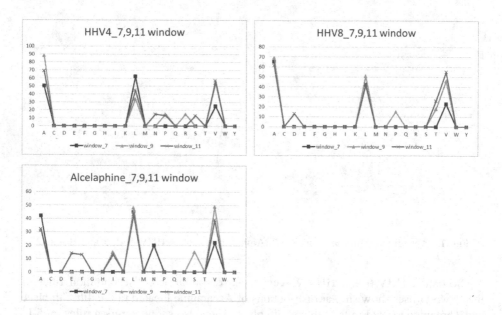

Fig. 3. Apriori algorithm analysis result. Gammapesvirinae, HHV4, 8 and Alcelaphine

different from other herpesviruses. Therefore we concluded that HHV 6, 7 are unique and can be classified differently with other betaherpesvirinae.

Lastly, the comparison of the species and the possibility of classification were found. Based on the distribution of the rules, we concluded that alphaherpesvirus subfamily is more similar with gammaherpesvirus subfamily and betaherpesvirus subfamily can be separated from them. The difference between human and animal herpes virus is not that clear in alphaherpesvirus subfamily but in beta and gamma-herpesvirus subfamily, the distribution and detailed rules showed a possibility of the classification. For instance, HHV4 and 8 showed different distribution of the rules with the Alcelaphine which is visible in the diagram.

3.2   Decision Tree Algorithm Analysis

Table 2 is the result of the decision tree algorithm experiment and it is automatically drawn by C 5.0 program after the experiment. This table ensembles an 11-by-11 matrix. For example, (1, 4) entry of the matrix, which is 10, indicates that class 4 was classified as class 1 for 10 times. The classification of the decision tree algorithm experiment is based on the obtained rule from the divided sequences. High value of the matrix can be explained as the similarities between the experimental classes. We found high similarities between HHV-6 and HHV-7, HHV-1, 2 and Bovine2.

Table 2. The result of the 7 window decision tree algorithm experiment. Program, C5.0 Release 2.07 GPL Edition, was used. Class 1 to 8 are the human herpes virus 1 to 8 respectively and class 9 to 11 are Alcelaphine(gamma subfamily), Bovine2(alpha subfamily) and Elphantid1(beta subfamily). Values above 30 and peak value from each classes were marked bold.

1	2	3	4	5	6	7	8	9	10	11	Classified as
37	**35**	**23**	10	11	14	7	15	7	**26**	12	class1
29	**35**	**23**	14	16	9	12	15	13	22	8	class2
27	29	19	11	**21**	20	13	12	16	19	13	class3
27	29	13	16	15	20	21	13	9	17	**18**	class4
25	22	25	14	11	26	12	14	15	18	14	class5
9	21	5	13	19	23	**54**	15	8	17	9	class6
12	14	19	12	12	**48**	26	17	12	10	11	class7
26	**31**	19	8	14	14	22	8	25	20	10	class8
25	25	20	15	8	10	23	**23**	13	17	17	class9
37	**43**	19	13	9	16	10	7	10	**26**	8	class10
18	21	14	**18**	7	**31**	20	20	13	14	17	class11

The 10 cross validation was held and the average classification error percentages was 89.3 %, standard error 0.5 % in 7window experiment. 9 and 11 window experiment also had similar results.

It is noticeable that the diagonal entries of the matrix are not the always biggest value. And some results like class 4 and 8 do not completely match with the apriori algorithm experiment. So considering the errors, this matrix cannot supply a solid value. Rather, it can be used to see the tendency of the large datasets.

4 Conclusion

Data mining is the process of finding and analyzing the relationship among the data. Through this project we used the data mining method to analyze the Herpes Virus configuration, and used the found patterns to investigate the similarities of the viruses and their distinguable characteristics. We were not only able to investigate the virus itself, but considered this investigation to be helpful in medical research and symptom analysis.

In the results of analyzing the general data, the rules of Alanine, Valine, and Leucine were the main components. Meticulously analyzing the results, we were able to conclude that these amino acids are distributed in patterns. We were also able to make a prediction that in the case that further investigation of Alanine, Valine, and Leucine is conducted, analysis of the Herpes Viruses and their cure centered around this investigation will develop.

HHV6 and HHV7 showed similar characteristics in several windows, but there were also unique patterns that could only be found in their boundaries. We discovered that, unlike other viruses, HHV6 and HHV7 have liver-related traits, and that this is connected to the fact that HHV6 and HHV7 are deeply related. This also can be supported by the decision tree algorithm experiment (Table 2, (6, 7) and (7, 6)). The patterns that commonly appear for these viruses will assist in identifying the uniqueness of the viruses.

Other than the relationships that showed the major patterns, the minor patterns that were found in the analysis process appeared in various forms. In other words, general patterns could also be seen including the observations that Threonine and Asparagine were located mainly under 4, and that Valine was located in HHV5-9 and its high frequency. We couldn't found the effects and the exact influence of these minor rules. So, the more understanding with the sequence pattern is necessary.

The three top main amino acids, Alanine, Valine and Leucine, were nonpolar amino acids. It is quite remarkable that rules regarding Leucine and Alanine were always extracted from all of the experiment regardless of window. Since amino acids sequence determines function and structure of the protein, we assumed that Alanine and Leucine are the key amino acid forming the major capsid protein. Even though we found a lot of other large and small rules, however, we couldn't utilize them in a solid knowledge. Therefore, a close research regarding the features of repeated amino acid should be held in near future.

The limitation of our research is that the size of the data was not big enough to have sufficient credibility and could only find vague information. Also, other kinds of protein should be analyzed to check the validity of our conclusion. So conducting a deeper research with more dataset is necessary in near future.

References

1. Jones, C.: Bovine herpes virus 1 (BHV-1) and herpes simplex virus type 1 (HSV-1) promote survival of latently infected sensory neurons, in part by inhibiting apoptosis. J. Cell Death 6 (2013)
2. Fatahzadeh, M., Schwartz, R.A.: Human herpes simplex virus infections: epidemiology, pathogenesis, symptomatology, diagnosis, and management. J. Am. Acad. Dermatol. 57(5), 737–763 (2007)

3. Anzivino, E., et al.: Herpes simplex virus infection in pregnancy and in neonate: status of art of epidemiology, diagnosis, therapy and prevention. Virol. J. **6** (2009)
4. Edelman, D.C.: Human herpesvirus 8 – a novel human pathogen. Virol. J. **2** (2005)
5. Arvin, A., Campadelli-Fiume, G., Mocarski, E., et al. (eds.): Human Herpesviruses: Biology, Therapy, and Immunoprophylaxis. Cambridge University Press, Cambridge (2007)
6. Ykihiro, N.: Herpesvirus genes: molecular basis of viral replication and pathogenicity. Nagoya J. Med. Sci. **59**, 107–119 (1996)
7. Bowman, B.R., et al.: Structure of the herpesvirus major capsid protein. EMBO J. **22**(4), 757–765 (2003). PMC. Web. 1 March 2015
8. Bayardo, Jr., R.J.: Efficiently mining long patterns from databases. In: SIGMOD 1998 Proceedings of the 1998 ACM SIGMOD International Conference on Management of Data, pp. 85–93 (2) (1998)
9. RuleQuest Research (2008). http://www.rulequest.com/see5-unix.html

Comparative Studies Based on a 3-D Graphical Representation of Protein Sequences

Yingzhao Liu[1(✉)], Yan-chun Yang[2], and Tian-ming Wang[2]

[1] Department of Mathematics,
Luoyang Normal University, Luoyang 471022, China
liuyzyz@yahoo.com
[2] School of Mathematical Sciences,
Dalian University of Technology, Dalian 116024, Liaoning, China
wangtm@dlut.edu.cn

Abstract. We perform comparisons among protein sequences based on a 3-D graphic representation of proteins [Bai and Wang, J. Biomol. Struc. Dyn., 23, 537–545 (2006)]. In detail, we make analysis of similarities among nine ATP6 (ATP synthase F0 subunit 6) proteins by comparing their corresponding 3-D curves and select the geometric center and the correlation matrix from the 3-D curve, as well as a angle metric for the construction of the phylogenetic tree. In comparison with the traditional alignments of sequences, the proposed method does not require multiple alignment and conceptually and computationally it is rather simple.

Keywords: Protein · Graphical representation · Sequences analysis · Phylogenetic tree

1 Introduction

The explosive growth of DNA and protein sequences data have made the sequence analysis be a discipline that grew enormously in recent years. One of the schemes for sequence analysis is to consider graphical representations for DNA and protein primary sequences. Some researchers have outlined different 2-D,3-D,4-D or higher dimensional graphical representation of DNA sequences [1–16]. The advantage of such representations is that they allow visual inspection of data, helping in recognizing major differences among similar DNA sequences, but also to derive numerical characterization for DNA primary sequences. So, once the graphical representation of a DNA sequence is obtained, we can use it to extract useful knowledge from the sequence, such as long-range correlation information, sequence periodicities, local nucleotide composition and other sequence characteristics [3, 5, 11, 17, 18]; furthermore, according to the representation, some mathematical characterizations are selected as invariants of sequence for comparisons of DNA primary sequences and constructions of phylogenetic trees [10, 19–21].

In contrast to DNA, the graphical representations of protein sequences emerged just very recently [22–25, 26–29]. The reason for the delay in emergence of the graphical representations of protein sequences is the increased complexity of biological strings

D.-S. Huang et al. (Eds.): ICIC 2015, Part II, LNCS 9226, pp. 436–444, 2015.
DOI: 10.1007/978-3-319-22186-1_43

built on a 20-letter alphabet (representing the 20 natural amino acids) in comparison with strings built from only four letters (representing DNA or RNA) (26) and consequently the research on applications of the graphical representations of protein sequences is insufficient. In additions, the currently available graphical representations of proteins are almost set in the 2-D space [22–25, 26–28]. The main advantage of the 2-D graphical representations of proteins is that these methods simplify and visualize the protein sequences, providing the hopes of gaining an understanding of the underlying proteome language. However, because the 20-letter alphabet are assigned to twenty directions in the Cartesian coordinates, it is very hard to arrive at unique graphical by the 2-D curve representing protein so that many biological information contained in the protein primary sequences may be lost. In order to overcome this difficult, Bai and Wang [29] proposed a 3-D representation of proteins with the assignment of twenty amino acids to the twenty vertices of the regular dodecahedron in the 3D space.

In this paper, based on the 3-D graphic representation of proteins, we pay our attentions to the study on the application of the graphic representation of proteins in this Letter. Since the 3-D curve contains the information of its corresponding protein sequence, multiple proteins can be compared by comparing their 3-D curves. We make analysis of similarities among the 3-D curves of ATP6(ATP synthase F0 subunit 6) proteins encoded in the H strand of mtDNA belonging to nine species and extract the geometric center and the correlation matrix from the 3-D curve to construct phylogentic trees for two protein sets shown in Tables 1 and 2, respectively. Unlike most existing phylogeny construction methods, the proposed method does not require multiple alignment. All of protein sequences used in this Letter were downloaded from GenBank.

Table 1. The accession number, abbreviation, name, and length for each of nine ATP6 proteins

No.	Accession	Abbreviation	Protein	Length (nt)
1	AAU02571	human	Human ATP synthase F0 subunit 6	226
2	NP_008204	chimpan	Pygmy chimpanzee ATP synthase F0 subunit 6	226
3	NP_008230	orang	Orangutan ATP synthase F0 subunit	226
4	NP_007165	horse	Horse ATP synthase F0 subunit 6	226
5	NP_007438	w rhino	White rhinoceros ATP synthase F0 subunit 6	226
6	NP_007386	donkey	Donkey ATP synthase F0 subunit 6	226
7	NP_006894	f whale	Finback whale ATP synthase F0 subunit 6	226
8	NP_007100	opossum	North American opossum ATP synthase F0 subunit 6	226
9	NP_008795	hippopo	Hippopotamus ATP synthase F0 subunit 6	226

2 A 3-D Graphic Representation of Protein Sequences

In 2006, a 3-D graphic representation of protein sequences was proposed by Bai and Wang [29]. A protein sequence, of length n, can be viewed as a linear sequence of n symbols from a finite alphabet set $\Omega = \{A, C, D, E, F, G,$

Table 2. The accession number, abbreviation, name, and length for each of nine protein sequences

No	Accession	Abbreviation	Protein	Length(nt)
1	CAB01581	r-caenor	Caenorhabditis elegans RAD50	1,298
2	Q9W252	r-droso	Drosophila melanogaster RAD50	1,318
3	NP_005723	r-homo	Homo sapiens RAD50 homolog isoform 1	1,312
4	NP_033038	r-mus	Mus musculus RAD50 homolog	1,312
5	Q9JIL8	r-ratt	Rattus norvegicus RAD50	1,312
6	NP_078900	s6-homo	Homo sapiens SMC6 protein	1,091
7	NP_079971	s6-mus	Mus musculus SMC6 protein	1,097
8	AAB34405	s1-homo	Homo sapiens SMC1	1,233
9	NP_062684	s1-mus	Mus musculus SMC1	1,233

$H, I, K, L, M, N, P, Q, R, S, T, V, W, Y\}$, consisting of twenty amino acids. We put each amino acid into a vertex of the regular dodecahedron and obtain the unit vectors representing 20 amino acids as follows

$$A \to (0,0,1), \quad M \to (-\frac{\sqrt{5}}{3}, \frac{\sqrt{3}}{3}, -\frac{1}{3}),$$

$$C \to (\frac{2}{3}, 0, \frac{\sqrt{5}}{3}), \quad N \to (-\frac{\sqrt{5}}{3}, -\frac{\sqrt{3}}{3}, -\frac{1}{3}),$$

$$D \to (-\frac{1}{3}, \frac{\sqrt{3}}{3}, \frac{\sqrt{5}}{3}), \quad P \to (\frac{\sqrt{5}-3}{6}, -\frac{\sqrt{3}+\sqrt{15}}{6}, -\frac{1}{3}),$$

$$E \to (-\frac{1}{3}, -\frac{\sqrt{3}}{3}, \frac{\sqrt{5}}{3}), \quad Q \to (\frac{3+\sqrt{5}}{6}, \frac{\sqrt{3}-\sqrt{15}}{6}, -\frac{1}{3}),$$

$$F \to (-\frac{3+\sqrt{5}}{6}, \frac{\sqrt{15}-\sqrt{3}}{6}, \frac{1}{3}), \quad R \to (\frac{3+\sqrt{5}}{6}, \frac{\sqrt{15}-\sqrt{3}}{6}, -\frac{1}{3}),$$

$$G \to (-\frac{3+\sqrt{5}}{6}, \frac{\sqrt{3}-\sqrt{15}}{6}, \frac{1}{3}), \quad S \to (\frac{\sqrt{5}-3}{6}, \frac{\sqrt{3}+\sqrt{15}}{6}, -\frac{1}{3}),$$

$$H \to (\frac{3-\sqrt{5}}{6}, -\frac{\sqrt{3}+\sqrt{15}}{6}, \frac{1}{3}), \quad T \to (\frac{1}{3}, \frac{\sqrt{3}}{3}, -\frac{\sqrt{5}}{3}),$$

$$I \to (\frac{\sqrt{5}}{3}, -\frac{\sqrt{3}}{3}, \frac{1}{3}), \quad V \to (-\frac{2}{3}, 0, -\frac{\sqrt{5}}{3}),$$

$$K \to (\frac{\sqrt{5}}{3}, \frac{\sqrt{3}}{3}, \frac{1}{3}), \quad W \to (\frac{1}{3}, -\frac{\sqrt{3}}{3}, -\frac{\sqrt{5}}{3}),$$

$$L \to (\frac{3-\sqrt{5}}{6}, -\frac{\sqrt{3}+\sqrt{15}}{6}, \frac{1}{3}), \quad Y \to (0, 0, -1).$$

In this way, we can convert a protein sequence into a series of vectors $P_0, P_1, P_2, \cdots, P_n$, whose coordinates $x_i, y_i, z_i (i = 0, 1, 2, \cdots, n)$ satisfy

$$
\begin{pmatrix} x_i \\ y_i \\ z_i \end{pmatrix} = \begin{pmatrix} 0 & \frac{2}{3} & -\frac{1}{3} & \cdots & \frac{1}{3} & 0 \\ 0 & 0 & \frac{\sqrt{3}}{3} & \cdots & -\frac{\sqrt{3}}{3} & 0 \\ 1 & \frac{\sqrt{5}}{3} & \frac{\sqrt{5}}{3} & \cdots & -\frac{\sqrt{5}}{3} & -1 \end{pmatrix} \begin{pmatrix} A_i \\ C_i \\ D_i \\ \vdots \\ W_i \\ Y_i \end{pmatrix}, \tag{1}
$$

where $A_i, C_i, \cdots Y_i$ are the cumulative occurrence numbers of $A, C, \cdots, Y$ in the subsequence from the 1st amino acid to the ith amino acid of the sequence, respectively. We define $A_0 = C_0 = \cdots = Y_0 = 0$, therefore $x_0 = y_0 = z_0 = 0$. When index i runs from 1 to n, we in turn connect the points $P_0, P_1, P_2, \cdots, P_n$ and get the 3-D curve corresponding to the protein.

3 Application of the 3-D Graphic Representation of Protein Sequences

3.1 Analysis of Similarities of Proteins Based on a Visual Inspection of the 3-D Curves Involved

We denote the vector pointing to the point P_i from the origin O as $r_i, i = 1, 2, \cdots, n$, and let $\Delta r_i = r_i - r_{i-1}$, of which the coordinates are $\Delta x_i, \Delta y_i$ and Δz_i. By the coordinates of the vectors representing 20 amino acids and Eq. (1), the vector Δr_i has at most twenty possible directions and the direction number (that is $\Delta x_i, \Delta y_i, \Delta z_i$) of Δr_i for each possible amino acid at the ith position are virtually equal to the coordinates of the corresponding amino acid, respectively (see Property and Table II in Ref. [29]). Furthermore, we have the value of Δz_i is negative if the amino acid at the ith position is one of $M, N, P, Q, R, S, T, V, W$ and Y, that is, the projection on the z-axis of Δr_i points to the negative orientation of z-axis, while the projection points to the positive orientation if the ith amino acid is one of $A, C, D, E, F, G, H, I, K$ and L.

As shown in the Fig. 1, the curve of opossum contains a segment of which projection on the z-axis distributes on the negative z-semi-axis and according to the analysis above, we conclude that in the ATP6 protein of opossum, there is a subsequence in which the amino acids are almost belonging to the group $\{M, N, P, Q, R, S, T, V, W, Y\}$ or $\{A, C, D, E, F, G, H, I, K, L\}$. We are here unable to make sure the amino acids belong to the first or second group because the direction of the curve with the index i running from 1 to n is not labeled in Fig. 1. As a result, we find there are nine amino acids belonging to the first set in the subsequence 90–100 bp of the ATP6 protein of opossum. This shows that based on the 3-D curve, some useful information can be extracted from the curve for the protein sequence.

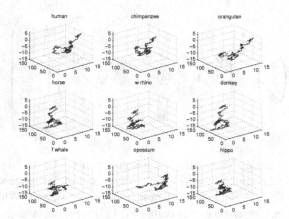

Fig. 1. The 3-D curves of ATP6(ATP synthase F0 subunit 6) proteins encoded in the H strand of mtDNA belonging to nine species in Table 1.

3.2   The Construction of Phylogenetic Tree of Protein Sequences

Let $P = p_1 p_2 \cdots p_n$ be an arbitrary protein sequence, where n is the length of this sequence. By Eq. (1), we can transform the protein sequence into a series of vectors $P_0, P_1, P_2 \cdots P_n$, whose coordinates are x_i, y_i, z_i $(i = 0, 1, 2, \cdots, n)$ and further get the 3-D curve of this protein by connecting the vectors in turn. The coordinates of the geometrical center of the curve, denoted by $\bar{x}, \bar{y}$ and $\bar{z}$, can be calculated by

$$\bar{x} = \frac{1}{n} \sum_{i=1}^{n} x_i, \quad \bar{y} = \frac{1}{n} \sum_{i=1}^{n} y_i, \quad \bar{z} = \frac{1}{n} \sum_{i=1}^{n} z_i. \tag{2}$$

Then, the correlation matrix which describes the global distribution pattern of the three-dimensional space curve is calculated as follows,

$$\Sigma = \begin{pmatrix} \sigma_{xx} & \sigma_{xy} & \sigma_{xz} \\ \sigma_{yx} & \sigma_{yy} & \sigma_{yz} \\ \sigma_{zx} & \sigma_{zy} & \sigma_{zz} \end{pmatrix} = (\sigma_{pq}), \tag{3}$$

where

$$\sigma_{pq} = \frac{n-1}{n} \frac{\sum_{i=1}^{n} (p_n - \bar{p})(q_n - \bar{q})}{\sqrt{\sum_{i=1}^{n} (p_n - \bar{p})^2} \sqrt{\sum_{i=1}^{n} (q_n - \bar{q})^2}}, \quad p, q = x, y, z, \tag{4}$$

Obviously, the matrix is a real symmetric 3×3 one. So there are three eigenvalues and their associated eigenvectors will be selected to characterize the distribution pattern of the corresponding curve. The eigenvalues and their associated eigenvectors are defined as follows,

$$\sum \mathbf{C}_k = \lambda_k \mathbf{C}_k, \quad \mathbf{C}_k = (C_{k,1}, C_{k,2}, C_{k,3})^T, \quad k = 1, 2, 3.$$

Corresponding to each eigenvalue λ_k, there's an eigenvector $\mathbf{C}_k$. Corresponding to $\lambda_1 < \lambda_2 < \lambda_3$, the three eigenvectors are denoted by $\mathbf{C}_1, \mathbf{C}_2, \mathbf{C}_3$, respectively. We will derive the distance matrix from the above parameters.

Suppose that the number of the protein sequences studied is M and the parameters of the uth protein are $\bar{x}_u, \bar{y}_u, \bar{z}_u; \mathbf{C}_k^u, u = 1, 2, \cdots, M; k = 1, 2, 3$, respectively. Firstly, the Euclid distance is used to reflect the diversity between two points,

$$d_{uv} = \sqrt{(\bar{x}_u - \bar{x}_v)^2 + (\bar{y}_u - \bar{y}_v)^2 + (\bar{z}_u - \bar{z}_v)^2}, \quad u, v = 1, 2, \cdots, M, \tag{5}$$

where d_{uv} denotes the distance between the geometric centers of the uth and the vth proteins. Secondly, to reflect the differences between the trends of every two three-dimensional curves, the eigenvectors of every two proteins are used. As pointed out by Stuart et al. [34], the cosine function of the angle between two vectors has been justified to measure relatedness of the vectors. Hence, we will employ the metric proposed by Stuart et al. [34] to measure the differences between the two curves. The cosine between any two eigenvectors of every two proteins can be computed as follows,

$$\cos \theta_{uv}^k = \frac{\mathbf{C}_k^u \cdot \mathbf{C}_k^v}{|\mathbf{C}_k^u| \cdot |\mathbf{C}_k^v|}, \quad u, v = 1, 2, \cdots, M, \quad k = 1, 2, 3,$$

These pairwise cosine values are then converted into pairwise distances measures using the formula

$$f_{uv}^k = -\ln \frac{1 + \cos \theta_{uv}^k}{2}, \quad u, v = 1, 2, \cdots, M, \quad k = 1, 2, 3, \tag{6}$$

The sum of f_{uv}^k over k for given u, v can be used to reflect the trend information of the eigenvectors involved

$$\Gamma_{uv} = f_{uv}^1 + f_{uv}^2 + f_{uv}^3, \quad u, v = 1, 2, \cdots, M. \tag{7}$$

Consequently, two sets of parameters are obtained. The first reflects the difference of center positions represented by the Euclid distance between the geometric centers; the second indicates the difference of the trends of the 3-D curves represented by the related eigenvectors. The overall distance D_{uv} between the proteins u and v is defined by

$$D_{uv} = d_{uv} \times \Gamma_{uv}, \quad u, v = 1, 2, \cdots, M. \tag{8}$$

Once the pairwise distance matrix is obtained, it is straightforward to generate a phylogenetic tree using the UPGMA method in the PHYLIP package (http://evolution. genetics.washington.edu/phylip.html). The final phylogenetic tree is viewed using the TreeView program (http://taxonomy.zoology.gla.ac.uk/rod/treeview.html).The branch

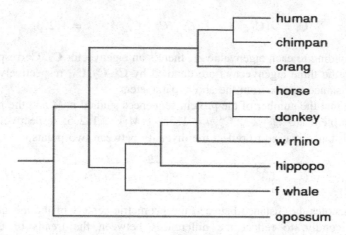

Fig. 2. Topological UPGMA species tree for nine ATP6 proteins in Table 1.

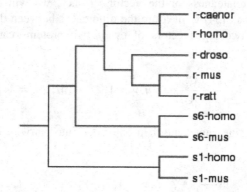

Fig. 3. Topological UPGMA tree for nine proteins in Table 2

lengths are not scaled according to the distances and only the topology of the tree is concerned. In Fig. 2, we show the phylogenetic tree containing nine ATP6 proteins in Table 1 based on our method (Fig. 3).

4   Concluding Remarks

The graphical representations provide a simple way of viewing, sorting, and comparing various sequences and more attentions have been paid to this research. However, in contrast to DNA sequences, the study on the graphical representations of protein sequences and its applications is insufficient. So, we explored the application of a 3-D graphic representation of proteins in this Letter. We compared nine ATP6 proteins based on their 3-D curves and then employed the geometric center and the correlation matrix of theses curves as well as a angle metric for the construction of the phylogenetic tree. In the same way, we also made the phylogenetic analysis for another nine proteins which are comprised of five RAD50 proteins, two SMC6 and two SMC1 proteins. The coincidence between the final evolutionary tree by us and that by other researchers corroborate the utility of our method.

In this Letter, the study of the application of the graphical representations is concentrated on primary structure of protein. As is well know, the higher-level structures of a protein plays an instrumental role in determining its functions and the higher-level, especially three-dimensional structure of protein is determined by its primary structure. On the other hand, the 3-D curve of a protein sequence contain mainly information of the protein, so a natural question occurs that does there exist a relationship between the spatial structure of the 3-D cure of the protein and its real three-dimensional structure? Further, can we predict its real three-dimensional structure directly by the spatial structure of corresponding 3-D cure? We think these questions are of great challenge as well as significance and hope to explore these pending questions in our future research. We would also welcome comments, inputs, modifications, etc. of other interested scientists which may facilitate further refinements of the outlined methodology.

Acknowledgments. We are indebted to both referees for their many valuable suggestions and comments, which are critical for improving the quality of the manuscript. This work is supported by the Science and Technology Development Program of Henan Province, China (Grant No. 102300410216).

References

1. Hamori, E., Ruskin, J.: H curves, a novel method of representation of nucleotide series especially suited for long DNA sequences. J. Biol. Chem. **258**, 1318–1327 (1983)
2. Gates, M.A.: A simple way to look at DNA. J. Theor. Biol. **119**, 319–328 (1986)
3. Peng, C.K., Buldyrev, S.V., Goldberger, A.L., Havlin, S., Sciortino, F., Simons, M., Stanley, H.E.: Long-range correlations in nucleotide sequences. Nature **356**, 168–170 (1992)
4. Ghosh, A., Nandy, A.: Adv. Protein Chem. Struct. Biol. **83**, 1–42 (2011)
5. Nandy, A.: Two-dimensional graphical representation of DNA sequences and intron-9exon discrimination in intron-rich sequences. Comput. Appl. Biosci. **12**, 55–62 (1996)
6. Roy, A., Raychaudhury, C., Nandy, A.: Novel techniques of graphical representation and analysis of DNA sequences - a review. J. Biosci. **23**, 55–71 (1998)
7. Randic, M., Vracko, M., Nandy, A., Basak, S.C.: On 3-D graphical representation of DNA primary sequences and their numerical characterization. J. Chem. Inf. Comput. Sci. **40**, 1235–1244 (2000)
8. Randic, M., Vracko, M., Lers, N., Plavsic, D.: Novel 2-D graphical representation of DNA sequences and their numerical characterization. Chem. Phys. Lett. **368**, 1–6 (2003)
9. Randic, M., Vracko, M., Lers, N., Plavsic, D.: Analysis of similarity/dissimilarity of DNA sequences based on novel 2-D graphical representation. Chem. Phys. Lett. **371**, 202–207 (2003)
10. Liao, B., Tan, M., Ding, K.: Application of 2-D graphical representation of DNA sequence. Chem. Phys. Lett. **414**, 296–300 (2005)
11. Zhang, C.T., Zhang, R., Ou, H.Y.: The Z curvedatabase: a graphic representation of genome sequences. Bioinformatics **19**, 593–599 (2003)

12. Yuan, C.X., Liao, B., Wang, T.M.: New 3D graphical representation of DNA sequences and their numerical characterization. Chem. Phys. Lett. **379**, 412–417 (2003)
13. Randic, M., Balaban, A.T.: On A Four-Dimensional Representation of DNA Primary Sequences. J. Chem. Inf. Comput. Sci. **43**, 532–539 (2003)
14. Chi, R., Ding, K.: Novel 4D numerical representation of DNA sequences. Chem. Phys. Lett. **407**, 63–67 (2005)
15. Liao, B., Wang, T.M.: Analysis of similarity/dissimilarity of DNA sequences based on nonoverlapping triplets of nucleotide bases. J. Chem. Inf. Comput. Sci. **44**, 1666–1670 (2004)
16. Liu, Y.Z., Wang, T.M.: Related matrices of DNA primary sequences based on triplets of nucleic acid bases. Chem. Phys. Lett. **417**, 173–178 (2006)
17. Berger, J.A., Mitra, S.K., Carli, M., Neri, A.: Visualization and analysis of DNA sequences using DNA walks. J. Franklin I. **341**, 37–53 (2004)
18. Nandy, A., Basak, S.C., Gute, B.D.: Graphical representation and numerical characterization of H5N1 avian flu neuraminidase gene sequence. J. Chem. Inf. Model. **47**, 945–951 (2007)
19. Zheng, W.X., Chen, L.L., Ou, H.Y., Gao, F., Zhang, C.T.: Coronavirus phylogeny based on a geometric approach. Mol. Phylogenet. Evol. **36**, 224–232 (2005)
20. Liao, B., Shan, X., Zhu, W., Li, R.: Phylogenetic tree construction based on 2D graphical representation. Chem. Phys. Lett. **422**, 282–288 (2006)
21. Wang, W.P., Liao, B., Wang, T.M., Zhu, W.: A graphical method to construct a phylogenetic tree. Int. J. Quantum Chem. **106**, 1998–2005 (2006)
22. Randic, M., Novic, M., Roy Choudhury, A., Plavsic, D.: Graphical representation of trans-membrane proteins. SAR QSAR Environ. Res. **23**, 327–343 (2012)
23. Randic, M., Zupan, J., Balaban, A.T.: Unique graphical representation of protein sequences based on nucleotide triplet codons. Chem. Phys. Lett. **397**, 247–252 (2004)
24. Randic, M., Novic, M., Vracko, M., Plavsic, D.: Study of proteome maps using partial ordering. J. Theor. Biol. **266**, 21–28 (2010)
25. Randic, M., Butina, D., Zupan, J.: Novel 2-D graphical representation of proteins. Chem. Phys. Lett. **419**, 528–532 (2006)

A Graph Theoretic Approach for the Feature Extraction of Transcription Factor Binding Sites

Yinglei Song[1(✉)], Albert Y. Chi[2], and Junfeng Qu[3]

[1] School of Electronics and Information Science,
Jiangsu University of Science and Technology,
Zhenjiang 212003, Jiangsu, China
syinglei2013@163.com
[2] Department of Mathematics and Computer Science,
University of Maryland Eastern Shore, Princess Anne, MD 21853, USA
[3] Department of Computer Science and Information Technology,
Clayton State University, Morrow, GA 30260, USA

Abstract. Recent work in molecular biology has revealed that transcription factors are biologically important in gene regulation. A transcription factor regulates the expression level of a gene by binding to the promoter region of the gene. A model that can accurately describe the binding sites of a transcription factor is thus crucial for understanding the biological mechanisms of gene regulation. In this paper, we develop a new feature extraction algorithm that can accurately obtain features of the binding sites of a transcription factor. The obtained features describe the pair-wise correlations of different positions in a binding site. Based on these features, pair-wise correlations can be integrated into a statistical model that describes the binding sites of a transcriptional factor. Our testing results show that, this approach is able to identify important features for transcription factor binding sites and statistical models based on these features can achieve prediction accuracy that is higher than or comparable with that of other feature extraction methods.

Keywords: Transcription factor binding sites · Graph theoretical approach, feature extraction · Statistical model

1 Introduction

Recent research in molecular biology has revealed that gene regulation is a fundamental biological process at molecular level. The expression levels of genes are precisely regulated by gene regulation processes and many other biological processes are then controlled by the expression levels of genes [9]. At the level of transcriptional regulation, a transcription factor, which is a certain type of protein molecule, binds to the promoter region of the gene whose expression level needs to be regulated. The binding often occurs between a subsequence of nucleotides in the promoter region of the gene and the transcription factor. This subsequence is also called

© Springer International Publishing Switzerland 2015
D.-S. Huang et al. (Eds.): ICIC 2015, Part II, LNCS 9226, pp. 445–455, 2015.
DOI: 10.1007/978-3-319-22186-1_44

a binding site. The features associated with the binding sites of a transcription factor are often closely related to the binding process and thus are crucial for gene regulation.

A number of experimental approaches, such as footprinting [7], in vitro and vivo methods [3, 15] and Southwestern blotting [6] etc., have been designed to study the binding process [6, 8]. Although these approaches can accurately recognize the binding sites, they are expensive and time-consuming. In addition, these approaches cannot be applied to accurately identify the features of the binding sites of a transcription factor. It is therefore highly desirable to develop approaches that can efficiently identify the binding sites from a large amount of sequence data with high accuracy. To this end, a number of computational models [5, 13, 20, 27, 28] have been developed to describe transcription factor binding sites. Based on these models, a few algorithms have been designed to predict their locations in a set of homologous sequences. Most of these models use a statistical profile to describe each individual position in a binding site and the corresponding prediction algorithm is based on the assumption that different positions in a binding site are statistically independent. In general, the statistical profile of a position contains the probability for the position to contain each individual nucleotide. When positions in a binding site are mutually independent, a matrix can thus be used to fully describe the binding sites of a transcription factor. Such a matrix is also called a position specific scoring matrix (PSSM).

One important advantage of the PSSM representation is its simplicity. Based on the independent assumption, the affinity of each subsequence to the model that describes the binding sites can be efficiently computed and a prediction on whether the subsequence is a binding site or not can then be made based on the statistical significance of this affinity value. Previous work has shown that the PSSM representation can lead to accurate prediction results on a few transcription factors [5, 13, 20, 27, 28]. However, recent work has demonstrated that the binding process is often complex and may require the collaboration from nucleotides in different positions [15]. The independent assumption thus may miss biological mechanisms that are important for binding to occur and thus leads to inaccurate prediction results. Models that are able to accurately describe the dependencies among nucleotides in a binding site are thus needed to further improve the accuracy of transcription factor binding site prediction.

A few approaches have been developed to model the correlations among multiple positions in a binding site [1, 19]. For example, a statistical approach is developed in [19] to select features that can accurately describe the correlations among multiple positions in a binding site. Testing results show that the features important for the binding process can be effectively selected by the approach. However, the approach must exhaustively search the space of all possible features to determine the features that are important for the binding process. The computational efficiency of the approach thus may drop significantly when the length of the binding site is moderate. In [1], higher order correlations among multiple positions are modeled by a Bayesian network approach. This approach can be used to process ChIP-chip data and improve the prediction accuracy for putative transcription factor binding sites. However, a Bayesian network only contains a limited set of topological features and is thus unable to describe all types of dependencies among the positions in a binding site.

Recently, research in molecular biology and bioinformatics has revealed the importance of pair-wise correlations of different positions in a binding site during the binding process [18, 29]. In [18], the in vivo ChiPSeq data of fly and mouse is analyzed and it is shown that the accuracy of PSSM based approaches is significantly lower than that of an approach that considers pair-wise correlations among positions in a binding site. Pair-wise correlations are thus important for an accurate description of the binding sites of certain genes. In this paper, we develop a new approach that can extract crucial features that describe pair-wise correlations among the positions of a transcription factor binding site. Specifically, we use a graph to model the pair-wise mutual correlations among all positions in a binding site. Each position in the binding site is represented by a vertex in the graph and the weight of an edge in the graph represents the correlation between the two positions that correspond to the two endpoints of the edge in the graph. The correlation between two positions in a binding site can be computed based on the alignment of the binding sites. In order to recognize the set of pair-wise correlations that are most significant without introducing duplicate information, we compute the maximum spanning forest in the graph and the edges in the tree represent the crucial features associated with pair-wise correlations for the binding sites.

Based on this feature extraction algorithm, we develop a new software tool BSPP that can iteratively predict the binding sites from the promoter regions of a set of homologous sequences. Specifically, the approach starts by selecting the two sequences that have the lowest similarity and perform a local alignment between the two sequences. Two subsequences that have the highest similarity value are then selected to be the initial set of binding sites. The feature extraction algorithm is then used to extract the features associated with pair-wise correlations in the binding sites. The rest of the sequences in the set are then processed one by one. For a given sequence that needs to be processed, the features are used to evaluate the affinity of each subsequence in the sequence to the model that describes the binding sites and the subsequence that has the highest affinity value is selected as the binding set in the sequence. Once a new binding site is identified, the feature extraction procedure is applied again to update the set of features for the binding sites. This procedure is repeated until all sequences in the set have been processed. We have tested the prediction accuracy of BSPP and compare it with both a few existing tools for binding site prediction. Our testing results have shown that including features extracted by our approach in a binding site model can significantly improve the prediction accuracy. In addition, we also compare the prediction accuracy of BSPP with that of two other feature extraction methods, including a Bayesian network based approach [1] and FMM [18], our results show that BSPP can achieve prediction accuracy that are higher than or comparable with that of the two other approaches on all testing datasets.

It is also worth pointing out that the binding pattern for different transcription factors may vary dramatically. For simple cases, PSSM based approaches can also reflect the pair-wise correlations to a certain extent. Our approach in fact provides a generalized framework that can be used to include the contributions from pair-wise interactions during the binding process to predict the transcription factor binding sites.

2 The Feature Extraction Algorithm

Given a transcription factor, its binding site is a subsequence in the promoter region of the gene that it regulates. In general, a binding site contains from 6 to 20 nucleotides. We use P to denote the set of the promoter regions of n homologous genes, each sequence in P contains a binding site. In general, the binding sites in the sequences in P are similar in their sequence content.

We use $S = \{B_1, B_2, \ldots, B_n\}$ to denote a set that contains n aligned binding sites, the objective is to extract pair-wise features that represent the binding site sequences in S. A pair-wise feature contains two positions in the binding site that are correlated to certain extent. The pair-wise correlation of the two positions can be evaluated based on the alignment of sequences in S. Given two positions p_1 and p_2 in the binding site, the dependency $C(p_1, p_2)$ of p_1 and p_2 is computed as follows.

$$C(p_1, p_2) = \sum_{t_1, t_2} P_{p_1, p_2}(t_1, t_2) \log_2 \frac{P_{p_1, p_2}(t_1, t_2)}{P_U(t_1, t_2)} \tag{1}$$

where the summation is over all possible pairs of nucleotides, $P_{p_1, p_2}(t_1, t_2)$ is the probability that nucleotides t_1 and t_2 appear in p_1 and p_2 respectively, and $P_U(t_1, t_2)$ is the probability for t_1 and t_2 to appear in p_1 and p_2 in a uniform distribution. It is not difficult to see that the value of $C(p_1, p_2)$ reflects the correlation between p_1 and p_2. A larger value of $C(p_1, p_2)$ suggests a stronger correlation between p_1 and p_2. The value of $C(p_1, p_2)$ is 0 if p_1 and p_2 are two independent positions.

We then use a graph model G to describe the pair-wise correlations among all positions in the binding sites in S. Specifically, each position in is represented by a vertex in G. Two vertices v_1, v_2 in G are joined by an edge if the correlation value of the two positions that correspond to v_1, v_2 is larger than a threshold value T_c. To select features that contain as much information as possible without introducing duplicate information, we compute a maximum spanning forest T_m in G, each edge in T_m represents a pair-wise feature for the binding sites and is included in a set F. Given a graph G, the maximum spanning forest in G can be efficiently computed with the Prim's algorithm [4] or the Kruskal's algorithm [4] in $O(|E| \log_2 |V|)$ time, where $|V|$ is the number of vertices in G and $|E|$ is the number of edges in G. Figure 1(a) and

(a)   (b)

Fig. 1. (a) A graph constructed to model a binding site that contains five positions a, b, c, d, and e, each edge in the graph is associated with a weight value which is the correlation between the two positions between the two endpoints of the edge; (b) a maximum spanning forest computed in the graph in (a).

(b) provide an example of a graph that represents the pair-wise corrections between positions in a binding site and a maximum spanning forest in the graph.

3 Statistical Model for Binding Sites

We construct a statistical profile for each pair-wise feature in F. For $f \in F$, we assume p_1 and p_2 are the two positions in f, the statistical profile for f contains the probability for each possible pair of nucleotides to appear in p_1 and p_2 in a binding site.

Based on the statistical profiles for all pair-wise features, we construct a statistical model to describe the binding sites of a transcription factor. We assume the binding sites contain a set of pair-wise features $F = \{f_1, f_2, \ldots, f_l\}$, where each feature contains a pair of highly correlated positions in the binding set. Given a subsequence S, the probability that S is a binding site of the transcription factor can be computed as follows.

$$P(S) = P(f_1, S)P(f_2, S)\ldots P(f_l, S) \tag{2}$$

where $P(f_i, S)$ is the probability for the nucleotides in the positions in f_i assigned by S to appear in the positions. Since the product of a large number of numbers that are all less than 1.0 is a small number, we use the logarithm of $P(S)$ for convenience. We have

$$\log_2 P(S) = \sum_{i=1}^{l} \log_2 P(f_i, S) \tag{3}$$

It is not difficult to see that the logarithm of $P(S)$ can be efficiently computed based on S and the statistical profile of each pair-wise feature in F. This new statistical model in fact provides a new approach to evaluating the affinity of a subsequence to the binding sites of a transcription factor. It thus can be integrated into any algorithm that has been developed for transcription factor prediction to improve the prediction accuracy.

4 An Algorithm for Binding Site Prediction

We develop a simple algorithm for binding site prediction based on the statistical model we have developed. Given a set of n sequences that are the promoter regions of n homologous genes, the algorithm starts by computing the mutual similarity between each pair of sequences in the set and select the pair of sequences that have the lowest mutual similarity. The similarity between two sequences is evaluated by a global alignment of the two sequences with the Needleman-Wunsch algorithm. A local alignment is then performed between the two selected sequences with the Smith-Waterman algorithm. The two subsequences that have the highest similarity value are included in a set B as the two binding sites that have been determined. The length of the binding site is also set to be the length of these two subsequences. The two sequences that have been processed are also included in a sequence set H, which stores the sequences that have been processed by the algorithm.

The similarity of each sequence that is not in H to each sequence in H is computed by global alignment and the average similarity is computed. The sequence that has the lowest similarity value is selected to be the one that is processed next. The algorithm then uses the feature extraction algorithm to extract pair-wise features based on the binding sites in B and process the next sequence based on this statistical model. Specifically, the affinity of each subsequence of the same length as the ones in B is computed based on the statistical model and the one that has the highest affinity value is included in B as a binding site that has been identified. The sequence is then included in H as a sequence that has been processed. The procedure is repeated until all sequences have been processed and the subsequences in B are output as the set of predicted binding sites. It is straightforward to see that this simple algorithm identifies a single binding site in each sequence in the sequence set.

5   Testing Results

We have implemented the transcription factor prediction approach together with the feature extraction algorithm into a software tool BSPP. We test the prediction accuracy of BSPP and compare it with that of a few other existing tools including Gibbs Sampler [12], BioProspector [13] and MDGA [5]. A threshold value T_c is used to construct the graph that models the pair-wise correlations among positions in a binding site and a statistical approach is used to obtain an appropriate threshold value for T_c. Specifically, we downloaded the aligned sequence sets of 200 transcription factor binding sites for 11 different species from the JASPAR database at http://jaspar.genereg.net. Different values of T_c are then used to construct statistical models to describe each of the 200 transcription factor binding sites. For each given threshold value T_c, we align each sequence in the sequence set of a transcription factor binding site to the corresponding statistical model and an alignment score can be computed based on the model. The statistical significance of this alignment score can be evaluated by computing its Z-score with respect to the random background. The average of Z-scores computed on all sequences in the sequence set is then used as a measure of the statistical significance of the model. Finally, the average value of the average Z-scores computed on all transcription factor binding sites is used to evaluate the effectiveness of the threshold T_c that corresponds to the model. A higher value of this measure suggests higher recognition ability on average. Table 1 shows the values of this measure on a few different values for T_c.

Table 1 clearly shows that the recognition ability of the approach is maximized when T_c is 0.5. We thus set the threshold T_c used in the feature extraction algorithm to be 0.5 in all of our experiments. We use two different approaches to evaluate the accuracy of a prediction result. First, we evaluate how close BSSP can reach the correct positions of transcription factor binding sites. We compare the predicted position of a binding site with the accurate position that has been reported in the data set. A predicted binding site is considered to be correct if both of its predicted left and right boundaries deviate from the correct positions by at most 3

Table 1. The Z-score average evaluated on a few different values for T_c.

T_c	0.1	0.3	0.5	0.7	0.9
Z-score average	2.732	3.125	4.263	3.672	3.137

positions. However, this evaluation metric only reflects the accuracy of the location of a predicted binding site.

In addition to comparing the prediction accuracy of BSPP with that of a few existing tools for transcription factor prediction, we also integrate two other feature selection methods, including the Bayesian network based approach [1] and the feature motif model (FMM) [19], into our simple algorithm for binding site prediction. We then compare the prediction accuracy of PSSM with that obtained with both the Bayesian network based approach and FMM.

5.1   On Yeast Datasets

We used the data set provided in [14] to test the prediction accuracy of our approach. There are in total 16371 regulatory interactions between transcription factors and binding sites in the data set. In addition, the data set includes data for the binding sites of 69 transcription factors. We use BSPP, Gibbs Sampler, BioProspector and MDGA to predict the binding sites of all 69 transcription factors and evaluate the average accuracy of each tool. The prediction accuracy for a transcription factor is computed by dividing the number of correctly predicted binding sites by the total number of sequences that contain the binding site in the data set. Table 2 shows the distribution of prediction accuracy for each of the four tools. It can be seen from the table that BSPP achieves prediction accuracy above 90 % in 51 of the 69 transcription factors, which significantly outperforms the other three tools. In addition, BSPP also significantly outperforms the other three tools in terms of the average prediction accuracy.

Table 3 further compares the prediction accuracy of BSPP with that of three other tools. It shows the number of transcription factors where BSPP outperforms three, two,

Table 2. The distribution of the prediction accuracy of BSPP, Gibbs Sampler, BioProspector, and MDGA on all 69 transcription factors in the yeast data set.

Prediction tool	>=90 %	80 %–90 %	70 %–80 %	<70 %	Average
BSPP	51	7	6	5	87.67 %
Gibbs sampler	37	12	13	7	82.62 %
BioProspector	41	10	12	6	83.71 %
MDGA	44	9	10	6	85.13 %

Table 3. The number of transcription factors where BSSP outperforms three, two, one, and none of the three other tools in the yeast data set.

	Outperforms 3	Outperforms 2	Outperforms 1	Outperforms 0
Number of TFs	43	15	4	7

one and none of the three other tools. It can be seen from the table that BSPP achieves higher prediction accuracy than all three other tools in 43 of the 69 transcription factors and achieves higher prediction accuracy than at least one of the three tools in 62 of the 69 transcription factors. It is therefore clear that BSPP significantly outperforms the other three tools in the yeast data set.

Table 3 also shows that at least one of the three other tools outperforms BSPP in 26 of the 69 transcription factors. This suggests that BSSP is unable to accurately capture the pair-wise interactions among positions in the binding sites of these transcription factors. Since BSSP determines the pair-wise interactions by computing the maximum spanning tree in the graph that models all possible pair-wise interactions, some pair-wise interactions that are biologically important may not be considered in the statistical model constructed in BSSP and this may adversely affect its prediction accuracy.

For each transcription factor, we compare the sequence content of each predicted binding site with its correct sequence content and determine the percentage of correctly predicted nucleotides. We then compute the average percentage of correctly predicted nucleotides for all binding sites of a transcription factor. Table 4 shows the distribution of the average percentage of nucleotides that can be correctly recognized for each transcription factor in the yeast data set.

5.2   On Other Datasets

In addition to the yeast data set, we also compare the prediction accuracy of BSSP with that of three other tools on three other datasets, including HSF1 [17], FOXO1 [2], BATF [16], and EGR1 [10]. In this experiment, we evaluate the accuracy of a prediction tool on each single sequence by computing the percentage of nucleotides that are correctly identified to be in a binding site. The prediction accuracy on a data set is the prediction accuracy averaged over all sequences in the data set. Table 5 shows the

Table 4. The distribution of average percentage of correctly predicted nucleotides for the yeast data set.

Prediction tool	>=90 %	80 %–90 %	70 %–80 %	<70 %	Average
BSPP	40	11	10	8	85.53 %
Gibbs sampler	25	17	19	8	81.34 %
BioProspector	31	14	16	9	83.27 %
MDGA	33	15	11	10	84.02 %

Table 5. The prediction accuracy of all four prediction tools on data sets HSF1, FOXO1, BATF, and EGR1.

Dataset	BSSP	Gibbs sampler	BioProspector	MDGA
HSF1	86.67 %	80.41 %	81.32 %	77.44 %
FOXO1	88.25 %	93.18 %	88.89 %	90.32 %
BATF	92.13 %	88.33 %	87.21 %	90.19 %
EGR1	95.20 %	86.47 %	84.62 %	86.08 %

Table 6. The prediction accuracy of all six different feature extraction methods on the Yeast datasets, HSF1, FOXO1, BATF, and EGR1.

Dataset	Bayesian network	FMM	Gibbs sampler	BioP	MDGA	BSSP
Yeast datasets	83.21 %	82.15 %	82.62 %	83.71 %	85.13 %	**87.67 %**
HSF1	**87.32 %**	85.52 %	83.32 %	83.27 %	80.36 %	86.67 %
FOXO1	84.61 %	85.67 %	**94.24 %**	90.13 %	91.14 %	88.25 %
BATF	82.35 %	86.72 %	90.16 %	89.12 %	91.06 %	**92.13 %**
EGR1	94.56 %	**97.18 %**	88.25 %	86.53 %	89.17 %	95.20 %

average prediction accuracy of all four tools on the four datasets. It can be seen from Table 5 that BSSP achieves higher prediction accuracy than all of the other three tools in data sets HSF1, BATF, and EGR1.

5.3   Comparison with Other Feature Extraction Methods

In [1], a Bayesian network based approach is developed to extract features that model the correlations among positions in a binding site. FMM is an approach that extracts features to describe the dependencies among positions in a binding site with Markov networks [19]. We integrate both approaches with our simple binding site prediction algorithm and compare their prediction accuracy with that of BSSP. The prediction accuracy of a predicted binding site is evaluated based on its position. In other words, a predicted binding site is considered to be correct if both of its predicted left and right boundaries deviate from the correct positions by at most 3 positions. Table 4 shows a unified comparison among Bayesian network, FMM, Gibbs Sampler, BioProspector, MDGA and BSSP on the Yeast datasets, HSF1, FOXO1, BATF, and EGR1.

It can be seen from the table that BSSP achieves the highest prediction accuracy on two of the five datasets and outperforms both Bayesian network and FMM in three data sets. In addition, we observe that the accuracy achieved by BSSP on the other three data sets is not significantly lower than that of the other methods. Both of the two other methods for feature extraction consider the correlations among multiple positions and our testing results show that pair-wise interactions are probably more important than correlations among multiple positions during the binding process. This observation is also consistent with the facts that are discovered recently in [18] (Table 6).

6   Conclusions

In this paper, we develop a graph algorithm to extract features for pair-wise correlations of positions in a transcription factor binding site. This approach uses a graph model to describe the pair-wise correlations between each pair of positions in a binding site. A maximum spanning forest in the graph can then be computed to obtain the pair-wise features that are important for the binding process between the transcription factor and the binding site. Based on the obtained pair-wise features, a new statistical model can be constructed to accurately describe the binding sites of a transcription factor and the affinity value of a sequence segment to the binding sites of the transcription factor can be efficiently computed based on this statistical model.

We have integrated this new approach for feature extraction into a simple algorithm for transcription factor binding site prediction and implemented it into a software tool BSSP. Our testing on yeast transcription factor binding site data set and a few other data sets have shown that including features for pair-wise correlations can significantly improve the prediction accuracy in most of the testing data sets. By comparing the prediction accuracy of BSSP with that of two other feature selection methods, we observe that BSSP can achieve comparable or higher prediction accuracy on all testing datasets.

Although BSSP is able to achieve higher prediction accuracy than a few existing approaches on the tested data sets, it needs to perform a large number of sequence alignments to process high-throughput datasets that are often of large scale. The computational efficiency of BSSP thus may decline significantly in this scenario. In addition, since we have not evaluated the accuracy of BSSP on whole genome sequences and real data with higher noise, such as ChIP-Seq data, it remains unknown whether the testing we have performed so far may lead to over estimation of its prediction accuracy or not. Additional testing that is more comprehensive is thus needed in the future to evaluate the prediction accuracy of BSSP and compare its accuracy with that of other tools. In addition, we have developed parameterized algorithms that can efficiently solve a few bioinformatics problems [11, 20–26]. We expect that this new feature extraction algorithm can be combined with the parameterized approaches we have developed in our previous work to further improve the prediction accuracy. This would constitute an important direction of our future work.

References

1. Barash, Y., Elidan, G., Friedman, N., Kaplan T.: Modeling dependencies in protein-DNA binding sites. In: Proceedings of the Seventh Annual International Conference on Computational Biology, pp. 28–37 (2003)
2. Brent, M.M., Anand, R., Marmorstein, R.: Structural basis for DNA recognition by Foxo1 and its regulation by posttranslational modification. Structure 16, 1407–1416 (2008)
3. Bulyk, M.: DNA microarray technologies for measuring protein-DNA interactions. Curr. Opin. Biotechnol. 17, 1–9 (2006)
4. Cormen, T., Leiserson, C., Rivest, R., Stein, C.: Introduction to Algorithms, 2nd edn. MIT Press, Cambridge (2001)
5. Che, D., Song, Y., Zhao, H.: MDGA: motif discovery using a genetic algorithm. In: Proceedings of the 2005 Conference on Genetic and Evolutionary Computation, pp. 447–452 (2005)
6. Elnitski, L., Jin, V.X., Farnham, P.J., Jones, S.J.M.: Locating mammalian transcription factor binding sites: a survey of computational and experimental techniques. Genome Res. 16(12), 1455–1464 (2006)
7. Gallas, D., Schmitz, A.: DNA footprinting: a simple method for the detection of protein-dna binding specificity. Nucleic Acids Res. 5(9), 3157–3170 (1978)
8. Garner, M., Revzin, A.: A gel electrophoresis method for quantifying the binding site of proteins to specific DNA regions: application to components of the escherichia coli lactose operon regulatory systems. Nucleic Acids Res. 9(13), 3047–3060 (1981)

9. Harbinson, C.T., et al.: Transcriptional regulatory code of a eukaryotic genome. Nature **431** (7004), 99–104 (2004)
10. Hu, T.C., et al.: Snail associates with EGR-1 and SP-1 to upregulate transcriptional activation of P15ink14b. FEBS J. **227**, 1202–1218 (2010)
11. Liu, C., Song, Y.: Parameterized complexity and inapproximability of dominating set problem in chordal and near chordal graphs. J. Comb. Optim. **22**(4), 684–698 (2011)
12. Liu, J., Neuwald, A., Lawrence, C.: Bayesian models for local alignment and gibbs sampling strategies. J. Am. Stat. Assoc. **90**(432), 1156–1170 (1995)
13. Liu, X., Brutlag, D., Liu, J.: Bioprospector: discovering conserved DNA motifs in upstream regulatory regions of coexpressed genes. In: Proceedings of 2001 Pacific Symposium on Biocomputing, pp. 127–138 (2001)
14. MacIssac, K., Wang, T., Gordon, D., Gifford, D., Stormo, G., Fraenkel, E.: An Improved map of conserved regulatory sites for saccharomyces cerevisiae. BMC Bioinf. **7**, 113 (2006)
15. Maerkl, S., Quake, S.: A systems approach to measuring the binding energy landscapes of transcription factors. Science **315**(5809), 233–236 (2007)
16. Quigley, M., et al.: Transcriptional analysis of HIV-specific CD8+ T cells shows that PD-1 inhibits T cell function by upregulating BATF. Nat. Med. **16**, 1147–1151 (2010)
17. Rigbolt, K.T., et al.: System-wide temporal characterization of the proteome and phosphoproteome of human embryonic stem cell differentiation. Sci. Signal. **4**, RS3 (2011)
18. Santolini, M., Mora, T., Hakim, V.: A general pair-wise interaction model provides an accurate description of in vivo transcription factor binding sites. PLoS One **9**(6), e99015 (2014)
19. Sharon, E., Segal, E.: A feature-based approach to modeling protein-DNA interactions. In: Speed, T., Huang, H. (eds.) RECOMB 2007. LNCS (LNBI), vol. 4453, pp. 77–91. Springer, Heidelberg (2007)
20. Song, Y., Wang, C., Qu, J.: A parameterized algorithm for predicting transcription factor binding sites. In: Huang, D.-S., Han, K., Gromiha, M. (eds.) ICIC 2014. LNCS, vol. 8590, pp. 339–350. Springer, Heidelberg (2014)
21. Song, Y., Yu, M.: On finding the longest antisymmetric path in directed acyclic graphs. Inf. Process. Lett. **115**(2), 377–381 (2015)
22. Song, Y., Chi, A.Y.: Peptide sequencing via graph path decomposition. Inf. Sci. **301**, 262–270 (2015)
23. Song, Y., Chi, A.Y.: A new approach for parameter estimation in the sequence-structure alignment of non-coding RNAs. J. Inf. Sci. Eng. **31**(2), 593–607 (2015)
24. Song, Y.: An improved parameterized algorithm for the independent feedback vertex set problem. Theoret. Comput. Sci. **535**, 25–30 (2014)
25. Song, Y.: A new parameterized algorithm for rapid peptide sequencing. PLoS One **9**(2), e87476 (2014)
26. Song, Y., Yu, M.: On the treewidths of graphs of bounded degree. PLoS One **10**(4), e0120880 (2015)
27. Stormo, G.: Computer methods for analyzing sequence recognition of nucleic acids. Annu. Rev. Biochem. **17**, 241–263 (1988)
28. Stormo, G., Hartzell, G.: Identifying protein-binding sites from unaligned DNA fragments. Proc. Natl. Acad. Sci. **86**(4), 1183–1187 (1989)
29. Wang, L.-S., Jensen, S.T., Hannenhalli, S.: An interaction-dependent model for transcription factor binding. In: Eskin, E., Ideker, T., Raphael, B., Workman, C. (eds.) RECOMB 2005. LNCS (LNBI), vol. 4023, pp. 225–234. Springer, Heidelberg (2007)

The λ-Turn: A New Structural Motif in Ribosomal RNA

Huizhu Ren[1], Ying Shen[2,3(✉)], and Lin Zhang[2]

[1] Key Laboratory of Hormones and Development (Ministry of Health), Metabolic Disease Hospital, 2011 Collaborative Innovation Center of Tianjin for Medical Epigenetics, Tianjin Institute of Endocrinology, Tianjin Medical University, Tianjin 300070, China
rose92316@163.com
[2] School of Software Engineering, Tongji University, Shanghai, China
{yingshen,cslinzhang}@tongji.edu.cn
[3] Key Laboratory of Intelligent Perception and Systems for High-Dimensional Information, Ministry of Education, Nanjing University of Science and Technology, Nanjing 210094, People's Republic of China

Abstract. RNA structural motifs are recurrent structural elements occurring in RNA molecules. They play essential roles in consolidating RNA tertiary structures and in binding proteins. Recently, we identified a new type of RNA structural motif, namely λ-turn, from ribosomal RNAs. This motif has a helix-internal loop-helix structure. The directions of its two helices are changed ~90° due to the existence of the internal loop. A guanine from the 3′-end of the internal loop extrudes out and forms a base triple with a G-C WC base pair from one helix of the motif. From the global perspective, the λ-turn is often capped by a helix and a tetraloop. A nucleotide between the capped helix and the tetraloop forms a consecutive base triple next to the first one with a G-C pair from the same helix that the first G-C pair resides. The λ-turn motif has a consensus sequence pattern and its 3D structure is conserved across different species. All the identified λ-turns are located on surfaces of ribosomal RNAs. Structures of ribosomes reveal direct interactions between λ-turns and ribosomal proteins. All these observations indicate that λ-turns have an important role in binding with ribosomal proteins.

Keywords: λ-Turn · RNA structural motif · Ribosomal RNA

1 Introduction

RNA structural motifs are recurrent structural elements occurring in RNA tertiary structures and they play essential roles when RNAs performing their functions in various biological processes (Moore 1999; Hendrix et al. 2005; Leontis et al. 2006; François et al. 2005). Furthermore, they are key components to consolidate RNA tertiary structures. In view of their importance, researchers endeavour to understand RNA motifs with their structures, characteristics, and functions. Currently, more than twenty types of RNA structural motifs have been identified (Woese et al. 1990; Szewczak et al. 1993; Cate et al. 1996; Chang and Tinoco 1994; Klein et al. 2001;

© Springer International Publishing Switzerland 2015
D.-S. Huang et al. (Eds.): ICIC 2015, Part II, LNCS 9226, pp. 456–466, 2015.
DOI: 10.1007/978-3-319-22186-1_45

Nissen et al. 2001; Wadley and Pyle 2004; Leontis and Westhof 2003). However, the steps of discovering new motifs will never stop. For instance, motifs like G-ribo and adenosine wedge, to name a few, have been recently identified. (Steinberg and Boutorine 2007; Jaeger et al. 2009; Gagnon and Steinberg 2010; Shen et al. 2013). These newly identified motifs greatly enrich our knowledge about RNA structures and their functions.

In this paper, we present a new type of RNA structural motif, namely λ-turn, the name of which is given by the shape of its characteristic strand. We successfully identified nine λ-turns from ribosomal RNAs (rRNAs). Based on observations on these nine instances, we find that the λ-turn motif has a consensus sequence pattern and its 3D structure is conserved across different species. In addition, the structure of the λ-turn has several distinct characteristics. Firstly, the λ-turn contains two helices which are arranged orthogonally. Secondly, the two helices are connected by a bulge which is composed by 2–4 nucleotides. Thirdly, a guanine located at the 3'-end of the bulge extrudes out and forms a triple base pair with the first G-C pair on a helix of the motif. All the identified λ-turns are located on surfaces of ribosomal RNAs, which implies that they may be involved in binding with other molecules, such as ribosomal proteins. The observations on four structures of ribosomes further support this assumption. In these ribosomes, λ-turns have direct interactions with ribosomal protein S17 or S11, which suggests that λ-turns play an important role when rRNAs try to bind with ribosomal proteins. Any mutation in positions of λ-turns may affect the stableness of ribosome structures.

2 Results

2.1 Definition of the λ-Turn

Similar to the kink-turn motif, the λ-turn is a helix-internal loop-helix-structure motif. It is composed of two strands containing 14–17 nucleotides (see Fig. 1(A)). The longer strand, which contains an unpaired region, is regarded as the characteristic strand. The other strand in the motif is called the complementary strand. Two strands form two segments of helices which are arranged orthogonally in the λ-turn. The first helix (Helix 1 in Fig. 1(A)), which is called the "UG-stem", is composed of two base pairs, a C-G WC pair and a U•G wobble base pair. The second helix (Helix 2 in Fig. 1(A)), which is called "AU-stem", is composed of four base pairs, two A-U (or U-A) pairs followed by two G-C pairs. Between the two helices, there is a bulge which is formed by the unpaired region of the characteristic strand. The bulge contains 2–4 nucleotides. At its 3'-end, there is always a guanine which interacts with the first G-C pair on the AU-stem and they together form a base triple (see Fig. 1(A)). The other residues in the bulge extrude out to form remote interactions with residues from other parts of the RNA to consolidate the tertiary structure of the RNA molecule.

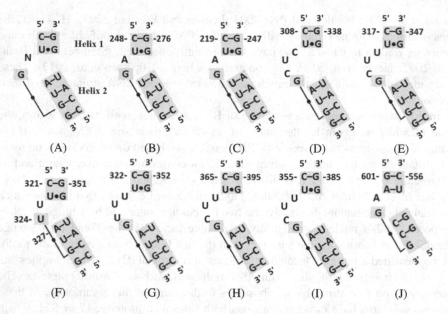

Fig. 1. Diagrams of secondary structure of λ-turns found in ribosomal RNAs. Solid lines represent Watson-Crick (WC) base pairings and dots represent non-WC base pairings. Yellow shading indicates the consensus sequence pattern. (A) The consensus sequence pattern of λ-turn. N represents one or three nucleotides of any type; (B) A:C248-G255/C271-G276 (PDB:2AW7); (C) A:C219-G226/C242-G247 (PDB:3BBN); (D) A:C308-G317/C333-G338 (PDB:2XZM); (E) 6:C317-G326/C342-G347 (PDB:3U5F); (F) A:C321-G330/C346-G351 (PDB:3IZ7); (G) 2: C322-G331/C347-G352 (PDB:3J3C); (H) 2:C365-G374/C390-G395 (PDB:3J3D); (I) i: C355-G364/C380-G385 (PDB:4KZX); (J) 3:G601-G607/G553-C556 (PDB:1S1I) (Color figure online).

2.2   Identification of λ-Turns

We have searched 550 RNA complexes (their PDB ID have been listed in Table S1 in the supplementary data) and identified nine instances of the λ-turn from nine rRNAs, which include two 16S rRNAs (PDB:2AW7, PDB:3BBN), six 18S rRNAs (PDB:2XZM, PDB:3U5F, PDB:3IZ7, PDB:3J3C, PDB:3J3D, PDB:4KZX), and a 5.8S/25S rRNA (PDB:1S1I). Since 16S and 18S rRNAs are homologous gene products, λ-turns mostly reside in 16S/18S rRNAs. The sequences of nine λ-turns have been listed in Table S2 in the supplementary data. It needs to be stressed that, we didn't find λ-turn-like structures from non-ribosomal RNAs.

2.3   Consensus Pattern of λ-Turns

The diagrams of secondary structure of nine identified λ-turns have been shown in Fig. 1(B)–(J). These diagrams clearly reveal that the λ-turn is composed of two strands which form two segments of helices and a bulge connecting the two helices. The first helix (Helix 1) contains a WC base pair (C-G) and a wobble base pair (U•G). It can be observed that eight λ-turns from 16S/18S rRNAs share the same components of Helix 1 (see Fig. 1(B)–(I)). It indicates that, in the evolution of 16S/18S rRNAs, no substitution takes place in Helix 1. However, in the λ-turn from 5.8S/25S rRNA, the C-G pair is replaced by a G-C pair and the U•G pair is replaced by an A-U pair (see Fig. 1(J)). The difference is possibly because that 16S/18S rRNAs are not homologous with 5.8S/25S rRNAs.

The second helix (Helix 2) contains four WC base pairs: two A-U (or U-A) pairs and two G-C pairs. In 16S rRNAs, Helix 2 starts with a U-A pair, followed by an A-U (Fig. 1(B)) or a U-A pair (Fig. 1(C)). As to cases of 18S rRNAs, Helix 2 always starts with an A-U pair, followed by a U-A pair (see Fig. 1(D)–(I)). The bases in the position of two G-C pairs remain the same across all λ-turns from 16S and 18S rRNAs. It indicates that, in the evolution from prokaryotes to eukaryotes, the substitution takes place in the position of two A-U/U-A base pairs in Helix 2. But it doesn't happen in the position of the following two G-C pairs. However, such pattern does not exist in the instance from 5.8S/25S rRNA, where A-U and U-A pairs are substituted by two C-G pairs (see Fig. 1(J)). In addition, the following two G-C pairs are replaced by a single guanine. It can be seen that the base-pair composition of Helix 2 in 5.8S/25S is quite different from those in 16S/18S rRNAs.

Between two helices, there is a bulge connecting them. The bulge contains two or four nucleotides (see Fig. 1(B)–(J)). Compared with the conserveness of two helices, the components of the bulge vary a lot across nine λ-turns, which suggests that substitutions occurring in the bulge are more frequent than those occurring in helices. The content of the bulge can be AG, CUCG, UUCG, UU, and UUUG. The nucleotide at the 3'-end of the bulge is often a guanine except the instance in PDB:3IZ7, where the guanine is replaced by a uracil. The base of the guanine has a remote interaction with bases of the first G-C pair in Helix 2. They together form a base triple which helps to consolidate the structure of the λ-turn. Three bases in the base triple are coplanar and their interaction diagrams have been shown in Fig. 2. There are two ways of interactions occurring in the base triple. The first one is a cis-WC-WC (cWW) interaction between the G-C WC pair in Helix 2, which connects WC edges of the guanine and the cytosine. The second one is a trans-Hoogsteen-WC (tHW) interaction, which connects the Hoogsteen edge of the guanine (or the cytosine) from the G-C pair in Helix 2 and the WC edge of the inserted guanine (or uracil) (see Fig. 2(A)–(C)). Therefore, the base triple in the λ-turn can be represented as a combination of cWW and tHW (cWW/tHW) interactions.

Since G/U-GC base triples appear in all nine λ-turns, they seem indispensable for λ-turns' formation. We also notice that, in all the base triples the G-C WC pair remains the same. Therefore, we reach to the conclusion that the first G-C pair in Helix 2 has an important role in forming the base triple and in consolidating the λ-turn's structure. At the same time, we are wondering what makes the first G-C pair in Helix 2 non-substituted. To answer this question, we looked into the RNA Base Triples Database (Petrov et al. 2013) and investigated the family of cWW/tHW, the type of which G/U-GC base triples in λ-turns belong to. In the cWW/tHW family, the total number of RNA fragments discovered by FR3D (Sarver et al. 2008) is 80, among which there are 25 G-GC triples. In other families involving cWW interactions (such as cWW/cHW, cWW/cHS), G-GC and U-GC triples also account for a large proportion.

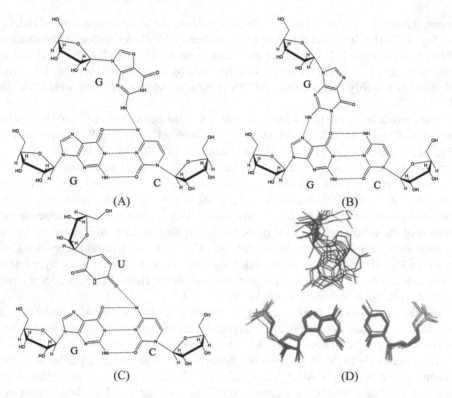

Fig. 2. Interaction diagrams of the G-GC and U-GC base triples occurring in λ-turns. (A) The G-GC base triple with interactions occurring between the inserted guanine and the cytosine from G-C WC pair; (B) the G-GC base triple with interactions occurring between the inserted guanine and the guanine from G-C WC pair; (C) the U-GC base triple with interactions occurring between the inserted uracil and the cytosine from G-C WC pair; (D) structure alignment of nine triple base pairs.

The great number of G/U-GC triples indicates that their structures are more stable than other types of base triples. It further suggests that, as long as the nucleotide at the 3′-end of the bulge is a guanine, the G-C pair in Helix 2 prefers to be non-substituted. Otherwise the stableness of the base triple will be weakened. Interestingly, in the case of PDB:3IZ7, the guanine in the bulge has been replaced by a uracil. We can't help wondering that, in this circumstance, whether the G-C pair could be substituted by an A-U pair without affecting the stableness of the base triple and the whole RNA structure[1]. However, the consequence of such substitution remains unclear.

The tertiary structure of the λ-turn has some distinct characteristics. Firstly, the directions of two helices are changed ∼90° due to the existence of the bulge (Fig. 3

---

[1] The replacement of G-C pair to A-U pair is possible. First, the cytosine is substituted by a uracil which consequently forms a G•U wobble pair. Then, the guanine is substituted by an adenosine and A-U pair will be formed finally. The U-AU triples are also abundant in the RNA Base Triples Database.

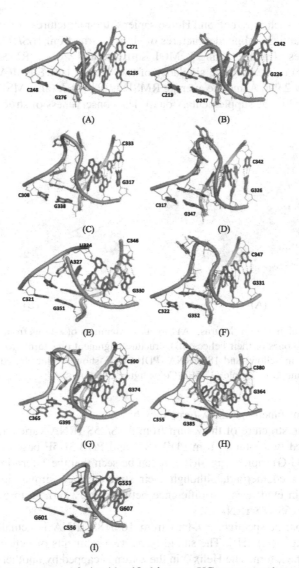

Fig. 3. The tertiary structures of nine identified λ-turns. UG stems are shown in magenta, AU stems are shown in blue, and the internal loops are shown in teal. (A) A:C248-G255/C271-G276 (PDB:2AW7); (B) A:C219-G226/C242-G247 (PDB:3BBN); (C) A:C308-G317/C333-G338 (PDB:2XZM); (D) 6:C317-G326/C342-G347 (PDB:3U5F); (E) A:C321-G330/C346-G351 (PDB:3IZ7); (F) 2:C322-G331/C347-G352 (PDB:3J3C); (G) 2:C365-G374/C390-G395 (PDB:3J3D); (H) i:C355-G364/C380-G385 (PDB:4KZX); (I) 3:G553-C556/G601-G607 (PDB:1S1I) (Color figure online).

(A)–(I)). Secondly, the nucleotides in the bulge extrude out and have remote interactions with other parts of RNA molecule. Despite that nine λ-turns are identified from different species, such as E. coli and Homo sapiens, their structures are quite conserved. To verify our claim, we aligned structures of eight λ-turns from 16S/18S rRNAs based on the backbones of helices using PyMOL software and we use RMSD (Root Mean Square Distance) to measure the similarity of two λ-turns (see Fig. 4(A)). The largest RMSD is merely 2.996 Å and the average RMSD is 0.8944 Å (all RMSD values can be found in Table S3 in the supplementary data). The conserveness of structures of λ-turns

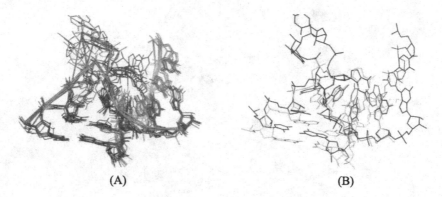

(A)   (B)

Fig. 4. Structure alignment of λ-turns. (A) Structure alignment of λ-turns from 16S/18S rRNAs based on the backbones of their helices; (B) structure alignment of λ-turns from 5.8/25S rRNA (PDB:1S1I, show in yellow) and 18S rRNA (PDB:3U5F, show in blue) based on positions of G-G pair (1S1I) and G-GC triple (3U5F) (Color figure online).

implies important functions that λ-turns may have.

However, the structure of the λ-turn from 5.8S/25S rRNA varies a bit from other ones. We aligned two λ-turns from PDB:1S1I and PDB:3U5F based on positions of their G-GC (or G-G) triples (Fig. 4(B)). It can be seen that the λ-turn in 3U5F is more compact than its counterpart, although their structures are similar in general. This observation again evidences the difference between the λ-turn from 5.8S/25S rRNA and those from 16S/18S rRNAs.

From a global perspective, a λ-turn from 16S/18S rRNA is actually folded by a single strand (Fig. 5(A)–(H)). The strand forms two segments of helices and bends at the position of the λ-turn. The Helix 2 in the λ-turn is capped by another helix and then a tetraloop. In this global structure, an interesting pattern is observed: between the capped helix and the tetraloop there is always a guanine stretching inside and interacting with the second G-C pair in Helix 2 (see nucleotides coloured blue in Fig. 5). The new G-GC base triple is also formed by cWW/tHW interactions, the same way by which the first G-GC base triple in Helix 2 is formed. The new base triples can be found in all λ-turns from 16S/18S rRNAs. Due to the existence of the inserted guanine, the second G-C pair in Helix 2 prefers not to be substituted for the same reason as the first G-C pair in Helix 2. It further explains why the second G-C pair in Helix 2 remains the same across 16S/18S rRNAs.

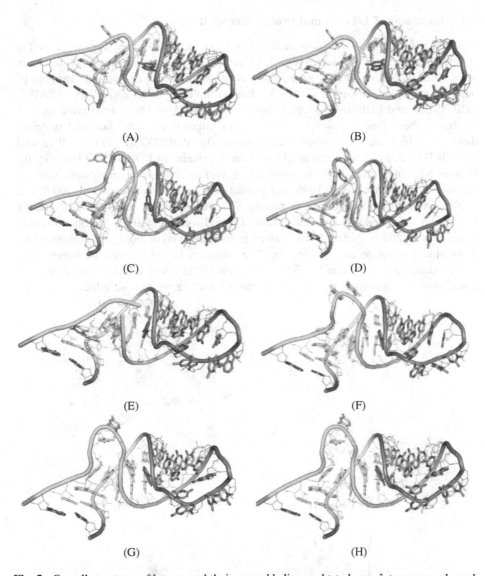

Fig. 5. Overall structures of λ-turns and their capped helices and tetraloops. λ-turns are coloured yellow and the guanines located between the capped helix and the tetraloop are coloured blue. (A) A:C248-G255/C271-G276 (PDB:2AW7); (B) A:C219-G226/C242-G247 PDB:3BBN); (C) A:C308-G317/C333-G338 (PDB:2XZM); (D) 6:C317-G326/C342-G347 (PDB:3U5F); (E) A:C321-G330/C346-G351 (PDB:3IZ7); (F) 2:C322-G331/C347-G352 (PDB:3J3C); (G) 2: C365-G374/C390-G395 (PDB:3J3D); (H) i:C355-G364/C380-G385 (PDB:4KZX) (Color figure online).

2.4   Locations of λ-Turns and Protein Recognition

Nine identified λ-turns are all located on surfaces of rRNAs (see Figure S1 in the supplementary data), which indicates that λ-turns may have significant functions, such as binding with ribosomal proteins. The structures of ribosomes provide insights of possible functions of λ-turns. There are four complexes (PDB:2XZM, PDB:2AW7, PDB:3BBN, and PDB:4KZX) which contain both rRNA and ribosomal proteins. The λ-turns in these four complexes all have direct interactions with ribosomal proteins, three of which are S17 ribosomal proteins (in PDB:2XZM, PDB:2AW7, and PDB:3BBN) and one of which is S11 ribosomal protein (in PDB:4KZX) (see Fig. 6). Protein S17 in human body is involved in various biological processes, such as translation (Vladimirov et al. 1996) and translational initiation (Cmejla et al. 2007), etc. Studies indicate that a shortage of protein S17 may result in the disease of anemia because of the increasing self-destruction of blood-forming cells in the bone marrow (Cmejla et al. 2007). Protein S11 in human is also involved in biological processes like translational initiation and termination. The mutation in the location of λ-turns may cause a deficiency of protein S11/S17 binding for rRNAs, which hinders the process of translation. Consequently a similar syndrome of anemia may be introduced.

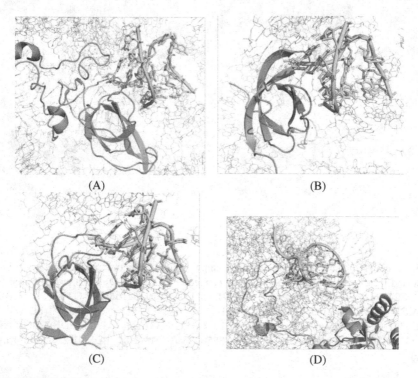

(A)

(B)

(C)

(D)

Fig. 6. Interaction diagrams between λ-turns and ribosomal proteins. λ-turns are coloured yellow and proteins are represented by green ribbons. (A) The λ-turn and protein S17 in PDB:2XZM; (B) the λ-turn and protein S17 in PDB:2AW7; (C) the λ-turn and protein S17 in PDB:3BBN; (D) the λ-turn and protein S11 in PDB:4KZX (Color figure online).

3 Discussion

In this paper, we describe a new type of RNA structural motif, namely λ-turn, which is only identified from ribosomal RNAs. The λ-turn is a helix-internal loop-helix-structure motif. It bends at the position of the internal loop, which leads to a ~ 90° change in the direction of its two helices. The region of the internal loop contains unpaired nucleotides which have remote interactions with nucleotides from other parts of the RNA molecule. At the 3'-end of the internal loop, there is a guanine which extrudes out and forms a base triple with the first G-C pair in Helix 2. From the global perspective, the λ-turn is capped by another helix followed by a tetraloop. Between the capped helix and the tetraloop, there is a guanine stretching inside and forming a base triple with the second G-C pair from Helix 2. The two base triples help to consolidate the structure of the λ-turn.

All the identified λ-turns are located on surfaces of rRNAs. In structures of ribosomes, it can be seen that λ-turns have direct interactions with ribosomal proteins. Since proteins S11 and S17 are involved in the translational initiation and termination, we speculate that λ-turns may be involved the process of translation and mutations in λ-turns may lead to the disease of anemia.

Supplementary Data
Supplementary data are available at: http://sse.tongji.edu.cn/yingshen/LTurn/supp.docx.

Acknowledgement. The work described in this paper is supported partially by the Natural Science Foundation of China under grants no. 61303112 and no. 81373864, and partially by the Key Laboratory of Intelligent Perception and Systems for High-Dimensional Information, Ministry of Education, Nanjing University of Science and Technology, Nanjing, China, under grant no. 30920140122006.

References

Cate, J.H., Gooding, A.R., Podell, E., Zhou, K., Golden, B.L., Kundrot, C.E., Cech, T.R., Doudna, J.A.: Crystal structure of a group I ribozyme domain: principle of RNA packing. Science **273**, 1678–1685 (1996)

Chang, K., Tinoco, I.: Characterization of a "kissing" hairpin complex derived from the human immunodeficiency virus genome. Proc. Natl. Acad. Sci. U.S.A. **91**, 8705–8709 (1994)

Cmejla, R., Cmejlova, J., Handrkova, H., Petrak, J., Pospisilova, D.: Ribosomal protein S17 gene (RPS17) is mutated in Diamond-Blackfan anemia. Hum. Mutat. **28**, 1178–1182 (2007)

François, B., Russell, R.J., Murray, J.B., Aboul-ela, F., Masquida, B., Vicens, Q., Westhof, E.: Crystal structures of complexes between aminoglycosides and decoding A site oligonucleotides: role of the number of rings and positive charges in the specific binding leading to miscoding. Nucleic Acids Res. **33**, 5677–5690 (2005)

Gagnon, M.G., Steinberg, S.V.: The adenosine wedge: a new structural motif in ribosomal RNA. RNA **16**, 375–381 (2010)

Hendrix, D.K., Brenner, S.E., Holbrook, S.R.: RNA structural motifs: building blocks of a modular biomolecule. Q. Rev. Biophys. **38**, 221–243 (2005)

Jaeger, L., Verzemnieks, E.J., Geary, C.: The UA_handle: a versatile submotif in stable RNA architectures. Nucleic Acids Res. **37**, 215–230 (2009)

Klein, D.J., Schmeing, T.M., Moore, P.B., Steitz, T.A.: The kink-turn: A new RNA secondary structure motif. EMBO J. **20**, 4214–4221 (2001)

Leontis, N.B., Westhof, E.: Analysis of RNA motifs. Curr. Opin. Struct. Biol. **16**, 300–308 (2003)

Leontis, N.B., Lescoute, A., Westhof, E.: The building blocks and motifs of RNA architecture. Curr. Opin. Struct. Biol. **16**, 279–287 (2006)

Moore, P.B.: Structural motifs in RNA. Annu. Rev. Biochem. **68**, 287–300 (1999)

Nissen, P., Ippolito, J.A., Ban, N., Moore, P.B., Steitz, T.A.: RNA tertiary interactions in the large ribosomal subunit: The A-minor motif. Proc. Natl. Acad. Sci. U.S.A. **98**, 4899–4903 (2001)

Petrov, A.I., Zirbel, C.L., Leontis, N.B.: Automated classification of RNA 3D motifs and the RNA 3D Motif Atlas. RNA **19**, 1327–1340 (2013)

Sarver, M., Zirbel, C.L., Stombaugh, J., Mokdad, A., Leontis, N.B.: FR3D: finding local and composite recurrent structural motifs in RNA 3D structures. J. Math. Biol. **56**, 215–252 (2008)

Shen, Y., Wong, H.-S., Zhang, S., Zhang, L.: RNA structural motif recognition based on least-squares distance. RNA **19**, 1183–1191 (2013)

Steinberg, S.V., Boutorine, Y.I.: G-ribo: a new structural motif in ribosomal RNA. RNA **13**, 549–554 (2007)

Szewczak, A.A., Moore, P.B., Chang, Y.L., Wool, I.G.: The conformation of the sarcin/ricin loop from 28S ribosomal RNA. Proc. Natl. Acad. Sci. U.S.A. **90**, 9581–9585 (1993)

Vladimirov, S.N., Ivanov, A.V., Karpova, G.G., Musolyamov, A.K., Egorov, T.A., Thiede, B., Wittmann-Liebold, B., Otto, A.: Characterization of the human small-ribosomal-subunit proteins by N-terminal and internal sequencing, and mass spectrometry. Eur. J. Biochem. **239**, 144–149 (1996)

Wadley, L.M., Pyle, A.M.: The identification of novel RNA structural motifs using COMPADRES: an automated approach to structural discovery. Nucleic Acids Res. **32**, 6650–6659 (2004)

Woese, C.R., Winker, S., Gutell, R.R.: Architecture of ribosomal RNA: constraints on the sequence of "tetra-loops". Proc. Natl. Acad. Sci. U.S.A. **87**, 8467–8471 (1990)

Discovery of Ovarian Cancer Candidate Genes Using Protein Interaction Information

Di Zhang[1], Qingbao Wang[2], RongRong Zhu[1], Hai-Tao Li[3],
Chun-Hou Zheng[3], and Junfeng Xia[1(✉)]

[1] Key Laboratory of Intelligent Computing and Signal Processing,
Ministry of Education, and Institute of Health Sciences,
School of Computer Science and Technology,
Anhui University, Hefei 230039, Anhui, China
[2] Clinical Laboratory Center, The First Affiliated Hospital,
Anhui University of Traditional Chinese Medicine,
Hefei 230012, Anhui, China
[3] College of Electrical Engineering and Automation,
Anhui University, Hefei 230601, Anhui, China

Abstract. Ovarian cancer (OVC) is one of the deadliest cancers in women. At present, effective clinical therapy for OVC is still limited. We adopted a computational method to identify OVC candidate genes based on subnetwork extraction algorithm in the protein interaction network. Of the final identified five genes, *MSH3* and *PIK3R3* genes have been reported to be implicated in OVC tumorigenesis, indicating that the method is effective, whereas the other three genes (*MAPK8IP1*, *PICK1*, and *IQGAP3*) might have potential roles in OVC development. The results provided a novel insight into identifying the OVC-related genes and exploring new methods for OVC therapy.

Keywords: Ovarian cancer · Protein interaction network · Degree · Betweeness

1 Introduction

In 2015, 21,290 patients will suffer from ovarian cancer (OVC) and 14,180 cases will die of the disease in American [1]. Early detection of OVC is crucial to improve the effectiveness of OVC treatment. Therefore, it is very important to identify candidate genes associated with OVC, which may become potential biomarkers for further clinical application.

Recently, with the rapid development of next-generation sequencing technology, some genetic markers have been identified in OVC, such as *TP53* and *RB1* [2]. However, the genes in networks rather than the single gene drive the cancer initiation and progression. Protein-protein interaction (PPI) data makes it possible to extract the genes/proteins that involve in the same or similar biological processes or signaling pathways. Lots of PPI network-based algorithms have been developed to identify cancer associated genes, such as NeAT [3] and Genes2Networks [4]. However, their applications are limited by only providing the online service.

© Springer International Publishing Switzerland 2015
D.-S. Huang et al. (Eds.): ICIC 2015, Part II, LNCS 9226, pp. 467–472, 2015.
DOI: 10.1007/978-3-319-22186-1_46

In contrast, GenRev [5] makes up for their shortcomings and exhibits a series of advantages.

In this paper, we applied a computational method for identifying OVC related genes by combining the existing OVC genes and the human PPI network. A subnetwork extraction algorithm in GenRev was then applied to construct the paths between each pair of genes. Finally, 51 candidate genes were identified. Among them, two high-degree genes (*MAPK8IP1*, *MSH3*) and three high-betweenness genes (*PICK1*, *IQGAP3*, and *PIK3R3*) were analyzed. These predictions may be significant for future efforts to unravel the role of genes in tumorigenesis and the design of gene targeted therapeutics in OVC.

2 Methods

2.1 Collection of Known OVC Related Genes

To collect the known OVC-related genes, we curated five data sources, including the top 20 genes with the highest frequency of mutations in COSMIC (Catalog Of Somatic Mutations in Cancer) [6], OMIM (Online Mendelian Inheritance in Man) [7], DriverDB (a database for tumor driver gene identification) [8], KEGG (Kyoto Encyclopedia of Genes and Genomes) [9], and six candidate genes extracted from a literature [10]. Finally, 38 non-redundant OVC associated genes were curated for further study.

2.2 Construction of PPI Network

In order to construct a PPI network, we acquired the PPI data from HPRD [11]. HPRD provides a resource of experimentally verified protein interaction for Homo sapiens. In PPI networks, each node stands for an encoded gene (protein) and each edge stands for a protein interaction.

2.3 Selection of Candidate Genes

GenRev [5] was used to explore functional relevance of genes in a defined molecular network (PPI network). 38 non-redundant OVC related genes were used as the input seed genes to construct a subnetwork module through the limited kWalks algorithm [12] in GenRev [5].

2.4 GO Enrichment Analysis

To evaluate the function of the gene set, we carried on the functional analysis utilizing the online tool DAVID [13]. We selected the GO terms with the adjusted p-value smaller than 0.05 followed by the Benjamini-Hochberg approach [14], which was completed in DAVID website.

3   Results and Discussions

3.1   Biological Analysis of Significant OVC-Related Genes

Using GenRev [5], we identified 51 significant linker genes with a p-value smaller than 0.05. The GO enrichment analysis of these genes in DAVID was shown in Table 1. We can find that many of these genes are mainly related to nucleotide binding activity, including ATP binding, adenyl ribonucleotide binding, adenyl nucleotide binding, purine nucleoside binding and so on. Altered nucleotide binding activity might contribute to OVC progression. The GO enrichment analysis suggests that the detected candidate genes may implicate in OVC development.

3.2   Five Linker Genes (Two Linker Genes from the Top 20-High Degree Genes and Three Linker Genes from the Top 20 High-Betweenness Genes in the Subnetwork) Related to OVC Progression

In this study, we have obtained 89 OVC candidate genes, including 38 seed genes and 51 linker genes. We focus on the top 20 genes with high-degree (which represents a protein with many interacting partners) and high-betweeness (which represents that many "shortest paths" going through them), respectively. Apart from the seed genes, we obtained 5 novel linker genes, including two high-degree genes (MAPK8IP1, MSH3) and three high-betweenness genes (PIK3R3, PICK1, and IQGAP3) (Table 2). Below, we will briefly discuss the relationship between these five linker genes and OVC.

The MSH3 (mutS homolog 3) gene is a component of DNA mismatch repair (MMR) system. Loss of MMR function contributes to genetic instability. There is a high relevance between the MMR deficiency and several types of cancer [15–17], including ovarian cancer [17]. Whether the loss of MSH3 expression contributes to OVC need receive further study.

Table 1. GO analysis for identified OVC related genes

GO-ID	Description	Count	p-value
GO:0030554	Adenyl nucleotide binding	12	0.0054
GO:0004428	Inositol or phosphatidylinositol kinase activity	3	0.0055
GO:0001883	Purine nucleoside binding	12	0.0061
GO:0001882	Nucleoside binding	12	0.0064
GO:0005524	ATP binding	11	0.0101
GO:0032559	Adenyl ribonucleotide binding	11	0.0111
GO:0017076	Purine nucleotide binding	12	0.0224
GO:0030971	Receptor tyrosine kinase binding	2	0.0238
GO:0019207	Kinase regulator activity	3	0.0317
GO:0032555	Purine ribonucleotide binding	11	0.0409
GO:0032553	Ribonucleotide binding	11	0.0409

Table 2. (a) List the top 20 genes ranked according to their degree in the subnetwork. (b) List the top 20 genes ranked according to their betweenness in the subnetwork

(a)		(b)	
Gene	Degree	Gene	Betweenness
TP53	19	TP53	0.221912437
BRCA1	18	MYC	0.203638463
MYC	17	EGFR	0.183340673
EGFR	17	BRCA1	0.172483315
SMARCA4	15	SMARCA4	0.163066073
AKT1	14	CTNNB1	0.153191445
CREBBP	12	AKT1	0.151770847
CTNNB1	12	CREBBP	0.072440259
CHEK2	10	BRCA2	0.067136886
MSH2	10	PARK2	0.060032344
ERBB2	9	ERBB2	0.058544783
PARK2	8	*PICK1	0.04586479
CDH1	8	CHEK2	0.04228114
RB1	7	PPP2R1A	0.042152486
PPP2R1A	7	*IQGAP3	0.041007337
MLH1	7	*PIK3R3	0.03796305
*MAPK8IP1	6	RB1	0.03296284
CDKN2A	5	MLH1	0.031270584
PTEN	5	PTEN	0.03023639
*MSH3	5	PIK3CA	0.030107417

*represents the linker gene (the identified candidate gene

MAPK8IP1 (mitogen-activated protein kinase 8 interacting protein 1) enhances JNK signaling by placing JNK next to upstream kinase, which is important in oncogenic transformation during apoptosis in many cell death paradigms, cell survival and proliferation [15, 16]. A study has reported that SNPs in *MAPK8IP1* was associated with overall survival in pancreatic cancer [17]. Our results suggested that it is important to further investigate the relationship between *MAPK8IP1* and OVC.

PIK3R3 (phosphoinositide 3-kinases) is a member of phosphatidylinositol 3'-kinase (PI3 K) family, which plays a crucial role in many cancer-associated signal transduction pathways. Knockdown of *PIK3R3* gene expression have been found to significantly contribute to the apoptosis in OVC cell lines, which suggest that *PIK3R3* might act as a potential therapeutic target in OVC [18].

PICK1 (protein interacting with C α kinase 1) is a peripheral membrane protein coding gene, which is thought to play vital roles in multiple pathologies including cancer. Besides, it can promote cancer cell proliferation [19]. *PICK1* has been proposed that it may serve not only as an indicator for poor prognosis, but also as a therapeutic target in breast tumor [19]. Our results implied that *PICK1* might be a promising marker for OVC therapy.

IQGAP3 (IQ motif containing GTPase activating protein 3) is special because its expression is just confined to proliferating cells. What's more, it is essential and enough for driving cell proliferation through activating Ras-dependent ERK [20]. Previous analyses have indicated that *IQGAP3* was overexpressed in various types of cancers, including ovary, lung, gastric, and breast malignancies [21]. Therefore, *IQGAP3* was expected to be an OVC candidate gene.

4 Conclusion

We applied a subnetwork extraction method, i.e. GenRev, on OVC and identified five liner genes with the most potential to be biomarkers for OVC. Among them, *MSH3* and *PIK3R3* genes have direct evidence for their roles in OVC, whereas the other three genes (*MAPK8IP1*, *PICK1*, and *IQGAP3*) may play potential roles in OVC development. These results might provide new insights into discovering candidate genes associated with OVC development and new strategies of clinical therapy.

Acknowledgments. This work was supported by National Natural Science Foundation of China (31301101 and 61272339), the Anhui Provincial Natural Science Foundation (1408085QF106), the Specialized Research Fund for the Doctoral Program of Higher Education (20133401120011), and the Technology Foundation for Selected Overseas Chinese Scholars from Department of Human Resources and Social Security of Anhui Province (No. [2014]-243).

References

1. Siegel, R.L., Miller, K.D., Jemal, A.: Cancer statistics, 2015. CA Cancer J. Clin. **65**(1), 5–29 (2015)
2. Flesken-Nikitin, A., et al.: Induction of carcinogenesis by concurrent inactivation of p53 and Rb1 in the mouse ovarian surface epithelium. Cancer Res. **63**(13), 3459–3463 (2003)
3. Brohée, S., et al.: Network analysis tools: from biological networks to clusters and pathways. Nat. Protoc. **3**(10), 1616–1629 (2008)
4. Berger, S.I., Posner, J.M., Ma'ayan, A.: Genes2Networks: connecting lists of gene symbols using mammalian protein interactions databases. BMC Bioinf. **8**(1), 372 (2007)
5. Zheng, S., Zhao, Z.: GenRev: exploring functional relevance of genes in molecular networks. Genomics **99**(3), 183–188 (2012)
6. Forbes, S.A., et al.: COSMIC: mining complete cancer genomes in the catalogue of somatic mutations in cancer. Nucleic Acids Res. **39**, gkq929 (2010)
7. Hamosh, A., et al.: Online mendelian inheritance in man (OMIM), a knowledgebase of human genes and genetic disorders. Nucleic Acids Res. **33**(suppl. 1), D514–D517 (2005)
8. Cheng, W.-C., et al.: DriverDB: an exome sequencing database for cancer driver gene identification. Nucleic Acids Res. **42**(D1), D1048–D1054 (2014)
9. Kanehisa, M., Goto, S.: KEGG: kyoto encyclopedia of genes and genomes. Nucleic Acids Res. **28**(1), 27–30 (2000)
10. Network, C.G.A.R.: Integrated genomic analyses of ovarian carcinoma. Nature **474**(7353), 609–615 (2011)

11. Goel, R., et al.: Human protein reference database and human proteinpedia as resources for phosphoproteome analysis. Mol. BioSyst. **8**(2), 453–463 (2012)
12. Dupont, P., et al.: Relevant subgraph extraction from random walks in a graph. Universite catholique de Louvain, UCL/INGI, Number RR 7 (2006)
13. Alvord, G., et al.: The DAVID gene functional classification tool: a novel biological module-centric algorithm to functionally analyze large gene lists. Genome Biol. **8**(9), 183 (2007)
14. Benjamini, Y., Hochberg, Y.: Controlling the false discovery rate: a practical and powerful approach to multiple testing. J. Roy. Stat. Soc. Ser. B (Methodol.) **57**, 289–300 (1995)
15. Stebbins, J.L., et al.: Identification of a new JNK inhibitor targeting the JNK-JIP interaction site. Proc. Natl. Acad. Sci. **105**(43), 16809–16813 (2008)
16. Cui, J., Zhang, M., ZHANG, Y.Q.: JNK pathway: diseases and therapeutic potential1. Acta Pharmacol. Sin. **28**(5), 601–608 (2007)
17. Reid-Lombardo, K.M., et al.: Survival is associated with genetic variation in inflammatory pathway genes among patients with resected and unresected pancreatic cancer. Ann. Surg. **257**(6), 1096 (2013)
18. Zhang, L., et al.: Integrative genomic analysis of phosphatidylinositol 3′-kinase family identifies PIK3R3 as a potential therapeutic target in epithelial ovarian cancer. Clin. Cancer Res. **13**(18), 5314–5321 (2007)
19. Zhang, B., et al.: Protein interacting with C α kinase 1 (PICK1) is involved in promoting tumor growth and correlates with poor prognosis of human breast cancer. Cancer Sci. **101**(6), 1536–1542 (2010)
20. Nojima, H., et al.: IQGAP3 regulates cell proliferation through the Ras/ERK signalling cascade. Nat. Cell Biol. **10**(8), 971–978 (2008)
21. Yang, Y., et al.: IQGAP3 promotes EGFR-ERK signaling and the growth and metastasis of lung cancer cells. PLoS ONE **9**(5), e97578 (2014)

Oscillatory Behavior of the Solutions for a Coupled van der Pol-Duffing Oscillator with Delay

Yuanhua Lin[(⊠)] and Zongyi Hou

School of Mathematics and Statistics,
Hechi University, Yizhou, Guangxi 546300, China
{lyh4473,hcxyhouzongyi}@163.com

Abstract. This paper studies the oscillatory behavior for a coupled van der Pol-Duffing oscillator with delay. Two theorems are provided to guarantee the oscillation of the system. Computer simulations are given to support the theoretical analysis.

Keywords: Coupled van der Pol-Duffing oscillator · Delay · Equilibrium point · Oscillation

1 Introduction

Dynamics of coupled van der Pol oscillators and coupled van der Pol-Duffing oscillators have many applications in various areas of science and technology, such as physical, chemical, biological systems and so on [1–14]. Noting that there is a relatively large number of parameter affecting dynamical phenomena in these systems, the dynamics behavior is more complex than the problem concerning the dynamics of one oscillator. In [1], Li et al. have discussed the following coupled van der Pol oscillators with time delays as following:

$$
\begin{cases}
y_1'' + w_1^2 y_1 - \varepsilon(1 - y_1^2)y_1' = \varepsilon\alpha\left[y_2(t-\tau) + y_2'(t-\tau)\right] \\
y_2'' + w_2^2 y_2 - \varepsilon(1 - y_1^2)y_1' = \varepsilon\alpha\left[y_1(t-\tau) + y_1'(t-\tau)\right]
\end{cases}
\tag{1}
$$

By means of the method of averaging together with truncation of Taylor expansions, the Hopf bifurcations for symmetric modes were determined. Zhang and Gu extend the self-excited system to two delays as follows [2]:

$$
\begin{cases}
x_1''(t) + \varepsilon\left(x_1^2(t) - 1\right)x_1' + x_1(t) = \alpha[y_1(t-\tau_2) + x_1(t)] \\
y_1''(t) + \varepsilon\left(y_1^2(t) - 1\right)y_1' + x_1(t) = \alpha[x_1(t-\tau_1) + y_1(t)]
\end{cases}
\tag{2}
$$

Using the theory of normal form and the center manifold theorem, the authors have provided the explicit expression for determining the direction of the Hopf bifurcations and the stability of the bifurcating periodic solutions. Sabarathinam et al. [3] considered the Duffing oscillator which is one of the typical and widely explored examples of nonlinear oscillator without time delays as follows:

© Springer International Publishing Switzerland 2015
D.-S. Huang et al. (Eds.): ICIC 2015, Part II, LNCS 9226, pp. 473–479, 2015.
DOI: 10.1007/978-3-319-22186-1_47

$$\begin{cases} z_1''(t) + bz_1' - \alpha z_1 + \beta z_1^3 + k(x_2 - x_1) = 0 \\ z_2''(t) + bz_2' - \alpha z_2 + \beta z_2^3 + k(x_1 - x_2) = 0 \end{cases} \tag{3}$$

The authors provided numerical and experimental evidence in the case of small positive and negative damping. Two different types of transient chaos in the neighborhood of unstable tori appeared. For another kind of coupled van der Pol-Duffing oscillators without time delays of the following [4]:

$$\begin{cases} x'' - (\lambda_1 - x^2)x' + \left(1 - \frac{\Delta}{2}\right)x + \beta x^3 + \varepsilon(x - y) + \mu(x' - y') = 0 \\ y'' - (\lambda_2 - y^2)y' + \left(1 + \frac{\Delta}{2}\right)y + \beta y^3 + \varepsilon(y - x) + \mu(y' - x') = 0 \end{cases} \tag{4}$$

where λ_1 and λ_2 are parameters responsible Hopf bifurcation, Δ is the detuning parameter, β is the nonisochronism parameter, ε and μ are coupling parameters. By using the method of charts of dynamic regimes in the parameter planes, synchronization in the system of coupled van der Pol-Duffing oscillators with inertial and dissipative coupling has been studied. Numerical study of the parameters space of the system has been also carried out. Motivated by the above models, in this paper, we extend (4) to time delay system as follows:

$$\begin{cases} x'' - (\lambda_1 - x^2)x' + \left(1 - \frac{\Delta}{2}\right)x + \beta_1 x^3 + \varepsilon_1(x(t-\tau) - y(t-\tau)) + \mu_1(x'(t-\tau) - y'(t-\tau)) = 0 \\ y'' - (\lambda_2 - y^2)y' + \left(1 + \frac{\Delta}{2}\right)y + \beta_2 y^3 + \varepsilon_2(y(t-\tau) - x(t-\tau)) + \mu_2(y'(t-\tau) - x'(t-\tau)) = 0 \end{cases} \tag{5}$$

where the parameters $\lambda_1, \lambda_2, \beta_1, \beta_2, \varepsilon_1, \varepsilon_2, \mu_1$ and μ_2 may be different values. The oscillatory behavior of system (5) is considered.

2 Preliminaries

System (5) can be rewritten as a four first order delay differential equations as follows:

$$\begin{cases} u_1' = -u_2 \\ u_2' = (\lambda_1 - u_1^2)u_2 + \left(1 - \frac{\Delta}{2}\right)u_1 + \beta_1 u_1^3 + \varepsilon_1(u_1(t-\tau) - u_3(t-\tau)) - \mu_1(u_2(t-\tau) - u_4(t-\tau)) \\ u_3' = -u_4 \\ u_4' = (\lambda_2 - u_3^2)u_4 + \left(1 + \frac{\Delta}{2}\right)u_3 + \beta_2 u_3^3 + \varepsilon_2(u_3(t-\tau) - u_1(t-\tau)) - \mu_2(u_4(t-\tau) - u_2(t-\tau)) \end{cases} \tag{6}$$

The nonlinear system (6) can be expressed in the following matrix form:

$$U'(t) = AU(t) + BU(t - \tau) + M(U) \tag{7}$$

where

$$U(t) = [u_1, u_2, u_3, u_4]^T,$$

$$U(t - \tau) = [u_1(t - \tau), u_2(t - \tau), u_3(t - \tau), u_4(t - \tau)]^T,$$

$$A = \left(a_{ij}\right)_{4\times4} = \begin{pmatrix} 0 & -1 & 0 & 0 \\ \delta_1 & \lambda_1 & 0 & 0 \\ 0 & 0 & 0 & -1 \\ 0 & 0 & \delta_2 & \lambda_2 \end{pmatrix}, \quad B = \left(b_{ij}\right)_{4\times4} = \begin{pmatrix} 0 & 0 & 0 & 0 \\ \varepsilon_1 & -\mu_1 & -\varepsilon_1 & \mu_1 \\ 0 & 0 & 0 & 0 \\ \varepsilon_2 & -\mu_2 & -\varepsilon_2 & \mu_2 \end{pmatrix}$$

and $M(U) = \left(0, \ -u_1^2 u_2 + \beta_1 u_1^3, \ 0, \ -u_3^2 u_4 + \beta_2 u_3^3\right)^T$, where $\delta_1 = 1 - \frac{A}{2}$, and $\delta_2 = 1 + \frac{A}{2}$.

The linearization of system (7) about zero is the following:

$$U'(t) = AU(t) + BU(t - \tau) \tag{8}$$

We adopt the following norms of vectors and matrices [15]: $\|U(t)\| = \sum_{i=1}^{4} |u_i(t)|$, $\|A\| = \max_i \sum_{i=1}^{4} |a_{ij}|$, $\|B\| = \max_i \sum_{i=1}^{4} |b_{ij}|$.

The measure of matrix A is defined by $\mu(A) = \max_{1 \le i \le 4} \left(a_{ii} + \sum_{j=1, j\neq i} |a_{ij}|\right)$.

3 Oscillation Analysis

Theorem 1. Assume that the following inequality holds,

$$\mu(A) + \|B\| \neq 0, \quad (\|B\|\tau e)e^{-|\mu(A)|\tau} > 1 \tag{9}$$

Then the trivial solution of system (6) (or (7)) is unstable, there exists an oscillatory solution of system (6).

Proof. Noting that system (7) and (8) both have zero equilibrium point. According to the property of delay differential equation, the instability of the zero point of system (8) implies the instability of the zero point of system (7). Therefore, we will focus on the instability of the zero point of system (8). Let $y(t) = |u_1| + |u_2| + |u_3| + |u_4|$, consider the scalar delay equation

$$y'(t) = \mu(A)y(t) + \|B\|y(t - \tau) \tag{10}$$

If the trivial solution is stable, then the characteristic equation associated with Eq. (10) given by

$$\rho = \mu(A) + \|B\|e^{-\rho\tau} \tag{11}$$

should have a real non-positive root say ρ^*. Obviously, $\rho^* \neq 0$. Otherwise, we will have $\mu(A) + \|B\| = 0$. From (11), we get

$$|\rho^*| + |\mu(A)| \ge \|B\|e^{-\rho^*\tau} = \|B\|e^{|\rho^*\tau|} \tag{12}$$

Therefore,

$$|\rho^*| + |\mu(A)| \geq \|B\|e^{|\rho^*\tau|}, \text{ or } 1 \geq \frac{\|B\|e^{|\rho^*\tau|}}{|\rho^*| + |\mu(A)|} \tag{13}$$

By using inequality $e^x \geq ex$, we have

$$1 \geq \frac{\tau\|B\|e^{|\rho^*\tau|}}{|\rho^*|\tau + |\mu(A)|\tau} = \frac{\tau\|B\|e^{|\rho^*\tau| + |\mu(A)|\tau}e^{-|\mu(A)|\tau}}{|\rho^*|\tau + |\mu(A)|\tau} \geq (\|B\|\tau e)e^{-|\mu(A)|\tau} \tag{14}$$

Inequality (14) contradicts condition (9). Hence, our claim regarding the unstable of the trivial solution is valid. It follows that system (6) (or (7)) has an oscillatory solution.   □

Theorem 2. Suppose that the parameters $\lambda_1 > 0$ and $\lambda_2 > 0$, there exists a positive real number L such that the following inequality holds,

$$L - \mu(A) - \|B\|e^{-L\tau} > 0 \tag{15}$$

Then the trivial solution of system (6) (or (7)) is unstable, system (6) generates an oscillatory solution.

Proof. We shall prove that the trivial solution of system (6) (or (7)) is unstable. Similar to Theorem 1, we only need to discuss the instability of the trivial solution of system (8). It is sufficiently to show that the characteristic equation associated with (8) given by (11) has a real positive root. Noting that Eq. (11) is a transcendental equation, the eigenvalues may be complex numbers. We show that there is a real positive root under restrictive condition (15). Set

$$f(\rho) = \rho - \mu(A) - \|B\|e^{-\rho\tau} \tag{16}$$

then $f(\rho)$ is a continuous function of ρ. Since $\lambda_1 > 0$ and $\lambda_2 > 0$, based on the definition of the measure of matrix A we know that $\mu(A) > 0$, and the value of $\|B\|$ is bounded. This implies that $f(0) = 0 - \mu(A) - \|B\| = -\mu(A) - \|B\| < 0$. From condition (16), we know that there exists a suitably large real number L such that

$$f(L) = L - \mu(A) - \|B\|e^{-L\tau} > 0 \tag{17}$$

By the continuity of function $f(\rho)$, there exists a number ρ^{**} such that $f(\rho^{**}) = 0$, where $\rho^{**} \in (0, L)$. In other words, ρ^{**} is a real positive eigenvalue. The trivial solution of system (8) is unstable. It implies that system (6) (or (7)) has an oscillatory solution. The proof is complete.   □

4 Simulation Results

In system (6) (or (7)), the parameter values are selected as $\lambda_1 = 0.1$, $\lambda_2 = 0.2$, $\beta_1 = 0.45$, $\beta_2 = 0.3$, $\varepsilon_1 = 0.4$, $\varepsilon_2 = 0.5$, $\mu_1 = 0.7$, $\mu_2 = 0.8$, $\Delta = 0.6$ then $\mu(A) = 1.2$, $\|B\| = 1.5$. When the time delays are fixed to 0.8 and 1.5 respectively, it is easy to see that the conditions of Theorem 1 are satisfied. There exist oscillatory solutions (see Figs. 1 and 2). In Figs. 3 and 4, the parameter values are selected as $\lambda_1 = 0.85$, $\lambda_2 = 0.95$, $\beta_1 = 1.5$, $\beta_2 = 1.8$, $\varepsilon_1 = 1.4$, $\varepsilon_2 = 1.4$, $\mu_1 = 0.75$, $\mu_2 = 0.75$, $\Delta = 0.8$. The time delays are fixed to 1.5 and 2.0 respectively. One can see the conditions of Theorem 2 are satisfied. System (6) (or (7)) generates oscillatory solutions (see Figs. 3 and 4).

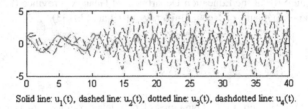

Solid line: $u_1(t)$, dashed line: $u_2(t)$, dotted line: $u_3(t)$, dashdotted line: $u_4(t)$.

Fig. 1. Oscillatory behavior of the solution, delay: 0.8.

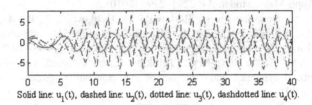

Solid line: $u_1(t)$, dashed line: $u_2(t)$, dotted line: $u_3(t)$, dashdotted line: $u_4(t)$.

Fig. 2. Oscillatory behavior of the solution, delay: 1.5.

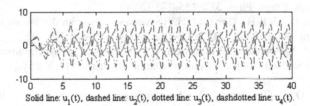

Solid line: $u_1(t)$, dashed line: $u_2(t)$, dotted line: $u_3(t)$, dashdotted line: $u_4(t)$.

Fig. 3. Oscillatory behavior of the solutions, delay: 1.5.

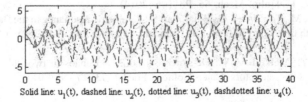

Solid line: $u_1(t)$, dashed line: $u_2(t)$, dotted line: $u_3(t)$, dashdotted line: $u_4(t)$.

Fig. 4. Oscillatory behavior of the solutions, delay: 2.0.

5   Conclusion

This paper investigates the oscillation of a coupled van der Pol-Duffing oscillators with delay. Two simple criteria are provided to ensure the oscillation of the solutions. Our method to determine the parameters in order to guarantee the existence of oscillatory solutions of the system is different from the approach of charts of dynamical regimes in computations from initial differential equations which used by [4]. Computer simulations are given to support the effectiveness of the results.

Acknowledgements. This research was supported by NNSF of China (11361010), the high school specialty and curriculum integration project of Guangxi Zhuang Autonomous Region (GXTSZY2220), the SRF of the Education Department of Guangxi Province (LX2014330), and the Key Discipline of Statistics of Hechi University of China (20133).

References

1. Li, X.Y., Ji, J.C., Hansen, C.H.: Dynamics of two delay coupled van Der Pol oscillators. Mech. Res. Commun. **33**, 614–627 (2006)
2. Zhang, J.M., Gu, X.S.: Stability and bifurcation analysis in the delay-coupled van Der Pol oscillators. Appl. Math. Model. **34**, 2291–2299 (2010)
3. Sabarathinam, S., Thamilmaran, K., Borkowski, L., Perlikowski, P., Bzeski, P., Stefanski, A., Kapitaniak, T.: Transient chaos in two coupled, dissipatively perturbed Hamiltonian duffing oscillators. Commun. Nonlinear. Sci. Numer. Simul. **18**, 3098–3107 (2013)
4. Kuznetsov, A.P., Stankevich, N.V., Turukina, L.V.: Coupled van der Pol-Duffing oscillators: phase dynamics and structure of synchronization tongues. Physica D **238**, 1203–1215 (2009)
5. Kuznetsov, A.P., Paksyutov, V.I.: Feature of the parameter space structure of two non - identical coupled van der Pol-Duffing oscillators. Appl. Nonlin. Dynam. **13**, 3–19 (2005)
6. Peano, V., Thorwart, M.: Dynamics of the quantum duffing oscillator in the driving induced bistable regime. Chem. Phys. **322**, 135–143 (2006)
7. Camacho, E., Rand, R.H., Howland, H.: Dynamics of two van der Pol oscillators coupled via a bath. Int. Solids Struct. **41**, 2133–2143 (2004)
8. Kuznetsov, A.P., Paksyutov, V.I., Roman, Y.P.: Features of the synchronization of coupled van der Pol oscillators with nonidentical control parameters. Tech. Phys. Lett. **8**, 636–638 (2007)
9. Gendelman, O.V., Starosvetsky, Y.: Quasiperiodic response regimes of linear oscillator coupled to nonlinear energy sink under periodic forcing. Trans. ASME J. Appl. Mech. **74**, 325–331 (2007)
10. Rusinek, R., Weremczuk, A., Kecik, K., Warminski, J.: Dynamics of a time delayed Duffing oscillator. Int. J. Nonlin. Mech. **65**, 98–106 (2014)
11. Siewe Siewe, M., Tchawoua, C., Rajasekar, S.: Parametric resonance in the Rayleigh-Duffing oscillator with time-delayed feedback. Commun. Nonlinear Sci. Num. Simul. **17**, 4485–4493 (2012)
12. Beregov, R.Y., Melkikh, A.V.: De-synchronization and chaos in two inductively coupled Van der Pol auto-generators. Chaos Solitons Fractals **73**, 17–28 (2015)

13. Sun, Y., Xu, J.: Experiments and analysis for a controlled mechanical absorber considering delay effect. J. Sound Vibr. **339**, 25–37 (2015)
14. Zhang, C.M., Li, W.X., Wang, K.: Boundedness for network of stochastic coupled van Der Pol oscillators with time-varying delayed coupling. Appl. Math. Model. **37**, 5394–5402 (2013)
15. Horn, R.A., Johnso, C.R.: Matrix Analysis. Cambridge Press, Cambridge (1990)

Prediction of Clinical Drug Response Based on Differential Gene Expression Levels

Zhenyu Yue[1], Yan Chen[1], and Junfeng Xia[2(✉)]

[1] School of Life Sciences, Anhui University, Hefei 230601, Anhui, China
[2] Institute of Health Sciences, Anhui University, Hefei 230601, Anhui, China
371381220@qq.com

Abstract. We demonstrated a new method for the prediction of in vivo drug sensitivity using before-treatment baseline tumor gene expression data. First, we fitted ridge regression models for differential gene expression against drug sensitivity in a large panel of cell lines. Following data homogenization and filtering, drug response was predicted based on baseline expression levels from primary tumor biopsies. We validated this approach on two clinical trial data-sets, and obtained predictions better than those from whole-genome gene expression. The findings may point out new directions for the prediction of anticancer drug sensitivity and the development of personalized medicine.

Keywords: Gene expression · Drug response · Ridge regression · Machine learning

1 Introduction

Acquisition of knowledge of clinical drug response of a specific cancer type is a major challenge in cancer therapy that can ultimately lead to an option for personalized drug treatment. To address this challenge, molecular and cellular biology approaches have been broadly applied in the last few decades and have made some achievements. However, experimental methods suffered from the drawback of expensive and time-consuming.

High-throughput screenings of drugs against a panel of genetically heterogeneous cancer cell lines have uncovered multiple relationships between genomic alterations and drug sensitivity [1, 2]. Additionally, numerous studies have explored large-scale, publicly available pharmacogenomic data, which have accelerated the search for new cancer therapies [3–6].

The quantification of gene expression has been applied to drug response prediction for many years [7, 8]. However, many of the results reported are limited due to their methodological and computational challenges. In an effort to complement these approaches, we developed machine learning regression models to predict the clinical response to drug treatment, based on differential gene expression levels between sensitive and resistant cancer cells. We demonstrated that it is possible to capture a significant proportion of the variability in drug response in patients with differential gene expression levels by building regression models, and obtained predictions better

© Springer International Publishing Switzerland 2015
D.-S. Huang et al. (Eds.): ICIC 2015, Part II, LNCS 9226, pp. 480–488, 2015.
DOI: 10.1007/978-3-319-22186-1_48

than those based on whole-genome gene expression. The findings may help guide the investigation of potential cancer therapy for humans.

2 Materials and Methods

2.1 Data

For model development, our approach was applied to the data from Geeleher et al. [9], who recently constructed predicted models using the GDSC (The Genomics of Drug Sensitivity in Cancer) cell line data [10], with the aim of predicting IC50 value (the concentration required for 50 % of cellular growth inhibition) of clinical drug response based on whole-genome gene expression data. The database (GDSC) consists of baseline (i.e. before drug treatment) gene expression microarray data and sensitivity to 138 drugs in a panel of almost 800 cell lines.

To test our approach, we also used clinical trial datasets (both baseline tumor expression and post-treatment drug response) from Geeleher et al. [9]. The authors applied their method to four trials for three different types of cancer. The drug (Erlotinib) used in the fourth trial was a targeted agent, which IC_{50} values for most cell lines were derived using extrapolated data in GDSC, and thus has very large associated confidence intervals. The other drugs in the test sets were cytotoxic agents (docetaxel and cisplatin in breast cancer, bortezomib in myeloma). Here, we used the docetaxel and bortezomib clinical trials dataset [11, 12] as the test sets, which were denoted as Data1 and Data2, respectively.

2.2 Models Building

In all cases, both training (cell lines) and test (clinical trial) sets were generated on different microarray platforms, thus we used a subset of genes come from both platforms, which led to approximately 10,000 genes. Once the data were prepared, a linear ridge regression model was fitted for in vitro drug sensitivity (IC50 values) depended on the differential expression levels. To do this, the linearRidge() function from the ridge [13] package in R [14] was used. This function chooses the ridge regression tuning parameter automatically. After the model was fitted, it was applied to the same differential gene expression data from the clinical trial. The predict.linearRidge() function from the ridge package in R was used to obtain a drug sensitivity estimation for each patient.

3 Results and Discussions

3.1 Workflow

Our goal was to predict patients' response to drugs based on differential gene expression data, which includes three key steps.

Clustering of IC50 Values. Clustering of drug IC50 values from GDSC was carried out by using SimpleKMeans in WEKA [15] (Waikato Environment For Knowledge Analysis), which is a popular algorithm for cluster analysis in data mining. Here K was set 2 or 3, which means that the cell lines were divided into 2 (the sensitive cell lines and the resistant cell lines) or 3 groups. In the case K = 3, the sensitive cell lines referred to the most sensitive group, the resistant cell lines referred to the most resistant group, and the intermediate group was ignored.

Feature Selection. In this step, the TopN genes with the most differentially expressed between the sensitive cell lines and the resistant cell lines were chosen as the features of our training set. These genes were selected using t-tests with the rowttests() function in the R library genefilter [16]. In the case K = 3, 2108 and 4305 genes that had statistical significant differences (P value < 0.05), which were denoted as TopSig genes, were chosen for the docetaxel-treated and the bortezomib-treated trial, respectively. In the case K = 2, 1314 and 2069 genes were chosen, respectively. In addition, we chose Top50, Top100, Top500 and Top1000 genes with the most differentially expressed in this feature selection step. Finally, we obtained the optimal TopN genes which are highlighted with triangles in Fig. 2. The detailed results were shown in Table 1.

Machine Learning Models. A linear ridge regression model was fitted and then was applied to the tumor expression data from the clinical trial, with the purpose of generating drug sensitivity estimates (IC50).

The prediction workflow for in vivo sensitivity model is shown in Fig. 1.

3.2   Selection of TopN Genes

In the two test sets, we chose Top50, Top100, Top500, Top1000 and TopSig genes that were most significantly differential expressed in this feature selection step. In the case K = 2, the Top1000 and Top500 groups have the highest AUC (area under the curve) values for Data1 (0.871) and Data2 (0.586), respectively. In the case K = 3, the Top1000 and Top100 groups have the highest AUC values for Data1 (0.857) and

Table 1. Detailed regression results (AUC). The best result in each row except for the Optimal is highlighted in red. TopSig genes refer to gene numbers that have statistical significant differences (P value < 0.05). N was the TopN gene number. Optimal referred to the AUC when N was chosen as the feature number of ridge regression model.

Data	Type	Top50	Top100	Top500	Top1000	TopSig (N)	Optimal (N)
Data1	Original						0.814 (8239)
Data1	Cluster2	0.600	0.800	0.800	**0.871**	0.850 (1314)	0.886 (1140)
Data1	Cluster3	0.550	0.179	0.700	**0.857**	0.836 (2108)	0.886 (1150)
Data2	Original						0.635 (11952)
Data2	Cluster2	0.494	0.484	**0.586**	0.542	0.532 (2069)	0.598 (450)
Data2	Cluster3	0.542	**0.620**	0.574	0.525	0.507 (4305)	0.637 (115)

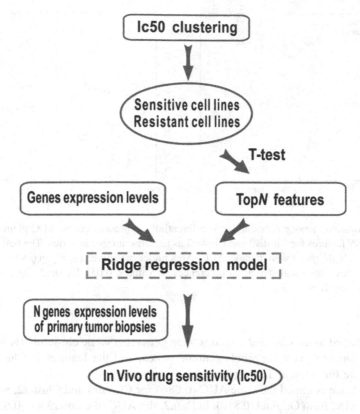

Fig. 1. In vivo sensitivity prediction workflow. This flow diagram demonstrates the simulation workflow for creation and optimization of drug sensitivity prediction model, which includes three key steps: (1) clustering of drug IC50 values from the GDSC, then the cell lines were divided into 2 or 3 groups, and the sensitive cell lines and the resistant cell lines were generated. (2) TopN genes that were most differentially expressed between the sensitive and resistant cell lines were chosen as the features of our training set. (3) Machine learning models were fitted in R and can then be applied to the tumor expression data from the clinical trials, with the purpose of generating drug sensitivity estimates.

Data2 (0.620), respectively. In addition, we search the optimal TopN genes (Fig. 2) around the highest dots of each curve. The final optimal gene number for Data1 and Data2 are 1140 and 450 when K = 2, and the final optimal gene number for Data1 and Data2 are 1150 and 115 when K = 3. The detailed results were shown in Table 1.

3.3   Comparison of AUC Among Three Cases

Here, ROC curves in three cases for Data1 in Fig. 3(A) and Data2 in Fig. 3(B) were generated using R. Original referred to the case in the work of Geeleher et al. [9]. Cluster2 referred to the case that 2 groups were generated while clustering (K = 2).

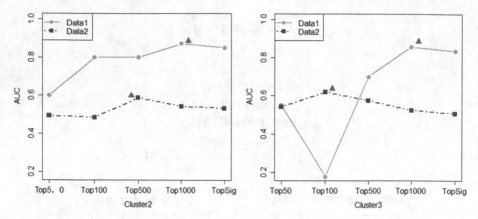

Fig. 2. Comparison among different Top*N* differentially expressed genes. AUC plotted against different Top*N* features for Cluster2 and Cluster3 as features number is varied. The best group of each test set is highlighted with triangle. AUC, Area under the curve. TopSig genes refer to gene numbers that have statistical significant differences (P value < 0.05). Cluster2, the case K = 2. Cluster3, the case K = 3.

Cluster3 referred to the case that 3 groups were generated while clustering (K = 3). The features in Original were the whole-genome genes, and the features in Cluster2 and Cluster3 were the optimal Top*N* genes.

In Data1, we obtained the same AUC (0.886) for Cluster3 and Cluster2, which are better than AUC from Original (0.814). In Data2, the AUC of Cluster3 was 0.637 better than Cluster2 (0.598) and Original (0.635). So, Cluster3 was chosen as standard method in the following analysis.

3.4   Prediction of Sensitivity by Cluster3

Strip charts and box plots (Fig. 4) shown that the predicted drug sensitivity value was lower in the patients who were defined (by the trial) to be sensitive to docetaxel and bortezomib, compared to the patients defined as resistant. There was a statistically significant difference between the two groups for either Data1 (P = 7.2×10^{-4}, t-test) or Data2 (P = 4.3×10^{-3}, t-test).

3.5   Detailed Regression Results

Table 1 listed detailed results of the prediction models by AUC. According to Geeleher et al., the Data2 clinical dataset presented some obstacles, such as the raw microarray data were not available and clinical samples were collected from various sites, which may lead to a decreased predictive ability of the model.

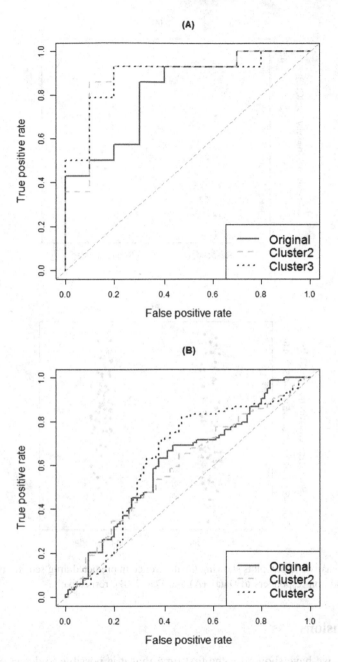

Fig. 3. ROC curve showing the proportion of true positives against the proportion of false positives as the classification threshold is varied for Data1 (A) and Data2 (B). ROC, receiver operating characteristic. Original, the case in the work of Geeleher et al. Cluster2, the case K = 2. Cluster3, the case K = 3.

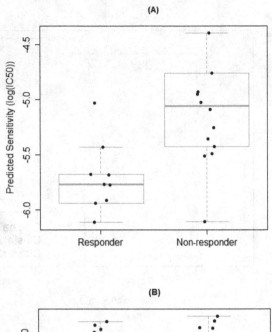

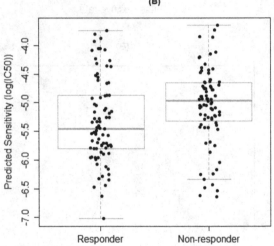

Fig. 4. Strip charts and box plots showing the difference in predicted drug sensitivity for in vivo responders and non-responders to Data1 (A) and Data2 (B), respectively.

4   Conclusions

In summary, we have shown for the first time that it is possible to detect cancer drug response using differential gene expression levels, by applying models generated from a large panel of cell lines. We have demonstrated that our approach performed well in the test sets compared with the original whole-genome gene expression data. In the future, we will investigate how to improve our method with RNA-Seq data and advanced intelligent computing methodologies such as deep learning [17].

Acknowledgments. This work was supported by National Natural Science Foundation of China (31301101), the Anhui Provincial Natural Science Foundation (1408085QF106), the Specialized Research Fund for the Doctoral Program of Higher Education (20133401120011), and the Technology Foundation for Selected Overseas Chinese Scholars from Department of Human Resources and Social Security of Anhui Province (No. [2014]-243).

References

1. Barretina, J., Caponigro, G., Stransky, N., et al.: The cancer cell line encyclopedia enables predictive modelling of anticancer drug sensitivity. Nature **483**(7391), 603–607 (2012)
2. Garnett, M.J., Edelman, E.J., Heidorn, S.J., et al.: Systematic identification of genomic markers of drug sensitivity in cancer cells. Nature **483**(7391), 570–575 (2012)
3. Kim, N., He, N., Yoon, S.: Cell line modeling for systems medicine in cancers (review). Int. J. Oncol. **44**(2), 371–376 (2014)
4. Masica, D.L., Karchin, R.: Collections of simultaneously altered genes as biomarkers of cancer cell drug response. Cancer Res. **73**(6), 1699–1708 (2013)
5. Menden, M.P., Iorio, F., Garnett, M., et al.: Machine learning prediction of cancer cell sensitivity to drugs based on genomic and chemical properties. PLoS One **8**(4), e61318 (2013)
6. Papillon-Cavanagh, S., De Jay, N., Hachem, N., et al.: Comparison and validation of genomic predictors for anticancer drug sensitivity. J. Am. Med. Inf. Assoc. **20**(4), 597–602 (2013)
7. Van De Vijver, M.J., He, Y.D., van't Veer, L.J., et al.: A gene-expression signature as a predictor of survival in breast cancer. N. Engl. J. Med. **347**(25), 1999–2009 (2002)
8. Sotiriou, C., Wirapati, P., Loi, S., et al.: Gene expression profiling in breast cancer: understanding the molecular basis of histologic grade to improve prognosis. J. Natl. Cancer Inst. **98**(4), 262–272 (2006)
9. Geeleher, P., Cox, N.J., Huang, R.S.: Clinical drug response can be predicted using baseline gene expression levels and in vitro drug sensitivity in cell lines. Genome Biol. **15**(3), R47 (2014)
10. Yang, W., Soares, J., Greninger, P., et al.: Genomics of drug sensitivity in cancer (GDSC): a resource for therapeutic biomarker discovery in cancer cells. Nucleic Acids Res. **41** (Database issue), D955–D961 (2013)
11. Chang, J.C., Wooten, E.C., Tsimelzon, A., et al.: Gene expression profiling for the prediction of therapeutic response to docetaxel in patients with breast cancer. Lancet **362** (9381), 362–369 (2003)
12. Mulligan, G., Mitsiades, C., Bryant, B., et al.: Gene expression profiling and correlation with outcome in clinical trials of the proteasome inhibitor bortezomib. Blood **109**(8), 3177–3188 (2007)
13. Cule, E., De Iorio, M.: Ridge regression in prediction problems: automatic choice of the ridge parameter. Genet. Epidemiol. **37**(7), 704–714 (2013)
14. Ihaka, R., Gentleman, R.R.: A language for data analysis and graphics. J. Comput. Graph. Stat. **5**(3), 299–314 (1996)

15. Hall, M., Frank, E., Holmes, G., Pfahringer, B., et al.: The WEKA data mining software: an update. ACM SIGKDD Explor. Newslett. **11**(1), 10–18 (2009)
16. Gentleman, R., Carey, V., Huber, W., Hahne, F.: Genefilter: methods for filtering genes from microarray experiments. R Package Version **1**(0) (2011)
17. Arel, I., Rose, D.C., Karnowski, T.P.: Deep machine learning-a new frontier in artificial intelligence research [research frontier]. IEEE Comput. Intell. Mag. **5**, 13–18 (2010)

SVM-Based Classification of Diffusion Tensor Imaging Data for Diagnosing Alzheimer's Disease and Mild Cognitive Impairment

Wook Lee, Byungkyu Park, and Kyungsook Han[✉]

Department of Computer Science and Engineering,
Inha University, Incheon, South Korea
wooklee@inha.edu, {bpark,khan}@inha.ac.kr

Abstract. Alzheimer's disease (AD) is the most common cause of dementia. Early detection of AD is important since treatment is more efficacious if introduced earlier. Mild cognitive impairment (MCI) is often a precursory stage of AD, so is considered to be a good target for early detection of AD. However, MCI is not easy to diagnose due to the subtlety of cognitive impairment. In this study, we developed a method to automate the diagnosis of AD and MCI using support vector machines (SVMs) and diffusion tensor imaging (DTI) data. We implemented two SVM models: one for classifying AD and MCI and another for classifying AD and normal control (NC). In both SVM models, the fractional anisotropy (FA) and the mode of anisotropy (MO) values of DTI were used as features. MO values resulted in a better performance than FA values in both models. In independent testing, the AD-MCI classifier showed a sensitivity of 69.2 %, a specificity of 100 % and an accuracy of 89.7 %, and the AD-NC classifier showed a sensitivity of 84.6 %, a specificity of 90.9 % and an accuracy of 87.5 %. These results are encouraging and suggest that SVM-based classification of DTI data is potentially powerful in early detection of MCI and AD.

Keywords: Alzheimer's disease · Mild cognitive impairment · Diffusion tensor imaging · Tract-based spatial statistics · Relief · Support vector machine

1 Introduction

Alzheimer's disease (AD) is one of most common degenerative brain diseases, but essential treatment is not available yet [1, 2]. Since early detection of AD is very important [3], many studies were carried out for diagnosis of AD [4, 5]. Mild cognitive impairment (MCI) is known as a precursory stage of AD and the rate of MCI to AD (10–15 %) is higher than normal to AD (1–2 %) [6]. Hence, the classification of AD and MCI is as important as classification of AD and healthy elderly.

Recently diffusion tensor imaging (DTI) has received increasing attention from many studies of Alzheimer's disease (AD) since it can reveal the microscopic tissue structure of the brain white matter [7–9]. The diffusion of water molecules shows anisotropy by axon (body of the neuron) in the brain. Water diffusion in the direction parallel to the fiber bundles is rapid whereas water diffusion in the orthogonal direction

© Springer International Publishing Switzerland 2015
D.-S. Huang et al. (Eds.): ICIC 2015, Part II, LNCS 9226, pp. 489–499, 2015.
DOI: 10.1007/978-3-319-22186-1_49

is slow. DTI is an imaging technique that can estimate the proportion or direction of diffusion in the white matter. We can also obtain the information on the property of white matter and connectivity of nerve fibers [10–13]. Studying DTI data from multiple subjects requires all DTI data to be aligned in a single space, which is not easy. We use tract-based spatial statistics (TBSS) to solve the alignment problem. TBSS align all DTI data of multiple subjects using non-linear registration and project them in the skeleton region [14].

The fractional anisotropy (FA) is the most commonly used value of DTI data. FA represents anisotropy of diffusion with a range from 0 to 1 [15]. The mode of anisotropy (MO) is closely related to crossing fiber regions. Studies that consider the diffusion in the primary orientation only cannot explain crossing regions of two fiber bundles. MO defines the shape of anisotropy using the diffusion of the secondary orientation (orthogonal to the primary direction), and has a value from -1 to $+1$ [16]. The diffusion tensor can be decomposed into three components: planar anisotropy (mode = -1), orthotropy (mode = 0) and linear anisotropy (mode = $+$ 1). The planar anisotropy means that the second eigenvalue is similar to the first one ($\lambda 1 \sim \lambda 2 > \lambda 3$) and the linear anisotropy means that the second eigenvalue is similar to the third one ($\lambda 1 > \lambda 2 \sim \lambda 3$) [17].

In this study, we developed support vector machines (SVMs) [18] that use FA and MO values along with TBSS and the Relief feature selection algorithm [19]. The rest of this paper presents the details of our method and its experimental results with DTI data of AD, MCI and healthy elderly subjects.

2 Materials and Methods

2.1 Dataset

A total of 141 DTI data from the LONI Image Data Archive (http://ida.loni.ucla.edu) were used in our study (Table 1). Ninety one DTI data of 22 AD, 47 MCI and 22 normal control (NC) subjects were selected for training and cross validation of our SVM models. The training dataset comprised of AD, MCI, and NC subjects with matched ages and imaging parameters; the average ages of 22 AD, 47 MCI and NC subjects are 76.62 years, 75.00 years and 75.19 years, respectively. The imaging parameters of the training data are as follows: field strength = 3.0 T; flip angle = 90°; b = 1000 s/mm^2; gradient directions = 41; pixel size = 1.3672 × 1.3672 mm^2; repetition time (TR) = 9050 ms; slice thickness = 2.7 mm; DTI matrix = 256 × 256.

Table 1. DTI data. avg: average

	AD	MCI	NC
#Subjects	35	73	33
#Training data/#test data	22/13	47/26	22/11
Avg. age of training data	76.62 ± 8.72	75.00 ± 7.42	75.19 ± 4.60
Age range of test data	67.87 ~ 92.41	57.78 ~ 84.21	63.56 ~ 80.61
MMSE score	20.34 ± 4.81	26.74 ± 1.87	28.27 ± 1.31

The remaining 50 DTI data of 13 AD, 26 MCI and 11 NC subjects were used to construct an independent test dataset of SVM models. Age and imaging parameters were not considered in constructing the test dataset because we wanted to examine the performance of SVM models with new data of various types. The imaging parameters of the test dataset are as follows: field strength = 3.0 T; flip angle = 90°; b = 1000 s/mm^2; gradient directions = 41; pixel size = 1.3672 × 1.3672 mm^2; repetition time (TR) = 9050–14200 ms; slice thickness = 2.7 mm; DTI matrix = 256 × 256.

2.2   Data Processing of DTI

All DTI data were processed using FMRIB software library (FSL) [20] and Tract-Bbased Spatial Statistics (TBSS) [2]. FSL provides many functions that can process DTI data. The time series data of DTI were corrected to remove distortions caused by eddy currents and simple head motion. From the corrected DTI data, a standalone 3D image with no diffusion was generated and the skull of brain was removed to obtain pure brain images using Brain Extraction Tool (BET) [21]. Pure brain mask images, eigenvectors and eigenvalues, FA and MO images were also generated from the corrected DTI data [22].

Figure 1 shows the whole process of TBSS. TBSS aligned FA images of each subject to a space of a target image using non-linear registration. As the target image, we used 1 x 1 x 1 mm standard FA map (FMRIB58_FA) provided by FSL. After registration, an all_FA image was made from the aligned FA images of all subjects and a mean_FA 3D image with the average of FA values was generated. In the next step, a mean_FA_skeleton image was generated from the mean_FA image. Voxels of this skeleton have higher FA values than neighbor voxels in the white matter. This skeleton is a common white matter region of all subjects. The TBSS process of MO images was performed in the same way.

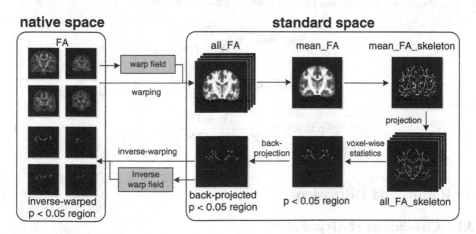

Fig. 1. The Tract-Based Spatial Statistics (TBSS) process of FA images. TBSS aligns FA images of all subjects to a single space and projects the aligned FA images (all_FA) to a common white matter shaped skeleton (mean_FA_skeleton). The TBSS process of MO images is performed in the same way.

2.3   SVM Classification

We implemented a SVM model to classify AD and MCI and performed a cross validation. We extracted FA values of each subject by aligning the FA image (all_FA) to the mean_FA_skeleton image and encoded the FA values into a feature vector for the SVM model. A typical DTI of a brain contains a huge number of voxelx, so selecting voxels that are likely to be relevant to classification would reduce the computational burden and help us find relevant information. In several studies [23, 24], the best performance was achieved when 1,000 or fewer voxels of DTI were analyzed. In this study, we applied the Relief feature selection algorithm [19] to the FA values, and obtained 10 datasets with different numbers of features (100, 200, 300, 400, 500, 600, 700, 800, 900 and 1000 features each). The 10 datasets were used in cross validations and training the SVM model.

Since the class label of the data in the test dataset is unknown, the Relief algorithm could not be used in testing the DTI data. So, we used the coordinate values of voxels in training datasets. We found corresponding voxels in all_FA image of test DTI data, such as same coordinate with training voxels and made test data sets. The class label is not required until the step of all_FA_skeleton in TBSS, so we obtained the all_FA image of new DTI data. We prepared 10 test datasets, which is the same number of training data sets. The training and test datasets of MO images were generated in the same way as those of FA images.

For the AD-NC classifier, we prepared additional 20 training and 20 test datasets. We used a library for support vector machines (LIBSVM) (http://www.csie.ntu.edu.tw/~cjlin/libsvm) with the radial basis function (RBF) kernel [25]. The parameter values of the RBF kernel were set as follows: C = 100 and GAMMA = 1/#feature vectors in the dataset. The Relief feature selection algorithm was executed using WEKA (http://www.cs.waikato.ac.nz/ml/weka) [26].

The performance of the SVM models was evaluated using three measures (sensitivity, specificity and accuracy), which are defined as follows.

$$\text{Sensitivity} = \frac{TP}{TP + FN} \tag{1}$$

$$\text{Specificity} = \frac{TN}{TN + FP} \tag{2}$$

$$\text{Accuracy} = \frac{TP + TN}{TP + FP + TN + FN} \tag{3}$$

3   Results and Discussion

3.1   Classification of AD and MCI

Table 2 shows the performance of the AD-MCI classifier in the 10-fold cross validation with 22 AD and 47 MCI subjects. Specificity was higher than sensitivity, and FA

Table 2. The performance of the AD-MCI classifier in a 10-fold cross validation

#Voxels	FA			MO		
	SN (%)	SP (%)	AC (%)	SN (%)	SP (%)	AC (%)
100	77.3	89.4	85.5	90.9	93.6	92.8
200	77.3	91.5	87.0	**95.5**	**100.0**	**98.6**
300	72.7	91.5	85.5	**95.5**	**100.0**	**98.6**
400	77.3	91.5	87.0	95.5	97.9	97.1
500	72.7	91.5	85.5	95.5	95.7	95.7
600	77.3	91.5	87.0	95.5	97.9	97.1
700	68.2	93.6	85.5	95.5	97.9	97.1
800	72.7	91.5	85.5	90.9	100.0	97.1
900	72.7	93.6	87.0	**95.5**	**100.0**	**98.6**
1000	**77.3**	**93.6**	**88.4**	95.5	100.0	98.6

images showed a bigger difference between specificity and sensitivity than MO images. The best performance of the AD-NC classifier with FA images was a sensitivity of 77.3 %, a specificity of 93.6 % and an accuracy of 88.4 %; the best performance of the AD-NC classifier with MO images was a sensitivity of 95.5 %, a specificity of 100 % and an accuracy of 98.6 %.

Table 3 shows the result of independent testing the AD-NC classifier with 13 AD and 26 MCI subjects. Its performance with FA images was similar to that of the 10-fold cross validation, but slightly lower especially in specificity. Its sensitivity with MO images was decreased, ranging from 53.8 % to 69.2 %. However, it still showed a high specificity. The best performance with FA images was a sensitivity of 76.9 %, a specificity of 80.8 % and an accuracy of 79.5 %; The best performance with MO images was a sensitivity of 69.2 %, a specificity of 100 % and an accuracy of 89.7 % in independent testing.

Table 3. The performance of of the AD-MCI classifier in independent testing

#Voxels	FA			MO		
	SN (%)	SP (%)	AC (%)	SN (%)	SP (%)	AC (%)
100	69.2	73.1	71.8	53.8	96.2	82.1
200	76.9	69.2	71.8	69.2	96.2	87.2
300	69.2	69.2	69.2	61.5	96.2	84.6
400	84.6	69.2	74.4	**69.2**	**100.0**	**89.7**
500	76.9	76.9	76.9	61.5	100.0	87.2
600	76.9	76.9	76.9	61.5	96.2	84.6
700	69.2	80.8	76.9	53.8	96.2	82.1
800	**76.9**	**80.8**	**79.5**	53.8	96.2	82.1
900	69.2	80.8	76.9	53.8	96.2	82.1
1000	61.5	80.8	74.4	53.8	96.2	82.1

3.2   Classification of AD and NC

Tables 4 and 5 show the performance of the AD-NC classifier in a 10-fold cross validation and independent testing, respectively. In the cross validation with 22 AD and 22 NC subjects, the classifier achieved a high performance in all three measures with both FA and MO images (Table 4). In independent testing with 13 AD and 11 NC subjects, it showed a slightly lower performance than the cross validation. While the performance of the AD-NC classifier with FA images increases with more voxels, its performance with MO images showed the best performance with 700 voxels. In independent testing, the best performance of the classifier with FA images was a sensitivity of 92.3 %, a specificity of 81.8 % and an accuracy of 87.5 %, and the best performance with MO images was a sensitivity of 84.6 %, a specificity of 90.9 % and an accuracy of 87.5 %.

Table 4. The performance of the AD-NC classifier in a 10-fold cross validation

#Voxels	FA			MO		
	SN (%)	SP (%)	AC (%)	SN (%)	SP (%)	AC (%)
100	90.9	86.4	88.6	90.9	90.9	90.9
200	90.9	90.9	90.9	95.5	95.5	95.5
300	90.9	90.9	90.9	95.5	95.5	95.5
400	90.9	90.9	90.9	100.0	95.5	97.7
500	90.9	90.9	90.9	100.0	95.5	97.7
600	90.9	90.9	90.9	100.0	95.5	97.7
700	90.9	90.9	90.9	100.0	95.5	97.7
800	90.9	90.9	90.9	100.0	95.5	97.7
900	90.9	90.9	90.9	100.0	95.5	97.7
1000	90.9	90.9	90.9	100.0	95.5	97.7

Table 5. The performance of the AD-NC classifier in independent testing

#Voxels	FA			MO		
	SN (%)	SP (%)	AC (%)	SN (%)	SP (%)	AC (%)
100	76.9	81.8	79.2	76.9	72.7	75.0
200	84.6	90.9	87.5	76.9	63.6	70.8
300	76.9	90.9	83.3	84.6	81.8	83.3
400	69.2	90.9	79.2	84.6	81.8	83.3
500	61.5	90.9	75.0	84.6	81.8	83.3
600	76.9	100.0	87.5	84.6	81.8	83.3
700	76.9	81.8	79.2	**84.6**	**90.9**	**87.5**
800	84.6	81.8	83.3	84.6	81.8	83.3
900	**92.3**	**81.8**	**87.5**	84.6	81.8	83.3
1000	**92.3**	**81.8**	**87.5**	84.6	81.8	83.3

3.3   Discussion

The most interesting result found in this study is that the classifiers showed a better performance with MO images than FA images. So far, many studies of AD with DTI data reported that FA images were better than other types of images.

The mode of anisotropy defines the shape of diffusion tensor (type of anisotropy) by the second eigenvalue ($\lambda 2$). The linear anisotropy increases as the mode becomes close to +1 because of $\lambda 2$ is dominated by the first eigenvalue ($\lambda 1$) and the planar anisotropy increases as the mode becomes close to −1 [16]. The planar anisotropy in which $\lambda 1$ and $\lambda 2$ have the almost same value indicates that two fiber bundles are crossing each other [17]. We made two fiber crossing models using the Bayesian Estimation of Diffusion Parameters Obtained using Sampling Techniques (BEDPOSTX) of FSL [27] to determine the entire crossing region of fiber bundles.

Figure 2 is an example of the two fiber crossing models for AD and NC. The background image shows the estimated proportion of diffuse toward the first axis orientation. The nested red image shows the information of the same type toward the second orientation. Voxels in the red image exist in the crossing fiber region of two fiber bundles. Most crossing voxels are found in the white matter of AD and NC, but their distribution in the AD white matter is obviously wider than in the NC white matter. MO values used in our classifiers were obtained in a common white matter of all subjects. Thus, the difference of MO values between AD and NC was reflected in SVM model. We think that this is the main cause of the good performance of the classifiers with MO values.

Figure 3 shows ROC curves of the two classifiers in independent testing with MO images. The AD-MCI classifier achieved the highest area under the curve (AUC) of

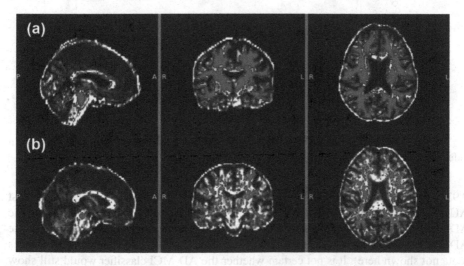

Fig. 2. Crossing region of two fiber bundles in AD and NC. The background image shows the proportion of the diffusion toward the primary axis orientation. The red area shows the same information toward the secondary orientation. Voxels in red area indicate crossing of two fiber bundles. Most voxels in crossing regions were found in white matter. (a) AD, (b) NC (Color figure online).

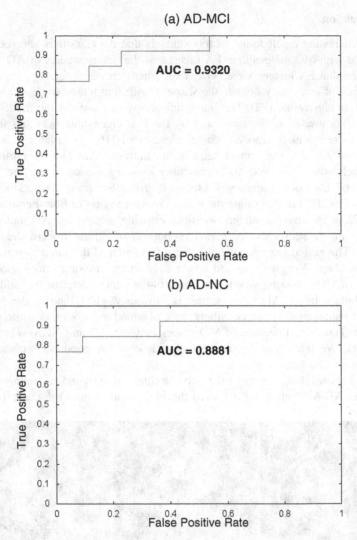

Fig. 3. The ROC curves of the AD-MCI classifier (top) and the AD-NC classifier (bottom) in independent testing with MO images

0.932 with 400 voxels (Fig. 3a), and the AD-NC classifier showed the highest AUC AUC of 0.8881 with 700 voxels (Fig. 3b). Although the best AUC of the AD-MCI classifier is larger than that of the AD-NC classifier with MO images, the AD-MCI classifier showed a smaller AUC than the AD-NC classifier with FA images (data not shown here). It is not certain whether the AD-MCI classifier would still show a larger AUC than the AD-NC classifier with MO images from other subjects, and this requires further investigation when more extensive DTI data become available for analysis.

4 Conclusion

Using the tract-based spatial statistics of diffusion tensor imaging (DTI) and the Relief feature selection algorithm, we developed two support vector machine (SVM) models that classify Alzheimer's disease (AD) from mild cognitive impairment (MCI) and normal control (NC). The fractional anisotropy (FA) and the mode of anisotropy (MO) values of DTI were used as features of the SVM models. On average, SVM models using MO values showed a better performance than those using FA values. In a 10-fold cross validation, the best performance of the AD-MCI classifier showed a sensitivity of 95.5 %, a specificity of 100 % and an accuracy of 98.6 %, and the AD-NC classifier achieved a sensitivity of 100 %, a specificity of 95.5 % and an accuracy of 97.7 %. In independent testing with datasets which was not used in training the SVM models, the AD-MCI classifier showed a sensitivity of 69.2 %, a specificity of 100 % and an accuracy of 89.7 %, and the AD-NC classifier showed a sensitivity of 84.6 %, a specificity of 90.9 % and an accuracy of 87.5 %. Although preliminary, these results are promising and indicate that SVM-based classification of DTI data can be useful in early detection of MCI and AD.

Acknowledgments. This research was supported by the Ministry of Education (2010-0020163) and in part by the Basic Science Research Program through the National Research Foundation (NRF) funded by the Ministry of Science, ICT & Future Planning (2015R1A1A3A04001243).

References

1. Vetrivel, K.S., Thinakaran, G.: Amyloidogenic processing of beta-amyloid precursor protein in intracellular compartments. Neurology **66**(2 Suppl. 1), S69–S73 (2006)
2. Jonsson, L., Lindgren, P., Wimo, A., Jonsson, B., Winblad, B.: The cost-effectiveness of donepezil therapy in Swedish patients with Alzheimer's disease: a Markov model. Clin. Ther. **21**(7), 1230–1240 (1999)
3. Sperling, R.A., Aisen, P.S., Beckett, L.A., Bennett, D.A., Craft, S., Fagan, A.M., Iwatsubo, T., Jack, Jr, C.R., Kaye, J., Montine, T.J., Park, D.C., Reiman, E.M., Rowe, C.C., Siemers, E., Stern, Y., Yaffe, K., Carrillo, M.C., Thies, B., Morrison-Bogorad, M., Wagster, M.V., Phelps, C.H.: Toward defining the preclinical stages of Alzheimer's disease: recommendations from the National Institute on Aging-Alzheimer's Association workgroups on diagnostic guidelines for Alzheimer's disease. Alzheimers Dement. **7**(3), 280–292 (2011)
4. Kloppel, S., Stonnington, C.M., Chu, C., Draganski, B., Scahill, R.I., Rohrer, J.D., Fox, N.C., Jack, Jr., C.R., Ashburner, J., Frackowiak, R.S.: Automatic classification of MR scans in Alzheimer's disease. Brain **131**(Pt. 3), 681–689 (2008)
5. Magnin, B., Mesrob, L., Kinkingnehun, S., Pelegrini-Issac, M., Colliot, O., Sarazin, M., Dubois, B., Lehericy, S., Benali, H.: Support vector machine-based classification of Alzheimer's disease from whole-brain anatomical MRI. Neuroradiology **51**(2), 73–83 (2009)
6. Bischkopf, J., Busse, A., Angermeyer, M.C.: Mild cognitive impairment—a review of prevalence, incidence and outcome according to current approaches. Acta Psychiatr. Scand. **106**, 403–414 (2002)

7. Medina, D.A., Gaviria, M.: Diffusion tensor imaging investigations in Alzheimer's disease: the resurgence of white matter compromise in the cortical dysfunction of the aging brain. Neuropsychiatr. Dis. Treat. **4**(4), 737–742 (2008)

8. Oishi, K., Mielke, M.M., Albert, M., Lyketsos, C.G., Mori, S.: DTI analyses and clinical applications in Alzheimer's disease. J. Alzheimers Dis. **26**(Suppl. 3), 287–296 (2011)

9. Hess, C.P.: Update on diffusion tensor imaging in Alzheimer's disease. Magn. Reson. Imaging Clin. N. Am. **17**(2), 215–224 (2009)

10. Bastin, M.E., Le Roux, P.: On the application of a non-CPMG single-shot fast spin-echo sequence to diffusion tensor MRI of the human brain. Magn. Reson. Med. **48**(1), 6–14 (2002)

11. Ito, R., Mori, S., Melhem, E.R.: Diffusion tensor brain imaging and tractography. Neuroimaging Clin. N. Am. **12**(1), 1–19 (2002)

12. Melhem, E.R., Itoh, R., Jones, L., Barker, P.B.: Diffusion tensor MR imaging of the brain: effect of diffusion weighting on trace and anisotropy measurements. AJNR Am. J. Neuroradiol. **21**(10), 1813–1820 (2000)

13. Pierpaoli, C., Jezzard, P., Basser, P.J., Barnett, A., Di Chiro, G.: Diffusion tensor MR imaging of the human brain. Radiology **201**(3), 637–648 (1996)

14. Smith, S.M., Jenkinson, M., Johansen-Berg, H., Rueckert, D., Nichols, T.E., Mackay, C.E., Watkins, K.E., Ciccarelli, O., Cader, M.Z., Matthews, P.M., Behrens, T.E.: Tract-based spatial statistics: voxelwise analysis of multi-subject diffusion data. Neuroimage **31**(4), 1487–1505 (2006)

15. Basser, P.J., Pierpaoli, C.: Microstructural and physiological features of tissues elucidated by quantitative-diffusion-tensor MRI. 1996. J. Magn. Reson. **213**(2), 560–570 (2011)

16. Ennis, D.B., Kindlmann, G.: Orthogonal tensor invariants and the analysis of diffusion tensor magnetic resonance images. Magn. Reson. Med. **55**(1), 136–146 (2006)

17. Douaud, G., Jbabdi, S., Behrens, T.E., Menke, R.A., Gass, A., Monsch, A.U., Rao, A., Whitcher, B., Kindlmann, G., Matthews, P.M., Smith, S.: DTI measures in crossing-fiber areas: increased diffusion anisotropy reveals early white matter alteration in MCI and mild Alzheimer's disease. Neuroimage **55**(3), 880–890 (2011)

18. Noble, W.S.: What is a support vector machine? Nat. Biotechnol. **24**(12), 1565–1567 (2006)

19. Robnik-Šikonja, M., Kononenko, I.: Theoretical and empirical analysis of ReliefF and RReliefF. Mach. Learn. **53**(1–2), 23–69 (2003)

20. Jenkinson, M., Beckmann, C.F., Behrens, T.E., Woolrich, M.W., Smith, S.M.: FSL. Neuroimage **62**(2), 782–790 (2012)

21. Smith, S.M.: Fast robust automated brain extraction. Hum. Brain Mapp. **17**(3), 143–155 (2002)

22. Smith, S.M., Jenkinson, M., Woolrich, M.W., Beckmann, C.F., Behrens, T.E., Johansen-Berg, H., Bannister, P.R., De Luca, M., Drobnjak, I., Flitney, D.E., Niazy, R.K., Saunders, J., Vickers, J., Zhang, Y., De Stefano, N., Brady, J.M., Matthews, P.M.: Advances in functional and structural MR image analysis and implementation as FSL. Neuroimage **23** (Suppl. 1), S208–S219 (2004)

23. Haller, S., Nguyen, D., Rodriguez, C., Emch, J., Gold, G., Bartsch, A., Lovblad, K.O., Giannakopoulos, P.: Individual prediction of cognitive decline in mild cognitive impairment using support vector machine-based analysis of diffusion tensor imaging data. J. Alzheimers Dis. **22**(1), 315–327 (2010)

24. O'Dwyer, L., Lamberton, F., Bokde, A.L.W., Ewers, M., Faluyi, Y.O., Tanner, C., Mazoyer, B., O'Neill, D., Bartley, M., Collins, D.R., Coughlan, T., Prvulovic, D., Hampel, H.: Using support vector machines with multiple indices of diffusion for automated classification of mild cognitive impairment. PLoS ONE **7**(2), e32441 (2012)

25. Scholkopf, B., Sung, K.K., Burges, C.J.C., Girosi, F., Niyogi, P., Poggio, T., Vapnik, V.: Comparing support vector machines with Gaussian kernels to radial basis function classifiers. IEEE Trans. Sig. Process. **45**(11), 2758–2765 (1997)
26. Frank, E., Hall, M., Trigg, L., Holmes, G., Witten, I.H.: Data mining in bioinformatics using Weka. Bioinformatics **20**(15), 2479–2481 (2004)
27. Behrens, T.E., Berg, H.J., Jbabdi, S., Rushworth, M.F., Woolrich, M.W.: Probabilistic diffusion tractography with multiple fibre orientations: what can we gain? Neuroimage **34** (1), 144–155 (2007)

Extraction of Features from Patch Based Graphs for the Prediction of Disease Progression in AD

Tong Tong[1] and Qinquan Gao[2(✉)]

[1] Department of Computing, Imperial College London, London, UK
[2] Province Key Lab of Medical Instrument and Pharmaceutical Technology,
Fuzhou University, Fuzhou, China
gqinquan@fzu.edu.cn

Abstract. There has been significant interest in approaches that utilize local intensity patterns within patches to derive features for disease classification. Several existing methods explore the patch relationship between different subjects for classification. Other methods utilize the patch relationship within the same subject to aid classification. In this paper, we proposed a new approach to extract different types of features by utilizing the patch relationships within and between subjects. Specifically, features are first extracted by exploiting the patch relationship between subjects. Then, the relationship among patches within the same subject is modeled as a network and features are derived from the constructed network. Finally, these two different types of features are integrated into one framework for identifying mild cognitive impairment (MCI) subjects who will progress to Alzheimer's disease. Using the standardized ADNI database, the proposed method can achieve an area under the receiver operating characteristic curve (AUC) of 81.3 % in discriminating patients with stable MCI and progressive MCI in a 10-fold cross validation, demonstrating that the integration of patch relationship between and within subjects can aid the prediction of MCI to AD conversion.

Keywords: Brain MR imaging · Alzheimer's disease · MCI conversion · Patch-based technique · Graph

1 Introduction

Alzheimer's disease (AD), the most common type of dementia, is a progressive brain disease, leading to gradual disorders that can affect patients' everyday activities. Approximately 30 million people worldwide have AD, and this number is predicted to triple in the next four decades [1]. In spite of the severity of AD, no drug or treatment has so far been reported to be able to reverse or stop the progress of AD. This has led to intensive research in academia and industry to fight this disease. Mild cognitive impairment (MCI) is a transitional stage between normal cognition and clinical dementia. Studies have shown that individuals with MCI progress to AD at a rate of 10 %–15 % per year [6]. Identifying MCI subjects who will progress to AD is crucial in clinical practice because the early diagnosis would allow doctors to

© Springer International Publishing Switzerland 2015
D.-S. Huang et al. (Eds.): ICIC 2015, Part II, LNCS 9226, pp. 500–509, 2015.
DOI: 10.1007/978-3-319-22186-1_50

perform clinical trials sooner and may slow down the progression to AD. However, the anatomical changes between stable MCI (sMCI) and progressive MCI (pMCI) are subtle, which makes the prediction of MCI conversion quite challenging. For example, in a recent comprehensive study [4], ten methods were evaluated for the prediction of MCI conversion and only four of these methods can perform the prediction of MCI to AD conversion more accurately than a random classifier. Therefore, more advanced methods need to be developed for accurate prediction of MCI to AD conversion.

Many researchers have tried to identify subjects with AD or MCI by using different types of imaging modalities, such as structural magnetic resonance imaging (MRI) [18], functional MRI [16], fluorodeoxyglucose positron emission tomography (FDG-PET) [5] and diffusion tensor imaging (DTI) [9]. Among these neuroimaging modalities, structural MRI has attracted a significant interest in the studies of neurodegenerative diseases since it is widely available and offers good diagnostic accuracy to cost ratio. Brain MR images have a high dimensionality of the voxel-wise intensity features and many of these features may not be related to pathological changes. If these intensity features are directly used for classification, it may result in low performance due to a large number of irrelevant features. Therefore, it is necessary to extract discriminative and representative features from the original MR images for classification. In previous studies [17, 22], features derived from MR images such as volumes or cortical thickness are extracted for multiple regions of interests (ROI) for disease classification. The feature dimensionality is significantly reduced after extraction, which allows the use of conventional classifiers. However, ROI-based features are not independent of each other. It is observed that the pathological changes of AD occur in several inter-related regions rather than at isolated voxels [22]. The pairwise correlations between ROIs within the same subject can provide further information about the disease-induced changes and may be helpful for disease classification. In [17, 22], correlated features are derived from ROI-based features. The correlated features are then integrated with ROI-based features, which can significantly improve the classification performance [17, 22].

Although promising classification results can be achieved using ROI-based features, these features may ignore subtle morphological changes between sMCI and pMCI. For example, the morphological changes between sMCI and pMCI may occur in subfields of hippocampus rather than the whole hippocampus as indicated in [3]. However, the ROI-based features characterize the average changes within each ROI, which may include the less affected region within each ROI. In recent approaches [3, 10, 15], local intensity patterns within patches are utilized for brain disease classification. These local patches are very small regions, which can capture subtle structural changes for disease classification. In addition, the correlations among patches may also be associated with pathological changes and provide complementary information for disease classification. In [15], the correlations among patches within the same subject are modeled as a network for a more accurate characterization of pathology. However, the derived correlated features only consider the ROI or patch interactions within the same subject. It is possible to find similar pathological changes of one subject in other subjects. Therefore, the pathological status of patches extracted from training subjects can be propagated to the patches of MCI subjects by investigating the patch similarities across subjects. In [3], the SNIPE (Scoring by Nonlocal Image Patch Estimator) method was proposed to investigate the patch relationship between different subjects and has shown a good prediction accuracy of MCI conversion.

In this paper, we proposed a novel classification method to integrate the patch relationship within and between subjects for the prediction of MCI conversion. In particular, two types of features are extracted for each MCI subjects as illustrated in Fig. 1. First, grading features are extracted to investigate patch relationship across subjects. A new approach based on sparse representation techniques is also proposed to improve the discriminative ability of the grading features. Second, network features are extracted to represent the relationship among patches within the same subject. Specifically, patches are modeled as nodes in a graph and the relationship among patches are modeled as edges. This yields a brain network for each subject. Then, features can be extracted from the network for classification. Finally, the grading features and the network features are integrated via multi-kernel Support Vector Machines (SVM) to perform the final classification. The classification performance of the proposed method was evaluated on the standardized list [20] of baseline scans from the Alzheimer's Disease Neuroimaging Initiative (ADNI) and compared with those of other state-of-the-art methods [3, 11, 15, 18]. Moreover, the influence of the combination parameter was studied.

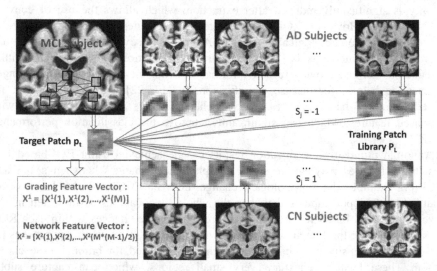

Fig. 1. Feature extraction of the proposed method. Two types of features are extracted for each MCI subject: network features and grading features. The red graph (on the top left) is constructed by modeling the patch relationship within the MCI subject and network features are extracted from the red graph. The blue graph (in the middle) is constructed for each target patch p_t extracted from the MCI subject. The blue graph is based on the patch similarities across subjects. A grading value is estimated for p_t via the blue graph (Color figure online).

2 Method

As illustrated in Fig. 1, two types of graphs are constructed for extracting the grading features and the network features respectively. The blue graph as shown in Fig. 1, which is based on patch similarities across subjects, is constructed to propagate the pathological status of cognitively normal (CN) and AD subjects to the target patch pt, which is extracted from the MCI subject. A disease score can be estimated for each target patch pt. If M patches are extracted from the MCI subject, M grading features will be obtained. Furthermore, the red graph as shown in Fig. 1 is also constructed by modeling the relationship of the M patches within the MCI subject. Network features with a dimensionality of M × (M − 1)/2 are then extracted from the red graph for classification.

2.1 Patch Based Grading Features

The CN and AD were used as the training populations to calculate the grading features for MCI subjects as suggested in [3, 21]. To calculate a grading value for each target patch pt in the MCI subject, all corresponding patches across the training populations are extracted to form a training patch library PL. The patch library PL typically contains thousands of patches. The relationship between the target patch pt and the patches in the training library PL is modeled by a weighting function in [2]. Each training patch receives a weight wt(j) via this weighting function:

$$w_t(j) = e^{\frac{-\|p_t - P_L(j)\|_2^2}{h^2}} \qquad (1)$$

where PL(j) represents the jth training patch in the patch library. The weight wt(j) reflects the similarity between the target patch pt and the training patch PL(j). If the training patch is similar to the target patch, a weight close to one is assigned; otherwise, a weight close to zero is assigned. The pathological status of the training patch PL(j) is denoted as sj. If the training patch is extracted from CN subjects, sj is set to 1. sj = −1 is used for patches extracted from AD subjects. Since we know the pathological status of the training patches and the relationship between the target patch and the training patches, a grading value can be estimated for the target patch using the weighted fusion method as proposed in [2]. Finally, this patch based grading (PBG) method can estimate grading values for all M patches in each MCI subject.

The PBG method uses all the training patches to estimate the grading value according to their similarities to the target patch. However, not all training patches with high appearance similarities have the same pathological status with that of the target patch. If misleading training patches are used, this undermines the final grading estimation result. To overcome this limitation, we propose a sparse representation patch based grading (SRPBG) method, inspired by the patch based segmentation studies in [14, 19]. The proposed SRPBG method uses a sparse representation technique [14], which seeks for the best representation of the target patch pt from the patch library PL and then estimates a grading value for pt using the selected training patches. The Elastic Net [14] sparse regression model is used in this paper. The sparse representation of the target patch pt is calculated as:

$$\hat{a}_t = \min_{a_t} \frac{1}{2} \|p_t - P_L(j)\|_2^2 + \beta_1 \|a_t\|_1 + \beta_2 \|a_t\|_2^2 \tag{2}$$

where at are the coding coefficients for the target patch pt. Most of the coefficients in at are zero due to the sparsity constraint. If the coefficient in at is not zero, it indicates that the corresponding training patch has been selected to propagate their pathological information to the target patch. After the sparse solution is obtained, the scoring of the target patch is based on the coding coefficients at and the pathological status of the selected training patches. The grading value of pt can be estimated as:

$$g_t = \frac{\sum_{j=1}^{N} \hat{a}_t(j) s_j}{\sum_{j=1}^{N} \hat{a}_t(j)} \tag{3}$$

where N is the number of the training patches in the patch library PL. at(j) is the coding coefficient corresponding to the training patch PL. If gt is close to -1, it indicates that pt is more characteristic of pMCI than sMCI.

2.2   Patch Based Network Features

Each MCI subject is represented with M patches. The relationship among the M patches within the same subject can be modeled as a network [15]. In the constructed network, each node denotes a patch and each edge represents the correlation between a pair of patches. The correlation between two patches can be computed using different measures, such as correlation coefficient and mutual information. To simplify the problem, we use the correlation coefficient in our work. If the intensity values within a patch are affected by MCI due to factors such as atrophy, the correlation of this patch with other patches will be affected.

It should be noted that the constructed network for each MCI subject is a dense matrix, which may not have the same network properties that are observed in structural or functional brain networks. In addition, the edges in the constructed network do not reflect the real neuronal connections, but imply the similarities of intensity patterns between patches. Commonly used network features, such as local clustering coefficients, can be extracted from sparse networks. However, these network features may not work efficiently in our case because the constructed networks are dense. Since the correlations between patches may reflect the structural changes, we keep all the correlations as features to capture possible structural changes for classification. As there are M patches within each MCI subject, the dimensionality of the network features is $M \times (M - 1)/2$.

2.3 Classification Using Multi-kernel SVM

After feature extraction, feature selection is performed to select useful features for classification because the extracted grading and network features may contain irrelevant or redundant features. For example, if a patch is in a non-affected region, the grading value of this patch is close to zero but will provide little discriminative information. In addition, the correlation between this patch and other patches may not reflect the structural changes caused by pathology. The learned classifier may lead to overfitting and become less generalizable if noisy features are present. Therefore, feature selection is utilized to identify informative and discriminative features for classification. In our study, one of the most widely used feature selection methods called Least Absolute Shrinkage and Selection Operator (LASSO) [7] is used.

To integrate different types of features, two categories of methods can be used: feature concatenation approaches and kernel based approaches. In our study, a multi-kernel SVM was utilized to integrate information from the grading and network features. Specifically, a kernel matrix was first computed for the grading and network features respectively (denoted as kGrading and kNetwork), and a radial basis function (RBF) kernel was adopted. Then, different subkernels were integrated to form a mixed kernel matrix K with weights given to each subkernel. The mixed kernel k(xi, xj) is then computed by using the grading and network features of the ith and jth subject:

$$k\left(x_i, x_j\right) = \lambda_{Grading} k_{Grading}\left(x_i^1, x_j^1\right) + \lambda_{Network} k_{Network}\left(x_i^2, x_j^2\right)$$
$$= \lambda_{Grading} e^{-\frac{\left\|x_i^1 - x_j^1\right\|_2^2}{\delta_1}} + \lambda_{Network} e^{-\frac{\left\|x_i^2 - x_j^2\right\|_2^2}{\delta_2}} \tag{4}$$

where x1 i and x2 i respectively represent the grading and network feature vector of the ith subject. λGrading and λNetwork are the weights given to each feature subkernel, whose values are constrained to be positive and sum to one as in [21]. The above equation can then be formulated as:

$$k\left(x_i, x_j\right) = \lambda e^{-\frac{\left\|x_i^1 - x_j^1\right\|_2^2}{\delta_1}} + (1 - \lambda) e^{-\frac{\left\|x_i^2 - x_j^2\right\|_2^2}{\delta_2}} \tag{5}$$

After the mixed kernel matrix is computed, a kernel SVM is used for training and testing.

3 Experiments and Results

Baseline MR images acquired at 1.5T from the Alzheimer's Disease Neuroimaging Initiative (ADNI) study (www.loni.ucla.edu/ADNI) were downloaded for evaluation in this paper. To enable future comparison, we used the standardized database from ADNI [20], which consists of 818 subjects and was classified into five different groups as shown in Table 1. The AD and CN groups were determined according to the diagnosis at baseline. Subjects in the MCI group were classified as pMCI if the subjects converted to AD during a 3 year follow-up, but without reversion to MCI or CN at any

Table 1. Demographic information of the dataset used in this study. MMSE: Mini-Mental State Examination; CDR-SB: Clinical Dementia Rating-Sum of Boxes.

Group	Number	Age	MMSE	CDR-SB
NC	229	75.9 ± 5.0	29.1 ± 1.0	0.03 ± 0.12
sMCI	100	74.6 ± 7.5	27.6 ± 1.7	1.34 ± 0.65
pMCI	164	74.5 ± 7.0	26.6 ± 1.7	1.88 ± 0.97
uMCI	134	75.2 ± 8.0	27.0 ± 1.9	1.48 ± 0.87
AD	191	75.3 ± 7.5	23.3 ± 2.0	4.31 ± 1.63

available follow-up (0–96 months). Subjects were classified as sMCI if the diagnosis was MCI at all available time points (0–36 months). Those whose diagnosis was missing at 36 months or whose diagnosis was not stable at all available time points were grouped as unknow MCI (uMCI). The definition of these different groups is the same as that in [11]. All images were preprocessed by the standard ADNI pipeline as described in [8]. After that, non-rigid registration, based on B-spline free-form deformation [13] with a final control point spacing of 10 mm, was performed to align all images to the MNI152 template space. The intensity was then normalized between each image and the template using the approach proposed in [12]. In order to compare disease-specific differences between groups, we also performed an age correction step as proposed in [11]. Then, an ROI around the hippocampus is defined in template space followed by a dilation of 5 mm and all non-overlapping patches within this defined ROI are extracted. This produces 227 patches within each MCI subject for extracting features.

Experiments were performed using 10-fold cross validation. This validation was repeated 100 runs to remove the randomness due to sampling. The classification performance in terms of accuracy (ACC), sensitivity (SEN), specificity (SPE) and area under the receiver operating characteristic curve (AUC) are averages over 100 runs. The patch size was fixed to $7 \times 7 \times 7$ voxels as suggested in [2, 3]. For calculating grading features using sparse representation, the parameters $\beta1$ and $\beta2$ in Eq. (2) were set to 0.15 as suggested in [14]. The parameter λ in Eq. (5) was studied. The other parameters including $\delta1$, $\delta2$ in Eq. (5) and the cost parameter C in SVM were determined via inner 10-fold cross validations on the training samples.

3.1   Results Using Different Features

Experiments were conducted to compare the discriminative power of different features. The single kernel SVM was applied to the network features and the grading features for classification. The multi-kernel SVM was applied to the integrated features. The classification results are presented in Table 2. We can see that the classification results using the grading features are slightly more accurate than the results using the network features. When the network features and the grading features are combined, there is a significant improvement in the classification between sMCI and pMCI (p < 0.01 using two-tailed t-tests over accuracy), indicating that there are complementary information between these two types of features for the prediction of MCI to AD conversion.

Table 2. Comparison of classification results of sMCI vs pMCI using different features.

Features	AUC	ACC	SPE	SPE
Network	76.6 ± 1.8	72.4 ± 1.8	80.9 ± 2.1	58.4 ± 3.1
Grading	78.3 ± 1.0	72.6 ± 1.3	81.2 ± 1.6	58.5 ± 2.4
Network + grading	81.3 ± 1.0	76.3 ± 1.3	83.6 ± 1.8	64.4 ± 2.7

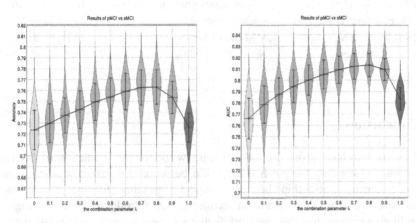

Fig. 2. The effect of λ over the classification performance of sMCI versus pMCI. The parameter space of λ was searched between 0 and 1 with an increment of 0.1. $\lambda = 0$ represents the classification results only using the network features, while $\lambda = 1$ represents the classification results only using the grading features.

We further evaluated the classification performance using different settings of the combination parameter λ in Eq. (5). The classification accuracy and AUC with varying settings of λ are shown in Fig. 2.

3.2 Comparison with State-of-the-Art Methods

Table 3 shows the classification results of other state-of-the-art methods [3, 11, 15, 18] for comparison. The comparison with previous methods using the ADNI database is hampered by sereral factors, such as different subsets for study, different cross validation strategy, different denifition criteria of sMCI and pMCI and so on. Here, to enble a fair comparison, we used the same subsets, the same definition of sMCI and pMCI as in [11]. We can see that our proposed method outperforms other state-of-the-art methods, which demonstrates the effectiveness of the proposed method in the prediction of MCI to AD conversion. It should be mentioned that we achieved a similar accuracy of about 73 % as in [3] using the grading features. When the network features were added, a significant improvement was achieved, demonstrating the benefit of combining these two types of features.

Table 3. Comparison of classification results of sMCI vs pMCI using different features.

Methods	Features	AUC	ACC	SEN	SPE
Multi-method (Wolz et al. [18])	Hippocampal volume	-	65	63	67
	Cortical thickness	-	56	63	45
	Manifold-based learning	-	65	64	66
	Tensor-based morphometry	-	64	65	62
	All	-	68	67	69
SNIPE (Coupé et al. [3])	Hippocampus grading + age	-	73	68	76
mi-Graph (Tong et al. [15])	Intensity patches	-	70	67	71
LDS (Moradi et al. [11])	MRI biomarker	76.6	74.7	88.9	51.6
Our proposed method	Network + grading	81.3	76.3	83.6	64.4

4   Conclusion

In this work, we have proposed a novel framework for the prediction of MCI conversion by investigating the patch relationship between and within subjects. Considering the patch relationship between subjects, grading features are extracted. Furthermore, the sparse representation technique was introduced for extracting more discriminative grading features. Network features are extracted to investigate the patch relationship within subjects. Finally, these two different types of features are integrated into one framework for the prediction of MCI conversion. The evaluation using the standardized database from the ADNI study demonstrates the efficiency of the proposed approach, comparing favourably with state-of-the-art classification methods.

References

1. Barnes, D.E., Yaffe, K.: The projected effect of risk factor reduction on Alzheimer's disease prevalence. Lancet Neurol. **10**(9), 819–828 (2011)
2. Coupé, P., Eskildsen, S.F., Manjón, J.V., Fonov, V., Collins, D.L.: Simultaneous segmentation and grading of hippocampus for patient classification with Alzheimer's disease. In: Fichtinger, G., Martel, A., Peters, T. (eds.) MICCAI 2011, Part III. LNCS, vol. 6893, pp. 149–157. Springer, Heidelberg (2011)
3. Coupé, P., Eskildsen, S.F., Manjon, J.V., Fonov, V.S., Pruessner, J.C., Allard, M., Collins, D.L.: Scoring by nonlocal image patch estimator for early detection of Alzheimer's disease. NeuroImage: Clin. **1**(1), 141–152 (2012)
4. Cuingnet, R., Gerardin, E., Tessieras, J., Auzias, G., Lehéricy, S., Habert, M.O., Chupin, M., Benali, H., Colliot, O.: Automatic classification of patients with Alzheimer's disease from structural MRI: a comparison of ten methods using the ADNI database. Neuroimage **56**(2), 766–781 (2011)
5. Gray, K.R., Wolz, R., Heckemann, R.A., Aljabar, P., Hammers, A., Rueckert, D.: Multi-region analysis of longitudinal FDG-PET for the classification of Alzheimer's disease. NeuroImage **60**(1), 221–229 (2012)

6. Grundman, M., Petersen, R.C., Ferris, S.H., Thomas, R.G., Aisen, P.S., Bennett, D.A., Foster, N.L., Jack, Jr., C.R., Galasko, D.R., Doody, R., et al.: Mild cognitive impairment can be distinguished from Alzheimer disease and normal aging for clinical trials. Arch. Neurol. 61(1), 59 (2004)

7. Haury, A.C., Gestraud, P., Vert, J.P.: The influence of feature selection methods on accuracy, stability and interpretability of molecular signatures. PLoS ONE 6(12), e28210 (2011)

8. Jack, C.R., Bernstein, M.A., Fox, N.C., Thompson, P., Alexander, G., Harvey, D., Borowski, B., Britson, P.J., Whitwell, J.L., Ward, C., et al.: The Alzheimer's disease neuroimaging initiative (ADNI): MRI methods. J. Magn. Reson. Imaging 27(4), 685–691 (2008)

9. Keihaninejad, S., Zhang, H., Ryan, N.S., Malone, I.B., Modat, M., Cardoso, M.J., Cash, D., Fox, N.C., Ourselin, S.: An unbiased longitudinal analysis framework for tracking white matter changes using diffusion tensor imaging with application to Alzheimer's disease. NeuroImage 72, 153–163 (2013)

10. Liu, M., Zhang, D., Shen, D.: Hierarchical fusion of features and classifier decisions for Alzheimer's disease diagnosis. Hum. Brain Mapp. 35, 1305–1319 (2013)

11. Moradi, E., Pepe, A., Gaser, C., Huttunen, H., Tohka, J.: ADNI: machine learning framework for early MRI-based Alzheimer's conversion prediction in MCI subjects. NeuroImage 104, 398–412 (2015)

12. Nyu, L.G., Udupa, J.K.: On standardizing the MR image intensity scale. Magn. Reson. Med. 42(6), 1072 (1999)

13. Rueckert, D., Sonoda, L.I., Hayes, C., Hill, D.L.G., Leach, M.O., Hawkes, D.J.: Nonrigid registration using free-form deformations: application to breast MR images. IEEE Trans. Med. Imaging 18(8), 712–721 (1999)

14. Tong, T., Wolz, R., Coupé, P., Hajnal, J.V., Rueckert, D.: Segmentation of MR images via discriminative dictionary learning and sparse coding: application to hippocampus labeling. NeuroImage 76, 11–23 (2013)

15. Tong, T., Wolz, R., Gao, Q., Hajnal, J.V., Rueckert, D.: Multiple instance learning for classification of dementia in brain MRI. Med. Image Anal. 18(5), 808–818 (2014)

16. Wee, C.Y., Yap, P.T., Li, W., Denny, K., Browndyke, J.N., Potter, G.G., Welsh-Bohmer, K.A., Wang, L., Shen, D.: Enriched white matter connectivity networks for accurate identification of MCI patients. Neuroimage 54(3), 1812–1822 (2011)

17. Wee, C.Y., Yap, P.T., Shen, D.: Prediction of Alzheimer's disease and mild cognitive impairment using cortical morphological patterns. Hum. Brain Mapp. 34(12), 3411–3425 (2013)

18. Wolz, R., Julkunen, V., Koikkalainen, J., Niskanen, E., Zhang, D.P., Rueckert, D., Soininen, H., Lotjonen, J.: Multi-method analysis of MRI images in early diagnostics of Alzheimer's disease. PLoS ONE 6(10), e25446 (2011)

19. Wu, G., Wang, Q., Zhang, D., Nie, F., Huang, H., Shen, D.: A generative probability model of joint label fusion for multi-atlas based brain segmentation. Med. Image Anal. 18(6), 881–890 (2014)

20. Wyman, B.T., Harvey, D.J., et al.: Standardization of analysis sets for reporting results from ADNI MRI data. Alzheimer's Dement. 9(3), 332–337 (2013)

21. Young, J., Modat, M., Cardoso, M.J., Mendelson, A., Cash, D., Ourselin, S.: Accurate multimodal probabilistic prediction of conversion to Alzheimer's disease in patients with mild cognitive impairment. NeuroImage Clin. 2, 735–745 (2013)

22. Zhou, L., Wang, Y., Li, Y., Yap, P.T., Shen, D.: Hierarchical anatomical brain networks for MCI prediction: revisiting volumetric measures. PLoS ONE 6(7), e21935 (2011)

Group Sparse Representation for Prediction of MCI Conversion to AD

Xiaoying Chen[✉], Kaifeng Wei, and Manhua Liu

Department of Instrument Science and Engineering, School of EIEE,
Shanghai Jiao Tong University, Shanghai 200240, China
xiaoying.chen2009@gmail.com, mhliu@sjtu.edu.cn

Abstract. Early Diagnose of Alzheimer's disease (AD) is a problem which scientists are committed to solving for a long time. As the prodromal stage of AD, the mild cognitive impairment (MCI) patients have high risk of conversion to AD. Thus, for early diagnosis and possible early treatment of AD, it is important for accurate prediction of MCI conversion to AD, i.e., classification between MCI non-converter (MCI-NC) and MCI converter (MCI-C). In this paper, we propose a group discriminative sparse representation algorithm for prediction of MCI conversion to AD. Unlike the previous researches which are based on l_1-norm sparse representation classification (SRC), we focus on how to mining the group label information which will help us to do the classification more correctly and efficiently. We apply the group label restricted condition as well as the sparse condition when doing the sparse coding procedure, which makes the sparse coding coefficients discriminative. The Moreau-Yosida regularization method is utilized to help us solving this convex optimization problem. In our experiments on magnetic resonance brain images of 403 MCI patients (167 MCI-C and 236MCI-NC) from ADNI database, we demonstrate that the proposed method performs better than the traditional classification methods such as SRC with l_1-norm and group sparse representation with l_2-norm.

Keywords: AD/MCI diagnosis · Group sparse representation · Sparse representation classification

1 Introduction

Alzheimer's disease (AD) is a chronic neurodegenerative disease that usually starts slowly and gets worse over time to interfere with daily tasks. Nowadays, there has no effective cure for AD, and current treatments can only temporarily slow the worsening of dementia symptoms and improve quality of life for those with AD. One effective way to deal with the AD is to diagnose it as early as possible and then take proper treatment to slow down the disease progression. However in its early stages od elderly persons, the symptoms of AD are difficult to distinguish from those of normal aging. Fortunately, recent research has identified that neuroimaging features of AD obtained by Magnetic Resonance Imaging (MRI) or Positron emission tomography (PET) are also observed in mild cognitive impairment (MCI), which causes a slight but noticeable

© Springer International Publishing Switzerland 2015
D.-S. Huang et al. (Eds.): ICIC 2015, Part II, LNCS 9226, pp. 510–519, 2015.
DOI: 10.1007/978-3-319-22186-1_51

and measurable decline in cognitive abilities, including memory and thinking skills. So we can regard MCI as the prodromal stage of AD naturally and correct classification of MCI-C (MCI converter) and MCI-NC (MCI non-converter) becomes a research highlight in AD research area.

For decades, many laboratories have concentrated their efforts on AD diagnosis. A Work Group on the Diagnosis of Alzheimer's Disease was established by the National Institute of Neurological and Communicative Disorders and Stroke (NIN-CDS) and the Alzheimer's disease and Related Disorders Association (ADRDA), which successfully established and described uniform clinical criteria for the diagnosis of Alzheimer's disease. They are useful for comparative studies of patients in different kinds of researches, including case control studies, therapeutic trials and so on [1]. Also the American Psychiatric Association's Diagnostic and Statistical Manual of Mental Disorders (DSM-IV-TR) is another prevailing diagnostic standard of Alzheimer's disease in researches. However, the above two standards have now fallen behind the dramatic growth of scientific knowledge and technology. Nowadays, with the development of structural MRI, molecular neuroimaging with PET, and cerebrospinal fluid analyses, we can obtain numbers of distinctive and reliable biomarkers of AD. Dubois, et al. said that it is necessary to set up a brand new framework to capture both the earliest stages, before full-blown dementia, as well as the full spectrum of illness [2].

Although there have been many inspiring research achievements about the diagnosis of Alzheimer's disease, it still remains challenging for the early diagnosis of AD, and relatively a small number of works have been performed in the literature. For instance, Davatzikos said that it is helpful to predict the short-term conversion from MCI to AD by analyzing magnetic resonance imaging (MRI) and cerebrospinal fluid (CSF) biomarkers together [3]. Risacher SL analyzed hundreds of participants from the ADNI database by baseline diagnosis and one-year MCI to probable AD conversion status to identify neuroimaging phenotypes associated with MCI and AD, and to identify the potential predictive markers of imminent conversion [4]. Cheng, et al. proposed a sparse semi-supervised manifold-regularized transfer learning classification method to fully exploit the auxiliary domain information (AD/Normal Control) and unlabeled data to do the MCI conversion prediction [5].

Besides the deeper understanding of the neuroimaging features of Alzheimer's disease as well as the Mild Cognitive Impairment, the rapid development of the theory and algorithms of pattern recognition and machine learning have significantly improved the performance of their computer aided diagnosis. Recently sparse representation techniques have gotten surprising results in image classification area, including face recognition [6], etc. Researchers attribute the success of sparse representation based classification to two factors as follows. One is that the model of sparse representation (sparse coding) imitates the structure of human vision system, the learned bases resemble the receptive fields of neurons in the visual cortex [7]. So a high-dimensional image can be represented or coded by some low-dimensional images which belong to the same class. The other factor is the progression convex optimization algorithms, especially the l_0-norm and l_1-norm minimization problems.

As for the sparse representation based classification, there are two key procedures. One is dictionary learning, which is used to train an over-completed space where the given signal/image could be well represented for classification or restoration. The other critical procedure is sparse coding, which converts the feature vector of original signal/image into a combination of the dictionary atoms, then the classification is performed based on the coding coefficients. The aim of sparse coding is to use as little atoms of over-complete dictionary as possible to represent the signal/image, which will obtain the concise way to describe the signal/image. Mallat and Zhang (1993) proposed a greedy solution known as Matching Pursuit, which is the first effective way to do the sparse factorization with l_1-norm regularized term to guarantee the sparsity [8]. After that, series of algorithm including Orthogonal Matching Pursuit (OMP) [9], Regularized Orthogonal Matching Pursuit (ROMP) [10] were proposed to solve the sparse coding problem with l_1-norm more efficiently. The sparse representation with l_1-norm is utilized in image denoising [11, 12], image restoration [13] and Super-Resolution [14] etc. However, the researchers found that although using the l_1-norm will guarantee the sparsity, it didn't take full advantages of the label information which might be valuable in image classification problem. Therefore, they proposed group Euclidean and Hilbert norms (i.e., l_2-norms) to do the sparse representation. This scheme was well developed with theory and algorithms [15]. As we can see, lots of researchers concentrated themselves on performing nice and efficient algorithm on sparse coding.

In this paper, we propose a new discriminative sparse representation framework which employs the group prior information to obtain a structured coding coefficients, for prediction of MCI conversion to AD. The idea can be briefly described as follows. Suppose that there is a testing subject that belongs to class j, for the sparse coding coefficients which come from the same class j, we hope them to be similar with each other because this will help to eliminate the noise and corruption of the dictionary atom itself. For the sparse coding coefficients which come from other class k, they should be totally zero theoretically, or in the other word, the absolute value of difference between sparse coding coefficients belong to class j and class k should be as large as possible for improving the discriminability. These two issues can be both converted to the problem of combination of l_1-norm and l_2-norm minimization. With the proposed group sparse coding method, the coding coefficients will be discriminative, which will be helpful for our subsequent classification tasks. Compared with the SRC method, our proposed method has competitive performance in the pattern recognition tasks of the neuroimaging data.

The rest of this paper is organized as follows. In next section, we describe the neuroimaging materials which we used in the experiment. The pre-processing process and specific group sparse coding method which we proposed to do the classification is described in Sect. 3. The experimental results and comparisons on the ADNI dataset are demonstrated in Sect. 4. The conclusion and future work are discussed in Sect. 5.

2   Materials

The imaging data used in this paper were obtained from the Alzheimer's Disease Neuroimaging Initiative (ADNI) database (http://adni.loni.ucla.edu/). ADNI began in October 2004 and was designed to find more sensitive and accurate methods to detect Alzheimer's disease at earlier stages and mark its progress through biomarkers. The study gathered and analyzed thousands of brain scans, genetic profiles, and biomarkers in blood and cerebrospinal fluid that are used to measure the progress of disease or the effects of treatment. ADNI researchers collect, validate and utilize data such as MRI and PET images, cognitive tests and so on as predictors for the disease. All the data were collected from the North American ADNI's study participants, including Alzheimer's disease patients, mild cognitive impairment subjects and healthy subjects, which were divided by some criteria such as Mini-Mental State Examination (MMSE) scores. The initial goal of ADNI was to recruit 800 adults, aged from 55 to 90, to participate in the research, approximately 200 cognitively normal older individuals to be followed for 3 years, 400 people with MCI to be followed for 3 years, and 200 people with early AD to be followed for 2 years.

Although the proposed method makes no assumption on a specific neuroimaging modality, MR images are widely available, non-invasive and often used as the first biomarker in the diagnostics of AD. In ADNI, the MRI datasets included standard T1-weighted MR images acquired sagittally using volumetric 3D MPRAGE with 1.25×1.25 mm^2 in-plane spatial resolution and 1.2 mm thick sagittal slices. Most of these images were obtained with 1.5T scanners, while a few were acquired using 3T scanners. Detailed information about MR acquisition procedures is available at the ADNI Web site. For testing our proposed method, in this paper, we use the T1-weighted MR brain image data from 403 MCI subjects' data, including 167 MCI-C and 236 MCI-NC. These subjects have different numbers of time-point imaging data. The maximum number of time points is 5, which include the baseline, 6-month, 12-month, 18-month, and 24-month brain images. The demographic characteristics of the studied subjects are summarized in Table 1.

Table 1. Mean age, education, and MMSE by diagnosis

Diagnosis	Number	Age	Gender (M/F)	MMSE
MCI-C	167	74.9 ± 6.8	102/65	26.6 ± 1.7
MCI-NC	236	74.9 ± 7.7	158/78	27.3 ± 1.8

3 Method

3.1 Image Preprocessing and Feature Extraction

To facilitate the following image feature extraction and classification, the MR brain images of multiple time points were preprocessed using some techniques. Specifically, a correction of image intensity inhomogeneity was first performed on the T1-weighted MR brain images using nonparametric nonuniform intensity normalization (N3) algorithm [16]. Secondly, a robust and automated skull stripping method [17] was applied for brain extraction and the MR brain images were cerebellum-removed. Thirdly, each brain image is segmented into three kinds of tissue volumes, e.g., gray matter (GM), white matter (WM), and cerebrospinal fluid (CSF) volumes. Fourthly, a fully automatic 4-dimensional atlas warping method called HAMMER [18] is used to register the three tissue volumes of all different time-point images of each subject to a template with 93 manually labeled ROIs [19]. After registration, we can label all images based on the 93 ROIs in the template. Finally, we compute the total GM, WM and CSF tissue volumes of each ROI. In this work, since it is known that the GM is more related to AD and/or MCI than other tissues, the GM volumes of 93 ROIs are used as the representation features for each MR brain image.

3.2   Classification by Group Sparse Representation

One of the key ideas of the sparse representation based classification (SRC) method is that we consider the representation of a target neuroimaging feature vector based on the training data. Suppose that there are N training samples $\{s_1, \ldots, s_n, \ldots, s_N\}$, represented by $X = [X_1 \ldots, X_l \ldots, X_C] \in \Re^{M \times N}$, belonging to C classes (C = 2 in this study), where $N = N_1 + \ldots N_l + \ldots N_C$ and $X_l \in \Re^{M \times N_l}$ consists of N_l training samples from the l-th class. Unless specially noted, all feature vectors in this paper are represented by the column vectors, and $\|\cdot\|_2$ denotes the standard Euclidean norm while $\|\cdot\|_1$ represents the standard l_1-norm. Instead of using the sparsity to identify relevant features, the sparse representation-based classifier (SRC) exploits the discriminative nature of sparse representation for classification. Specifically, it constructs a nonparametric dictionary using all training dataset across all classes and seeks for the sparse representation of a test sample in the nonparametric dictionary by l_1-norm minimization. The classification is performed by evaluating which class produces the minimum reconstruction error. Given a test sample y, SRC seeks to solve the following l_1-norm minimization problem [6]

$$\hat{\alpha} = arg \min_{\alpha} \|X\alpha - y\|_2^2 + \lambda \|\alpha\|_1 \tag{1}$$

The above l_1-norm minimization can be efficiently solved by using the l_1-regularized sparse coding methods such as those proposed in [20, 21]. Ideally, the sparse coefficients of $\hat{\alpha}$ are associated with the training samples from a single class, so that the test sample can be easily assigned to that class. However, noise and modeling error may also lead to small nonzero sparse coefficients associated with other multiple classes. Instead of classifying test sample based on the sparse coefficients, SRC makes classification by evaluating how well the sparse coefficients $\hat{\alpha}_l$ associated with the training data X_l in the l-th class can be used to reconstruct the test sample. This classification can better harness the subspace structure of each class. Thus, for a test sample y, its final classification is made by:

$$\text{Label}(y) = arg \min_{l} \|X_l \hat{\alpha}_l - y\|_2 \tag{2}$$

The above l_1-norm based SRC model assumes that the nonzero coefficients appear randomly. However, the sparse coefficients often exhibit intrinsic structure in form of groups. Therefore, introducing the intrinsic group structure into sparse representation model is a reasonable strategy to improve the performance of sparse representation for classification. In addition, sharing of dictionary atoms for data in the same group had been shown to increase the discriminative power of the sparse representation.

To consider intrinsic group structure of sparse signal, group sparse representation was proposed to include the class group information of training data for sparse representation. In this method, the l_1-norm regularization is replaced by l_2-norm combined with l_1-norm regularization as below [22]:

$$\hat{\alpha} = arg \min_{\alpha} \|X\alpha - y\|_2^2 + \lambda \sum_{i=1}^{c} \omega_i \|\alpha_{G_i}\|_2 \tag{3}$$

Where α_{G_i} denotes the sparse coefficients associated with the class/group i. With the group information, the test data can be approximated by a union of the training data in each group. While the l_1-norm based SRC model induces sparsity in the representation coefficients, it selects training data in a subject dependent manner. But it has a limitation in inducing the group-wise information. We combine a group analysis with a class-discriminative feature extraction by extending the group lasso:

$$\hat{\alpha} = arg \min_{\alpha} \|X\alpha - y\|_2^2 + \lambda_1 \|\alpha\|_1 + \lambda_2 \sum_{i=1}^{c} \omega_i \|\alpha_{G_i}\|_2 \tag{4}$$

In this work, all the training data X is partitioned into two groups, one is MCI-C group X_1 and the other one is MCI-NC group X_2. α_{G_1} represents the sparse coefficients corresponding to the MCI-C group subjects while α_{G_2} represents the sparse coefficients corresponding to the MCI-NC group subjects. This proposed sparse representation scheme can make use of the l_1-norm based SRC and the group discriminative information for better classification.

For optimization, the objective function of group discriminative sparse representation in Eq. (4) can be solved by the Moreau-Yosida regularization associated with the sparse group Lasso penalty [23]. The parameter λ_1 and λ_2 are adjusted to balance the tradeoff between the l_1-norm regularization and the group regularization.

Finally, with the sparse coefficients computed with Eq. (4), we can perform the classification based on minimization of the reconstruction residues as in Eq. (2). The class of the test subject is assigned as the one with the minimum of reconstruction residues/errors. For the test subject with the imaging data of multiple time points, the classification decision can be made by sum of all reconstruction residues/errors at multiple time points.

4   Experimental Results

To evaluate the performance of the proposed method, we apply it to predict the MCI conversion to AD, i.e., classification of MCI non-converter (MCI-NC) and MCI converter (MCI-C). Data used in this study were the T1-weighted MR brain images of 403 MCI subjects, including 236 MCI-NC and 167 MCI-C subjects, at multiple time points obtained from the ADNI database. The time points examined in this study include the baseline, 6 months, 12 months, 18 months and 24 months. The subjects with either all or some of these time points are used. The demographic characteristics of the studied subjects are shown in Table 1.

Before making classification, we perform the image preprocessing and feature extraction as described in Sect. 3. For each subject, we compute the GM volumes of 93 ROIs from the MR brain images at multiple time points as one type of imaging features. To evaluate the classification performance, the ten-folds cross validation was performed to compute the classification accuracy (ACC), i.e., the proportion of correctly classified subjects among the test dataset, as well as the sensitivity (SEN), i.e., the proportion of MCI-C patients correctly classified, and the specificity (SPE), i.e., the proportion of MCI-NC subjects correctly classified. Each time, the multiple time-point data of one fold was used for testing, while the baseline imaging data of the other nine folds with the remaining subjects were used for training.

The first experiment is to test the effectiveness of the regularization parameter λ on the classification performance. For simplicity, the regularization parameter λ_1 is set same as the λ_2. Table 2 shows the results with various regularization parameter λ.

The second experiment is to compare our proposed method with some state-of-the-art classification methods, for instance, the l_1-norm regularized SRC, and l_2-norm regularized group sparse representation, and DTSVM (Domain Transfer SVM) [24] method for classification of MCI-NC and MCI-C patients. The experimental results of these three methods as well as the proposed method are summarized in Table 3. For the SRC and group sparse representation method, we can change the regularization parameter λ in Eqs. (1) and (3) to adjust the sparsity and obtain the optimal classification performance. The regularization parameter of SRC with l_1-norm is empirically set as 0.0005 while the regularization parameter of l_2-norm regularized group sparse representation is empirically set as 0.01. For the DTSVM, we simply implement it using LIBSVM toolbox with a linear kernel and a default value for the parameter C (i.e., C = 1). The experimental results for MRI modality are ACC = 63.3, SEN = 59.8, SPEC = 66.0 respectively. In our proposed method, one more parameter ω_i in Eq. (4) should be clarified. Since the number of subjects from two class are not

Table 2. The experiment results of the proposed method with various regularization parameter on the ADNI data

$\lambda_1 = \lambda_2$	0.01	0.005	0.001	0.0005	0.0001	0.00005	0.00001
ACC	0.685	**0.715**	0.688	0.685	0.690	0.698	0.680
SEN	0.634	0.563	0.491	0.491	0.497	0.521	0.480
SPEC	0.721	0.823	0.827	0.823	0.827	0.823	0.822

Table 3. The experiment results of various methods on the ADNI data

		ACC	SEN	SPEC	AUC
State-of-the-art methods	SRC with l_1-norm	0.668	0.557	0.746	0.687
	SRC with l_2-norm	0.661	0.647	0.670	0.713
	DTSVM	0.633	0.598	0.660	0.700
Proposed method	Baseline(B)	0.688	0.581	0.763	0.718
	B + 6 M + 12 M + 18 M + 24 M	**0.715**	0.563	0.823	0.729

equal, so we defined parameter ω_i to balance the weight of coefficients from different classes. Suppose the number of subjects from class MCI-C and MCI-NC are N_1 and N_2 respectively, then the parameter ω_i in our experiment can be set as $\frac{\omega_1}{\omega_2} = \sqrt{\frac{N_2}{N_1}}$. In our proposed method, the ACC is about 0.688 for baseline data. We can see that our proposed method on the same dataset have better performance than the other two methods.

Furthermore, we also apply the proposed method on the test data of multiple time points for ensemble of the results. Every testing subject has at most five time point data. The experimental results are shown in Table 3. We can see that ensemble of the data at multiple time points can improve the classification accuracy by more than 2 % (The accuracy is increased from 0.688 to 0.715).

Finally, we also draw the Receiver Operating Characteristic Curve (ROC) and calculate the AUC (Area under curve) to evaluate the performance of each binary classifier system in Fig. 1. ROC is created by plotting the true positive rate against the false positive rate at various threshold settings. ROC analysis provides tools to select possibly optimal models and to discard suboptimal ones independently from the thresholds or the class distribution. From the comparisons of ROCs and the AUC value, we can see that the proposed method is effective for prediction of the MCI conversion.

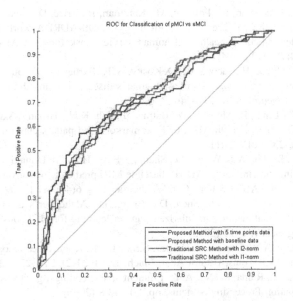

Fig. 1. ROC curves of classification results with different methods

5 Conclusion

In this paper, a discriminative group sparse representation scheme was proposed to predict the conversion of MCI to AD. To improve the representative and discriminative power of the sparse representation, the proposed method combines the strengths of the l1-norm regularized SRC and group sparse representation. The SRC can guarantee the sparsity of coding parameters, while group sparse representation with l2-norm fully utilizes the group label information of training data. Our preliminary experimental results on ADNI database show that the group sparse representation based classifier achieves better performance for prediction of MCI conversion to AD than the traditional SRC and group SRC methods. In the future works, we continue to improve and extend the group sparse representation based classifier by including corresponding dictionary learning scheme and fully utilize the longitudinal information of features.

Acknowledgment. This work was supported by National Natural Science Foundation of China grants (No. 61005024, No. 61375112) and Medical and Engineering Foundation of Shanghai Jiao Tong University grants (No. YG2012MS12, No. YG2014MS39).

References

1. McKhann, G., Drachman, D., Folstein, M., Katzman, R., Price, D., Stadlan, E.M.: Clinical diagnosis of Alzheimer's disease report of the NINCDS-ADRDA work group* under the auspices of department of health and human services task force on Alzheimer's disease. Neurology **34**(7), 939 (1984)
2. Dubois, B., Feldman, H., Jacova, C., DeKosky, S.T., Barberger-Gateau, P., Cummings, J.: Research criteria for the diagnosis of Alzheimer's disease: revising the NINCDS–ADRDA criteria. Lancet Neurol. **6**(8), 734–746 (2007)
3. Davatzikos, C., Bhatt, P., Shaw, L.M., Batmanghelich, K.N., Trojanowski, J.Q.: Prediction of MCI to AD conversion, via MRI, CSF biomarkers, and pattern classification. Neurobiol. Aging **32**(12), 2322-e19 (2011)
4. Risacher, S.L., Saykin, A.J., West, J.D., Shen, L., Firpi, H.A., McDonald, B.C.: Alzheimer's disease neuroimaging initiative (ADNI): baseline MRI predictors of conversion from MCI to probable AD in the ADNI cohort. Curr. Alzheimer Res. **6**(4), 347 (2009)
5. Cheng, B., Liu, M., Suk, H.I., Shen, D., Zhang, D.: Alzheimer's disease neuroimaging initiative: multimodal manifold-regularized transfer learning for MCI conversion prediction. Brain Imaging Behav. 1–14 (2015)
6. Wright, J., Yang, A.Y., Ganesh, A., Sastry, S.S., Ma, Y.: Robust face recognition via sparse representation. IEEE Trans. Pattern Anal. Mach. Intell. **31**(2), 210–227 (2009)
7. Lee, H., Battle, A., Raina, R., Ng, A.Y.: Efficient sparse coding algorithms. In: Advances in Neural Information Processing Systems, pp. 801–808 (2006)
8. Mallat, S.G., Zhang, Z.: Matching pursuits with time-frequency dictionaries. IEEE Trans. Sig. Process. **41**(12), 3397–3415 (1993)
9. Pati, Y.C., Rezaiifar, R., Krishnaprasad, P.S.: Orthogonal matching pursuit: recursive function approximation with applications to wavelet decomposition. In: 1993 Conference Record of the Twenty-Seventh Asilomar Conference on Signals, Systems and Computers, pp. 40–44. IEEE, November 1993

10. Needell, D., Vershynin, R.: Uniform uncertainty principle and signal recovery via regularized orthogonal matching pursuit. Found. Comput. Math. **9**(3), 317–334 (2009)
11. Elad, M., Aharon, M.: Image denoising via sparse and redundant representations over learned dictionaries. IEEE Trans. Image Process. **15**(12), 3736–3745 (2006)
12. Protter, M., Elad, M.: Image sequence denoising via sparse and redundant representations. IEEE Trans. Image Process. **18**(1), 27–35 (2009)
13. Dong, W., Zhang, D., Shi, G.: Centralized sparse representation for image restoration. In: 2011 IEEE International Conference on Computer Vision (ICCV), pp. 1259–1266. IEEE, November 2011
14. Yang, J., Wright, J., Huang, T.S., Ma, Y.: Image super-resolution via sparse representation. IEEE Trans. Image Process. **19**(11), 2861–2873 (2010)
15. Shawe-Taylor, J., Cristianini, N.: Kernel Methods for Pattern Analysis. Cambridge University Press, Cambridge (2004)
16. Sled, J.G., Zijdenbos, A.P., Evans, A.C.: A nonparametric method for automatic correction of intensity nonuniformity in MRI data. IEEE Trans. Med. Imaging **17**(1), 87–97 (1998)
17. Wang, Y., Nie, J., Yap, P.-T., Shi, F., Guo, L., Shen, D.: Robust deformable-surface-based skull-stripping for large-scale studies. In: Fichtinger, G., Martel, A., Peters, T. (eds.) MICCAI 2011, Part III. LNCS, vol. 6893, pp. 635–642. Springer, Heidelberg (2011)
18. Shen, D., Davatzikos, C.: Very high-resolution morphometry using mass-preserving deformations and HAMMER elastic registration. NeuroImage **18**(1), 28–41 (2003)
19. Thompson, P.M., Schwartz, C., Toga, A.W.: High-resolution random mesh algorithms for creating a probabilistic 3D surface atlas of the human brain. NeuroImage **3**(1), 19–34 (1996)
20. Koh, K., Kim, S.J., Boyd, S.P.: An interior-point method for large-scale l1-regularized logistic regression. J. Mach. Learn. Res. **8**(8), 1519–1555 (2007)
21. Liu, J., Ji, S., Ye, J.: SLEP: Sparse Learning with Efficient Projections, vol. 6, p. 491. Arizona State University, Arizona (2009)
22. Chi, Y.T., Ali, M., Rajwade, A., Ho, J.: Block and group regularized sparse modeling for dictionary learning. In: 2013 IEEE Conference on Computer Vision and Pattern Recognition (CVPR), pp. 377–382. IEEE, June 2013
23. Liu, J., Ye, J.: Moreau-Yosida regularization for grouped tree structure learning. In: Advances in Neural Information Processing Systems, pp. 1459–1467 (2010)
24. Cheng, B., Zhang, D., Shen, D.: Domain transfer learning for MCI conversion prediction. In: Ayache, N., Delingette, H., Golland, P., Mori, K. (eds.) MICCAI 2012, Part I. LNCS, vol. 7510, pp. 82–90. Springer, Heidelberg (2012)

Prediction of Molecular Substructure Using Mass Spectral Data Based on Deep Learning

Zhi-Shui Zhang, Li-Li Cao, Jun Zhang, Peng Chen, and Chun-hou Zheng[✉]

School of Electronic Engineering and Automation,
Anhui University, Hefei 230601, Anhui, China
wwwzhangjun@163.com

Abstract. The success of machine learning algorithms generally depends on effective features. Prediction of molecular substructure based on mass spectral data is try to extract more useful information or feature expression. In this paper, deep learning (DBN) was used to extract mass spectral features automatically. A large dataset consisting 11 molecular substructure is extracted from NIST mass spectral library. The experimental results show that deep learning (DBN) achieve best classification performance in 11 molecular substructure contrasting traditional classification methods.

Keywords: Machine learning · Molecular substructure · Feature representation · Deep learning

1 Introduction

Development of computer-assisted methods for automatic elucidation of chemical structures of organic compounds is a central theme in chemometrics [1]. There are two classes of methods to identify compound: one is library search methods based on spectra similarities, another is classification method [2]. Usually, the original mass spectral data are transformed to some mass spectral features, many classification methods have been applied in the analysis of mass spectral features. Linear discriminant analysis (LDA) and discriminant partial least squares (DPLS) are common and simple methods that have been used to mass spectra classification [3]. Since the mass spectral features is high dimensional data, some dimension deduction methods like principal component analysis (PCA) and partial least squares (PLS) [4] and some feature selection methods such as Fisher ratios and genetic algorithm (GA) [5] have been used in mass spectral classification. Some software tools has been developed for these purpose such as PaDEL-Descriptor [6], OpenBabel. However, the relation between mass spectral data and molecular structure are so complicated that make the prediction is difficult. Deep learning (learning representations) [8–12] is a method which can extract features from original data. In fact, the mass spectral features calculated by some devised algorithms can not reveal the potential relation between mass spectral data and molecular structure. In this paper, in order to get optimal prediction results, we use deep learning to extract effective mass spectral features automatically, then some methods are used to investigate the relationships between mass spectral data

© Springer International Publishing Switzerland 2015
D.-S. Huang et al. (Eds.): ICIC 2015, Part II, LNCS 9226, pp. 520–529, 2015.
DOI: 10.1007/978-3-319-22186-1_52

and molecular substructures. In this work, 11 substructures are selected to construct prediction model, AdaBoost-CART based on mass spectral features are employed to compare with our proposed method, the experimental results show that deep learning achieve better predictive performance.

2 Materials and Methods

2.1 Mass Spectra and Substructures Dataset

Mass Spectral Library 2005 (NIST05) main library containing 163,195 mass spectra is used [7] in this study. In order to obtain the structures of every compounds, we first extract molecular structure file (mol format) for each spectrum, then OpenBabel has been used to calculate the PubChem molecular fingerprint from structure file. In this work, we select 14000 molecules from NIST MS library, software MassFeatGen was employed to transform the original mass peak data into an 862 mass spectral features. The corresponding 11 molecular substructures have been chosen from PubChem molecular fingerprint. The details of the molecular substructure are listed in Table 1. For each substructure, if the substructures present the corresponding class set as 1, otherwise, the class set as 2.

2.2 Algorithm

In probabilistic models, a good expression is summarizing the distribution of the feature for the observed input. From the probabilistic modeling perspective, the question of feature learning can be seen as an attempt to recover a set of latent random variables that describe a distribution over the observed data. We can express any probabilistic model over the joint space of the latent variables h, and observed or visible variables x, as $p(x; h)$. Feature values are conceived as the result of an inference process to determine the probability distribution of the latent variables given the data, i.e. $p(h|x)$, directed latent factor models are parameterized through a decomposition of the joint distribution, $p(x; h) = p(h|x)p(h)$, involving a prior $p(h)$, and a likelihood $p(h|x)$ that describes the observed data xx in terms of the latent factors h. Unsupervised feature learning models that can be interpreted with this decomposition include: principal components analysis (PCA), undirected graphical models, also called Markov random fields, parametrize the joint $p(h|x)$ through a factorization in terms of unnormalized clique potentials:

$$p(x, h) = \frac{1}{Z_\theta} \prod_i \varphi_i(x) \prod_j \mu_j(h) \prod_k v_k(x, h) \tag{1}$$

where $\varphi_i(x)$, $\mu_j(h)$ and $v_k(x, h)$ describe the interactions between the visible elements, between the hidden variables, and those interaction between the visible and hidden variables respectively. The partition function Z_θ ensures that the distribution is normalized. Within the context of unsupervised feature learning, we generally see a

Table 1. The 11 substructure categories and corresponding chemical structures

Structure no.	Chemical structure	Substructure name
1		4-methylbenzenethiol
2		3-methylbenzenethiol
3		o-cresol
4		1-chloro-2-methylbenzene
5		4-methylcyclohexanol
6		4-methylcyclohexanethiol

(*Continued*)

Table 1. (*Continued*)

Structure no.	Chemical structure	Substructure name
7		3-methylcyclohexanol
8		3-methylcyclohexanethiol
9		2-methylcyclohexanol
10		2-methylcyclohexanethiol
11		1-chloro-2-methylcyclohexane

particular form of Markov random field called a Boltzmann distribution with clique potentials constrained to be positive:

$$p(x, h) = \frac{1}{Z_\theta} \exp(-\varepsilon_\theta(x, h)) \tag{2}$$

where $-\varepsilon_\theta(x, h)$ is the energy function and contains the interactions described by the MRF clique potentials and θ are the model parameters that characterize these interactions. A Boltzmann machine is defined as a network of symmetrically-coupled binary random variables or units. These stochastic units can be divided into two groups: (1) the visible units $x \in \{0,1\}^{d_z}$ that represent the data, and (2) the hidden or latent units $h \in \{0,1\}^{d_h}$ that mediate dependencies between the visible units through their mutual interactions. The pattern of interaction is specified through the energy function:

$$\varepsilon_\theta^{BM}(x,h) = -\frac{1}{2}x^T U x - \frac{1}{2}h^T V h - x^T W h - b^T x - d^T h \tag{3}$$

where $\theta = \{U, V, W, b, d\}$ are the model parameters which respectively encode the visible-to-visible interactions, the hidden-to-hidden interactions, the visible-to-hidden interactions, the visible self-connections, and the hidden self-connections (also known as biases). To avoid over-parametrization, the diagonals of U and V are set to zero. The Boltzmann machine energy function specifies the probability distribution over the joint space [x, h]. This joint probability distribution gives rise to the set of conditional distributions of the form:

$$p(h_i|x, h_i) = sig \bmod \left(\sum_j W_{ji}x_j + \sum_{i' \neq i} V_{ii'}h_{i'} + d_i \right) \tag{4}$$

$$p(h_j|x, h_j) = sig \bmod \left(\sum_j W_{ji}x_j + \sum_{j' \neq j} V_{ij'}h_{j'} + d_j \right) \tag{5}$$

In general, inference in the Boltzmann machine is intractable. The restricted Boltzmann machine (RBM) is likely the most popular subclass of Boltzmann machine. It is defined by restricting the interactions in the Boltzmann energy function, in Eq. 3, to only those between h and x, i.e. ε_θ^{BM} is ε_θ^{BM} with U = 0 and V = 0. As such, the RBM can be said to form a bipartite graph with the visibles and the hiddens forming two layers of vertices in the graph (and no connection between units of the same layer). With this restriction, the RBM possesses the useful property that the conditional distribution over the hidden units factorizes given the visible:

$$p(x|h) = \prod_j p(x_j|x) \tag{6}$$

$$p(h_i = 1|x) = sig \bmod \left(\sum_i W_{ji}x_j + d_j \right) \tag{7}$$

Likewise, the conditional distribution over the visible units given the hiddens also factorizes:

$$p(x|h) = \prod_j p(x_j|h) \tag{8}$$

$$p(h_i = 1|x) = sig \bmod \left(\sum_i W_{ji}h_i + b_i \right) \tag{9}$$

This conditional factorization property of the RBM immediately implies that most inferences we would like make are readily tractable. For example, the RBM feature representation is taken to be the set of posterior marginals $p(h_i|x)$ which are, given the conditional independence described in Eq. 7, are immediately available. Note that this is in stark contrast to the situation with popular directed graphical models for unsupervised feature extraction, where computing the posterior probability is intractable.

Hierarchical neural networks are motivated by the hierarchical structure of the ventral stream of the visual cortex, where a sequence of retinotopic representations is computed that represents the field-of-view with decreasing spatial resolution, but increasing feature complexity. This network structure reflects the typical hierarchical structure of samples.

To ease the following discussion, let us first define the terms. A deep architecture (as shown in Fig. 1) with depth L is a function $h^{(L)}$ with

$$h^{(L)}(x) = sig \bmod (W^{(l)}h^{(l-1)}(x) + b^{(l)}), \quad h^{(0)}(x) = x \tag{10}$$

Here, x is the input vector, The weight matrices $W^{(l)}$ and the biases $b^{(l)}$ constitute the layer parameters $\theta^{(l)}$ Intermediate $h^{(l)}(x)$ are referred to as hidden layers. When unambiguous, we omit the dependency on x and write $h^{(l)}$. While Eq. (10) is based on dot products of input and weight vectors, it is also possible to rely on differences between the two vectors instead and set sigmod function to a radial basis function, our features correspond to the rows of $W^{(l)}$ and can be determined by learning. We first formalize the task using a loss function which is minimal when the task is solved. Learning is then to find parameters such that the loss function is minimal on some training data D. For example, we might choose the mean square loss

$$l_{MSE}(\theta^{(1)}, \theta^{(2)}, \theta^{(3)}, \ldots, \theta^{(L)}, D) = \sum_{d=1}^{D} \sum_{n=1}^{N} (h_n^{(L)}(x^d) - x_n^d)^2 \tag{11}$$

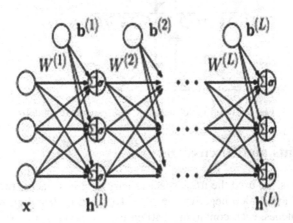

Fig. 1. Visualization of a deep architecture

for an i.i.d. dataset $D = \{x^1, x^2, \ldots, x^D\}$ with Gaussian noise model. This is an unsupervised task where the input is reconstructed from the features. Supervised tasks provide some desired output or label in addition to the input data, and we learn a classification task by minimizing the cross-entropy loss

$$l_{CE}(\theta^{(1)}, \theta^{(2)}, \theta^{(3)}, \ldots, \theta^{(L)}, D) = \sum_{d=1}^{D} [y_m^d \bullet \log \frac{\exp(h_m^{(L)}(x^d))}{\sum_m^M \exp(h_m^{(L)}(x^d))}] \qquad (12)$$

In networks with at least one non-linear hidden layer, both losses are non-convex in the parameters. They are typically minimized by gradient descent, where the gradients are efficiently (linearly in the number of parameters) computed with the backpropagation algorithm.

A deep belief network with l layers models the joint distribution between observed vector x and l hidden layers $h^{(k)}$ as follows:

$$p(x, h^1, h^2, \ldots, h^l) = (\prod_{k=1}^{l=2} p(h^k | h^{k+1})) p(h^{l-1}, h^l) \qquad (13)$$

where $x = h^0$, $p(h^{k-1} | h^k)$ is a conditional distribution for visible hidden units in an RBM associated with level k of the DBN, and $p(h^{l-1} | h^l)$ is the visible-hidden joint distribution in the top-level RBM. This is illustrated in Fig. 2.

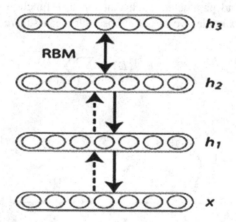

Fig. 2. Deep belief network as a generative model

3 Experiments and Discussions

In our experiments, we used the mass spectrometry data sets, each structure has 14000 positive samples and 14000 negative samples. To speed up learning, we divided data sets into mini-batches, each containing 140 cases, and updated the weights after each mini-batch. The number of sample particles used for approximating the model's expected sufficient statistics was set as 100. For the stochastic approximation algorithm, we always used 3 full Gibbs updates of the particles, we build a 1659-100-200-862 deep network (1659 is original dimensions of mass spectral data). For comparison, MassFeatGen was employed to transform the original mass peak data into an 862 mass features, the comparison experiment is set the top layer as 862 nodes. For all experiments, the initial learning rate was set as 0.6, learning rate for biases of visible units and hidden units were set as 0.1, weight cost was set as 0.0002, in order to explain the results fairly, every experiment is run 5 times and take the average results, we also compare the results to some classical methods, such as PCA, ISOMAP, AdaBoost is a popular machine learning algorithm [13, 14]. In this paper, AdaBoost-CART was used to classify the mass spectral features, the boosting iterations were set as 100. The results are shown in Fig. 3. (See Table 2 for a detailed listing of results).

Experimental results were obtained by averaging over multiple runs on 5-fold cross-validation of each data set. The results on the 11 data sets are shown in Fig. 3. (See Table 2 for a detailed listing of results). We can see that DBN achieve best classification performance in 11 substructure. DBN transform the original mass spectral data into more effective features contrasting the artificial features generated by MassFeatGen.

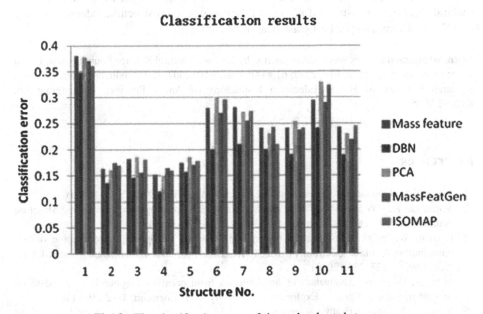

Fig. 3. The classification error of the molecules substructure

Table 2. The detailed classification error of the molecules substructure

Structure no.	Mass feature	DBN	PCA	MassFeatGen	ISOMAP
1	0.3804	0.3486	0.3779	0.3706	0.3601
2	0.1629	0.1357	0.1611	0.1745	0.1702
3	0.1825	0.1466	0.1856	0.1564	0.1811
4	0.1524	0.1201	0.1511	0.1651	0.1602
5	0.1751	0.1566	0.1856	0.1712	0.1789
6	0.2803	0.2012	0.3011	0.2701	0.2956
7	0.281	0.2103	0.2723	0.2551	0.2741
8	0.2412	0.2001	0.2301	0.2423	0.2103
9	0.2411	0.1902	0.2536	0.2382	0.2412
10	0.2952	0.2411	0.3301	0.2902	0.3233
11	0.2423	0.1897	0.2301	0.2189	0.2451

4 Conclusion

Improving the accuracy of classification based on mass spectra is important for chemical structure elucidation. In this paper, deep learning (DBN) was used to extract effective features from original mass spectral data, then adaboost-cart is employed as classifier to predict the molecular sub-structure. Contrasting to traditional classification method, experimental results show that the deep learning was used for learning mass spectral features is successful, in 11 molecular sub-structure, deep learning (DBN) algorithm achieve best classification accuracy.

Acknowledgments. This work was supported by National Natural Science Foundation of China under grant nos. 61271098, 61472282, 61300058 and 61032007, and Provincial Natural Science Research Program of Higher Education Institutions of Anhui Province under grant no. KJ2012A005.

References

1. Gray, N.A.: Computer-Assisted Structure Elucidation. Wiley, New York (1986)
2. Varmuza, K., Werther, W.: Mass spectral classifiers for supporting systematic structure elucidation. J. Chem. Inf. Comput. Sci. **36**(2), 323–333 (1996)
3. Cynkar, W., et al.: Classification of tempranillo wines according to geographic origin: combination of mass spectrometry based electronic nose and chemometrics. Anal. Chim. Acta **660**(1), 227–231 (2010)
4. Werther, W., et al.: Evaluation of mass spectra from organic compounds assumed to be present in cometary grains. Exploratory data analysis. J. Chemom. **16**(2), 99–110 (2002)
5. Yoshida, H., et al.: Feature selection by genetic algorithms for mass spectral classifiers. Anal. Chim. Acta **446**(1), 483–492 (2001)
6. Yap, C.W.: PaDEL-descriptor: an open source software to calculate molecular descriptors and fingerprints. J. Comput. Chem. **32**(7), 1466–1474 (2011)

7. Boger, Z.: Selection of quasi-optimal inputs in chemometrics modeling by artificial neural network analysis. Anal. Chim. Acta **490**(1), 31–40 (2003)
8. Lee, H., et al.: Unsupervised feature learning for audio classification using convolutional deep belief networks. Adv. Neural Inf. Process. Syst. **22**, 1096–1104 (2009)
9. Erhan, D., et al.: Why does unsupervised pre-training help deep learning? J. Mach. Learn. Res. **11**, 625–660 (2010)
10. Bengio, Y., Aaron C.C., Pascal, V.: Unsupervised feature learning and deep learning: a review and new perspectives. CoRR, abs/1206.5538 1 (2012)
11. Ngiam, J., et al.: On optimization methods for deep learning. In: Proceedings of the 28th International Conference on Machine Learning (ICML-11) (2011)
12. Ranzato, M., et al.: Unsupervised learning of invariant feature hierarchies with applications to object recognition. In: IEEE Conference on Computer Vision and Pattern Recognition, CVPR 2007 (2007)
13. Freund, Y.:Boosting a weak learning algorithm by majority. In: COLT (1997)
14. Breiman, L.: Bagging predictors. Mach. Learn. **24**(2), 123–140 (1996)

An Enhanced Algorithm for Reconstructing a Phylogenetic Tree Based on the Tree Rearrangement and Maximum Likelihood Method

Sun-Yuan Hsieh[1,2(✉)], I-Pien Tsai[2], Hao-Che Hung[1],
Yi-Chun Chen[1], Hsin-Hung Chou[3], and Chia-Wei Lee[1]

[1] Department of Computer Science and Information Engineering,
National Cheng Kung University, No. 1, University Road, Tainan 701, Taiwan
hsiehsy@mail.ncku.edu.tw, hwdhung@gmail.com,
alex111880@hotmail.com.tw, cwlee@csie.ncku.edu.tw
[2] Institute of Medical Informatics,
National Cheng Kung University, No. 1, University Road, Tainan 701, Taiwan
hsiehsy@mail.ncku.edu.tw, ipientsai@gmail.com
[3] Department of Information Management,
Chang Jung Christian University, No.1, Changda Road,
Gueiren District, Tainan 711, Taiwan
chouhh@mail.cjcu.edu.tw

Abstract. The *phylogeny reconstruction problem* is a fundamental problem in computational molecular biology and biochemical physics. Since the number of data sets has grown substantially in recent years, the accuracy and speed of constructing phylogenies become increasingly critical. Numerous studies have demonstrated that the maximum likelihood (ML) method is the most effective method for reconstructing a phylogenetic tree from sequence data. Conversely, tree bisection and reconnection (TBR) is a tree topology rearrangement method that can generate an extensive tree space. In this paper, we propose an enhanced method for reconstructing phylogenetic trees in which the TBR operation is modified and combined with the minimum evolution principle to filter out some unnecessary reconnected positions to reduce the search time. The experiment results demonstrate that the proposed method can assist other algorithms in constructing more accurate trees within a reasonable time.

Keywords: Bioinformatic algorithms · Computational biology · Maximum likelihood · Phylogeny reconstruction

1 Introduction

In this section, we introduce the background of the *phylogeny reconstruction problem* and survey its related works.

© Springer International Publishing Switzerland 2015
D.-S. Huang et al. (Eds.): ICIC 2015, Part II, LNCS 9226, pp. 530–541, 2015.
DOI: 10.1007/978-3-319-22186-1_53

1.1   Background

A phylogenetic tree (also called evolutionary tree) is a tree with leaves labeled by taxa or species, which can be used to infer the evolutionary relationships among a group of organisms, or among a family of related nucleic acids. The phylogenetic tree shows the similarities among organisms during the course of their evolution, and this information can help biologists to understand the changes of organisms as they evolve further. The phylogenetic tree also plays an important role in many biological problems, such as protein structure (or function) prediction [15, 23, 36], and drug design [22, 29, 40].

Phylogenetic tree reconstruction is a classical computational biology problem that has been a scientific challenge for decades. As the number of known organisms has grown, the accuracy and speed of reconstructing phylogenetic trees has become increasingly important. In the past, biologists expended a great deal of time and efforts on conducting tests to determine the relationships of species by their morphological, behavioral and anatomical characteristics. Over time, there have a number of theories, such as probabilistic sequence evolution models and the rate variation among sites, been proposed to increase the accuracy and speed of reconstructing phylogenetic trees [10, 16, 20, 24].

There were some heuristic algorithms that utilized a tree rearrangement strategy for reconstructing phylogenetic trees [12, 47]. The simplest one is the so-called *Nearest Neighbor Interchange* (NNI) [6, 7] that is applied in PHYML (*Phylogenetic Maximum Likelihood*) [14]. Each internal branch of a hypothetical phylogenetic tree (a binary unrooted tree) has four subtrees connected to it. NNI exchanges these four subtrees to obtain three possible trees, including the original tree. The method is fast because the exchange of every move can be calculated locally. Nevertheless, it would often be trapped in a local optima, which happens frequently with data sets obtained from multiple genes with conflicting signals [21]. An alternative, more wide-ranging tree rearrangement method is called *Subtree Pruning and Regrafting* (SPR) [3, 17, 42], which is implemented in RAxML (*Randomized Axelerated Maximum Likelihood*) [42, 43]. In SPR, a subtree is pruned and then regrafted to a different location in the remaining tree. Although the method can avoid being trapped in bad local optima, it is more expensive in terms of computation time. Moreover, once the SPR operation begins, the likelihood of the whole tree has to be recalculated.

A more elaborated tree rearrangement method, called *Tree Bisection and Reconnection* (TBR) [5, 30, 46], cuts a given tree into two parts by breaking an internal branch. Then, two branches, each from one subtree, are chosen and reconnected to form a new tree. This is quite a rigorous and complicated method of rearranging trees, but it does generate more neighbors than NNI or SPR. Although the method is not guaranteed to find a true tree in all cases, it can reconstruct a true phylogenetic tree from data more frequently than the previous two methods. However, TBR is much more time-consuming because of its extensive tree space and the fact that the likelihood has to be re-evaluated for each possible rearrangement. Under TBR, there are many potential reconnected positions, but not all of them can yield a good result. To reduce the TBR search space and improve the likelihood-based approach further, we modify the standard TBR method to generate neighboring topologies and utilize the *minimum evolution* principle [9] to select possible reconnected positions that can speed up the search time. The results of experiments demonstrate that our method can help other currently known algorithms to reconstruct a more accurate phylogenetic tree within a reasonable time.

1.2 Related Works

The methods used to infer phylogenetic trees can be classified into two categories: *distance-based* methods and *character-based* methods. The former count the distances (e.g., Hamming distance) between sequences, and then store the distances in an $N \times N$ matrix to construct a phylogenetic tree. This approach has been adopted in Neighbor-Joining (NJ) [34] and the Unweighted Pair Group Method with an Arithmetic mean (UPGMA) [39]. Although distance-based methods are fast, some biological information may be lost when characters are transformed into pairwise distances. The methods can calculate the distance between sequences from the characters, but not *vice versa*. In contrast, character-based methods align the sequences of many different species, and then use probabilistic or statistical methods to count the differences between characters as the basis for constructing trees. This approach has been adopted in *Bayesian analysis* [28, 33, 45], *maximum-likelihood* (ML) estimation [10, 13, 18] and *maximum-parsimony* (MP) estimation [1, 27, 43].

Numerous studies have shown that ML can reconstruct a true phylogenetic tree from data more frequently than other methods [14, 19, 25, 37, 38]. The advantage of ML is that it can compare different trees within a statistical framework. However, ML must deal with several computational difficulties, which make it almost impossible to obtain a nearly true phylogenetic tree in a reasonable computation time. For example, let us assume that there are totally $\frac{(2n-5)!}{2^{n-3}(n-3)!}$ trees, where n is the number of leaf nodes [12]. Because the number of possible tree topologies grows rapidly with the number of taxa, a practical solution must rely on a heuristic method to reduce the search space when the tree contains more than 10–15 organisms. MetaPIGA [31] combines a genetic algorithm and the stochastic approach to reconstruct phylogenetic trees. This method navigates in the tree space, randomly perturbs a population of trees by modifying the lengths of their branches and their topologies and then selects the best tree. DPRml [26] creates a tree by stepwise insertion of taxa to enlarge the topological structure. It estimates the likelihood of all the potential insertion positions and then chooses the most appropriate one to grow the tree. In contrast, several ML-based phylogenetic tree reconstruction methods that incorporate some heuristics have been proposed, e.g., PHYML [14], DNAML [10], and RAxML [41, 42].

The remainder of this paper is organized as follows. In the next section, we provide some basic definitions and notations used throughout this paper. In Sect. 3, we present our method for reconstructing phylogenetic trees. Experiment results are reported in Sect. 4. Then, some concluding remarks are given in Sect. 5.

2 Preliminaries

In this section, we define some basic graph terms, and introduce ML together with the minimum evolution principle.

2.1 Basic Definitions and Notations

A *graph* $G = (V, E)$ comprises a *node set* V and an *edge set* E, where V is a finite set and E is a subset of $\{(u, v)|\ (u, v)$ is an unordered pair of $V\}$. We also use $V(G)$ and $E(G)$ to denote the node set and edge set of G, respectively. A *subgraph* of $G = (V, E)$ is a graph (V', E') such that $V' \subseteq V$ and $E' \subseteq E$. Given $U \subseteq V(G)$, the *subgraph of G induced by U* is defined as $G[U] = (U, \{(u, v) \in E(G)|\ u, v \in U\})$. A *rooted tree T* (*tree* for short) is a connected and acyclic graph in which one node is specified as the root. We use $V(T)$ and $E(T)$ to denote the node set and the edge set of T, respectively. In $V(T)$, the nodes with degree 1 are called *leaves*, and nodes that are not leaves are called *internal nodes*. For two arbitrary nodes u and v in T, where u is on the path from the root of T to v, we say (1) u is the *ancestor* of v, and (2) v is the *descendant* of u. Moreover, if $(u, v) \in E(T)$, then u is the *parent* of v, and v is the *child* of u. If two nodes in T have the same parent, they are called *siblings*. A *subtree* of T rooted at node $u \in V(T)$ is the tree induced by u and all its descendants. For convenience, we use U to denote the subtree of T rooted at $u \in V(T)$. The phylogenetic tree we consider here is a *binary tree* in which each node v has at most two children, the *left child* and the *right child* of v.

2.2 Maximum Likelihood

Maximum likelihood (ML) is a statistical estimation method that is widely used to fit a mathematical model to a data set. It can also be used to estimate the phylogeny of organisms. ML for phylogeny was introduced by [8] for gene frequency data. The first application of ML for molecular sequences was proposed in [35], and another application of ML for nucleotide sequences was first proposed in [10].

Given an input data comprising a set of k-sites corresponding to DNA aligned sequences, a phylogenetic tree associated with branch lengths and a substitution model, we can compute the probability of changes between different states on the tree. Suppose that D is the sequence data, M is the substitution model of characters, and T is the tree topology. The conditional likelihood can be estimated by: $L_T = \mathrm{Prob}(D|M, T)$.

In particular, we need a model to compute the *transition probability*, denoted by $P_{ij}(e)$, which is the probability of the change at a branch e from the state i to the state j. For every initial state, the *priori probability* is denoted by $\{\pi_A, \pi_T, \pi_C, \pi_G\}$. Two assumptions are made when using models for ML estimation:

Sites evolve independently.

Lineages evolve independently.

The first assumption allows us to estimate the likelihood of a tree and decompose it into the product of all the characters: $L_T = \mathrm{Prob}(D|M, T) = \prod_{i=1}^{k} \mathrm{Prob}(D_i|M, T)$, where D_i is the data at the ith site. Therefore, if we know how to compute the likelihood at each site, we can get the likelihood of the tree. However, a new problem arises in that we do not know the states at two internal nodes x and y. There are several solutions that may apply to this problem; for example, try all combinations of states and use the one that can maximize the likelihood. In phylogeny inference, we sum over all probabilities of each state at the internal nodes to get the likelihood for observing data instead.

2.3   The Minimum Evolution Principle

There are many ways to estimate the edge-length when we modify the tree topology [4, 11]. When estimating the length of two nodes u and v, denoted by $l(u, v)$, the iterative estimation methods proposed in [4, 11] usually overestimate or underestimate the edge-lengths. To speed up the estimation process, we utilize a faster way based on the minimum evolution principle [9].

2.4   Moving Distance

According to our method, the tree topology needs to be modified. Specifically, the modification involves separating a chosen subtree from the original tree temporarily, "moving" it to a specific position by the way described in Sect. 3.3, and then recon- necting it to the main tree. Figure 1 illustrates how the subtree is moved. The subtree U_2 is separated from the original tree by deleting edges (u1, u2) and (u2, u3); then u1 and u3 are connected by a new edge. After the subtree has been separated, all edges in the remaining tree, except for (u_1, u_3), can be viewed as reconnected positions. If the moving distance is 1, we move U2 to the reconnected position at edge (u_3, u_4), which means that it passes through one node. However, if the moving distance is k, we move U_2 to the reconnected position at edge (u_k, u_{k+1}), which means that it passes through k nodes. The subtree can move along the tree until the reconnected position is at a *terminal edge* (the edge on which one end-node is a leaf).

3   The Proposed Method

The steps of our proposed algorithm, called TRMLE (Tree Rearrangement + Maximum Likelihood + Minimum Evolution), are described as follows. First, using the method adapted from TBR, we modify the topology of a given tree to generate neighboring

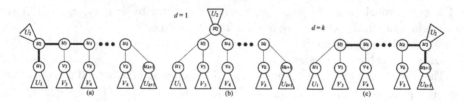

Fig. 1. Illustration of moving a subtree U_2.

topologies. Second, the branch lengths are estimated and the likelihood of the resulting tree is computed. Third, based on the likelihood, if the new topology is better than the present one, we use it instead; otherwise, the present one remains unchanged. The above steps are repeated until the likelihood that could not be improved further or a termination condition is satisfied. Next, we describe each step in detail.

Based on standard TBR procedure, we first select and delete an edge $e = (u, v)$ in a tree T to divide T into two subtrees, U and V, where $u \in U$ and $v \in V$. Then, we delete node u from U and link the remaining components of U to make it connected. The same operation is applied to V and node v. Furthermore, we randomly choose edges (c, x) and (d, y) as the reconnection positions, and let the edges, (c, x) and (d, y), be subdivided by u and v[1], respectively. Finally, we add a new edge $e' = (u, v)$ to form the tree T', to complete the TBR procedure.

3.1 The Modified TBR Method

Because the standard TBR method chooses the reconnected positions randomly, the likelihood of the resulting tree may be far from that of the true phylogenetic tree. To resolve the problem, we modify the standard TBR method and combine it with the minimum evolution principle to select the correct reconnected positions.

Let us consider the phylogenetic tree shown in Fig. 2. The modified TBR method is decomposed into two phases.

Phase 1: All edges in a given tree T are sorted in descending order of their lengths, and the largest one, say (i_2, j_2), is selected (Fig. 2(a)). The end nodes of the selected edge are then moved to another position by applying the following operations to nodes i_2 and j_2. The subtree of T, which is induced by $\{i_2, j_2\}$ and the descendants of j_2, is separated from T, and a new edge (i_1, i_3) is added. Then, we determine the moving distance d by the method described in Sect. 3.3, and move the subtree based on the moving distance d to the new reconnected position at (i_d, i_{d+1}). Let T' be the resulting tree (Fig. 2(b)). The new average distance of T' is computed. As we will show in the next section, the new average distance can be obtained by accurately adjusting the average distance of the original tree.

Phase 2: The subtree J_2 of T', which is induced by $\{i_2, j_2\}$ and the descendants of i_2, is separated from T', and a new edge (h_1, h_3) is added. Then, we determine another moving distance s by the method described in Sect. 3.3, and move the subtree based on the moving distance s to the new reconnected position at (h_s, h_{s+1}) (Fig. 2(c)). Let T'' be the resulting tree (Fig. 2(d)). The new average distance of T'' can be obtained by accurately adjusting that of T'.

Phases 1–2 yield the modified TBR. We then calculate the likelihood of T'', denoted by $L_{T''}$. If $L_{T''}$ is larger than the likelihood of the original tree T, the topology of T' is closer to the true topology than T. The larger the gap between L_T and $L_{T''}$, the more confidence we have in the rearrangement from T to T''.

---

[1] In a graph G, the *subdivision* of an edge (x, y) by a node z involves replacing (x, y) with a path $\langle x, z, y \rangle$ through a new node z.

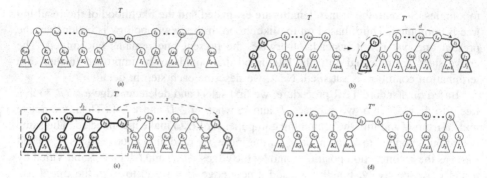

Fig. 2. Illustration of the modification of TBR

3.2   Edge-Length Estimation

Applying the modified TBR will change the tree topology significantly; hence, the likelihood and average distances computed previously should be updated. In this section, we derive a new property and demonstrate that the new average distance can be computed efficiently. The following proposition is useful to our method.

Proposition 1. Let v be an internal node in a tree T with three neighbors u_1, u_2, and u_3. Then, $l(ui, uj) = l(ui, v) + l(uj, v)$, where $i, j \in \{1,2,3\}$ and $i \neq j$.

Under the modified TBR procedure illustrated in Fig. 2, because node i_2 and the subtree rooted at i_2 are pruned, edges (i_2, i_3) and (i_1, i_2) must also be deleted. As a result, nodes i_1 and i_3 will now be connected by a new edge (see Fig. 2(b)). A general and straightforward way to estimate $l(i_1, i_3)$ is to sum the lengths of (i_2, i_3) and (i_1, i_2) before deleting i_2, that is, $l(i_1, i_3) = l(i_1, i_2) + l(i_2, i_3)$ according to Proposition 1.

Next, let us consider the subtree rooted at i_2 (induced by i_2, j_2, and the descendants of j_2), which is separated from the original tree T and moved based on the moving distance d to the reconnected position at (id, id + 1) to form a new tree T', as shown in Fig. 2(b). The relevant nodes and edges around the reconnected position are shown in Fig. 3. The edge-lengths, which need to be estimated, are (i_2, id), $(i_2, \text{id} + 1)$ and (i_2, j_2). Therefore, we can only recompute the above three edge-lengths for the resulting tree after applying the modified TBR operation.

When the new distances has been updated, the operations in Phase 2 of the modified TBR (Fig. 2(c)) are executed in a similar manner. The operations for dealing with the subtree of T', which is induced by $\{i_2, j_2\}$ and the descendants of i_2, are similar to the operations in Phase 1. The computation of the required edge-lengths is also similar to that described in Phase 1.

3.3   Determining the Moving Distance

Let the *length of T*, denoted by *len(T)*, equal $\sum_{(u,v) \in E(T)} l(u, v)$ (i.e., *len(T)* is the summation of the edge-lengths of a tree T). The minimum evolution requires that *len(T)* should be minimized. Suppose we move a subtree based on the moving distance $d = 1$.

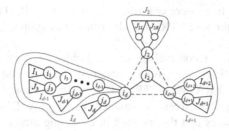

Fig. 3. The relevant nodes and edges around the reconnected position.

Clearly, the resulting tree, denoted by T_1, is similar to the tree T' obtained by applying the nearest neighbor interchange (NNI)[2]. It has been demonstrated that the $len(T')$ can be estimated by the following equation [2, 9]: $len(T') = len(T) - \frac{1}{4}\big((\Delta_{I_1|J_2} + \Delta_{I_4|J_3}) - (\Delta_{I_1|J_3} + \Delta_{I_4|J_2})\big)$.

The tree reconstruction algorithm implemented by TRMLE is shown in Algorithm 1.

Algorithm 1: TRMLE

input : sequence data, a model, a phylogenetic tree T construted using a
         maximum-likelihood-based algorithm
output: an improved phylogenetic tree
1 Compute the average distance for any pair of non-intersecting subtrees in the
   phylogenetic tree T based on Equation (9)
2 *counter* := 0
3 m:=the number of edges in T
4 **while** *counter* $< 0.3m$ **do**
5   Sort the edges according to their lengths in an array $E = (e_1, e_2, \cdots, e_m)$, where
        $l(e_1) \le l(e_2) \le \cdots \le l(e_m)$
6      $i := 0$
7   Select the ith edge $e_i = (u, v)$ to start the modified TBR
8   Identify the subtree S in T, where S is the subtree of T induced by $\{u, v\}$ and the
        descendants of u
9   Determine the moving distance d using the method described in Section 3.3
10   Move S using the moving distance d and reconnect it to the tree
11   Update the information associated with the resulting tree using the method described
        in Section 3.2
12   **if** *Phase 2 has been complete* **then**
13   **if** *the likelihood of the current tree is larger than that of the original tree* **then**
14   Store the new likelihood and topology of the tree
15   **else**
16             $i := i + 1$
17   *counter* := *counter* $+ 1$
18   **if** *counter* $< 0.3m$ and $i < m$ **then**
19   go to line 7
20   **else**
21   break
22   **else**
23   Identify the subtree S in T, where S is the subtree of T induced by $\{u, v\}$ and the
            descendants of v
24   go to line 9

---

4 Experiment Results

In the experiment, the TRMLE method was applied to four real data sets comprising 25 taxa, 35 taxa, 40 taxa, and 100 taxa. These benchmark data sets were obtained from [32, 44, 47].

---

[2] NNI is a local tree rearrangement method that generates two alternative trees by swapping a subtree
on one side of the branch with a subtree on the other side.

First, phylogenetic trees were constructed using DMAML, PhyML, and RAxML, and subsequently the proposed TRMLE algorithm was applied to these phylogenetic trees.

As indicated in Fig. 4 comparison of similarities of three algorithms with and without using TRMLE, the trees obtained using DMAML + TRMLE, PhyML + TRMLE, and RAxML + TRMLE were more similar to true phylogenetic trees than the phylogenetic trees constructed using only DMAML, PhyML, and RAxML. Table 1 indicates that the increase in executing time caused by applying the TRMLE algorithm to the three algorithms were reasonably small, such that the total executing time did not substantially increase.

The results demonstrated that TRMLE is highly effective and can assist numerous established algorithms in reconstructing phylogenetic trees.

Table 1. The increased executing time (seconds) by executing TRMLE after running DNAML, PhyML, and RAxML.

# of taxa	25	35	40	100
DNAML + TRMLE	3.740	4.612	12.977	153.740
PhyML + TRMLE	3.676	4.604	12.615	149.565
RAxML + TRMLE	3.763	4.601	12.620	153.978

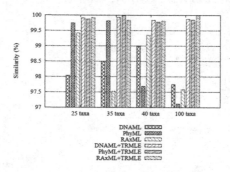

Fig. 4. Comparison of similarities of three algorithms with and without using TRMLE.

5   Concluding Remarks

In this paper, we propose a TRMLE method that combines a modified tree bisection and reconnection (TBR) with the minimum evolution principle to reconstruct a phylogenetic tree, based on the ML condition. By computing the changes in the tree length and updating new average distances of the original and reconnected positions, the proposed method avoids having to calculate the ML for every modified TBR. The optimal position can be selected only by considering changes in the tree length to recalculate the likelihood of the resulting tree. The proposed method demonstrates that the minimum evolution principle can assist in selecting the candidate reconnected positions efficiently.

The experiment results demonstrate that TRMLE can determine a near optimal ML tree. The method is highly effective and efficient, and it can be used to help assist numerous established algorithms in reconstructing phylogenetic trees within a reasonable time.

6 Availability and Supplemental Material

The TRMLE method is implemented in a web-based program, which is available at 140://140.116.82.73/TRMLE/index.php.

References

1. Baba, M.L., Darga, L.L., Goodman, M., Czelusniak, J.: Evolution of cytochrome C investigated by the maximum parsimony method. J. Mol. Evol. **17**, 197–213 (1981)
2. Bordewich, M., Gascuel, O., Huber, K.T., Moulton, V.: Consistency of topological moves based on the balanced minimum evolution principle of phylogenetic inference. IEEE/ACM Trans. Comput. Biol. Bioinf. **6**, 110–117 (2009)
3. Bordewich, M., Semple, C.: On the computational complexity of the rooted subtree prune and regraft distance. Ann. Comb. **8**, 409–423 (2004)
4. Brent, R.: Algorithms for Minimization Without Derivatives. Prentice Hall Inc., Englewood Cliffs (1973)
5. Bryant, D.: The splits in the neighborhood of a tree. Ann. Comb. **8**, 1–11 (2004)
6. Culik II, K., Wood, D.: A note on some tree similarity measures. Inf. Process. Lett. **15**, 39–42 (1982)
7. Day, W.H.E.: Properties of the nearest neighbor interchange metric for trees of small size. J. Theor. Biol. **101**, 275–288 (1983)
8. Edwards, A.W.F., Cavalli-Sforza, L.L.: The reconstruction of evolution. Ann. Hum. Genet. **27**, 105–106 (1963)
9. Desper, R., Gascuel, O.: Fast and accurate phylogeny reconstruction algorithms based on the minimum-evolution principle. J. Comput. Biol. **9**, 687–705 (2002)
10. Felsenstein, J.: Evolutionary trees from DNA sequences: a maximum likelihood approach. J. Mol. Evol. **17**, 368–376 (1981)
11. Fauvel, J.: Algorithms in the pre-calculus classroom: who was Newton- Raphson? Math. Sch. **27**, 45–47 (1998)
12. Felsenstein, J.: Inferring Phylogenies. Sinauer, Sunderland (2004)
13. Gaut, B.S., Lewis, P.O.: Success of maximum likelihood phylogeny inference in the four-taxon case. Mol. Biol. Evol. **12**, 152–162 (1995)
14. Guindon, S., Gascuel, O.: A simple, fast, and accurate algorithm to estimate large phylogenies by maximum likelihood. Syst. Biol. **52**, 696–704 (2003)
15. Harper, J.T., Waanders, E., Keeling, P.J.: On the monophyly of chromalveolates using a six-protein phylogeny of eukaryotes. Int. J. Syst. Evol. Microbiol. **55**, 487–496 (2005)
16. Hasegawa, M., Kishino, H., Yano, T.: Dating of the human-ape splitting by a molecular clock of mitochondrial DNA. J. Mol. Evol. **21**, 160–174 (1985)
17. Hordijk, W., Gascuel, O.: Improving the efficiency of SPR moves in phylogenetic tree search methods based on maximum likelihood. Bioinformatics **21**, 4338–4347 (2005)

18. Huelsenbeck, J.P., Crandall, K.A.: Phylogeny estimation and hypothesis testing using maximum likelihood. Ann. Rev. Ecol. Syst. **28**, 437–466 (1997)
19. Huelsenbeck, J.F., Hillis, D.M.: Success of phylogenetic methods in the four-taxon Case. Syst. Biol. **42**, 247–264 (1993)
20. Jukes, T.H., Cantor, C.R.: Evolution of protein molecules. In: Munro, H.N. (ed.) Mammalian Protein Metabolism. Academy Press, New York (1969)
21. Jones, N.C., Pevzner, P.A.: An Introduction to Bioinformatics Algorithms. The MIT Press, Cambridge (2004)
22. Ho, C.K., Shuman, S.: Trypanosoma brucei RNA triphosphatase: antiprotozoal drug target and guide to eukaryotic phylogeny. J. Biol. Chem. **276**, 46182–46186 (2001)
23. Jiang, H., Blouin, C.: Insertions and the emergence of novel protein structure: a structure-based phylogenetic study of insertions. BMC Bioinf. **8**, 444–458 (2007)
24. Kimura, M.: A simple method for estimating evolutionary rate of base substitutions through comparative studies of nucleotide sequences. J. Mol. Evol. **16**, 111–120 (1980)
25. Kuhner, M.K., Felsenstein, J.: A simulation comparison of phylogeny algorithms under equal and unequal evolutionary rates. Mol. Biol. Evol. **11**, 459–468 (1994)
26. Keane, T.M., Naughton, T.J., Travers, S.A.A., McInerney, J.O., McCormack, G.P.: DPRml: distributed phylogeny reconstruction by maximum likelihood. Bioinformatics **21**, 969–974 (2005)
27. Lamboy, W.F.: The accuracy of the maximum parsimony method for phylogeny reconstruction with morphological characters. Syst. Bot. **19**, 189–505 (1994)
28. Larget, B., Simon, D.L.: Markov chain Monte Carlo algorithms for the Bayesian analysis of phylogenetic trees. Mol. Biol. Evol. **16**, 750–759 (1999)
29. Ledford, R.M.: VP1 sequencing of all human rhinovirus serotypes: insights into genus phylogeny and susceptibility to antiviral capsid-binding compounds. J. Virol. **78**, 3663–3674 (2004)
30. Lemey, P., Pybus, O.G., Wang, B., Saksena, N.K., Salemi, M., Vandamme, A.M.: Tracing the origin and history of the HIV-2 epidemic. Nat. Acad. Sci. **100**, 6588–6592 (2003)
31. Lemmon, A., Milinkovitch, M.: The metapopulation genetic algorithm: an efficient solution for the problem of large phylogeny estimation. Nat. Acad. Sci. US Am. **99**, 10516–10521 (2002)
32. Ludwig, W.: ARB: a software environment for sequence data. Nucleic Acids Res. **32**, 1363–1371 (2004)
33. Mau, B., Newton, M.A., Larget, B.: Bayesian phylogenetic inference via Markov Chain Monte Carlo methods. Biometrics **55**, 1–12 (1999)
34. Michener, C., Sokal, R.: A quantitative approach to a problem in classification. Evolution **11**, 130–162 (1957)
35. Neyman, J.: Statistical Decision Theory and Related Topics. Academy Press, New York (1971)
36. Ohkuma, M., Saita, S., Inoue, T., Kudo, T.: Comparison of four protein phylogeny of parabasalian symbionts in termite guts. Mol. Phylogenet. Evol. **42**, 847–853 (2007)
37. Ranwez, V., Gascuel, O.: Improvement of distance-based phylogenetic methods by a local maximum likelihood approach using triplets. Mol. Biol. Evol. **19**, 1952–1963 (2002)
38. Rosenberg, M., Kumar, S.: Traditional phylogenetic reconstruction methods reconstruct shallow and deep evolutionary relationships equally well. Mol. Biol. Evol. **18**, 1823–1827 (2001)
39. Saitou, N., Nei, M.: The neighbor-joining method: a new method for reconstructing phylogenetic trees. Mol. Biol. Evol. **4**, 406–425 (1987)
40. Song, Y.S.: Properties of subtree-prune-and-regraft operations on totally-ordered phylogenetic trees. Ann. Comb. **10**, 147–163 (2006)

41. Stamatakis, A., Ludwig, T.: RAxML-III: a fast program for maximum likelihood-based inference of large phylogenetic trees. Bioinformatics **21**, 456–463 (2005)
42. Stamatakis, A., Hoover, P., Rougemont, J.: A rapid bootstrap algorithm for the RAxML web servers. Syst. Biol. **75**, 758–771 (2008)
43. Takezaki, N., Nei, M.: Inconsistency of the maximum parsimony method when the rate of nucleotide substitution is constant. J. Mol. Evol. **39**, 210–218 (1994)
44. Winkworeth, R.C., Bryant, D., Lockhart, P.J., Havell, D., Moulton, V.: Biogeographic interpretation of splits graph: least squares optimization of branch length. Syst. Biol. **54**, 56–65 (2005)
45. Yang, Z., Rannala, B.: Bayesian phylogenetic Inference using DNA sequences: a Markov Chain Monte Carlo method. Mol. Biol. Evol. **14**, 717–724 (1997)
46. Yang, Z.: Computational Molecular Evolution. Oxford University Press, Oxford (2006)
47. http://www.ncbi.nlm.nih.gov/genbank/

Rapid Annotation of Non-coding RNA Structures with a Parameterized Filtering Approach

Yinglei Song[1]([✉]), Junfeng Qu[2], and Chunmei Liu[3]

[1] School of Electronics and Information Science,
Jiangsu University of Science and Technology,
Zhenjiang 212003, Jiangsu, China
syinglei2013@163.com
[2] Department of Computer Science and Information Technology,
Clayton State University, Morrow, GA 30260, USA
[3] Department of Systems and Computer Sciences,
Howard University, Washington, DC 20059, USA

Abstract. An important problem in structural bioinformatics is to search genomes for RNA sequences with known secondary structures. Most of the existing approaches use sequence-structure alignment to evaluate the probability for a sequence segment to be a member of the searched RNA family. Due to the large amount of computation time needed by accurate sequence-structure alignments, filtering approaches have been developed to rapidly eliminate most part of the genome that are unlikely to contain sequences from the desired family. Most of the existing filtering tools construct and select filters based on the recognition ability of filters and do not consider the computation time needed by the filtering process. In this paper, we develop a parameterized filter selection approach that considers both the recognition ability and the computational efficiency of filters. As the first step, the approach constructs a set of filters that contain highly conserved parts in a searched family. A dynamic programming approach is then used to evaluate the recognition ability of each filter constructed in the first step. Finally, a set of filters are selected by solving a variant of the 0-1 knapsack problem that considers both their sizes and recognition ability. Our testing results showed that this new filtering approach can significantly speed up the search procedure without adversely affecting the search accuracy. It therefore can be combined with most of the existing search tools to significantly reduce the computation time needed for search.

Keywords: Noncoding RNA search · Sequence-Structure alignment · Filter selection · Dynamic programming

1 Introduction

In molecular biology and bioinformatics, the relationships between the structure of a molecule and its biological functions have been under intensive study for many years. As a result, a large number of models and tools have been developed to utilize the

© Springer International Publishing Switzerland 2015
D.-S. Huang et al. (Eds.): ICIC 2015, Part II, LNCS 9226, pp. 542–553, 2015.
DOI: 10.1007/978-3-319-22186-1_54

relationships that have been discovered to improve the accuracy of structure prediction or recognition of these molecules [4, 11, 13, 21, 22, 26]. So far, proteins and noncoding RNAs (ncRNAs) are the two most important types of such molecules.

ncRNAs are RNA molecules that do not encode proteins. Recent research in molecular biology has revealed that ncRNAs are of fundamental importance in many biological processes, such as gene regulation, chromosome replication and RNA modification [3, 12, 25]. Recently, since the amount of fully sequenced data has grown explosively, computational search tools have become an important alternative for searching homologous RNA structures in genomic sequences and recognizing new ncRNAs [6, 7, 11, 13]. In general, a genome is scanned through by a computational search tool during the search procedure and each sequence segment in the genome is aligned to a structural model that describes the secondary structure of the searched family. A segment is reported as a potential member of the family if its alignment score is statistically significant. The structure model generally consists of information from both the primary sequence content and secondary structure.

A covariance model (CM) is a structural model that can accurately describe the secondary structure of an ncRNA family. Compared with profiling models based on Hidden Markov Models (HMMs), a CM [11] contains additional emission states that can generate base pairs in the stems of secondary structure. CMs can thus be used as structural models to model RNA families. A well known fact about CM is that the computational efficiency of a CM-based search tool drops significantly when the searched sequence is only of moderate length.

Our previous work has developed a new parameterized approach for efficiently computing the optimal sequence-structure alignment [13]. However, the computation time needed by the approach may rise sharply when the parameters become large integers. In [21], we developed a machine learning based approach that can accurately extract features to represent sequences in a family and a classifier-based approach is used to determine whether a sequence segment is a member of the searched family or not. Although this approach is able to achieve higher search accuracy than those based on CM, the computation time needed for feature extraction is of the same asymptotic order as that of the optimal sequence-structure alignment. In practice the approach may need more computation time than CM-based search tools.

An important approach that can significantly improve the computational efficiency of search is the filtering method [5]. A filtering method employs a simpler sequence or structural model to describe the family and use it to efficiently screen out genome segments that are unlikely to contain the searched sequence. So far, a number of filtering methods [1, 9, 23] have been developed to improve the search efficiency. For example, tRNAscan-SE [9] uses two efficient tRNA detection algorithms as filters to preprocess a genome and remove most parts that are unlikely to contain the searched tRNA structure. The remaining part of the genome is processed with a CM to determine the location of tRNA. FastR [1] considers an RNA structure as a collection of structural units and evaluates the specificity of each structural unit. A set of filters can be constructed with the specificities of these structural units. In [23], base pairs in an RNA structure are safely broken by an algorithm and a set of filters can be automatically generated from the resulting HMM models. TDFilter uses a graph theoretic approach to construct a set of filters that reflect the most functionally important parts in the secondary structure of a family [8].

These approaches have significantly improved the computational efficiency of ncRNA search. However, most of them only select filters based on their specificity or recognition ability and thus are not guaranteed to achieve the optimal computational efficiency for filtering. For tasks where the genomes are of large sizes, the computation time needed by filtering may significantly affect the speed of searching. Therefore, new approaches that can effectively obtain the most functionally important parts in the searched family while optimizing the computational efficiency of filtering are highly desirable for rapid search of ncRNA structures in genomes of large sizes.

In this paper, we develop a parameterized approach that can construct and select filters based on the expected filtration ratio of filtering, which is the probability that a random sequence segment can pass the filtering process. This ratio is a parameter that can be determined based on the length of the genomes that need to be searched. The approach starts by evaluating the degree of conservation for each position in the structural alignment of the family and uses a dynamic programming method to find the parts that are of appropriate lengths and have the highest degree of conservation. Each part contains a set of aligned subsequences and a profile HMM can be constructed to describe the subsequences. Each of these profile HMMs is considered as a potential filter. We then employ a dynamic programming approach to evaluate the recognition ability of each potential filter. Finally, we select a set of filters from the set of all potential filters to minimize the amount of computation time of filtering by solving a variant of the 0-1 knapsack problem.

We have implemented this filter selection algorithm and compare its performance with that of TDFilter [8] and the filter implemented in Infernal 1.1.1 [11, 24]. We tested the performance of the three filtering approaches on benchmark dataset RMARK3 and compared both the search accuracy and computational efficiency among the three filtering approaches. Our testing results showed that this filter approach is significantly faster and can achieve accuracy comparable with the other two filtering approaches. Specifically, it is around 6 times faster than TDFilter and 9 times faster than the filter implemented in Infernal 1.1.1.

2   The Approach

Given the structural alignment of a set of training sequences in a noncoding RNA family, the goal of the search problem is to locate the potential member of the family in a given DNA genome. A filtering approach processes the sequence segments in the genome with one or a set of filters and only segments that can be aligned to the filters with statistically significant scores are considered to be a candidate for a member of the family. Each candidate is further processed by a more sophisticated structural model to determine whether it is indeed a member of the family or not.

As the first step of our filter generation approach, we use the information content to evaluate the degree of conservation for each position in the structural alignment of training sequences and use a dynamic programming approach to compute the average information content for each subsequence of certain lengths in the structural alignment. A set of potential filters can be selected by choosing a set of subsequences that have the highest values of average information content.

In the second step, we use a dynamic programming approach to compute the mean and standard deviation of the alignment scores of all possible subsequences on each potential filter. Since a set of subsequences are used to construct a potential filter, we align each subsequence in the set to the potential filter and a Z-score can be computed based on the alignment score. We evaluate the recognition ability of each filter by computing the average Z-score of the alignment scores obtained with all subsequences in the set. From the average Z-score, an expected p-value can be obtained to represent the probability for a random sequence to be aligned to a potential filter with a statistically significant score.

In the last step, we select a set of filters that satisfy the expected filtration ratio while minimizing the amount of computation time needed for filtering. We formulate this optimization problem into a variant of the 0-1 knapsack problem and use a dynamic programming approach to efficiently solve the problem.

2.1   Generating the Potential Filters

For each position in the structural alignment of a set of homologous sequences in an ncRNA family, we use information content as a measure of the degree of conservation for nucleotides in the position. The information content $I(l)$ of a position l in the alignment can be computed as follows.

$$I(l) = \sum_{t=0}^{4} p_{l,t} \log_2 p_{l,t} \tag{1}$$

where $p_{l,0}$ is the frequency of a gap to appear in l, and $p_{l,t}$ $(1 \leq t \leq 4)$ represent the frequencies of nucleotides A, C, G, U to appear in l respectively. It is straightforward to see that the value of $I(l)$ is minimized when the gap and each possible nucleotide appears with equal probability in l. On the other hand, $I(l)$ reaches its maximum when l only contains one of the four possible nucleotides. $I(l)$ is therefore a measure of the degree of conservation for l in the alignment.

A region in an alignment is comprised of a set of positions that are consecutive in the alignment. It is not difficult to see that the average information content of a region reflects the degree of conservation of the region. Recent research has revealed that regions of higher degree of conservation are generally more important for the biological functionality of the ncRNA. A set of regions with high values of average information content can thus be selected to construct the potential filters.

The average information content of all regions in the alignment can be computed with a dynamic programming approach. Specifically, we use a two dimensional table $M[i,j]$ to represent the average information content for the region that contains the part between positions i and j, where $i \leq j$. It is straightforward to see that $M[i,j]$ can be determined with the following recursion relationship.

$$M[i,i] = I(i) \tag{2}$$

$$M[i,j] = \frac{(j-i)M[i,j-1] + I(j)}{j-i+1} \tag{3}$$

where Eq. (2) is the case where the region contains only one nucleotide and Eq. (3) relates $M[i,j]$ with $M[i,j-1]$ when $i<j$.

After the value of each element in M has been determined, we use the following algorithm to select regions that have high average information content and are of lengths between a pair of given integers L_1 and L_2.

1. Set F to be an empty set.
2. Compute the mean of the average information content of all regions that are of lengths between L_1 and L_2 based on M and store it in A.
3. Find all regions that have an average information content value larger than A and do not overlap with any region in F. All these regions are included in a set R.
4. If R is empty, go to step 6, otherwise find the region T that has the largest average information content in R and include T in F.
5. Go to step 3.
6. Output the regions in F.

A profile HMM can be constructed based on each region selected by the above algorithm. Such a profile HMM is a potential filter. A profile HMM models each position in a region with two states. A matching state is able to generate a nucleotide while a deletion state models the gaps that may appear in the position. Given a selected region that includes all positions between positions m and n in the alignment, we use the following algorithm to construct a potential filter.

1. Include two states S and E in state set H. Set the set T of transitions to be empty.
2. Generate two states D_m and M_m for position m and set the transition probability from S to D_m to be $p_{m,0}$ and the transition probability from S to M_m to be $1 - p_{m,0}$. Include both transitions in T and D_m and M_m in H.
3. Set the emission probability of nucleotide t in M_m to be $p_{m,t} / \sum_{q=1}^{4} p_{m,q}$.
4. Set i to be $m+1$.
5. Generate two states D_i and M_i for position i and set the transition probability from both D_{i-1} and M_{i-1} to D_i to be $p_{i,0}$ and the transition probability from both D_{i-1} and M_{i-1} to M_i to be $1 - p_{i,0}$. Include all four transitions in T and D_i and M_i in H.
6. Set the emission probability of nucleotide t in M_i to be $p_{i,t} / \sum_{q=1}^{4} p_{i,q}$.
7. Increment i by one, if $i \leq n$, go back to step 5.
8. Set the transition probability from both D_n and M_n to E to be 1.
9. Output H and T as the potential filter.

Figure 1 provides an illustration on the potential filter constructed for a region in the structural alignment of a set of homologous sequences. The rectangle in (a) contains

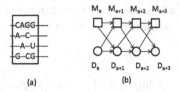

Fig. 1. (a) Columns $a, a+1, a+2, a+3$ in the rectangle are selected by the algorithm; (b) the states and transitions among them in the profile HMM constructed for the region, where a square represents a matching state and a circle is a deletion state.

four columns in a selected region and (b) shows the corresponding states and transitions that describe the columns in the profile HMM constructed for the region.

2.2 Evaluating the Recognition Ability of Potential Filters

To evaluate the recognition probability of a potential filter, we consider the distribution of alignment scores of all sequences randomly generated with the base composition of the genome that needs to be searched. Based on the distribution, the statistical significance of each alignment score obtained on the potential filter can be evaluated by computing the Z-score associated with the alignment score. Specifically, given a distribution with mean m and standard deviation σ, the Z-score associated with an alignment score A can be computed as follows.

$$Z = \frac{A - m}{\sigma} \tag{4}$$

We thus consider each subsequence the region contains in the alignment and align the subsequence to the potential filter to obtain an alignment score. The average Z-score of the alignment scores of all such subsequences is therefore a reliable measure of the recognition ability of the potential filter.

Equation (4) suggests that the mean m and standard deviation σ of the distribution need to be computed to evaluate the recognition ability of each potential filter. One approach that has been extensively used to determine m and σ is based on sampling. Specifically, a large number of random sequences are generated with the base composition of the genome and each of them is aligned to the potential filter to obtain an alignment score. The distribution is then approximated by the obtained alignment scores. To address this problem, we develop a dynamic programming approach that can compute the values of m and σ both accurately and efficiently.

Specifically, given a potential filter constructed from a region that contains positions between positions s and t in an alignment. It is straightforward to see that a profile HMM can also be constructed with positions between s and r, where r is an integer between s and t. We denote this partial profile HMM with Q_r. We use a one dimensional table $N(r)$ to store the natural logarithm of the expected value of the alignment scores obtained by aligning all random sequences generated with the base composition of the genome to Q_r. It is not difficult to see that $N(r)$ can be computed with the following recursion relation.

$$N[s] = \ln(p_{s,0} + (1 - p_{s,0}) \sum_{t=1}^{4} p_{s,t} q_t) \tag{5}$$

$$N[r] = N[r-1] + \ln(p_{r,0} + (1 - p_{r,0}) \sum_{t=1}^{4} p_{r,t} q_t) \tag{6}$$

where q_t's $(1 \leq t \leq 4)$ are the probabilities for nucleotides A, C, G, T to appear in the genome that needs to be searched. After every element in N is determined, the value of m can be computed with $m = \exp(N[t])$.

Based on the same principle, we are able to compute the expected value of the square of the alignment scores obtained by aligning all random sequences generated with the base composition of the genome. We use a one dimensional table $S[r]$ to store the natural logarithm of the expected value of the square of the alignments scores by aligning all random sequences to Q_r.

$$S[s] = \ln(p_{s,0}^2 + (1 - p_{s,0})^2 \sum_{t=1}^{4} p_{s,t}^2 q_t) \tag{7}$$

$$S[r] = S[r-1] + \ln(p_{r,0}^2 + (1 - p_{r,0})^2 \sum_{t=1}^{4} p_{r,t}^2 q_t) \tag{8}$$

After every element in S is determined, the value of σ can be computed with Eq. (9).

$$\sigma = \sqrt{\exp(S[t]) - m^2} \tag{9}$$

2.3   Selecting Filters

After a list of potential filters has been generated and their recognition ability is evaluated, we are ready to select filters from all potential filters such that the computation time needed by filtering can be minimized. Based on the average Z-score we have computed for each potential filter, we estimate the probability for a random sequence to achieve an alignment score comparable with those in the training dataset. We use $p(F)$ to denote this probability for a potential filter F. The estimation of $p(F)$ is based on the assumption that the distribution of the alignment scores on F is approximately a Gaussian distribution. In other words, $p(F)$ is approximately computed based on its average Z-score $Z(F)$ with Eq. (10).

$$p(F) \approx 1 - erc(Z(F)) = 1 - \frac{2}{\sqrt{\pi}} \int_0^{Z(F)} e^{-t^2} dt \qquad (10)$$

where the value of the error function can be obtained by looking up a table.

Since all potential filters are selected from regions that do not overlap in the alignment, the probability that a random sequence can achieve statistically significant alignment scores on a list of potential filters can be computed with Eq. (11).

$$p(F_1, F_2, \ldots, F_k) = \prod_{c=1}^{k} p(F_c) \qquad (11)$$

By taking logarithm with base 2 on both sides of Eq. (11), we obtain the following equation.

$$-\log_2 p(F_1, F_2, \ldots, F_k) = \sum_{c=1}^{k} (-\log_2 p(F_c)) \qquad (12)$$

On the other hand, the computation time needed to optimally align a sequence segment to a potential filter that contains b states is $O(b^2)$. We thus can assign a value $W(F)$ to each potential filter F based on the number of states in F. $W(F)$ can be computed with Eq. (13).

$$W(F) = b(F)^2 \qquad (13)$$

where $b(F)$ is the number of states in F.

The filter selection problem can then be formulated as a variant of the 0-1 knapsack problem, where the objective is to select a set of filters from all potential filters such that the logarithmic of the filtration ratio evaluated by Eq. (12) is approximately equal to the logarithm of a filtration ratio provided by the user and the sum of the values of all selected filters is minimum. This problem is similar to the 0-1 knapsack problem. The only difference is that the objective here is to minimize the total values of all "objects" that can be placed into the "knapsack".

Based on the above observation, we assign each potential filter F a capacity value $c(F) = -\log_2 p(F)$ and a value $W(F)$ that can be computed with Eq. (13). The total capacity of the knapsack is defined to be $C = -\log_2 f_r$, where f_r is the filtration ratio provided by the user. All capacity values will be rounded to integers. The well known dynamic programming for the 0-1 knapsack problem can be slightly changed to solve this problem. Specifically, assume we have n potential filters and we assign each filter an integer in $\{1, 2, 3, \ldots, n\}$ such that no two filters are assigned the same integer. We use a two dimensional table $V(D, j)$ to store the minimum total value of a maximal set of filters such that they are selected from potential filters with integer identities from $\{1, 2, 3, \ldots, j\}$ and the sum of their capacities is at most D. It is not difficult to see that Eq. (14) holds for any $1 \leq j \leq n$ if $0 \leq D < C_{m,j}$, where $C_{m,j}$ is the minimum capacity of all potential filters with integer identities from $\{1, 2, 3, \ldots, j\}$.

$$V(D,j) = 0 \qquad (14)$$

We use $F_1, F_2, \ldots, F_n$ to denote the potential filters with integer identity numbers $1, 2, 3, \ldots, n$. It is not difficult to see that Eq. (15) holds if $D < c(F_j)$.

$$V(D,j) = V(D,j-1) \qquad (15)$$

For $D \geq c(F_j)$, if $V(D,j-1) > 0$, the value of $V(D,j)$ can be computed with Eq. (16). Otherwise, its value can be computed with Eq. (17).

$$V(D,j) = \min\{V(D,j-1), V(D-c(F_j),j-1) + W(F_j)\} \qquad (16)$$

$$V(D,j) = W(F_j) \qquad (17)$$

It is not difficult to see that the above dynamic programming approach can efficiently select the filters that satisfy the required filtration ratio and minimizes the total values of the selected filters. A traceback table can be used to store the decision made in each step of the dynamic programming procedure and the selected filters can be obtained with a standard traceback procedure.

3 Experimental Results

We have implemented this filter selection approach into a computer program New-Filter. The program is freely available and it can be downloaded at the following link: http://www.rworldpublishing.org/software.html. We have tested its performance and compared its accuracy and computational efficiency with those of other filtering tools, including TDFilter [8] and the filters of Infernal 1.1.1 [11, 24]. TDFilter is a filter selection approach we have developed based on graph tree decomposition. TDFilter selects structure units that are functionally important with a tree decomposition based dynamic programming approach and the filters are constructed based on these selected structure units. Infernal 1.1.1 is a CM-based search tool and it uses four separate filter stages where each filter stage is implemented with a set of profile HMMs. The accuracy of a filter stage is successively higher than the previous stage while more computation time is needed for alignments.

The testing is performed on the RMARK3 benchmark dataset [11]. The dataset contains a set of 106 ncRNA families. The structural alignment of the training sequences for each family is constructed based on the seed alignments of Rfam 10.0 database [2]. The mutual identity between any pair of sequences in a structural alignment is below 70 % and the mutual identity between any pair of training and testing sequences is not higher than 60 %. The dataset contains 780 testing sequences that are inserted into ten genome-like sequences and each such sequence contains 10^6 nucleotides and the dataset contains around 1.016×10^7 nucleotides in total. All testing sequences are also available in the Rfam 10.0 database.

We have integrated all three filters with RNASC [21], a search tool developed based on a machine learning approach. The accuracy and efficiency of the three combined tools on the RMARK3 benchmark dataset. For NewFilter, we generate potential filters based on regions with lengths from 5 to 10 and the filtration ratio is set to be 1.0×10^{-4}. For each tested family, we compute the sensitivity and specificity based on the search results returned by all three filters. We use the average of the sensitivity and specificity to evaluate the search accuracy of a combined search tool. Table 1 shows distribution of the accuracy of the three combined tools and that of RNASC when no filters are used. It can be seen from the table that the combined search tool with NewFilter integrated achieves higher average search accuracy than two other combined search tools. It is also clear from the table that the average search accuracy of RNASC does not drop significantly after the NewFilter is integrated into it. We believe selecting highly conserved regions to construct filters is an important reason why NewFilter is able to achieve higher accuracy than both TDFilter and Infernal Filter.

In addition to the search accuracy, we also evaluate the computational efficiency of all four tools. Table 2 shows the average amount of computation time needed by each of the four search tools to search in a sequence that contains 10^5 nucleotides. It can be seen from the table that NewFilter is around 6 times faster than TDFilter and 9 times faster than Infernal Filter. This is not surprising since NewFiler selects short regions to construct filters and does not include information on base pairs in the secondary structure, which significantly reduces the computation time needed for alignments. In addition, NewFilter selects filters that can optimize the computational efficiency of the filtering process, which contributes further to its advantage in computational efficiency.

Table 1. The distribution of search accuracy of RNASC (without a filter) and three other combined search tools on the RMARK3 benchmark.

Search tools	RNASC (without a filter)	NewFilter	TDFilter	Infernal filter
Above 90 %	65	60	54	61
80 %–90 %	38	32	30	29
70 %–80 %	3	12	17	13
Below 70 %	0	2	5	3
Average	92.7 %	88.6 %	85.4 %	87.8 %

Table 2. The average amount of computation time needed by each search tool to search a sequence of 10^5 nucleotides.

Search tools	RNASC (with no filter)	NewFilter	TDFilter	Infernal filter
Time (s)	439.23	8.72	52.78	73.62

4 Conclusions

In this paper, we develop a new approach to improve the computational efficiency for the annotation of non-coding RNAs. This approach develops a dynamic programming approach to accurately evaluate the recognition ability of a potential filter and selects a set of filters to minimize the computation time needed for filtering while maintaining the given filtration ratio. Our testing results also show that the filters selected by this approach can significantly improve the computational efficiency of search without adversely affecting the accuracy. However, evaluations have been only performed on a few ncRNA families and the results thus could be biased. More comprehensive testing is needed to further evaluate its accuracy. In [15–19], we have developed parameterized approaches that can solve a few NP-hard Bioinformatics Problems. Combining NewFilter with these approaches can probably further improve its search accuracy.

References

1. Bafna, V., Zhang, S.: FastR: fast database search tool for non-coding RNA. In: Proceedings of the 3rd IEEE Computational Systems Bioinformatics Conference, pp. 52–61 (2004)
2. Burge, S.W., et al.: Rfam 11.0: 10 years of RNA families. Nucleic Acids Res. **41**, D226–D232 (2013)
3. Frank, D.N., Pace, N.R.: Ribonuclease P: unity and diversity in a tRNA processing ribozyme. Annu. Rev. Biochem. **67**, 153–180 (1998)
4. Huang, D.S., Yu, H.: Normalized feature vectors: a novel alignment-free sequence comparison method based on the numbers of adjacent amino acids. IEEE/ACM Trans. Comput. Biol. Bioinform. **10**(2), 457–467 (2013)
5. Kolbe, D.L., Eddy, S.R.: Fast filtering for RNA homology search. Bioinformatics **27**, 3 102–3109 (2011)
6. Meyer, F., Kurtz, S., Backofen, R., Will, S., Beckstette, M.: Structator: fast index-based search for RNA sequence-structure patterns. BMC Bioinformatics **12**, 214 (2011)
7. Liu, C., Song, Y., Malmberg, R.L., Cai, L.: Profiling and searching for RNA pseudoknot structures in genomes. In: Sunderam, V.S., van Albada, G.D., Sloot, P.M.A., Dongarra, J. (eds.) ICCS 2005. LNCS, vol. 3515, pp. 968–975. Springer, Heidelberg (2005)
8. Liu, C., Song, Y., Hu, P., Malmberg, R.L., Cai, L.: Efficient annotation of non-coding RNA structures including pseudoknots via automated filters. In: Proceedings of the 2006 Computational Systems Bioinformatics Conference, pp. 99–110 (2006)
9. Lowe, T.M., Eddy, S.R.: tRNAscan-SE: a program for improved detection of transfer RNA genes in genomic sequence. Nucleic Acids Res. **25**, 955–964 (1997)
10. Mistry, J., Finn, R.D., Eddy, S.R., Bateman, A., Puna, M.: Challenges in homology search: HMMER3 and convergent evolution of coiled-coil regions. Nucleic Acids Res. **41**, e123 (2013)
11. Nawrocki, E.P., Eddy, S.R.: Infernal 1.1: 100-fold faster RNA homology searches. Bioinformatics **29**, 2933–2935 (2013)
12. Nguyen, V.T., Kiss, T., Michels, A.A., Bensaude, O.: 7SK small nuclear RNA binds to and inhibits the activity of CDK9/cyclin T complexes. Nature **414**, 322–325 (2001)

13. Song, Y., Liu, C., Malmberg, R.L., Pan, F., Cai, L.: Tree decomposition based fast search of rna structures including pseudoknots in genomes. In: Proceedings of the 2005 IEEE Computational Systems Bioinformatics Conference, pp. 223–234 (2005)
14. Song, Y., Yu, M.: On finding the longest antisymmetric path in directed acyclic graphs. Inf. Process. Lett. 115(2), 377–381 (2015)
15. Song, Y., Chi, A.Y.: Peptide sequencing via graph path decomposition. Inf. Sci. 301, 262–270 (2015)
16. Song, Y.: An improved parameterized algorithm for the independent feedback vertex set problem. Theor. Comput. Sci. 535, 25–30 (2014)
17. Song, Y.: A new parameterized algorithm for rapid peptide sequencing. PLoS ONE 9(2), e87476 (2014)
18. Song, Y., Yu, M.: On the treewidths of graphs of bounded degree. PLoS ONE 10(4), e0120880 (2015)
19. Song, Y., Chi, A.Y.: A new approach for parameter estimation in the sequence-structure alignment of non-coding RNAs. J. Inf. Sci. Eng. 31(2), 593–607 (2015)
20. Song, Y., Liu, C., Huang, X., Malmberg, R.L., Xu, Y., Cai, L.: Efficient parameterized algorithms for biopolymer structure-sequence alignment. IEEE/ACM Trans. Comput. Biol. Bioinform. 3(4), 423–432 (2006)
21. Song, Y., Liu, C., Wang, Z.: A machine learning based approach for accurate annotation of noncoding RNAs. IEEE/ACM Trans. Comput. Biol. Bioinform. (to appear)
22. Wang, B., Huang, D.S., Jiang, C.: A new strategy for protein interface identification using manifold learning method. IEEE Trans. Nanobiosci. 13(2), 118–123 (2014)
23. Weinberg, Z., Ruzzo, W.L.: Faster genome annotation of non-coding RNA families without loss of accuracy. In: Proceedings of the Eighth Annual International Conference on Computational Molecular Biology, pp. 243–251 (2004)
24. Wheeler, T.J., Eddy, S.R.: NHMMER: DNA homology search with profile HMMs. Bioinformatics 29, 2487–2489 (2013)
25. Yang, Z., Zhu, Q., Luo, K., Zhou, Q.: The 7SK small nuclear RNA inhibits the Cdk9/cyclin T1 kinase to control transcription. Nature 414, 317–322 (2001)
26. Zhu, L., You, Z.H., Huang, D.S., Wang, B.: t-LSE: a novel robust geometric approach for modeling protein-protein interaction networks. PLoS ONE 8(4), e58368 (2013)

Prediction of Pre-miRNA with Multiple Stem-Loops Using Feedforward Neural Network

Gaoqiang Yu, Dong Wang, and Yuehui Chen[✉]

School of Information Science and Engineering,
University of Jinan, Jinan, China
yhchen@ujn.edu.cn

Abstract. miRNA is a kind of single non-coding RNA that plays a pivotal regulated role in gene expression and has a very important influence in disease occurrence, growth and development, cell proliferation and so on. Therefore prediction miRNA has become the most important task in understanding miRNA regulation mechanism. Existing computational prediction methods are usually good at recognition pre-miRNA with multiple stem-loops. In this study, in order to further improve predictive precision of pre-miRNA, we quoted a set of new biologically multiple stem and loop secondary structure features based on the previous research work, then handled the imbalance problem of dataset, combined feedforward neural network. Finally, the new classifier system was constructed successfully with the proposed approach to separate the real pre-miRNA from datasest. By using the dataset of human pre-miRNAs and employing systematic 5-fold cross-validation methods for evaluating the classifier performance. We discover that the new classifier improved predictive precision effectively.

Keywords: Multiple stem-loops · miRBase · Feedforward neural network · Particle swarm optimization

1 Introduction

MiRNA synthesis begins with the generation of pri-miRNA that is produced by gene transcription of nucleus encoded miRNA. Then it is processed into about 60–70 nt pre-miRNA by RNA incision enzyme Drosha. Subsequently, pre-miRNA is transported into cytoplasm by transfer protein: Exportin-5. In cytoplasm pre-miRNA is sheared into double-chains complex by another RNA incision enzyme Dicer. Finally mature miRNA originates from one arm of double-chains complex. Figure 1 is the miRNA generation process.

MiRNA recognises target mRNA according to complementary base pairing, then degrades or prevents mRNA expression according to different degree of complementary base pairing [1]. Recent research suggests that miRNA regulates various human physiological activities [2].

Most of the pseudo pre-miRNA and about 4.76 % real pre-miRNA have multiple stem-loops in their secondary structure [3]. However existing computational prediction

© Springer International Publishing Switzerland 2015
D.-S. Huang et al. (Eds.): ICIC 2015, Part II, LNCS 9226, pp. 554–562, 2015.
DOI: 10.1007/978-3-319-22186-1_55

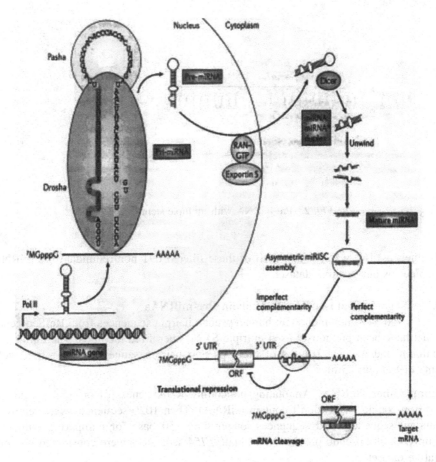

Fig. 1. MiRNA generation process

methods are usually good at recognition pre-miRNA with multiple stem-loops as shown in Fig. 2. Part of the machine learning methods, such as triple-SVM [4] even excludes this pre-miRNA with multiple stem-loops when selection samples. This inevitably reduces the performance of classifier. So we quoted a set of new biologically multiple stems and loops secondary structure features for getting better classification performance.

2 Materials and Methods

2.1 Biological Datasets

2.1.1 Positive Dataset-Human Pre-miRNAs
We used microPred [5] dataset for comparison of experimental result. 695 human pre-miRNA sequences were downloaded from miRBase12 (http://microrna.sanger.ac.

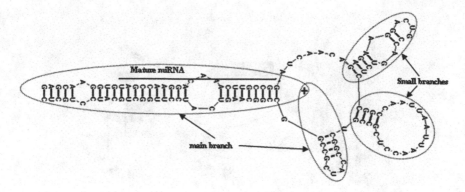

Fig. 2. Pre-miRNA with multiple stem-loops

uk/sequences) [6], we considered all of these filtered 691 non-redundant pre-miRNA sequences as our positive dataset.

2.1.2 Negative Dataset-Pseudo Hairpin Pre-miRNAs

We retrieved 8494 non redundant human pseudo hairpin sequences from RefSeq genes [7] that have been previously used in triple-SVM, microPred methods. The minimum, maximum and average lengths of these pseudo hairpin sequences are respectively 62 nt, 119 nt and 85 nt.

Human Other NcRNA. Annotating noncoding RNA genes [8] originally contains 1020 non-coding RNA (don't contain miRNA). Then 1020 sequences were removed redundant sequences and sequences longer than 150 base for comparing with real pre-miRNA and pseudo pre-miRNA. Finally, 754 sequences were considered into our negative dataset.

2.2 Features

We need to extract more specific and more representative features of pre-miRNA with multiple stem-loops for effectively improving performance of our classifier. First, we extracted 17 pre-miRNA sequence features that includes (AA %, AC %, ..., UU %), and (%C + G). 6-folding measures (MFEI1, MFEI2, dG, dP, dQ, dD) and one topological descriptor (dF) that are calculated from pre-miRNA secondary structure. At last, zF, zD, zQ, zP and zG is five normalized variants of dF, dD, dQ, dP and dG.

On the other hand, we adopted the following 19 features except above characteristics. L is the length of pre-miRNA sequence.

- Get following related features according to Minimum Free Energy [9].
  MFEI3 = dG/n_loops; n_loops is the number of pre-miRNA loops, dG = MFE/L.
  MFEI4 = MFE/tot_bases; tot_bases is the number of pairing base.
- Get following related features by using RNAfold [10] (default temperature 37°C).

Normalized Ensemble Free Energy (*NEFE*).

The frequency of the MFE structure (*Freq*).

The structural diversity (*Diversity*).

Diff = |MFE − EFE|/L; where EFE is the ensemble free energy.

- Get following related features by using UNAfold [11].

  dS, and dS/L; dS is Structure Entropy.

  dH, and dH/L; dH is Structure Enthalpy.

  Tm, and Tm/L; Tm is Melting Energy of the structure.

- Pairing base features.

  $|A - U|/L, |G - C|/L, |G - U|/L; |A - U|, |G - C|$ and $|G - U|$ are the number of pairing base.

Average base pairs per stem: $Avg\_BP\_Stem = tot\_bases/n\_stems$; tot_bases is the total number of pairing base. n_stems is the number of stem.

$$\%(A - U)/n\_stems, \%(G - C)/n\_stems, \%(G - U)/n\_stems.$$

2.3 Evaluation Criterion

We use SE, SP, Gm [12] as our classifier evaluation index. Gm is the final evaluation standard. Evaluation index is defined as follows.

$$\text{Sensitivity: } SE = \frac{TP}{TP + FN} \tag{1}$$

$$\text{Specificity: } SP = \frac{TN}{TN + FP} \tag{2}$$

$$\text{Geometric mean: } Gm = \sqrt{SE \times SP} \tag{3}$$

TP is the number of correctly predicted positive sample; FN is the number of wrong predicted negative sample; TN is the number of correctly predicted negative sample; FP is the number of wrong predicted positive sample.

2.4 Feedforward Neural Network

In 1986, Rumelhart and MeClelland research team put forward error bask propagation [13] of multilayer network that can approximate any nonlinear function and is research foundation of multilayer feedforward neural network. Feedforward neural network [14] is composed of input layer, hidden layer and output layer.

Node between hidden layer and output layer generally use the sigmoid excitation function As shown in the formula (4).

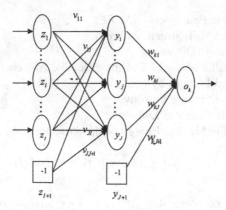

Fig. 3. Feedforward neural network

$$f(x) = \frac{1}{1 + e^{-x}} \tag{4}$$

We used the feedforward neural network that belongs to network with tuor training. A training sample is composed of input vector and corresponding target vector. Input vector is 48 dimension, target vector is 1 or 0 that respectively represents real or pseudo pre-miRNA.

It is a typical feedforward neural network that Fig. 3 includes input layer, hidden layer and output layer. Input layer dimension I is 48, the number of hidden layer neuron are 10, input is 1 or 0.

For any given input sample Zp, output result as follows:

$$
\begin{aligned}
O_{k,p} &= f_{O_k}\left(net_{O_{k,p}}\right) \\
&= f_{O_k}\left(\sum_{j=1}^{J+1} W_{kj} f_{yj}\left(net_{y_{j,p}}\right)\right) \\
&= f_{O_k}\left(\sum_{j=1}^{J+1} W_{kj} f_{yj}\left(\sum_{i=1}^{I+1} V_{ji} Z_{i,p}\right)\right)
\end{aligned}
\tag{5}
$$

f_{oi} and f_{yj} are activation function of output unit o_k and hidden unit y_j. W_{ki} is weight between output unit o_k and hidden unit y_i. $z_{i,p}$ is the value of input sample z_p from output unit z_i. The input unit $(I + 1)$ and the hidden unit $(J + 1)$ are 1 that respectively represents bias unit of next layer neuron threshold.

$t_k(X)$ reprensents ideal output of input sample X, then square error is E(X).

$$E(X) = \frac{1}{2}\sum_{k} \left(t_k(X) - o_k(X)\right)^2 \tag{6}$$

We need to train FNN for diminishing square error. In the training process, FNN weights are adjusted by particle swarm optimization (PSO) [15] algorithm.

2.5 Particle Swarm Optimization

It reflects inferential knowledge of artificial neural network that neural network weights are optimized by PSO. PSO is an optimal search method, bases on swarm intelligence, and was proposed by Kennedy and Russell Eberhart in 1995. Each particle in the swarm is associated with a position vector x^i and a velocity vector v^i in the solution space, each particle has a fitness value by the fitness function measure. The iterative process is toward the direction of optimal fitness value.

a. initialize particles with arbitrary positions and velocities.
b. each particle's fitness value is calculated by fitness function.
c. if particle's fitness value is better than pbest, then pbest and its position are updated.
d. if any particle's fitness value is better than gbest, then gbest and its position are updated.
e. particle's position and velocity are updated according to following two formula.

$$V_{id} = w \times V_{id} + c_1 \times rand() \times (P_{id} - X_{id}) + c_2 \times Rand() \times (P_{gd} - X_{id}) \qquad (7)$$

$$X_{id} = X_{id} + V_{id} \qquad (8)$$

f. repeat these steps from step b until reaching stop criterion. Usually stopping criterion will be set to reach maximum execution times or achieve desired fitness value.

c_1 and c_2 are learning factor, ω is inertia factor that adjusts search range of solution space. p^i is historical optimal solution of each particle. p^g is optimal solution of entire population.

3 Dataset Class Imbalance Problem

We will find that the ratio of positive to negative samples is 1:13.38 through analysis of positive and negative dataset. Studies have shown that training the classifier with an imbalance dataset can lead to poor classification performance with inclined to minority class [16]. In our method, it inclined to negative sample. FNN has its unique advantages in solving classification, but it's classification accuracy of less samples is far below large number of samples, thus leading to a serious decline in the performance of our classifier.

Improved Random_SMOTE [17] is used for solving above problem. Random_S-MOTE is a new novel over-sampling method, which generates synthetic examples for minority class.

The basic idea of Random_SMOTE sampling method is that randomly selection minority samples y_1, y_2 for each minority sample x, then construction a triangle taking x, y_1, y_2 as vertexes, finally, N new minority samples are randomly generated according to over-sampling rate N in triangle region. Specific process is as follows.

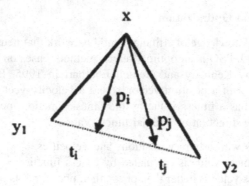

Fig. 4. The show of how Random_SMOTE generates synthetic examples

a. stochastic linear interpolation between y_1, y_2, generate N temporary samples $t_j(j = 1, 2, \ldots, N)$;

$$t_j = y_1 + rand(0, 1) \times (y_2 - y_1), j = 1, 2, \ldots N \tag{9}$$

b. stochastic linear interpolation between t_j, x, generate minority samples $p_j(j = 1, 2, \ldots, N)$;

$$p_j = x + rand(0, 1) \times (t_j - x), j = 1, 2, \ldots, N \tag{10}$$

Rand (0, 1) is a random number between 0 and 1. As shown in Fig. 4.

4 Results and Discussion

In order to ensure reliability of our experiment, 5-fold cross-validation [18] is used to test our classifier. Whole dataset is divided into five equally sized parts, imbalaned datasets are processed by Random_SMOTE until reaching sample balance. We randomly take four parts as training set to train FNN classifier, and make remaining part as our test set. Above procedure was repeated five times with different combinations of

Table 1. Experimental comparison results

	SE	SP	Gm
triplet-SVM	93.30	88.10	90.66
MiPred [19]	89.35	93.21	91.26
MiPred [20]	84.55	97.97	91.01
miRabela [21]	71.00	97.00	82.99
microPred [22]	90.02	97.28	93.58
FNN	94.7	97.9	95.8

The experiment with 48 features produced following average classification results: SE = 94.7, SP = 97.9 and Gm = 95.8. Gm is increased by 2.22 % when comparison with microPred. This shows that our proposed approach has a stronger predictive recognition ability, and can effectively predict pre-miRNA.

training (four parts) and test (the remaining part) dataset. Final result is average values of five experiments as shown in Table 1.

Acknowledgement. Gaoqiang Yu and Dong Wang contributed equally to this work and should be considered co-first authors. This research was partially supported by Program for Scientific research innovation team in Colleges and universities of Shandong Province 2012–2015, the Key Project of Natural Science Foundation of Shandong Province (ZR2011FZ001), the Natural Science Foundation of Shandong Province (ZR2011FL022, ZR2013FL002), the Key Subject Research Foundation of Shandong Province and the Shandong Provincial Key Laboratory of Network Based Intelligent Computing. This work was also supported by the National Natural Science Foundation of China (Grant No. 61302128, 61201428, 61203105).The scientific research foundation of University of Jinan(xky1109).

References

1. Song, X., Wang, M., Chen, Y.P.P., et al.: Prediction of pre-miRNA with multiple stem-loops using pruning algorithm. Comput. Biol. Med. **43**(5), 409–416 (2013)
2. Denli, A.M., Tops, B., Plasterk, R., Ketting, R.F., Hannon, G.J.: Processing of primary microRNAs by the microprocessor complex. Nature **432**, 231–235 (2004)
3. Xue, C.H., Li, F., He, T., Liu, G.P., Li, Y.D., Zhang, X.G.: Classification of real and pseudo microRNA precursors using local structure-sequence features and support vector machine. BMC Bioinformatics **6**, 310 (2005)
4. Xue, C., Li, F., He, T., et al.: Classification of real and pseudo microRNA precursors using local structure-sequence features and support vector machine. BMC Bioinformatics **6**, 310–316 (2005)
5. Batuwita, R., Palade, V.: microPred: effective classification of pre-miRNAs for human miRNA gene prediction. Bioinformatics **25**, 989–995 (2009)
6. Griffiths-Jones, S.: The microRNA Registry. Nucleic Acids Res. **32**(1), 109–111 (2004)
7. Pruitt, K.D., Maglott, D.R.: RefSeq and LocusLink: NCBI gene-centered resources. Nucleic Acids Res. **29**, 137–140 (2001)
8. Griffiths-Jones, S.: Annotating noncoding RNA genes. Annu. Rev. Genomics Hum. Genet. **8**, 279–298 (2007)
9. Kaffe-Abramovich, T., Unger, R.: A simple model for evolution of proteins towards the global minimum of free energy. Fold. Des. **3**(5), 389–399 (1998)
10. Hofacker, I.L.: Vienna RNA secondary structure server. Nucleic Acids Res. **31**(13), 3429–3431 (2003)
11. Zuker, M.: Mfold web server for nucleic acid folding and hybridization prediction. Nucleic Acids Res. **31**(13), 3406–3415 (2003)
12. Akbani, R., Kwek, S.S., Japkowicz, N.: Applying support vector machines to imbalanced datasets. In: Boulicaut, J.-F., Esposito, F., Giannotti, F., Pedreschi, D. (eds.) ECML 2004. LNCS (LNAI), vol. 3201, pp. 39–50. Springer, Heidelberg (2004)
13. McClelland, H., Davis, W.H.: Discrimination against atheists. Free Inquiry **24**(3), 11 (2004)
14. Jin, W., Li, Z.J., Wei, L.S., et al.: The improvements of BP neural network learning algorithm. In: 5th International Conference on Signal Processing Proceedings, 2000, WCCC-ICSP 2000, vol. 3, pp. 1647–1649. IEEE (2000)
15. Kennedy, J., Eberhart, R.: Particle swarm optimization. In: Proceedings of the IEEE International Conference on Neural Networks, vol. 4, pp. 1942–1948 (1995)

16. Weiss, G.M.: Mining with rarity: a unifying framework. ACM SIGKDD Explor. Newsl. **6** (1), 7–19 (2004)
17. Dong, Y., Wang, X.: A new over-sampling approach: random-SMOTE for learning from imbalanced data sets. In: Xiong, H., Lee, W.B. (eds.) KSEM 2011. LNCS, vol. 7091, pp. 343–352. Springer, Heidelberg (2011)
18. Kohavi R. A study of cross-validation and bootstrap for accuracy estimation and model selection. In: IJCAI, vol. 14, no. 2, pp. 1137–1145 (1995)
19. Jiang, P., et al.: MiPred: classification of real and pseudo microRNA precursors using random forest prediction model with combined features. Nucleic Acids Res. **35**, 339–344 (2007)
20. Loong, K., Mishra, S.: De novo SVM classification of precursor microRNAs from genomic pseudo hairpins using global and intrinsic folding measures. Bioinformatics **23**, 1321–1330 (2007)
21. Sewer, A., et al.: Identification of clustered microRNAs using an ab initio prediction method. BMC Bioinformatics **6**, 267–282 (2005)
22. Batuwita, R., Palade, V.: microPred: effective classification of pre-miRNAs for human miRNA gene prediction. Bioinformatics **25**(8), 989–995 (2009)

Gait Image Segmentation Based Background Subtraction

Shanwen Zhang[✉], Ping Li, Le Wang, and Yunlong Zhang

SIAS International University of Zhengzhou University,
Zhengzhou, Henan 451150, China
wjdw716@163.com

Abstract. Gait represents a potential biometric as humans can perceive it. It is attractive because it requires no contact and less likely to be concealed. In gait recognition processes, gait image segmentation is important to final recognition result. In this paper, based on background subtraction, a segmentation method is proposed. The experiment results show the effectiveness of the proposed method.

Keywords: Biometrics characteristics · Gait recognition · Background subtraction · Image segmentation

1 Introduction

Biometrics technologies can verify a person's identity by analyzing human characteristics such as fingerprints, facial and plamprint images, irises, gait, heat patterns, keystroke rhythms, and voice record. Among them, gait as a biometric can be performed at a distance and a low resolution, while other biometric need higher resolution [1–3]. And it is difficult to disguise the gait information by body-invading equipment. Medical studies indicate that gait is unique if all gait movements are considered. In these cases, gait recognition is an attractive biometric and becoming increasingly important for surveillance, control area and medical treatment, etc. More and more researchers have devoted to this field. Because gait is different from other physical biometrics such as face and plamprint in that gait mainly captures the dynamic aspect of human activity instead of the static appearance of human [4]. By extracting profile, a large part of physical appearance features have been removed from the image representation of human. Nevertheless, a profile still contains information about the shape and stance of human body [5–8]. Han and Bhanu et al. [9] argued that selecting the most relevant gait features that are invariant to changes in gait covariate conditions is the key to develop a gait recognition system that works without subject cooperation. Lee and Grimpson [10] proposes an automatic gait recognition approach for analyzing and classifying human gait by computer vision techniques. The approach attempts to incorporate knowledge of the static and dynamics of human gait into the feature extraction process. The gait recognition approaches are evaluated without the cooperative subjects, i.e. both the gallery and the probe sets consist of a mixture of gait sequences under different and unknown covariate conditions. Specifically, Gait

© Springer International Publishing Switzerland 2015
D.-S. Huang et al. (Eds.): ICIC 2015, Part II, LNCS 9226, pp. 563–569, 2015.
DOI: 10.1007/978-3-319-22186-1_56

Entropy Image (GEI) can be used to measure the relevance of gait features [11–13], which represents a gait sequence using a single image; it is thus a compact representation which is an ideal starting point for feature selection. Current approaches to gait recognition can be divided into two categories: Appearance-based ones that deal directly with image statistics and Model-based ones that first model the image data and then analyze the variation of its parameters. The majority of current approaches are the appearance-based approaches which are simple and fast. But the silhouette appearance information is only indirectly linked to gait dynamics. Constructed by computing Shannon entropy for the profile extracted from a gait sequence [13], a GEI can be readily used to distinguish dynamic gait information and static shape information contained in a GEI with the former being selected as features that are invariant to appearance changes. The steps of the gait based identity recognition methods are as, (1) for every gait image, some preprocessing steps must be implemented, i.e., cutting, reducing noise, segmentation, etc.; (2) feature extraction and selection or dimensionality reduction; (3) image classification. Among them, gait image segmentation is the basis for recognition. The segmentation result directly influences the final recognition rate. In this paper, we discuss the gait image segmentation method based background subtraction.

2 Background Subtraction

Generally the interest regions of gait image are human in its foreground. The following analyses make use of the function of G(x, y, t) as a video gait sequence where t is the time dimension, x and y are the pixel location variables, i.e. G(x, y, t) is the pixel intensity at (x, y) pixel location of the image at t in the video sequence. The simplest way to implement the gait detection and segmentation is to take an gait image as background and take the frames obtained at the time t, denoted by G(x, y, t) to compare with the background image denoted by B(x, y, t). This difference gait image would only show some intensity for the pixel locations which have changed in the two frames. By a simple calculation, we can segment the gaits simply by using image subtraction technique of computer vision meaning for each pixels in G(x, y, t), take the pixel value denoted by P[G(x, y, t)] and subtract it with the corresponding pixels at the same position on the background image denoted as P[B(x, y, t)], denoted as follows,

$$P[F(t)] = P[G(x, y, t)] - P[B(t)] \tag{1}$$

By Eq. (1), we have seemingly removed the background, but this approach may be only well for the cases where all foreground pixels are moving and all background pixels are static. A threshold is put on this difference image to improve the subtraction.

$$|P[F(t)] - P[F(t+1)]| > \delta \tag{2}$$

where δ is a threshold.

To extract the gait image containing only the background, a series of preceding gait images are averaged, which likely GEI of gait images. The averaging results refer to averaging corresponding pixels in the given gait images. For calculating the background image at the instant t,

$$B(x,y) = \frac{1}{N} \sum_{i=1}^{N} G(x,y,t-i) \tag{3}$$

where N is the number of preceding images taken for averaging, which depends on the video speed (number of images per second in the video) and the amount of movement in the video.

We can simply use median instead of mean to compute B(x, y). After calculating the background B(x, y), we can subtract it from the image G(x, y, t) at time t and threshold it. The foreground is

$$\left| G(x,y,t) - B(x,y) \right| > \lambda \tag{4}$$

where λ is threshold.

Usage of global and time-independent Thresholds (same λ value for all pixels in the image) may limit the accuracy of the above two approaches.

The probabilistic density function of every pixel is characterized by mean μ_t and variance σ_t^2. The following is a possible initial condition (assuming that initially every pixel is background):

$$\mu_0 = G_0, \sigma_0^2 = \,<\text{some default value}> \tag{5}$$

where G_t is the value of the pixel's intensity at time t.

In order to initialize variance, we can use the variance in (x, y) from a small window around each pixel. In fact, background may change over time because of the illumination changes or non-static background objects. To accommodate for that change, at every frame t, every pixel's mean and variance must be updated, as follows:

$$\begin{aligned} \mu_t &= \rho G_t + (1 - \rho)\mu_{t-1} \\ \sigma_t^2 &= d^2 G_t + (1 - \rho)\sigma_{t-1}^2 \\ d &= |G_t - \mu_t| \end{aligned} \tag{6}$$

where ρ determines the size of the temporal window that is used to fit the probabilistic density function (default value $\rho = 0.01$) and d is the Euclidean distance between the mean and the value of the pixel. A pixel is classified as background if its current intensity lies within some confidence interval of its distribution's mean:

$$\frac{|G_t - \mu_t|}{\sigma_t} > k \rightarrow Foreground, \frac{|G_t - \mu_t|}{\sigma_t} \leq k \rightarrow Background \tag{7}$$

where k is a free threshold (default value k = 2.5).

A larger value for k allows for more dynamic background, while a smaller k increases the probability of a transition from background to foreground due to more subtle changes. A pixel's distribution is only updated if it is classified as background. This is to prevent newly introduced foreground objects from fading into the background. The update formula for the mean is changed accordingly:

$$\mu_t = M\mu_{t-1} + (1 - M)(G_t + (1 - \rho)\mu_{t-1}) \tag{8}$$

where M = 1 when G_t is considered foreground and otherwise M = 0.

Above segmentation processing only works if all pixels are initially background pixels, and it cannot cope with gradual background changes: If a pixel is categorized as foreground for a too long period of time, the background intensity in that location might have changed because the illumination has changed etc.

At any time t, a particular history of pixel (x_0, y_0) is

$$X_1, X_2, \ldots, X_t = \{V(x_0, y_0, t) | 1 \le i \le t\} \tag{9}$$

This history is modeled by a mixture of K Gaussian distributions:

$$P(X_t) = \sum_{i=1}^{K} \varpi_{i,t} N(X_t | \mu_{i,t}, \sum_{i,t}) \tag{10}$$

where

$$N(X_t | \mu_{i,t}, \sum_{i,t}) = \frac{1}{(2\pi)^{D/2}} \frac{1}{\left|\sum_{i,t}\right|^{1/2}} \exp(-\frac{1}{2}(X_t - \mu_{i,t})^T \frac{1}{\sum_{i,t}} (X_t - \mu_{i,t}))$$

First, each pixel is characterized by its intensity in RGB color space. Then probability of observing the current pixel is given by the following formula.

Once the parameters initialization is made, a first foreground detection can be made, and then the parameters are updated. The first B Gaussian distribution which exceeds the threshold T is retained for a background distribution

$$B = \arg \min(\sum_{i-1}^{B} \varpi_{i,t} > T) \tag{11}$$

The other distributions are considered to represent a foreground distribution. Then, when the new frame incomes at times t + 1, a match test is made of each pixel. A pixel matches a Gaussian distribution if the Mahalanobis distance

$$((X_{t+1} - \mu_{t+1})^T \sum_{i-1}^{b} (X_{t+1} - \mu_{t+1}))^{1/2} < k\sigma_{i,t} \tag{12}$$

where k is a constant threshold equal to 2.5.

There are two cases as follows,
Case 1: A match is found with one of the K Gaussians. For the matched component, the update is done as follows

$$\sigma_{i,t+1} = (1 - \rho)\sigma_{i,t}^2 + \rho(X_{t+1} - \mu_{t+1})(X_{t+1} - \mu_{t+1})^T$$

Then use the same algorithm to segment the foreground of the image

$$\sigma_{i,t+1} = (1 - \alpha)\varpi_{i,t} + \alpha P(k|X_t, \phi)$$

The essential approximation to $P(k|X_t, \phi)$ is given by $M(k,t)$.
When $M(k,t) = 1$; match, otherwise $M(k,t) = 0$
Case 2: No match is found with any of the k Gaussians. In this case, the least probable distribution k is replaced with a new one with parameters

$$k_{i,t} = \text{lowPriorWeight}, \mu_{i,t+1} = X_{t+1}, k_{i,t+1} = \text{LargeInitialWeight}$$

Given a human gait sequence, a human profile is extracted from each frame. After normalizing the size and adjusting the horizontal alignment of each extracted profile image, gait cycles are evaluated by estimating gait frequency using a maximum entropy estimation technique presented. Gait energy image (GEI) is then computed by

$$G(x,y) = \frac{1}{T}\sum_{t=1}^{T} G(x,y,t) \tag{13}$$

where T is the number of frames in a complete gait cycle, I is a profile image whose pixel coordinates are given by x and y, and t is the frame number in the gait cycle.
In gait segmentation, we input $G(x,y)$ an input data $G(x,y,t)$.

3 Experiments

A typical gait analysis laboratory has several cameras (video and/or infrared) placed around a walkway or a treadmill or in natural environment, which are linked to one or more computer. The humans walk naturally at any angles and are shot by computer in three dimensions. The gait sequence is obtained for identity recognition. We use a background subtraction method to segment the body's profile from the gait sequence. Firstly, we model the background image. In fact, the static background pixels include non-background pixels mainly due to the shadows, and the profile can contain holes while directly subtracting the foreground image from original image. To solve this problem, some researchers amend the deficient profile. Figure 1 shows an example of profile extraction.

From Fig. 1, we can find that pixels with high intensity values in a GEI correspond to body parts that move little during a walking cycle, and whilst pixels with low intensity values correspond to body parts that move constantly.

Fig. 1. The result of silhouette extraction

4 Conclusions

In this paper, we applied background subtraction to gait segmentation. The results show that the proposed method is effective. There is great scope for future research effort, both in application and development. The developments could be improvements in recognition procedure or in automated technique. As such, future work will establish more precisely the results that can be achieved by this new biometric.

Acknowledgment. This work was supported by the grant of the basic and frontier technology research projects of Henan Province (Nos. 142300410309 & 142400410853), Science and Technology projects of the Henan Department of Education (No. 15A520101), and the grants of the introduction of talent project of SIAS University (2012YJRC01, 2012YJRC02).

References

1. Akita, K.: Image sequence analysis of real world human motion. Pattern Recogn. **17**(1), 73–83 (1984)
2. Murary, M.P., Drought, A.B., Kory, R.C.: Walking pattern of movement. Am. J. Med. **46** (1), 290–332 (1967)
3. Veeraraghavan, A., Chowdhury, A., Chellappa, R.: Role of shape and kinematics in human movement analysis. In: IEEE Conference on Computer Vision and Pattern Recognition, pp. 730–737 (2004)
4. Cunado, D., Nixon, M.S., Carter, J.N.: Automatic extraction and description of human gait models for recognition purposes. Comput. Vis. Image Underst. **90**(1), 1–41 (2003)
5. Yam, C. Y., Nixon, M. S., Carter, J.N.: Gait recognition by walking and running: a model-based approach. In: Proceedings of ACCV2002, pp. 1–6 (2002)
6. Yoo, J.-H., Nixon, M.S., Harris, C. J.: Model-driven statistical analysis of human gait motion. In: Proceedings of ICIP2002, Rochester, pp. 285–288 (2002)
7. Su, H., Huang, F.-G.:. Human gait recognition based on motion analysis. In: Proceedings of the Fourth International Conference on Machine Learning and Cybernetics, pp. 4464–4468 (2005)
8. Bashir, K., Xiang, T., Gong, S.: Gait recognition without subject cooperation. Pattern Recogn. Lett. **31**, 2052–2060 (2010)
9. Han, J., Bhanu, B.: Individual recognition using gait energy image. IEEE Trans. PAMI **28** (2), 316–322 (2006)
10. Lee, L., Grimson, W.: Gait analysis for recognition and classification. In: Proceedings of the Fifth IEEE International Conference on Automatic Face and Gesture Recognition, pp. 155–162, May 2002

11. Collins, R., Gross, R., Shi, J.: Silhouette-based human identification from body shape and gait. In: Proceedings of International Conference on Automatic Face and Gesture Recognition, pp. 366–371, Oct 2002
12. Wang, L., Tan, T.N., Hu, W.M., Ning, H.Z.: Automatic gait recognition based on statistical shape analysis. IEEE TIP **12**(9), 1120–1131 (2003)
13. Sarkar, S., Phillips, P., Liu, Z., Vega, I., Grother, P., Bowyer, K.: The human ID gait challenge problem: data sets, performance, and analysis. IEEE Trans. PAMI **27**(2), 62–177 (2005)

Multi-objective Optimization Method
for Identifying Mutated Driver Pathways
in Cancer

Wu Yang[1], Junfeng Xia[2(✉)], Yan Zhang[1], and Chun-Hou Zheng[1]

[1] College of Electrical Engineering and Automation, Anhui University,
Hefei, Anhui 230601, China
[2] Institute of Health Sciences, Anhui University, Hefei, Anhui 230601, China
jfxia@ahu.edu.cn

Abstract. Genome aberrations in cancer cells can be divided into two types as random 'passenger mutation' and functional 'driver mutation'. Identifying mutated driver genes and driver pathways from cancer genome sequencing data is one of the greatest challenges. In this paper, we introduced a Multi-Objective optimization model based on Genetic Algorithm (MOGA) to solve the so-called maximum weight submatrix problem, which can be used to identify driver genes and driver pathways in cancer. The maximum weight submatrix problem is built on two specific properties, i.e., high coverage and high exclusivity. Those two properties are considered as two inconsistent objectives in our MOGA algorithm. The results show that our MOGA method is effective in real biological data.

Keywords: Mutated driver pathways · Multi-objective optimization · Genetic algorithm

1 Introduction

Cancer is a very complex disease which is largely driven by somatic mutations that include single nucleotide mutations, larger copy-number aberrations, and structural aberrations [1]. And even worse, the cancer cells can invade other tissues and organs through lymphatic system or blood circulation [2]. Hence, cancer has been treated as one of the most serious diseases to human being.

Researchers should understand the molecular mechanisms of cancer before exploring the effective treatment for cancer patients. Therefore, an open challenge is to identify functional mutations, named "driver mutations", which are important for the development in cancer, and filter out the random "passenger mutations", which are neutral to the proliferation of cancer cells [3]. There is no doubt that it is important to design an effective method to identify the driver mutations and pathways in carcinogenesis procession, which is helpful to personalized medicine programs [4].

It is well recognized that pathways rather than individual genes govern the cancer progression. Therefore, shifting the point of view from gene level to pathway level is a key to capture the heterogeneous phenomenon in cancer [5]. Enrichment of somatic

© Springer International Publishing Switzerland 2015
D.-S. Huang et al. (Eds.): ICIC 2015, Part II, LNCS 9226, pp. 570–576, 2015.
DOI: 10.1007/978-3-319-22186-1_57

mutations is a routine method [6]. The disadvantage of this method is that background knowledge of pathways remains incomplete, and those known pathway databases have obvious limitation because they often contain overlapped and unavailable data [7]. Taking these factors into consideration, it's indispensable to design a new algorithm to identify mutated genes as well as driver pathways without relying on prior knowledge absolutely.

In this paper, we proposed a novel Multi-Objective optimization model based on Genetic Algorithm (MOGA) to adjust the trade-off between with higher coverage versus higher overlap in finding gene sets for solving the so-called maximum weight submatrix problem. We applied our MOGA on real biological datasets, and the results show that the proposed MOGA method is effective and accurate in distinguishing the functional "driver mutations" from random "passenger mutations" in cancer pathway.

2 Methods

2.1 Maximum Weight Submatrix Problem

In recent years, detection of mutated driver pathways has been transformed into a maximum weight submatrix problem with two constraints, i.e., 'high coverage', which means a pathway should cover a large number of the patient samples, and 'high exclusivity', that is, the mutations in one pathway should be heterogeneous [8]. In particular, to solve this problem, the Markov Chain Monte Carlo (MCMC) method, which is based on these two properties, was used to detect the driver genes. MCMC method combined these two constraints by using a scoring function. A binary mutation matrix A with m rows (patients) and n columns (genes) based on the somatic mutation data is first constructed, where matrix element $A_{ij} = 1$ indicates gene j is mutated in patient i and $A_{ij} = 0$ indicates the absence of mutation. The scoring function is defined as:

$$W(M) = |\Gamma(M)| - \omega(M) = 2|\Gamma(M)| - \sum_{g \in M} |\Gamma(g)| \tag{1}$$

where M is an $m \times k$ submatrix of mutation matrix A, $\omega(M) = \sum |\Gamma(g) - \Gamma(M)|$ is defined to measure the coverage overlap in a set of M genes. $\Gamma(M) = \bigcup \Gamma(g)$ denotes the coverage of submatrix M. $\Gamma(g) = \{i : A_{ig} = 1\}$ means the gene g is mutated in the set of sample i. For example, we can detect gene set B and C as driver genes when $k = 2$ and $k = 3$ in Fig. 1. The MCMC method didn't evaluate the trade-off between high coverage and high exclusivity. Gene set D may be identified by using GA and MCMC while the heterogeneity of D is low. In order to take both two properties into consideration, a multi-objective optimization method is designed in this work.

2.2 Multi-objective Optimization Method

In practical application, many multi-optimization problems (MOPs) can be characterized by multiple objectives which conflict with each other sometimes. In general, we can formulate a simple single-objective optimization problem as min $f(x), x \in S$, where

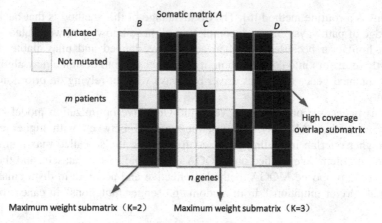

Fig. 1. Illustration of the somatic mutation matrix A. Gene sets B and C with high coverage and high exclusivity, and gene set D with high coverage but no significant heterogeneity.

$f(x)$ is a scalar function and S is the constraints on the variables in the function $f(x)$ which can be defined as $S = \{x \in R, h(x) = 0, g(x) \geq 0\}$. For the same reason, we often describe the multi-objective optimization problem as $\min[f_1(x)f_2(x), \ldots, f_n(x)]$, where n is an integer greater than 1, and S is the set of constraints on the variables in the functions $f_i(x)$.

If we can get an optimal value of $f_i^* = \min\limits_{x \in R} f_j(x)$, $j = 1, 2, \ldots, p$, which is defined to measure each single-objective and named as an ideal point. Then we can construct the ideal point model to calculate the degree that actual result deviates from the optimal solution in theory [9]. The new model is defined as follows:

$$\min \psi[f(x)] = \{[(f_c(x) - f_{cm})/(f_{ch} - f_{cl})]^2 + [(f_o(x) - f_{om})/(f_{oh} - f_{ol})]^2\}^{1/2} \quad (2)$$

where the function $f_c(x)$ is the mathematical quantification of the constraint of high coverage. Similarly, $f_o(x)$ represents the coverage overlap, f_{cm} and f_{om} are the optimal value for the two constraints of high coverage and high exclusivity, respectively. Specifically, normalization method is used to unify the different dimensional data, $(f_{ch} - f_{cl})$ is the range of values of function $f_c(x)$, and $(f_{oh} - f_{ol})$ is the range of values of function $f_o(x)$. The maximum weight submatrix problem is translated as minimum the function $\psi[f_c(x)]$. That is,

$$\min \psi = \{[(|\Gamma(M)| - \Gamma(M)_{\max})/|\Gamma(M)|_{\max}]^2 + [\omega(M)/\omega(M)_{\max}]^2\}^{1/2} \quad (3)$$

The details of GA are described in the reference [10]. Note that, a local search strategy was used to avoid premature convergence and improve the accuracy of our MOGA algorithm.

2.3   Permutation Test

A permutation test method was used to evaluate the significance as follows: the mutation matrix datasets are permuted $N = 1000$ times, and the MOGA method was run on the N randomly generated matrix. We tested the significance of both coverage and exclusivity of the result by measuring the scoring function ψ, which is defined as the degree that the result deviates from the optimal solution in theory. The p-value is calculated as below:

$$pvalue = \frac{\sum_{i=1}^{N} (\psi_{Mi} < \psi_M)}{1000} \qquad (4)$$

where ψ_M is the optimal value obtained by MOGA, M_i is the submatrix selected randomly.

3   Results

In this section, we applied our MOGA method on real biological data to solve the maximum weight submatrix problem. A permutation test method was used to calculate the significance of the result. In order to detect the cooperative pathways, we checked the second optimal patterns when the 'optimal' submatrix was removed. The pipeline of the MOGA algorithm is shown in Fig. 2.

Our MOGA method was used to analyze a dataset of 1013 somatic mutations identified in 623 genes in 188 lung adenocarcinoma samples from the tumor sequencing project [11]. Note that, our MOGA can get the optimal solutions in less than 10 s, while the MCMC method gets the result in more than 60 s.

Firstly, we applied our MOGA method and got the gene pair (EGFR, KRAS) as the optimal result when $k = 2$. When $k = 3$, the triplet (EGFR, KRAS, NTRK3) was obtained as the optimal solution. The pairs (EGFR, KRAS) and the (EGFR, NTRK3) are reported as the significant pairs in mutual exclusivity test in Ding et al. [11], which shows the effectiveness of our model. The p-value of the gene pair (EGFR, KRAS) and

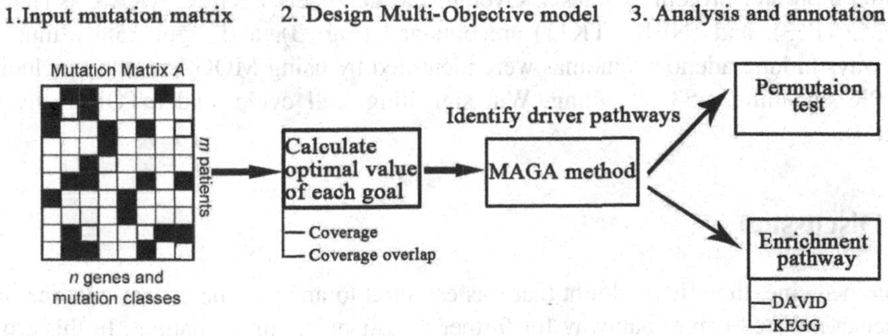

Fig. 2. The MOGA pipeline. The optimal gene set obtained with MOGA is regarded as mutated driver pathway.

Table 1. Results for lung adenocarcinoma when $2 \leq k \leq 4$, the MCMC method leads to the same gene sets with GA method, $\psi 1$ and $\psi 2$ is the fitness value of the sets obtained by GA and MOGA, respectively.

k	GA (MCMC)	MOGA	$\psi 1$	$\psi 2$
k = 2	*EGFR KRAS*	*EGFR KRAS*	0.1	0.1
k = 3	*EGFR KRAS STK11*	*EGFR KRAS NTRK3*	0.3136	0.2063
k = 4	*EGFR KRAS STK11 PRKCG*	*EGFR KRAS NTRK3 NF1*	0.2637	0.2426

the triplet (EGFR, KRAS, NTRK3) are both less than 0.01. Thus, it may be a novel discovery of the coverage and mutual exclusivity of the triplet (EGFR, KRAS, NTRK3). Table 1 reports the results obtained with GA, MCMC and MOGA algorithm, When $k = 3$, the most significant triplet is (EGFR, KRAS, STK11) with GA and MCMC. The KRAS mutated in 60 samples and STK11 mutated in 34 samples. Although the coverage of this gene pair is high, the pair is not mutual exclusive significant, and STK11 and KRAS mutated in 13 patients together. It's obvious that MCMC method can't adjust the trade-off between the high coverage and the high exclusivity well, and $\psi 1$ and $\psi 2$ is the fitness value of the sets obtained by MCMC and MOGA. To assess the quality of our results, we expect to minimize the fitness function ψ.

Note that, the EGFR and the NTRK receptor families show significant exclusivity. NTRK3 is known as one of three high-affinity neurotrophin receptors which regulate cell growth and necrosis of neurons. Recently, it is reported that mutations of NTRK3 may play an important role in neoplasia of lung cancer [12].

Further, we applied the MOGA algorithm to identify more gene sets after removing the gene pair (EGFR, KRAS). On the remaining matrix, the optimal solution is the gene pair (ATM TP53), which is mutated in 76 samples. The p-value of this pair is less than 0.01, so this pair is significant. Previous studies have shown that both ATM and TP53 genes play an important role in the cell cycle progression, and they are involved in the cell cycle checkpoint control and interact directly [13].

To identify more gene sets, we deleted the gene set (EGFR, KRAS, NTRK3, ATM, TP53).The gene set (NF1, STK11) was identified. Note that, STK11 and NF1 are tumor suppressor genes in lung adenocarcinoma [14]. Both of them play an important role in proliferation and protein synthesis. Overall, the gene sets (EGFR, KRAS, NTRK3), (ATM, TP53), and (NF1, STK11) are obtained (Fig. 3), and significantly mutated pathways in lung adenocarcinomas were identified by using MOGA method including MAPK signalling, p53 signalling, Wnt signalling, cell cycle, and mTOR pathways (Fig. 4).

4   Discussion

In biomedicine, there is no doubt that it is essential to analyze the cancer genome data to detect mutated driver pathway for further treatment of cancer patient. In this work, we try to distinguish the driver gene from passenger genes by using an appropriate algorithm named MOGA without relying on the prior knowledge.

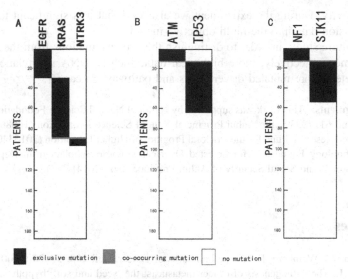

exclusive mutation   co-occurring mutation   no mutation

Fig. 3. Results for lung adenocarcinoma. Black bars mean exclusive mutations, and the gray bars mean co-occurring mutations.

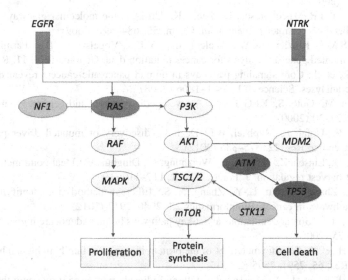

Fig. 4. Illustration of the relevant lung adenocarcinomas signaling pathways of the three gene sets. The blue nodes denote the gene set identified when $k = 3$. The red nodes denote the gene set detected when $k = 2$ after removing gene set (*EGFR, KRAS, NTRK3*). The gray nodes denote the relationship between the gene pair (*NF1, STK11*) (Color figure online).

We applied MOGA method to lung adenocarcinoma data. Compared with MCMC and GA method, our MOGA method shows reliable performance. Note that, multi-objective model focuses on adjusting the trade-off between high coverage and high exclusivity. During the experiment we also find that it is significant to integrate gene expression data into the multi-objective model.

In the future study, in order to distinguish the driver mutations from the passenger mutations more effectively, more biological data, such as DNA methylation, can be added to detect more mutated driver genes and pathways in cancer.

Acknowledgments. This work was supported by National Natural Science Foundation of China (31301101 and 61272339), the Anhui Provincial Natural Science Foundation (1408085QF106), the Specialized Research Fund for the Doctoral Program of Higher Education (20133401120011), and the Technology Foundation for Selected Overseas Chinese Scholars from Department of Human Resources and Social Security of Anhui Province (No. [2014]-243).

References

1. Hanahan, D., Weinberg, R.A.: The hallmarks of cancer. Cell **100**, 57–70 (2000)
2. Fidler, I.J.: The pathogenesis of cancer metastasis: the 'seed and soil 'hypothesis revisited. Nat. Rev. Cancer **3**, 453–458 (2003)
3. Greenman, C., et al.: Patterns of somatic mutation in human cancer genomes. Nature **446**, 153–158 (2007)
4. Overdevest, J.B., Theodorescu, D., Lee, J.K.: Utilizing the molecular gateway: the path to personalized cancer management. Clin. Chem. **55**, 684–697 (2009)
5. Boca, S.M., Kinzler, K.W., Velculescu, V.E., Vogelstein, B., Parmigiani, G.: Patient-oriented gene set analysis for cancer mutation data. Genome Biol. **11**, R112 (2010)
6. Jones, S., et al.: Core signaling pathways in human pancreatic cancers revealed by global genomic analyses. Science **321**, 1801–1806 (2008)
7. Kanehisa, M., Goto, S.: KEGG: kyoto encyclopedia of genes and genomes. Nucleic Acids Res. **28**, 27–30 (2000)
8. Vandin, F., Upfal, E., Raphael, B.J.: De novo discovery of mutated driver pathways in cancer. Genome Res. **22**, 375–385 (2012)
9. Anchen, Q., Linsen, Z., Jianguo, L., Weizhong, L., Dinghua, C.: Ideal point method applied in forest harvest regulation. J. Forest. Res. **8**, 117–119 (1997)
10. Zhao, J., Zhang, S., Wu, L.-Y., Zhang, X.-S.: Efficient methods for identifying mutated driver pathways in cancer. Bioinformatics **28**, 2940–2947 (2012)
11. Ding, L., et al.: Somatic mutations affect key pathways in lung adenocarcinoma. Nature **455**, 1069–1075 (2008)
12. Davies, H., et al.: Somatic mutations of the protein kinase gene family in human lung cancer. Cancer Res. **65**, 7591–7595 (2005)
13. Khanna, K.K., et al.: ATM associates with and phosphorylates p53: mapping the region of interaction. Nat. Genet. **20**, 398–400 (1998)
14. Sanchez-Cespedes, M., et al.: Inactivation of LKB1/STK11 is a common event in adenocarcinomas of the lung. Cancer Res. **62**, 3659–3662 (2002)

A Two-Stage Sparse Selection Method
for Extracting Characteristic Genes

Ying-Lian Gao[1], Jin-Xing Liu[2,3(✉)], Chun-Hou Zheng[4],
Sheng-Jun Li[2], and Yu-Xia Lei[2]

[1] Library of Qufu Normal University, Qufu Normal University, Rizhao, China
yinliangao@126.com
[2] School of Information Science and Engineering, Qufu Normal University,
Rizhao, China
{sdcavell,yx_lei}@126.com, qfnulsj@163.com
[3] Bio-Computing Research Center, Shenzhen Graduate School,
Harbin Institute of Technology, Shenzhen, China
sdcavell@126.com
[4] College of Electrical Engineering and Automation, Anhui University,
Hefei, Anhui, China
zhengch99@126.com

Abstract. In this paper, we view gene selection as a feature selection issue, evaluate the response to the abiotic stresses of the genes in a holistic way and propose the Two-Stage Sparse Selection (TSSS) method for gene data analysis. When performing feature selection, our method properly takes into account the interactions and correlations among thousands of genes. Moreover, the two stages of our method have the following rationale: the first stage discards the genes that have very weak relationships with the abiotic stresses and preserves the others that are likely related to the abiotic stresses. The second stage further enables the remaining genes to show their relationships with the abiotic stresses in a competitive way. The experiments show that our method has promising performance in catching the genes that have the strong relationship with the abiotic stresses.

Keywords: Sparse selection method · Gene selection · Abiotic stresses · Plant genes

1 Introduction

Abiotic stress responses are important for plants because the organisms of this type cannot survive unless they are able to cope with environmental changes [1]. Plants have evolved a number of defending mechanisms that function to increase tolerance to the abiotic factors. To adapt to the changed environment, plants may have a specific set of interacting genes responding to the abiotic stresses.

To obtain the expression pattern of the characteristic genes, many conventional experimental methods were introduced, such as RT-PCR and Northern blotting [2, 3]. These methods are very useful, e.g., RT-PCR can accurately position the genes in tissue or cell and Northern blotting can display the information of detected genes.

© Springer International Publishing Switzerland 2015
D.-S. Huang et al. (Eds.): ICIC 2015, Part II, LNCS 9226, pp. 577–588, 2015.
DOI: 10.1007/978-3-319-22186-1_58

However, these two methods have a common flaw that only a limited amount of genes can be studied at one time. To overcome the flaw, people developed the gene micro-array technology, which makes it possible to monitor gene expression levels on a genomic scale [4].

However, there is another hot issue posed to scientists on how to mine the interesting knowledge from the large number of gene expression data. Feature selection from gene expression data has been studied extensively in the last few years. The most common methods based on feature selection used the following scheme: First, a score for each gene is independently calculated. Then the genes with high scores [5], which are often denoted as univariate feature selection (UFS), are selected. The main virtues of UFS are its simplicity, interpretability and efficiency.

However, a large part of the information contained in the data is lost when genes are selected solely from gene expression data. Unlike UFS, the methods of multivariate feature selection (MFS), also denoted as dimension reduction, use all the gene expression data simultaneously to select the genes. So far, many mathematical methods for MFS, such as principal component analysis (PCA) [6, 7] and partial least squares (PLS) regression, have been demonstrated to be used for gene data analysis. For example, PCA was used to identify differential gene pathways by Ma [8]. De Haan et al. used PCA to analyze microarray data [9]. Cao et al. gave a comparison of PCA and ICA for dimensionality reduction in [10]. PLS was used to identified genes for new diagnostics by Musumarra et al. [11]. Boulesteix used PLS method for feature selection [12]. According to Boulesteix [12], the only well established supervised linear dimension reduction method working even for small-sample-size problem is the Partial Least Squares method (PLS). Feature selection is sometimes criticized for its lack of interpretability, especially for the applied scientists who often need more concrete answers about individual genes.

Recently, a new method for feature selection has been proposed from a novel viewpoint. Such method addresses feature selection by evaluating representation ability on a test sample of each class. The methods in [13, 14] first used a sparse linear combination of the training samples to represent the test sample. Xu et al. adopted a two-phase method to obtain the sparse representation of the test sample [15]. It is noticeable that this method provided a very simple and computationally efficient way to implement the idea of sparse representation and achieved a very good performance in biometrics. We think that the method proposed in [15] is a big progress of image-based biometrics. However, it is not very clearly known whether or not the method can be used in microarray gene expression data.

In order to extract the characteristic genes, we view gene selection as a feature selection issue. Our approach is to evaluate the response to the abiotic stress of the genes in a holistic way, i.e. the following Two-Stage Sparse Selection (TSSS): The first stage acts as a coarse procedure of feature selection, which selects a large number of genes that probably have direct relationships with the abiotic stress and discards the others. The second stage acts as a refining procedure of feature selection and finally determines the genes that are the most related to the abiotic stress. The experimental results show that our method achieves good performance in extracting the plant characteristic genes responding to abiotic stresses. Our work has the following con-tributions: first, it proposes, for the first time, the idea and method of TSSS method.

Second, it clearly presents the rationale explanations for the proposed method. Third, it provides a large number of experiments and the experimental results show that the proposed method is competitive.

2 Methods

In this section, the details of the TSSS method will be presented. Let $\mathbf{X}$ denote an $n \times p$ matrix of real-valued data, which consists of p genes in n samples. In the case of microarray data, x_{ij} is the expression level of the j-th gene in the i-th sample. The elements of the i-th row of $\mathbf{X}$ form the p-dimensional vector $\mathbf{s}_i$, which is referred to as the expression profile of the i-th sample. Correspondingly, the elements of the j-th column of $\mathbf{X}$ form the n-dimensional vector r_j, which is referred to as the transcriptional response of the j-th gene. Usually $p \gg n$ and we face with a classical small-sample-size problem. Our TSSS method employs a two-stage scheme: The first stage of the TSSS acts as a coarse procedure of feature selection and selects a large number of genes that probably have direct relationships with the abiotic stress and discards the others. The second stage acts as a refining procedure of feature selection and finally determines the genes that are the most related to the abiotic stress.

2.1 The First Stage of the TSSS

The first stage of the TSSS uses all of the features to represent each sample and uses sparse principal component analysis (SPCA) to select the features probably associated with the abiotic stress to construct a new feature space. Principal component analysis (PCA) is a well established unsupervised tool for making sense of high dimensional data by reducing it to a smaller dimension. Principal components can be viewed as weighted sum of all the input variables. In the case of gene expression data, each variable represents the expression level of a particular gene. A good analysis tool for biological interpretation should be capable to highlight "simple" structures in the genome-structures. The highlighted "simple" structure had better include a few genes and can explain a significant amount of the specific biological processes encoded in the data. Components that are linear combinations of a small number of variables are quite naturally and usually easier to interpret. It is clear, however, that with this additional goal, some of the explained variance has to be sacrificed. The objective of SPCA is to find a reasonable trade-off between these conflicting goals. For the merits of SPCA, SPCA is used to perform the first sparse selection. Here, we introduce the SPCA via L_0-penalty proposed by Journée et al. in [16].

Let us consider the optimization problem

$$\varphi_{l_0}(\gamma) \stackrel{def}{=} \max_{z \in B^p} z^T \Sigma z - \gamma \|z\|_0 \tag{1}$$

with sparsity-controlling parameter $\gamma > 0$ and sample covariance matrix $\Sigma = X^T X$, $\|\bullet\|_0$ denotes the L_0-norm, that is, the number of non-zero components (cardinality), $B^p = \{z \in R^p | z^T z \leq 1\}$.

According to [16], Eq. (1) can be reformed as:

$$\varphi_{l_0}(\gamma) = \max_{v \in B^n} \max_{z \in B^p} \left(v^T X z\right)^2 - \gamma \|z\|_0, \tag{2}$$

where the maximization with respect to $z \in B^p$ for a fixed $v \in B^n$ has the closed form solution.

$$z_i^* = z_i^*(\gamma) = \frac{[\text{sign}((a_i^T v)^2 - \gamma)]_+ a_i^T v}{\sqrt{\sum_{k=1}^{p} [\text{sign}((a_k^T v)^2 - \gamma)]_+ (a_k^T v)^2}}, \quad i = 1, \ldots, p. \tag{3}$$

In [16], it has been inferred that Eq. (2) can be cast in the following form

$$\varphi_{l_0}(\gamma) = \max_{v \in B^n} \sum_{i=1}^{p} \left[\left(a_i^T v\right)^2 - \gamma\right]_+. \tag{4}$$

Here, for $t \in R$, sign(t) is written for the sign of the argument and $t_+ = \max\{0, t\}$.

Given a solution v^* of Eq. (4), the set of active indices of z^* is given by $I = \{i | (a_i^T v^*)^2 > \gamma\}$.

So, we have

$$\gamma \geq \|a_i\|_2^2 \quad \Rightarrow \quad z_i^*(\gamma) = 0, \quad i = 1, \cdots, p. \tag{5}$$

By choosing an appropriate γ, we can get the first sparse principal components (FPC) C with many entries being zeros. Supposing that the m entries in C are non-zero, we sort the elements of C in descending order in terms of their absolute values and get a new vector $\hat{C}$.

$$\hat{C} = [c_1, \cdots, c_m, \underbrace{0, \ldots, 0}_{p-m}]^T. \tag{6}$$

As shown in Eq. (6), only the genes corresponding to the first m entries of $\hat{C}$ are related to the abiotic stress. So, we utilize $\hat{C}$ to select the genes that probably have direct relationships with the abiotic stress and discard the others. The first m elements of $\hat{C}$ can be selected into a new feature set Φ_{SPCA}. Then to obtain the new data set $\hat{X}$, the data X are projected into Φ_{SPCA}. It is clear that $\hat{X}$ consists of the m genes selected.

The steps of the first stage are shown in Algorithm 1.

Algorithm 1. The steps of the first stage.
1. Given data set $D=\{(X_i,Y_i)\}$, $i=1,\cdots,n$, **features Set**
$\Phi_{all}=\{\Phi_j\}$, $j=1,\cdots,p$, **and** $\gamma>0$.
2. Get the first sparse principal components C **of SPCA.**
3. By sorting the elements of C **in descending order in terms of their absolute values, a new vector** $\hat{C}$ **is gotten.**
4. Select the first m **elements of** $\hat{C}$ **as a new feature set** Φ_{SPCA}.
5. By projecting the data X into Φ_{SPCA}, **the data** $\hat{X}$ **are obtained.**

2.2   The Second Stage of the TSSS

The second stage acts as a refining procedure of feature selection and finally determines the genes that are most related to the abiotic stresses. After the first stage of the TSSS, the feature space has been reduced. Let m and p denote the numbers of selected features and all the features, respectively. Usually, the number of samples, n, is less than m. As a result, we have $n<m<p$. The partial least squares (PLS) method is able to analyze the importance of individual variables. So the PLS regression is utilized to identify the closest related features. In the case of PLS feature selection applied to binary classification problems, the weight vector $\mathbf{w}_1 = (w_{11},\ldots,w_{m1})^T$ defining the first latent component may be used to order the m features in terms of their relevance for the classification problem. Here, we introduce the PLS proposed by Dejong in [17].

Let $\hat{X}$ denote an $n \times m$ matrix, which consists of the n observations of the m features obtained using the first stage of TSSS. In other words, $\hat{X}$ is composed of the elements in matrix $\mathbf{X}$ corresponding to the selected m features. Let $\mathbf{Y}$ be an $n \times 1$ vector to denote the classification label. PLS method decomposes $\hat{X}$ and $\mathbf{Y}$ into the form:

$$\hat{X} = \mathbf{TP}^T + \mathbf{E}, \tag{7}$$

$$\mathbf{Y} = \mathbf{UQ}^T + \mathbf{F}, \tag{8}$$

where the $\mathbf{T}$, $\mathbf{U}$ are $n \times k$ matrices of the k extracted score vectors (components). The $m \times k$ matrix $\mathbf{P}$ and the $1 \times k$ matrix $\mathbf{Q}$ represent matrices of loadings and the $n \times m$ matrix $\mathbf{E}$ and the $n \times 1$ matrix $\mathbf{F}$ are the matrices of residuals. According to [17], $\mathbf{U}$ can be expressed as follows:

$$\mathbf{U} = \mathbf{TD} + \mathbf{H}, \tag{9}$$

where $\mathbf{D}$ is the $k \times k$ diagonal matrix and $\mathbf{H}$ denotes the matrix of residuals. Substitute Eq. (9) into it, Eq. (8) can be rewritten as follows:

$$\mathbf{Y} = \mathbf{TDQ}^T + (\mathbf{HQ}^T + \mathbf{F}).$$ (10)

According to [17], Eq. (10) can be written as:

$$\mathbf{Y} = \hat{\mathbf{X}}\mathbf{B} + \mathbf{F}^*,$$ (11)

where $\mathbf{F}^* = \mathbf{HQ}^T + \mathbf{F}$ is the residual matrix, and $\mathbf{B}$ represents the matrix of regression coefficients

$$\mathbf{B} = \hat{\mathbf{X}}^T \mathbf{U} \left(\mathbf{T}^T \hat{\mathbf{X}} \hat{\mathbf{X}}^T \mathbf{U} \right)^{-1} \mathbf{T}^T \mathbf{Y}.$$ (12)

For $\hat{\mathbf{X}}$ and $\mathbf{Y}$ are an $n \times m$ and $n \times 1$ matrix, respectively, $\mathbf{B}$ must an $m \times 1$ vector. Without lose generality, let

$$\mathbf{B} = [b_1, \ldots, b_m]^T.$$ (13)

Given a test sample $\hat{X}_t = [x_{t1}, \ldots, x_{tm}]$, according to Eq. (11), the following classification result can be obtained.

$$Y_t = \hat{X}_t \mathbf{B} = [x_{t1}, \ldots, x_{tm}][b_1, \ldots, b_m]^T = \sum_{i=1}^{m} x_{ti} b_i.$$ (14)

If omit b_i which is small enough (such as $|b_i| < \varepsilon$), the following prediction result can be obtained.

$$\hat{Y}_t = \sum_{|b_i| \geq \varepsilon} x_{ti} b_i = \sum_{i=1}^{m} x_{ti} b_i - \sum_{|b_i| < \varepsilon} x_{ti} b_i = Y_t - \sum_{|b_i| < \varepsilon} x_{ti} b_i.$$ (15)

The prediction result of Eq. (15) is approximately in agreement with the Eq. (14).

If the elements of $\mathbf{B}$ in descending order in terms of their absolute values are sorted, a new vector of regression coefficients $\tilde{\mathbf{B}}$ is gotten.

$$\tilde{\mathbf{B}} = \left[\tilde{b}_1, \ldots, \tilde{b}_m \right]^T.$$ (16)

If use only the first part of the elements of $\tilde{\mathbf{B}}$ to predict the classification, we can get the nearly same classification result with the Eq. (14). It does mean the corresponding genes to the first part of the elements of $\tilde{\mathbf{B}}$ are most related to the abiotic stresses.

In summary, a scheme of the TSSS method is shown in Algorithm 2.

Algorithm 2. The main steps of the TSSS method.
1. Given data set $D = \{(X_i, Y_i)\}$, $i = 1, \cdots, n$, **features Set**
$\Phi_{all} = \{\Phi_j\}$, $j = 1, \cdots, p$, $\gamma > 0$, $0 < \kappa < 1$, **and** $Num > 0$.

2. By executing Algorithm 1 on the features set Φ_{all}, **the new feature set** Φ_{SPCA} **and data** $\hat{X}$ **are obtained.**

3. By executing PLS algorithm on $\hat{X}$, **the regress coefficients B is obtained.**

4. By sorting the elements of B in descending order in terms of their absolute values, we get a new vector of regression coefficients $\tilde{B}$.

5. Selecting the closest related NUM features according to $\tilde{B}$.

2.3 Rational Explanation

The first stage of our method can greatly reduce the number of features. For example, it can select hundreds of genes from the original hundreds of thousands genes. Its rationale is based upon the following assumption that data sets with thousands of features often contain many irrelevant features. We wish to eliminate the majority of these irrelevant ones before we search for the relevant ones. Therefore, the feature space reduction step is useful for eliminating a large number of irrelevant features.

If the reasonable feature subset is obtained, the data may be much easier to be processed on the reduced feature space than on the complete feature space. Even though this improvement is sometimes small, the result is still impressive since we use a much smaller set of features than the complete set.

3 Results

In this section the TSSS is evaluated by extracting the characteristic genes responding to abiotic stresses.

3.1 Data Source

The raw data were downloaded from NASCArrays [http://affy.arabidopsis.info/], reference numbers are: control, NASCArrays-137; cold stress, NASCArrays-138; osmotic stress, NASCArrays-139; salt stress, NASCArrays-140; drought stress, NASCArrays-141; UV-B light stress, NASCArrays-144; heat stress, NASCArrays-146 [18]. The arrays are adjusted for background of optical noise using the GC-RMA software by Wu et al. [19] and normalized using quantile normalization. The results of GC-RMA are gathered in a matrix to be processed by TSSS.

3.2 Selection of the Parameters

Here, we refer to the sample corresponding to each stress as the positive one label and the control sample as negative one label. We compare our method with the PLS method, since the first PLS component can be used for gene selection and the proposed procedure proved by Boulesteix was equivalent to a well-known gene selection procedure found in the literature [12]. Here PLS is used for gene selection, so the number of components is set to 1, that is, only the first component is obtained.

For our method, it is a crucial problem how to decide the ratio κ of the first sparse selection. We test the different κ on the data set of cold-, drought-, and salt- stresses. As listed in Table 1, while $\kappa \approx 0.1$, the TSSS method achieves the best results for gene selection. According to Sect. 2, $\kappa \approx 0.1$ means that the first stage selects around 10 % genes from the complete features. The number of features has been significantly reduced among three data sets. In first stage of TSSS, we take L_0-norm penalty and set $\gamma = 0.01$, $\kappa = 0.1$. In second stage of TSSS, we roughly set $num = 200$.

3.3 Gene Ontology (GO) Analysis

The Gene Ontology (GO) Term Enrichment tool can be used to find significant shared GO terms, and to describe the genes in the query/input set to help discover what those genes may have in common [20]. The genes selected using TSSS and PLS are checked by GO, its threshold parameters are set as follows: maximum p-value = 0.01 and minimum number of gene products = 2. The remarkable results are listed in Tables 2 and 3.

Table 1. Testing result of the parameter κ

κ	Cold-stress			Drought-stress			Salt-stress		
	Cold	Stimulus[?]	To stress	Water	Stimulus	Stress	Salt	Stimulus	Stress
1	32/312[*]	142/312	90/312	24/336	146/336	103/336	32/371	163/371	115/371
0.9	32/312	142/312	90/312	24/338	148/338	103/338	32/371	163/371	115/371
0.7	32/312	142/312	90/312	24/339	151/339	106/339	32/371	163/371	115/371
0.5	33/312	143/312	91/312	25/340	152/340	107/340	32/371	164/371	116/371
0.4	33/313	143/313	91/313	25/339	151/339	106/339	32/371	164/371	116/371
0.3	33/313	143/313	91/313	27/342	155/342	109/342	32/371	164/371	116/371
0.2	**33/316**	**145/316**	**91/316**	27/345	157/345	109/345	32/371	164/371	116/371
0.1	32/318	143/318	90/318	32/354	166/354	114/354	**32/371**	**164/371**	**116/371**
0.05	32/318	140/318	87/318	**33/355**	**175/355**	**115/355**	32/371	163/371	115/371
0.04	32/318	142/318	89/318	29/357	168/357	108/357	32/371	164/371	116/371
0.03	32/320	143/320	89/320	26/360	167/360	109/360	32/371	164/371	116/371
0.02	34/320	141/320	92/320	23/367	166/367	111/367	32/371	165/372	116/371
0.01	32/320	131/320	92/320	27/381	166/381	120/381	29/378	171/378	114/378

[?]In Table 1, stimulus denotes the response to stimulus of Gene Ontology (GO) Term. Stress denotes the response to stress term.

[*]The data item, e.g. 32/312 denotes that the algorithm selects 312 genes, in which there are 32 genes having been with the response to cold.

Table 2. Response to stimulus (GO:0050896)*

Stress	TSSS		PLS	
	Sample frequency	p-value	Sample frequency	p-value
Cold	145/316, 45.9 %	1.30E-31	140/312, 44.9 %	2.57E-32
Drought	175/355,49.3 %	5.54E-39	144/336, 42.9 %	2.19E-30
Salt	164/371, 44.2 %	2.66E-36	163/371, 43.9 %	2.31E-36
UV-B	157/357, 44.0 %	2.22E-34	156/357, 43.7 %	3.02E-34
Heat	100/306, 32.7 %	2.22E-08	103/308, 33.4 %	2.47E-12
Osmotic	116/331, 35.0 %	3.39E-15	116/331, 35.0 %	8.87E-16

*The response to stimulus (GO:0050896) on characteristic genes obtained by GO Term
Enrichment Analysis are shown, whose background frequency in TAIR set is 4570/29887
(15.3 %), where 4570/29887 denotes having 4570 genes to respond to stimulus in whole 29887
genes set.

As listed in Table 2, according to sample frequency and the genes number of the
response to stimulus, there are 4 out of 6 data sets (cold, drought, salt and UV-B) the
TSSS method exceeds the PLS; Both of them is equal on 1 out of 6 data set (osmotic);
And the TSSS is inferior to PLS on 1 out of 6 data set (heat).

The genes number of the response to stress (GO:0006950) by GO Term Enrichment
Analysis are listed in Table 3. According to sample frequency and genes number, there
are 5 out of 6 data sets (cold, drought, salt, UV-B and heat) TSSS exceeds PLS;
And TSSS is equal to PLS on only 1 out of 6 data set (osmotic).

Moreover, we investigate the characteristic terms concerning straight with the
abiotic stresses. The results are given in Table 4. As listed in Table 4, there are 3 out of
6 data sets (cold, drought and osmotic) TSSS is superior to PLS on the characteristic
terms concerning straight with the stresses. For drought stress, 33 genes are selected
responding to water deprivation with the p-value = 2.63E-23 using TSSS method,
whereas PLS only select 24 genes. For the cold stress, TSSS method can select 33
genes responding to cold with the p-value = 6.77E-20, and PLS select 32 genes. There

Table 3. Response to stress (GO:0006950)*

Stress	TSSS		PLS	
	Sample frequency	p-value	Sample frequency	p-value
Cold	91/316, 28.8 %	4.88E-26	89/312, 28.5 %	1.83E-25
Drought	115/355, 32.4 %	1.12E-34	101/336, 30.1 %	4.65E-31
Salt	116/371, 31.3 %	2.74E-54	115/371, 31.0 %	4.11E-37
UV-B	96/357, 26.9 %	4.84E-26	95/357, 26.6 %	1.34E-24
Heat	83/306, 27.1 %	2.58E-19	82/308, 26.6 %	4.26E-21
Osmotic	89/331, 26.9 %	3.11E-23	89/331, 26.9 %	2.42E-23

*The response to stress (GO:0006950) obtained by GO Term Enrichment Analysis are shown,
whose background frequency in TAIR set is 2351/29887 (7.9 %), where 2351/29887 denotes
having 2351 genes to respond to stress in whole 29887 genes set.

Table 4. Characteristic terms selected from GO by algorithms

Stress	GO Terms	Background frequency	TSSS		PLS	
			Sample frequency	p-value	Sample frequency	p-value
cold	GO:0009410 response to cold	276/29887 (0.9 %)	**33/316** **(10.4 %)**	6.77E-20	32/312 (10.3 %)	3.66E-20
drought	GO:0009414 response to water deprivation	207/29887 (0.7 %)	**33/355** **(9.3 %)**	2.63E-23	24/336 (7.1 %)	7.21E-14
salt	GO:0009651 response to salt stress	395/29887 (1.3 %)	**32/371** **(8.6 %)**	1.80E-13	**32/371** **(8.6 %)**	1.77E-13
UV-B	GO:0009642 response to light intensity	88/29887 (0.3 %)	**8/357** **(2.2 %)**	4.34E-3	**8/357** **(2.2 %)**	4.31E-3
heat	GO:0009408 response to heat	140/29887 (0.5 %)	**42/306** **(13.7 %)**	9.18E-46	42/308 (13.6 %)	1.18E-45
osmotic	GO:0006950 response to osmotic stress	2221/29887 (7.4 %)	**33/331** **(10.0 %)**	9.64E-15	32/332 (9.6 %)	7.83E-14

are 33 genes responding to osmotic stress with p-value = 9.64E-15 selected using TSSS method, and PLS select 32 genes. On the other data sets, both of them give the same results.

Based on the above experimental results and analyses, we can conclude that our proposed method is superior to PLS method.

4   Conclusions

In this paper, the two-stage sparse selection method (TSSS) is proposed. The first and second stages of TSSS play a collaborative role in features selection. The first stage of the method selects the features associated with specific conditions to construct a new feature space and reduces its complexity. The second stage refines the closest related features subset on the new feature space. For processing of gene expression data, the method can achieve more comprehensible and interpretable results and extract the characteristic genes to respond the abiotic stresses.

Furthermore, the genes selected by the methods are checked by GO terms enrichment analysis. The results indicate that the proposed TSSS method is superior to PLS on extracting the characteristic terms concerning straight with the stresses. In the end, the characteristic genes verifications from literatures demonstrate that the proposed TSSS method should be potentially effective.

In future, we will focus on the biological interpretation of the characteristic genes.

Acknowledgements. This work was supported in part by the NSFC under grant Nos. 61370163 and 61272339; China Postdoctoral Science Foundation funded project, No.2014M560264; Shandong Provincial Natural Science Foundation, under grant Nos. ZR2013FL016 and BS2014DX004; Shenzhen Municipal Science and Technology Innovation Council (Nos. JCYJ20140417172417174, CXZZ20140904154910774 and JCYJ20140904154645958).

References

1. Hirayama, T., Shinozaki, K.: Research on plant abiotic stress responses in the post-genome era: past, present and future. Plant J. **61**, 1041–1052 (2010)
2. Tichopad, A., Dzidic, A., Pfaffl, M.W.: Improving quantitative real-time RT-PCR reproducibility by boosting primer-linked amplification efficiency. Biotechnol. Lett. **24**, 2053–2056 (2002)
3. Solanas, M., Moral, R., Escrich, E.: Improved non-radioactive Northern blot protocol for detecting low abundance mRNAs from mammalian tissues. Biotechnol. Lett. **23**, 263–266 (2001)
4. Kilian, J., Whitehead, D., Horak, J., Wanke, D., Weinl, S., Batistic, O., D'Angelo, C., Bornberg-Bauer, E., Kudla, J., Harter, K.: The AtGenExpress global stress expression data set: protocols, evaluation and model data analysis of UV-B light, drought and cold stress responses. Plant J. **50**, 347–363 (2007)
5. Dudoit, S., Shaffer, J.P., Boldrick, J.C.: Multiple hypothesis testing in microarray experiments. Stat. Sci. **18**, 71–103 (2003)
6. Madu, E.A., Uguru, M.I.: Inter-relations of growth and disease expression in pepper using principal component analysis (PCA). Afr. J. Biotechnol. **5**, 1054–1057 (2006)
7. Genet, T., Labuschagne, M.T., Hugo, A.: Genetic relationships among Ethiopian mustard genotypes based on oil content and fatty acid composition. Afr. J. Biotechnol. **4**, 1256–1268 (2005)
8. Ma, S., Kosorok, M.R.: Identification of differential gene pathways with principal component analysis. Bioinformatics **25**, 882–889 (2009)
9. De Haan, J.R., Piek, E., van Schaik, R.C., de Vlieg, J., Bauerschmidt, S., Buydens, L.M.C., Wehrens, R.: Integrating gene expression and GO classification for PCA by preclustering. BMC Bioinformatics **11**, 158 (2010)
10. Cao, L.J., Chua, K.S., Chong, W.K., Lee, H.P., Gu, Q.M.: A comparison of PCA, KPCA and ICA for dimensionality reduction in support vector machine. Neurocomputing **55**, 321–336 (2003)
11. Musumarra, G., Barresi, V., Condorelli, D.F., Fortuna, C.G., Scire, S.: Potentialities of multivariate approaches in genome-based cancer research: identification of candidate genes for new diagnostics by PLS discriminant analysis. J. Chemom. **18**, 125–132 (2004)
12. Boulesteix, A.L., Strimmer, K.: Partial least squares: a versatile tool for the analysis of high-dimensional genomic data. Brief Bioinform. **8**, 32–44 (2007)
13. Wright, J., Yang, A.Y., Ganesh, A., Sastry, S.S., Ma, Y.: Robust face recognition via sparse representation. IEEE Trans. Pattern Anal. Mach. Intell. **31**, 210–227 (2009)
14. Shi, Y., Dai, D.Q., Liu, C.C., Yan, H.: Sparse discriminant analysis for breast cancer biomarker identification and classification. Prog. Nat. Sci. **19**, 1635–1641 (2009)
15. Xu, Y., Zhang, D., Yang, J., Yang, J.Y., Shenzhen, C.: A two-phase test sample sparse representation method for use with face recognition. IEEE Trans. Circuits Syst. Video Technol. **99**, 1–12 (2011)

16. Journée, M., Nesterov, Y., Richtarik, P., Sepulchre, R.: Generalized power method for sparse principal component analysis. J. Mach. Learn. Res. **11**, 517–553 (2010)
17. Dejong, S.: Simpls - an alternative approach to partial least-squares regression. Chemometr. Intell. Lab. **18**, 251–263 (1993)
18. Craigon, D.J., James, N., Okyere, J., Higgins, J., Jotham, J., May, S.: NASCArrays: a repository for microarray data generated by NASC's transcriptomics service. Nucleic Acids Res. **32**, D575–D577 (2004)
19. Wu, Z., Irizarry, R.A., Gentleman, R., Martinez-Murillo, F., Spencer, F.: A model-based background adjustment for oligonucleotide expression arrays. J. Am. Stat. Assoc. **99**, 909–917 (2004)
20. Ashburner, M., Ball, C.A., Blake, J.A., Botstein, D., Butler, H., Cherry, J.M., Davis, A.P., Dolinski, K., Dwight, S.S., Eppig, J.T.: Gene Ontology: tool for the unification of biology. Nat. Genet. **25**, 25–29 (2000)

Identification of Mild Cognitive Impairment Using Extreme Learning Machines Model

Wen Zhang[1,2,3], Hao Shen[1,2,3], Zhiwei Ji[1,2,3], Guanmin Meng[4],
and Bing Wang[1,2,3(✉)]

[1] School of Electronics and Information Engineering,
Tongji University, Shanghai 201804, China
wangbing@ustc.edu
[2] The Advanced Research Institute of Intelligent Sensing Network,
Tongji University, Shanghai 201804, China
[3] The Key Laboratory of Embedded System and Service Computing,
Tongji University, Shanghai 201804, China [4] Department of Clinical Laboratory,
Tongde Hospital of Zhejiang Province,
234th Gucui Road, Hangzhou 310012, China

Abstract. Alzheimer's disease (AD) is very common disorder among the older people. Predicting mild cognitive impairment (MCI), an intermediate stage between normal cognition and dementia, in individuals with some symptoms of cognitive decline may have great influence on treatment choice and disease progression. In this work, a novel prediction model based on extreme learning machines algorithm was proposed to identify mild cognitive impairment using the information of some biomarkers, i.e., Position Emission Tomography (PET), Magnetic Resonance Imaging (MRI), and the cerebrospinal fluid (CSF). Firstly, each subject will be represented by three biomarkers using a vector with 201 dimensions after some essential pre-processing. Then, a prediction model based on extreme learning machine algorithm was trained to catch the difference between the patients and healthy persons. Finally, the model had been employed to identify identifying MCI participants from the normal people. Based on a dataset, including 99 MCI patients and 52 health controls, our proposed method achieved 70.86 % prediction accuracy with 67.68 % sensitivity at the precision of 76.92 %. Extensive experiments are performed to compare our method with state-of-the-art techniques, i.e., support vector machine (SVM) and SVM with fusion kernels. Experimental results demonstrate that proposed extreme learning machine is a powerful tool for predicting MCI with excellent performance and less time.

Keywords: Alzheimer's disease · Mild cognitive impairment · Extreme learning machines · Clinical biomarker · Computational model

© Springer International Publishing Switzerland 2015
D.-S. Huang et al. (Eds.): ICIC 2015, Part II, LNCS 9226, pp. 589–600, 2015.
DOI: 10.1007/978-3-319-22186-1_59

1 Introduction

Alzheimer's disease (AD) is the most common forms of neurodegenerative disorders characterized by a gradual loss of cognitive functions such as episodic memory [1, 2]. The disease is related to pathological amyloid depositions and hyperphosphorylation of structural proteins which leads to progressive loss of function, metabolic alterations and structural changes in the brain [3]. According to the 2013 World Alzheimer report, between 2010 and 2050, the total number of older people with care needs will nearly treble from 101 to 277 million, and around half of all older people who need personal care have dementia. However, there is currently no known treatment that slows the progression of this disorder.

In the past two decades, the development of Position Emission Tomography (PET) and Magnetic Resonance Imaging (MRI), has allowed the non-invasive investigation of the structure and function of the human brain in health and pathology [4]. These techniques have been applied to identify possible biomarkers which could be used for early diagnosis, treatment planning and monitoring of disease progression [5]. MRI captures disease-related structural patterns by measuring loss of brain volume and decreases in cortical thickness. A number of studies, covering region of interest (ROI), volume of interest, voxel-based morphometry and shape analysis, have reported that the degree of atrophy in several brain regions, such as the hippocampus, entorhinal cortex and medial temporal cortex, are sensitive to disease progression and predict MCI conversion [6–11]. Biochemical changes in the brain are reflected in the cerebrospinal fluid (CSF). CSF concentrations of total tau (t-tau), amyloid-β1 to 42 peptide (Aβ1–42) and tau phosphorylated at the threonine 181 (p-tau181p) are considered to be CSF biomarkers which are diagnostic for AD [12–14]. An increase in levels of CSF t-tau and a decline in Aβ1–42 have been identified as being amongst the most promising and informative AD biomarkers [15]. This has revealed structural and functional alterations in several disorders including, amongst others, mild cognitive impairment, probable dementia of Alzheimer type, major depression, bipolar disorder, schizophrenia and generalized anxiety disorder [16–21].

Mild cognitive impairment (MCI), a disorder situated in the spectrum between normal cognition and dementia, is very important for clinical because some patients with MCI will progress to AD, and others not. Identification of MCI participants from normal individuals can offers the opportunity for the enrichment of clinical trails to slow or prevent AD [22, 23]. Some cognitive measurements have shown statistically significant differences between MCI progressors and nonprogressors over the course of 12 months [24]. Currently, there are many efforts had been presented for the early identification of MCI, and different modalities of disease indicators have been studied for AD progression including neuroimaging biomarkers [25–29], biomedical biomarkers [12], and neuropsychological assessments [30, 31].

While most research focuses on a single modality of data, different modalities of data may provide complementary information. Zhang et al. showed that a combination of MRI, CSF and fluorodeoxyglucose positron emission tomography (FDG-PET) predicted MCI converters within 18 months with a sensitivity of 91.5 % and a specificity of 73.4 % (total 99 individuals) [32]. Davatzikos and colleagues analyzed MRI and CSF biomarkers and correctly classified 55.8 % (sensitivity, 94.7 %; specificity, 37.8 %) of 239 individuals using SPARE-AD (Spatial Pattern of Abnormalities for Recognition of Early AD) index [11]. Ewers et al. obtained accuracies from 64 % to 68.5 % for 130 MCI participants with different markers: MRI, CSF, neuropsychological tests, and their combinations [33]. However, the diagnosis accuracy of MCI patients from the healthy people is still low, which will defer the timely treatment of patients. Meanwhile, some methods, such as SVM with the combined kernels, are too complicate to optimize the parameters, and therefore more time needed for achieving satisfyingly computational results [32].

In consideration of both precision and computing time, we proposed a novel computational model in this work for the identification of MCI participants from the normal individuals using extreme learning machine (ELM) method. Here, three biomarkers, i.e., MRI image, PET image and CSF information, had been extracted from each subject after some necessary pre-processing steps. An ELM model will be subsequently trained to capture the difference between the MCI participants and healthy people. To evaluate the performance, the proposed method was applied to a dataset, which including 99 MCIs and 52 health controls (HCs), downloaded from the Alzheimer's Disease Neuroimaging Initiative (ADNI) database. The experiment results show that our method achieved 70.86 % prediction accuracy with 67.68 % sensitivity at the precision of 76.92 % Promisingly, as can be seen that the developed ELM-based MCI prediction system has achieved high accuracy and runs very fast as well.

2 Experiments and Algorithm

2.1 Dataset Preparation

The data used in this work were obtained from the Alzheimer's Disease Neuroimaging Initiative (ADNI) database (adni.loni.usc.edu). As a global research effort that actively supports the investigation and development of treatments that slow or stop the progression of AD, the overall goal of ADNI is to validate biomarkers for use in Alzheimer's disease clinical treatment trials. Data from the North American ADNI's study participants, including Alzheimer's disease patients, mild cognitive impairment subjects and elderly controls, are available from this site. The ADNI study began in 2004 and included 400 subjects diagnosed with mild cognitive impairment (MCI), 200 subjects with early AD and 200 elderly control subjects. In 2009, ADNI1 was extended with ADNI GO which assessed the existing ADNI1 cohort and added 200 participants identified as having early mild cognitive impairment (EMCI). The objective of this phase was to examine biomarkers in an earlier stage of disease. In 2011, as ADNI GO was ending, ADNI2 began with another $67 million in funding. The ADNI general eligibility criteria are described at www.adni-info.org.

In this paper, the dataset extracted by the previous work by Zhang et al. had been used, where only healthy and MCI subjects in ADNI database with all corresponding MRI, CSF and PET baseline data are included [32]. This yields a total of 151 subjects including 99 MCI patients (43 MCI converters who had converted to AD within 18 months and 56 MCI non-converters who had not converted to AD within 18 months), and 52 healthy controls. The description in details of MRI, PET, CSF data, and Image analysis can also find in Zhang's work. For each subject in our dataset, it can be represented by a vector with 189 dimensions, where 99 of them are MRI information, 99 are corresponding PET information and 3 are CSF information.

2.2   Extreme Learning Machines

Feed-forward neural networks (FNN) are ideal classifiers due to their approximation capabilities for nonlinear mappings. However, the dominant algorithms in training FNN are back-propogation (BP) learning algorithm and its variants, which are well known for several challenging issues such as local minima, trivial human intervention and time consuming in learning [34]. The slow learning speed of FNN has been a major bottleneck in different applications. The input weights and hidden layer biases of FNN had to be adjusted using some parameter tuning approach such as gradient descent based methods, which are generally time-consuming due to inappropriate learning steps with significantly large latency to converge to a local maxima. Another important machine learning method, support vector machine (SVM), and its variants achieve surprising performance in both research and classification related applications. In several previous works, SVM algorithms had been employed as classification method to differentiate MCI participants from healthy people [1, 32].

In this work, Extreme learning machines (ELMs), a machine learning method which become attractive to more and more researchers recently, has been used to fix our problem in consideration of both classification precision and computing time [35]. ELM was first proposed for the single-hidden layer feedforward neural networks (SLFNNs) and was then extended to the generalized single-hidden layer feedforward networks where the hidden layer need not be neuron alike [36, 37]. In previous works, Huang et al. proved that SLFNN could exactly learn N distinct observations for almost any non-linear activation function with almost N hidden nods [37–39]. Though ELM's architecture is similar to that of a SLFNN, the input weights and first hidden layer biases need not be adjusted but they are randomly assigned. The ELM algorithm has been proven to perform learning at an extremely fast speed, and obtains good generalization performance for activation functions that are infinitely differentiable in hidden layers.

Give a training dataset $X = \{(x_i, t_i)|x_i \in R^n, x \in R^m, i = 1, \ldots, N\}$, activation function g(x), and hidden neuron number L, the output function of ELM for generalized SLFNNs is

$$o_j = \sum_{i=1}^{L} \beta_i g(w_i \cdot x_j + b_j), \; j = 1, \ldots, N \tag{1}$$

where $w_i = [w_{i1}, w_{i2}, \ldots, w_{in}]^T$ represents the weight vector connecting the ith hidden node and the input nodes, $\beta_i = [\beta_{i1}, \beta_{i2}, \ldots, \beta_{in}]^T$ represents the weight vector connecting the ith hidden neuron and the output neurons, and b_i denotes the threshold of the ith hidden neuron. $w_i \cdot x_j$ denotes the inner product of w_i and x_j.

From the learning point of view, ELM theory aims to reach the smallest training error, which means that $\sum_{j=1}^{N} \|o_j - t_j\| = 0$, i.e., there exist β_i, w_i and b_i such that:

$$\sum_{i=1}^{L} \beta_i g(w_i \cdot x_j + b_i) = \sum_{i=1}^{L} \beta_i h_i(x_j) = t_j, j = 1, \ldots, N \tag{2}$$

Therefore, the training problem has been transformed to find the solution of the equation:

$$H\beta = T \tag{3}$$

where H is the hidden layer output matrix:

$$H = \begin{bmatrix} h(x_1) \\ \vdots \\ h(x_N) \end{bmatrix} = \begin{bmatrix} h_1(x_1) & \cdots & h_L(x_1) \\ \vdots & \ddots & \vdots \\ h_1(x_N) & \cdots & h_L(x_N) \end{bmatrix} \tag{4}$$

and T is the training data target matrix:

$$T = \begin{bmatrix} t_1^T \\ \vdots \\ t_N^T \end{bmatrix} \tag{5}$$

The basic implementation of ELM uses the minimal norm least square method instead of the standard optimization method in the solution:

$$\beta = H^\dagger T \tag{6}$$

where $H^\dagger$ is the Moore-Penrose inverse method for obtaining good generalization performance with extremely fast learning speed.

Unlike the back-propagation or conjugate gradient descent training algorithm, the ELM employs a completely different algorithm for calculating weights and biases. The input weights w_i, and the hidden layer bias are randomly chosen and the output weights β_i are analytically determined based on the Moore-Penrose generalized inverse of the hidden layer output matrix. The algorithm is implemented easily and tends to produce a small training error. It also produces the smallest weights norm, performs well and the learning speed can be thousands of times faster than traditional feed-forward network learning algorithms like back-propagation (BP) algorithm while obtaining better generalization performance.

2.3   10-Fold Cross Validation

In this work, the classifiers to identify the MCI participants from the health controls are constructed based on the ELM algorithm. The ELM classifiers are implemented in two stages: the first one is to train the ELM structure using the training data, and the second is to test the trained ELM structure with the test data. Therefore, to evaluate our proposed method, a 10-fold cross validation strategy has been adopted. Specifically, the dataset we used has been partitioned into 10 subsets with the almost same size. One of these subsets will be selected as test data, and the others as training data. The process will run 10 times in the strategy that each subset can be select once as test data. Here, the activation function in ELM algorithm is Sigmoid, and the parameters are not optimized for comparison with other methods.

3   Results

In this section, we discuss the evaluation strategies used in this work, and present the results using our proposed ELM model in identifying the MCI participants from normal controls. Also, a comparison with other computational methods has been taken to evaluate the effectiveness of our proposed ELM model.

3.1   Evaluation Measures

Four measures have been adopted to evaluate our method of predicting MCIs: sensitivity (SN), specificity (SP), accuracy (Acc) and correlation coefficient (CC). Let TP be the number of true positives, i.e. predicted MCIs that actually are MCIs, and FP be the number of false positives, i.e. predicted MCIs that are in fact health controls. In addition, let TN be the number of true negatives, and FN the number of false negatives. Then the evaluation measures can be computed as follows [40, 41]:

$$\begin{cases} SN = \frac{TP}{TP+FN} \\ SP = \frac{TN}{TN+FP} \\ Acc = \frac{TP+TN}{TP+FN+TN+FP} \\ CC = \frac{TP \times TN - FP \times FN}{\sqrt{(TP+FN)(FP+TN)(TP+FP)(FN+TN)}} \end{cases} \tag{7}$$

The correlation coefficient (CC) is a measure of how well the predicted class labels correlate with the actual class labels. Its range is from -1 to 1, where a correlation coefficient of 1 corresponds to perfect prediction and -1 to the worst possible prediction; a correlation coefficient of 0 corresponds to random guessing.

3.2   Prediction Performance

We evaluated the performance of the proposed ELM model using the data published in ADNI project, where 99 MCIs and 52 HCs included. Three biomarkers, i.e., MRI, PET and CSF, had been used to represent each subject in the dataset. After a series of pre-processing, each subject therefore is represented by a 99 dimensional vector that denote MRI information, 99 dimensional vector for PET information and 3 dimensional vector for CSF information.

To figure out whether the three biomarkers can be complimentary for each other, four ELM predictors had been investigated in this work, i.e., MRI image based-ELM predictor (ELMm), PET image based-ELM predictor (ELMp), CSF information based-ELM predictor (ELMc), and combined information based predictor (ELMmpc). All of the input values were normalized in the range of [−1, 1]. To reduce the bias of training and testing data, a 10-fold cross-validation technique is adopted. More specifically, the dataset is divided into 10 subsets, and the holdout method is reiterated 10 times. Each time nine of the ten subsets are put together as the training dataset, and the other one subset is utilized for testing the model. Thus ten experiments were generated for the ten sets of data. Because the number of MCIs and HCs subjects are very imbalance in our dataset, i.e., the number of HCs is almost half of that of MCIs, there will introduce some overfitting issue into model training if all the samples are used. Therefore, only the same number of MCIs with that of HCs had been selected randomly for model training. Owing to the stochastic method used for selecting MCI samples, the results could rarely be reproduced exactly for the same subject with another cross-validation run. Therefore, the entire cross-validation procedure was repeated five times, and the resulting performances were used to evaluate our method.

The overall prediction performance of our proposed ELM models can be found in the Table 1. It can seen that, for the predictor based on single biomarkers, the prediction capability of MRI and CSF is very close, i.e., the same level of accuracy, very close level of other three measures. Though the SN value of ELMm is a little bit higher than that of ELMc, the SP value is lower. Comparing with MRI and CSF information, PET got the worst measures, i.e., almost 0.1 lower in Acc, SN and SP, as well as 0.2 lower in CC. Even so, the result denoted that the PET information is still a biomarker which can be used for differentiate the MCIs from Healthy people because ELMp can get a CC value of 0.2086.

From Table 1, it can be also found that when the three biomarkers combined together, the best prediction performance can be achieved, i.e., a prediction accuracy value of 0.7086, and a correlation coefficient of 0.4243. Though the SP value, i.e., 0.7692, is slight lower than that of ELMc, the overall performance of ELMmpc is the best among the four predictors, which means the three biomarkers we used in this work can be complimentary when they are used to differentiate the MCIs from HCs. Furthermore, the relative high value of CC denotes that the ELM model can capture the

Table 1. The overall prediction performance of four ELM-based predictors

	Acc	SN	SP	CC
ELM_M	0.6954	0.6768	0.7308	0.3883
ELM_P	0.5960	0.5657	0.6538	0.2086
ELM_C	0.6954	0.6465	0.7885	0.4133
ELM_{MPC}	0.7086	0.6768	0.7692	0.4243

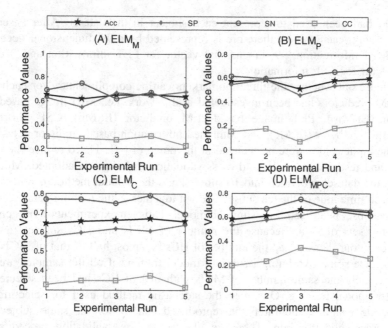

Fig. 1. The four performance measures in different ELM-based predictors.

difference between MCIs and HCs, which demonstrated the effectiveness of our proposed method.

In the training process of every ELM model, the MCI sample was selected stochastically, it is therefore important to infer the influence of this randomness. In this work, the randomness was studied by computing the variance of each performance measure through five repetitions across the 99 MCIs and 52 HCs. The values of four performance measures of the four predictors can be found in the Fig. 1. It can be seen that fluctuations of measure are small. The biggest differences of Acc, SN, SP and CC are 0.08, 0.14, 0.13, and 0.15 respectively among the five experimental runs for the predictor ELMmpc. The very small standard deviations can be got from our experiments, i.e., 0.0365, 0.0578, 0.0490 and 0.0603 for Acc, SN, SP and CC respectively. Obviously, the relative stable prediction performance demonstrate that the ELM model can certainly find the difference between the MCIs and HCs which contained inside of the biomarkers information we used in this work. Furthermore, the small changes among the five repeats of our experiments indicate the robustness of the ELM algorithm, even the training samples are different in each time.

4 Discussion

To infer the effectiveness of our proposed method, we compare the prediction performance with other two approaches, i.e., SVM and SVM with fusion kernel (SVM-fusion) [32]. As a popular machine learning algorithm, SVM had been adopted to solve classification and regression problems in medical or biological studies [42, 43].

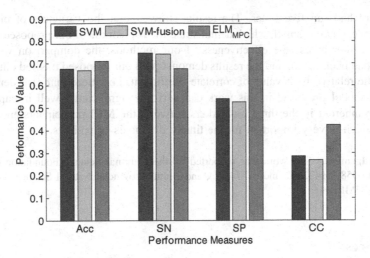

Fig. 2. The comparison of our proposed ELM model with the SVM-based method

To make the methods more comparable, all the SVMs employ the same linear kernel function used in the previous works. Based on the same dataset we used in this work, the prediction performance of different methods can be found in the Fig. 2, where ELMmpc is our proposed classifier. As may be expected, it can be found that our method outperform the SVM predictor, though SN is a little bit lower than SVM. Our proposed method get a 0.03 in Acc, 0.23 in SP, and 0.14 in CC higher than SVM, and 0.03 in Acc, 0.24 in SP, and 0.16 in CC higher than SVM-fusion predictor. Very interesting, the SVM-fusion model did not get the same prediction performance as what Zhang reported, though the same best combination (0.5:0.4:0.1) of three information, i.e., MRI, PET and CSF, had been used here [32]. The difference should be brought from the different toolbox of support vector machine. Zhang et al. implemented their algorithm using lib-svm toolbox. The relative higher value of correlation coefficient is very important for prediction model because it present how fit between the predictive values and the really value, and the more than 0.4 of CC show that ELM model can actually to identify the MCIs from HCs, which is very significant for the clinical diagnosis of patients.

5 Conclusions

In this study, an efficient computational model based extreme learning machine algorithm has been proposed to identify MCIs from health control samples, where three biomarker information, i.e., MRI, PET and CSF, had been extracted from each subject in our dataset as the input features in the model. A 10-fold cross validation strategy had been employed in the training and test process of the ELM model. In the consideration of the imbalance composition of the MCIs and HCs in the original dataset, a same number of MCIs as the HCs had been selected randomly when the ELM model was trained. In order to avoid the influence of random selection of MCI samples, the experiment had run five times. The results showed that the variances of prediction performance are very small, which demonstrate the robustness of our proposed method. Also, to further assess the effectiveness of our methods, the comparison with other methods had been taken, and the results demonstrated our proposed method outperform others. The relative high value of correlate coefficient, i.e., more than 0.4, denote that the ELM model proposed in this work can perform significantly well in capture the difference inherent in the three biomarkers between the MCI participants and healthy people, which is very important for the timely diagnosis of patients.

Acknowledgement. This work was supported by the National Science Foundation of China (Nos. 61300058, 61472282 and 61374181), and Anhui Provincial Natural Science Foundation (No.1508085MF129).

References

1. Westman, E., et al.: Combining MRI and CSF measures for classification of Alzheimer's disease and prediction of mild cognitive impairment conversion. Neuroimage **62**, 229–238 (2012)
2. Abasolo, D., et al.: Approximate entropy and auto mutual information analysis of the electroencephalogram in Alzheimer's disease patients. Med. Biol. Eng. Comput. **46**, 1019–1028 (2008)
3. Weiner, M.W., et al.: The Alzheimer's disease neuroimaging initiative: a review of papers published since its inception. Alzheimers Dement. **9**, e111–e194 (2013)
4. Friston, K.J.: Modalities, modes, and models in functional neuroimaging. Science **326**, 399–403 (2009)
5. Orru, G., et al.: Using support vector machine to identify imaging biomarkers of neurological and psychiatric disease: a critical review. Neurosci. Biobehav. Rev. **36**, 1140–1152 (2012)
6. Costafreda, S.G., et al.: Automated hippocampal shape analysis predicts the onset of dementia in mild cognitive impairment. Neuroimage **56**, 212–219 (2011)
7. Misra, C., et al.: Baseline and longitudinal patterns of brain atrophy in MCI patients, and their use in prediction of short-term conversion to AD: results from ADNI. Neuroimage **44**, 1415–1422 (2009)
8. Li, Y., et al.: Discriminant analysis of longitudinal cortical thickness changes in Alzheimer's disease using dynamic and network features. Neurobiol. Aging **33**(2), 427 (2012)
9. McEvoy, L.K., et al.: Alzheimer disease: quantitative structural neuroimaging for detection and prediction of clinical and structural changes in mild cognitive impairment. Radiology **251**, 195–205 (2009)
10. Cuingnet, R., et al.: Automatic classification of patients with Alzheimer's disease from structural MRI: a comparison of ten methods using the ADNI database. Neuroimage **56**, 766–781 (2010)
11. Davatzikos, C., et al.: Prediction of MCI to AD conversion, via MRI, CSF biomarkers, and pattern classification. Neurobiol. Aging **32**, 2322.e19–2322.e27 (2010)
12. Shaw, L.M., et al.: Cerebrospinal fluid biomarker signature in Alzheimer's disease neuroimaging initiative subjects. Ann Neurol. **65**, 403–413 (2009)
13. Mattsson, N., et al.: CSF biomarkers and incipient Alzheimer disease in patients with mild cognitive impairment. JAMA **302**, 385–393 (2009)

14. Minati, L., et al.: Current concepts in Alzheimer's disease: a multidisciplinary review. Am. J. Alzheimers Dis. Other Demen. **24**, 95–121 (2009)
15. Mattsson, N., et al.: CSF biomarkers and incipient Alzheimer disease in patients with mild cognitive impairment. JAMA-J. Am. Med. Assoc. **302**, 385–393 (2009)
16. Arnone, D., et al.: Magnetic resonance imaging studies in unipolar depression: systematic review and meta-regression analyses. Eur. Neuropsychopharmacol. **22**, 1–16 (2011)
17. Davatzikos, C., Resnick, S.M.: Degenerative age changes in white matter connectivity visualized in vivo using magnetic resonance imaging. Cereb. Cortex **12**, 767–771 (2002)
18. Ellison-Wright, I., Bullmore, E.: Anatomy of bipolar disorder and schizophrenia: a meta-analysis. Schizophr. Res. **117**, 1–12 (2010)
19. Etkin, A., Wager, T.D.: Functional neuroimaging of anxiety: a meta-analysis of emotional processing in PTSD, social anxiety disorder, and specific phobia. Am. J. Psychiatry **164**, 1476–1488 (2007)
20. Smieskova, R., et al.: Neuroimaging predictors of transition to psychosis–a systematic review and meta-analysis. Neurosci. Biobehav. Rev. **34**, 1207–1222 (2010)
21. Zakzanis, K.K., et al.: A meta-analysis of structural and functional brain imaging in dementia of the Alzheimer's type: a neuroimaging profile. Neuropsychol. Rev. **13**, 1–18 (2003)
22. Cui, Y., et al.: Identification of conversion from mild cognitive impairment to Alzheimer's disease using multivariate predictors. PLoS ONE **6**, e21896 (2011)
23. Marmarelis, V.Z., et al.: Model-based physiomarkers of cerebral hemodynamics in patients with mild cognitive impairment. Med. Eng. Phys. **36**, 628–637 (2014)
24. Petersen, R.C., et al.: Alzheimer's disease neuroimaging initiative (ADNI): clinical characterization. Neurology **74**, 201–209 (2010)
25. Risacher, S.L., et al.: Baseline MRI predictors of conversion from MCI to probable AD in the ADNI cohort. Curr. Alzheimer Res. **6**, 347–361 (2009)
26. Davatzikos, C., et al.: Detection of prodromal Alzheimer's disease via pattern classification of magnetic resonance imaging. Neurobiol. Aging **29**, 514–523 (2008)
27. Qiu, A., et al.: Regional shape abnormalities in mild cognitive impairment and Alzheimer's disease. Neuroimage **45**, 656–661 (2009)
28. Vemuri, P., et al.: Alzheimer's disease diagnosis in individual subjects using structural MR images: validation studies. Neuroimage **39**, 1186–1197 (2008)
29. Fan, Y., et al.: Structural and functional biomarkers of prodromal Alzheimer's disease: a high-dimensional pattern classification study. Neuroimage **41**, 277–285 (2008)
30. Chapman, R.M., et al.: Predicting conversion from mild cognitive impairment to Alzheimer's disease using neuropsychological tests and multivariate methods. J. Clin. Exp. Neuropsychol. **33**, 187–199 (2010)
31. Perri, R., et al.: Amnestic mild cognitive impairment: difference of memory profile in subjects who converted or did not convert to Alzheimer's disease. Neuropsychology **21**, 549–558 (2007)
32. Zhang, D., et al.: Multimodal classification of Alzheimer's disease and mild cognitive impairment. Neuroimage **55**, 856–867 (2011)
33. Ewers, M., et al.: Prediction of conversion from mild cognitive impairment to Alzheimer's disease dementia based upon biomarkers and neuropsychological test performance. Neurobiol. Aging **33**, 1203–1214 (2010)
34. Huang, G.B.: An insight into extreme learning machines: random neurons, random features and kernels. Cogn. Comput. **6**(3), 376–390 (2014)
35. Shi, J., et al.: SEMG-based hand motion recognition using cumulative residual entropy and extreme learning machine. Med. Biol. Eng. Comput. **51**, 417–427 (2013)

36. Huang, G.B., Chen, L.: Convex incremental extreme learning machine. Neurocomputing **70**, 3056–3062 (2007)
37. Huang, G.B., et al.: Extreme learning machine: theory and applications. Neurocomputing **70**, 489–501 (2006)
38. Huang, G.B., et al.: Extreme learning machine for regression and multiclass classification. IEEE Trans. Syst. Man Cybern. B Cybern. **42**, 513–529 (2012)
39. You, Z.H., et al.: Prediction of protein-protein interactions from amino acid sequences with ensemble extreme learning machines and principal component analysis. BMC Bioinformatics **14**(Suppl 8), S10 (2013)
40. Wang, B., et al.: Radial basis function neural network ensemble for predicting protein-protein interaction sites in heterocomplexes. Protein Pept. Lett. **17**, 1111–1116 (2010)
41. Wang, B., et al.: Predicting protein interaction sites from residue spatial sequence profile and evolution rate. FEBS Lett. **580**, 380–384 (2006)
42. Hidalgo-Munoz, A.R., et al.: EEG study on affective valence elicited by novel and familiar pictures using ERD/ERS and SVM-RFE. Med. Biol. Eng. Comput. **52**, 149–158 (2014)
43. Kim, K.A., et al.: Mortality prediction of rats in acute hemorrhagic shock using machine learning techniques. Med. Biol. Eng. Comput. **51**, 1059–1067 (2013)

Application of Graph Regularized Non-negative Matrix Factorization in Characteristic Gene Selection

Dong Wang[1], Ying-Lian Gao[2], Jin-Xing Liu[1,3(✉)], Ji-Guo Yu[1],
and Chang-Gang Wen[1]

[1] School of Information Science and Engineering,
Qufu Normal University, Rizhao, China
{dongwshark, sdcavell, chgwenqrnu}@126.com,
jiguoyu@sina.com
[2] Library of Qufu Normal University, Qufu Normal University, Rizhao, China
yinliangao@126.com
[3] Bio-Computing Research Center, Shenzhen Graduate School, Harbin Institute
of Technology, Shenzhen, Guangdong, China
sdcavell@126.com

Abstract. Nonnegative matrix factorization (NMF) has become a popular method and widely used in many fields, for the reason that NMF algorithm can deal with many high dimension, non-negative problems. However, in real gene expression data applications, we often have to deal with the geometric structure problems. Thus a Graph Regularized version of NMF is needed. In this paper, we propose a Graph Regularized Non-negative Matrix Factorization (GRNMF) with emphasizing graph regularized on error function to extract characteristic gene set. This method considers the samples in low-dimensional manifold which embedded in a high-dimensional ambient space, and reveals the data geometric structure embedded in the original data. Experiment results on tumor datasets and plants gene expression data demonstrate that our GRNMF model can extract more differential genes than other existing state-of-the-art methods.

Keywords: Abiotic stresses · Tumor gene expression data · Gene selection · Manifold embed · Non-negative matrix factorization

1 Introduction

With the development of microarray technologies many problems of biomedical will be smoothly done or easily solved. For example, environmental abiotic stresses, that is to say cold, drought, heat and high salinity negatively impact plant growth and development [1, 2]. In order to reduce the adverse effects of these conditions, plants have developed multifaceted strategies, including morphological, physiological, and biochemical adaptations. During the response and adaptation to diverse abiotic stresses, many stress-related genes are induced. Furthermore, a precise identification of the type of tumor is important for effective treatment of cancer. As a result, how to extract the interacting genes responding to abiotic stress and tumor in gene expression data is

© Springer International Publishing Switzerland 2015
D.-S. Huang et al. (Eds.): ICIC 2015, Part II, LNCS 9226, pp. 601–611, 2015.
DOI: 10.1007/978-3-319-22186-1_60

becoming a new challenge. Facing with this problem, lots of traditional methods have been used to analyze gene expression data [3]. Such as, Journée et al. proposed a Sparse principal component analysis (SPCA) method by using generalized power method [4]. Witten et al. proposed penalized matrix decomposition (PMD) [5] method which has been shown to be useful for microarray analysis via imposing penalization on factor matrices [6].

Although those methods have been widely applied, there also existing some drawbacks:

(i) The microarray data typically contain thousands of genes on each sample, so how to reduce the data dimension poses some challenges to the study [7, 8].
(ii) In many real data applications, we usually consider the samples in low-dimensional manifold which embedded in a high-dimensional ambient space, and thus, it is important and necessary to consider the data geometric structure embedded in the original data.

Previous study have shown that the problem in (i) is usually solved by nonnegative matrix factorization (NMF) which has been universally used in many fields and can decipher the high dimension problem. But how to preserve the data geometric structure in learning subspace is another problem.

In this paper, we propose a novel method, Graph Regularized Non-negative Matrix Factorization (GRNMF) algorithm, via imposing the Manifold Regularization on error function [9] to solve the problems simultaneously. GRNMF has been widely used in clustering problem, while it is a new trial in gene expression data for further study. This is the first time to quote this method in gene expression data for gene selection.

Motivated by previous research [10], NMF algorithm is used to solve the non-negative and high dimension problems. The Graph Regularized aims to consider the data geometric structure to obtain parts-based representation [11]. Extensive experiments have been performed on real gene expression data sets and have shown that our method has the advantage in the extracting of genes.

The advantages of this paper are described as follows:

Firstly, in order to avoid the negative data and the high dimension problem, we use the non-negative factorization which can make a low rank data and nonnegative values.

Secondly, the addition of manifold regularization is to find the low dimensional manifold in the high dimensional feature space [12].

The rest of the paper is organized as follows. In Sect. 2, we propose the GRNMF method and demonstrate the efficient algorithm of our method. In Sect. 3, we compare our method with other two methods (PMD and SPCA). Finally, the conclusions are given in Sect. 4.

2 Methodology

2.1 Mathematical Definition of Manifold Regularization

Further studies in graph theory [13] and manifold learning theory [14] have displayed that the data in high-dimension geometric structure can be modeled effectively through a K nearest neighbor graph in the low-dimension data. For each data point X_j, we find its K nearest neighbors and put weight between X_j and its neighbors. There are many methods to define the weight matrix $\mathbf{W}$ on the graph. Three of the most commonly used methods are listed as follows [15]:

(1) 0–1 weighting $\mathbf{W}_{jc} = 1$ if and only if nodes j and c are connected by an edge. This is the simplest weighting method and is very easy to compute.
(2) Heat kernel weighting. If nodes j and c are connected, put

$$\mathbf{W}_{jc} = e^{-\frac{\left\| x_j - x_c \right\|^2}{\sigma}} \tag{1}$$

Heat kernel has an intrinsic connection to the Laplace Beltrami operator on differentiable functions on a manifold [16].
(3) Dot-product weighting. If nodes j and c are connected, put

$$\mathbf{W}_{jc} = \mathbf{x}_j^T \mathbf{x}_c. \tag{2}$$

2.2 Graph Regularization NMF

Given data matrix $\mathbf{X}$, let $\mathbf{X} = (\mathbf{x}_1, \mathbf{x}_2, \ldots, \mathbf{x}_j) \in R^{i \times j}$, $\mathbf{Y} = (\mathbf{y}_1, \mathbf{y}_2, \ldots \mathbf{y}_k)^T \in R^{j \times k}$. The error function of the GRNMF formulation is [17]:

$$\min_{\mathbf{A}, \mathbf{Y}} \left\| \mathbf{X} - \mathbf{A}\mathbf{Y}^T \right\|^2 + \alpha Tr(\mathbf{Y}^T \mathbf{L} \mathbf{Y}), \quad st. \mathbf{Y} > 0, \ \mathbf{A} > 0. \tag{3}$$

where $Tr(\cdot)$ denotes the trace of a matrix and the regularization parameter α controls the smoothness of the new representation [18]. We define $\mathbf{D}$ is a diagonal matrix whose entries are column (or row, since $\mathbf{W}$ is symmetric) sums of $\mathbf{W}$, $\mathbf{D}_{jj} = \sum_c \mathbf{W}_{jc}$. $\mathbf{L} = \mathbf{D} - \mathbf{W}$, which is called graph Laplacian [19].

2.3 An Efficient Algorithm of GRNMF

To solve the constrained optimization problem in Eq. (3), we first introducing Lagrange multiplier ψ_{ik} and ϕ_{jk} by constraint $a_{ik} \geq 0$ and $y_{jk} \geq 0$ respectively [20], $\Psi = [\psi_{ik}]$, $\Phi = [\phi_{jk}]$ the Lagrange L is

$$L = Tr(\mathbf{X}\mathbf{X}^T) - 2Tr(\mathbf{X}\mathbf{Y}\mathbf{A}^T) + Tr(\mathbf{A}\mathbf{Y}^T\mathbf{Y}\mathbf{A}^T) \\ + \alpha Tr(\mathbf{Y}^T \mathbf{L} \mathbf{Y}) + Tr(\Psi \mathbf{A}^T) + Tr(\Phi \mathbf{Y}^T). \tag{4}$$

Using the KKT conditions [21] $\psi_{ik}a_{ik} = 0, \phi_{jk}y_{jk} = 0$, left multiplying the two sides of the derivatives of L with respect to $\mathbf{A}$ and $\mathbf{Y}$ by a_{ik} and y_{jk}, then we obtain the updating roles.

Then the updating roles are listed in the follow:

$$a_{ik} \leftarrow a_{ik} \frac{(\mathbf{XY})_{ik}}{(\mathbf{AY}^T\mathbf{Y})_{ik}} \tag{5}$$

$$y_{jk} \leftarrow y_{jk} \frac{(\mathbf{X}^T\mathbf{A} + \alpha\mathbf{WY})_{jk}}{(\mathbf{YA}^T\mathbf{A} + \alpha\mathbf{DY})_{jk}}. \tag{6}$$

The detail of the method is summarized in Algorithm 1. The iteration procedure is repeated until the algorithm convergence.

Algorithm 1: GRNMF

Input: $\mathbf{X} \in R^{i \times j}$ **and parameter** α .

Output: $\mathbf{Y} \in R^{j \times k}$, $\mathbf{A} \in R^{i \times k}$, $\mathbf{W} \in R^{j \times j}$

1: **Initialize** $\mathbf{A}_0 \in R^{i \times k}$ **and** $\mathbf{Y}_0 \in R^{j \times k}$ **as non-negative matrix. Set r=0.**

2: **repeat**

   Update $\mathbf{A}_{r+1}$ **as** $\mathbf{A}_{r+1} \leftarrow \mathbf{A}_r \dfrac{\mathbf{XY}_r}{\mathbf{A}_r \mathbf{Y}_r{}^T \mathbf{Y}_r}$

   Update $\mathbf{Y}_{r+1}$ **as** $\mathbf{Y}_{r+1} \leftarrow \mathbf{Y}_r \dfrac{(\mathbf{X}^T\mathbf{A}_{r+1} + \alpha\mathbf{WY}_r)_{jk}}{(\mathbf{Y}_r\mathbf{A}_{r+1}{}^T\mathbf{A} + \alpha\mathbf{DY}_r)_{jk}}$

   r=r+1

Until convergence

---

3   Results and Discussion

In order to validate the performance of our method, several experiments on plant gene expression data and tumor expression data based on three methods are carried out to display the performance of our proposed GRNMF method.

In the first subsection, we give the source of gene expression data. How to set the parameter α are demonstrated in the next section. Finally, our method is compared with the following methods and represents a better effect: (a) PMD method (proposed by Witten et al. [22]); (b) SPCA method (proposed by Nyamundanda et al. [23]).

3.1 Results on Gene Expression Data of Plants Responding to Abiotic Stresses

Data Source. The first gene expression data used in our experiment are downloaded from the NASC Arrays [http://affy.arabidopsis.info/] which contain 22810 genes and two classes: shoot and root in each sample, while different stress has different sample numbers. The reference numbers and sample numbers of each stress are listed in Table 1 [24].

The Selection of Parameter α. In this subsection, in order to get the most efficient effort, we use the simulation data to obtain the parameter α. The simulation data are generated with $j = 3000$ genes and $n = 8$ samples. The simulation data are generated as $\mathbf{X} \sim (0, \Sigma_4)$. Let $\tilde{\mathbf{v}}_1 \sim \tilde{\mathbf{v}}_4$ be four 3000-dimensional vectors, such that $\tilde{\mathbf{v}}_{1k} = 1, k = 1, \ldots, 50$, and $\tilde{\mathbf{v}}_{1k} = 0, k = 51, \ldots, 3000$; $\tilde{\mathbf{v}}_{2k} = 1, k = 51, \ldots, 100$, and $\tilde{\mathbf{v}}_{2k} = 0, k \neq 51, \ldots, 100$; $\tilde{\mathbf{v}}_{3k} = 1, k = 101, \ldots, 150$, and $\tilde{\mathbf{v}}_{3k} = 0, k \neq 101, \ldots, 150$; $\tilde{\mathbf{v}}_{4k} = 1, k = 151, \ldots, 200$, and $\tilde{\mathbf{v}}_{4k} = 0, k \neq 151, \ldots, 200$. Let $\mathbf{E}$ be a 3000-dimensional noise matrix, $\mathbf{E} \sim N(0, 1)$. Then the noise matrix with different Signal-to-Noise Ratio (SNR) is added into $\tilde{\mathbf{v}}$ The four eigenvectors of Σ_4 are normalized to be $\tilde{\mathbf{v}}_k = \tilde{\mathbf{v}}_k / \|\tilde{\mathbf{v}}_k\|$, $k = 1, 2, 3, 4$. With these four eigenvectors dominate, we let the eigenvalues be $c_1 = 400$, $c_2 = 300$, $c_3 = 200$, $c_4 = 100$ and $c_k = 1$ for $k = 5, \ldots, 3000$ [25].

The parameter varies from 1 to 10000 in each iterate, after each iterate we calculate the accuracy and select α with the highest accurate. The results are described in Table 2, from the result we can see that when parameter α is 100, we can obtain the best result, so we chose the parameter α as 100.

Gene Ontology (GO) Analysis. In this paper, the tool to evaluate the genes response to the plant abiotic stresses is GO Term [26] which can provides profound information for the biological interpretation of high-through put experiments. The Term Enrichment tool of GO can find the corresponding shared common genes which is available publicly at [http://go.princeton.edu/cgi-bin/GOTermFinderS] [27].

Table 1. The reference No. and sample numbers of each stress type in the data

Stress type	Drought	Salt	UV-B	Cold	Heat	Osmotic	Control
Reference	141	140	144	138	146	139	137
Number	6	7	6	7	8	6	9

Table 2. The selection of parameter α

α	1	10	100	1000	10000
Accuracy (%)	13 %	17 %	31.90 %	28 %	20.60 %

For a fair comparison, 500 genes are selected from gene expression data by GRNMF, PMD and SPCA methods. The threshold parameters are set in listed below: maximum p-value = 0.01 and minimum number of gene sample = 2.

Response to Abiotic Stimulus. In this subsection, we performance the genes responding to abiotic stimulus whose background frequency in TAIR set is 1539/29556 (5.2 %). The results of sample frequency and p-values are shown in Table 3, the p-value is calculated via using the hyper-geometric distribution (details can be seen in [28]). Sample frequency denotes the numbers of core genes selected in the total genes data, for example, 124/500 denotes 124 genes corresponding to the abiotic stimulus in 500 genes selected by the methods. The details are described in Table 3.

There are 12 samples in our experiment which are listed in Table 3 and we compare the three methods with the sample frequency and p-values. From the result we can see

Table 3. Response to abiotic stimulus (GO: 0009628)

Stress type	GRNMF		PMD		SPCA	
	p-value	Sample frequency	p-value	Sample frequency	p-value	Sample frequency
Drought s	6.32E-59	141/500 28.2 %	1.95E-30	107/500 21.40 %	7.50E-21	87/500 17.00 %
Drought r	1.82E-22	107/500 21.4 %	2.37E-13	68/500 13.60 %	4.14E-08	63/500 12.60 %
Salt s	2.71E-50	124/500 24.80 %	1.96E-21	113/500 22.60 %	9.83E-33	105/500 21.00 %
Salt r	7.32E-36	111/500 22.2 %	5.11E-15	78/500 15.60 %	6.18E-12	71/500 14.00 %
UV-B s	9.11E-46	98/500 19.6 %	1.34E-09	74/500 14.80 %	7.84E-23	90/500 18.00 %
UV-B r	1.17E-14	64/500 12.8 %	5.30E-10	67/500 13.40 %	8.00 E-4	52/500 10.00 %
Cold s	5.78E-64	138/500 27.60 %	2.56E-28	106/500 21.60 %	1.17E-19	85/500 17.00 %
Cold r	1.05E-53	125/500 25.00 %	1.50E-20	91/500 18.20 %	4.10E-19	84/500 16.80 %
Heat s	7.16E-22	125/500 25.00 %	1.44E-24	93/500 18.60 %	4.64E-22	89/500 17.80 %
Heat r	9.27E-34	145/500 29.00 %	1.41E-15	78/500 15.60 %	1.35E-08	64/500 12.80 %
Osmotic s	6.35E-56	128/500 25.60 %	3.79E-15	112/500 22.40 %	2.02E-18	83/500 16.60 %
Osmotic r	1.32E-49	119/500 23.80 %	1.40E-16	76/500 15.20 %	2.87E-17	81/500 16.20 %

's' denotes the shoot samples; 'r' denotes the root samples; 'none' denotes that the algorithm cannot give the GO terms

that in eleven of twelve samples, our method performs better than the other two methods, for example, on the drought samples of shoot, whose sample frequency is 28.2 % in our method, 21.4 % in PMD and 17 % in SPCA, which shows that our method gives 7 percent higher than PMD and almost 11 percent higher than SPCA. From Table 3, it demonstrates that our method is far superior to PMD and SPCA ones. Only in one sample set, our method performs worse than PMD method.

The Characteristic Terms. As listed the characteristic terms in Table 4, in all 12 terms, our method outperforms SPCA and PMD in nine items and our method performs

Table 4. Characteristic terms selected from GO by algorithms

Stress type	GO terms	Background frequency	GRNMF	PMD	SPCA
Drought s	GO: 0009414 response to water deprivation	207/29887 0.70 %	**56/500** **11.20 %**	47/500 9.40 %	23/500 4.60 %
Drought r	GO: 0009415 response to water deprivation	207/29887 0.70 %	**50/500** **10 %**	26/500 5.20 %	24/500 4.80 %
Salt s	GO: 0009651 response to salt stress	395/29887 1.30 %	**60/500** **12 %**	41/500 8.20 %	28/500 5.60 %
Salt r	GO: 0009651 response to salt stress	395/29887 1.30 %	**64/500** **12.80 %**	33/500 6.60 %	22/500 4.40 %
UV-B s	GO: 0009416 response to light stimulus	557/29887 1.90 %	**43/500** **8.60 %**	23/500 4.60 %	30/500 6.00 %
UV-B r	GO: 0009416 response to light stimulus	557/29887 1.90 %	none	24/500 4.80 %	none
Cold s	GO: 0009409 response to cold	276/29887 0.90 %	**50/500** **10 %**	44/500 8.80 %	34/500 6.80 %
Cold r	GO: 0009409 response to cold	276/29887 0.90 %	28/500 5.60 %	**43/500** **8.60 %**	33/500 6.60 %
Heat s	GO: 0009408 response to heat	140/29887 0.50 %	**42/500** **8.40 %**	45/500 8.00 %	30/500 6.00 %
Heat r	GO: 0009408 response to heat	140/29887 0.50 %	**45/500** **9 %**	43/500 8.60 %	28/500 5.60 %
Osmotic s	GO: 0006970 response to osmotic stress	474/29887 1.60 %	34/500 6.80 %	**55/500** **11.00 %**	29/500 5.80 %
Osmotic r	GO: 0006970 response to osmotic stress	474/29887 1.60 %	32/500 6.40 %	**39/500** **7.80 %**	27/500 5.40 %

'none' denotes that the algorithm cannot give the GO terms

far beyond the two methods. From the results of experiments, it can be concluded that our method is efficient and effective.

Response to water deprivation (GO: 0009414) only in shoot samples is also analyzed in Table 4. The background frequency of response to water deprivation (GO:

Table 5. The terms of P-values in DLBCL dataset

ID	Name	PMD p-value	SPCA p-value	GRNMF p-value
M1865	Up-regulated genes in the subpopulation of invasive PyMT cells (breast cancer) compared to the general population of PyMT cells	3.67E-21	1.03E-15	**1.68E-22**
M16858	Genes down-regulated in blood samples from bladder cancer patients	5.32E-25	7.25E-13	**2.29E-26**
M5138	Genes up-regulated in comparison of peripheral blood mononuclear cells (PBMC) from healthy donors versus PBMCs from infant with acute influenza infection	5.59E-25	8.51E-13	**1.36E-26**
M19666	Genes down-regulated in Caco-2 cells (intestinal epithelium) after coculture with the probiotic bacteria L. casei for 24 h	1.19E-29	8.47E-14	**2.69E-31**
M2535	The '3/3 signature': genes consistently down-regulated in all three pools of normal mammary stem cells (defined by their ability to retain the dye PKH26)	1.58E-29	1.78E-24	**2.57E-31**
M15555	Protein biosynthesis, transport or catabolism genes up-regulated in hyperploid multiple myeloma (MM) compared to the non-hyperploid MM samples.	7.16E-30	4.50E-14	**4.11E-32**
GO: 0003723	RNA binding	5.59E-38	3.09E-19	**2.56E-39**
M19251	Genes up-regulated in Caco-2 cells (intestinal epithelium) after coculture with the probiotic bacteria L. casei for 6 h	1.37E-41	6.10E-24	**5.74E-44**
M2328	Genes translationally regulated in MEF cells (embryonic fibroblasts) in response to serum starvation and by rapamycin (sirolimus)	2.12E-54	1.56E-18	**3.75E-54**
M11197	Housekeeping genes identified as expressed across 19 normal tissues	6.86E-95	1.96E-87	**1.09E-96**

0009414) is 1.4 %. It is obvious to see that, GRNMF method can extract more feature genes than the others, such as the sample frequency of GRNMF is 11.2 %, 9.4 % in PMD, 4.6 % in SPCA, which demonstrates our method gives 2 percent higher than PMD and almost 8 percent higher than SPCA.

3.2   The Results on Tumor Dataset

In our experiment, we use the Diffuse Large B-Cell Lymphoma (DLBCL) dataset, which is provided in Shipp's data (www.genome.wi.edu/MPR/lymphoma). Shipp et al. applied a supervised learning method on an expression profiling dataset of 7139 genes on 58 tumor specimens, and identified 13 genes that are highly predictive to the outcomes [29].

For a fair comparison, 100 genes are extracted as the differentially expressed genes on the tumor datasets based on the three methods. The Gene Ontology (GO) enrichment of functional annotation of the extracted genes by three methods is detected by ToppFun which is publicly available at [http://toppgene.cchmc.org/enrichment.jsp].

Table 5 lists the top 10 closely related terms of P-value corresponding to different methods on DLBCL tumor datasets. In these tables, it can be found that GRNMF outperforms the other two methods in all the terms.

In summary, after a comparison of abiotic stimulus and the characteristic terms, we can conclude that GRNMF is more superior to the other two methods. It demonstrates that our method is more effective and interpretable in feature selection, especially in the gene expression data.

4   Conclusions

In this paper, we applied a more comprehensible and interpretable core gene extraction method with emphasizing Graph Regularized on error function. Accordingly, by previous work, the manifold regularization can find the inherent law of data from the observations. The non-negative factorization method is used to avoid the high dimension, non-negative problems. To the summary, our method can handle high-dimension, non-negative and manifold embed problems simultaneously. By extensive experiments on the plant gene expression data and tumor expression data, we can conclude that our method performs better than other state-of-the-art methods.

Acknowledgement. This work was supported in part by the NSFC under grant Nos. 61370163, 61373027 and 61272339; China Postdoctoral Science Foundation funded project, No. 2014M560264; Shandong Provincial Natural Science Foundation, under grant Nos. ZR2013FL016 and ZR2012FM023; Shenzhen Municipal Science and Technology Innovation Council (Nos. JCYJ20140417172417174, CXZZ20140904154910774 and JCYJ20140904154645958); the Scientific Research Reward Foundation for Excellent Young and Middle-age Scientists of Shandong Province (BS2014DX004).

References

1. Liu, J.-X., Zheng, C.-H., Xu, Y.: Extracting plants core genes responding to abiotic stresses by penalized matrix decomposition. Comput. Biol. Med. **42**, 582–589 (2012)
2. Liu, J.-X., Liu, J., Gao, Y.-L., Mi, J.-X., Ma, C.-X., Wang, D.: A class-information-based penalized matrix decomposition for identifying plants core genes responding to abiotic stresses. PloS one **9**, e106097 (2014)
3. Livak, K.J., Schmittgen, T.D.: Analysis of relative gene expression data using real-time quantitative pcr and the $2^{-\Delta\Delta CT}$ method. Methods **25**, 402–408 (2001)
4. Journée, M., Nesterov, Y., Richtárik, P., Sepulchre, R.: Generalized power method for sparse principal component analysis. J. Mach. Learn. Res. **11**, 517–553 (2010)
5. Liu, J.-X., Gao, Y.-L., Xu, Y., Zheng, C.-H., You, J.: Differential expression analysis on RNA-Seq count data based on penalized matrix decomposition. IEEE Trans. Nanobiosci. **13**, 12–18 (2014)
6. Yalavarthy, P.K., Pogue, B.W., Dehghani, H., Paulsen, K.D.: Weight-matrix structured regularization provides optimal generalized least-squares estimate in diffuse optical tomography. Med. Phys. **34**, 2085–2098 (2007)
7. Chen, D., Cao, X., Wen, F., Sun, J.: Blessing of dimensionality: high-dimensional feature and its efficient compression for face verification. In: 2013 IEEE Conference on Computer Vision and Pattern Recognition (CVPR), pp. 3025–3032. IEEE (2013)
8. Hall, P., Marron, J., Neeman, A.: Geometric representation of high dimension, low sample size data. J. R. Stat. Soc. Ser. B (Stat. Method.) **67**, 427–444 (2005)
9. Geng, B., Tao, D., Xu, C., Yang, Y., Hua, X.-S.: Ensemble manifold regularization. IEEE Trans. Pattern Anal. Mach. Intell. **34**, 1227–1233 (2012)
10. Lee, D.-C., Wuest, M., McEwan, A., Jans, H.-S.: Dynamic FDG PET images of mice analyzed with a novel non-negative matrix factorization (NMF) technique. In: Society of Nuclear Medicine Annual Meeting Abstracts, p. 1422. Soc Nuclear Med (2008)
11. Li, S.Z., Hou, X., Zhang, H., Cheng, Q.: Learning spatially localized, parts-based representation. In: Proceedings of the 2001 IEEE Computer Society Conference on Computer Vision and Pattern Recognition (CVPR 2001), vol. 201, pp. I-207–I-212. IEEE (2001)
12. Huang, Y., Pei, J., Yang, J., Wang, T., Yang, H., Wang, B.: Kernel generalized neighbor discriminant embedding for SAR automatic target recognition. EURASIP J. Adv. Signal Process. **2014**, 72 (2014)
13. Hammond, D.K., Vandergheynst, P., Gribonval, R.: Wavelets on graphs via spectral graph theory. Appl. Comput. Harmonic Anal. **30**, 129–150 (2011)
14. Carr, J.: Applications of Centre Manifold Theory. Springer, Berlin (1981)
15. Deng, C., Xiaofei, H., Jiawei, H., Huang, T.S.: Graph regularized nonnegative matrix factorization for data representation. IEEE Trans. Pattern Anal. Mach. Intell. **33**, 1548–1560 (2011)
16. Izenman, A.J.: Modern Multivariate Statistical Techniques: Regression, Classification, and Manifold Learning. Springer, New York (2009)
17. Liu, J., Wu, Z., Sun, L., Wei, Z., Xiao, L.: Hyperspectral image classification using kernel sparse representation and semilocal spatial graph regularization. IEEE Geosci. Remote Sens. Lett. **11**, 1320–1324 (2014)
18. Roughgarden, T., Schoppmann, F.: Local smoothness and the price of anarchy in splittable congestion games. J. Econ. Theory **156**, 317–342 (2014)

19. Liu, X., Zhai, D., Zhao, D., Zhai, G., Gao, W.: Progressive image denoising through hybrid graph laplacian regularization: a unified framework. IEEE Trans. Image Process. Publ. IEEE Signal Process. Soc. **23**, 1491–1503 (2014)
20. Wu, G.-C., Baleanu, D.: Variational iteration method for the burgers' flow with fractional derivatives—new lagrange multipliers. Appl. Math. Model. **37**, 6183–6190 (2013)
21. Facchinei, F., Kanzow, C., Sagratella, S.: Solving quasi-variational inequalities via their KKT conditions. Math. Program. **144**, 369–412 (2014)
22. Witten, D.M., Tibshirani, R., Hastie, T.: A penalized matrix decomposition, with applications to sparse principal components and canonical correlation analysis. Biostatistics (2009). doi:10.1093/biostatistics/kxp008
23. Nyamundanda, G., Gormley, I.C., Brennan, L.: A dynamic probabilistic principal components model for the analysis of longitudinal metabolomics data. J. R. Stat. Soc. Ser. C (Appl. Stat.) **63**(5), 763–782 (2014)
24. Craigon, D.J., James, N., Okyere, J., Higgins, J., Jotham, J., May, S.: NASCArrays: a repository for microarray data generated by NASC's transcriptomics service. Nucleic Acids Res. **32**, D575–D577 (2004)
25. Shen, H., Huang, J.Z.: Sparse principal component analysis via regularized low rank matrix approximation. J. Multivar. Anal. **99**, 1015–1034 (2008)
26. Jenks, M.A., Hasegawa, P.M.: Plant Abiotic Stress. Wiley, Hoboken (2008)
27. Feigelman, J., Theis, F.J., Marr, C.: MCA: multiresolution correlation analysis, a graphical tool for subpopulation identification in single-cell gene expression data. arXiv preprint. arXiv:1407.2112 (2014)
28. Dinkla, K., El-Kebir, M., Bucur, C.-I., Siderius, M., Smit, M.J., Westenberg, M.A., Klau, G.W.: eXamine: exploring annotated modules in networks. BMC Bioinformatics **15**, 201 (2014)
29. Tembhare, P.R., Subramanian, P.G., Sehgal, K., Yajamanam, B., Kumar, A., Gujral, S.: Hypergranular precursor B-cell acute lymphoblastic leukemia in a 16-year-old boy. Indian J. Pathol. Microbiol. **52**(3), 421 (2009)

Graph Regularized Non-negative Matrix with L₀-Constraints for Selecting Characteristic Genes

Chun-Xia Ma[1], Ying-Lian Gao[2], Dong Wang[1], Jian Liu[3],
and Jin-Xing Liu[1,4(✉)]

[1] School of Information Science and Engineering,
Qufu Normal University, Rizhao, China
{mcxia87,dongwshark,sdcavell}@126.com
[2] Library of Qufu Normal University,
Qufu Normal University, Rizhao, China
yinliangao@126.com
[3] School of Communication,
Qufu Normal University, Rizhao, China
liujiansqjxt@126.com
[4] Bio-Computing Research Center, Shenzhen Graduate School,
Harbin Institute of Technology, Shenzhen, Guangdong, China
sdcavell@126.com

Abstract. Non-negative Matrix Factorization (NMF) has been widely concerned in computer vision and data representation. However, the penalized and restriction L_0-norm measure are imposed on the NMF model in traditional NMF methods. In this paper, we propose a novel graph regularized non-negative matrix with L_0-constraints (GL₀NMF) method which comprises the geometrical structure and a more interpretation sparseness measure. In order to extract the characteristic gene effectively, the steps are shown as follows. Firstly, the original data $\mathbf{Q}$ is decomposed into two non-negative matrices $\mathbf{F}$ and $\mathbf{P}$ by utilizing GL₀NMF method. Secondly, characteristic genes are extracted by the sparse matrix $\mathbf{F}$. Finally, the extracted characteristic genes are validated by using Gene Ontology. In conclusion, the results demonstrate that our method can extract more genes than other conventional gene selection methods.

Keywords: Non-negative matrix factorization · Gene expression data · Gene selection · Characteristic genes

1 Introduction

With the development of DNA microarray technology, tens of thousands of genes expression data can be measured at the same time and they are shown in a matrix. In the matrix, the numbers of samples are much lower than the number of genes. In these massive data, how to extract the characteristic gene efficiently is still a challenge in the biological treatment process.

© Springer International Publishing Switzerland 2015
D.-S. Huang et al. (Eds.): ICIC 2015, Part II, LNCS 9226, pp. 612–622, 2015.
DOI: 10.1007/978-3-319-22186-1_61

Many mathematical methods have been used to analyze gene expression data [1–8], such as, principal component analysis (PCA), independent component analysis (ICA), singular value decomposition (SVD) and non-negative matrix factorization (NMF). High dimensional data are mapped into a low dimensional space by these matrices decomposition methods. In gene expression data analysis, PCA is an unsupervised method, which has been extensively applied to various fields [1, 9]. The penalized discriminant method based on ICA is proposed for tumor classification by Huang and Zheng [4]. ICA [3] is used as an extension of PCA. SVD was introduced by Alter et al. for modeling and processing gene expression data [10]. However, these methods have some common defects: They allow the negative component exists and need to standardize the original data.

To seek a better solution, the Non-negative Matrix Factorization (NMF) method was proposed by Lee and Seung, which has been used to decompose the image matrix [11]. NMF has the characteristics of nonnegative dimensionality reduction and it usually involves some simple operations. So NMF has been extensively applied in various fields. The corresponding versions of NMF have been proposed [12–14], such as, Sparse NMF [15], Fisher NMF [16], Non-negative Matrix Factorization with Sparse Constraints (NMFSC) [17] and Graph Regularized Non-negative Matrix Factorization (GNMF) [18]. These methods have been widely used in gene expression data analysis [19], images processing [20], genes selection [21, 22] and so on. Among them, GNMF takes into account the manifold learning on the basis of NMF (GNMF) for data representation. It has been demonstrated that it has certain advantages over PCA and SVD methods [18]. However, these methods impose sparsity on $\mathbf{F}$ and $\mathbf{P}$ by penalized and restriction L_1-norm measure. Indeed, L_0-norm permits specified number of non-zero entries, and it is more natural sparseness measure than others [23, 24]. In addition, a handful of work has been done using L_0-norm. So we put forward a graph regularized non-negative matrix with L_0-constraints (GL$_0$NMF) method, which introduces manifold learning and constraint L_0-norm. The scheme of GL$_0$NMF is given as follows: Firstly, the sparse matrix $\mathbf{F}$ is obtained by GL$_0$NMF method; Secondly, the characteristic genes are extracted via the sparse matrix $\mathbf{F}$; Thirdly, the characteristic genes are evaluated by Gene Ontology.

This paper is organized as follows. In Sect. 2, we introduce the methodology of GL$_0$NMF. Section 3 provides experimental results and discussion. Section 4 concludes the paper.

2 Methods

2.1 Mathematical Definition of GL$_0$NMF

Given a data matrix $\mathbf{Q} = [\mathbf{q}_1, \mathbf{q}_2, \ldots, \mathbf{q}_n] \in \mathrm{R}^{m \times n}$, $\mathbf{Q}$ is an $m \times n$ non-negative matrix, each column of $\mathbf{Q}$ contains m genes, each row of $\mathbf{Q}$ contains n samples. Similar to GNMF, our GL$_0$NMF aims to find two non-negative matrices $\mathbf{F} = [f_{ik}] \in R^{m \times k}$ and $\mathbf{P} = [p_{jk}] \in \mathrm{R}^{n \times k}$. Here we use the Euclidean distance, GL$_0$NMF model is shown as the following objective function:

$$\mathbf{Z} = \left\| \mathbf{Q} - \mathbf{FP}^T \right\|_F^2 + \xi Tr(\mathbf{P}^T \mathbf{LP}), \forall \mathbf{F} \geq 0, \mathbf{P} \geq 0, \|\mathbf{F}\|_0 \leq s, \tag{1}$$

where ξ is the non-negative regularization parameter, $\mathbf{L} = \mathbf{B} - \mathbf{C}$, $\mathbf{C} \in \mathbf{R}^{n \times n}$ is a weight matrix on the graph with '0–1 weighting'. $\mathbf{B} \in \mathbf{R}^{n \times n}$ is a diagonal matrix, $\mathbf{B}_{ii} = \sum_{j=1}^{n} \mathbf{C}_{ij}$, $\mathbf{B}_{ii}$ is the row sum of the matrix $\mathbf{C}$. $\mathbf{L} \in \mathbf{R}^{n \times n}$ is named graph Laplacian matrix. Parameter s is the maximal permitted number of non-zero entries of $\mathbf{F}$.

The optimization problem can be described as the following:

$$\min_{F,P} \left\| \mathbf{Q} - \mathbf{FP}^T \right\|_F^2 + \xi Tr(\mathbf{P}^T \mathbf{LP}), \forall \mathbf{F} \geq 0, \mathbf{P} \geq 0, \tag{2}$$

s. t. optional constraint:

$$\|\mathbf{F}\|_0 \leq s, \forall i. \tag{3}$$

2.2 The Update Rules of GL$_0$NMF

Similar to GNMF, the objective function with L_0-constraints in Eq. (1) can be described as the following:

$$\begin{aligned} Y &= Tr((\mathbf{Q} - \mathbf{FP}^T)(\mathbf{Q} - \mathbf{FP}^T)^T) + \xi Tr(\mathbf{P}^T \mathbf{LP}) \\ &= Tr(\mathbf{Q}^T \mathbf{Q}) - 2Tr(\mathbf{QPF}^T) + Tr(\mathbf{FP}^T \mathbf{PF}^T) + \xi Tr(\mathbf{P}^T \mathbf{LP}), \end{aligned} \tag{4}$$

where the equality applies the matrix properties $Tr(\mathbf{AB}) = Tr(\mathbf{BA})$ and $Tr(\mathbf{B}) = Tr(\mathbf{B}^T)$. Let γ_{ik} and ϕ_{jk} be the Lagrange multiplier for constraint $\mathbf{f}_{ik}$ and $\mathbf{p}_{jk}$ respectively. Let $\mathbf{F} = [\mathbf{f}_{ik}] \geq 0$, $\mathbf{P} = [\mathbf{p}_{jk}] \geq 0$, $\Upsilon = [\gamma_{ik}]$ and $\Phi = [\phi_{jk}]$. The Lagrange $\mathbf{O}$ can be rewritten as:

$$\begin{aligned} \mathbf{O} &= Tr(\mathbf{Q}^T \mathbf{Q}) - 2Tr(\mathbf{QPF}^T) + Tr(\mathbf{FP}^T \mathbf{PF}^T) \\ &\quad + \xi Tr(\mathbf{P}^T \mathbf{LP}) + Tr(\Upsilon \mathbf{F}^T) + Tr(\Phi \mathbf{P}^T), \end{aligned} \tag{5}$$

The partial derivatives of $\mathbf{O}$ with respect to $\mathbf{F}$ and $\mathbf{P}$ are set as follows:

$$\frac{\partial \mathbf{O}}{\delta \mathbf{F}} = -2\mathbf{QP} + 2\mathbf{FP}^T \mathbf{P} + \Upsilon, \tag{6}$$

$$\begin{aligned} \frac{\partial \mathbf{O}}{\delta P} &= -2\mathbf{Q}^T \mathbf{F} + 2\mathbf{PF}^T \mathbf{F} + 2\xi \mathbf{LP} + \Phi \\ &= -2\mathbf{Q}^T \mathbf{F} + 2\mathbf{PF}^T \mathbf{F} + 2\xi (\mathbf{B} - \mathbf{C})\mathbf{P} + \Phi \\ &= -2(\mathbf{Q}^T \mathbf{F} + \xi \mathbf{CP}) + 2\mathbf{PF}^T \mathbf{F} + 2\xi \mathbf{BP} + \Phi \end{aligned} \tag{7}$$

Using the conditions of Karush-Kuhn-Tucker (KKT): $\gamma_{ik}\mathbf{f}_{ik} = 0$ and $\phi_{jk}\mathbf{p}_{jk} = 0$. At the same time, let the Eqs. (6) and (7) be equal to zero. On both sides of these gotten equation are multiplied by $\mathbf{f}_{ik}$ and $\mathbf{p}_{jk}$, respectively. We get the following equations for $\mathbf{f}_{ik}$ and $\mathbf{p}_{jk}$:

$$-(\mathbf{QP})_{ik}\mathbf{f}_{ik} + (\mathbf{FP}^T\mathbf{P})_{ik}\mathbf{f}_{ik} = 0, \tag{8}$$

$$-(\mathbf{Q}^T\mathbf{F} + \xi\mathbf{CP})_{jk}\mathbf{p}_{jk} + (\mathbf{PF}^T\mathbf{F} + \xi\mathbf{BP})_{jk}\mathbf{p}_{jk} = 0 \tag{9}$$

By using the Lagrange function and the Karush-Kuhn-Tucker (KKT) conditions, the corresponding updating rules for $\mathbf{F}$ and $\mathbf{P}$ can be obtained:

$$\mathbf{f}_{ij} \leftarrow \mathbf{f}_{ij}\frac{(\mathbf{QP})_{ik}}{(\mathbf{FP}^T\mathbf{P})_{ik}}, \tag{10}$$

$$\mathbf{p}_{jk} \leftarrow \mathbf{p}_{jk}\frac{(\mathbf{Q}^T\mathbf{F} + \xi\mathbf{CP})_{jk}}{(\mathbf{PF}^T\mathbf{F} + \xi\mathbf{BP})_{jk}}. \tag{11}$$

2.3   The Algorithm

The details of the GL$_0$NMF algorithm are listed as follows:

1. **Initialize P randomly**
2. **For** $i =1$**: numIter do**
      $F^T = NNLS(\mathbf{Q}^T\mathbf{P})$
   End
3. **Normalize F and P:**
      [F, P] = NormalizeFP (F, P, 2, 0)
4. **For** $j =1$**:**k **do**
         Set $m-s$ **smallest values in F to zero**
      End
5. **Repeat**
         1). Update F by rule (10)
         2). Update P by rule (11)
      Convergence
6. **Extracting characteristic genes by GL$_0$NMF**

The gene expression data matrix $\mathbf{Q}$ is an $m \times n$ matrix, rows represent the expression level of the m genes in n samples, each column of $\mathbf{Q}$ represents the expression level of m genes across one sample. The gene expression data matrix $\mathbf{Q}$ can be written as: $\mathbf{Q} \sim \mathbf{FP}^T$, where $\mathbf{F}$ is an $m \times k$ non-negative basis matrix. $\mathbf{P}$ is an $n \times k$ non-negative coefficient matrix. Here, $m > > n$ and $k < min(m,n)$. The optimization problem is convex in $\mathbf{F}$ and $\mathbf{P}$ separately and various optimizations have been testified in [18, 25]. The sample expression profile $\mathbf{l}_i$ can be denoted as:

$$\mathbf{l}_i = \sum_{j=1}^{k} \mathbf{f}_i \mathbf{p}_{ji}^T, i = 1, 2, \ldots, n, \tag{12}$$

where $\mathbf{l}_i$ is the row of $\mathbf{Q}$ and it is a linear combination of the metasamples $\{\mathbf{f}_i\}$. $\mathbf{p}_{ji}^T$ is the entry of $\mathbf{P}^T$. Rows of $\mathbf{P}$ are considered as coefficient vectors. The matrix $\mathbf{F}$ is called basis vectors (metasample). The relationships between these matrices are presented in Fig. 1.

In the experiment, the goal is to extract characteristic genes responding to the abiotic stresses from the matrix with the high dimension small sample. For saving the computational complexity, we use the part of sample characteristics to replace $\mathbf{Q}$. As Fig. 1 shows the matrix $\mathbf{F}$ contains all genes and it is one subset of metasample of $\mathbf{Q}$. So we can extract characteristic genes from the basis matrix $\mathbf{F}$. In Fig. 1, $\mathbf{r}_j$ and $\mathbf{l}_i$ represent the samples characteristic and characteristic vectors of the matrix $\mathbf{Q}$, a_{ij} shows the expression level of the i-th gene in the j-th sample. $\hat{\mathbf{l}}_i$ and $\hat{\mathbf{r}}_j$ are the i-th characteristic vector and j-th sample of $\mathbf{F}$, respectively. $\tilde{s}_i$ and $\tilde{r}_j$ refer to the i-th characteristic vector and j-th sample of $\mathbf{P}^T$, respectively, which consists of k genes in n samples. In a word, $\mathbf{l}_i$ can be replaced by $\mathbf{f}_i$. By giving the parameter s of GL_0NMF, the sparse matrix $\mathbf{F}$ can be gained. So, the characteristic genes can be extracted from the non-zero entries of $\mathbf{F}$.

In the end, the steps of obtaining characteristic genes by GL_0NMF method are listed as the following:

(1) Randomly initialize the non-negative matrix $\mathbf{P}$.
(2) The initialize matrix $\mathbf{F}$ is obtained by least square method (NNLS).
(3) Normalize $\mathbf{F}$ and $\mathbf{P}$.
(4) The matrix $\mathbf{F}$ is imposed on L_0-constraints.
(5) Iterative multiplication of matrix $\mathbf{F}$ and $\mathbf{P}$.
(6) Extract the characteristic genes in the non-zero entries of $\mathbf{F}$.
(7) The characteristic genes are checked by Gene Ontology tool.

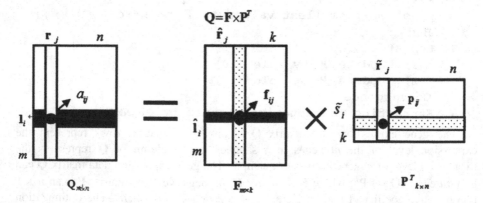

Fig. 1. The graph description of the matrix $\mathbf{Q}$ with the factors $\mathbf{F}$ and $\mathbf{P}$.

3 Results and Discussion

In this section, the results on simulation data and gene expression data sets are indicated by GL_0NMF method. In addition, our method will compare with GNMF [18], PMD [6] and SPCA [2] methods.

3.1 Data Source

In our simulation experiment, the size of simulation data X is 5000×30. The simulation data contains 30 samples and 5000-dimensional vectors. The simulation data is generated as $X \sim (0, \sum_4)$. $\tilde{v}_1 \sim \tilde{v}_4$ are considered as four 5000-dimensional vectors, such that $\tilde{v}_{1k} = 1, k = 1, \ldots, 125$, and $\tilde{v}_{1k} = 0, k = 126, \ldots, 5000; \tilde{v}_{2k} = 1, k = 126, \ldots, 250$, and $\tilde{v}_{2k} = 0, k \neq 126, \ldots, 250; \tilde{v}_{3k} = 1, k = 251, \ldots, 375$, and $\tilde{v}_{3k} = 0, k \neq 251, \ldots, 375;$ $\tilde{v}_{4k} = 1, k = 376, \ldots, 500$, and $\tilde{v}_{4k} = 0, k \neq 376, \ldots, 500$. The noise matrix is added into $\tilde{v}$ with different Signal-to-Noise Ratio (SNR) [26]. The first four eigenvectors of $\sum_4$ are chosen to be $v_k = \tilde{v}_k / \|\tilde{v}_k\|$, $k = 1, 2, 3, 4$. The eigenvalues $c_1 = 400$, $c_2 = 300$, $c_3 = 200, c_4 = 100$ and $c_k = 1$ for $k = 2, \ldots, 5000$ are set to make the four eigenvectors dominate. Then, the simulation scheme in [26] is serves to generate the simulation data.

The NASCArrays [http://affy.arabidopsis.info/] offers the gene expression data. We can download the data from this website. The size of data is 22810×136 and six stress types and the reference numbers are listed in [27]. In our experiment, the corresponding background light noise of these data can be fitted by using the GC-RMA method which was proposed by Wu et al. [28]. The GC-RMA results are collected in a matrix to be further processing the Gene Ontology (GO) TermFinder tool [29] is used as assessment tool and the website is freely used at <http://go.princeton.edu/cgi-bin/GOTermFinder>.

3.2 Selection of the Parameters

In [18], the parameter ξ was set to 10^0, 10^1, 10^2, 10^3, 10^4. So in our simulation experiment, the parameter $\xi \in \{1, 10, 100, 1000, 10000\}$ is set to search the beat result. The minimum number of gene products is set to 2, the maximum p-value is set to 0.01.

3.3 Simulation Results

In this experiment, the parameter ξ is set to 1, 10, 100, 1000, 10000. And we iterate 100 times to randomly generate the simulation data to test the parameter in the best results. The extraction accuracies with different parameters while $P = 5000$ and $n = 30$ are shown in Fig. 2. As we see, GNMF obtains best result when ξ is set from 100 to 10000. GL_0NMF obtains good results when ξ varies from 10 to 10000. So in our gene expression experiment, the parameter is set to 100.

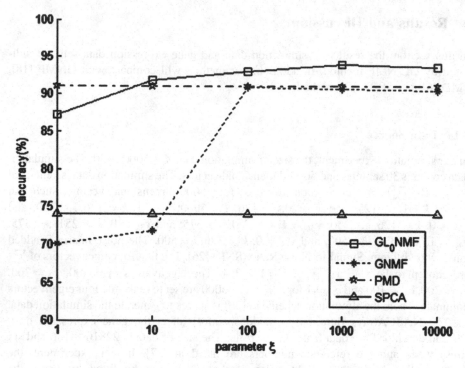

Fig. 2. The performance of GL_0NMF with the parameter ξ.

3.4   Gene Ontology (GO) Analysis

In this subsection, the results of real gene expression data are introduced. For comparison, 500 genes are selected by GL_0NMF, GNMF, PMD and SPCA methods. Then these characteristic genes are verified by Gene Ontology (GO) analysis.

(1) Response to stimulus (GO: 0050896) in root and shoot samples

In TAIR set, the background frequency of response to stimulus (GO: 0050896) is 22.1 % (6690/30323). In this experiment, 500 genes are selected by GL_0NMF, GNMF, PMD and SPCA methods. The detailed results are presented in Tables 1 and 2. In root sample, the results of characteristic genes response to stimulus (GO: 0050896) are listed in Table 1. This table lists the p-value and sample frequency of various stresses. Obviously, bold fonts in the Table 1 show that only one stress (drought stress) that these methods are superior to our method. In other stresses, our method outperforms the three methods. On the whole, our method is far superior to GNMF, PMD and SPCA methods.

Table 1. Response to stimulus (GO: 0050896) in root samples

Stress type	GL_0NMF		GNMF		PMD		SPCA	
	PV	SF	PV	SF	PV	SF	PV	SF
Drought	2.33E-46	260*, 52.0 %	5.98E-49	264,52.9 %	3.67E-65	287,57.4 %	1.39E-66	**289, 57.8 %**
Salt	7.88E-84	**314, 62.8 %**	1.81E-16	199,39.8 %	1.19E-84	313,62.6 %	1.42E-34	237, 47.4 %
uv-b	1.39E-34	**239, 47.8 %**	6.22E-13	189,37.8 %	8.78E-38	237,47.6 %	9.56E-22	210, 42.0 %
Cold	9.27E-64	**287, 57.4 %**	1.76E-41	251,50.4 %	5.06E-68	286,57.0 %	1.54E-61	281, 56.4 %
Heat	1.76E-20	**208, 41.9 %**	3.44E-07	170,34.1 %	1.18E-19	205,41.0 %	1.23E-17	200, 40.0 %
Osmotic	1.52E-31	**233, 46.6 %**	3.33E-20	208,41.7 %	1.10E-26	221,44.2 %	1.49E-34	232, 46.0 %

In this table, the background frequency of response to stimulus (GO: 0050896) in TAIR set is 22.1 % (6690/30323). And in the sample frequency, 260* denotes having 260 genes to response to stimulus in the 500 selection genes. PV: p-value, and SF: sample frequency.

In shoot samples, the sample frequencies of response to stimulus (GO: 0050896) are shown in Table 2. Table 2 describes the p-value and sample frequency of numerous stresses in detail. It can be observed that our method is superior to GNMF. Only in salt stress data set, PMD method is better than our method. In the five remaining data sets, our method is far beyond PMD method. Compared with SPCA, only in salt stress and heat stress data sets, SPCA method is superior to our method. In the four remaining data points, our method is better than SPCA method. The bold fonts in the Table 1 show that GL_0NMF has a certain advantage in the extraction of genes.

Table 2. Response to stimulus (GO: 0050896) in shoot samples

Stress type	GL_0NMF		GNMF		PMD		SPCA	
	PV	SF	PV	SF	PV	SF	PV	SF
Drought	1.93E-92	**325, 64.9 %**	1.10E-52	270,54.1 %	8.09E-77	303,60.6 %	5.45E-54	270,54.0 %
Salt	3.94E-41	251, 50.2 %	3.39E-22	213,42.6 %	1.43E-45	256,51.2 %	9.59E-49	**262,52.4 %**
uv-b	2.37E-129	**364, 73.1 %**	2.32E-37	244,48.9 %	6.4E-128	361,72.2 %	3.12E-101	332,66.4 %
Cold	1.97E-69	**295, 59.0 %**	2.89E-70	295,59.0 %	1.82E-70	294,58.8 %	3.54E-60	279,55.8 %
Heat	2.62E-46	260, 52.0 %	1.10E-03	156,31.2 %	3.69E-26	220,44.0 %	8.04E-51	**265,53.0 %**
Osmotic	8.06E-74	**301, 60.2 %**	2.76E-47	261,52.4 %	3.67E-70	294,58.8 %	2.30E-49	263,52.6 %

(2) Response to stress (GO: 0006950) in root and shoot samples

Tables 3 and 4 describe the corresponding sample frequency of response to stress (GO: 0006950) in root and shoot samples, respectively. The numbers we select and p-value of response to stress in root and shoot samples are described in the two tables. And the background frequency of response to stress is 13.3 % (4044/30323) in TAIR. In Table 3, bold fonts show that only cold stress point of PMD method and drought stress point of SPCA method have an advantage over ours. In other data points, our method is superior to the two methods. In shoot sample, PMD method excels our method in the salt stress data set, and the detailed contents are listed in Table 4. In other data sets, our method outperforms PMD method. In addition, 218 of characteristic genes are extracted by SPCA in cold stress set; however, 216 of characteristic genes are extracted by our method. SPCA method is higher than our method in drought stress set. In other stress data sets, our method exceeds SPCA. In addition, from the two tables, it easy to see that our method extracts more genes than GNMF method in the six data sets.

Table 3. Response to stress (GO: 0006950) in root samples.

Stress type	GL$_0$NMF		GNMF		PMD		SPCA	
	PV	SF	PV	SF	PV	SF	PV	SF
Drought	1.98E-52	207,41.4 %	5.38E-44	194,38.9 %	9.89E-64	222,44.4 %	1.49E-70	**231,46.2 %**
Salt	6.09E-81	**245,49.0 %**	8.79E-12	131,26.2 %	1.48E-82	244,48.9 %	3.65E-34	177,35.4 %
uv-b	1.04E-28	**168,33.6 %**	2.51E-08	121,24.2 %	1.36E-27	165,33.0 %	1.38E-21	153,30.6 %
Cold	1.59E-64	224,44.8 %	4.98E-48	200,40.2 %	4.03E-72	**233,46.6 %**	7.11E-65	223,44.7 %
Heat	6.58E-33	**175,35.2 %**	1.38E-05	112,22.5 %	2.98E-30	170,34.0 %	1.47E-21	153,30.6 %
Osmotic	6.69E-31	**172,34.4 %**	1.29E-17	145,29.1 %	1.55E-24	159,31.8 %	1.07E-29	169,33.8 %

Table 4. Response to stress (GO: 0006950) in shoot samples.

Stress type	GL$_0$NMF		GNMF		PMD		SPCA	
	PV	SF	PV	SF	PV	SF	PV	SF
Drought	6.41E-90	**256, 51.1 %**	3.11E-53	208,41.7 %	2.43E-83	247, 49.4 %	7.15E-46	196,39.2 %
Salt	5.36E-32	174,34.8 %	4.48E-25	161,32.2 %	2.77E-33	**175, 35.0 %**	3.66E-29	168,33.6 %
uv-b	4.85E-127	**295, 59.2 %**	1.00E-33	177,35.5 %	9.65E-128	295, 59.0 %	1.56E-85	249,49.8 %
Cold	1.05E-58	216,43.2 %	2.11E-56	212,42.7 %	2.01E-57	213, 42.6 %	3.48E-61	**218,43.6 %**
Heat	1.06E-51	**206,41.2 %**	3.02E-05	111,22.2 %	1.97E-32	174, 34.8 %	1.34E-39	186,37.2 %
Osmotic	2.33E-71	**233,46.6 %**	1.59E-38	185,37.1 %	1.00E-66	226, 45.2 %	1.52E-42	191,38.2 %

(3) Characteristic genes response to stresses

In order to further study, the osmotic stress (GO: 0006970) in root sample is set out in Table 5. The background frequency of response to osmotic stress is 2.8 %. The corresponding sample frequency of response to osmotic stress and p-value are shown in the table. From the table, we can see that GL$_0$NMF method extracts more characteristic genes than GNMF, PMD and SPCA methods.

Table 5. Response to osmotic stress (GO: 0006970) in root sample.

Stress type	GL$_0$NMF		GNMF		PMD		SPCA	
	PV	SF	PV	SF	PV	SF	PV	SF
Root	1.72E-22	**65,13.0 %**	9.85E-03	32,6.4 %	8.07E-22	64,12.8 %	1.62E-7	42,8.4 %

Sum up, these experiments and analyses show our method can extract more characteristic genes than other methods. Therefore, our method has more advantages than other methods, especially in extracting characteristic genes.

4   Conclusions

A new method (GL_0NMF) is proposed to extract characteristic genes in this paper. GL_0NMF method incorporates into the NMF model both constraint L_0-norm measure and geometrical structure. So it can obtain more comprehensible and interpretable results. The simulation and real experimental results show that GL_0NMF can extract more characteristic genes than other methods. These experiments affirm that our method is efficient and suitable for selecting characteristic genes.

In future, we will be focused on the biological interpretation of the characteristic genes.

Acknowledgements. This work was supported in part by the NSFC under grant Nos. 61370163, 61373027 and 61272339; China Postdoctoral Science Foundation funded project, No. 2014M560264; Shandong Provincial Natural Science Foundation, under grant Nos. ZR2013-FL016 and ZR2012FM023; Shenzhen Municipal Science and Technology Innovation Council (Nos. JCYJ20140417172417174, CXZZ20140904154910774 and JCYJ20140904154645958); the Scientific Research Reward Foundation for Excellent Young and Middle-age Scientists of Shandong Province (BS2014DX004).

References

1. Alter, O., Brown, P.O., Botstein, D.: Singular value decomposition for genome-wide expression data processing and modeling. Proc. Natl. Acad. Sci. **97**, 10101–10106 (2000)
2. Journée, M., Nesterov, Y., Richtárik, P., Sepulchre, R.: Generalized power method for sparse principal component analysis. J. Mach. Learn. Res. **11**, 517–553 (2010)
3. Bartlett, M.S., Movellan, J.R., Sejnowski, T.J.: Face recognition by independent component analysis. IEEE Trans. Neural Netw. **13**, 1450–1464 (2002)
4. Huang, D.-S., Zheng, C.-H.: Independent component analysis-based penalized discriminant method for tumor classification using gene expression data. Bioinformatics **22**, 1855–1862 (2006)
5. Witten, D.M., Tibshirani, R., Hastie, T.: A penalized matrix decomposition, with applications to sparse principal components and canonical correlation analysis. Biostatistics **10**, 515–534 (2009)
6. Liu, J.-X., Zheng, C.-H., Xu, Y.: Extracting plants core genes responding to abiotic stresses by penalized matrix decomposition. Comput. Biol. Med. **42**, 582–589 (2012)
7. Liu, J., Zheng, C., Xu, Y.: Lasso logistic regression based approach for extracting plants coregenes responding to abiotic stresses. In: 2011 Fourth International Workshop on Advanced Computational Intelligence (IWACI), pp. 461–464 (2011)
8. Liu, J.-X., Xu, Y., Zheng, C.-H., Wang, Y., Yang, J.-Y.: Characteristic gene selection via weighting principal components by singular values. PLoS ONE **7**, e38873 (2012)
9. Liu, J.-X., Wang, Y.-T., Zheng, C.-H., Sha, W., Mi, J.-X., Xu, Y.: Robust PCA based method for discovering differentially expressed genes. BMC Bioinformatics **14**, 1–10 (2013)
10. Aradhya, V.N.M., Masulli, F., Rovetta, S.: A novel approach for biclustering gene expression data using modular singular value decomposition. In: Masulli, F., Peterson, L.E., Tagliaferri, R. (eds.) CIBB 2009. LNCS, vol. 6160, pp. 254–265. Springer, Heidelberg (2010)

11. Lee, D.D., Seung, H.S.: Learning the parts of objects by non-negative matrix factorization. Nature **401**, 788–791 (1999)
12. Li, Y., Ngom, A.: The non-negative matrix factorization toolbox for biological data mining. Source Code Biol. Med. **8**, 1–15 (2013)
13. Yang, S., Ye, M.: Global minima analysis of Lee and Seung's NMF algorithms. Neural Process. Lett. **38**, 29–51 (2013)
14. Lee, D.D., Seung, H.S.: Algorithms for non-negative matrix factorization. In: Advances in Neural Information Processing Systems, pp. 556–562 (2000)
15. Gao, Y., Church, G.: Improving molecular cancer class discovery through sparse non-negative matrix factorization. Bioinformatics **21**, 3970–3975 (2005)
16. Wang, Y., Jia, Y.: Fisher non-negative matrix factorization for learning local features. In: Proceedings of Asian Conference on Computer Vision. Citeseer (2004)
17. Hoyer, P.O.: Non-negative matrix factorization with sparseness constraints. J. Mach. Learn. Res. **5**, 1457–1469 (2004)
18. Deng, C., Xiaofei, H., Jiawei, H., Huang, T.S.: Graph regularized nonnegative matrix factorization for data representation. IEEE Trans. Pattern Anal. Mach. Intell. **33**, 1548–1560 (2011)
19. Zheng, C.-H., Ng, T.-Y., Zhang, D., Shiu, C.-K., Wang, H.-Q.: Tumor classification based on non-negative matrix factorization using gene expression data. IEEE Trans. Nanobiosci. **10**, 86–93 (2011)
20. Du, J.-X., Zhai, C.-M., Ye, Y.-Q.: Face aging simulation based on NMF algorithm with sparseness constraints. In: Huang, D.-S., Gan, Y., Gupta, P., Gromiha, M.M. (eds.) ICIC 2011. LNCS, vol. 6839, pp. 516–522. Springer, Heidelberg (2012)
21. Kong, X., Zheng, C.-H., Wu, Y., Shang, L.: Molecular cancer class discovery using non-negative matrix factorization with sparseness constraint. In: Huang, D.-S., Heutte, L., Loog, M. (eds.) ICIC 2007. LNCS, vol. 4681, pp. 792–802. Springer, Heidelberg (2007)
22. Tang, Z., Ding, S.: Nonnegative dictionary learning by nonnegative matrix factorization with a sparsity constraint. In: Wang, J., Yen, G.G., Polycarpou, M.M. (eds.) ISNN 2012, Part II. LNCS, vol. 7368, pp. 92–101. Springer, Heidelberg (2012)
23. Peharz, R., Stark, M., Pernkopf, F.: Sparse nonnegative matrix factorization with ℓ^0-constraints. Neurocomputing **80**, 38–46 (2010)
24. Morup, M., Madsen, K.H., Hansen, L.K.: Approximate L_0 constrained non-negative matrix and tensor factorization. In: IEEE International Symposium on Circuits and Systems, ISCAS 2008, pp. 1328–1331 (2008)
25. Long, X., Lu, H., Peng, Y., Li, W.: Graph regularized discriminative non-negative matrix factorization for face recognition. Multimedia Tools Appl. **72**, 2679–2699 (2014)
26. Shen, H., Huang, J.Z.: Sparse principal component analysis via regularized low rank matrix approximation. J. Multivar. Anal. **99**, 1015–1034 (2008)
27. Craigon, D.J., James, N., Okyere, J., Higgins, J., Jotham, J., May, S.: NASCArrays: a repository for microarray data generated by NASC's transcriptomics service. Nucleic Acids Res. **32**, D575–D577 (2004)
28. Wu, Z., Irizarry, R.A., Gentleman, R., Murillo, F.M., Spencer, F.: A model based background adjustment for oligonucleotide expression arrays (2004)
29. Consortium, G.O.: The gene ontology in 2010: extensions and refinements. Nucleic Acids Res. **38**, D331–D335 (2010)

Hypergraph Supervised Search for Inferring Multiple Epistatic Interactions with Different Orders

Junliang Shang[1(✉)], Yan Sun[1], Yun Fang[1], Shengjun Li[1],
Jin-Xing Liu[1,2], and Yuanke Zhang[1]

[1] School of Information Science and Engineering, Qufu Normal University,
Rizhao 276826, Shandong, China
{shangjunliang110,yunfang6278,qfnulsj,
yuankezhang}@163.com
[2] Bio-Computing Research Center, Shenzhen Graduate School,
Harbin Institute of Technology, Shenzhen 518055, Guangdong, China
{sunyan225,sdcavell}@126.com

Abstract. Nonlinear interactive effects of Single Nucleotide Polymorphisms (SNPs), namely, epistatic interactions, have been receiving increasing attention in understanding the mechanism underlying susceptibility to complex diseases. Though many works have been done for their detection, most only focus on the detection of pairwise epistatic interactions. In this study, a Hypergraph Supervised Search (HgSS) is developed based on the co-information measure for inferring multiple epistatic interactions with different orders at a substantially reduced time cost. The co-information measure is employed to exhaustively quantify the interaction effects of low order SNP combinations, as well as the main effects of SNPs. Then, highly suspected SNP combinations and SNPs are used to construct a hypergraph. By deeply analyzing the hypergraph, some clues for better understanding the genetic architecture of complex diseases could be revealed. Experiments are performed on both simulation and real data sets. Results show that HgSS is promising in inferring multiple epistatic interactions with different orders.

Keywords: Epistatic interactions · Single nucleotide polymorphisms (SNPs) · Genome-wide association study · Hypergraph · Genetic interaction network

1 Introduction

Complex diseases, such as cancer, heart disease, Alzheimer's disease, diabetes and so on, are supposed to be caused by multiple Single Nucleotide Polymorphisms (SNPs), their nonlinear interactive effects, and/or their interactions with environmental factors [1]. The nonlinear interactive effects of multiple SNPs that underlie complex diseases are often referred to as epistatic interactions or epistasis, which are widely believed to unveil a large portion of missing heritability of complex diseases. Furthermore, recent advances made it widely be accepted that a complex disease is not caused by only a SNP or an epistatic interaction, but multiple epistatic interactions with different orders

© Springer International Publishing Switzerland 2015
D.-S. Huang et al. (Eds.): ICIC 2015, Part II, LNCS 9226, pp. 623–633, 2015.
DOI: 10.1007/978-3-319-22186-1_62

interacting with other causative factors [2]. Hence, the detection of multiple epistatic interactions with different orders is a compelling step in Genome-Wide Association Studies (GWAS).

Though many works for detecting epistatic interactions have been performed, most were constrained to one or few pairwise epistatic interactions, and algorithms are still being developed, mainly because of the mathematical and computational challenges that are involved. The first challenge is the intensive computational burden mainly imposed by the enormous search space, which has significant implications for GWAS with hundreds of thousands or millions of SNPs. For instance, search space of a 100 K SNP data set with maximum order of 3 is an astronomical number 1.67×10^{14}, where the order refers to the number of SNPs in a SNP combination. The second challenge is the complexity of genetic architecture of a complex disease. It might involve multiple epistatic interactions and each epistasis might consist of several SNPs with small or even no main effects. No prior knowledge available for a disease, such as the number of epistatic interactions, the order and the effect magnitude of each epistatic interaction, makes their detection difficult. Existing methods often specify the order of epistatic interactions beforehand, which ignore the broader epistasis landscape [3]. The third is the association measure that determines how well a SNP combination contributes to the phenotype. A suitable association measure is required to be efficient in time cost, and more importantly, it can be truly capture causative epistatic interactions. Hence, how to select or develop a suitable measure is still a direction.

Among existing association measures, the co-information based measures appear promising in quantifying epistatic interactions since they have well-developed theory, and can measure multivariate dependence without a complex modeling. For instance, Chanda et al. [4] developed a co-information based metric called the interaction index to prioritize interacting SNPs. They also developed another two co-information based measures, which have been applied to three epistasis detection methods: AMBIENCE [5] with the Phenotype Associated Information (PAI) measure and KWII [6] with the Modified Co-Information (MCI) measure for identifying associations with the binary phenotype, CHORUS [7] combining both PAI and MCI measures together to detect epistatic interactions associated with quantitative traits. Sucheston et al. [8] said that the co-information based measures are flexible and have perfect power for detecting epistatic interactions under a variety of conditions that characterize complex diseases.

Besides association measures, search strategies are of great importance. They can be classified into three categories [9]: exhaustive search, stochastic search and heuristic search. The exhaustive search enumerates all SNP combinations with all considered orders. It becomes no longer feasible for large scale data sets like those in GWAS. The stochastic search performs a random investigation of search space and its performance relies on random chance to select phenotype associated SNPs or SNP combinations. With the number of SNPs or SNP combinations growing, it is believed that the chances of correct guess dramatically drop [9]. Recently, the heuristic search is gaining increasing favor since it performs well on detection power while largely reducing computational complexity. Graph (or network) supervised search belongs to the heuristic search, and shows huge potential in detecting epistatic interactions. Moore et al. [10] built an interaction graph to detect epistasis. McKinney et al. [11]

constructed a genetic association interaction network to identify epistatic interactions, whose edges quantify the synergy between pair SNPs with respect to the phenotype. Hu et al. [12] proposed a statistical epistasis network, which has been proven to be able to discover pairwise epistatic interactions of bladder cancer and prostate cancer. They also demonstrated that graph supervised search is able to infer several 3 order epistatic interactions with significantly high associations at a substantially reduced computational cost [13]. Though these graph supervised search can provide a global map of pairwise epistatic interactions, and can indirectly capture higher order epistatic interactions, they are limited in directly detecting and visualizing high order epistatic interactions, for example, 3 or larger.

In this paper, we present a Hypergraph Supervised Search (HgSS) based on the co-information measure for inferring multiple epistatic interactions with different orders. At first, HgSS employs the co-information measure to exhaustively quantify the main effects of all SNPs and the interaction effects of lower order SNP combinations. Then, highly suspected SNPs and SNP combinations are applied to construct a hypergraph, where two types of vertices are introduced: a real vertex representing the main effect of a SNP, and a virtual vertex denoting the interaction effect of a SNP combination with an arbitrary order. By deeply analyzing the constructed hypergraph, multiple epistatic interactions with different orders could be inferred simultaneously; for a SNP combination, interaction effects of itself and its subsets, as well as main effects of the involved SNPs, could be distinguished precisely; higher order epistatic interactions, even several topology structures, could be captured; high order epistatic interactions could be inferred at a substantially reduced time cost. These discoveries might play an important role on understanding the genetic architectures of complex diseases. Experiments of HgSS are performed on both simulation data set and a real Age-related Macular Degeneration (AMD) data set. Results demonstrate that HgSS is promising in inferring multiple epistatic interactions with different orders, and is an alternative to existing methods.

2 Methods

2.1 Co-information Measure

Co-information [14] is one of several generalizations of the mutual information, and expresses the amount information bound up in a set of variables, beyond that which is present in any subset of those variables. In other words, it is a parsimonious measure that can quantify multivariable dependence that cannot be obtained without observing all variables at the same time. Therefore, it seems promising in measuring epistatic interactions [3, 7]. In this study, the co-information is introduced to HgSS as the association measure.

The definition of co-information among n SNPs S_1, ..., S_n and the phenotype C is that it is the alternating sum of the joint entropies of all possible subsets T of V using the difference operator notation of Han [15], namely,

$$CI(S_1; \ldots; S_n; C) = -\sum_{T \subseteq V} (-1)^{n+1-|T|} H(T),$$

where $V = \{S_1, \ldots, S_n, C\}$ and $H(T)$ is the joint entropy of T, which is written as

$$H(T) = -\sum_{t \in T} p(t) log p(t),$$

where $p(t)$ is the probability mass function.

It is seen that co-information is equal to mutual information and its value is always positive in the bivariate case, however, in the multivariate case, co-information value can be positive or negative. The interpretation of this property is generally intuitive: a positive value is an evidence of interactions among variables; a negative value implies the presence of redundancy; and zero denotes that variables are independent or, more likely, interact with a mixture of synergy and redundancy. In HgSS, only those SNP combinations are considered as candidate epistatic interactions that their values of co-information are positive. Among them, top m candidate epistatic interactions with high co-information values are selected as the detected low order epistatic interactions, where m is the user-specified number.

2.2   Hypergraph Supervised Search

A hypergraph is a generalization of a graph in which an edge can connect any number of vertices. Therefore, the hypergraph is of great benefit to represent the interaction effects of multiple SNPs. For a fair and intuitive comparison of effects with different orders, a generalization of hypergraph is introduced into the study to supervise the search of multiple epistatic interactions.

The list of highly suspected SNPs and SNP combinations that have already been screened out by the co-information measure is used to construct a hypergraph, which is composed of weighted vertices and unweighted edges. For weighted vertices, two types of them are designed in the hypergraph, that is, real vertices and virtual vertices. Each real vertex represents a SNP, and its weight is the co-information value between the SNP and the phenotype, corresponding to main effect of the SNP to the phenotype. Each virtual vertex represents the interaction effect of the combination of the linking SNPs to the phenotype, and its weight is the co-information value among the involved SNPs and the phenotype. In order to facilitate the visualization of hypergraph, HgSS uses red circles to represent the real vertices, and non-red circles to represent the virtual vertices. The sizes of circles are respectively in proportion to their weights. The number in the red circle is the SNP index, which can be replaced easily by the SNP name. For unweighted edges, each of them links a SNP and a related interaction effect of this SNP to the phenotype. By analyzing the constructed hypergraph, some hidden clues for better understanding the underlying architecture of complex diseases could be revealed.

3 Results and Discussion

3.1 Simulation Data

To evaluate the performance of HgSS, a simulation data set, *dat*100, is used, which has been developed previously [16]. Data set *dat*100 records genotypes of 100 SNPs of 2000 cases and 2000 controls as a matrix, where a row denotes genotypes of an individual and a column denotes a SNP. Genotypes of a SNP are coded as {0, 1, 2}, corresponding to homozygous common genotype, heterozygous genotype, and homozygous minor genotype. The label of an individual is a binary phenotype being either 0 (control) or 1 (case). In *dat*100, 7 models are added that jointly influence the phenotype, including 2 single-SNP models, 2 models of 2-order epistatic interactions, 2 models of 3-order epistatic interactions and a model of 5-order epistatic interactions. These models are generated by complex inheritance patterns. Details of first 6 models are summarized in Tables 1, 2 and 3, respectively. Model 7 assumes a 5-order epistatic interaction under a dominant genetic model for the minor allele (Minor Allele Frequency or MAF of each SNP is set to 0.30) at each SNP [16]. Its penetrance function is 0 if the minor allele is not present at each SNP and 0.90 if the minor allele is present at each SNP. Indices of ground-truth SNPs of each model are listed in Table 4, where ground-truth SNPs refer to the causative SNPs that truly associated with the phenotype, in other words, the SNPs in models added into *dat*100.

Table 1. Details of 2 single-SNP models.

Model	MAF (*a*)	Penetrance function		
		AA	*Aa*	*aa*
Model 1	0.10	0.00	0.50	0.50
Model 2	0.40	0.00	0.00	0.50

Table 2. Details of 2 models of 2-order epistatic interactions.

Model	MAF (*a*)	MAF (*b*)	Penetrance function	Genotypes (SNP *B*)		
			Genotypes (SNP *A*)	*BB*	*Bb*	*bb*
Model 3	0.25	0.25	*AA*	0.10	0.10	0.00
			Aa	0.10	0.10	0.00
			aa	0.00	0.00	0.00
Model 4	0.20	0.30	*AA*	0.00	0.00	0.00
			Aa	0.00	0.75	0.75
			aa	0.00	0.75	0.75

Table 3. Details of 2 models of 3-order epistatic interactions.

Model	MAF (a)	MAF (b)	MAF (c)	Penetrance function									
				A	BB			Bb			bb		
					CC	Cc	cc	CC	Cc	cc	CC	Cc	cc
Model 5	0.25	0.20	0.20	AA	0.00	0.00	0.00	0.00	0.00	0.00	0.00	0.00	0.00
				Aa	0.00	0.00	0.00	0.00	0.00	0.00	0.00	0.00	0.00
				aa	0.00	1.00	1.00	1.00	1.00	1.00	1.00	1.00	1.00
Model 6	0.40	0.25	0.25	AA	0.00	0.00	0.00	0.00	0.00	0.00	0.00	0.00	0.00
				Aa	0.00	0.00	0.00	0.00	0.50	0.50	0.00	0.50	1.00
				aa	0.00	0.00	0.00	0.00	0.50	1.00	0.00	1.00	1.00

Table 4. Indices of ground-truth SNPs of each model in *dat*100. In the table, the underlined ground-truth SNPs can be inferred by the hypergraph Fig. 1A and the bolded ground-truth SNPs can be inferred by the hypergraph Fig. 1B.

Model	Indices				
Model 1	<u>98</u>				
Model 2	<u>78</u>				
Model 3	<u>60</u>	93			
Model 4	<u>44</u>	75			
Model 5	<u>20</u>	<u>47</u>	<u>97</u>		
Model 6	**83**	**85**	**100**		
Model 7	25	**81**	**87**	92	**99**

3.2   Experiments on Simulation Data

We apply HgSS on *dat*100 2 times with the specified maximum order being 2 and 3, respectively, hypergraphs of which are shown in Fig. 1. Observations from the figure demonstrate that HgSS is promising in inferring multiple epistatic interactions with different orders at a substantially reduced computational cost.

Firstly, HgSS is capable of distinguishing interaction effects from main effects precisely. Specifically, SNPs 98, 60, 78 and 93 have far stronger main effects than other SNPs, illustrating that each of them could independently and significantly modify the phenotype; nevertheless, SNPs 60 and 93 also have the strongest 2-order interaction effect among all displayed 2-order interaction effects, indicating that the combination of them also modify the phenotype significantly; though SNP 98 has the strongest main effect among all displayed main effects, it also shows a strong 2-order interaction effect with SNP 60, suggesting that its total effect consists of main effect and interaction effect; SNP 78 only displays a strong main effect, implying that its total effect is mainly composed of main effect; the combination of SNPs 20, 47, and 97 has both 2-order interaction effects and a 3-order interaction effect, suggesting that they might modify the phenotype jointly. Observations of other SNPs in the figure are consistent with observations of above mentioned SNPs.

Secondly, HgSS is able to detect multiple epistatic interactions simultaneously. From the Table 4, it is seen that 12 and 14 out of 17 ground-truth SNPs are detected in

Fig. 1A and B, respectively, and 5 out of 7 models are identified accurately in both subfigures, which implies that HgSS is suitable for detecting multiple epistatic interactions at the same time. Two models (Model 4 and Model 7) cannot be detected, though several ground-truth SNPs added into them have been identified, which is mainly because of their total effects to the phenotype becoming small in $dat100$, and being not as expected as those in their respective models. This is understandable since the phenotypic status is controlled by total effects of all added models in $dat100$, and these models interact with each other in a complicated way.

Thirdly, HgSS can discover epistatic interactions with different orders. Either in Fig. 1A or B, 5 out of 7 models are detected perfectly, including 2 single-SNP models, one model of 2-order epistatic interaction, and 2 models of 3-order epistatic interactions, which implying that HgSS is indeed developed for capturing multiple epistatic interactions with different orders. This is one of the highlights of HgSS since existing methods often specify the order of epistatic interactions beforehand, which obviously ignore the broader epistasis landscape.

Fourthly, HgSS hypergraph constructed by main effects and low order interaction effects is a promising guide map for capturing higher order epistatic interactions, even graph structures, such as connected subgraphs or density subgraphs, which might play an important role on understanding the genetic architectures of complex diseases. It is clear that such topology structures are difficult to be inferred by existing non-graph methods. For instance, combinations {83, 85} and {85, 100} have strong interaction effects, and the SNP 85 displays strong main effect, implying that superset of them, i.e., {83, 85, 100}, might be a 3-order epistatic interaction. In Fig. 1B, SNPs 20, 47, and 97 forms a density subgraphs, meaning that it is a 3-order epistatic interaction. There are in total 10 and 12 maximal connected subgraphs in respective Fig. 1A and B. Among them, {98, 93, 60} and {20, 47, 97} appear in both hypergraphs, and display strong total effects to the phenotype, implying that both of them are really causative factors. The largest maximal connected subgraph consists of 12 SNPs, none of which

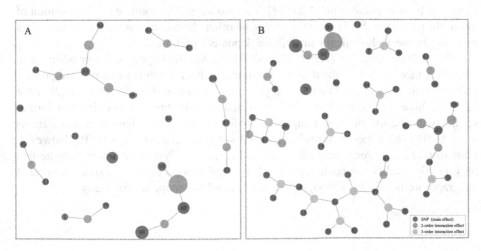

Fig. 1. Hypergraphs of $dat100$ constructed by HgSS. Figure 1A is constructed with the maximum order specified to 2 and Fig. 1B is constructed with the maximum order specified to 3.

has strong effect, either main effect or interaction effect; however, their total effects might be reinforced through connectivity structures in the hypergraph.

Fifthly, the bright spot of HgSS is its computational complexity in inferring high order epistatic interactions. It is seen that HgSS is capable of searching high order epistatic interactions at a substantially reduced time cost. For example, hypergraph of Fig. 1A is built by main effects and 2-order interaction effects, which needs only 5050 steps. Its time cost is far less than that of the exhaustive search of 3-order epistatic interactions; even though time cost of building hypergraph is also considered together. This reduction of time cost is more encouraging in the era of GWAS.

3.3   Application to Real AMD Data

HgSS is also applied on a real data set of AMD (Age-related Macular Degeneration), which is the most important cause of irreversible visual loss in elderly populations, and is considered as a complex disease whereby multiple epistatic interactions interact with other factors to the disease [9, 17, 18]. The AMD data set contains 103611 SNPs genotyped with 96 cases and 50 controls. Because its small sample size is insufficient for the search of higher order (>3) epistatic interactions, hypergraph is constructed by real vertices (top 10 SNPs with high main effects, and other SNPs linking 2-order or 3-order interaction effects), and virtual vertices (top 10 2-order interaction effects and top 10 3-order interaction effects), which is shown in Fig. 2A.

It is seen that rs380390 and rs1329428 mainly modify the phenotype according to their strong main effects. They are believed to be significantly associated with AMD [17], and HgSS confirms them successfully. They are in the CFH gene, and there are biologically plausible mechanisms for the involvement of CFH in AMD. CFH is a regulator that activates the alternative pathway of the complement cascade, the mutations in which can lead to an imbalance. This phenomenon is thought to account for substantial tissue damage in AMD [3, 19]. SNP rs1394608, residing within SGCD gene, has been implicated in AMD [17]. Variants of SGCD regulate the degradation of extracellular matrix by facilitating access of other degradative matrix enzymes, thus resulting in the pathological extracellular deposits in retinal.

The SNP combination {rs994542, rs9298846} has the largest co-information value (Fig. 2B), though each of them shows small main effect, implying that they might be a pure epistatic interaction. Its estimated penetrance function (Fig. 2C) and the reference [20] reinforce our view. Other SNPs reported in both Fig. 2A and B need further studies with the use of large samples to confirm whether they have true associations with AMD. The largest maximal connected subgraph consists of 8 SNPs. However, none of them has strong main effect and/or interaction effect, which may indicate that they jointly modify the phenotype. Further function tests of this maximal connected subgraph are necessary, although they are beyond the scope of this study.

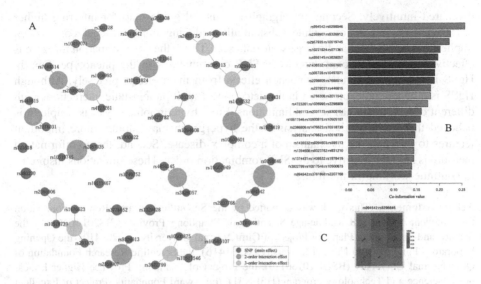

Fig. 2. Experimental results of HgSS on AMD data set. Figure 2A is the constructed hypergraph where SNP names are labeled in their corresponding vertices, and sizes of vertices are in proportion to their effects to the phenotype. Figure 2B shows the top 20 detected epistatic interactions sorted by their descending co-information values. Figure 2C is the estimated penetrance function of the combination of SNPs rs994542 and rs9298846, where penetrance is the probability of occurrence of a complex disease given a particular genotype.

4 Conclusions

Epistatic interactions have been widely believed to play an increasing important role in unraveling the mystery of missing heritability, and detection of them has become a compelling step of GWAS. Though many works have been done for their detection, most only focus on the detection of one or several pairwise epistatic interactions, and the algorithmic development is still ongoing. In this study, a hypergraph supervised search, HgSS, is proposed based on co-information measure to infer multiple epistatic interactions with different orders. At first, the co-information measure is employed to exhaustively quantify the interaction effects of low order SNP combinations (≤ 3) to the phenotype, as well as main effects of all considered SNPs. Then, all detected low order epistatic interactions and the highly suspected SNPs are applied to construct a hypergraph, where real vertex represents the main effect of a SNP and a virtual vertex represents the interaction effect of a SNP combination with an arbitrary order. By deeply analyzing the HgSS hypergraph, some hidden clues for better understanding the underlying architecture of complex diseases could be revealed. Experiments are performed on both simulation and real data sets. Results demonstrate that HgSS is promising in inferring multiple epistatic interactions with different orders, and is an alternative to existing methods.

Besides, HgSS also has several attractive advantages. First, HgSS creates a global interaction map that shows all orders of effects simultaneously, where effects can be compared intuitively. Second, hypergraph is a useful guide map for inferring higher order epistatic interactions while substantially reducing computational cost, which implies that HgSS can handle large scale data sets. Third, the co-information measure is effective in measuring contributions of SNP combinations to the phenotype. Fourth, HgSS is able to distinguish interaction effects from main effects precisely. Although HgSS is a beneficial exploration in the detection of multiple epistatic interactions with different orders, it still has several limitations. First, hypergraph analysis is simple. It is believed that more detailed analysis of the hypergraph may capture more important features to interpret the mechanism of a complex disease. Second, the co-information measure is slightly sensitive to the SNP combination order. These limitations inspire us to continue working in the future.

Acknowledgments. This work was supported by the Scientific Research Reward Foundation for Excellent Young and Middle-age Scientists of Shandong Province (BS2014DX004), the Science and Technology Planning Project of Qufu Normal University (xkj201410), the Opening Laboratory Fund of Qufu Normal University (sk201416), the Scientific Research Foundation of Qufu Normal University (BSQD20130119), the Project of Shandong Province Higher Educational Science and Technology Program (J13LN31), the Award Foundation Project of Excellent Young Scientists in Shandong Province (BS2014DX005), the Shenzhen Municipal Science and Technology Innovation Council (JCYJ20140417172417174), the Shandong Provincial Natural Science Foundation (ZR2013FL016), the China Postdoctoral Science Foundation Funded Project (2014M560264).

Conflict of Interests The authors declare that there is no conflict of interests regarding the publication of this paper.

References

1. Maher, B.: Personal genomes: the case of the missing heritability. Nature **456**(7218), 18–21 (2008)
2. Maher, B.: The case of the missing heritability. Nature **456**(7218), 18–21 (2008)
3. Shang, J., Zhang, J., Sun, Y., Zhang, Y.: EpiMiner: a three-stage co-information based method for detecting and visualizing epistatic interactions. Digit. Signal Process. **24**, 1–13 (2014)
4. Chanda, P., Sucheston, L., Zhang, A., Ramanathan, M.: The interaction index, a novel information-theoretic metric for prioritizing interacting genetic variations and environmental factors. Eur. J. Hum. Genet. **17**(10), 1274–1286 (2009)
5. Chanda, P., Sucheston, L., Zhang, A., Brazeau, D., Freudenheim, J.L., Ambrosone, C., Ramanathan, M.: AMBIENCE: a novel approach and efficient algorithm for identifying informative genetic and environmental associations with complex phenotypes. Genetics **180**(2), 1191–1210 (2008)
6. Chanda, P., Zhang, A., Brazeau, D., Sucheston, L., Freudenheim, J.L., Ambrosone, C., Ramanathan, M.: Information-theoretic metrics for visualizing gene-environment interactions. Am. J. Hum. Genet. **81**(5), 939–963 (2007)
7. Chanda, P., Sucheston, L., Liu, S., Zhang, A., Ramanathan, M.: Information-theoretic gene-gene and gene-environment interaction analysis of quantitative traits. BMC Genom. **10**(1), 509 (2009)

8. Sucheston, L., Chanda, P., Zhang, A., Tritchler, D., Ramanathan, M.: Comparison of information-theoretic to statistical methods for gene-gene interactions in the presence of genetic heterogeneity. BMC Genom. **11**(1), 487 (2010)
9. Shang, J., Zhang, J., Sun, Y., Liu, D., Ye, D., Yin, Y.: Performance analysis of novel methods for detecting epistasis. BMC Bioinformatics **12**(1), 475 (2011)
10. Moore, J.H., Gilbert, J.C., Tsai, C.-T., Chiang, F.-T., Holden, T., Barney, N., White, B.C.: A flexible computational framework for detecting, characterizing, and interpreting statistical patterns of epistasis in genetic studies of human disease susceptibility. J. Theor. Biol. **241**(2), 252–261 (2006)
11. McKinney, B.A., Crowe, J.E., Guo, J., Tian, D.: Capturing the spectrum of interaction effects in genetic association studies by simulated evaporative cooling network analysis. PLoS Genet. **5**(3), e1000432 (2009)
12. Hu, T., Sinnott-Armstrong, N.A., Kiralis, J.W., Andrew, A.S., Karagas, M.R., Moore, J.H.: Characterizing genetic interactions in human disease association studies using statistical epistasis networks. BMC Bioinformatics **12**, 364 (2011)
13. Hu, T., Andrew, A.S., Karagas, M.R., Moore, J.H.: Statistical epistasis networks reduce the computational complexity of searching three-locus genetic models. In: Pacific Symposium on Biocomputing, pp. 397–408. World Scientific, Singapore (2013)
14. Bell, A.J.: The co-information lattice. In: The 4th International Symposium on Independent Component Analysis and Blind Signal Separation, pp. 921–926 (2003)
15. Sun Han, T.: Multiple mutual informations and multiple interactions in frequency data. Inf. Control **46**(1), 26–45 (1980)
16. Miller, D.J., Zhang, Y., Yu, G., Liu, Y., Chen, L., Langefeld, C.D., Herrington, D., Wang, Y.: An algorithm for learning maximum entropy probability models of disease risk that efficiently searches and sparingly encodes multilocus genomic interactions. Bioinformatics **25**(19), 2478–2485 (2009)
17. Tang, W., Wu, X., Jiang, R., Li, Y.: Epistatic module detection for case-control studies: a Bayesian model with a Gibbs sampling strategy. PLoS Genet. **5**(5), e1000464 (2009)
18. Klein, R.J., Zeiss, C., Chew, E.Y., Tsai, J.Y., Sackler, R.S., Haynes, C., Henning, A.K., SanGiovanni, J.P., Mane, S.M., Mayne, S.T.: Complement factor H polymorphism in age-related macular degeneration. Science **308**(5720), 385–389 (2005)
19. Adams, M.K., Simpson, J.A., Richardson, A.J., Guymer, R.H., Williamson, E., Cantsilieris, S., English, D.R., Aung, K.Z., Makeyeva, G.A., Giles, G.G.: Can genetic associations change with age? CFH and age-related macular degeneration. Hum. Mol. Genet. **21**(23), 5229–5236 (2012)
20. Wan, X., Yang, C., Yang, Q., Xue, H., Tang, N.L., Yu, W.: Detecting two-locus associations allowing for interactions in genome-wide association studies. Bioinformatics **26**(20), 2517–2525 (2010)

Predicting Protein-Protein Interactions from Amino Acid Sequences Using SaE-ELM Combined with Continuous Wavelet Descriptor and PseAA Composition

Yu-An Huang, Zhu-Hong You[(⊠)], Jianqiang Li, Leon Wong,
and Shubin Cai

College of Computer Science and Software Engineering, Shenzhen University,
Shenzhen 518060 Guangdong, China
zhuhongyou@gmail.com

Abstract. Protein-protein interactions (PPIs) are known for its crucial role in almost all cellular processes. Although many innovative techniques for detecting PPIs have been developed, these methods are still both time-consuming and costly. Therefore, it is significant to develop computational approaches for predicting PPIs. In this paper, we propose a novel method to identify new PPIs in ways of self-adaptive evolutionary extreme learning machine (SaE-ELM) combined with a novel representation using continuous wavelet (CW) transform and Chou's pseudo amino acid feature vector. We apply Meyer continuous wavelet transform to extracting wavelet power spectrums from a protein sequence representing a protein as an image, which allows us to use well-known image texture descriptors for extracting protein features. Chou's pseudo-amino-acid composition (PseAAC) expands the simple amino-acid composition (AAC) by retaining information embedded in protein sequence. SaE-ELM, a variant of extreme learning machine (ELM), optimizes the single hidden layer feedforward network (SLFN) hidden node parameters using self-adaptive different evolution algorithms. When performed on the PPI data of yeast, the proposed method achieved 87.87 % prediction accuracy with 91.19 % sensitivity at the precision of 82.62 %. Extensive experiments are performed to compare our method with the method base on state-of-the-art classifier, support vector machine (SVM). It is observed from the achieved results that the proposed method is very promising for predicting PPI.

Keywords: Protein-protein interaction · Self-adaptive evolutionary extreme learning machine · Protein sequence

1 Introduction

Proteins are major component of organisms and play an important role in nearly all cell functions including DNA transcription and replication, metabolic cycles, and signaling cascades. Usually, proteins cooperate with each other by forming a big network of protein-protein interactions (PPIs) rather than carry out particular biological functions alone. Thus, the elucidation of the interactions in cellular systems is one of the

© Springer International Publishing Switzerland 2015
D.-S. Huang et al. (Eds.): ICIC 2015, Part II, LNCS 9226, pp. 634–645, 2015.
DOI: 10.1007/978-3-319-22186-1_63

indispensable objectives of the post-genomic era. In the past decades, there have been many advanced techniques for detecting PPIs developed [1–3]. Due to the progress in large-scale experimental technologies such as yeast two-hybrid (Y2H) screens [2, 4], tandem affinity purification (TAP) [1], mass spectrometric protein complex identification (MS-PCI) [3] and other high-throughput biological techniques for PPIs detection, the last decade has seen an explosion in the collection of PPIs data for different species, which provides the wealth of data for further investigations. However, those experiment-based methods for detecting PPIs are both expensive and tedious, and the PPI pairs obtained only cover a very small fraction of the whole networks. In addition, experiment-based methods usually fail to achieve low rates of both false positive and false negative predictions [5]. For this reason, it is of great benefit to develop computational methods capable of identifying PPIs.

Extracting features from proteins is an indispensable first step in building machine systems for predicting PPIs. A number of methods have been proposed for protein representation [28, 29, 32]. Because of proteins' extremely complicated variation in both sequence order and length, it is hardly to build a feasible predictor by using the sequential mode to express a protein. For this reason, the Chou's pseudo amino acid composition, a method based on discrete mode, has been proposed and becomes one of the most widely used feature extractors for peptides and proteins. In Chou's method [10], the first 20 factors represent the components of its conventional amino acid (AA) composition. In order to avoid losing the information of order and length in an original sequence, λ additional factors would be added and incorporate some sequence order information using various modes, such like a series of rank-different correlation factors.

Recently, some methods for extracting features from the proteins using wavelets have been proposed [20–22]. Employing graphic approaches to study biological systems is of great novelty and sufficient to reveal the complicated relations in biological systems [16]. In this study, a novel approach - the continuous wavelet images of protein sequence - is introduced to extracting protein features. This method is based on the calculation of texture descriptors derived from a wavelet representation of the protein. More specifically, we first encode the protein sequence as a numerical sequence by substituting every amino acid with proteins' hydrophobicity index value, and then apply Meyer continuous wavelet to the encoding process representing a protein as an image CW. In the last step, we choose Local Binary Pattern Histogram Fourier (LBP-HF) as the texture descriptors to extract feature from the wavelet images.

It is drawing increasing attentions to use computation methods the mining biological knowledge [27, 30, 31]. Among them, the popular computational methods for predicting protein- protein interaction are mainly machine learning approaches, such as neural networks (NNs) [6], support vector machines (SVM) [7, 8, 14, 26], and random forests [9]. The general trend in current study for predicting PPIs usually focuses on high accuracy with little attention to the time taken to train the classification models which is a significant factor because of the tremendous scale of the PPIs network [11, 12]. Therefore, some classification models with high accuracy may have to go through a tedious training process as a cost. Here, considering the trade-off between the classification accuracy and the time consuming, we investigate the use of novel paradigm of learning machine called self-adaptive evolutionary extreme learning machine (SaE-ELM) [13]. Extreme learning machine (ELM), a quick and popular offline classifier, randomly assigns the hidden node parameters of generalized single-hidden layer feed-forward networks (SLFNs) and calculates the output weights with least square algorithm. SaE-ELM, as one of the variations of basic ELM, optimizes the hidden node parameters by the self-adaptive differential evolution algorithm. Different strategies would be employed in different generations and controlled by associated parameters which are self-adapted in a strategy pool by learning from their previous experiences in generating promising solutions.

Much efforts have been devoted to build a PPIs prediction system [24, 25]. This paper proposes a SaE-ELM based model combing continuous wavelet descriptor and PseAA composition for predicting PPIs. The remainder of this paper is arranged as follows. In Sect. 2, we introduce the feature extraction methods and classifier employed in this work. In Sect. 3, we report the experimental results obtained on interaction protein pairs problem. The conclusion is drawn in Sect. 4.

2   Materials and Methodology

In this section, we describe the proposed approach for predicting protein interactions from protein sequences. Our method to predict the PPIs by the following steps: (1) Represent protein as a CW image by using the Meyer continuous wavelet transform. (2) Local Binary Pattern Histogram Fourier would be used as the image texture descriptor for feature extraction. (3) In addition, we use Chou's pseudo amino acid composition as another feature extraction method to generate another 40-dimensional feature vector. (4) SaE-ELM is used to perform the protein interaction prediction tasks.

2.1   Generation of the Data Set

We evaluated the proposed method with the real PPIs dataset of human. It was collected from the Human Protein References Database (HPRD). After the redundant protein pairs which have ≥25 % sequence identity were deleted, the remaining 3,899 protein-protein pairs of experimentally verified PPIs from 2,502 different human proteins comprise the golden standard positive dataset.

The selection of golden standard negative dataset has an important impact on the prediction performance, and it can be artificially inflated by a bias towards dominant samples in the positive data. For golden standard negative dataset, we followed the previous work [23] assuming that the proteins in different subcellular compartments do not interact with each other. After strictly following the steps in *You*'s work, we finally obtained 4,262 protein pairs from 661 different human proteins as the golden standard negative dataset. By combining the above two golden standard positive and negative PPI datasets, the final whole PPI dataset consists of 8,161 protein pairs, where nearly half are from the positive dataset and half are from the negative dataset.

2.2   Protein Sequence Representation

To apply machine learning machine methods to mining PPIs information from protein sequences, one of the most important computational challenges is to extract feature vectors expressing corresponding proteins. In this work, a protein would be presented by a 216-dimensional vector which is constituted by a 176-dimensional vector from CW feature and a 40-dimensional vector from PseAAC feature.

Continuous Wavelet Transform. Since introduced in the early 1980s, wavelets transform (WT) has become an important signal analysis tool due to its ability to contain both spectral and temporal information within the signal. We can use different scales with CWT to decompose a digital signal into many groups of coefficients which can represent characteristics in both time domain and frequency domain. In our work, features are extracted by considering 100 decomposition scales using Meyer continuous transform. CWT can be described as follows:

$$W_f(a,b) = \frac{1}{\sqrt{a}} \int f(t) \psi(\frac{t-b}{a}) dt \tag{1}$$

where a is the scale parameter and b is the shift factor; $\psi(t)$ is wavelet core; $f(t)$ is the digital signal sequence; and $W_f(a,b)$ is the result of inner product operation between $f(t)$ and $\psi(t)$.

Local Binary Pattern Histogram Fourier (LBP-HF). The Local Binary Patterns (LBP) is a texture operator assigning each pixel a specific value based on the level of its neighborhood [17]. Compared to the neighbors (i_n), the value of the centered pixel (i_c) is computed based on the following equations:

$$LBP(x_c, y_c) = \sum_{m=0}^{p-1} s(i_n - i_c) \times 2^n \tag{2}$$

$$s(i_n - i_c) = \begin{cases} 1 & if \quad i_n - i_c \geq 0 \\ 0 & if \quad i_n - i_c < 0 \end{cases} \tag{3}$$

where p is the number of neighboring pixels. LBP-HF, a rotation invariant image descriptor, is computed globally from the discrete fourier transforms of local binary pattern histograms. This descriptor use discrete fourier transform to construct rotationally features which are invariant to cyclic shifts in the input vector. LBP-HF has a lower overhead than LBP histogram because it only computes $P - 1$ fast fourier transforms of P points from the LBP histogram. Here $(P = 16; R = 2)$ and $(P = 8; R = 1)$.

2.2.1   Pseudo Amino Acid Composition (PseAAC)

The PseAAC approach is a popular technique for protein's feature extraction due to its ability to express proteins merely according to the information of their sequences. This method allows us to deal with the complication of protein structure with a discrete model without completely losing the sequence-order and sequence-length effects like the case treated by the conventional amino acid composition. In the PseAA composition, a protein sequence P should be defined in a $20 + \lambda$ dimensional space, which can be formulated as follow:

$$p = [p_1 \quad p_2 \quad \cdots \quad p_{20} p_{21} \quad \cdots \quad p_{20+\lambda}]^T \tag{4}$$

where p_1, p_2, ... p_{20} represent the 20 values of the conventional amino acid composition, while the rest components are obtained by adopting autocovariance approach (AC) method:

$$AC^d(i) = \sum_{k=1}^{N-i+20} \frac{(index(p_k, d) - \mu_d) \cdot (index(p_{k+i-20}, d) - \mu_d)}{\sigma_d \cdot (N - i + 20)} \quad i \in [21, \ldots, 20 + \lambda] \tag{5}$$

$$\mu_d = \frac{1}{20} \sum_{i=1}^{20} index(i, d), \tag{6}$$

$$\sigma_d = \frac{1}{20} \sum_{i=1}^{20} (index(i, d) - \mu_d)^2 \tag{7}$$

where $index(i,d)$ is a function returning the value of the property d for the amino acid i and here we choose hydrophobicity index as property d; μ_d and σ_d denote the normalized mean and the variance of d on the 20 amino acids.

2.3   Self-adaptive Evolutionary Extreme Learning Machine

Feed-forward neural networks (FFN) have been widely used in many fields due to their strong approximation capabilities [15]. But the main methods for training FNN including Gradient-descent, standard optimization and least-square have a low learning speed which goes against the big scale of training data. Extreme learning machine, proposed by Huang et al., is a new learning algorithm for single-hidden layer feed-forward neural networks. This learning scheme randomly chooses hidden node parameters and uses the Moore-Penrose (MP) generalized inverse to compute the output weights of SLFNs. Due to the fact that the hidden layer of SLFNs need not be tuned, ELM can achieve a speed thousands of times faster than conventional learning methods [18]. SaE-ELM, an improved algorithm based on ELM, uses an adaptive differential evolution algorithm to optimize the hidden node parameters, which generates a more compact nodes network than ELM.

Training a data set consisting of N arbitrary distinct samples $s = \{(x_j, t_j) | x_j \in R^n, t_j \in R^m, j = 1, 2, \ldots, N\}$, the SaE-ELM algorithm includes following steps.

Step 1. Initialization. The population of the first generation where each one includes all the hidden node parameters would be initialized forming a set of NP vectors as:

$$\theta_{k,G} = [w_1^{T,(k,G)}, \ldots, w_L^{T,(k,G)}, b_{1,(k,G)}, \ldots b_{L,(k,G)}] \tag{8}$$

where W_j and b_j $(j = 1 \ldots L)$ are randomly generated, G represents the generation and $k = 1, 2, \ldots, NP$.

Step 2. Calculate the output weight. A standard SLFNN with L hidden neurons and an activation function g(x) can be modeled as:

$$\sum_{i=1}^{L} \beta_i g(x_j) = \sum_{i=1}^{L} \beta_i g(w_i \cdot x_j + b_i) = o_j, j = 1, \ldots, N \tag{9}$$

where $w_i = [w_{i1}, w_{i2}, \ldots, w_{in}]^T$ is the weight vector connecting the *ith* hidden neuron and the input neurons; $\beta_i = [\beta_{i1}, \beta_{i2}, \ldots, \beta_{im}]^T$ is the weight vector connecting the *ith* hidden neuron and the ouput neurons; b_j presents the bias of the *ith* hidden neurons. The above N equations can be written compactly as:

$$H_{k,G}\beta_{k,G} = T \tag{10}$$

where $H = \begin{bmatrix} g(w_{1,(k,G)} \cdot x_1 + b_{1,(k,G)}) & \cdots & g(w_{L,(k,G)} \cdot x_1 + b_{L,(k,G)}) \\ \vdots & \cdots & \vdots \\ g(w_{1,(k,G)} \cdot x_N + b_{1,(k,G)}) & \cdots & g(w_{L,(k,G)} \cdot x_N + b_{L,(k,G)}) \end{bmatrix}_{N \times L}$,

$$\beta = \begin{bmatrix} \beta_1^T \\ \vdots \\ \beta_L^T \end{bmatrix}_{L \times m} \quad \text{and} \quad T = \begin{bmatrix} t_1^T \\ \vdots \\ t_N^T \end{bmatrix}_{N \times m}$$

The *ith* column of H is the *ith* hidden node output with respect to the inputs $x_1, x_2, \ldots x_N$ of *kth* individual in *Gth* generation; T denotes the label matrix. In SaE-ELM algorism, the output weight β is obtained by the Moor-Penrose generalized inverse as:

$$\hat{\beta} = H^\dagger T \tag{11}$$

where $H^\dagger = (H^T H)^{-1} H^T$.

Step 3. Mutation and crossover. The new population would be produced through tree operations including mutation, crossover and selection. The success of the algorisms based on differential evolution (DE) is highly dependent on the choice of correct trial vector generation strategies [19]. In SaE-ELM algorism, DE employs a self-organizing scheme based on randomly generated parameters in mutating process. For each target vector, the values of crossover rate *CR* and control parameters *F* are randomly generated based on the Gaussian distributions.

Here, four mutation strategies we used are presented as follows:

1. $V_{k,G} = \theta_{r1,G} + F\left(\theta_{r2,G} - \theta_{r3,G}\right)$
2. $V_{k,G} = \theta_{r1,G} + F\left(\theta_{best,G} - \theta_{r1,G}\right) + F\left(\theta_{r2,G} - \theta_{r3,G}\right) + F\left(\theta_{r4,G} - \theta_{r5,G}\right)$
3. $V_{k,G} = \theta_{r1,G} + F\left(\theta_{r2,G} - \theta_{r3,G}\right) + F\left(\theta_{r4,G} - \theta_{r5,G}\right)$
4. $V_{k,G} = \theta_{r1,G} + F\left(\theta_{best,G} - \theta_{k,G}\right) + F\left(\theta_{r2,G} - \theta_{r3,G}\right)$

In these equations, the indices $r1$, $r2$, $r3$, $r4$, $r5$ are generated in the range $[1, 2 ..., NP]$; k is a control parameter generated randomly within the region $[0, 1]$; and in our work, F is set to be generated according to the normal distributions $N(0.5, 0.5)$. After generating all the mutant vectors by different strategies described above, a trial vector $u_{k,G}(j)$ would be created through a crossover procedure. The crossover equation is as follow:

$$u_{k,G} = \begin{cases} v_{k,G}(j) & if\,(rand_j \leq CR)\ or\ (j = j_{rand}) \\ \theta_{k,G}(j) & otherwise \end{cases} \qquad (12)$$

where CR, the crossover rate to control the replication from the mutant vector, is set to be generated according to the normal distributions $N(0.7, 0.1)$ in our experiment; $rand_j$ is a uniform random parameter; j_{rand} is an integer between 1 and L ensuring at least one parameter in $u_{k,G}$ different from $\theta_{k,G}$.

Step 4. Evaluation. In this step, the root mean square error (RMSE) is computed as an evaluation value for each vector. The next generation $\theta_{k,G+1}$ is generated based on RMSE which can be calculated as:

$$RMSE_{k,G} = \sqrt{\frac{\sum_{i=1}^{N} \left\| \sum_{j=1}^{L} \beta_j g(w_{j,(k,G)} \cdot x_i + b_{j,(k,G)}) - t_i \right\|}{m \cdot N}} \qquad (13)$$

Then generate the new best population vector $\theta_{k,G+1}$ using the following equations.

$$\theta_{k,G+1} = \begin{cases} u_{k,G+1} & if\ \ RMSE_{\theta_{k,G}} - RMSE_{\theta_{k,G+1}} > \lambda \cdot RMSE_{\theta_{k,G}} \\ u_{k,G+1} & if\ \ \left|RMSE_{\theta_{k,G}} - RMSE_{\theta_{k,G+1}}\right| < \lambda \cdot RMSE_{\theta_{k,G}}\ and\ \|\beta_{u_k,G+1}\| < \|\beta_{u_k,G}\| \\ \theta_{k,G} & else \end{cases}$$
$$(14)$$

where $u_{k,G+1}$ denotes the kth individual vector at the $G+1th$ generation; λ is the preset small positive tolerance rate. After this evaluation and selection procedure, mutation and crossover would be operated again until the goal is achieved or the maximum iterations are reached.

3 Experiments and Results

3.1 Evaluation Measures

To measure the prediction performance of the proposed method, the overall prediction accuracy (Accu.), sensitivity (Sens.), precision (Prec.) and Matthews correlation coefficient (MCC) were calculated. They are defined as follows:

$$Accuracy = \frac{TP + TN}{TP + FP + TN + FN} \tag{15}$$

$$Sensitivity = \frac{TP}{TP + FN} \tag{16}$$

$$PE = \frac{TP}{TP + FP} \tag{17}$$

$$MCC = \frac{TP \times TN - FP \times FN}{\sqrt{(TP + FN) \times (TN + FP) \times (TP + FP) \times (TN + FN)}} \tag{18}$$

where true positive (TP) denotes the number of true samples which are predicted correctly; false negative (FN) is the number of true samples predicted to be non-interacting pairs incorrectly; false positive (FP) is the number of true non-interacting pairs predicted to be PPIs falsely, and true negative (TN) is the number of true non-interacting pairs predicted correctly.

3.2 Parameter Selection

In our work, we set $NP = 15$, $CR = 0.7$ and $F = 0.5$. It should be noticed is that the prediction performance of the model will be significantly different if we choose different number of hidden nodes. As observed form Fig. 1, with the increase of hidden nodes, the testing accuracy increases at first and then has a downward trend while training accuracy keeps increasing. When the number of hidden nodes is 2400, the testing accuracy is 87.87 % reaching the maximum. At last, we choose sigmoidal function and use 2400 nodes to train the SaE-ELM model. In addition, we have compared proposed approach against methods utilizing SVM with $c = 0.5$, $g = 3$.

3.3 Prediction Performance of Proposed Model

We evaluated the performance of the proposed method using yeast PPIs data set. In order to reduce the bias of training and testing process, a 5-fold cross-validation is adopted. More specifically, we divide the dataset into five subsets which four of them are used for training and the rest subset is used for testing. Thus five models were generated from the original data set. The prediction results of SVM prediction models with the same data set are exhibited in Table 1.

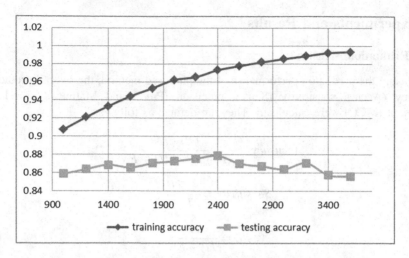

Fig. 1. Comparison among different numbers of hidden nodes

Table 1. The experimental result of the PPIs using proposed method compared with SVM.

Methods	Testing set	Accu. (%)	Prec. (%)	Sen. (%)	MCC (%)
Proposed method	1	87.87	82.85	90.97	78.48
	2	87.22	81.61	90.83	77.48
	3	87.41	83.50	89.21	77.81
	4	89.06	83.69	92.77	80.32
	5	87.79	81.45	92.21	78.30
	Average	**87.87 ± 0.72**	**82.62 ± 1.04**	**91.19 ± 1.38**	**78.48 ± 1.10**
SVM	1	80.88	71.80	96.97	67.98
	2	83.21	74.24	98.57	71.13
	3	84.31	77.54	96.16	72.87
	4	82.17	73.56	96.87	69.81
	5	84.63	77.39	96.73	73.36
	Average	**83.04 ± 1.55**	**74.94 ± 2.50**	**97.06 ± 0.90**	**71.03 ± 2.21**

It can be observed from Table 1, for all five models the precisions are ≥81.45 %, the sensitivities are ≥89.21 %,the prediction accuracies are ≥87.22 %, and the Matthews correlation coefficients are ≥77.48 %. On average, proposed method reach a high accuracy as 87.79 ± 0.72 %. Furthermore, the Receiver Operating Characteristic curve (ROC) in Fig. 2 shows that the proposed method can reach a high true positive rate at the cost of low false positive rate. Above all, we can see that proposed method gives a good prediction performance. In addition, it can noticed that the standard deviation of accuracy, precision, sensitivity and MCC are as low as 0.72 %, 1.04 %, 1.38 % and 1.10 % respectively (Fig. 3).

Recently, many researchers have proposed other computational methods for predicting PPIs. In order to evaluate the forecasting ability of SaE-ELM prediction model

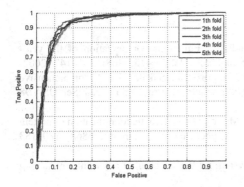

Fig. 2. ROC from proposed method result

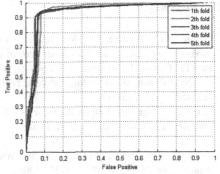

Fig. 3. ROC from SVM-based method result

combining continuous wavelet features and Chou's features, SVM, as a state-of-the-art technique, is performed to compare with our method. From Table 1, we can see that the model based on SVM gives relative poor results with the average accuracy, precision and MMC of 83.04 %, 74.94 % and 71.03 % respectively. The results show that our method is superior to the SVM-based method.

4   Discussion and Conclusions

The last decade has seen an explosion in the collection of protein data. There is a growing demand for developing computational methods utilizing this wealth of data to predict potential PPIs. In this study, we proposed a novel approach based on SaE-ELM combining features extracted from continuous wavelet and PseAAC. Image texture features were extracted from the protein's CW image and connected with Chou's features, and then we employ SaE-ELM algorithm to establish the prediction model. It is a good attempt to use graphic approaches to mine biological massage, which can provide useful insights into the complicated biological system. The proposed feature extraction method retains the protein sequence in both continuous and discontinuous ways, and enables us to mine more PPIs information from the protein sequences. When performed on the PPI data of yeast the proposed method achieved 87.87 % prediction accuracy with 91.19 % sensitivity at the precision of 82.62 % on average. All the analysis shows that our model is an accurate and promising method for the prediction of PPIs.

Acknowledgments. This work is supported in part by the National Science Foundation of China, under Grants 61373086, 61202347, and 61401385, in part by the Guangdong Natural Science Foundation under Grant 2014A030313555, and in part by the Shenzhen Scientific Research and Development Funding Program under grants JCYJ20140418095735569.

References

1. Gavin, A.C., Bosche, M., Krause, R., Grandi, P.: Functional organization of the yeast proteome by systematic analysis of protein complexes. Nature **415**(6868), 141–147 (2002)
2. Ito, T., Chiba, T., Ozawa, R., Yoshida, M., Hattori, M., Sakaki, Y.: A comprehensive two-hybrid analysis to explore the yeast protein interactome. Proc. Natl. Acad. Sci. USA **98**(8), 4569–4574 (2001)
3. Ho, Y., Gruhler, A., Heilbut, A., Bader, G.D., Moore, L.: Systematic identification of protein complexes in Saccharomyces cerevisiae by mass spectrometry. Nature **415**(6868), 180–183 (2002)
4. Pazos, F., Valencia, A.: In silico two-hybrid system for the selection of physically interacting protein pairs. Proteins **47**, 219–227 (2002)
5. You, Z.H., Lei, Y.K., Gui, J., Huang, D.S., Zhou, X.: Using manifold embedding for assessing and predictingprotein interactions from high-throughput experimental data. Bioinformatics **26**(21), 2744–2751 (2010)
6. Chen, H.L., Zhou, H.X.: Prediction of interface residues in proteinprotein complexes by a consensus neural network method: test against NMR data. Proteins Struct. Funct. Bioinform. **61**(3), 21–35 (2005)
7. Koike, A., Takagi, T.: Prediction of protein–protein interaction sites using support vector machines. Protein Eng. Des. Sel. **17**(2), 165–173 (2004)
8. Dong, Q.W., Wang, X.L., Lin, L., Guan, Y.: Exploiting residue-level and profile-level interface propensities for usage in binding sites prediction of proteins. BMC Bioinformatics **8**, 147–159 (2007)
9. Sikic, M., Tomic, S., Vlahovicek, K.: Prediction of protein–protein interaction sites in sequences and 3D structures by random forests. Comput. Biol. **5**(1), e1000278 (2009)
10. Chou, K.C.: Prediction of protein cellular attributes using pseudo-amino acid composition. Proteins **43**(3), 246–255 (2001)
11. Shoemaker, B., Panchenko, A.: Deciphering proteinprotein interactions - part ii: computational methods to predict protein and domain interaction partners. PLoS Comput. Biol. **3**(4), 595–601 (2007)
12. Walker-Taylor, A., Jones, D.: Proteomics and protein-protein interactions: biology, chemistry, bioinformatics, and drug design. In: Waksman, G. (ed.) Computational Methods for Predicting Protein-Protein Interactions, pp. 89–114. Springer, Berlin (2005)
13. Cao, J., Lin, Z., Huang, G.B.: Self-adaptive evolutionary extreme learning machine. Neural Process. Lett. **36**(3), 285–305 (2012)
14. Vapnik, V.: The Nature of Statistical Learning Theory. Springer, Berlin (1995)
15. Huang, G.B., Wang, D.H., Lan, Y.: Extreme learning machines: a survey. Int. J. Mach. Learn. Cybern. **2**(2), 107–122 (2011)
16. Nanni, L., Brahnam, S., Lumini, A.: Wavelet images and Chou's pseudo amino acid composition for protein classification. Amino Acids **43**(2), 657–665 (2012)
17. Ahonen, T., Matas, J., He, C., Pietikäinen, M.: Rotation invariant image description with local binary pattern histogram fourier features. In: Salberg, A.-B., Hardeberg, J.Y., Jenssen, R. (eds.) SCIA 2009. LNCS, vol. 5575, pp. 61–70. Springer, Heidelberg (2009)
18. Huang, G.B., Zhou, H., Ding, X., Zhang, R.: Extreme learning machine for regression and multiclass classification. IEEE Trans. Syst. Man Cybern. Part B-Cybern. **42**, 513–529 (2012)
19. Subudhi, B., Jena, D.: Differential evolution and levenberg marquardt trained neural network scheme for nonlinear system identification. Neural Process. Lett. **27**, 285–296 (2008)

20. Qiu, J.D., Huang, J.H., Liang, R.P., Lu, X.Q.: Prediction of G-protein-coupled receptor classes based on the concept of Chou's pseudo amino acid composition: an approach from discrete wavelet transform. Anal. Biochem. **390**, 68–73 (2009)
21. Shi, S.P., Qiu, J.D., Sun, X.Y., Huang, J.H., Huang, S.Y., Suo, S.B., Liang, R.P., Zhang, L.: Identify submitochondria and subchloroplast locations with pseudo amino acid composition: approach from the strategy of discrete wavelet transform feature extraction. Biochim. Biophys. Acta **1813**, 424–430 (2011)
22. Wen, Z.N., Wang, K.L., Li, M.L., Nie, F.S., Yang, Y.: Analyzing functional similarity of protein sequences with discrete wavelet transform. Comput. Biol. Chem. **29**, 220–228 (2005)
23. You, Z.H., Yu, J.Z., Zhu, L., Li, S., Wen, Z.K.: A MapReduce based parallel SVM for large-scale predicting protein-protein interactions. Neurocomputing **145**(5), 37–43 (2014)
24. You, Z.H., Lei, Y.K., Huang, D.S., Zhou, X.B.: Using manifold embedding for assessing and predicting protein interactions from high-throughput experimental data. Bioinformatics **26**(21), 2744–2751 (2010)
25. Luo, X., You, Z.H., Zhou, M.C., Li, S., Leung, H.: A highly efficient approach to protein interactome mapping based on collaborative filtering framework. Sci. Rep. **5**, Article 7702 (2015)
26. You, Z.H., Yu, J.Z., Zhu, L., Li, S., Wen, Z.K.: A MapReduce based parallel SVM for large-scale predicting protein-protein interactions. Neurocomputing **145**(5), 37–43 (2014)
27. You, Z.H., Yin, Z., Han, K., Huang, D.S., Zhou, X.B.: A semi-supervised learning approach to predict synthetic genetic interactions by combining functional and topological properties of functional gene network. BMC Bioinformatics **11**, 343 (2010)
28. You, Z.H., Lei, Y.K., Zhu, L., Xia, J.F., Wang, B.: Prediction of protein-protein interactions from amino acid sequences with ensemble extreme learning machines and principal component analysis. BMC Bioinformatics **14**(8), 10 (2013)
29. You, Z.H., Zhu, L., Zheng, C.H., Yu, H.J., Deng, S.P., Ji, Z.: Prediction of protein-protein interactions from amino acid sequences using a novel multi-scale continuous and discontinuous feature set. BMC Bioinformatics **15**(S15), Article S9 (2014)
30. You, Z.H., Li, S., Gao, X., Luo, X., Ji, Z.: Large-scale protein-protein interactions detection by integrating big biosensing data with computational model. BioMed Res. Int. **2014**, Article. 598129, p. 9 (2014)
31. Zhu, L., You, Z.H., Wang, B., Huang, D.S.: t-LSE: a novel robust geometric approach for modeling protein-protein interaction networks. PLOS One, **8**(4), Article e58368 (2013)
32. Lei, Y.K., You, Z.H., Ji, Z., Zhu, L., Huang, D.S.: Assessing and predicting protein interactions by combining manifold embedding with multiple information integration. BMC Bioinformatics **13**, Article. S3 (2012)

News Event Detection Based Web Big Data

Bohai Yu, Xia Zhang, and Zhengyou Xia[✉]

College of Computer Science and Technology,
Nanjing University of Aeronautics and Astronautics, Nanjing, China
zhengyou_xia@nuaa.edu.cn

Abstract. News event detection is also called TDT (Topic Detection and Tracking), which is hot research field. Previous studies about TDT are general based on news headline. However, the headline in certain situations can not accurately reflect the actual content of the news. In this paper, our approaches focus on news event that extract from the websites. We propose an approach to generate the feature vectors that represent by the text summary of news rather than the whole news content or news headlines. This method is called Incremental clustering algorithm. Experimental results demonstrate the reliability and effectiveness of our method.

1 Introduction

Topic Detection and Tracking (TDT) [1–3] refers to a variety of automatic techniques for discovering and threading together topically related material in streams of data such as newswire and broadcast news. Topic detection and tracking is a DARPA-sponsored [4, 5] initiative to investigate the state of the art in finding and following new events in a stream of broadcast news stories. And the problems that TDT will deal with are the tasks [7–9]: (1) to find the boundary of each story in the streams; (2) identifying new event stories; (3) finding all related stories when given a small number of sample news stories.

The first research of TDT (Topic Detection and Tracking) was implemented in 1996. When the Defense Advanced Research Projects Agency (DARPA) [1] want to automatically detect the structure of news streams without human intervention. In 1997 a pilot study [1] laid the essential groundwork, producing a small corpus and establishing feasibility. During 1998 and 1999, TDT research blossomed, with new and more challenging tasks, many more participating sites, and considerably larger multilingual corpora (adding ASR data in 1998 and Chinese data in 1999). TDT research is continuing under the new DARPA program known as TIDES (Translingual Information Detection, Extraction, and Summarization). New sites are welcome to participate in the annual TDT evaluations conducted by the National Institute of Standards and Technology (NITS) [10–12]. The research of TDT is based on news headline, which are generalization of news contents [13–20]. Readers often can know the news content through this headline. This is because news headline contain highly relevant keywords about news content. For news headline from different news websites, if the news contents are similar, they commonly exist the same keywords in headline.

News headline are high summary of news contents, so we can understand news contents according to headline. However, in practice, we find that judging news contents

D.-S. Huang et al. (Eds.): ICIC 2015, Part II, LNCS 9226, pp. 646–653, 2015.
DOI: 10.1007/978-3-319-22186-1_64

through headline is not accurate enough in some cases. The news headlines are basically the same. However, the two news contents are not similar. This also means that the headline in certain situations can not accurately reflect the actual content of the news. However, We propose an approach to generate the feature vectors that represent by the summary of news rather than the whole news content or news headlines. Experimental results demonstrate the reliability and effectiveness of our method. In the incremental clustering algorithm the data vector is important to experiment.

2   Our Model Based on Text Summary

As we all know, the content of news is about 200–400 words or even more. If we can extract the key words or key sentences from the content to represent the feature vector, which is more efficiency than the vector forming by the whole corpus. The approach we use to extract the summary is TextRank algorithm, which is derived from iterative idea of PageRank. Chinese text according to the following steps to construct the candidate keyword map:

Divide the news text by sentence, the text represent by T then $T = [s_1, s_2, \cdots s_m]$.

For each sentence $s_i \in T$ is segmented by Chinese words, then remove the stop words that are rarely represented the idea of news content. After that we can obtain the collection $s_i = [t_{i,1}, t_{i,2}, \cdots t_{i,m}]$ where $t_{i,1}$ are the candidate words of the sentence.

Construct the graph of candidate words $G = (V, E)$, where V represent the nodes consist of the candidate words generated by step2 and $E \subseteq V \times V$ for each $t_{i,j} \in s_i$, $t_{i,j+1} \in s_i$ then $< t_{i,j}, t_{i,j+1} > \in E$.

Let $G = (V, E)$ be directed graph consist of collection of nodes V and collection of edges E. For each node V_i let:

$$In(V_i) = \{V_j | <V_j, V_i > \in E\}$$

$$Out(V_i) = \{V_j | <V_i, V_j > \in E\}$$

Where $In(V_i)$ represent the set pointing the V_i, $Out(V_i)$ represent the edges start at V_i.

Equation 1 is the recursion formula of TextRank that completely consistent with PageRank. We sort the words through inter node polling and recommendation mechanism. A node chains into a node set that represents voter. If voter is more important, the more the number nodes of voters have. This suggests that voters ranking will be top. In Eq. 1 d is the Damping Factor that general equals to 0.85.

$$WS(V_i) = (1 - d) + d * \sum_{V_j \in In(V_i)} \frac{\omega_{ji}}{\sum_{V_k \in Out(V_j)} \omega_{jk}} WS(V_j) \tag{1}$$

When we get the summary of the news, the next we should do is to particle the summary. we statistic the each words number and choose top 100 words as the feature. If the word is less then 100 we regard it as 0. Vector = [$word_1$, $word_2$, ..., $word_{100}$] where the word belong to summary.

To calculate the value of each word, we use the approach of TF-IDF (term frequency–inverse document frequency). The main idea of TF-IDF is: if the frequency is TF, a 'word or phrase in an article appearing in the high, and is rarely seen in other articles. We think that this word or phrase has good ability of distinguishing categories, suitable for classification. TF-IDF actually is: TF* IDF, frequency TF, IDF inverse document frequency. TF is the frequency of entry appears in document d. The main idea of IDF is: if the document contains T entry is less, the smaller n, larger IDF, indicates that the entry of T has good ability to distinguish category.

We use Incremental clustering algorithm to deal the corpus and the training corpus are labeled by human. First, we use the training data to generate the cluster when new story comes we compare new story with each cluster. We calculate the similarity and find the maximum then we compare the maximum value with the threshold that we pre-define. If the maximum value is less then, the threshold indicates the story is a new cluster. We put it in the new cluster. If the maximum value is larger, the threshold indicates the story belong to the cluster with biggest similarity. The similarity of two stories is defined as the cosine value of the corresponding story vectors. Figure 1 shows the approach of Incremental clustering algorithm.

```
Input the cluster that already exist C
While (story) {
        String [] collect = news summary of story;
        String [] vector= calculate IF-IDF;
        For (all cluster vector cvector) {
                value= calculate (vector,cvector);
                if ( value > threshold) {
                        add cvector to the cluster;
                }
                Else {
                        Add to new cluster
                        Add to new event;
                }
        }
}
```

Fig. 1. Approach of incremental algorithm based on text summary of news

3 Experiment and Analysis

3.1 TDT Corpus and Our Data Set

A corpus of text and transcribed speech has been developed to support the TDT study effort. This study corpus spans the period from July 1, 1994 to June 30, 1995 and includes nearly 16,000 stories, with about half taken from Reuters newswire and half from CNN broadcast news transcripts. The transcripts were produced by the Journal of Graphics Institute (JGI). The stories in this corpus are arranged in chronological order, are structured in SGML format, and are available from the Linguistic Data Consortium (LDC).

The data set in this paper is collect form Chinese news websites, the data set also the training set to cluster. For our approach we collect 500 stories about human label 30 events. Table 1 shows the data set of our approach.

Table 1. Data set of our approach

Website name	URL	Numbers of stories
163	http://news.163.com	120
SINA	http://news.sina.com.cn	115
SOHU	http://news.sohu.com	105
IFENG	http://news.ifeng.com	115
QQ	http://news.qq.com	30
XINHUA	http://news.xinhuanet.com	15

3.2   Evaluation

In the TDT setting, we have chosen to measure a system's effectiveness primarily by the miss (false negative) and false alarm (false positive or fallout) rates. The major reason for choosing these is a perception of the problem as being a detection task rather than a ranking task: a system needs to indicate whether or not a story is new or is on the event being tracked, not provide a ranked list of stories that might discuss the event. Unfortunately, although it is fairly straightforward for systems to generate ranked lists of stories and to provide a score that creates that ranking, it is more difficult to determine a good score that can be used as a threshold. An ideal system might output a score that corresponds to the probability that the story discusses the event.

In our approach we choose the number of features of vector base on the segmentation of news summary rather than the segmentation of the whole corpus. According to Fig. 2 we choose 100 as the number of features. If we use the segmentation of the whole corpus the length of vector the incremental clustering algorithm cost too much time.

In the TDT approach it chooses the threshold value as 0.23, however the value in our approach the miss rate and false alarm rate is too high. Figures 3 and 4 shows the miss rate and false alarm rate when the threshold is 0.23. Due to we use the summary of news content as our feature rather then the whole corpus so the threshold that the threshold is 0.23 in TDT pilot study dose not apply to our approach.

According to find a better threshold we do the experiment a lot of times with different threshold, after the experiment we find the threshold about 3.75 the miss rate and false alarm rate is acceptable. Figures 5 and 6 show the miss rate and false alarm rate at threshold 3.75.

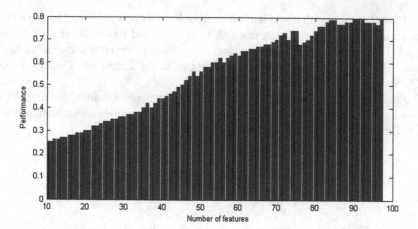

Fig. 2. Performance of different features

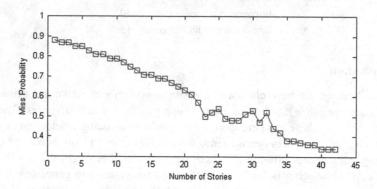

Fig. 3. Miss probability at threshold is 0.23

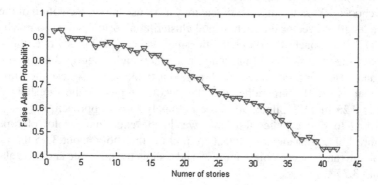

Fig. 4. False alarm probability at threshold is 0.23

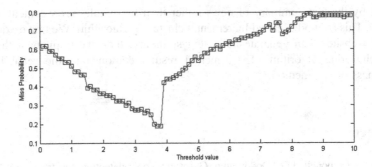

Fig. 5. Miss probability of different threshold value

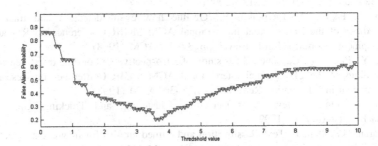

Fig. 6. False alarm probability of different threshold value

Table 2. Comparison with two methods

Approach	Miss rate (%)	False alarm rate (%)	Average cost(ms)
News summary	25.3 %	21.3 %	341
UMass	20.5 %	17.7 %	1103

At last we will compare our approach with the UMass [1] method that use the whole corpus as features. And Table 2 shows the details. The difference between the two method are the corpus and the features. In the news summary method we use data only contain the news extract from the website, UMass use the TDT corpus that contain different type of data. Through Table 2 we find the news summary method has advantage than UMass method because of the number of features is smaller than UMass.

4 Conclusion

Topic Detection and Tracking (TDT) refers to a variety of automatic techniques for discovering and threading together topically related material in streams of data such as newswire and broadcast news. In this paper, our approaches focus on news event that extract from the websites. We propose an approach to generate the feature vectors that represent by the text summary of news rather than the whole news content or news headlines. This method is called Incremental clustering algorithm. We first extract news form the websites then generate news text summary, then we implement the incremental clustering algorithm. Experimental results demonstrate the reliability and effectiveness of our method.

References

1. Allan, J., Carbonell, J.G., Doddington, G., et al.: Topic detection and tracking pilot study final report (1998)
2. Wayne, C.L.: Multilingual topic detection and tracking: successful research enabled by corpora and evaluation. In: LREC (2000)
3. Allan, J., Papka, R., Lavrenko, V.: On-line new event detection and tracking. In: Proceedings of the 21st Annual International ACM SIGIR Conference on Research and Development in Information Retrieval, pp. 37–45. ACM (1998)
4. Yang, Y., Pierce, T., Carbonell, J.: A study of retrospective and on-line event detection. In: Proceedings of the 21st Annual International ACM SIGIR Conference on Research and Development in Information Retrieval, pp. 28–36. ACM (1998)
5. Papka, R.: On-line New Event Detection, Clustering, and Tracking. University of Massachusetts Amherst (1999)
6. Kumaran, G., Allan, J.: Text classification and named entities for new event detection. In: Proceedings of the 27th Annual International ACM SIGIR Conference on Research and Development in Information Retrieval, pp. 297–304. ACM (2004)
7. Brants, T., Chen, F., Farahat, A.: A system for new event detection. In: Proceedings of the 26th Annual International ACM SIGIR Conference on Research and Development in Information Retrieval, pp. 330–337. ACM (2003)
8. Sayyadi, H., Hurst, M., Maykov, A.: Event detection and tracking in social streams. In: ICWSM (2009)
9. Zhang, K., Zi, J., Wu, L.G.: New event detection based on indexing-tree and named entity. In: Proceedings of the 30th Annual International ACM SIGIR Conference on Research and Development in Information Retrieval, pp. 215–222. ACM (2007)
10. Allan, J.: Introduction to topic detection and tracking. In: Allan, J. (ed.) Topic Detection and Tracking, pp. 1–16. Springer, New York (2002)
11. Kleinberg, J.: Temporal dynamics of on-line information streams. In: Data Stream Management: Processing High-Speed Data Streams (2006)
12. Makkonen, J., Ahonen-Myka, H., Salmenkivi, M.: Applying semantic classes in event detection and tracking. In: Proceedings of International Conference on Natural Language Processing (ICON 2002), pp. 175–183 (2002)
13. Li, Z., Wang, B., Li, M., et al.: A probabilistic model for retrospective news event detection. In: Proceedings of the 28th Annual International ACM SIGIR Conference on Research and Development in Information Retrieval, pp. 106–113. ACM (2005)
14. Makkonen, J.: Investigations on event evolution in TDT. In: Proceedings of the 2003 Conference of the North American Chapter of the Association for Computational Linguistics on Human Language Technology: Proceedings of the HLT-NAACL 2003 Student Research Workshop, vol. 3, pp. 43–48. Association for Computational Linguistics (2003)
15. Allan, J. (ed.): Topic Detection and Tracking: Event-Based Information Organization. Springer Science and Business Media, New York (2002)

16. Makkonen, J., Ahonen-Myka, H., Salmenkivi, M.: Topic detection and tracking with spatio-temporal evidence. In: Sebastiani, F. (ed.) ECIR 2003. LNCS, vol. 2633, pp. 251–265. Springer, Heidelberg (2003)
17. Wayne, C.L.: Topic detection and tracking (TDT). In: On Workshop Held at the University of Maryland, vol. 27, p. 28 (1997)
18. Allan, J., Lavrenko, V., Jin, H.: First story detection in TDT is hard. In: Proceedings of the Ninth International Conference on Information and Knowledge Management, pp. 374–381. ACM (2000)
19. Kurt, H.: On-line New Event Detection and Tracking in a Multi-resource Environment. Bilkent University (2001)
20. Zhang, K., Zi, J., Wu, L.G.: New event detection based on indexing-tree and named entity. In: Proceedings of the 30th Annual International ACM SIGIR Conference on Research and Development in Information Retrieval, pp. 215–222. ACM (2007)
21. Xia, Z., Bu, Z.: Community detection based on a semantic network. Knowl. Based Syst. 26, 30–39 (2012)

Self-organized Neural Network Inspired by the Immune Algorithm for the Prediction of Speech Signals

D. Al-Jumeily[1(✉)], A.J. Hussain[1], P. Fergus[1], and N. Radi[2]

[1] Applied Computing Research Group, Faculty of Technology and Environment,
Liverpool John Moores University, Byrom Street, Liverpool L3 3AF, UK
{d.aljumeily,a.hussain,p.fergus}@LJMU.ac.uk
[2] Alkhawarizmi International College, Abu Dhabi, UAE
n.radi@khawarizmi.com

Abstract. This paper presents the use of self-organized neural network inspired by the immune algorithm for the prediction of speech signal. Two speech signals are utilized, woman and man voices counting from one to ten in Arabic. The simulation results were compared with the multilayer perceptrons neural network. A new training algorithm was used with the self-organised multilayer perceptrons neural network that is inspired by the immune using weight decay. The simulation results indicated slight improvement of the use of regularisation technique for the multilayer perceptrons and no improvement when using the self-organized neural network inspired by the immune algorithm for the two speech signals.

Keywords: Speech time series · Immune algorithm · Regularised SMIA

1 Introduction

Speech signals have nonlinear and nonstationary statistical properties. The prediction of the speech signals requires the use of neural network models with temporal capabilities and the memory should be embedded in some form in the network.

Kim [1] showed that there are several methods for embedding memory within the structure of the neural networks which include: (1) creating a spatial representation of the temporal pattern; (2) putting time delays into the neurons or their connection; (3) employing recurrent connections; (4) using neurons with activations summing inputs over time; and (5) using combinations of the previous methods.

The highly popularised multi layer perceptrons (MLPs) models have been successfully applied in various applications. MLPs utilise computationally intensive training algorithms such as the error back-propagation [2] and can get stuck in local minima. In addition, these networks have problems in dealing with large amounts of training data, while demonstrating poor interpolation properties, when using reduced training sets.

In order to improve the recognition and generalisation capability of the back-propagation neural networks, Widyanto et al. [3] used a self organized hidden layer

© Springer International Publishing Switzerland 2015
D.-S. Huang et al. (Eds.): ICIC 2015, Part II, LNCS 9226, pp. 654–664, 2015.
DOI: 10.1007/978-3-319-22186-1_65

inspired by immune algorithm for the prediction of sinusoidal signal and time temperature based quality food data. The simulation results for the self-organized multilayer network inspired by the immune algorithm (SMIA) indicated that the prediction of sinusoidal signal showed an improvement of 1/17 in the approximation error in comparison to the backpropagation and 18 % improvement in the recognition capability for the prediction of time temperature based quality food data. The learning of the SMIA network concentrats on the local properties of the signal and the aim of the network is to adapt to the local properties of the observed signal using the self-organised hidden layer inspired by the immune algorithm. Thus the self organized hidden layer inspired by immune algorithm networks have a more detailed mapping of the underlying structure within the data and are able to respond more readily to any greater changes or regime shifts which are common in nonstationary signals.

In this paper, we propose the use of the SMIA as a prediction tool for the speech signals. Furthermore, we propose an improvement in the training algorithm of the SMIA in which weight decay was introduced. Hanson and Pratt [4] compare weight decay training with standard backpropogation BP using the 4-bit parity problem. The simulation results indicated that training improves the performance achieved by standard BP. Weigend et al. [5] utilized weight decay training for time series forecasting, their simulation results indicated that weight decay training reliably extracts a signal that accounts for between (2.5 % and 4.0 %) of the variance, in which it has been shown that the network will produce better results than networks trained using the standard BP.

The remainder of this paper is as follows: the neural network and structure of the SMIA will be discussed in Sect. 2. Section 3 is dedicated for speech time series prediction, while Sect. 4 shows the simulation results. Conclusion and further works can be found in Sect. 5.

2 The Networks

Although most neural network models share a common goal in performing functional mapping, different network architectures may vary significantly in their ability to handle different types of problems. For some tasks, higher order combinations of some of the inputs or activations may be appropriate to help form good representation for solving the problems.

This section is concerned with introducing functional link neural network, and the immune based neural networks.

2.1 Functional Link Neural Network (FLNN)

FLNN was first introduced by Giles and Maxwell [10]. It naturally extends the family of theoretical feedforward network structure by introducing nonlinearities in inputs patterns enhancements [12]. These enhancement nodes act as supplementary inputs to the network. FLNN calculates the product of the network inputs at the input layer, while at the output layer the summations of the weighted inputs are calculated.

FLNN can use higher order correlations of the input components to perform nonlinear mappings using only a single layer of units. Since the architecture is simpler, it is suppose to reduce computational cost in the training stage, whilst maintaining good approximation performance [13]. A single node in FLNN model could receive information from more than one node by one weighted link. The higher order weights, which connect the high order terms of the input products to the upper nodes, have simulated the interaction among several weighted links. For that reason, FLNN could greatly enhance the information capacity and complex data could be learnt [11, 13, 15]. Fei and Yu [14] showed that FLNN has a powerful approximation capability than conventional backpropagation network, and it is a good model for system identification [13]. Cass and Radl [15] used FLNN in the optimisation process and found that FLNN can be trained much faster than MLP network without scarifying computational capability. FLNN has the properties of invariant under geometric transformations [12]. The model has the advantage of inherent invariance, and only learns the desired signal. Figure 1 shows an example of third order FLNN with three external inputs x_1, x_2, and x_3, and four high order inputs which act as supplementary inputs to the network. The output of FLNN is determined as follows:

$$Y = \sigma\left(W_0 + \sum_j W_j X_j + \sum_{j,k} W_{jk} X_j X_k + \sum_{j,k,l} W_{jkl} X_j X_k X_l + \cdots\right) \quad (1)$$

where σ is a nonlinear transfer function, and w_o is the adjustable threshold. Unfortunately, FLNN suffers from the explosion of weights which increase exponentially with the number of inputs. As a result, second or third order functional link networks are considered in practice [6].

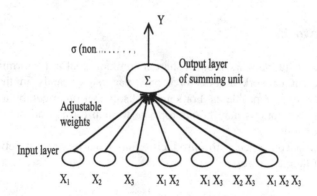

Fig. 1. Functional link neural network.

2.2   The Self-organised Network Inspired by the Immune Algorithm

The immune algorithm which was first introduced by [16] has attracted many interests. Widyanto et al. [3] introduced a method to improve recognition as well as generalisation capability of the backpropagation by suggesting a self-organisation hidden layer inspired by immune algorithm (SMIA) network. The input vector and hidden layer of SMIA network are considered as antigen and recognition ball, respectively. The recognition ball which is the generation of the immune system is used for hidden unit creation.

In time series prediction, the recognition balls are used to solve overfitting problem. In the immune system, the recognition ball has a single epitope and many paratopes. In which, the epitope is attached to B cell and paratopes are attached to antigen, where there is a single B cell that represents several antigens.

For SMIA network, each hidden unit has a centre that represents the number of connections of the input vectors that are attached to it. To avoid the overfitting problem, each centre has a value which represents the strength of the connections between input units and their corresponding hidden units. The SMIA network consists of three layers which are input, self-organised and output layers as shown in Fig. 2.

In what follows the dynamic equations of SMIA network are considered. The ith input unit receives normalised external input S_i where $i = 1....N_I$ and N_I represents the number of inputs. The output of the hidden units is determined by the Euclidean distance between the outputs of input units and the connection strength of input units and the jth hidden unit. The use of the Euclidian distance enables the SMIA network to exploit locality information of input data. This can lead to improve the recognition capability. The output of the jth hidden unit is determined as follows:

$$X_{Hj} = f\left(\sqrt{\sum_{i=1}^{N_I}(w_{Hij} - x_{Ii})^2}\right)$$

$$j = 1\cdots\cdots, N_H$$

(2)

where W_{Hij} represents the strength of the connection from the ith input unit to the jth hidden unit, and f is a nonlinear transfer function.

The outputs of the hidden units represent the inputs to the output layer. The network output can be determined as follows:

$$y_k = g\left(\sum_{j=1}^{N_H}w_{ojk}X_{Hj} + b_{ok}\right)$$

$$k = 1, \cdots, N_o$$

(3)

where w_{ojk} represents the strength of the connection from the jth hidden unit to the kth output unit and b_{ok} is the bias associated with the kth output unit, while g is the nonlinear transfer function.

2.3   Training the SMIA Network

In this subsection, the training algorithm of the SMIA network will be shown. Furthermore, a B cell construction based hidden unit creation will be described.

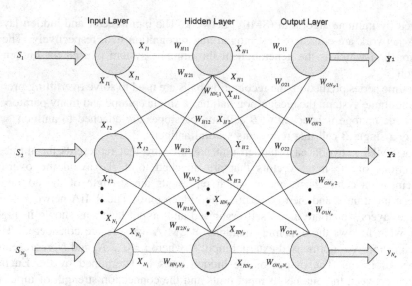

Fig. 2. The structure of the SMIA network [3, 25].

For the immune algorithm, inside the recognition ball, there is a single *B* cell which represents several antigens. In this case the hidden unit is considered as the recognition ball of immune algorithm. Let $d(t + 1)$ represents the desired response of the network at time $t + 1$. The error of the network at time $t + 1$ is defined as:

$$e(t+1) = d(t+1) - y(t+1) \qquad (4)$$

The cost function of the network is the squared error between the original and the predicted value, that is:

$$J(t+1) = \frac{1}{2}[e(t+1)]^2 \qquad (5)$$

The aim of the learning algorithm is to minimise the squared error by a gradient descent procedure. Therefore, the change for any specified element wo_{ij} of the weights matrix is determined according to the following equation:

$$\Delta wo_{ij}(t+1) = -\eta \frac{\partial J(t+1)}{\partial wo_{ij}} \qquad (6)$$

where ($i = 1...., N_H, j = 1...,N_o$) and η is a positive real number representing the learning rate. The change for any specified element b_{ok} of the bias matrix can is determined as follows:

$$\Delta b_{oj}(t+1) = -\eta \frac{\partial J(t+1)}{\partial b_{oj}} \qquad (7)$$

where $(j = 1...,N_o)$. The initial values of w_{oij} are set to zero and the initial values of b_{oj} are given randomly.

2.4   Regularised SMIA Network (R-SMIA)

In this section, the regularisation technique has been introduced in order to improve the performance of the SMIA network. Regularisation is the technique of adding a penalty term Ω to the error function which can help obtaining a smoother network mappings. It is given by:

$$A = E + \lambda\Omega \tag{8}$$

where E represents one of the standard error functions such as the sum-of-squares error and the parameter λ controls the range of the penalty term Ω in which it can influence the form of the solution.

The network training should be implemented by minimising the total error function Ã [18]. One form of regularisation is called weight decay. This form is based on the sum of the squares of the adaptive parameter in the network.

$$\Omega = \frac{1}{2}\sum_i w_i^2 \tag{9}$$

Although the use of weight decay in some cases leads to degraded performance of the network, it has been proven in most cases that it can avoid the overfitting problem and as a result enhance the network performance [17].

The reason behind the popularity of weight decay approach is the simplicity of using this method. The idea is that every weight once updated, is simply decayed or shrunk as follows:

$$w^{new} = w^{old}(1 - \lambda) \tag{10}$$

where $0 < \lambda < 1$. The weight decay is performed by adding a bias term to the original objective function E, thus the weight decay cost function is determined as follows [18]:

$$Ewd = E + (\lambda/2)\,B \tag{11}$$

where λ is the weight decay rate, B represents the penalty term.

The simplest form of calculating the penalty term B is:

$$B = \sum W2ij \tag{12}$$

where w_{ij} is the weight connections between the ith units and j^{th} nodes in the next layer. In R-SMIA network the weight decay were used to adjust the weights between the hidden nodes and output units. The change of weights using weight decay method could be calculated as follows:

$$\Delta W_{ojk} = -\eta \frac{\partial E}{\partial W_{ojk}} = -\eta \frac{\partial}{\partial W_{ojk}} \left(E_{std} + \frac{\lambda}{2} \sum W_{ojk}^2 \right) \qquad (13)$$

$$\Delta W_{ojk} = \eta \left(\sum ef_{ot}'f_{ot} - \lambda \sum W_{ojk} \right) \qquad (14)$$

where Δw_{ojk} is the updated weights between hidden units and output unit. The R-SMIA network is used to examine the effect of the regularisation technique and to enhance the performance of the SMIA network in the prediction.

3 Speech Time Series Prediction

Time series forecasting takes an existing series of data $x_{t-n}, \ldots., x_{t-2}, x_{t-1}, x_t$ and forecasts the next incoming values of the time series $x_{t+1}, x_{t+2}, \ldots\ldots$ Time series forecasting is the process of predicting future values using current values. The prediction of a time series is synonymous with modeling of the underlying physical mechanism responsible for its generation. Most of the physical signals encountered in practice have two distinct properties: nonlinearity, and nonstationary. The speech signal is an example of a physical time series that exhibits those two properties. It should be pointed out that the use of prediction plays a key role in the modeling and coding of speech signals [9]. The production of a speech signal is known to be the result of a dynamic process that is both nonlinear and nonstationary.

In these experiments, male and female voices counting from one to ten using Arabic language are used.

The speech signal is divided into three data sets which are the training, the validation and the test data sets. Each network is trained by incremental

Metrics	Calculations
NMSE	$NMSE = \dfrac{1}{\sigma^2 n} \sum\limits_{i=1}^{n} (y_i - \hat{y}_i)^2$
	$\sigma^2 = \dfrac{1}{n-1} \sum\limits_{i=1}^{n} (y_i - \bar{y})^2$
	$\bar{y} = \sum\limits_{i=1}^{n} y_i$
SNR	$SNR = 10 * \log_{10}(sigma)$
	$sigma = \dfrac{m^2 * n}{SSE}$
	$SSE = \sum\limits_{i=1}^{n} (y_i - \hat{y}_i)$
	$m = max(y)$

Fig. 3. Performance metrics and their calculations, in this case n is the total number of data patterns y and ŷ represent the actual and predicted output values.

backpropagation learning algorithm. Early stopping with maximum number of 3000 epochs was utilized. An average performance of 50 trials is used. The learning rate was selected between 0.05 and 0.5 and the momentum term was experimentally selected between 0.4 and 0.9. Two sets of random weights initialization are employed which are in the range of [−0.5, 0.5] and [−1, 1]. The prediction performance of our networks is evaluated using two statistical metrics as shown in Fig. 3.

4   Simulation Results

In this section, the simulation results for the prediction of the speech signals will be presented using the SMIA. The SMIA using regularization technique such as the weight decays (R_SMIA), the MLP neural networks and the MLP network trained using weight decays (R_MLP).

Table 1 shows the average simulation results for 50 simulations using the various neural network architectures, while Figs. 4 and 5 demonstrate part of the predicted and the original speech signal.

As it can be noticed from Table 1, the simulation results indicated that the SMIA generates the highest value of the SNR and the lowest NMSE in comparison to the MLP, the R_MLP and the R_SMIA.

Table 1. Simulation results for the prediction of the speech signal.

Network	MLP	R_MLP	SMIA	R_SMIA
(a) *Female voice*				
SNR(dB)	21.23	22.22	27.02	26.04
NMSE	0.4524	0.3542	0.1169	0.1466
(b) *Male voice*				
SNR(dB)	22.02	22.34	24.61	22.71
NMSE	0.4758	0.4412	0.2624	0.330

Simulation results indicated that using the regularization technique such as the weight decay did not improve the performance of the SMIA network using the male and the female speech signals. However for the MLP network (using weight decay) has shown slight improvement. This is likely because the selection of hidden nodes in the SMIA network produce its own control of smoothing of the decision surface, which is the role of weight decay in the standard MLP network. For this reason, adding weight decay to the SMIA network does not improve its performance, although it does for the MLP.

5   Conclusion and Further Work

In this paper the simulation results for the prediction of speech signal time series using the SMIA models (self-organized neural network inspired by the immune algorithm) are presented. Results obtained from the experiments showed that SMIA outperformed multilayer perceptron (MLP) using the female and the male voices.

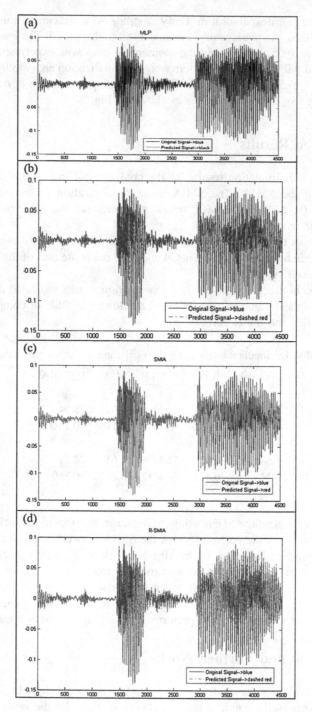

Fig. 4. Part of the predicted and the original signal for the female voice signal using (a) MLP (b) R_MLP and (c) SMIA and (d) R_SMIA neural networks.

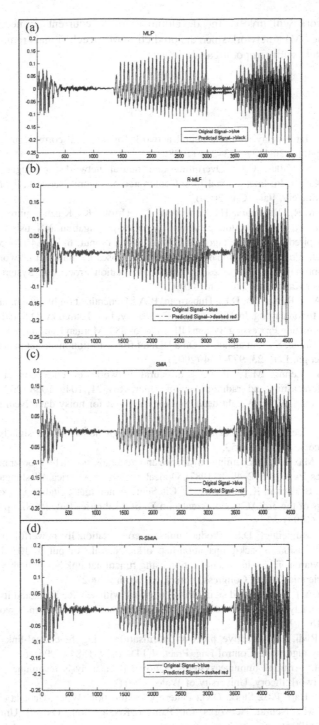

Fig. 5. Part of the predicted and the original signal for the male voice signal using (a) MLP (b) R_MLP and (c) SMIA and (d) R_SMIA neural networks.

Further work will involve the development of a recurrent connections into the structure of the SMIA network since the prediction of speech signals requires the use of recurrent models with temporal capabilities.

References

1. Kim, S.-S.: Time-delay recurrent neural network for temporal correlations and prediction. Neurocomputing **20**(1–3), 253–263 (1998)
2. Lawrence, S., Giles, C.L.: Overfitting and neural networks: conjugate gradient and backpropagation. In: International Joint Conference on Neural Network, pp 114–119. IEEE Computer Society, Italy, CA (2000)
3. Widyanto, M.R., Nobuhara, H., Kawamoto, K., Hirota, K., Kusumoputro, B.: Improving recognition and generalization capability of back-propagation NN using self-organized network inspired by immune algorithm. Appl. Soft Comput. **6**, 72–84 (2005)
4. Hanson, S.J., Pratt, L.: A comparison of different biases for minimal network construction with back-propagation. Advances in Neural Information Processing Systems II, pp. 177–185. Morgan Kaufmann, Los Altos (1990)
5. Weigend, A.S., Rumelhart, D.E., Huberman, B.A.: Generalization by weight elimination with application to forecasting. In: Lippmann, R.P., Moody, J.E., Touretzky, D.S. (eds.) Advances in Neural Information Processing Systems III, pp. 875–882. Morgan Kaufmann, Los Altos (1991)
6. Kaita, T., Tomita, S., Yamanaka, J.: On a HONN for distortion invariant pattern recognition. Pattern Recogn. Lett. **23**, 977–984 (2002)
7. Chen, A.S., Leung, M.T.: Regression neural network for error correction in foreign exchange forecasting and trading. Comput. Oper. Res. **31**, 1049–1068 (2004)
8. Gupta, A., Lam, S.M.: Weight decay backpropagation for noisy data. Neural Netw. **11**(6), 1127–1137 (1998)
9. Haykin, S., Lee, L.: Nonlinear adaptive prediction of nonstationary signals. IEEE Trans. Signal Process. **43**(2), 526–535 (1995)
10. Giles, C.L., Maxwell, T.: Learning, invariance and generalisation in high-order neural networks. Appl. Optics **26**(23), 4972–4978 (1987). Optical Society of America, Washington DC
11. Giles, C.L., Bollacker, K., Lawrence, S.: CiteSeer: an automatic citation indexing system. In: Proceedings of the 3rd ACM Conference on Digital Libraries (DL 1998), pp. 89–98. ACM, New York (1998)
12. Durbin, R., Rumelhart, D.E.: Product units: a computationally powerful and biologically plausible extension to back-propagation networks. Neural Comput. **1**, 133–142 (1989)
13. Mirea, L., Marcu, T.: System identification using functional-link NN with dynamic structure. In: 15th Triennial World Congress, Barcelona, Spain (2002)
14. Fei, G., Yu, Y.L.: A modified sigma-Pi BP network with self-feedback and its application in time series analysis. In: Proceedings of the 5th International Conference, vol. 2243–508F, pp. 508–515 (1994)
15. Cass, R., Radl, B.: Adaptive process optimisation using functional-link networks and evolutionary algorithm. Control Eng. Pract. **4**(11), 1579–1584 (1996)
16. Timmis, J.I.: Ariticial immune systems: a novel data analysis technique inspired by the immune network theory. University of Wales (2001)
17. Duda, R.O., Hart, P.E., Stork, D.G.: Pattern Classification. Wiley, Canada (2001)
18. Bishop, C.M.: Neural Networks for Pattern Recognition. Oxford University Press, Cambridge (1995)

A Framework to Support Ubiquitous Healthcare Monitoring and Diagnostic for Sickle Cell Disease

Mohammed Khalaf[1(⊠)], Abir Jaafar Hussain[1], Dhiya Al-Jumeily[1],
Paul Fergus[1], Russell Keenan[2], and Naeem Radi[3]

[1] Applied Computing Research Group,
School of Computing and Mathematical Sciences,
Liverpool John Moores University, Byrom Street, Liverpool L3 3AF, UK
M.I.Khalaf@2014.ljmu.ac.uk,
{a.hussain,d.aljumeily,P.Fergus}@ljmu.ac.uk
[2] Liverpool Paediatric Haemophilia Centre, Haematology Treatment Centre,
Alder Hey Children's Hospital,
Eaton Road, West Derby, Liverpool L12 2AP, UK
Russell.keenan@alderhey.nhs.uk
[3] Al Khawarizmi International College, Abu Dhabi, United Arab Emirates
n.radi@khawarizmi.com

Abstract. Recent technology advances based on smart devices have improved the medical facilities and become increasingly popular in association with real-time health monitoring and remote/personals health-care. Healthcare organisations are still required to pay more attention for some improvements in terms of cost-effectiveness and maintaining efficiency, and avoid patients to take admission at hospital. Sickle cell disease (SCD) is one of the most challenges chronic obtrusive disease that facing healthcare, affects a large numbers of people from early childhood. Currently, the vast majority of hospitals and healthcare sectors are using manual approach that depends completely on patient input, which can be slowly analysed, time consuming and stressful. This work proposes an alert system that could send instant information to the doctors once detects serious condition from the collected data of the patient. In addition, this work offers a system that can analyse datasets automatically in order to reduce error rate. A machine-learning algorithm was applied to perform the classification process. Two experiments were conducted to classify SCD patients from normal patients using machine learning algorithm in which 99 % classification accuracy was achieved using the Instance-based learning algorithm.

Keywords: Sickle cell disease · Mobile healthcare service · Real-time data · Self-care management system · E-Health

1 Introduction

Nowadays, the development in communication technologies and their implementation in medical/science data have helped the utilisation of new solutions to enhance healthcare services and outcomes. SCD is one of the most prevalent diseases, which

© Springer International Publishing Switzerland 2015
D.-S. Huang et al. (Eds.): ICIC 2015, Part II, LNCS 9226, pp. 665–675, 2015.
DOI: 10.1007/978-3-319-22186-1_66

could have an influence on patients' lives due to red blood cell (RBC) abnormality. SCD is caused by a genetic disorder that reduces life expectancy around 45 years in developed countries but it is rare for children with sickle cell in Africa and Asia continents to survive beyond their 5th birthday. The main reason behind this disease is that a group of ancestral disorders have an impact on the protein inside the RBC called haemoglobin. SCD can easily inherited to the child throughout genetic of sickle haemoglobin (Hb S) either from both parents or from one of them measured as abnormal haemoglobin [1].

According to the World Health Organisation (WHO), 7 million born each year suffer either from congenital anomaly or from an inherited disease [2]. Furthermore, 5 % of the population around the globe carries traits genes for haemoglobin disorders, primarily, thalassemia and SCD [3]. Within this context, there is still a significant need for constructing cooperative care environment to improve quality and safety of care, and increase caregivers' efficiency with the purpose of providing regular information for association healthcare providers and patients. In order to provide such facilities, this research focuses on the development of self-care monitoring system (personalised monitoring system) based on mobile devices. Therefore, continuous self-care monitoring of chronic diseases and medicine intake are vital for patients to mitigate the severity of the disease by taking the appropriate medicine at the proper time.

Over the past decade, mobile technologies have supported medical fields through intelligent data analytics. The state of the arts growth of communication technologies has raised the capacity to adopt smart devices health applications in clinics and medical environments [4]. The quick improvement of e-Health systems concentrate on developing the pattern of diagnosing as well as treating chronic diseases with the assistance of disease management centre or integrated care strategies [5]. The main reason behind using mobile devises is that they can offer real-time healthcare services efficiently, delivering information and more flexible tools for doctors-patients interaction, which is relevant to the patient's condition. They are also assist healthcare professionals in order to give accurate decision-making and store data for further investigation. These developments bring potential advantages to both medical expert's staff and patients; doctors can concentrate more on high priority responsibilities and patients could acquire a regular consultancy without having regular visits to hospitals, especially if they live in a remote area [6].

This paper provides a framework for monitoring and managements of patients' diary to provide better health-care and services for patients who suffer from SCD. An initial essential part of the proposed system is a machine learning algorithm which is used to classify patients who suffer from the disease and those who do not.

The remainder of this paper is organised as follows. Section 2 will discuss literature review about SCD while Sect. 3 shows the proposed monitoring system. The methodology for classifying sickle cell patients from normal patients is illustrated in Sect. 4, while Sect. 5 shows the conclusion section.

2 Related Work

Intelligent mobile medical care plays an important tool on providing accurate information for immediate healthcare services and diagnosis. Self-care management system provides patients and medical experts continuous monitoring to obtain instant response and save people lives in case of abnormal conditions. In recent years, healthcare organisations worldwide have faced many problems in order to meet the demands of enhanced medical sectors. There are a number of research projects developed for healthcare environments based on mobile computing. Tabish et al. [7] proposed a ubiquitous healthcare solution based on IPv6 over Low power Wireless Personal Area Networks (6LowPAN) that uses sensors to record electrocardiogram (ECG) and temperature data to a remote processing centre. This solution, named as U-health uses 3G/4G communication to provide real-time monitoring to the patients within a mile of coverage area.

Kim et al. [8] developed a new framework based on ubiquitous healthcare systems, which can work anywhere and at any time. In this case, their system provides real-time service based upon various biosensors measurements such as Electrocardiogram (ECG), blood pressure and temperature. They also have created Hadoop platform (Big Data Centre) in order to store medical data. Leijdekkers [9] proposed a new prototype system supported by wireless sensor network (biosensor) and smart phone to monitor heart fluctuation. The system itself provides analysis of the giving data and automatically alarms the ambulance to take action immediately. Furthermore, the system transmits sensor data to a healthcare management centre for further investigations. Chen et al. [10] presented pervasive healthcare monitoring system (PHMS) combined with cloud computing environment, mobile application, and planar super wideband (SWB). The PHMSs can deliver facilities for those who need long term and continuous collection of data, in particular disabled and elderly people for living independent life. The vital signs monitor (VSM) can deliver immediate information to the healthcare centre server with regard to analysis and storage of such data. The medical experts can seamlessly access the database and check final status.

A mobile phone diary platform for adolescent in association with SCD is proposed [11]. This solution can provide principal support for the utilisation of electronic diary to track symptoms and pain for young people. Their work based on utilising the internet communication capability of state-of-the-art smartphone based on 3G/4G [12]. Jung, et al. [13] established a new method based on IEEE802.15.4 with wireless sensor networking (WSN) for advanced healthcare monitoring using two techniques. First, wearable healthcare devices located in the human body. Second, a number of sensors placed in house celling with all information transmitted to the server in order to be stored and later analysed. The key important point behind applied like this system is to provide periodic monitoring system when situation required. Our literature review shows the current contributions are still limited for providing immediate information about SCD normality or abnormality [11, 13]. Therefore, this research work was focusing on the field of building robust system that can track disease and send instant information to the medical consultants in order to provide accurate decision, especially with critical conditions cases.

3   System Design of Healthcare Based on Mobile Healthcare (M-Healthcare)

In this research, we present a framework for applying smartphone device management system for the ubiquitous healthcare. The proposed system consists of mobile Platform, network coordinator and personal computers. In our framework, there are two important sides that have been proposed. The patient side is responsible for analysis, data storing and send feedback messages to the healthcare professionals in association with high-risk condition. The high-risk condition can be obtained when the number of heartbeat is significantly increasing or breathing becomes difficult also known as vaso-occlusive crisis. The medical side is responsible for providing accurate decision support to advice patient by taking appropriate type of medicine, particularly within crisis cases. Therefore, the proposed system has the ability of analysing and storing data without sending the information to the health professionals when detecting low-level condition with regards to reduce communication cost, workload, and encourage personalised patient's self-care system. The low-level condition can be obtained when the number of heartbeat is slightly decreasing. The proposed framework design is shown in Fig. 1.

The deployment of self-care management system consists of two main categories, the data collecting, and data analysis. In this paper, we also consider medical centre (located at hospital and clinics environments). The overall responsibility of medical centre is monitoring patients, gather data and sending reminder when the data indicates the patient is in critical condition.

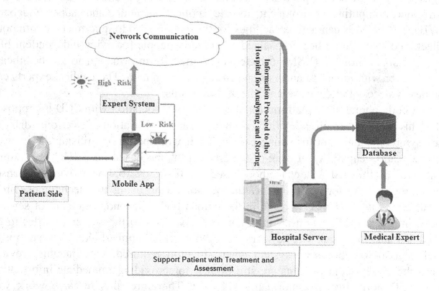

Fig. 1. System architecture of the proposed system.

These days, smart devices play such an important role to support people by accessing their condition from home. Within this context, mobile technology were used in this research to analyse all data that collected from patient's body in order to send that data to the medical consultants for diagnosing the patient's situation. The communications platform between patient and mobile device will be implemented through the application interface. In this aspect, we intend to store all the relevant information that come through patients input with the purpose of further investigations and analysis. It is essential that a patient who is suffering from SCD, is required to store relevant information, for instance personal details. This information can be used in the processing stage, which will be located in the device's platform. This would in turn, remind the patient of their current condition. Furthermore, the system then carries out analysis on the data that sent to determine whether the patient's condition is normal or abnormal. Afterward, the system will advise the patient of what medication is needed to correct their condition promptly and accurately. Otherwise, the system needs to send the whole details to a medical consultant which in turn to decide what action is required in order to save patient's life. The main motivation behind utilising mobile application (app) is to determine the patients' condition more promptly.

Based on the large volume of data that is generated in medical sectors, it has become vital to utilise expert system in the purpose of enhancing the quality of care. Expert system provides an effective and reliable way on improving the patients facilities within healthcare sectors [14]. The main significant factors that we used expert system in this paper is to deliver unlimited services. For example, managing, analysing and diagnosing patient's data to detect normal and abnormal patterns in order to save the patient's life.

3.1 Self-care Management System

There is no laboratory facilities available to test the blood for genetic blood disorder in the patients' home. The main reason behind this is because of the high financial costs of installing a blood test machine in the patient's home. In addition, the blood test machine requires a specialist nurse or medical expert to understand the main attributes that will be obtained after the blood has been tested. There are four types of SCD which is considered of an abnormal level, for instance (Hb SS, Hb SC, Hemoglobin S Beta + Thalassemia, Hemoglobin S Beta + 0 Thalassemia). The patient's diagnosis will be assessed according to the symptoms (e.g. severe pain in the bones, painful enlargement of the spleen and heart problems, headache, very pale skin, and chest pain) that could appear in various areas of the body. The data will be collected by the android mobile App, which will provide patients with a set of multi choice to select the proper symptoms. The proposed expert system will then analyse the input data and determine whether the patient's condition is normal or abnormal. This information will be transferred to the medical center in order to be analysed by a professional who makes any decisions needed in order to tackle patient's condition. This would be extremely beneficial, within crisis cases. In these circumstances, the healthcare professionals or specialist nurses could contact patients for professional advises on their condition.

4   Methodology

This paper has proposed a framework for monitoring and diagnosis SCD for the purpose of helping patients and healthcare organisation. The first component of the proposed system, is the classification of the sickle cell patients from normal patients using machine learning algorithms.

The original dataset was collected from a local hospital in Liverpool, UK for the purpose of classifying normal and sickle cell patients. Within this context, the local hospital has been selected according to the highest number of patients who actually suffer from SCD. The data represents 12 primary attributes using 100 patients checked for SCD of which 63 tested negative (patients has already carrying SCD traits), whereas 37 person tested positive (Non SCD traits affected). Table 1 illustrates the characteristics of SCD.

Table 1. Characteristics of SCD dataset.

No	Types of Attributes	Min	Max
1	Gender	Male, female	Male, female
2	Age	Between 15	Between 80
3	Hb	g/L 118	g/L 148
4	Hb F	0.8 %	2.0 %
5	PLTS	x109/L 150	x109/L 400
6	MCV	fL 80	fL 100
7	BILI	umol/L 21	umol/L 30
8	ALT	U/L 35	U/L 45
9	EGFR	ml/min/1.73 m^2 > 45	ml/min/1.73 m^2 > 60
10	Ferritin	ug/113	ug/150
11	Urea	mmol/L 2.5	mmol/L 7.8
12	Creatinine	umol/L 50	umol/L 130

4.1   Confusion Matrix Evaluation

In this section, there are two individual classification in the field of machine learning (including Zero-Rule and Instance-based learning algorithm) have examined to find the better value for Accuracy by class and the weight average. The evaluation technique was carried out in a confusion matrix (also known as contingency table). Figure 2 illustrates the confusion matrix. There are four denotes that are located in the contingency table. True Negative (TN) and True Positive (TP) donates are considered one of the most accurate classification of negative instance and the accurate classification of positive instance respectively. In addition, False Negatives (FN) illustrate the positive instance, which is incorrectly classified in terms of negative type, whereas False Positives (FP) show negative symbols, which is incorrectly classified in association with positive type.

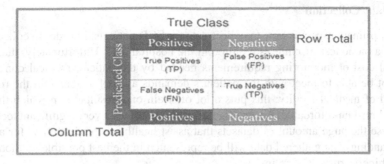

Fig. 2. Confusion matrix.

Based on the confusion matrix, there are a number of measurements that can be acquired to examine the model performance in terms of accuracy (also known as producer's accuracy) this can be determined using formula (1) below.

$$AC = \frac{TP + TN}{(TP + FP) + (FN + TN)} \tag{1}$$

The main purpose of applying Eq. (2) is to evaluate the proportion of positive instances that were correct.

$$TP = \frac{TN}{(FN + TN)} \tag{2}$$

The FP were classified incorrectly can be obtained from the following equation that illustrated in (3).

$$FP = \frac{TN}{(TP + FP)} \tag{3}$$

In order to classify the TN that were classified correctly, Eq. (4) were conducted.

$$TN = \frac{TP}{(TP + FP)} \tag{4}$$

The FN is belong positives instances were incorrectly classified. The following formula (5) illustrates how datasets can be calculated.

$$FN = \frac{FN}{(FN + TN)} \tag{5}$$

To evaluate the accuracy, it is typical practice to utilise confusion matrix. In this case, the confusion matrix mainly depends on the selection of datasets. Our motivation behind applying contingency table is to improve the accuracy and performance of benchmark datasets.

4.2   Data Collection

The first primary aspect is how to collect data for SCD patients. In order to collect SCD dataset, a blood test machine is required in the hospital side. Unfortunately, due to the financial cost of monitoring requirements needed by the patient, medical consultants were not be able to meet with the patient's demands as well as maintain the required standard of medical facility; this puts a lot of strain on consultants in which the total cost is almost un-affordable. An intelligent system becomes a very significant technique to analyse the huge amount of datasets that assist healthcare professionals for a better understanding. The gathered data will be represented in the best possible method based on machine learning algorithm.

This research proposed Zero-Rule based and Instance-based learning algorithm for diagnosing and managing SCD. Data was simulated using real-life data sets provided by local hospital when a patient takes blood test sample, which contains a set of attributes that used for checking patients' condition. In the medical centre, there are 12 attributes were used in this research as shown in Table 1 (e.g., Haemoglobin (Hb), Platelets (PLTS), Mean Corpuscular Volume (MCV), Neutrophils (white blood cell NEUT), Reticulocyte Count (RETIC), Reticulocyte Count (RETIC F), Body Bio Blood (BIO), Hb F, Bilirubin (BILI), Alanine aminotransferase (ALT), an Aspartate Aminotransferase (AST), Lactate dehydrogenase (LDH)). Furthermore, there are two features should be considered when analysing the blood test, which are gender and age. These features have some significant effects on the blood test.

Table 2. Classifier output for zero-rule algorithm.

No	Features	Classifier output
1	Correctly classified instances	62 %
2	Incorrectly classified instances	38 %
3	Kappa statistic	0
4	Mean absolute error	0.4722
5	Root mean squared error	0.4858
6	Relative absolute error	100 %
7	Root relative squared error	100 %
8	Total number of Instances	100

Table 3. Accuracy by class for Zero-Rule algorithm.

Measures	Accuracy by class	Weighted avg.
True positive rate	1	0.62
False positive rate	1	0.62
Precision	0.61	0.372
Recall	1	0.61
F-measure	0.758	0.462
ROC area	0.483	0.483

The Zero Rule algorithm with cross validation was utilised for the classification of the patients' data. It is a rule based classifier which is also called 0-R. Table 2 shows the accuracy and average weight classified by the proposed algorithm. The correct classified Instances that we obtained after utilising Zero-Rule algorithm is 62 %, whereas, incorrect classified Instances is 38 %. Further accuracy of the classifier is shown in Table 3.

In order to enhance our result with more achievable outcomes, we have used another simple classifier which is called the Instance-based learning algorithm developed by Aha et al. [15]. The outcomes with proposed classifier shows better performance than Zero-Rule algorithm. A performance of 99 % accuracy was achieved as illustrated in Tables 4 and 5.

Table 4. Classifier output for the Instance-based learning algorithm.

No	Features	Classifier output
1	Correctly classified instances	99 %
2	Incorrectly classified instances	1 %
3	Kappa statistic	0.9789
4	Mean absolute error	0.01
5	Root mean squared error	0.1
6	Relative absolute error	2.1179 %
7	Root relative squared error	20.5858 %
8	Total number of instances	100 patients

Table 5. Accuracy by class for instance-based learning algorithm.

Measures	Accuracy by class	Weighted avg.
True positive rate	1	0.99
False positive rate	0.016	0.006
Precision	0.974	0.99
Recall	1	0.99
F-measure	0.987	0.99
ROC area	0.992	0.992

The Instance-based learning algorithm achieved a better outcomes in comparison to Zero-Rule algorithm due to the optimal accessible heuristic for calculating the values of attributes. This algorithm can deal with large amount of datasets that need to be classified. In addition, this technique requires low storage and provides high performance in terms of classification accuracies. Our motivation behind using Instance-based learning algorithm is that it can save documents of completed training-set. Then, compares any new documents to be classified with the saved data. Furthermore, it can classify a large number of new datasets more rapidly. The performance method compare among other techniques are considerably much more beneficial in providing accurate outcomes. Thus, the main purpose of applying these two experiments is to examine all the attributes (excluding the class value) whether the patients carry SCD traits or otherwise.

5 Conclusion

This paper proposed a new framework for self-care management system in order to improve the idea of electronic healthcare for supporting patient's medical facilities. This will provide better satisfaction, quality of service, and lowering the costs of patient care. Self-care monitoring system is the most crucial function for patients' daily life. This could provide the ability to check and monitor their own health condition through smart devices technology. Based on this technology, personal monitoring system supports SCD patients to manage their health status instead of visiting healthcare professionals at regular time or being admitted into hospitals. Two machine learning algorithms were used to classify patients with SCD. The Instance-based learning algorithm offers an important improvement in terms of accuracy over Zero-Rule classifier with 99 % accuracy in comparison to 62 % using Zero-Rule algorithm. In this case, the possible way to improve the accuracy of these experiments is to increase the sample size of the datasets.

References

1. Weatherall, D.J.: The role of the inherited disorders of hemoglobin, the first "molecular diseases," in the future of human genetics. Annu. Rev. Genomics Hum. Genet. **14**, 1–24 (2013)
2. Weatherall, D.J.: The importance of micromapping the gene frequencies for the common inherited disorders of haemoglobin. Br. J. Haematol. **149**, 635–637 (2010)
3. Weatherall, D.J.: The inherited diseases of hemoglobin are an emerging global health burden. Blood **115**, 4331–4336 (2010)
4. Lin, M.K.: Evaluating the acceptance of mobile technology in healthcare: development of a prototype mobile ECG decision support system for monitoring cardiac patients remotely. University of Southern Queensland (2012)
5. Gillespie, G.: Deploying an IT cure for chronic diseases. Health Data Manage. **8**, 68 (2000)
6. Chan, V., Ray, P., Parameswaran, N.: Mobile e-Health monitoring: an agent-based approach. IET Commun. **2**, 223–230 (2008)
7. Tabish, R., Ghaleb, A.M., Hussein, R., Touati, F., Ben Mnaouer, A., Khriji, L., et al.: A 3G/WiFi-enabled 6LoWPAN-based U-healthcare system for ubiquitous real-time monitoring and data logging. In: 2014 Middle East Conference on Biomedical Engineering (MECBME), pp. 277–280 (2014)
8. Kim, T.W., Park, K.H., Yi, S.H., Kim, H.C.: A big data framework for u-Healthcare systems utilizing vital signs. In: 2014 International Symposium on Computer, Consumer and Control (IS3C), pp. 494–497 (2014)
9. Leijdekkers, P., Gay, V.: Personal heart monitoring and rehabilitation system using smart phones. In: International Conference on Mobile Business, ICMB 2006, pp. 29–29 (2006)

10. Chen, K.-R., Lin, Y.-L., Huang, M.-S.: A mobile biomedical device by novel antenna technology for cloud computing resource toward pervasive healthcare. In: 2011 IEEE 11th International Conference on Bioinformatics and Bioengineering (BIBE), pp. 133–136 (2011)
11. Yang, S., Jacob, E., Gerla, M.: Web-based mobile e-Diary for youth with sickle cell disease. In: 2012 IEEE Consumer Communications and Networking Conference (CCNC), pp. 385–389 (2012)
12. Venugopalan, J., Brown, C., Cheng, C., Stokes, T.H., Wang, M.D.: Activity and school attendance monitoring system for adolescents with sickle cell disease. In: 2012 Annual International Conference of the IEEE Engineering in Medicine and Biology Society (EMBC), pp. 2456–2459 (2012)
13. Jung, S.-J., Kwon, T.-H., Chung, W.-Y.: A new approach to design ambient sensor network for real time healthcare monitoring system. In: 2009 IEEE on Sensors, pp. 576–580 (2009)
14. Doherty, J.A., Reichley, R.M., Noirot, L.A., Resetar, E., Hodge, M.R., Sutter, R.D., et al.: Monitoring pharmacy expert system performance using statistical process control methodology. In: AMIA Annual Symposium Proceedings, p. 205 (2003)
15. Aha, D.W., Kibler, D., Albert, M.K.: Instance-based learning algorithms. Machine learning, 6, 37–66 (1991)

A Machine Learning Approach to Measure and Monitor Physical Activity in Children to Help Fight Overweight and Obesity

P. Fergus[1,4(✉)], A. Hussain[1,4], J. Hearty[2,4], S. Fairclough[2,4],
L. Boddy[1,4], K.A. Mackintosh[3,4], G. Stratton[3,4], N.D. Ridgers[2,4],
and Naeem Radi[4,5]

[1] Liverpool John Moores University, Byrom Street, Liverpool L3 3AF, UK
[2] Edge Hill University, St. Helens Road, Ormskirk, Lancashire L39 4QP, UK
[3] Swansea University, Singleton Park, Swansea SA2 8PP, UK
[4] Deakin University, Geelong, Victoria, Australia
p.fergus@ljmu.ac.uk
[5] Al-Khawarizmi International College, Abu Dhabi, UAE

Abstract. Physical Activity is important for maintaining healthy lifestyles. Recommendations for physical activity levels are issued by most governments as part of public health measures. As such, reliable measurement of physical activity for regulatory purposes is vital. This has lead research to explore standards for achieving this using wearable technology and artificial neural networks that produce classifications for specific physical activity events. Applied from a very early age, the ubiquitous capture of physical activity data using mobile and wearable technology may help us to understand how we can combat childhood obesity and the impact that this has in later life. A supervised machine learning approach is adopted in this paper that utilizes data obtained from accelerometer sensors worn by children in free-living environments. The paper presents a set of activities and features suitable for measuring physical activity and evaluates the use of a Multilayer Perceptron neural network to classify physical activities by activity type. A rigorous reproducible data science methodology is presented for subsequent use in physical activity research. Our results show that it was possible to obtain an overall accuracy of 96 % with 95 % for sensitivity, 99 % for specificity and a kappa value of 94 % when three and four feature combinations were used.

Keywords: Physical activity · Overweight · Obesity · Machine learning · Neural networks · Sensors

1 Introduction

According to the McKinsey Global Institute report[1], the global cost of obesity is comparable with smoking and armed conflict and greater than both alcoholism and climate change. The report claims that it costs £1.3tn, or 2.8 % of annual economic activity. In the UK, the cost is £47bn. The prevalence of overweight and obesity is

---

[1] Source: McKinsey Global Institute (2014).

© Springer International Publishing Switzerland 2015
D.-S. Huang et al. (Eds.): ICIC 2015, Part II, LNCS 9226, pp. 676–688, 2015.
DOI: 10.1007/978-3-319-22186-1_67

alarming, with 2.1bn people (30 % of the world's population) being overweight or obese. According to the World Health Organization (WHO), at least 2.8 million people worldwide die from being overweight or obese, and are the cause of a further 35.8 million of global Disability-Adjusted Life Years (DALY)[2]. In the UK, data extracted from the Health Survey for England (HSE) in 2012 (Ryley 2013) shows that the percentage of obese children between the age of 2 and 15 has increased since 1995. Fourteen percent of boys and girls were classified as obese and 28 % as either over-weight or obese. In addition, 19 % of children aged between 11 and 15 were more likely to be obese than children between 2–10 years.

The physical condition of adults is strongly influenced by the early stages of life. In this sense, the data provided by the HSE revealed that in 2012 almost a quarter of men (24 %), and a quarter of women (25 %) were obese, while 42 % of men and 32 % of women were overweight. Therefore, an increase in levels of obesity and overweight in the UK is evident, which depends, among other factors, on the lack of physical activity. This growing trend of obesity within the UK and other countries and its associated health risks are cause for national and international concern.

Consequently, the accurate measurement of physical activity (PA) in children (particularly in free-living environments) is of great importance to health researchers and policy-makers (Oyebode and Mindell 2013). One of the most reliable means of activity measurement in children is using accelerometers to measure movement intensity and frequency (Konstabel et al. 2014). This method offers the advantage of providing quantified values for activity intensity over time. However, methodological variations in data gathering and analysis methods used between studies have led to a growing saturation of conflicting cut-points (the quantitative boundaries between PA intensity classes) in the literature (Trost et al. 2010).

Consequently, several new research directions have been proposed that try to address this challenge. One such approach is the use of machine learning techniques for the prediction or detection of activity types and their associated intensity (Barshan and Yuksek 2014; Dalton and O'Laighin 2013; Trost 2012). This paper builds on existing research and previous works and presents a methodology for predicting physical activity types and intensity using a dataset obtained via the field-based protocol described in (Machkintosh et al. 2012). Exploratory data analysis is utilized to determine what activities and feature sets produce the best results when activity types are predicted using a Multilayer Perceptron (MLP).

2 Background

Physical activity is defined as any bodily movement produced by skeletal muscles that results in energy expenditure (Caspersen et al. 1985) and is measured in Kilojoules (Kj). The measurement of physical activity has become a fundamental component in healthy lifestyle management. Recommendations for physical activity levels are issued by most governments as part of public health measures (Pate, et al. 1995). However, they tend to

---

[2] http://www.who.int/.

be updated frequently or adjusted due to external circumstances, such as changes in diet and food pricing (Duffey et al. 2010), sedentary lifestyle (Martinez-Gonzalez 1999), technology (Kautiainen et al. 2005), the built environment (Saelens et al. 2003), family structure (Lissau and Sorensen 1994) and social influences (Mcferran et al. 2010). Consequently, it has become increasingly important, from a public health policy-makers perspective, to develop reliably measuring physical activity intensity to ground public health guidelines.

Artificial neural networks (ANN) have been used to classify physical activity (De Vries et al. 2011). In one study, Staudenmayer et al. developed an ANN, to classify activity type in adults, using time windows, with 88 % overall accuracy and a consistently low Root Mean Squared Error (rMSE) measure (Staudenmayer et al. 2009). In a study carried out by De Vries et al., a series of ANNs were developed to predict PA in children across a range of activity types. However, the results reported were significantly lower than those reported in (Staudenmayer et al. 2009) with classification accuracies between 57.2 % and 76.8 % (De Vries et al. 2011; De Vries et al. 2001).

While, Trost et al. conducted a rigorous study in which 90 ANN designs (different hidden layer and weight sizes) were developed and trained to predict PA type and intensity (Trost 2012). The best performing design was trained using features extracted from a range of time windows (10, 15, 20, 30 and 60 s), with the most successful network able to predict PA type with 88.4 % accuracy over a 60-s window, and PA intensity, with the network able to classify moderate to vigorous intensity activities 93 % of the time.

While Trost's study is one of the forerunners for artificial neural network usage in the classification of physical activity types and intensity, Trost points out their ANNs produce high error margins (as high as 44.6 % in the case of sedentary activity), and recommend that a combination of triaxial accelerometer use and different pattern recognition algorithms may help to generate more precise ANN outputs.

3 Methodology

The Mackintosh et al. dataset, used in this study, contains records for Twenty-eight children aged between 10 and 11 years of age from a North-West England primary school who participated in the study (Machkintosh et al. 2012). Children completed seven different physical activities performed in a randomized order, which took place in the school playground or classroom as appropriate with 5 min seated rest between each activity. To capture both the sporadic nature of children's activity (Orme et al. 2014) and locomotive movement best suited to accelerometers (Welk 2005), the activities incorporated both intermittent and continuous (i.e., walking and jogging) movements representative of culturally-relevant-free-play situations. Children who were performing sedentary activities were watching a DVD and drawing, which were consistent with those used previously (Evenson et al. 2008).

The dataset contains 28 records of children, age 11.4 ± 0.3 years, height 1.45 ± 0.09 m, body mass 42.4 ± 9.9 kg, and BMI 20.0 ± 4.7, where 46 % of the population was boys and the remainder girls. The dataset also contains physical activity codes from the System for Observing Fitness Instruction Time (SOFIT) (Mckenzie et al. 1992) to directly observe (DO) the children's physical activity behaviors during the activities. The physical activity coding element of SOFIT uses momentary time sampling to quantify health-related physical activity where codes 1 to 3 represent participants' body positions (lying down, sitting, standing), code 4 is walking, and code 5 (very active) is used for more intense activity than walking (Mckenzie et al. 1992). These DO physical activity codes have been validated with heart rate monitoring (Rowe et al. 2004), oxygen consumption (Row et al. 2004), (Honas et al. 2008), and accelerometry (Scruggs et al. 2005), (Sharma et al. 2011) with preschool to year 12 children, including those with development delays (McKenzie 2010). Throughout the protocol each child's activity was coded every 10-s by a trained observer.

The accelerometer and DO values obtained during the recording and observation period were processed to generate mean values per epoch. Some values were subsequently used to approximate values for an additional set of features. This approximation was achieved via the use of established calculations for mean hand accelerometer count (HAC), mean waist accelerometer count (WAC), direct observation values (DO), body mass index, heart rate count (HR), moderate physical activity percentage (MPA %), vigorous physical activity percentage (VPA%), indirect calorimetry oxygen consumption (V02), and energy expenditure (EE). For a complete description of the dataset the reader is referred to (Machkintosh et al. 2012).

3.1 Data Pre-processing

One notable concern with the dataset is that a significant number of values were missing. While the study involved 28 participants not all subjects performed every activity, and in some cases, values were missing for a number of features and/or activities. One subject performed no activities and two subjects performed only one activity and were consequently removed from the dataset. A further six subjects had a significant number of missing values for some or all of the features HR, MPA%, MPA time, VPA% and VPA time. Four of the six subjects had missing values for all activities and were therefore removed from the dataset. The remaining two subjects were missing values from three activities. Substitute values for these children were computed using cubic spline interpolation. For the features EE and V02, five subjects had missing values for some or all of the activities performed. As a result, these five records were removed, leaving a total dataset containing 16 cases per activity, with no missing or null values.

3.2 MLP Classification Trial for Activity Selection

MLP Classification Trial for Activity Selection: The classifier trial uses the mean hand and waist accelerometers, BMI and Direct Observation. A four-class classification problem was performed using combinations of four activities from the initial seven activities. Thirty iterations of 35 permutations of the 4-class classification problem were performed, and in each case, sensitivity and specificity data for each activity class was obtained.

Figure 1 shows that the activities Free Play and Playground demonstrate a greater spread of both specificity and sensitivity values across a range of classification problems, as well as having lower mean sensitivity and specificity values, than almost all other activities.

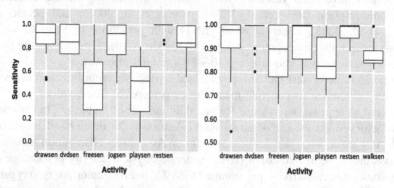

Fig. 1. Mean sensitivity and specificity values for each activity across 30 MLP classification trials per 4-activity combination.

Activity values surrounding Free Play and Playground (Jogging and Walking) show substantial variance in specificity, while, Jogging also shows variance in sensitivity. This is related to the significant overlap between feature values for the activities Free Play, Playground and other activities. Classification sensitivity (true positive rate) tends to be high for most activities, with mean values above 0.8 for all activities except Free Play and Playground. While mean classification specificity (false positive rate) is very high (>0.95) for all activities except Free Play, Playground and Walking, the variance in classification specificity values is significant for all activities except DVD watching. Observations for Drawing, Resting and DVD watching were confused for one another, which lowered the sensitivity of both features. DVD watching was often mistaken for Drawing or Resting and vice versa, which led to DVD watching having far higher specificities than either Drawing or Resting.

This analysis suggests that the activity set currently used is not appropriate for ANN classification analysis. This is due to the presence of multiple classes whose observation cases occupy the same region of values. For this reason, only one of the three sedentary activities (Drawing, DVD Watching and Resting) is used. Despite the excellent sensitivity values for the activity Resting showed, it was decided that the activity Drawing would be used. The Resting activity was initially intended for use in calibrating basal rates for various features, and was not intended for classification analysis. Furthermore, a significant number of features (MPA, MPA time, VPA and VPA time) are missing from the Resting data, which would significantly complicate MLP analyses using those features. Conversely, both Drawing and DVD watching possess a full complement of feature data and were intended for use in classification of sedentary activities. The final dataset following this analysis contains four activities,

Drawing, Free Play, Jogging, and Walking. This set covers a good breadth of activity intensities, while minimizing the risk of value overlap or classification error.

3.3   MLP Classification Trial for Activity Selection

Statistical comparison of features is performed on a per-activity basis. Figure 2 shows the plots for the statistical analysis of features. The feature BMI was retained for all four activities, although naturally the range and distribution of values is identical across all activities. This was done to establish a common scaling for all four plots. The stationary, sedentary nature of the Drawing activity was intended to provide a resting comparison to the more vigorous activities used. As such, all features have values at or around the minimum value of -1. For the most part, this suggests that Drawing may be easily distinguished from the other three activities.

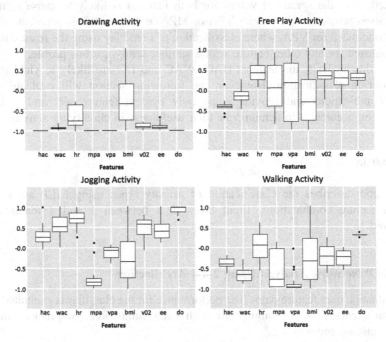

Fig. 2. Boxplots per activity for feature statistical analysis

Feature values show a broader spread for Free Play than for any other activity. The features HR, MPA, VPA, V02 and EE show significant coincidence between Free Play and Jogging, while some degree of coincidence between values for Free Play and Walking is present for almost every feature; suggesting that misclassification may occur at those class boundaries. The features HAC, WAC, and DO possess a small interquartile range, entailing that the majority of the data falls within a limited space of values. This is a positive finding for classification purposes, but one, which requires validation through MLP analysis.

Conversely, the interquartile range of the features MPA and VPA varies greatly. These features are measures of what proportion of the activity time was spent at vigorous or moderate levels of physical activity. In some activities this leads to an unusual distribution of values for both features; if an activity is vigorous, for instance, the VPA value may be at the maximum value for all subjects. If an activity is not vigorous, the values for all subjects performing the activity may be at the minimum of −1. However, activities which may or may not be vigorous, or which alternate between vigorous and non-vigorous activity states, tend to contain a range of MPA or VPA values.

In the case of Free Play, the range of values stretches between (approx.) 0.9 and −1, which implies that the activity was classed as vigorous for some participants, and not vigorous for others. In the case of MPA, the spread of values is less pronounced; with a mean value of −0.5, participants performing Free Play were classified as non-MPA more often than as MPA.

Nonetheless, the spread of values for both features is likely to cause significant classification problems when using VPA or MPA as features. This problem is particularly pronounced for VPA, where classification of Free Play using the feature is likely to be confused for any other feature with a similar, semi-vigorous profile.

From this analysis, the features HAC, WAC and DO are likely to yield the best results during classification. However, the following section will evaluate several feature combinations and provide empirically evident feature sets and associated classification accuracies to demonstrate their usefulness in classifying activity types.

4   Results

This section describes the classification of activity types using MLP analysis and different feature combinations. Input layer sizes between 1 and 4 features were considered.

4.1   MLP Network Analysis Using 2-4-4 Architecture

This evaluation uses feature pairs. The performance for the classifier is evaluated, using the mean accuracy of 30 simulations with each simulation comprising randomly selected training and test sets.

Classifier Performance: The first evaluation uses all the features in the data set to construct feature pairs. Table 1, shows the top 10 highest mean accuracies obtained over 30 simulations (the remainder were excluded because of their low accuracy values).

Table 1 shows that the mean classification accuracy rarely exceeded 70 % and in many cases was between 40 % and 60 %. Variance between classification accuracy during trials was also high, with some feature combinations. This combination of high variance and low classifier accuracy indicate that feature pairs are insufficiently consistent and insufficiently accurate for use in subsequent MLP analysis.

Table 1. Mean percentage classification correctness by feature pair

	Feature One	Feature Two	Accuracy
1	hr	hac	74
2	hac	hr	74
3	hr	ee	67
4	ee	hr	67
5	hac	ee	61
6	ee	hac	61
7	hr	v02	60
8	v02	hr	60
9	hac	do	59
10	do	hac	59

4.2 MLP Network Analysis Using 3-4-4 Architecture

The feature space was increased to triple feature combinations. The performance for the classifier is determined, using the mean accuracy obtained from 30 simulations. The metric includes Sensitivity, Specificity and Kappa estimates. Again, randomly selected training and test sets are used for each simulation.

Classifier Performance: Using the triple feature combinations, Table 2, shows the top 10 highest mean accuracies, sensitivity, specificity and kappa values obtained from 30 simulations.

Table 2 shows that the classification accuracy using triple feature combinations improves the results significantly. While mean classifier accuracies in the low 60th percentile were observed in a number of cases, several cases displayed mean classification accuracies >90 %. These findings are highly positive, suggesting that modification or sophistication of the classification techniques used may further improve classification accuracy.

Table 2. Mean percentage classification correctness by three features

	Features	Acc.	Sens	Spec	Kappa
1	Wac bmi do	96	0.95	0.99	0.94
2	Wac ee do	96	0.95	0.99	0.94
3	v02 ee do	96	0.95	0.99	0.94
4	hreedo	94	0.93	0.97	0.92
5	bmieedo	94	0.94	0.97	0.91
6	wacv02do	93	0.93	0.97	0.89
7	bmiv02do	93	0.93	0.97	0.89
8	hacv02do	92	0.95	0.97	0.89
9	hacv02ee	91	0.88	0.92	0.87
10	haceedo	90	0.91	0.94	0.86

4.3   MLP Network Analysis Using 4-4-4 Architecture

This set of results extends the feature space to four to determine whether further improvements can be made. Table 3 presents the results.

Classifier Performance: Using a combination of four features, Table 3 shows the top 10 highest mean accuracies, sensitivity, specificity and kappa values.

The results show that 4-feature combinations improve the results further. Of all the combinations empirically tested no feature combination showed mean classification accuracies below 87 %.

5   Discussion

The initial classifications on the dataset obtained a relatively low accuracy with HR and HAC providing the best pair of features. Heart rate displayed higher classification accuracy across a number of feature combinations. While MPA and VPA failed to

Table 3. Mean percentage classification correctness by four features

	Features	Acc	Sens	Spec	Kappa
1	bmiv02doee	96	0.95	0.99	0.94
2	bmiv02wachac	96	0.95	0.99	0.94
3	bmidoeewac	96	0.95	0.99	0.94
4	v02doeewac	96	0.95	0.99	0.94
5	v02doeehac	96	0.95	0.99	0.94
6	doeewachac	96	0.95	0.99	0.94
7	hrdoeewac	96	0.95	0.98	0.93
8	bmiv02eehac	95	0.94	0.98	0.93
9	bmieewachac	95	0.94	0.97	0.93
10	v02dowachac	95	0.94	0.97	0.93

perform sufficiently well to justify their inclusion in further trials. These features followed different trends to others. For example, MPA reached maximum values during the performance of activities such as walking and free play, where other features tended to show mid-range values. This is considered an advantage due to the potential additional information content of features with this pattern in conjunction with more normally distributed features. The preceding feature pair analysis demonstrates that neither MPA nor VPA provide useful classifications of activities by type and should thus be excluded from the feature set.

Extending the feature space to three showed a marked improvement in classifier performance. In particular, it should be observed that classification accuracy peaked at 96 %, with a maximum kappa value of 0.94. This value was seen consistently across all trials of a small number of feature combinations. This ceiling was due to the consistent misclassification of a single value; each of the network designs in question successfully classified all other values correctly across trials, but misclassified this single record on every occasion. Specifically, one record captured from participants performing the activity Jogging was consistently misclassified by the MLP networks as an instance of the activity Free play.

However, what these findings show is that larger input feature combinations produce higher classification accuracy and reduce variance between MLP trial iterations. A logical extension of the proceeding analysis, then, was to extend the input feature combination to a total of four features. The results showed that the top end classifiers as seen in Table 3 continue to fail to classify certain data values. Only one input feature combination (HR, DO, EE, V02) enabled perfect classification of the dataset, and perfect classification occurred in less than 7 % of the cases (2/30). However, these instances of perfect classification do demonstrate that improved classification accuracy is attainable although it may not be achievable without the use of new techniques.

6 Conclusions

This study used an existing dataset from recent research into physical activity in youth to classify data by the type of activity engaged upon. The dataset was analyzed using rigorous data science techniques, which led to an improved understanding of activity types and features. Data items whose properties impeded classification were removed. This did affect the size of the dataset. A series of machine learning analyses were performed. A range of classifier types, input feature combinations and architectural parameters were employed and refined to develop improved classification accuracy.

While the results show, specific activities and features tailored around a machine learning approach are promising, a great deal of research remains. Further development of Data Science techniques will help provide a varied and broad range of possibilities. The data exploration and analysis techniques used in this study produced results that suggest extensions and raise further questions. First, it would be useful to validate the preceding results using a substantial non-interpolated data set. Second, it would also be useful to explore other supervised machine learning methods, such as support vector machines and other advanced artificial neural network architectures. While this study focused on youth, it would be interesting to look at other population groups, such as adults and the elderly. Finally, one important point would be to standardize the use of activities and cut points in PA research that is underpinned with strong Data Science evidence and advanced machine learning techniques.

References

1. Barshan, B., Yuksek, M.C.: Recognizing daily and sports activities in two open source machine learning environments using body-worn sensor units. Comput. J. (2014)
2. Caspersen, C.J., Powell, K.E., Christenson, G.M.: Physical activity, exercise and physical fitness: definitions and distinctions for health-realted research. Public Health Rep. **100**(2), 126–131 (1985)
3. Dalton, A., O'Laighin, G.: Comparing supervised learning techniques on the task of physical activity recognition. Biomed. Health Inf. **17**(1), 46–52 (2013)
4. De Vries, S.I., Engels, M., Garre, F.G.: Identification of children's activity type with accelerometer-based neural networks. Med. Sci. Sports Exerc. **43**(10), 1994–1999 (2011)

5. De Vries, S.I., Garre, F.G., Engbers, L.H., Hildebrandt, V.H., Van Buuren, S.: Evaluation of neural networks to identify types of activity using accelerometers. Med. Sci. Sports Exerc. **43**(1), 101–107 (2001)
6. Duffey, K.J., Gordon-Larsen, P., Shikany, J.M., Guilkey, D., Jacobs, D.R., Popkin, B.M.: Food Price and Diet and Health Outcomes: 20 Years of the CARDIA Study. Arch. Intern. Med. 170(5), 420–426 (2010)
7. Evenson, K.R., Cattellier, D., Gill, K., Ondrak, K., McMurray, R.G.: Calibration of two objective measures of physical activity for children. J. Sports Sci. **26**(14), 1557–1565 (2008)
8. Honas, J.J., Washburn, R.A., Smith, B.K., Greene, J.L., Cook-Wiens, G., Donnelly, J.E.: The System for Observing Fitness Instruction Time (SOFIT) as a measure of energy expenditure during classroom based physical activity. Pediatr. Exerc. Sci. **20**, 439–445 (2008)
9. Kautiainen, S., Koivusilta, L., Lintonen, T., Virtanen, S.M., Rimpela, A.: Use of information and communication technology and prevalance of overweight and obesity among adolescents. Int. J. Obes. **29**(8), 925–933 (2005)
10. Konstabel, K., Veidebaum, T., Verbestel, V., Moreno, L.A., Bammann, K., Tornaritis, M., Eiben, G.: Objectively measured physical activity in European children: the IDEFICS study. Int. J. Obes. **38**, S135–S143 (2014)
11. Lissau, I., Sorensen, T.I.: Parental neglect during childhood and increased risk ofobesity in young children. Lancet. **343**(8893), 324–327 (1994)
12. Machkintosh, K.A., Fairclough, S.J., Stratton, G., Ridgers, N.D.: A calibration protocol for population-specific accelerometer cut-points in children. PloS One, 7(5), e36919 (2012)
13. Martinez-Gonzalez, M.A.: Physical inactivity, sedentary lifestyle and obesity in the European Union. Int. J. Obes. **23**(11), 1192–1201 (1999)
14. Mcferran, B., Dahl, D.W., Fitzsimons, G.J., Morales, A.C.: I'll have what she's having: effects of social influence and body type on the food choices of others. J. Consum. Res. **36** (6) (2010)
15. McKenzie, T.L.: 2009 C. H. McCloy Lecture. Seeing is beliving: observing physical activity and its contexts. Res. Q. Exerc. Sport. **8**(12), 113–122 (2010)
16. Mckenzie, T.L., Sallis, J.F., Nader, P.R.: SOFIT: System for observing fitness instruction time. J. Teach. Phys. Educ. **11**(2), 195–205 (1992)
17. Orme, M., Wijndaele, K., Sharp, S.J., Westgate, K., Ekelund, U., Brage, S.: Combined influence of epoch length, cut-point and bout duration on accelerometry-derived physical activity. Int. J. Behav. Nutrician Phys. Act. **11**(34), 11–34 (2014)
18. Oyebode, O., Mindell, J.: Use of data from the Health Survey for England in obesity policy making and monitoring. Obes. Rev. **14**(6), 463–476 (2013)
19. Pate, R.R., Pratt, M., Blair, S.N., Haskell, W.L., Macera, C.A., Bouchard, C., Buchner, D., Ettinger, W. et al.: Physical activity and public health: a recommendation from the centers for disease control and prevention and the Amercian College of Sports Medicine. JAMA **273** (5) (1995)
20. Rowe, P., van der Mars, H., Schuldheisz, J., Fox, S.: Measuring students' physical activity levels: validating SOFIT for use with high-school students. J. Teach. Phys. Educ. **23**(3), 235–251 (2004)
21. Ryley, A.: Health Survey for England 2012: Chapter 11. Children's BMI, overweight and obesity **1**, 1–22 (2013)
22. Saelens, B.E., Sallis, J.F., Black, J.B., Chen, D.: Neighborhood-based differences in physical activity: an environment scale evaluation. Am. J. Public Health **93**(9), 1552–1558 (2003)
23. Scruggs, P.W., Beveridge, S.K., Clocksin, B.D.: Tri-axial accelerometry and heart rate telemetry: relation and agreement with behavoural observation in elementarty physical education. Meas. Phys. Educ. Exerc. Sci. **9**(4), 203–218 (2005)

24. Sharma, S.V., Chuang, R.J., Skala, K.: Measuring physical activity in preschoolers: reliability and validity of the sysetm for observing fitness instruction time for preschoolers (SOFIT-P). Meas. Phys. Educ. Exerc. Sci. **15**(4), 257–273 (2011)
25. Staudenmayer, J., Prober, D., Crouter, S., Bassett, D., Freedson, P.S.: An artificial neural network to estimate phuysical activity energy expenditure and identify physical activity type from an accelerometer. J. Appl. Physiol. **107**(4), 1300–1307 (2009)
26. Trost, S.G.: Artificial neural networks to predict activity type and energy expenditure in youth. Med. Sci. Sports Exerc. **44**(5), 1801–1809 (2012)
27. Trost, S.G., Loprinzi, P.D., Moore, R., Pfeiffer, K.A.: Comparison of accelerometer cut points for predicting activity intensity in youth. Med. Sci. Sports Exerc. **43**(7), 1360–1368 (2010)
28. Welk, G.J.: Principles of design and analyses for the calibration of accelerometry-based activity monitors. Med Sci Sports Exerc. 37(11), 501–511 (2005)

A Position Paper on Predicting the Onset of Nocturnal Enuresis Using Advanced Machine Learning

Paul Fergus[1(✉)], Abir Hussain[1], Dhiya Al-Jumeily[1],
and Naeem Radi[2]

[1] Liverpool John Moores University, Byrom Street, Liverpool L3 3AF, UK
[2] Al-Khawarizmi International College, Abu Dhabi, UAE
n.radi@khawarizmi.com

Abstract. Bed-wetting during normal sleep in children and young people has a significant impact on the child and their parents. The condition is known as nocturnal enuresis and its underlying cause has been subject to different explanatory factors that include, neurological, urological, sleep, genetic and psychosocial influences. Several clinical and technological interventions for managing nocturnal enuresis exist that include the clinician's opinions, pharmacology interventions, and alarm systems. However, most have failed to produce any convincing results. Clinical information is often subjective and often inaccurate, the use of desmopressin and tricyclic antidepressants only report between 20 % and 40 % success, and alarms only a 50 % success fate. This position paper posits an alternative research idea concerned with the early detection of impending involuntary bladder release. The proposed framework is a measurement and prediction system that processes moisture and bladder volume data from sensors fitted into undergarments that are used by patients suffering with nocturnal enuresis. The proposed framework represents a level of sophistication in nocturnal enuresis treatment not previously considered.

Keywords: Nocturnal enuresis · Bedwetting · Machine learning · Classification · Neural networks · Sensors

1 Introduction

Although urination is a function performed effortlessly by healthy humans, it is an extremely complex process that involves the rapid and precise coordination of numerous muscles and nerves in the ureter, bladder, sphincter and urethra (Porth 2007). Nocturnal enuresis (incontinence or bedwetting) is an event that is commonly considered as a disruption to the normal process in achieving continence. Nocturnal enuresis occurs involuntary during sleep without any inherent suggestions of frequency or pathophysiology[1] and its etiology is complex. However, it is believed to be caused by three main non-exclusive pathogenic mechanisms: nocturnal polyuria (passing large amounts of urine at night and normal amounts during the day) (Fatah et al. 2009),

---

[1] http://www.nice.org.uk/.

© Springer International Publishing Switzerland 2015
D.-S. Huang et al. (Eds.): ICIC 2015, Part II, LNCS 9226, pp. 689–700, 2015.
DOI: 10.1007/978-3-319-22186-1_68

detrusor overactivity (involuntary contractions during the filling phase) (Prieto et al. 2012), and increased arousal thresholds (a patient's inability to waken in response to signals from a full bladder) (Dhondt et al. 2009). The underlying reason for these conditions has been subject to different explanatory factors that include, neurological, urological, sleep, genetic and psychosocial influences (Butler 2004; Campbell et al. 2009; Culbert and Banez 2008; Robson 2009). Furthermore, general parental knowledge of the causes and effective treatments for nocturnal enuresis (NE) is lacking. Only 55 % reported they would seek medical care for their child with NE and only 28 % reported awareness of effective treatments (Schlomer et al. 2013).

The condition is socially disruptive and stressful and is reported to affect 20–25 % of five year olds, 5 % at 10 years, and 1–2 % of 15 year olds (Kennea and Evans 2000). Data from UK Avon Longitudinal Study of Parents and Children (ALSPAC) put prevalence at 20 % at 10 years, 9 % at nine years and 1 % at 15 years (Darling 2010). Nocturnal enuresis can have profound effects on a child, low self-esteem (Thibodeau et al. 2013), social isolation (de Bruyne et al. 2009), and child abuse triggered by bedwetting (Can et al. 2004). Although, primary and secondary care interventions help, it has a major impact on the quality of life and healthcare resources. According to the British Association of Urological Surgeons (BAUS) between three and six million people in the UK suffer with urinary incontinence of which nocturnal enuresis is a subset of[2]. Urinary incontinence collectively has a significant cost implication, with conservative estimates suggesting that £424 million is spent annually on treatment in the UK.

In this position paper, we focus on using sensors and artificial neural networks to determine the onset of a voiding episode before it occurs. This is achieved by utilizing sensors to detect moisture and bladder volume and a personalized alarm system that utilizes a neural network to train the system and predict the onset of voiding episodes.

2 Nocturnal Enuresis: Diagnosis and Treatment

A child that is at least five years old and experiencing bed wetting episodes at least twice a week for a minimum of three months would be diagnosed as suffering with nocturnal enuresis (Bettina et al. 2010). Investigations are triggered by complaints from the parents or child following presentation to a General Practitioner (GP) or referral to a specialist provider (Nalbantoglu et al. 2013a). At the initial assessment, clinicians are mindful of the risk of parent intolerance towards their child's nocturnal enuresis since this can affect the treatments offered. During the initial assessment, a general history is completed to develop an understanding of the pattern of bedwetting over the previous few weeks. Further enquiry will include questions on urgency, frequency, and the type and amount of drinks consumed. Diaries are used as a self-reporting tool to record historical information (Bradley et al. 2011). A physical examination following an initial assessment is often required and laboratory tests requested to exclude other diseases, such as diabetes mellitus, urinary tract infection, and diabetes insipitus (Nalbantoglu

---

[2] http://www.baus.org.uk/.

et al. 2013b, Vande Walle et al. 2012). Physical examinations include abdominal/flank examination for masses, bladder distension, and relevant surgical scars; examination of the perineum and external genitalia and neurological testing (Abrams et al. 2010) is completed.

Lifestyle management is a first-line intervention that is useful during the exploratory stages of diagnosis, however, it is reported that only 20 % of patients suffering with nocturnal enuresis respond successfully, while 80 % will need additional treatment (Lottmann and Alova 2007). One approach is to use alarms, which have produced mixed results that range between 50 % and 80 % depending on the population (children and families need to be highly motivated) (Glazener and Evans 2007). In another approach pharmacology interventions have also been routinely used in the treatment of nocturnal enuresis. The use of desmopressin shows that one third of children remain dry; one third there is a partial effect, and in the remaining third it has no effect at all (Glazener and Evans 2002).

There have been several technological interventions for managing nocturnal enuresis that focus on detection and more recently on prediction, which have largely produced disappointing results. Arguably the earliest recorded attempt is the work of (Butler 1994) who used a urine alarm system to alert the clinical staff at a hospital when bedwetting episodes by patients occurred (the problem is that the event has already happened and requires sheets and cloths to be washed and changed). Nilsson et al. (2011) proposed a similar alarm system that uses a moisture sensing mechanism fitted into undergarments that alert patients and careers when fixed levels of moisture are detected (again the event has already happened).

3 Predicting the Onset of Nocturnal Enuresis (PRONE)

The discussion so far has highlighted a number of ways to treat nocturnal enuresis. Some of these include medication, alarms and the use of advanced sensor technology. Whilst these do benefit children and their careers, they do not stop the occurrence of bedwetting and in many instances, only alert the child or parent once the event has already occurred. This means that the child has to deal with the effects of bedwetting, which include, changing cloths and bedding, embarrassment and shame and the stigma this brings. This position paper describes a solution for a possible framework to treat nocturnal enuresis that goes far beyond any existing solutions to date (Fig. 1).

3.1   Sensor Platform

The proposed system, illustrated in Fig. 2, provides an overview of the logical architecture that comprises the framework. The system operates in two modes; training mode and prediction mode.

In the training mode, the system monitors the voiding habits of children during normal sleep. The training phase monitors the initial occurrence of moisture in the undergarment. This triggers a pulse generator in the electromagnetic sensor (also fitted into the undergarment above the navel), to start sending electromagnetic waves to

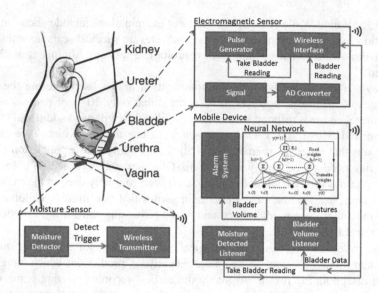

Fig. 1. Proposed prediction framework of nocturnal enuresis

decipher how full the bladder is. These values are wirelessly transmitted to the mobile device and in turn passed to the neural network to predict the volume of urine in the bladder. This value is used to configure an alarm that is specific to the patients voiding habits. The learning phase is configurable, but it is envisaged that several weeks of data will be required to provide an estimate of the bladder volume and voiding events. Following the completion of this training phase, an alarm will be configured to a percentage below the averaged bladder volume and voiding episodes learned by the system of the training period.

During prediction mode, electromagnetic data is continuously (or at set intermittent times), streamed wireless to the mobile device. This data is passed to the neural network and the predicted bladder volume level is passed to the alarm system. If the value is equal to or greater than the pre-configured alarm threshold as determined during the training phase, an alarm will be triggered that prompts the child to wake up and go to the toilet. The principle goal is to wake then just before the onset of a voiding episode.

3.2   Electromagnetic Wave Sensors

The communication platform is a Bluetooth low energy (BLE) network with supporting services for data processing, aggregation, storage and distribution. All sensors have a BLE interface connecting them to the network. Sensors transmit data using the Generic Attribute Profile (GATT) protocol in BLE (Gupta 2013) and all data is pre-processed using signal processing techniques. In the first instance, data is used to set the parameters of the algorithm and during runtime data is used for real-time prediction. The sensors are robust to variations in skin, fat, muscle thickness and bladder sizes.

Electromagnetic waves are attenuated while passing through the bladder. Different attenuation is encountered at different frequencies. This is primarily because different body tissue, such as muscle, fat and skin own different relative permittivity. This means that electromagnetic waves undergo numerous reflections from the boundary between two tissues/organs that have different relative permittivity. This method is proposed to detect the presence of water because the value of the relative permittivity of the water (in this instance urine) is much larger than the values of the relative permittivity of the surrounding organs and tissues.

A large reflection of the incident electromagnetic signal pulse occurs. The position, amount of water, and the water concentration in the reflecting organ, can be estimated since the electromagnetic signals own a fine range resolution and good penetration ability. The signals from the transmitting and receiving antenna are processed using classification and prediction algorithms, which represent the relation between the signal frequencies and bladder volume. Bladder volume prediction is triggered during training by moisture sensors fitted into undergarments that consist of two electricity conductors separated by an insulated moisture absorbing material. Figure 2 illustrates the two sensor systems.

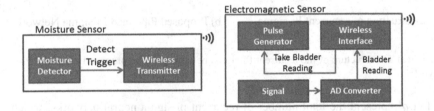

Fig. 2. Moisture and bladder volume sensor

In prediction mode, a similar process is followed. Data is streamed or sampled at set time intervals, as discussed above, vectors are created and transmitted for signal processing and bladder volume prediction.

3.3   Adaptive Neural Network Architecture

A system to predict the occurrence of nocturnal enuresis before it occurs is provided by the proposed framework. The purpose of the system is to raise an alarm to the user using a mobile device based on the information collected from the developed sensor technology. Since the bladder volume varies from one person to the other, an adaptive system will be designed to personalize prediction and alarm thresholds. A pipelined structure similar to the adaptive fully recurrent pipelined neural network architecture proposed by Haykin and Li (1995) for the prediction of future occurrences of bed-wetting, is introduced. The main structure of the proposed network is the recurrent pi-sigma unit (Shin and Ghosh 1991) as shown in Fig. 3(a), due to their simplicity and the high learning capabilities of higher order neural networks, which make them suitable for mobile device processing.

The network comprises a number of concatenated recurrent pi-sigma neural networks. It is designed to adaptively predict highly nonlinear and non-stationary signals, such as the bedwetting signal. Similar to the adaptive fully recurrent pipelined neural network architecture, the proposed network is designed using the principles of divide and conquer. This means that to solve a complex problem, it has to be broken into a number of smaller sub-problems. The proposed network is called pipelined recurrent pi-sigma neural network (PRPSN). Figure 3(b) shows the general structure of the proposed network, while Fig. 4 shows a specific structure of the proposed network with two modules for the recurrent pi-sigma units and m external inputs.

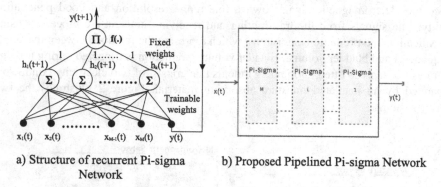

a) Structure of recurrent Pi-sigma Network

b) Proposed Pipelined Pi-sigma Network

Fig. 3. (a) The structure of the recurrent Pi-sigma network (b) The proposed pipelined Pi-sigma network

Let q represent the total number of recurrent pi-sigma neural networks, which are concatenated with each other. Each recurrent pi-sigma neural network is called a unit (or a module) and consists of $M-1$ external inputs, except for the last unit, which consists of M external inputs. All the units of the PRPSN receive the output from previous units as input, except the last module in the proposed pipeline.

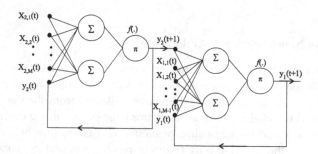

Fig. 4. Module recurrent pi-sigma pipelined neural network architecture with m external inputs and second order pi-sigma units.

In Fig. 4, the inputs and outputs, as well as, the processing equations of the pipelined recurrent pi-sigma neural network are presented.

If $S(t)$ represents the nonlinear and non-stationary signal at time t obtained from the sensor, then the external input vector presented to the ith module of the pipelined pi-sigma network is defined as follows:

$$X_i(t) = [S(t - i) \ S(t - (i + 1)), \ldots, \ S(t - (i + M - 2))]^T, \text{ for } i \neq q$$
$$X_i(t) = [S(t) \ S(t - i), \ S(t - (i + 1)) \ldots, \ S(t - (i + M - 1))]^T, \text{ for } i = q \tag{1}$$

Let y_i represent the output of module i, which is defined as follows:

$$y_i(t) = f(v_i(t)) \tag{2}$$

where f is a nonlinear transfer function and v_i is the net internal activation of module i. For the last module of the polynomial pipelined network:

$$v_q(t) = \prod_{L=1}^{k} h_{q,L}(t + 1)$$

$$h_{a,L}(t + 1) = \sum_{m=1}^{M} w_{q,Lm} \cdot s_{q,m}(t) + w_{q,L(M+1)} \cdot y_q(t - 1) \tag{3}$$

where k is the order of the recurrent pi-sigma unit.
For all other modules:

$$v_i(t) = \prod_{L=1}^{k} h_{i,L}(t + 1)$$

$$h_{i,L}(t + 1) = \sum_{m=1}^{M-1} w_{i,Lm} \cdot s_{i,m}(t) + w_{i,L(M+1)} \cdot y_i(t - 1) + w_{i,L(M+2)} \cdot y_{i+1}(t) \tag{4}$$

The proposed network can be trained using the real-time learning algorithm developed by Williams and Zipser [62]. Instead of assuming that the weights are constants during the whole trajectory, this condition is relaxed and the weights are updated for each input pattern presentation. The advantage of this learning algorithm is that the epoch boundaries are no longer required, making the implementation of the algorithm simpler and letting the network be trained for an indefinite period.

The learning algorithm starts by initializing the weights of one of the network units to small random values. Then, the weights of this unit are trained and used as the initial weights for the PRPSN. The PRPSN is trained adaptively in which the errors produced from each module are calculated and the overall cost function of the PPNN is defined as follows:

$$\varepsilon(t) = \sum_{i=1}^{q} \lambda^{i-1} e_i^2(t) \tag{5}$$

where λ is an exponential forgetting factor selected in the range (0, 1). At each time t, the output of each module $y_i(t)$ is determined and the error $e_i(t)$ is calculated as the difference between the actual value expected from each unit i and the predicted value $y_i(t)$. The weights are iteratively updated by gradient descent:

$$\Delta w_{ml}(t) = -\eta \frac{\partial \varepsilon(t)}{\partial w_{ml}} \tag{6}$$

where η is the manually adjusted gain and

$$\begin{aligned}
\frac{\partial \varepsilon(t)}{\partial w_{ml}} &= 2 \sum_{i=1}^{q} \lambda^{i-1} e_i(t) \frac{\partial e_i(t)}{\partial w_{ml}}, \\
&= -2 \sum_{i=1}^{q} \lambda^{i-1} e_i(t) \frac{\partial y_i(t)}{\partial w_{ml}}
\end{aligned} \tag{7}$$

Define

$$p_{ml}^i(t) = \frac{\partial y_i(t)}{\partial w_{ml}} \tag{8}$$

Then, the values of the triple $p_{ml}^i(t)$ matrix are updated by differentiating the processing equations as follows:

$$P_{ml}^i(t) = f'\left(\prod_{L=1}^{k} h_L\right) \prod_{\substack{L=1 \\ L \neq m}}^{k} h_L\left(Z_l(t) + w_{m(M+2)} P_{ml}^i(t-1)\right) \tag{9}$$

where

$$\begin{aligned}
z_l(t) &= \begin{cases} S(t-(l+L-1)), & 1 \leq L \leq M, \\ y_l(t) & L = M+1 \end{cases} \quad \text{for } i = q \\
z_l(t) &= \begin{cases} y_{l-1}(t) & L = 1 \\ S(t-(l+L-1)), & 1 < L \leq M-1, \quad \text{for } i \neq q \\ y_1(t) & L = M+1, \end{cases}
\end{aligned} \tag{10}$$

The proposed neural network architecture will be used to capture the properties of the sensor signals and its structure will be adapted according to the training period. The values will be predicted adaptively over time.

3.4   Mobile Platform

In the network configuration, a mobile device will act as the GATT server. All data will be pre-processed using signal processing techniques. Features are extracted from the data and passed to the neural network to predict the bladder volume level. This process is performed in both training and prediction mode – in training mode it is used to set the alarm thresholds in relation to the detection of moisture and the volume level in the bladder. In prediction mode it is used to determine whether an alarm threshold has been reached – if this is the case an alarm is raised to wake the child as shown in Fig. 5.

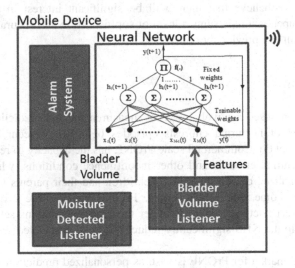

Fig. 5. Mobile platform

The proposed framework goes far beyond previous attempts to predict the likely time a child will wet the bed. It offers the means for the automated extraction of bladder measurements, by using advanced computer algorithms never attempted before in the prediction of nocturnal enuresis events that occur prior to bed-wetting. Overall, it is an adventurous medium risk/very-high impact approach, which will attempt to solve a significant real-world problem with advanced computer science methodologies.

4 Discussion

The timeliness of this work can be realistically judged in the context of large number of patients and the demographics of the population. The overall worldwide prevalence of urinary incontinence is said to be 200 million people. In the UK, the figure is between three and six million and in the US, it is said to be 25 million. The economic cost of treating adults over the age of 40 suffering with urinary incontinence annually was estimated to be £536 million in 1999/2000 prices. In addition, it is estimated to cost the individual £207 million for managing their symptoms (£29 million and £178 million for men and women, respectively). The conservative estimate for treating children and adolescents is £424 million annually in the UK. In 2005, the annual cost-of-illness estimates for urinary incontinence in Canada, Germany, Italy, Spain, Sweden, and the United Kingdom was 7 billion euros. A US cost-of-illness study reported a total cost of

$66 billion in 2007. These figures are likely to rise significantly over the next 20 years as awareness of the condition and the mean age of the population increases (Milsom et al. 2013).

There is no real "gold standard" for the diagnosis and treatment of nocturnal enuresis. Taking into account the important role of prediction in the clinical assessment of NHS patients, and the large volume of nocturnal enuresis research articles based on alarm systems, we believe that there will be significant interest in the framework posited in this paper as it represents a level of sophistication and accuracy in nocturnal enuresis treatment not previously considered.

5   Conclusion

This paper has proposed a novel, low cost measurement and prediction system that detects the onset of nocturnal enuresis episodes before they occur, which current technologies cannot do. Commercial scale PrONE has the capacity to revolutionize the treatment of nocturnal enuresis and other incontinence conditions, while offering an affordable personalized device that allows children and their parents to manage their condition in the home. In the UK alone, PrONE presents the NHS with a real opportunity to treat nocturnal enuresis under the recommendations set out by NICE, and will also help the NHS significantly reduce the costs associated with treating this condition.

The potential market for PrONE is vast, as personalized prediction systems can be used to offset the need for diagnostic and healthcare services. PrONE will improve the overall success rate of treatments (14 consecutive dry nights) while reducing relapses and help to gain a better and standardized understanding of the condition. It will reduce NHS costs and also costs incurred by parents (cleaning costs, mattress replacement, disposable products required). The anticipated efficiency of PrONE is such that it will standardize diagnosis and treatment compared with existing solutions and procedures as a result of its ease of use, data collection and communication capabilities between the child, parent and care practitioners.

There is no other product on the market that covers such technology that can help to treat and understand nocturnal enuresis personalized to an individual's condition and generalizable across all suffers of nocturnal enuresis.

References

Abrams, P., Andersson, K.E., Birder, L., Brubaker, L., Cardozo, L., et al.: Fourth international consultation on incontinence recommendations of the international scientific committee: evaluation and treatment of urinary incontinence, pelvic organ prolapse and fecal incontinence. Neurourol. Urodyn. **29**(1), 213–240 (2010)

Bettina, E., Shapira, E., Dahlen, P.: Therapeutic treatment protocol for enuressi using and enuresis alarm. J. Couns. Dev. **88**(2), 246–252 (2010)

Bradley, C.S., Brown, J.S., van Den Eeden, S.K., Schembri, M., Ragins, A., Thom, D.H.: Urinary incontinence self-report questions: reproducability and agreement with bladder diary. Int. J. Urogynedol, **22**(12), 1565–1571 (2011)

Butler, R.J.: Nocturnal Enuresis. the Child's Experience, p. 192, Butterworth Heinemann, Oxford (1994)

Butler, R.J.: Childhood nocturnal enuresis: developing a conceptual framework. Clin. Psychol. Rev. **24**(8), 909–931 (2004)

Campbell, L.K., Cox, D.L., Borowitz, S.M.: Elimination disorders. In: Handbook of Pediatric Psychology, p. 808, Guilford Press, New York (2009)

Can, G., Topbas, M., Okten, A., Kizil, M.: Child abuse as a result of enuresis. Pediatrics Int. **46**(1), 64–66 (2004)

Culbert, T.P., Banez, G.A.: Wetting the bed: integrative approaches to nocturnal enuresis. Explore **4**(3), 215–220 (2008)

Darling, J.C.: Management of nocturnal enuresis. Trends Urol. Gynaecol. Sexual Health **15**(3), 18–22 (2010)

De Bruyne, E., van Hoecke, E., van Gompel, K., Verbeken, S., Hoebeke, P., Vande Walle, J.: Problem behaviour, parental stress and enuresis. J. Urology **182**(4), 2015–2021 (2009)

Dhondt, K., Raes, A., Hoebeke, P., van Laecke, E., van Herzeele, C., Vande Walle, J.: Abnormal sleep architecture and refractory nocturnal enuresis. J. Urology, **182**(4), 1961–1965 (2009)

Fatah, D.A., Shaker, H., Ismail, M., Ezzat, M.: Nocturia polyuria and nocturnal arginine vasopression (AVP): a key factor in the pathophysiology of monosymptomatic nocturnal enuresis. Neurourol. Urodyn. **28**(6), 506–509 (2009)

Glazener, C.M., Evans, J.H.: Alarm interventions for nocturnal enuresis in children (Cochrane Review). The Cochrane Library (2007)

Glazener, C.M., Evans, J.M.: Desmopressin for nocturnal enuresis. cochrane database systematic review, CD002112 (2002)

Gupta, N.: Inside bluetooth low energy. IEEE Trans. Instrum. Measur. 210 (2013)

Haykin, L.: Nonlinear adaptive prediction of nonstationary signals. IEEE Trans. Signal Process. **43**(2), 526–535 (1995)

Kennea, N., Evans, J.: Drug treatment of nocturnal enuresis. Paediatr. Perinat. Drug Therapy **4**(1), 12–18 (2000)

Lottmann, H.B., Alova, I.: Primary monosymptomatic nocturnal enuresis in children and adolescents. Int. J. Clin. Pract. **61**(s155), 8–16 (2007)

Milsom, I., Coyne, K.S., Nicholson, S., Kvasz, M., Chen, C., Wein, A.J.: Global Prevalence and Economic Burden of Urgency Urinary Incontinence: A Systematic Review. Eur. Urol. (2013, in Press)

Nalbantoglu, B., Yazici, C.M., Nalbantoglu, A., Guzel, S., Topcu, B., Guzel, E.C et al.: Copeptin as a Novel Biomarker in Nocturnal Enuresis. Pediatr. Urol. (2013a, in Press)

Nalbantoglu, B., Yazici, C.M., Nalbantoglu, A., Guzel, S., Topcu, B., Guzel, E.C et al.: Copeptin as a Novel Biomarker in Nocturnal Enuresis. Pediatr. Urol. (2013b, in Press)

Nilsson, H., Gulliksson, J. An incontinence alarm solution utilizing RFID-based sensor technology. In: IEEE International Conference on RFID-Technologies and Applications, pp. 359–363 (2011)

Porth, C.M.:. Essentials of pathophysiology: concepts of altered health states, p. 1149 (2007)

Prieto, L., Castro, D., Esteban, M., Salinas, J., Jimenez, M., Mora, A.: Descriptive epidemiological study of the diagnosis of detrusor overactivity in urodynamic units in Spain. Actas Urologicas Espanolas, **36**(1), 21–28 (2012)

Robson, W.L.M.: Evaluation and management of enuresis. N. Engl. J. Med. **360**(14), 1429–1436 (2009)

Schlomer, B., Rodriquez, E., Weiss, D., Cropp, H.: Parental beliefs about nocturnal enuresis causes, treatments, and the need to seek professional medical care. J. Pediatr. Urol. (2013, In Press)

Shin, Y., Ghosh, J.: The pi-sigma network: an efficient higher-order neural network for pattern classification and function approximation. In: IEEE Int. Conf. Neural Computing, pp. 13–18 (1991)

Thibodeau, B.A., Metcalfe, P., Koop, P., Moore, K.: Urinary incontinence and quality of life in children. Pediatr. Urol. 9(1), 78–83 (2013)

Vande Walle, J., Rittig, S., Bauer, S., Eggert, P., Marshall-Kehrel, D., Tekgul, S.: Practical consensus guidelines for the management of enuresis. Eur. J. Pediatr. 171(6), 971–983 (2012)

Improved Bacterial Foraging Optimization Algorithm with Information Communication Mechanism for Nurse Scheduling

Ben Niu[1,2(✉)], Chao Wang[1], Jing Liu[1], Jianhou Gan[3(✉)],
and Lingyun Yuan[3]

[1] College of Management, Shenzhen University, Shenzhen, China
[2] Department of Industrial and System Engineering,
Hong Kong Polytechnic University, Hong Kong, China
Drniuben@gmail.com
[3] Key Laboratory of Educational Information for Nationalities,
Ministry of Education, Yunnan Normal University, Kunming, China
10800662@qq.com

Abstract. As a NP-hard combinatorial problem, nurse scheduling problem (NSP) is a well-known personnel scheduling task whose goal is to create a nurse schedule under a series of hard and soft constraints in a practical world. In this paper, a variant of structure-redesigned-based bacterial foraging optimization (SRBFO) with a dynamic topology structure (SRBFO-DN) is employed for solving nurse scheduling problem (NSP). In SRBFO-DN, each bacterium achieves cooperation by information exchange mechanism switching the topology structure between star topology and ring topology. A special encoding operation of bacteria in SRBFO-DN is adopted to transform position vectors into feasible solutions, which can make SRBFO-DN successfully dealing with this typical difficult and discrete NSP. Experiment results obtained by SRBFO-DN compared with SRBFO and SPSO demonstrated that the efficiency of the proposed SRBFO-DN algorithm is better than other two algorithms for dealing with NSP.

Keywords: Bacterial foraging optimization · Topology structure · Nurse scheduling problem

1 Introduction

Nurse scheduling problem (NSP) has been extensively studied in recent years, and is known as complexity NP-Hard [1]. In the past, the hospital used to arrange a nurse scheduling scheme based on accumulated personal experience. This would consume considerable time without even meeting most constraints. Therefore, a large variety of evolutionary computation approaches have been studied to find a better scheme for NSP, including tabu search [2], scatter search [3], Bayesian network [4], genetic algorithm [5], particle swarm optimization [6], variable neighborhood search [7] [8] and so on.

Recently, bacterial foraging optimization (BFO) has appeared as a powerful technique for optimization problems. SRBFO-DN an SRBFO are variants of BFO proposed in our previous research [9] to improve the performance of BFO. To expand the

© Springer International Publishing Switzerland 2015
D.-S. Huang et al. (Eds.): ICIC 2015, Part II, LNCS 9226, pp. 701–707, 2015.
DOI: 10.1007/978-3-319-22186-1_69

application area of BFO-based algorithms, this paper employed SRBFO-DN to specific nurse scheduling problem.

The rest of this paper is organized as follows. We will begin our paper by giving general information about the SRBFO-DN in Sect. 2. Section 3 emphasizes to extend a mathematical model of nurse scheduling problem. Section 4 presents experiment study and the results, and followed by conclusions in Sect. 5.

2   Improved BFO with Communication Mechanism

BFO method was invented by Passino [10] inspired by the natural selection which tends to eliminates the E coli. Bacteria with poor foraging abilities and remain those having stronger foraging abilities. In previous research [9], five variants of SRBFO have presented in detailed. In this section, taken the highest search precision and stability compared to all the other proposed algorithms into consideration, this paper only choose SRBFO-DN to solve nurse scheduling problems.

SRBFO-DN combines the advantages of both SRBFO-ST and SRBFO-Ri. In this algorithm, the condition of convergence (CoC) is referred to a criterion to judge whether the bacterial have been trapped into the local optima. Equation (1) is the calculation of CoC, where the $Tbest$ represents the fitness values of the current last m iterations. If CoC is equal to 1, the bacteria population become stagnated. However when it is larger than 1, the bacterial population is in the process of finding a better fitness value. The pseudo-code of SRBFO-DN is depicted in Table 1.

$$Coc = \left(\sum_{i=1}^{m-1} \frac{e^{Tbest(i)}}{e^{Tbest(i+1)}} \right) \Big/ m - 1 \tag{1}$$

Table 1. The pseudo-code of SRBFO-DN.

SRBFO-DN

```
Begin
Initialize variables
Randomize positions and the step size
Calculate CoC
While ChemIter <= MaxFEs
    While CoC
        Loop with Star topology
        Calculate CoC
    End While
    Loop with Ring topology
End While
```

3 Nurse Scheduling Problem

NSP can be defined as allocating the workload to the available staff nurses at healthcare organizations to meet the operational requirements [11]. More precisely, the aim is to assign each nurse to concrete pre-defined cycles satisfying some constrains (such as hard and soft constraints). Before describing the mathematical model, some parameters and definitions will be introduced to simplify the problem presentation in Table 2.

Table 2. Parameters and definitions of NSP.

Variables	Definitions
k	The shift type of each nurse; $k = 1, 2,..., ss$
j	The set of skill types; $j = 1, 2,..., sk$
i	Index for nurses; $\forall i = 1, 2, ..., nn$
d	The set of days in the current schedule period; $d = 1, 2,..., sd$
c	Punishment coefficients; $c = 1000$
lp	The lower limit of work shifts of each nurse in scheduling cycle
up	The upper limit of work shifts of each nurse in scheduling cycle
$w_{j,k}$	The wage of nurses in shift type k and skill type j
$m_{j,k,d}$	The number of nurses that did not meet demands on day d,in shift type k and skill type j
$x_{i,j,k,d}$	Nurse i is assigned on day d, in shift type k and skill type j
$h_{i,j,k,d}$	The hospital's minimum coverage demands for day d, shift k, and skill type j

In order to make the problem be simpler, we assume that each day has four working shifts denoted by $ss = 4$, composes of morning shift (A) of eight working hours(00:00–8:00), an afternoon shift (P) of eight working hours(8:00–16:00), a night shift (N) of eight working hours(16:00–24:00), and free shift(F). The mathematic model of NSP is described by Eq. (2–6), where the Eq. (3–5) are hard constraints of NSP and the Eq. (6) is soft constraint.

$$\min f(x) = \sum_{i=1}^{nn}\sum_{j=1}^{sk}\sum_{k=1}^{ss}\sum_{d=1}^{sd} x_{ijkd} * w_{jk} + c * \sum_{j=1}^{sk}\sum_{k=1}^{ss}\sum_{d=1}^{sd} m_{jkd} \qquad (2)$$

$$\sum_{k=1}^{ss} x_{i,j,k,d} = 1 \; i = 1,2,\ldots,nn; j = 1,2,\ldots,sk; d = 1,2,\ldots,sd \qquad (3)$$

$$x_{i,j,3,d} + x_{i,j,1,d+1} \le 1 \, i = 1,2,\ldots,nn; \; j = 1,2,\ldots,sk; \; d = 1,2,\ldots,sd-1 \qquad (4)$$

$$lp \le \sum_{d=1}^{sd} x_{i,j,k,d} \le up \, i = 1,2,\ldots,nn; j = 1,2,\ldots,sk; k = 1,2,\ldots,ss \qquad (5)$$

$$\sum_{i=1}^{nn} x_{i,j,k,d} \geq h_{i,j,k,d} j = 1, 2, \ldots, sk; k = 1, 2, \ldots, ss; d = 1, 2, \ldots, sd - 1 \qquad (6)$$

Equation (2) is the objective function of NSP consisted two parts, the first is the salary to be paid to nurse, the second is penalty when the hospital's minimum coverage demands is not satisfied. Equation (3) presents that each nurse must be scheduled for exactly on shift each day. Equation (4) represents that nurse can't be scheduled to work a night shift (N) followed immediately by a morning shift (A). Equation (5) means that each nurse may work no more than the large limit of work shifts and no less than lower limit of work shifts. Equation (6) shows that each schedule must satisfy the hospital's minimum coverage demands.

4 Experiment Result and Discussion

4.1 Encoding

In SRBFOs, each particle is defined as a cell array comprising a $1*nn$ array, and each element in the array is also a $ss*sd$ matrix, namely $Particle = \{ss*sd\ ss*sd\ldots\ ss*sd\ ss*sd\}$. An example assuming that the number of nurse nn is 5, shift type $ss = 4$ scheduling period $sd = 4$. The six element in the array represent six nurse's shift arrangements. The particle is regarded to a potential solution, rather than a feasible solution. Only by making some adjustments of the particle, the feasible solution can be obtained. Due to each nurse must be scheduled for exactly one shift each day, the value of each element in particle should be restricted to 0 or 1. The switching operation is defined as follows (Fig. 1).

Particle

0.7 0.6 0.5 0.7	0.4 0.6 0.6 0.6	0.2 0.6 0.4 0.1	0.8 0.8 0.9 0.9	0.8 0.1 0.8 0.1
0.7 0.1 0.3 0.2	0.9 0.1 0.7 0.1	0.1 0.3 0.3 0.4	0.9 0.2 0.9 0.4	0.9 0.2 0.2 0.2
0.2 0.1 0.8 0.5	0.7 0.8 0.7 0.7	0.1 0.9 0.7 0.4	0.1 0.2 0.3 0.8	0.5 0.8 0.9 0.8
0.6 0.4 0.2 0.8	0.9 0.9 0,3 0.5	0.8 0.2 0.7 0.8	0.9 0.5 0.9 0.1	0.1 0.2 0.3 0.4

1	1	0	1	0	0	0	0	0	0	0	0	0	1	1	1	0	1	1	1
0	0	0	0	1	0	1	0	0	0	0	0	1	0	0	0	1	0	1	0
0	0	1	0	0	0	0	1	0	1	1	0	0	0	0	0	0	0	0	1
0	0	0	0	0	1	0	0	0	0	0	0	0	0	0	0	0	0	0	0

Fig. 1. The witching operation

4.2   Experiment Settings

Experiment were implemented in MATLAB environment. SRBFO, SRBFO-DN and PSO are used to solve the e nurse scheduling model described in Sect. 3, besides, the population and other parameters in these three algorithms were all set the same. So, the comparison of the results is fair. In order to simplify the problem, the skill type of nurses is classified into three categories, namely junior, middle and senior. This paper choose an example of totally 23 nurses, including 10 juniors, 8 middles and 5 seniors. The demand of hospital's minimum coverage and shifts are randomly generated.

The procedures were tested with different number of iterations and the best trade-off between solution quality and computational effort was obtained with 1000 iterations and 20 particle numbers. In order to demonstrate the efficiency and good performance of SRBFO and SRBFO-DN, PSO algorithm is selected as comparable algorithm. In this paper, we set $c1 = c2 = c3, c4 = 1.5$ for SRBFO-DN, $v\_max = 0.8$, $w\_start = 0.9$, $w\_end = 0.4$ and inertia weight $w = 8$ for SPSO. In SRBFO, the length of chemotaxis step is 0.5, the swimming length is set to 5. The probability of elimination/dispersal is set to 0.2. The reproduction frequency is set as 3 and elimination-dispersal frequency is set as 7 in SRBFO and SRBFO-DN.

4.3   Results and Analyses

Numerical results of the experiment obtained by three algorithms are presented in Table 3, including the value of minimum, maximum, mean and standard deviation. Figure 2 demonstrates the convergence curve of all the three competitive algorithms on the nurse scheduling problems.

Table 3.   Numerical results obtained by the three algorithms.

Algorithm	Max	Min	Mean	Std.
SPSO	42650	40915	42042	6.6858e+002
SRBFO	53205	48770	51803	1.9685e+003
SRBFO-DN	41600	38940	40115	1.0994e+003

Table 3 shows part of 5 nurses' arrangement in two weeks as examples. A, P, N, F represent morning shift, afternoon shift, night shift and free shifts, respectively. Here, it chooses nurse No. 1 to give an example. Nurse No. 1 have ten working days in first two weeks, including three morning shifts, three afternoon shifts and four night shifts.

Analyzing the results, we concluded that the performance of SRBFO-DN algorithm are obviously better than PSO and SRBFO. SRBFO-DN maintains good global search capability as well as convergence speed and may be more effective for solving these difficult nurse scheduling problem (Table 4).

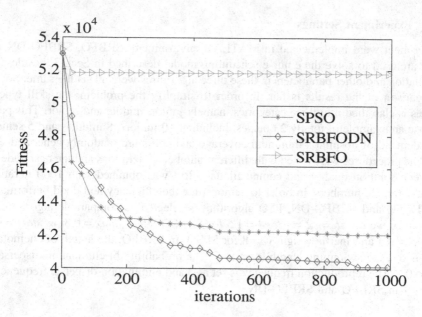

Fig. 2. Convergence curve of SRBFOs and SPSO.

Table 4. Results for nurse scheduling.

Nurse i	Day d													
	1	2	3	4	5	6	7	8	9	10	11	12	13	14
1	P	A	F	F	F	P	F	N	N	N	N	P	A	A
2	P	N	P	N	P	N	F	P	F	F	A	N	N	P
3	N	P	N	F	N	F	P	F	A	P	F	P	A	A
4	P	P	F	F	P	N	P	F	P	A	F	F	A	A
5	A	F	F	A	F	F	N	N	P	F	A	N	F	F

5   Conclusion

Nurse scheduling problem is studied in this paper and solved by SRBFO-DN. SRBFO is employed to decrease the complexity of computation. Moreover, SRBFO with information communication mechanism namely SRBFO-DN is improved to avoid premature convergence. Dynamic topology structure has been used in SRBFO-DN by combining the advantages of both SRBFO-ST and SRBFO-Ri algorithms.

Experimental results obtained verified the performance of the proposed SRBFO, SRBFO-DN, and indicated that SRBFO-DN algorithm is more effective in solving these difficult nurse scheduling problems than SPSO. In particular, the SRBFO-DN obtains a better performance than that SRBFO.

Future work will go on pay attention to optimize the efficiency of SRBFOs. It is a promising research area and a new application of BFO to solve a real-world problem with other discrete domains. Furthermore, more other optimization algorithms will be studied to solve nurse scheduling problem, such as multi-objective bacterial foraging optimization [12], bacterial colony optimization [13] and, etc.

Acknowledgements. This work is partially supported by The National Natural Science Foundation of China (Grants nos. 71001072, 71271140, 71471158, 71461027, 61262071), the Natural Science Foundation of Guangdong Province (Grant nos. S2012010008668, 945180600 1002294), Shenzhen Science and Technology Plan Project (Grant no. CXZZ2014041818 2638764).

References

1. Karp, R.M.: Reducibility Among Combinatorial Problems. Springer, US (1972)
2. Al-Betar, M.A., Khader, A.T.: A harmony search algorithm for university course timetabling. Ann. Oper. Res. **194**(1), 3–31 (2012)
3. Burke, E.K., Curtois, T., Qu, R.: A scatter search methodology for the nurse rostering problem. J. Oper. Res. Soc. **61**(11), 1667–1679 (2010)
4. Aickelin, U., Burke, E.K., Li, J.: An estimation of distribution algorithm with intelligent local search for rule-based nurse rostering. J. Oper. Res. Soc. **58**(12), 1574–1585 (2007)
5. Kim, S.J., Ko, Y.W., Uhmn, S.: A strategy to improve performance of genetic algorithm for nurse scheduling problem. Int. J. Soft. Eng. Appl. **8**(1), 53–62 (2014)
6. Rasip, N.M., Basari, A.S.H., Hussin, B.: A guided particle swarm optimization algorithm for nurse scheduling problem. Appl. Math. Sci. **8**(113), 5625–5632 (2014)
7. Al-Betar, M.A., Khader, A.T., Nadi, F.: Selection mechanisms in memory consideration for examination timetabling with harmony search. In: 12th Annual Conference on Genetic and Evolutionary Computation, pp. 1203–1210. ACM Press, New York (2010)
8. Alatas, B.: Chaotic harmony search algorithms. Appl. Math. Comput. **216**(9), 2687–2699 (2010)
9. Niu, B., Liu, J., Bi, Y.: Improved bacterial foraging optimization algorithm with information communication mechanism. In: 2014 Tenth International Conference on Computational Intelligence and Security, pp. 47–51. IEEE Press, New York (2014)
10. Passino, K.M.: Biomimicry of bacterial foraging for distributed optimization and control. IEEE Control Syst. **22**(3), 52–67 (2002)
11. Burke, E.K., Causmaecker, P.D., Berghe, G.V.: The state of the art of nurse rostering. J. Sched. **7**(6), 441–499 (2004)
12. Niu, B., Wang, H., Wang, J.W., Tan, L.J.: Multi-objective bacterial foraging optimization. Neurocomputing **116**, 336–345 (2012)
13. Niu, B., Wang, H., Chai, Y.J.: Bacterial colony optimization. Discrete Dyn. Nat. Soc. **2012**, 28 (2012)

Application of Disturbance of DNA Fragments in Swarm Intelligence Algorithm

Yanmin Liu[1](✉), Ben Niu[2], Felix T.S. Chan[3], Rui Liu[1],
and Sui changling[4]

[1] School of Mathematics and Computer Science, Zunyi Normal College,
Zunyi 563002, China
yanmin7813@gmail.com
[2] College of Management, Shenzhen University, Shenzhen 518060, China
drniuben@gmail.com
[3] Department of Industrial and System Engineering,
Hong Kong Polytechnic University, Kowloon, Hong Kong
[4] College of Life Science, Zunyi Normal College, Zunyi 563002, China

Abstract. Optimization problem is one of the most important problems encountered in the real world. In order to effectively deal with optimization problem, some intelligence algorithms have been put forward, for example, PSO, GA, etc. To effectively solve this kind of problem, in this paper, crossing strategy of DNA fragments is proposed to explore the effect on intelligence algorithms based on American genetic biologist Morgan theory. We mainly focus on DNA fragment decreasing strategy and DNA fragment increasing strategy based on disturbance in PSO. In order to test the role of the DNA mechanism, three test benchmarks were selected to conduct the analysis of convergence property and statistical property. The simulation results show that the PSO with DNA mechanism have an advantage on algorithm performance efficiency compared with the original proposed PSO. Therefore, DNA mechanism is an effective method for improving swarm Intelligence algorithm performance.

Keywords: DNA fragments · Swarm intelligence · Algorithm

1 Introduction

Optimization problem is one of the most important problems encountered in the real world. In order to effectively deal with optimization problem, some intelligence algorithms have been put forward, for example, PSO, GA, etc. These algorithms are easy to solve optimization problem and has been empirically shown to perform well. However, when the optimization problems are complicated, precocious phenomenon will appear in algorithm performance. To effectively solve the defect, some researchers investigated some new technology and novel algorithm to solve the problem.

In this paper, we propose a crossing strategy of DNA fragments to explore the effect on intelligence algorithms. According to the thought of the famous American genetic biologist Morgan "life = DNA + environment + interaction of environment and

© Springer International Publishing Switzerland 2015
D.-S. Huang et al. (Eds.): ICIC 2015, Part II, LNCS 9226, pp. 708–716, 2015.
DOI: 10.1007/978-3-319-22186-1_70

genetic", the environment interaction of environment and genetic will affect the development trend of the individual. In views of this, based on PSO we will discuss the effect of the environment on algorithm optimal process.

Usually, strengths and weaknesses of DNA will decide the life quality, so how to improve DNA to adapt environment and interaction with each other is a key problem for life evolution. DNA is composed of some of the pieces, in which the pieces with genetic effect is called gene. In this paper, we will explore the effect of DNA + environment + interaction of environment and genetic on various intelligence algorithm. As an individual, DNA genes and the ability to adapt to social environment will be absolutely decide itself quality and status in the existing society, thus the micro individual is equivalent to the microcosmic DNA respectively.

Because DNA importance in life, we regard an individual as a DNA molecule, and each dimension of an individual correspond with DNA fragment. Figure 2 gives the congruent relationship.

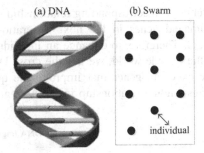

Fig. 1. DNA and swarm

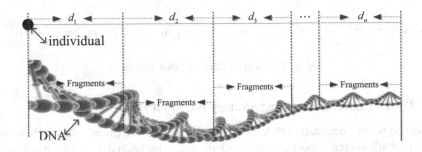

Fig. 2. Relationship DNA fragment and individual

2 Particle Swarm Optimization (PSO)

Particle swarm optimization (PSO) is a search algorithm based on population proposed by Kenney and Eberhart [1, 2]. It can effectively solve many optimization problems in the real world. But, when an optimization problem has many local optimal solutions, premature convergence will occur. To effectively use PSO to solve practical problems, some researches proposed some effective mechanisms [3–14]. In particle swarm optimization, each individual is called a particle composed of n-dimensional variables with the velocity $v_i \subset R^n$. The velocity and position of each individual will be updated as follows:

$$\overline{v}_i(t) = \overline{v}_i(t-1) + \phi_1 rand(\overline{p}_i - \overline{x}_i(t-1)) + \phi_2 rand(\overline{p}_g - \overline{x}_i(t-1)) \qquad (1)$$

$$\overline{x}_i(t) = \overline{v}_i(t) + \overline{x}_i(t-1) \qquad (2)$$

Literature [1] gives each parameter meaning. In this paper, the improved PSO proposed by [10] is adopted.

3 Crossing Strategy of DNA Fragments in PSO

The DNA traditional matching is a corresponding relationship showed in Fig. 3(a), but it can easily lead to damaged genes in the next generation, which will produce the lower quality individual. Therefore, to enhance an individual's adaptability to the environment to adapt to the outside world, we use the cross DNA fragments to make the individual inherit the excellent genes that improve the quality of the offspring. Figure 3(b) gives the corresponding relationship in DNA fragment.

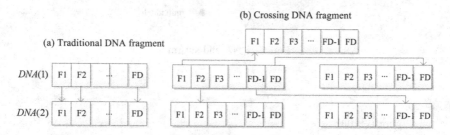

Fig. 3. Crossing strategy of DNA fragments

3.1 DNA Fragment Increasing Strategy Based on Disturbance

In order to better understand DNA disturbance strategies, we discuss the mechanism in PSO. In PSO search process, plenty and diversity individual is a key problem for swarm searching optima region, but in the existing various improvement PSO, it is always a worth exploring problem. In views of that, a strategy based on DNA fragment quality improvement is proposed in this paper. Here, population growth is equivalent to DNA disturbance that use produce new individual with better fitting, and its aim is to make sure plenty of particles (individual) to search unexplored areas of the space for more feasible solution. The strategy includes the following process.

(1) Determining the disturbing potential candidate DNA fragment

A disturbing DNA fragment selected in a group of DNAs has high probability convergence to the global optimal solution which can ensure that the individual have better heredity. In this condition, the most optimal DNA fragment for each individual (DNA) will have chance to generate the next for finding the best goal. Here, the optimal position (*Pbest*) for each particle in PSO is similar to the selected DNA fragment in which, *Pbest* for each particle will have chance to convergence to the global optima. Therefore, in each generation, some DNA fragment is selected to run disturbing operation to produce the promising DNA fragment. In this paper, we use $ns = \lfloor r \times N_{pbest} \rfloor$ to randomly choose the fragment to disturb it where r is random number in [0, 1]; ns is the chosen the number of DNA fragment; N_{pbest} is total fragment of DNA; $\lfloor \cdot \rfloor$ is *floor* function.

(2) The number of disturbance of candidate particle

Each candidate fragment of DNA is produced by disturbance. The disturbance number of candidate fragment uses the organization adjustment in each iteration. That is, we adopt a certain criteria showed in Eq. (4) to determine disturbance number. In the iteration t, the candidate fragment of disturbance number $np(t)$ is described by Eq. (3),

$$
np(t) = \begin{cases} (unp - \ln p) \times [1 - 2 \times \left(\dfrac{t}{T}\right)^2 + \dfrac{\ln p}{unp - \ln p}], & 0 \leq t \leq \dfrac{T}{2} \\[3mm] (unp - \ln p) \times [2 \times \left(\dfrac{t - T}{T}\right)^2 + \dfrac{\ln p}{unp - \ln p}], & \dfrac{T}{2} \leq t \leq T \end{cases} \tag{3}
$$

where, T is the max iteration number; lnp is the minimum disturbing number; unp is the max disturbing number. In this paper, $lnp = 1$, $unp = 5$. Figure 1 gives the number change of disturbance with the iteration.

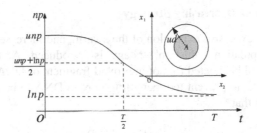

Fig. 4. Relationship between number of disturbance np and iterations

(3) New particles by disturbance for candidate particles

In order to avoid producing the particles far away from the candidate solution, we introduce inside neighborhood and outside neighborhood of the current individual. In the sub graph of Fig. 4, the gray area is the inside neighborhood of individual A, while

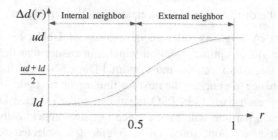

Fig. 5. Relationship between fragments added value of DNA and the random numbers

the white area is outside neighborhood of that. To make the new individual within inside neighborhood of the candidate solution by the disturbance operator, we adopt Eq. (4) to conduct the disturbance operator to produce new individual.

$$\Delta d(r) = \begin{cases} (ud - ld) \times (2 \times r^2 + ld/(ud - ld)), & 0 \le r \le 0.5 \\ (ud - ld) \times [1 - 2 \times (r - 1)^2 + ld/(ud - ld)], & 0 \le r \le 0.5 \end{cases} \quad (4)$$
$$x_{i,j} = x_{i,j} + \Delta d(r)$$

where, $r = abs(Gaussian(0, 1/9))$ is the absolute value of the gaussian distribution with the mean 0 and the variance 1/9; $ld = rld \times x_j^L$, $ud = rud \times x_j^U$; $\Delta d(r)$ is the fragment increasing value of DNA, and Fig. 5 gives the relationship with rand r; ld is the minimum added value; ud is the max added value; rud is the upper boundary rate based on hand problem; rld is the rate of boundary; x_j^U is the upper bound of the jth fragment in DNA; x_j^L is the lower bound of the jth fragment in DNA. In this paper, $rud = 0.9$, $rld = 0.02$.

3.2   DNA Fragment Decreasing Strategy

In order prevent the excessive expansion of the swarm, and increase the computational complexity, the population reduction strategy is introduced. At time t the deleted fragment of DNA will be deleted for not improved fragment of DNA of the continuous generation of L. Note: to prevent too many fragment in DNA to be dropped, we adopt Eq. (5) to deal with that.

$$N_{remove}(t) = R \times |S_L| \quad (5)$$

where, $N_{remove}(t)$ is at time t the deleted fragment of DNA; R is the delete ratio that is a random number in [0, 1]; $|S_L|$ is the number of fragment of DNA for the continuous generation of L.

3.3   PSO with DNA Mechanism

Based on Eqs. (1) and (2), in PSO, we added DNA disturbance strategies, DNA fragment decreasing strategy and DNA fragment increasing strategy. Equation (6) gives the positions updating equation.

$$\vec{x}_i(t+1) = \vec{x}_i(t) + \vec{v}_i(t+1) + \Delta D(r) \tag{6}$$

$\Delta D(r)$ is composed of $\Delta d(r)$. $\varphi_1 = \varphi_2 = \varphi_3 = 4/3$. Algorithm 1 gives PSO pseudocode with DNA mechanism.

Algorithm 1
Begin
Initialize positions and associated velocities of all particles.
Evaluate the fitness values of all particles.
Set the current position as *pbest*.
Set $V_{max}=0.25(X_{max}-X_{min})$.
While (fitcount < Max_FES) && (k<iteration)
For each particle (i=1:*ps*)
Updating particle velocity and position using Eq.(1) and (2)
Update *pbest* .
Evaluate the fitness values of the current particle *i*.
End for
Applying swarm-population growing strategy.
Applying swarm-population declining strategy.
Increase the generation count
End While
Output the result.
End Begin

4   Simulation Experiment and Analysis

4.1   DNA Mechanism Role in Algorithm Convergence Character

In order to test DNA mechanism role, we adopted the combination of the qualitative and quantitative analysis for different test function, namely the convergence property and statistical analysis to test performance of the algorithm.

DNA mechanism is introduced into PSO to test its role for intelligence algorithm showed in Algorithm 1. For convenience, PSO with DNA mechanism is call PSO with DNA (DNA-PSO for short). Sphere, Rosenbrock, Ackley and Griewank are selected to test the algorithm performance. Parameter settings of algorithm: maximum number of iterations of all algorithms (MAXT) is 1000; population size (PS) is 30; function dimension is 60; Fitness Evaluations of Rotated function is 6×104. Performance of DNA-PSO is compared with CF-LPSO [2], CF-GPSO [2]. Simulation platform is: Pentium Core Duo, 1.8 GB of RAM with 2 GB of RAM, CPU and Matlab R2008B. Figures 6 and 7 shows the algorithm convergence characteristics in non-rotated and rotated function.

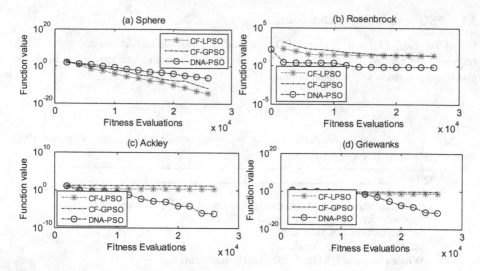

Fig. 6. Non-rotated function convergence figure

In non-rotated function, except for Sphere funciton, DNA-PSO achieved the best performance, especially on multimodal function, DNA mechanism has obvious merit. On rotated function, DNA-PSO has also the strong competitiveness, which.

4.2   Effect on Parameter L on Algorithm

In DNA fragment decreasing strategy, at iteration, no improvement performance of fragment of DNA for the continuous generation of L will be deleted. In order to test the effect of parameter L, Sphere, Rosenbrock, Ackley and Griewank functions are selected to independently run 30 algorithm, then averaging. Here, $L = 2, 6, 9, 12, 15, 40$. Table 1 gives the performance results for different parameter L. From Table 1, we can see parameter L will have some effect on algorithm performance where L with too big or too small will lead to poor performance. Only when $L = 6$, PSO with DNA mechanism achieve better results on all test functions.

Table 1. Effect of parameter L on algorithm performance

L	Sphere	Rosenbrock	Ackley	Griewank
2	6.9483e-007	3.1710e+000	9.5022e-006	7.4453e-012
6	3.1707e-007	1.1455e+000	5.1124e-006	2.8174e-012
9	6.7874e-007	3.9223e + 000	7.0605e-006	4.6137e-012
12	7.5774e-006	2.5548e+000	3.1833e-006	9.7143e-011
40	7.4313e-006	6.7119e+000	2.7692e-005	8.2346e-010

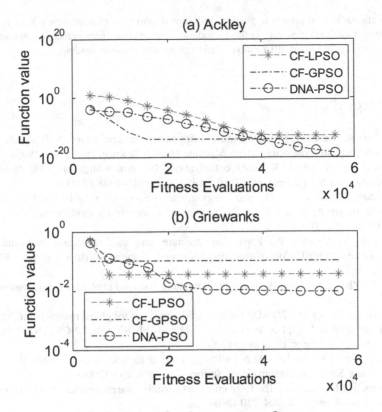

Fig. 7. Rotated function convergence figure

5 Conclusions and Future Work

In order to effectively deal with optimization problem, in this paper, crossing strategy of DNA fragments is proposed to explore the effect on intelligence algorithms based on American genetic biologist Morgan theory. We discuss role of DNA fragment decreasing strategy, and DNA fragment increasing strategy based on disturbance in PSO. In simulation experiment, we select Sphere, Rosenbrock, Ackley and Griewank to test algorithm performance in rotated and non-rotated function respectively. The results show that DNA mechanism havs active role on intelligence algorithm. Therefore, we may propose a novel intelligence algorithm based on DNA mechanism. In views of the thought, in the future, we will focus on: (i) proposing a novel intelligence algorithm based on DNA mechanism; (ii) exploring biology new role at life evolutionary.

Acknowledgments. This work is supported by the National Natural Science Foundation of China (Grants nos. 71461027, 71001072, 71271140, 71471158). Guizhou province science and technology fund (Qian Ke He J [2012] 2340 and [2012]2342, LKZS [2012]10 and [2012]22); Guizhou province natural science foundation in China (Qian Jiao He KY [2014]295);

The educational reform project in guizhou province department of education (Qian jiao gao fa [2013]446); Guizhou province college students' innovative entrepreneurial training plan (201410664004); 2013 and 2014 Zunyi 15851 talents elite project funding.

References

1. Eberhart, R., Kennedy, J.: New optimizer using particle swarm theory. In: Proceedings of the 6th International Symposium Micro Machine Human Science, pp. 39–43 (1995)
2. Kennedy, J., Eberhart, R.: PSO optimization. In: Proceedings of IEEE International Conference on Neural Networks, Perth, vol. 4, pp. 1941–1948 (1995)
3. Kennedy, J.: Small worlds and mega-minds: effects of neighborhood topology on particle swarm performance. In: Proceedings of Congress on Evolutionary Computation, pp. 1931–1938 (1999)
4. Kennedy J., Mendes, R.: Population structure and particle swarm performance. In: Proceedings of IEEE Congress on Evolutionary Computation, Honolulu, pp. 1671–1676 (2002)
5. Niu, B., Zhu, Y.: A lifecycle model for simulating bacterial evolution. Neurocomputing **72**, 142–148 (2008)
6. Niu, B., Li, L.: A novel PSO-DE-based hybrid algorithm for global optimization. In: Huang, D.-S., Wunsch II, D.C., Levine, D.S., Jo, K.-H. (eds.) ICIC 2008. LNCS (LNAI), vol. 5227, pp. 156–163. Springer, Heidelberg (2008)
7. Parsopoulos K.E., Vrahatis M.N.: UPSO-a unified particle swarm optimization scheme. In: Lecture Series on Computational Sciences, pp. 868–873 (2004)
8. Mendes, R., Kennedy, J.: The fully informed particle swarm: simpler, maybe better. IEEE Trans. Evol. Comput. **8**, 204–210 (2004)
9. Peram, T., Veeramachaneni, K.: Fitness-distance-ratio based particle swarm optimization. In: Proceedings of Swarm Intelligence Symposium, pp. 174–181 (2003)
10. Liang, J.J., Qin, A.K., Suganthan, P.N., Baskar, S.: Evaluation of comprehensive learning particle swarm optimizer. In: Pal, N.R., Kasabov, N., Mudi, R.K., Pal, S., Parui, S.K. (eds.) ICONIP 2004. LNCS, vol. 3316, pp. 230–235. Springer, Heidelberg (2004)
11. Liang, J.J., Qin, A.K.: Comprehensive learning particle swarm optimizer for global optimization of multimodal functions. IEEE Trans. Evol. Comput. **10**(3), 281–295 (2006)
12. Angeline, P.J.: Using selection to improve particle swarm optimization. In: Proceedings of IEEE Congress on Evolutionary Computation, Anchorage, pp. 84–89 (1998)
13. Lovbjerg, M., Rasmussen, T.K., Krink, T.: Hybrid particle swarm optimizer with breeding and subpopulations. In: Proceedings of Genetic Evolutionary Computation Conference pp. 469–476 (2001)
14. Miranda, V., Fonseca, N.: New evolutionary particle swarm algorithm (EPSO) applied to voltage/VAR control. In: Proceedings of 14th Power System Computation Conference, Seville (2002)

Multi-objective PSO Based on Grid Strategy

Yanmin Liu[1(✉)], Ben Niu[2], Felix T.S. Chan[3], Rui Liu[1],
and Changling Sui[4]

[1] School of Mathematics and Computer Science, Zunyi Normal College,
Zunyi 563002, China
yanmin7813@gmail.com
[2] College of Management, Shenzhen University, Shenzhen 518060, China
drniuben@gmail.com
[3] Department of Industrial and System Engineering,
Hong Kong Polytechnic University, Kowloon, Hong Kong
[4] College of Life Science, Zunyi Normal College, Zunyi 563002, China

Abstract. In multi-objective optimization problem (MOP), keeping solution diversity is key case for solution quality. To improve the MOP quality, the diversity maintenance threshold value ($\lambda\alpha$) is proposed to keep solutions diversity based on adaptive grid strategy. These strategies can adaptive maintain the non-inferior diversity to improve swarm individual fly to the global optimal. Four test problems are selected to test the proposed strategy compared with other classical methods, and three performance metrics are chosen to explore the algorithm effectiveness.

Keywords: Particle swarm optimizer · Grid strategy · Adaptive · Diversity maintenance · Multi-objective optimization

1 Introduction

Particle swarm optimization (PSO for short) [1] is a kind of random optimization algorithm based on birds foraging behavior which has simple operation and fast convergence. Based on these merits, many scholars put forward using PSO to solve multi-objective problem (Multi-objective PSO, MOPSO). The efficiency of evaluating an algorithm to deal with MOP is non inferior solution quality. Based on two goals, researchers have proposed many strategies, for example elite MOPSO (EM-MOPSO) [2], MOPSO of Coello [3]. In recent research, in [4, 5], the authors proposed dynamic population MOPSO with adaptive local external archive to enhance the diversity of each population which improve each particle's ability to converge to the true Pareto front. In view of above, in this paper we puts forward the adaptive distribution to maintain the multi-objective PSO (ADMMOPSO) to improve PSO to effectively to solve multi-objective problems.

© Springer International Publishing Switzerland 2015
D.-S. Huang et al. (Eds.): ICIC 2015, Part II, LNCS 9226, pp. 717–724, 2015.
DOI: 10.1007/978-3-319-22186-1_71

2 Problem Description

Due to minimize and maximization problem can be transformed each other, so we only select the minimum MOPs as the research object. The optimal solution for MOP is usually called Pareto solution. A MOPs with n decision variables and m objective functions can be described as follows,

$$\min \ y = F(x) = (f_1(x), f_2(x), \ldots, f_m(x))$$
$$s.t. \begin{cases} g_i(x) \geq 0 & i = 1, 2, \ldots, p \\ h_j(x) = 0 & j = 1, 2, \ldots, q \end{cases} \tag{1}$$

where $x = (x_1, x_2, \ldots, x_n) \in X \subset R^n$ is n decision vector; X is n decision space; $y = (y_1, y_2, \ldots, y_m) \in Y \subset R^m$ is m objective variable; Y is m objective space; $g_i(x) \geq 0 (i = 1, 2, \ldots, p)$ is p inequality constraints; $h_j(x) = 0 (j = 1, 2, \ldots, q)$ is q equality constraints. Based on MOP shown in Eq. (1), there are the following definitions.

Definition 1. Set of feasible solution. If $\overline{X}$ satisfy,

$$\overline{X} = \{x \in R^n | g_i(x) \geq 0, h_j(x) = 0 \ (i = 1, 2, \ldots, p; j = 1, 2, \ldots, q)\} \tag{2}$$

$\overline{X}$ is called the set of feasible solution.

Definition 2. Pareto domination. $x_u, x_v \in \overline{X}$ are two feasible solution for Eq. (1), and compared with x_u, x_v is Pareto domination, when and only when

$$\forall i = 1, 2, \ldots, m, \ f_i(x_v) \leq f_i(x_u) \wedge \exists j = 1, 2, \ldots, m, \ f_j(x_v) < f_j(x_u) \tag{3}$$

Definition 3. Pareto optimal solution set. For MOP shown in Eq. (1), its Pareto optimal solution set is defined as follows,

$$P^* = \{x \in \overline{X} | \exists x' \in \overline{X}, f_j(x') \leq f_j(x), (j = 1, 2, \ldots, m)\} \tag{4}$$

3 MOPSO with Adaptive Distribution Strategy

In the proposed algorithm, the external file stores the pareto solutions at each iteration and with iteration elapsed, crowded distance [3] is adopted to limit the size of the external file. For the global optimal selection of particles, when the pareto solutions distribution index is lower than the given threshold (λ_a), a non-inferior solutions with random method from external archive is selected to lead population flight. But when the distribution index is higher than λ_a, in order to improve the distribution of the pareto solutions, a global optimal particle selection strategy based on the grid is introduced. The following are discussed in turn.

3.1 Distribution Measurement

The distribution is uniformity and universality of the solution set, here we use distribution index of the literature [6] to evaluate the quality of the pareto solutions obtained, as shown in Eq. (5).

$$SP = \sqrt{\frac{1}{|Q|-1}\sum_{i=1}^{|Q|}(\bar{d}-d_i)^2}$$

$$d_i = \min_{q_j \in Q \wedge q_j \neq q_i}\sum_{m=1}^{M}|f_m(q_i)-f_m(q_j)|$$

(5)

where, di is euclidean distance of two consecutive vector in non-domination border of solution set; Q is the number of non-inferior in external archive; $\bar{d}$ is the average of the distance. SP is smaller which means that the distribution is more uniform.

3.2 Adaptive Strategy Based on Grid Strategy

When the distribution index is higher than the given threshold, the selection strategy of the global optimal particle selection strategy based on grid is invoked. The process of the construction of the grid is below: m dimension object space is divided into $K1 \times K2 \times \cdots \times Km$ grid, then using Eq. (6) gives the width di of ith dimension for each gird

$$d_i = \frac{\max\limits_{x \subset X} f_i(x) - \min\limits_{x \subset X} f_i(x)}{K_i}$$

(6)

where, d_i is objective width of ith dimension; $f_i(x)$ is the objective function value of ith dimension; K_i is objective number of ith dimension; x is decision variable. For convenience, let,

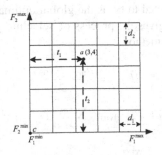

Fig. 1. The construction of the grid

$$\begin{cases} F_i^{\max} = \max_{x \subset X} f_i(x) \\ F_i^{\min} = \min_{x \subset X} f_i(x) \end{cases} \tag{7}$$

The commuting method of objective space position of new generator particles $a = (a_1, a_2, \ldots, a_m)$ at each iteration is given in Eq. (8),

$$\begin{cases} t_i = s_i - F_i^{\min} & i = 1, 2, \ldots, m \\ p_i = \mathrm{mod}(t_i, d_i) + 1 & i = 1, 2, \ldots, m \end{cases} \tag{8}$$

where, $\mathrm{mod}(x, y)$ is the integral part after x/y. Figure 1 gives the schematic the diagram with two targets. Here, particle a is in the grid (3, 4), and the coordinate of point c is $(F_1^{\max}, F_2^{\max})$. In order to discuss method expediently, the number of particles in each grid is called grid density.

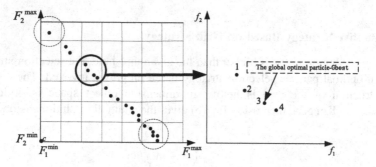

Fig. 2. The distribution of Pareto solutions

3.3 Selection of the Global Optimal Particle

In order to make non-inferior uniformly distribute in the Pareto front, the selected grid has the following characteristic: the number of non-inferior is equal to the average of that of the girds with max diversity and minimum diversity. After determining the grid, one in many particles is selected to be as the global optimal particle to guide flight for the whole swarm. Figure 2 gives the process of selecting global optimal particle, and Eqs. (9)–(11) give selection method.

$$H_{index} = \frac{n_{\max} + n_{\min}}{2} \tag{9}$$

$$index = \min_{i \in |H_{index}|} \sum_{j=1}^{m} (f_{ij} - \bar{f_j})^2 \tag{10}$$

$$\bar{f_j} = \frac{\sum_{i=1}^{n} f_{ij}}{n} \quad j = 1, 2, \ldots, m \tag{11}$$

where, n_{max} is in objective space the number of non-inferior in max diversity gird; n_{min} is in objective space the number of non-inferior in minimum diversity gird; $|H_{index}|$ is the grid with the number H_{index} of non-inferior; *index* is non-inferior corresponding to the global optimal particle; m is the number of objective; n is the number of particle; f_{ij} is the fitting function value of *jth* objective of *ith* particle.

3.4 ADMMOPSO

Combined with the above strategy, Eq. (11) gives the updating equation of velocity and position of particle.

$$\vec{v_i}(t) = w \cdot \vec{v_i}(t-1) + \varphi_1 \cdot r_1(\vec{p_i} - \vec{x_i}(t-1)) + \varphi_2 \cdot r_2(rep(index) - \vec{x_i}(t-1))$$
$$\vec{x_i}(t) = \vec{x_i}(t-1) + \vec{v_i}(t)$$
$$(11)$$

where, $\vec{v_i}(t)$ is the velocity of particle i at time t; $\vec{x_i}(t)$ is the position of particle i at time t; $\vec{p_i}(t)$ the history best position of particle i at time t; $rep(index)$ is the global optimal particle at time t; φ_1 and φ_2 are the acceleration constant of particle within [0, 2]; w is called the inertia weight; r_1 and r_2 are uniformly distributed random vector in [0, 1]. Usually, $w = (w_{ini} - w_{end})(T_{max} - t)/T_{max} + w_{end}$, here T_{max} is the maximum generation; w_{ini} is the initial weights; w_{end} is maximum generation at the end of evolution (Fig. 3).

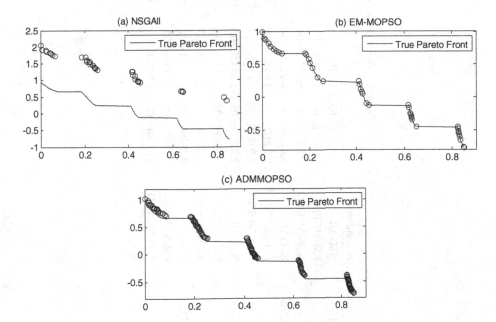

Fig. 3. Pareto solution for various algorithms on ZDT3 function

Table 1. The performance comparison for different algorithms

Problem	Algorithm	SP	D	T(s)	λ_a
ZDT1	NSGAII	$7.9713E-03 \pm 2.1183E-03$	$1.3901E+00 \pm 2.3761E-04$	$7.0943E+01 \pm 1.9375E-01$	8.3E−03
	EM-MOPSO	$5.3941E-03 \pm 3.8145E-03$	$2.2831E+00 \pm 4.1149E-03$	$6.2931E+01 \pm 3.8713E-01$	
	ADMMOPSO	$1.2081E-03 \pm 5.6624E-04$	$2.1038E+00 \pm 1.1391E-04$	$8.1143E+01 \pm 2.1392E-01$	
ZDT2	NSGAII	$6.6128E-03 \pm 4.1139E-04$	$2.0835E+00 \pm 1.3964E-03$	$7.9371E+01 \pm 2.9652E-01$	7.5E−03
	EM-MOPSO	$2.2291E-03 \pm 3.0982E-03$	$2.9871E+00 \pm 4.9326E-04$	$7.1532E+01 \pm 3.1749E-01$	
	ADMMOPSO	$5.9362E-04 \pm 1.1193E-04$	$3.7726E+00 \pm 6.0319E-04$	$6.2106E+01 \pm 1.1194E-02$	
ZDT3	NSGAII	$2.7763E-01 \pm 1.9642E-03$	$1.1036E+01 \pm 3.1284E-02$	$8.9218E+01 \pm 2.4399E-01$	9.2E−02
	EM-MOPSO	$9.6207E-02 \pm 1.1135E-03$	$2.6417E+01 \pm 1.3862E-03$	$8.7302E+01 \pm 4.1207E-02$	
	ADMMOPSO	$4.4173E-02 \pm 2.0749E-03$	$3.1037E+01 \pm 1.1372E-03$	$8.6302E+01 \pm 3.2904E-02$	
DTLZ2	NSGAII	$9.2846E-02 \pm 2.0751E-02$	$1.1139E+00 \pm 2.1643E-02$	$2.1108E+02 \pm 3.2901E-02$	7.9E−02
	EM-MOPSO	$4.2935E-02 \pm 3.9174E-02$	$3.9072E+00 \pm 1.0873E-02$	$1.8732E+02 \pm 1.0863E-02$	
	ADMMOPSO	$3.0384E-02 \pm 3.9281E-03$	$6.9311E+00 \pm 4.0932E-03$	$2.1937E+02 \pm 1.3857E-02$	

4 Analysis of Experimental Results

4.1 Function and Parameter Settings of Algorithm

ZDT1, ZDT2, ZDT3 and DTLZ2 [5] are selected to test algorithm performance where ZDT1, ZDT2, ZDT3and DTLZ2 are used to test the algorithm optimization ability for Pareto frontier of convex, non convex, discrete and high dimension. Two representative algorithms NSGAII [7] and EM-MOPSO [2] are selected to test the proposed algorithm performance. The iteration number for all algorithms is 200, the swarm size and the archive size are 100 respectively. In our proposed algorithm, $K = 20$, $\varphi_1 = \varphi_2 = 2$, $w_{ini} = 0.9$, $w_{end} = 0.4$, and different function λ_α is shown in Table 1. The simulation experiment platform is computer for ThinkPad-SL400 using software Matlab7.7.0. Each test function independently run 20 times, and the method for calculating the running time of the algorithm is the function (tic and toc).

4.2 Simulation Result

In this paper, we use the index SP [6] to evaluate the uniformity for different algorithm and, use D [8] to evaluate universality. Here, SP value that is the smaller means better uniformity, and D value that is the bigger means better universality. Table 1 gives the optimization results for different algorithm, where each value shows the mean and standard deviation for 20 independent run. For index SP, ADMMOPSO achieve the

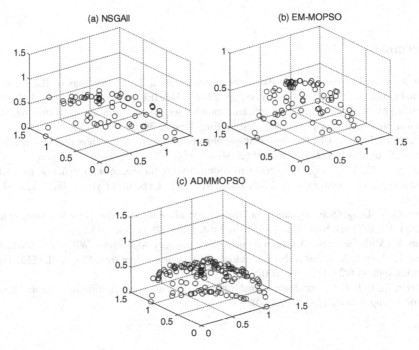

Fig. 4. Pareto solution for various algorithms on DTLZ2 function

best results for all test function, which the propose algorithm is effective for improving the uniformity quality of non-inferior. For index D, except ZDT1, ADMMOPSO also has the better performance compared with other two algorithms. At last, from the calculation time, the proposed algorithm and other algorithms of computing time are in the same order of magnitude, which means that the introduction of strategy did not increase the computational complexity (Fig. 4).

5 Conclusions

In this paper, we propose an adaptive distribution multi-objective particle swarm optimization (ADMMOPSO). By simulation experiment on ZDT1, ZDT2, ZDT3 and DTLZ2 function, ADMMOPSO shows better performance in dealing with the Pareto frontier for convex, concave, discontinuity and high dimension problem compared with algorithms. The results show that ADMMOPSO is an effective method for solving multi-objective optimization problem.

Acknowledgments. This work is supported by the National Natural Science Foundation of China (Grants nos. 71461027, 71001072, 71271140, 71471158). Guizhou province science and technology fund (Qian Ke He J [2012] 2340 and [2012]2342, LKZS [2012]10 and [2012]22); Guizhou province natural science foundation in China (Qian Jiao He KY [2014]295); The educational reform project in guizhou province department of education (Qian jiao gao fa[2013] 446); Guizhou province college students' innovative entrepreneurial training plan(2014106 64004); 2013 and 2014 Zunyi 15851 talents elite project funding.

References

1. Kennedy, J., Eberhart, R.C.: Particle swarm optimization. In: Proceedings of IEEE International Conference on Neural Networks, pp. 1942–1948. IEEE Press, Washington, DC (1995)
2. Reddy, M.J., Kumar, D.N.: An efficient multi-objective optimization algorithm based on swarm intelligence for engineering design. Eng. Optim. **39**(1), 49–68 (2007)
3. Coello, C., Pultdo, G.T.: Handling multiple objectives with particle swarm optimization. IEEE Trans. Evol. Comput. **8**(3), 256–279 (2004)
4. Leong, W.F., Yen, G.G.: PSO-based multiobjective optimization with dynamic population size and adaptive local archives. IEEE Trans. Syst. Man Cybern. B Cybern. **38**(5), 1270–1293 (2008)
5. Yen, G.G., Leng, W.F.: Dynamic multiple swarms in multiobjective particle swarm optimization. IEEE Trans. Syst. Man Cybern. A Syst. Hum. **39**(4), 890–911 (2009)
6. Deb, K.: Multi-objective Optimization using Evolutionary Algorithms. Wiley, London (2001)
7. Deb, K., Parpat, A.: A fast and elitist multiobjective genetic algorithm: NSGA–II. IEEE Trans. Evol. Comput. **6**(2), 182–197 (2002)
8. Zitzler, E., Deb, K.: Comparison of multiobjective evolutionary algorithms: empirical results. Evol. Comput. **8**(2), 173–195 (2000)

SRBFOs for Solving the Heterogeneous Fixed Fleet Vehicle Routing Problem

Xiaobing Gan[1], Lijiao Liu[1], Ben Niu[1,2,3(✉)], L.J. Tan[4],
F.F. Zhang[1], and J. Liu[1]

[1] College of Management, Shenzhen University, Shenzhen, China
[2] Hefei Institute of Intelligent Machine, Chinese Academy of Science,
Hefei, China
[3] Department of Industrial and System Engineering, Hong Kong Polytechnic
University, Kowloon, Hong Kong
Drniuben@gmail.com
[4] Department of Business Management, Shenzhen Institute of Information
Technology, Shenzhen 518172, China

Abstract. The purpose of this paper is to present a new method to solve the heterogeneous fixed fleet vehicle routing problem (HFFVRP) based on structure-redesign-based bacterial foraging optimization (SRBFO). The HFFVRP is a special case of the heterogeneous vehicle routing problem (HVRP), in which the number of each type of vehicles is fixed. To deal with this combinatorial optimization problem, two improved SRBFOs (SRBFOLDC and SRBFONDC) are presented by integrating the same time decreasing chemotaxis step size mechanism of BFOLDC and BFONDC into the optimization process of SRBFO. SRBFOLDC and SRBFONDC are successfully applied though encoding the position of bacteria by 2N dimensions. The first N dimensional vectors indicate the corresponding vehicle, and the next N dimensional vectors present the execution order of the corresponding vehicle routing. In the simulation experiments, it is demonstrated that SRBFONDC and SRBFOLDC are efficient to solve the heterogeneous fixed fleet vehicle routing problem with lower transportation cost and get better vehicle routing. Besides, SRBFONDC performs best compared with other five bacterial foraging optimization algorithms.

Keywords: Bacterial foraging optimization · Chemotaxis step · The heterogeneous fixed fleet vehicle routing problem

1 Introduction

The vehicle routing problem (VRP) is first proposed in 1959 by Dantzig and Ramser [1], which is an important problem in transportation, distribution and logistics areas. VRP is a combinatorial optimization and integer programming problem that calls for the determination of the optimal set of routes to be taken up by a fleet of vehicles serving customers subject to side constraints.

In the previous literature, a quite number of variants vehicle routing problem have been studied on either classical variants or the different versions, considering these factors: pickup and delivery, soft or hard time window, multi-depot, etc. such as the

© Springer International Publishing Switzerland 2015
D.-S. Huang et al. (Eds.): ICIC 2015, Part II, LNCS 9226, pp. 725–732, 2015.
DOI: 10.1007/978-3-319-22186-1_72

capacitated VRP (CVRP) [2], which takes the factor of vehicles having limited capacity into consideration. VRP with pick-up and delivery (VRPPD) [3], which considers the pick-up and delivery of goods. The multi-depot VRP (MDVRP) [4], which more than one depot is considered. VRP with time windows (VRPTW) [5], which adds the hard or soft service time windows of customers to research. The site-dependent VRP (SDVRP) [6] and the heterogeneous fleet vehicle routing problem (HVRP) [7] etc. The heterogeneous fixed fleet vehicle routing problem (HFFVRP) is one of the vehicle routing problems, which is much harder to solve. Taillard is the first researcher who has proposed an efficient and robust heuristic method to solve the HFFVRP in the past twenty years [8].

In reality, the phenomenon of various delivery types and different properties, or customer requirements is very common. Using the same type of vehicle involved in the distribution is likely to lead to significantly increased costs. And the use of heterogeneous fixed fleet vehicles may improve vehicle load factor, reduce logistics costs and improve economic efficiency of enterprises. Therefore, it has important theoretical and practical significance for research in the heterogeneous fixed fleet vehicle routing problem.

Bacterial foraging optimization (BFO) is a stochastic optimization technique that simulates the bacteria swarms' social foraging behavior, which was first proposed by Passino in 2002 [9]. It inspired by the foraging behavior of E. coli bacteria and it has attracted increasing attention as its strong optimization capabilities in solving optimization problems. In previous research, there are many improved BFO applied to VRP. In recent years, some research made great effort to settle the variants of VRP using BFO algorithms. Such as Niu et al. developed a novel bacterial foraging optimization with adaptive chemotaxis step in 2012 to solve Vehicle Routing Problem with Time Windows (VRPTW) [10]. Besides, a variant of the bacterial foraging optimization algorithm with time-varying chemotaxis step length and comprehensive learning strategy (ALCBFO) is proposed by Tan et al. to apply in vehicle routing problem with time windows in 2014 [11]. What's more, Seda et al. also presented BFO algorithm to solve the vehicle routing problem with simultaneous delivery and pick-up (VRPSDP) efficiently [12].

SRBFO is first proposed by Niu et al. [13], which is taking advantage of single-loop structure, and developed a new execution structure to improve the convergence rate as well as lower computational complexity. In pervious study, SRBFO is proved to solve portfolio selection such continuous problem efficiently. However, till now, SRBFO hasn't been applied to any other discrete optimization problems, such as VRP. In this paper, SRBFOs is presented to solve an improved kind of VRP model (HFFVRP). Furthermore, two improved SRBFOs with varying chemotaxis step (SRBFOLDC and SRBFONDC) are proposed, which imitate the same mechanism of BFOLDC and BFONDC by moderate the chemotaxis step length in the process of SRBFO.

This paper is structured as follows: Sect. 2 describes the heterogeneous fixed fleet vehicle routing problem (HFFVRP). Structure-redesign-based bacterial foraging optimizations for HFFVRP are presented in Sect. 3. Section 4 provides the experiment studies. Finally, the conclusion of this study and future research is presented in Sect. 5.

2 Description of HFFVRP

In real distribution management areas, the heterogeneous fixed fleet vehicle routing problem (HFFVRP) is more realistic compared with other standard VRP proposed by Taillard [8]. The HFFVRP is a design to seek a set of minimum cost routes, which is originated and terminated at a depot, and in which each types of vehicle have different fixed number and variable capacities and different unit running costs of servicing customers with known demands in the process of distribution. HFFVRP is a special case of the heterogeneous fleet vehicle routing problem (HVRP), for the number of each type of vehicles is fixed while in HVRP is unlimited [7, 14].

In HFFVRP, we assume that vehicles are start and end in the same depot. All the customers demand must be satisfied and only using one service time. Besides, all the goods can be mixed in vehicles and vehicles can't be overloaded. HFFVRP is described as follows: suppose that there are K vehicles in depot, including two types of vehicle. One with $K1$ vehicles, and another with $K2$ vehicles, and the speed are v_1, v_2 respectively. There are l customers to be serviced. For i_{th} customer, the demand is q_i, service time is S_i, and time window is $[ET_i, LT_i]$. In this model, there will generates a waiting cost pe if the vehicle reach customer i before ET_i and a late cost if the vehicle arrival at customer i over LT_i. The distance between customer i and customer j is d_{ij}. Assuming that n_k is the number of customers been serviced by the k_{th} vehicle, if $n_k = 0$, the k_{th} car is not involved in the delivery. R_k represents the k_{th} vehicle's routing path, where r_{ki} represent customer i in the path(not including the depot). r_{k0} represents the depot. $d_{r_{k(i-1)}r_{ki}}$ is the distance between $r_{k(i-1)}$ and r_{ki}. $d_{r_{kn_k}r_{k0}}$ is the distance between the customer of vehicle k last service and the depot. The structured Model is adopted in Ref. [15].

The objective function is:

$$
\min z = \sum_{k=1}^{K1} sign(n_k) * c_1 * \left[\sum_{i=1}^{n_k} d_{r_{k(i-1)}r_{ki}} + d_{r_{kn_k}r_{k0}} \right]
$$
$$
+ \sum_{k=1}^{K2} sign(n_k) * c_2 * \left[\sum_{i=1}^{n_k} d_{r_{k(i-1)}r_{ki}} + d_{r_{kn_k}r_{k0}} \right] \tag{1}
$$
$$
+ pe * \sum_{i=1}^{l} \max(ET_i - T_i, 0) + pl * \sum_{i=1}^{l} \max(T_i - LT_i, 0)
$$

where Eq. (1) represents the objective function for the sake of the total cost is minimized. Which consists of three parts, the first part is a first type of vehicles involved in the transport charges generated during transportation; the second part is the second type of vehicle transportation costs involved in the transport process produces; the third part is the total cost of late fees and waiting fees result from these does not meet the time window during transport procedure.

The constraints of the model are described as follows:

$$t_{r_{ki}} = t_{r_{k(i-1)}} + d_{r_{ki}r_{k(i-1)}}/v_k + s_{r_{k(i-1)}} \qquad (2)$$

$$\sum_{i=1}^{n_k} q_{r_{ki}} < Q_{K1} \qquad k = 1........K1 \qquad (3)$$

$$\sum_{i=1}^{n_k} q_{r_{ki}} < Q_{K2} \qquad k = 1........K2 \qquad (4)$$

$$0 \le n_k \le l \qquad (5)$$

$$\sum_{k=1}^{K} n_k = l \qquad (6)$$

$$sign(n_k) = \begin{cases} 1 & n_k \ge 1 \\ 0 & n_k = 0 \end{cases} \qquad (7)$$

$$R_{k1} \cap R_{k2} = \varphi \qquad \forall k_1 \ne k_2 \qquad (8)$$

where Eq. (2) is used to calculate the time of the vehicle reaches each customer. Equations (3) and (4) are used to ensure that transport cargo capacity on each route of the vehicle does not exceed the vehicle load; Eq. (5) ensures that the number of customer each vehicle serviced is less than the total number of customers. Equation (6) is used to assure that each customer is serviced. Equation (7) defines an variable, if vehicle k is not participate in the distribution, $sign(n_k) = 0$ else, $sign(n_k) = 1$. Finally, Eq. (8) indicates that each customer can only have a vehicle to make delivery.

3   SRBFOs for HFFVRP

3.1   SRBFONDC and SRBFOLDC

Original bacterial foraging algorithm (BFO) offers a constant chemotaxis step length C [9]. However, from the previous study, the proposed BFO-LDC [16] and BFO-NDC [17] has been demonstrated that chemotaxis C step length is one of the most important parameters to get a good balance between local and global search abilities. It proved that a small chemotaxis step length accelerate local search ability while a big one accelerates global search. SRBFO is making use of single-loop structure, and developed a new execution structure to improve the convergence rate as well as lower computational complexity [13]. According to those former studies, SRBFOLDC and SRBFONDC imitate the same mechanism through moderate the chemotaxis step length of SRBFO. Chemotaxis C step length is decided based on the following equation:

$$C_j = C_{min} + (1 - \frac{iter}{itermax}) \times (C_{max} - C_{min}) \qquad (9)$$

$$C_j = C_{min} + \exp(-a \times \left(\frac{iter}{itermax}\right)^n) \times (C_{max} - C_{min}) \qquad (10)$$

where Eqs. (9) and (10) respectively represent the SRBFOLDC and SRBFONDC chemotaxis step length change over the run process. C_{max} and C_{min} are the start value and end value of chemotaxis step. Different value of modulation index n will lead to different variants.

3.2 Design for HFFVRP

In previous VRP study, researchers often use $N + K - 1$ dimensional coding way to solve the problem with N customer and K vehicles, which is simple and efficient. However, in the research of HFFVRP, the number of vehicles involved in the distribution is uncertain. So we adopt 2N dimensional coding scheme for encoding, which corresponds to the vehicle routing problem with N customers. Each customer correspond two-dimension s. One dimension is the vehicle number k which service the customer, another is the k_{th} vehicle routing order r. That is to say, to express and to facilitate the calculation, each particle corresponding 2N dimensional vector X is divided into two N dimensional vectors: x_v indicate each task corresponding vehicle and x_r expressed execution order of the tasks in the corresponding path of the vehicle.

4 Experiments and Results

In this part, heterogeneous fixed fleet vehicle routing problem (HFFVRP) with 12 customers and 7 vehicles is developed. The depot coordination is set as (50, 60). And location of the customers is defined by coordinates and with each arc (x_i, y_i). The distance between customer i_{th} and customer j_{th} is associated with a Euclidean distance d_{ij}, which is approximately calculated: $d_{ij} = \sqrt{(x_i - x_j)^2 + (y_i - y_j)^2}$. The customer location coordinate parameters are shown in Table 1. There are two types of delivery vehicles, including 3 class A vehicles with 550 units capacity, 4 class B vehicles with 650 units capacity; the average speed was 75; the cost of the vehicle to wait for 25 yuan per hour, the delay cost 85 yuan per hour. In the experimental studies, six algorithms are used: SRBFOLDC, SRBFONDC, BFO [9], BFOLDC [16], BFONDC [17] and SRBFO [13]. The parameters of these algorithms are included in Table 2. Table 3 lists the customer demands and lay time and time windows. Table 4 gives the runs results including max value, min value, average value and variances of 20 runs. The best routes of vehicle to customers based on SRBFONDC are presented in Table 5. Finally, Fig. 1 shows the average values of all algorithms run 20 times.

From Table 4, SRBFONDC and SRBFOLDC get a better result regarding minimum, mean and maximum while BFO get the worst results. Besides, from the results,

SRBFONDC performs superior than SRBFOLDC. Meanwhile, SRBFONDC observed gain the fastest convergence to optimum and obviously performs better than other algorithms in Fig. 1.

Table 1. Customer location coordinate

Customer i	1	2	3	4	5	6	7	8	9	10	11	12
Xcoordinate	20	50	78	100	35	68	90	125	150	110	84	70
Ycoordinate	45	37	30	59	60	85	105	92	65	70	42	50

Table 2. Parameters of BFO algorithms

Algorithms	Nc	Nre	Ned	Cmax	Cmin	Ns	Popsize
BFO	400	5	2	0.1	0.1	5	100
BFO-LDC	400	5	2	0.2	0.01	5	100
BFO-NDC	400	5	2	0.5	0.01	5	100
SRBFO	4000	–	–	0.1	0.1	5	100
SRBFO-LDC	4000	–	–	0.2	0.01	5	100
SRBFO-NDC	4000	–	–	0.5	0.01	5	100

Table 3. The demands, lay-time and time window of customers

Customer i	Demands q_i	Lay-time (minutes)	Time window $[ET_i, LT_i]$
1	300	80	[9:00 12:00]
2	250	60	[9:00 17:00]
3	170	45	[12:00 17:00]
4	90	60	[8:00 16:30]
5	200	50	[8:00 12:00]
6	250	60	[9:00 17:00]
7	350	90	[9:00 17:30]
8	230	55	[9:00 18:00]
9	180	45	[9:00 16:00]
10	150	40	[9:00 17:00]
11	230	40	[9:00 18:00]
12	290	75	[9:00 17:00]

Table 4. The maximum, minimum, mean and variance results of all algorithms

Algorithms	Maximum	Minimum	Mean	Variance
BFO	3136	2691	2964	127.7
BFO-LDC	3212	2603	2883	154.3
BFO-NDC	3199	2637	2931	156.1
SRBFO	3035	2622	2820	91.9
SRBFO-LDC	2889	2626	2790	99.8
SRBFO-NDC	2964	2587	2786	87.6

Table 5. The best routes of the vehicle to each customer using SRBFONDC

Vehicle		Route of vehicle	Cost
Type A	1	$0 \rightarrow 11 \rightarrow 0$	2587
	2	$0 \rightarrow 5 \rightarrow 1 \rightarrow 0$	
	3	0	
Type B	4	$0 \rightarrow 10 \rightarrow 6 \rightarrow 2 \rightarrow 0$	
	5	$0 \rightarrow 9 \rightarrow 8 \rightarrow 12 \rightarrow 0$	
	6	$0 \rightarrow 4 \rightarrow 7 \rightarrow 3 \rightarrow 0$	
	7	0	

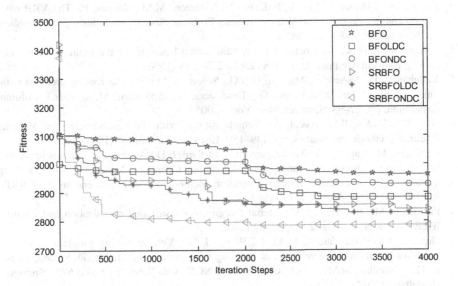

Fig. 1. The average values of all algorithms run 20 times.

5   Conclusions and Future Research

In this paper, a new way to solve HFFVRP based on BFO has been proposed. SRBFONDC and SRBFOLDC are improved SRBFO [13] with vary chemotaxis step change over the process of algorithm. From the results, we can conclude that SRBFONDC preforms best compared with other BFO algorithms, and get the best vehicle routing. SRBFONDC has the potential to applicate in other VRP variants. In further research, other vehicle routing problems, such as the multi-depot vehicle routing problem (MDVRP) [4] and the site-dependent vehicle routing problem (SDVRP) [6], will have a good capacity to solve by variant BFO algorithms.

Acknowledgements. This work is partially supported by the National Natural Science Foundation of China (Grants nos. 71001072, 71271140, 71471158), the Natural Science Foundation of Guangdong Province (Grant nos. S2012010008668, 9451806001002294), the Graduate Student's Innovation Project of School of Management Shenzhen University, the Project of Guangdong Province Promoting the Development of Science and Technology Service Industry

(Grant nos. 2013B040403005), and Shenzhen Science and Technology Plan Project (Grant no. CXZZ20140418182638764).

References

1. Dantzig, G.B., Ramser, J.H.: The truck dispatching problem. Manage. Sci. **6**(1), 80–91 (1959)
2. Augerat, P., Belenguer, J.M., Benavent, E., Corberin, A., Naddef, D.: Separating capacity constraints in the CVRP using tabu search. Europe J. Oper. Res. **106**, 546–557 (1998)
3. Esaulniers, G.D., Desrosiers, J., Erdman, A., Solomon, M.M., Soumis, F.: The VRP with pickup and delivery. In: The Vehicle Routing Problem. Society for Industrial and Applied Mathematics, Philadelphia (2001)
4. Renaud, J., Laporte, G., Boctor, F.F.: A tabu search heuristic for the multi-depot vehicle routing problem. Comput. Oper. Res. **23**(3), 229–235 (1996)
5. Kallehauge, B., Larsen, J., Madsen, O.B.G., Solomon, M.: Vehicle routing problem with time windows. In: Desaulniers, G., Desrosiers, J., Solomon, M.M. (eds.) Column Generation, pp. 67–98. Springer, New York (2005)
6. Nag, B., Golden, B.L., Assad, A.: Vehicle routing with site dependencies. In: Vehicle Routing: Methods and Studies, pp. 149–159. Elsevier, Amsterdam (1988)
7. Gendreau, M., Laporte, G., Musaraganyi, C., Taillard, E.D.: A tabu search heuristic for the heterogeneous fleet vehicle routing problem. Comput. Oper. Res. **26**(12), 1153–1173 (1999)
8. Taillard, E.D.: A heuristic column generation method for the heterogeneous fleet VRP. RAIRO Oper. Res. **33**(01), 1–14 (1999)
9. Passino, K.M.: Biomimicry of bacterial foraging for distributed optimization and control. IEEE Control Syst. Mag. **22**(3), 52–67 (2002)
10. Niu, B., Wang, H., Tan, L.-J., Li, L., Wang, J.-W.: Vehicle routing problem with time windows based on adaptive bacterial foraging optimization. In: Huang, D.-S., Ma, J., Jo, K.-H., Gromiha, M.M. (eds.) ICIC 2012. LNCS, vol. 7390, pp. 672–679. Springer, Heidelberg (2012)
11. Tan, L.J., Lin, F.Y.: Adaptive comprehensive learning bacterial foraging optimization and its application on vehicle routing problem with time windows. Neurocomputing **151**, 1208–1215 (2015)
12. Hezer, S., Kara, Y.: Solving vehicle routing problem with simultaneous delivery and pick up using bacterial foraging optimization algorithm. In: Proceedings of the 41st International Conference on Computers and Industrial Engineering, Los Angeles, pp. 380–385 (2011)
13. Niu, B., Bi, Y., Xie, T.: Structure-redesign-based bacterial foraging optimization for portfolio selection. In: Huang, D.-S., Han, K., Gromiha, M. (eds.) ICIC 2014. LNCS, vol. 8590, pp. 424–430. Springer, Heidelberg (2014)
14. Tutuncu, G.Y.: An interactive GRAMPS algorithm for the heterogeneous fixed fleet vehicle routing problem with and without backhauls. Eur. J. Oper. Res. **201**, 593–600 (2010)
15. Li, S.H.: The Model and Particle Swarm Optimization for Vehicle Routing Problem with Time Windows. 37–38
16. Niu, B., Fan, Y., Zhao, P., Xue, B., Li, L., Chai, Y.J.: A novel bacterial foraging optimizer with linear decreasing chemotaxis step. In: Second International Workshop on Intelligent Systems and Applications, pp. 1–4. IEEE Press, New York (2010)
17. Niu, B., Fan, Y., Wang, H.: Novel bacterial foraging optimization with time- varying chemotaxis step. Int. J. Artif. Intell. **7**(A11), 257–273 (2011)

SRBFO Algorithm for Production Scheduling with Mold and Machine Maintenance Consideration

Ben Niu[1,2,3]($\boxtimes$), Ying Bi[1], Felix T.S. Chan[3], and Z.X. Wang[3]($\boxtimes$)

[1] College of Management, Shenzhen University, Shenzhen, China
Drniuben@gmail.com
[2] Hefei Institute of Intelligent Machine,
Chinese Academy of Science, Hefei, China
[3] Department of Industrial and System Engineering,
Hong Kong Polytechnic University, Kowloon, Hong Kong
jeremy.wang@connect.polyu.hk

Abstract. A good production scheduling integrated with preventive maintenance scheduling scheme is significant for maintaining a higher reliability and stability for manufactory system. In this paper production scheduling problem with mold and machine maintenance (PS-MMS) consideration is studied and solved by structure-redesign-based bacterial foraging optimization (SRBFO) algorithm. PPS-MMS is a typical discrete combination optimization problem that allocating a certain number of jobs to available machine and mold and integrating maintenance activities for machine and mold with production activities. Unlike traditional maintenance activities operated with fixed duration, the maintenance duration in our PS-MMS model is varying with the usage age of machine/mold. To obtain a better solution for this difficult problem in acceptable time, SRBFO is adopted by encoding and decoding of bacteria on every dimension so that each bacterium can represent a potential solution. Five different scale instances were selected as test problems, experimental results demonstrated that SRBFO is more suitable than PSO to deal with PS-MMS problem in terms of the stability from the best solutions.

Keywords: Structure redesigned · Barterial foraging · Production scheduling · Machine maintenance · Mold maintenance

1 Introduction

In recent years, operation managers in production firms begin to focus on drawing up a good production scheduling scheme to lower the operational costs and deliver on time to improve competitive advantages under the current fierce competition environment. The issue has generated a lot of interest among researchers [1–4].

Preventive maintenance is an important action that can reduce deterioration with usage of machines to maintain a high reliability and stability for production system. The significance of preventive maintenance has attracted many practitioners and researchers [5–7].

© Springer International Publishing Switzerland 2015
D.-S. Huang et al. (Eds.): ICIC 2015, Part II, LNCS 9226, pp. 733–741, 2015.
DOI: 10.1007/978-3-319-22186-1_73

To address this problem, this paper explored production scheduling integrated with preventive maintenance scheduling (PS-MMS) problem under a plastic system, where jobs are allocated to an injection machine and a corresponding injection mold. The PS-MMS problem is a typical discrete combination optimization and difficult to be solved by traditional mathematical methods in acceptable time.

Bacteria foraging optimization (BFO) is a random search technology, has been received many attention by researchers from several aspects [8–11]. As a new variant of BFO, SRBFO was proposed by Niu et al. [12] to improve the performance of BFO in 2014.

In this paper, SRBFO is applied to find a better scheduling scheme for PS-MMS problem. The rest of paper is organized as follow: SRBFO is outlined in Sect. 2. Section 3 describes PS-MMS problem. In Sect. 4, four instances are selected as test problems and the experiment result is presented and analyzed. Finally, Sect. 5 gives a summary.

2 SRBFO

BFO was proposed by Passino (2002) [13] drawing inspiration from the foraging activities of Escherichia coli groups. Four essential behaviors of E. coli bacterium were simulated in BFO, namely chemotaxis, swarming, reproduction, elimination and dispersal, which demonstrate the social and cooperative feature when the bacteria survived in the area with sufficient or limited concentration of nutrients.

As a variant of BFO, SRBFO is put forward by redesigning the original BFO's implementation structure to improve convergence speed and lower time consumption. A single-loop execution structure is introduced into SRBFO instead of the triple nested ones in BFO, and the reproduction operation and elimination-dispersal operation are nested within chemotactic operation. It has been proved that SRBFO maintains the good global optimization ability of BFO and convergences faster for dealing with continue optimization problems. These related details can be referred to Ref. [11]. The Pseudo-code of SRBFO is presented in Table 1.

3 PS-MMS Problem

3.1 Maintenance Strategies

Production scheduling problem with machine and mold maintenance scheduling (PS-MMS) is studied in this section. Unlike traditional maintenance activities, which operate with fixed duration, in PS-MMS problem, maintenance duration varies with the usage age of machine/mold. We assumed that the relationship between maintenance time of machine/mold and the usage age of machine/mold can be fitted with a piece-wise linear function. The relationships are listed in Table 2 and shown in Fig. 1.

Table 1. Pseudo-code of SRBFO

SRBFO
Begin
For (Each run)
Calculate the fitness function J, and set $J_{last}= J$;
For (Each chemotaxis)
Update the position of bacteria by tumbling (Chemotaxis);
Calculate fitness function J;
If $(J<J_{last};$ and the maximum swimming steps are not met)
Swimming (Chemotaxis);
End If
If mod(the current number of chemotaxis, reproduction frequency Fre)==0
Reproduction operation;
End If
If mod(the current number of chemotaxis, dispersal frequency Fed)==0
Dispersal operation;
End If
End For
End For
End

Table 2. The relationships between machine/mold age and maintenance time

Machine age	Maintenance time	Mold age	Maintenance time
$0 < A_1 \leq 200$	150	$0 < A_2 \leq 160$	100
$200 < A_1 \leq 480$	$160 + A_1/4 - 40$	$160 < A_2 \leq 400$	$160 + A_2/3 - 40$
$480 < A_1 \leq 720$	$200 + A_1/3 - 50$	$400 < A_2 \leq 640$	$200 + A_2/2 - 50$
$720 < A_1$	720	$640 < A_2$	640

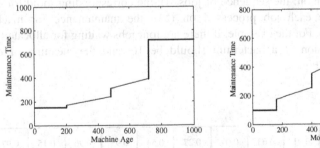

Fig. 1. The relationships between machine/mold age and maintenance time

3.2   Description of PS-MMS Problem

The PS-MMS problem can be described as allocate job j ($j = 1, 2, 3, …, J$) to machine m ($m = 1, 2, 3,…, M$) with a special mold d ($d = 1, 2, 3,…, D$). Each job j processed on machine m with mold d occupies a processing time P_{jmd}. Preventive maintenance activities for each machine (PM_{mt}) and mold (PMD_{dt}) are scheduled at some time point t $(t = 1, 2,..., T)$ in the whole makespan. The objective of PS-MMS problem is to minimize makespan $(Min(max(C_j)))$ by scheduling and integrating maintenance activities for machine and mold with production activities.

Some assumptions, conditions and constraints should be considered or satisfied are presented as follow:

(a) All jobs are available and independent at time zero.
(b) There are not suddenly interruption when machine processing a job.
(c) The set-up time of each job is included in processing time.
(d) The performances of machines and molds are very good at time zero, so the initial ages of them are zero.
(e) Not all molds can be operated on all machines.
(f) The completion time of each job equals the starting time plus the processing time of each job.
(g) Each job only can be processed on one machine with one mold.
(h) Each machine at each time unit can carry out one job with one mold, as well as each mold be carried out on one machine with one job at each time unit.
(i) The age of machine after a maintenance action is reset to 0.

4 Application

4.1 Encoding and Decoding

SRBFO is used to solve continuous domain problems, such as portfolio selection problem [11], while the PS-MMS problem is a typical discrete combinatorial optimization problem. Encoding and decoding of bacteria are necessary to make SRBFO successfully applied to PS-MMS problem.

Each bacterium represents a potential solution for PS-MMS problem and contains five aspects of information: the sequence of jobs (J), the corresponding mold for each job (D), the machine each job processed on (M), the maintenance for machine (PM) and mold (PMD). For the example, if there are four jobs waiting for allocated and processed, the dimension of a bacterium should be 16, and the meaning of each dimension is given in Fig. 2.

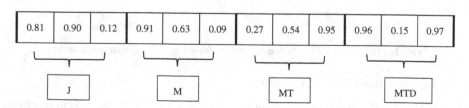

Fig. 2. An example of encoding a bacterium

The decoding schemes employed to transfer continuous position vector of bacteria in each dimension to a discrete scheduling solution are based on the random key representation [14] and the smallest position value (SPV) rule [15].

4.2   Case Studies

To demonstrate the optimization performance and effectiveness of SRBFO on PS-MMS problems, five representative instances with different scale are selected as test problems, as shown in Table 3.

Table 3. Parameters of test problems

Instances	J (job)	D (mold)	M (machine)	J × D × M
1	15	4	3	15 × 4 × 3
2	20	5	6	20 × 5 × 6
3	30	4	3	30 × 4 × 3
4	30	5	6	30 × 5 × 6
5	40	7	8	40 × 7 × 8

Some other parameters of test problems are listed below:

- The unit of jobs is a discrete random variable over the integers 1, 2, ..., 8.
- The processing time of molds on operable machine is randomly generated between 25 and 45 units of time.
- The initial age of each machine is zero.

PSO is chosen as comparable algorithm to reveal the optimization ability of SRBFO on the production scheduling problem in this paper. For the five instances, each test run 10 times, the initial population is 100 and the iteration is set as 1000.

Table 4 presents the min, max, mean and standard deviations obtained by SRBFO and PSO of 10 runs on the five cases. The average convergence characteristics of each algorithm for five test problems are shown in Figs. 3, 4 and 5. The best scheduling scheme obtained by SRBFO on case 2 are given in Figs. 6 and 7, where MT means the maintenance operation.

From Table 4, we found that PSO searched better minimum makespan on case 1, 2 and 4 than SRBFO, while the averages, standard deviations of fitness of 10 runs obtained by SRBFO are smaller than PSO on all instances. These results show that SRBFO's performance is more stable than PSO on PS-MMS problems. Figure 3 shows PSO converges faster than SRBFO because SRBFO need more CPU time to carry out four operations especially chemotaxis and swarming. From above results, it can be concluded that SRBFO is more suitable than PSO to solve these five different scale PS-MMS problems in terms of the stability from the best solutions.

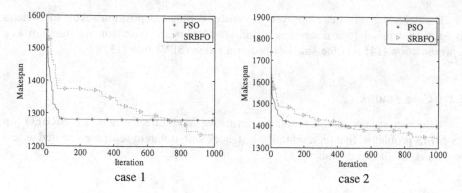

Fig. 3. Convergence curves of algorithms of case 1 and 2

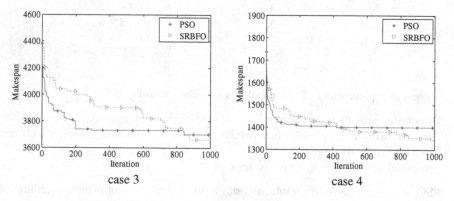

Fig. 4. Convergence curves of algorithms of case 3 and 4

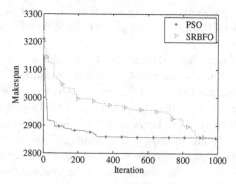

Fig. 5. Convergence curves of algorithms of case 3 and 4

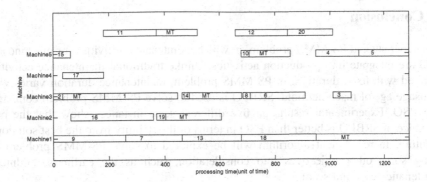

Fig. 6. The best machine scheduling scheme of instance 2 obtained by SRBFO

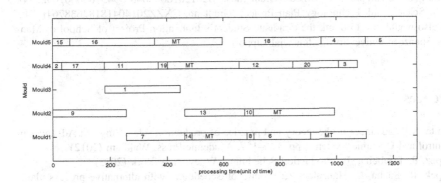

Fig. 7. The best mold scheduling scheme of instance 2 obtained by SRBFO

Table 4. Numerical results obtained by algorithms

Case	Algorithm	Min	Max	Average	Std.
1	PSO	**1178**	1378	1.2809e+003	7.5049e+001
	SRBFO	1198	1288	**1.2360e+003**	**3.5818e+001**
2	PSO	**1295**	1.4887e+003	1.4001e+003	6.0982e+001
	SRBFO	1305	1.4080e+003	**1.3459e+003**	**3.9256e+001**
3	PSO	3.5297e+003	3.8733e+003	3.6994e+003	1.4813e+002
	SRBFO	**3.5010e+003**	3.8140e+003	**3.6626e+003**	**1.3108e+002**
4	PSO	**1.7683e+003**	2.0741e+003	1.9285e+003	1.1841e+002
	SRBFO	1.7793e+003	1.9624e+003	**1.8790e+003**	**5.8283e+001**
5	PSO	2.6978e+003	2.9686e+003	2.8543e+003	7.8492e+001
	SRBFO	**2.7630e+003**	2.9590e+003	**2.8526e+003**	**7.4180e+001**

5 Conclusion

This paper studied PS-MMS problem, in which maintenance activities for machine and mold are integrate into production activities. Unlike traditional maintenance activities operated with fixed duration, in PS-MMS problem, maintenance duration varies with the usage age of machine/mold. SRBFO is used to solve PS-MMS problem compared with PSO. Experimental results on five different scale instances show that the performance of SRBFO is better than PSO in terms of the stability from the best solutions. In future, more efficient algorithm will be explored to solve PS-MMS problem by taking some other objectives into consideration, such as the earliness, tardiness, maintenance cost and so on.

Acknowledgements. This work is partially supported by The National Natural Science Foundation of China (Grants nos. 71001072, 71271140, 71471158, 71461027), the Natural Science Foundation of Guangdong Province (Grant nos. S2012010008668, 9451806001002294), Shenzhen Science and Technology Plan Project (Grant no. CXZZ20140418182638764). The authors also would like to thank the Graduate Student's Innovation Project of School of Management Shenzhen University for financial support.

References

1. White, K.P.: advances in the theory and practice of production scheduling. In: Advances in Control and Dynamic Systems, pp. 115–157. Academic Press, Waltham (2012)
2. Lopez, P., Roubellat, F.: Production Scheduling. Wiley, New York (2013)
3. Čapek, R., Šůcha, P., Hanzálek, Z.: Production scheduling with alternative process plans. Eur. J. Oper. Res. **217**(2), 300–311 (2012)
4. Karimi-Nasab, M., Ghomi, S.M.T.F.: Multi-objective production scheduling with controllable processing times and sequence-dependent setups for deteriorating items. Int. J. Prod. Res. **50**(24), 7378–7400 (2012)
5. Wong, C.S., Chan, F.T.S., Chung, S.H.: A joint production scheduling approach considering multiple resources and preventive maintenance tasks. Int. J. Prod. Res. **51**(3), 883–896 (2013)
6. Wang, S., Liu, M.: A branch and bound algorithm for single-machine production scheduling integrated with preventive maintenance planning. Int. J. Prod. Res. **51**(3), 847–868 (2013)
7. Fitouhi, M.C., Nourelfath, M.: Integrating noncyclical preventive maintenance scheduling and production planning for a single machine. Int. J. Prod. Econ. **136**(2), 344–351 (2012)
8. Qiuping, D.A.I., Liang, M.A., Ying, X.I.: Improved bacterial foraging algorithm for continuous optimization problem. J. Univ. Shanghai Sci. Technol. **2**, 002 (2013)
9. Elattar, E.E.: A hybrid genetic algorithm and bacterial foraging approach for dynamic economic dispatch problem. Int. J. Electr. Power Energy Syst. **69**, 18–26 (2015)
10. Niu, B., Fan, Y., Xiao, H., et al.: Bacterial foraging based approaches to portfolio optimization with liquidity risk. Neurocomputing **98**, 90–100 (2012)
11. Pandit, N., Tripathi, A., Tapaswi, S., et al.: An improved bacterial foraging algorithm for combined static/dynamic environmental economic dispatch. Appl. Soft Comput. **12**(11), 3500–3513 (2012)

12. Niu, B., Bi, Y., Xie, T.: Structure-redesign-based bacterial foraging optimization for portfolio selection. In: Huang, D.-S., Han, K., Gromiha, M. (eds.) ICIC 2014. LNCS, vol. 8590, pp. 424–430. Springer, Heidelberg (2014)
13. Passino, K.M.: Biomimicry of bacterial foraging for distributed optimization and control. IEEE Control Syst. **22**(3), 52–67 (2002)
14. Bean, J.C.: Genetic algorithms and random keys for sequencing and optimization. ORSA J. Comput. **6**(2), 154–160 (1994)
15. Tasgetiren, M.F., Liang, Y.C., Sevkli, M., et al.: A particle swarm optimization algorithm for makespan and total flowtime minimization in the permutation flowshop sequencing problem. Eur. J. Oper. Res. **177**(3), 1930–1947 (2007)

A Novel Branch-Leaf Growth Algorithm
for Numerical Optimization

Xiaoxian He[1,2]($\boxtimes$), Jie Wang[1], and Ying Bi[3]($\boxtimes$)

[1] College of Information Science and Engineering, Central South University,
Changsha, China
{xxhe, jwang}@csu.edu.cn
[2] College of Engineering, University of Tennessee, Knoxville, TN, USA
[3] College of Management, Shenzhen University, Shenzhen, China
1103979459@qq.com

Abstract. Inspired by branch and leaf growth behaviors of plants, a novel algorithm, named branch-leaf growth algorithm (BLGA), is presented for numerical optimization. In this algorithm, though branch and leaf implement different growth strategies, they cooperate closely to search the space for living resources. More specifically, branches grow into a stable self-similar architecture to support remote exploration, while leaves exploit local areas for better chances in each generation. An inhibition mechanism of plant hormones is applied to branches in case of overgrowth. In order to validate its efficiency, eight classic benchmark functions are adopted for test, and the results are compared with PSO, BFO and BCFO. The comparing results show that BLGA outperforms other evolutionary algorithms on most of benchmark functions.

Keywords: Branch-leaf growth · Plant-inspired algorithm · Self-similar propagation · Optimization

1 Introduction

In recent years, many bionic algorithms have achieved significant success in solving optimization problems. For example, Particle swarm optimizer (PSO) simulates swarm behaviors of birds and fish [1, 2]. Artificial bee colony (ABC) algorithm simulates cooperative foraging behaviors of a swarm of honeybees [3, 4]. Bacterial colony optimization algorithm simulates typical behaviors of bacteria during their lifecycle [5, 6]. Since there is no central control and all individuals cooperate to search for solutions simultaneously and spontaneously, partial failure has little effect on final performance of the population. This is a big advantage for heuristic algorithms to deal with complex multimodal problems.

Most of existing bionic algorithms imitate animal's behaviors. As another species of biology, however, plant has attracted only few scientists' attention until now [7–9]. Unlike animal that can move freely and forage in a large-scale range, plant chooses another way to adapt to the differing environments. Though anchored in fixed positions, they can grow iteratively to explore the surroundings to get nutrients and other living resources. Since they have also evolved for millions of years and shown

© Springer International Publishing Switzerland 2015
D.-S. Huang et al. (Eds.): ICIC 2015, Part II, LNCS 9226, pp. 742–750, 2015.
DOI: 10.1007/978-3-319-22186-1_74

surprising plasticity to the natural world, some biologists also regarded them as intelligent organisms[10, 11], which were only used to describe animals in the past.

Take the branch and leaf for example, plant can always steer their growth direction to get the most carbon dioxide and sunshine for photosynthesis. During the growth season, they continuously propagate some branches to occupy as much space with sunshine as possible, and then leaves are produced around these branches to get resources for photosynthesis, which improves the growth rate of branches in feedback. In the process, stems and branches grow gradually into a self-similar architecture to support exploring the whole surrounding space [12, 13], and leaves implement local search to get as much resources as possible. Though implement different growth strategies, branches and leaves cooperate so well that the plant becomes a very efficient resource collector.

The branch and leaf growth mechanism provides a perfect idea for designing optimization algorithms. Inspired by their growth behaviors, this paper presents a novel algorithm named branch-leaf growth algorithm for numerical optimization.

The remainder of this paper is organized as follows. Section 2 models the branch and leaf growth process. Section 3 presents the branch-leaf growth algorithm. Experiments and results are given in Sect. 4. Section 5 outlines the conclusions.

2 Artificial Branch-Leaf Growth

2.1 Basic Concept

In the growth model of artificial plant, an objective function is considered as growth environments, and the initial branches are considered as a homogeneous biomass. For the convenience of description, the growing process is defined as two phases separately, namely branch growth phase and leaf growth phase. In both growth phases, each growth apex represents a solution in the feasible region of the problem. All branches and leaves try to adjust their growth strategies in order to search for better growing conditions, which feed back to improve the growth rate of branches further.

In growing process, all branch apices can select their own growth strategies according to different conditions. However, all the strategies are composed of the following three basic actions [14]:

(1) Each branch apex may elongate forward (or sideways) in the environment.
(2) Each branch apex may produce daughter branch apices or leaf apices.
(3) Each branch apex may cease to function as above, and become an ordinary piece of branch mass.

The growth strategies of leaf apices consist of two basic actions:

(1) Each leaf apex may grow into a branch apex.
(2) Each leaf apex may cease to function and fall down.

In summary, a branch apex can produce many leaf apices, and each leaf apex belongs to a given branch apex. Though branch apex and leaf apex implement different growth strategies, they can transform into each other under some conditions.

Based on fitness values, the total initial branch mass are divided into two groups: the main branch group and the lateral branch group. The first group includes branches that have the best fitness values. Others are included in the second group. In these two groups, main branches and lateral branches implement different growth strategies.

2.2   The Growth Strategy of Main Branch: Monopodial Branching

Monopodial branching strategy means that the main branch itself regrows forward to form an axis, and produces branches in the lateral position in the meantime. In detail, the growth strategy contains three operators as follows.

(1) Regrowing. Firstly, the branch apex will elongate towards a local best position where there are better growth conditions. The operator is formulated as the following expression.

$$x_i^t = x_i^{t-1} + l \cdot rand() \cdot \left(x_{lbest} - x_i^{t-1} \right) \tag{1}$$

where x_i^{t-1} is the position of the ith branch apex in the last generation, and x_i^t is the new position in the current generation. l is the local learning constant. $rand()$ is a random number with uniform distribution in [0, 1]. x_{lbest} is the local best position in the last generation.

(2) Branching. Main branch will produces some lateral branches besides regrowing itself. The number of new-produced branch apices is calculated as follows.

$$w_i = \frac{f - f_{min}}{f_{max} - f_{min}} \cdot (s_{max} - s_{min}) + s_{min} \tag{2}$$

where f_{max} and f_{min} are the best and the worst fitness values in the current generation, respectively. f is the fitness value of the main branch apex. s_{max} and s_{min} are the maximal and the minimal branching number which are preset.

The positions of new-produced branch apices locate around the original branch apex with Gauss distribution $N(x_i^t, \sigma^2)$. The standard deviation σ is calculated as follows.

$$\sigma_i = ((i_{max} - i)/i_{max})^n \cdot (\sigma_{ini} - \sigma_{fin}) + \sigma_{fin} \tag{3}$$

where i_{max} and i are the maximal iteration number and current iteration number, respectively. σ_{ini} is the initial standard deviation which depends on the value of searching range, and σ_{fin} is the final standard deviation determined by expected accuracy standard in the applications.

From formula (3) we can see that when the value of i increases, the value of σ will become smaller and smaller. It is helpful to improve the convergence speed of the algorithm. Meanwhile, similar architectures will appear in variant scales according to these formulas. As a whole, it is an approximate self-similar architecture.

(3) Inhibition Mechanism of Plant Hormones. Because new-produced branch apices may be classified into the main branch group with high probability in the next generation if resources in the area are really rich enough, and all potential main branches will elongate and propagate again, the number of branches in this area may increase explosively, which we call overgrowth.

Overgrowth in some local area is absolutely harmful to the adaptability and sustainability of plants. From the view of optimization, it will make the algorithm plunge into local optima. In fact, this phenomenon is rarely seen from natural plants because plant hormones play an important role in making a balance.

Generally, the growth of plant is regulated by many plant hormones. Most of them can improve growth speed of plant organisms. However, if some part of a plant grows too quickly, one of plant hormones may increase rapidly and the balance between all plant hormones will be broken, then the growth will be inhibited [15, 16].

In order to imitate the mechanism of plant hormones to inhibit the overgrowth, we will calculate the local standard deviation $\sigma_{local}(f)$ of new-produced branch apices, and then get rid of some branch apices according to the calculating results by greedy principle. The operator is implemented as formula (4).

$$w_{-i} = \alpha \cdot \left(1 - \frac{\sigma_{local}(f)}{f_{max} - f_{min}} \right) \cdot w_i \qquad (4)$$

where w_{-i} is the number of branch apices which should be removed, and α is a control parameter.

From formula (4) we can see that the smaller $\sigma_{local}(f)$ is, the more branch apices will be removed in the next generation. On the one hand, rapid increase of branches is under control so that overgrowth can be avoided. On the other hand, essential diversity can be preserved to prevent the algorithm from pre-maturity in this way.

2.3   The Growth Strategy of Lateral Branch: Sympodial Branching

Sympodial branching strategy means that the branch will not elongate itself. Alternatively, it will produce a new branch apex at a lateral position, and the new-produced branch apex will grow into an axis by replacing the original one. The new branch apex may locate at a random position near the original branch with a random angle β. This strategy is formulated as follows.

$$x_i^t = rand() \cdot \beta \cdot x_i^{t-1} \qquad (5)$$

where $rand()$ is a random number with uniform distribution in [0, 1]. β is calculated as follows.

$$\beta = \lambda_i / \sqrt{\lambda_i^T \cdot \lambda_i} \qquad (6)$$

where λ_i is a random vector.

2.4   The Growth Strategy of Leaf: From Branch to Branch

Leaf is the lateral appendage of a branch which is specialized for photosynthesis. In the model, each branch apex will produce a cluster of leaf apices. Differing from branch apices, these leaf apices cannot form any static architecture. They just appear beside a branch seasonally. When they fall down at the end of the generation, the leaf apex that has the best fitness value will grow into a new branch apex at the next growth season.

In summary, they are born from branches at the start of the generation, and will grow into branches or fall down in the end.

The strategy contains two operators:

(1) Producing. After branch growth phase, each branch apex will produce some leaf apices around it. Simply, the number of new produced leaf apices will be set as s_{max}, which has been explained in formula (2). The positions of new produced leaf apices surround the branch apex with Gauss distribution $N(x_i^t, \varphi^2)$. The standard deviation φ is calculated as follows.

$$\varphi = \frac{\sigma_{ini} - \sigma_{fin}}{2} \tag{7}$$

(2) Falling Down. After a branch apex has produced leaf apices, the fitness values of all leaf apices will be estimated. The leaf apex that has the best fitness value will grow into a new branch apex to replace the original one, and then all leaves fall down.

3   Branch-Leaf Growth Algorithm

As another species of biology that has evolved for a long time, plant has shown strong vitality and is extremely adaptive to the differing environments in the natural world. Its growth mode is a wonderful inspiration for designing optimization algorithms.

As shown in the model presented above, all branches and leaves can cooperate closely to search the solution space. Branches grow into a stable self-similar architecture to support remote exploration tasks, and leaves exploit local areas for better results. Though the two growth phases are independent to some extent, this model has tried to describe the original appearance of natural plant with few assumptions. According to the model, corresponding algorithm for numerical optimization is designed. The pseudo code of BLGA is listed in Table 1.

4   Experiments and Results

In order to test the performance of BLGA, three bio-inspired algorithms including PSO, BFO and BCFO, are employed for comparison. In the experiments, eight classic benchmark functions are used to test its effectiveness.

Table 1. Pseudo code of BLGA

BLGA
1. Initialize the positions of branch apices
2. Evaluate the fitness values of branch apices
3. While not meet the terminal condition
4.   Divide the branch apices into main branches and lateral branches
5.   For each main branch apex
Regrow with the regrowing operator
Branch with the branching operator
Evaluate the fitness value of new branch apices
Calculate the deviations, implement inhibition mechanism of plant hormones
End for
6.   For each lateral branch apex
Produce a new apex to replace the original one
End for
7.   For each branch apex
Produce leaves with the producing operator
Evaluate fitness values and find the best one to replace the original branch apex
Leaves fall down
End for
8.   Rank branch apices and label elite branches
9. End while
10. Postprocess results

4.1 Experiment Sets and Benchmark Functions

The eight classic benchmark functions are widely employed to test evolutionary algorithms in many works. Among them, sphere is a unimodal function with separable variables which is easy to solve. Rosenbrock, Quadric and Sin are unimodal functions with non-separable variables. Among them, Rosenbrock function has a narrow valley sloping gently from local optima to the global optimum, thus can be treated as a multimodal function. Rastrigin and Schwefel are multimodal functions with separable variables. Ackley and Griewank are multimodal functions with non-separable variables. They all have a large number of local optima to make it difficult to reach the global optimum. All the benchmark functions are listed in Table 2.

In this paper, all functions use their standard ranges and variable data with 30 dimensions. The experiments are conducted by using Matlab 7.0 on a standard 2.5 GHZ desktop computer. The presented algorithm is run for 20 times and the mean values and standard deviation values are taken as final results. Related data of PSO, BFO and BCFO are picked up from [6]. In BLGA, the population size is set 100, and the number of branch apices in main branch group is thirty percent of the selected branch apices in each generation. s_{max} and s_{min} are set 3.0 and 1.0, respectively. α and l are all set 1.0.

Table 2. Test functions

Name	Function	Limits
Sphere	$f_1 = \sum\limits_{i-1}^{D} x_i^2$	$x \in [-5.12, 5.12]^D$
Rosenbrock	$f_2 = \sum\limits_{i-1}^{D} \left(\left(100(x_i^2 - x_{i+1})^2 + (1 - x_i)^2\right) \right)$	$x \in [-2.048, 2.048]^D$
Quadric	$f_3 = \sum\limits_{i=1}^{D} \left(\sum\limits_{j=1}^{i} x_j \right)^2$	$x \in [-30, 30]^D$
Sin	$f_4 = \frac{\pi}{D} \left(10 \sin^2 \pi x_1 + \sum\limits_{i=1}^{D-1} (x_i - 1)^2 (1 + 10 \sin^2 \pi x_{i+1}) + (x_n - 1)^2 \right)$	$x \in [-10, 10]^D$
Rastrigin	$f_5 = \sum\limits_{i=1}^{D} \left(x_i^2 - 10 \cos(2\pi x_i) + 10 \right)$	$x \in [-5.12, 5.12]^D$
Schwefel	$f_6 = 418.9829 * D + \sum\limits_{i=1}^{D} \left(-x_i \sin\left(\sqrt{|x_i|}\right) \right)$	$x \in [-500, 500]^D$
Ackley	$f_7 = 20 + e - 20 \exp\left(-0.2\sqrt{\frac{1}{D}\sum\limits_{i=1}^{D} x_i^2} \right) - \exp\left(\frac{1}{D}\sum\limits_{i=1}^{D} \cos(2\pi x_i)\right)$	$x \in [-32, 32]^D$
Griewank	$f_8 = \frac{1}{4000} \left(\sum\limits_{i=1}^{D} x_i^2 \right) - \left(\prod\limits_{i=1}^{D} \cos\left(\frac{x_i}{\sqrt{i}}\right) \right) + 1$	$x \in [-5.12, 5.12]^D$

4.2 Experiment Results and Analysis

The mean fitness values and standard deviation values obtained by the four algorithms are listed in Table 3. The best values obtained on each function are marked as bold.

Table 3. Results on eight benchmark functions with 30-D

Function		BLGA	PSO	BFO	BCFO
f_1	Mean	**3.95426E-12**	100.319	26.1113	12.3428
	Std	**5.13521E-12**	25.90979	32.6607	4.97197
f_2	Mean	3.50947E-14	1.42953	2.63748E-07	**5.26407E-16**
	Std	5.66421E-14	0.46578	5.87148E-07	**6.535E-17**
f_3	Mean	**6.28731E-04**	1441.36	36.3888	0.17389
	Std	**2.15421E-04**	1187.89	23.839	0.36929
f_4	Mean	3.24587E-14	2.73183	1.13101	**4.98689E-16**
	Std	7.12501E-15	1.13963	2.93271	**9.80152E-17**
f_5	Mean	**2.78054E-14**	144.859	9.95228	3.78956E-14
	Std	**6.10087E-15**	29.7319	2.48322	3.10747E-14
f_6	Mean	**0.54821**	1396.15	2248.02	3.9498
	Std	**0.06452**	455.262	420.689	21.6235
f_7	Mean	5.75265E-14	19.85	0.03607	4.23424E-13
	Std	3.52152E-14	0.45667	0.18014	1.60667E-13
f_8	Mean	6.12564E-11	5.4798	0.08760	8.26987E-04
	Std	5.01478E-11	1.40991	0.06669	3.21616E-03

As can be seen in this table, BLGA performs much better than others on function f_1, f_3, f_5, f_6, f_7 and f_8. PSO and BFO fail to hit the best results in all functions. BCFO gets the best results on function f_2 and f_4 as a newly improved algorithm. Though BLGA is slightly worse than BCFO on the two functions, it still performs quite well in terms of accuracy. The most notable is that BLGA hits the best results in all multimodal functions, which indicates its great potential for treating on complex multimodal optimization problems.

In view of the above comparison, we can see BLGA is a very promising algorithm. It has shown very strong optimizing ability on test functions.

5 Conclusions

Inspired by branch and leaf growth behaviors of plants, a novel numerical optimization algorithm (BLGA) is presented in this paper. Eight classic benchmark functions were used to validate its efficiency, and the results were compared with PSO, BFO and BCFO. The comparing results showed that BLGA outperforms other evolutionary algorithms on most of benchmark functions. Moreover, it has indicated great potential for solving complex multimodal optimization problems.

A further extension to the current BLGA algorithm may lead to even more effective optimization algorithms for solving high-dimension and multimodal problems. Future research efforts will be focused on how to improve the proposed algorithm and apply it to complex practical engineering problems.

Acknowledgements. This research is partially supported by the open fund of Key Laboratory of Networked Control System, Chinese Academy of Sciences, and National Natural Science Foundation of China under Grants Nos. WLHKZ2014004, 61202341 and 61202495.

References

1. Kennedy, J., Eberhart, R.C.: Particle swarm optimization. In: Proceedings of IEEE International Conference on Neural Networks, Piscataway, NJ, pp. 1942–1948. IEEE Press, New York (1995)
2. Liu, H., Cai, Z., Wang, Y.: Hybridizing particle swarm optimization with differential evolution for constrained numerical and engineering optimization. Appl. Soft Comput. **10** (2), 629–640 (2010)
3. Karaboga, D., Basurk, B.: A powerful and efficient algorithm for numerical function optimization: artificial bee colony (ABC) algorithm. J. Glob. Optim. **39**, 469–471 (2007)
4. Kiran, M.S., Hakli, H., Gunduz, M., Uguz, H.: Artificial bee colony algorithm with variable search strategy for continuous optimization. Inf. Sci. **300**, 140–157 (2005)
5. Niu, B., Wang, H.: Bacterial colony optimization. Discrete Dyn. Nat. Soc. 2012, 1–28 (2012)
6. Chen, H., Zhu, Y., Hu, K., Ma, L.: Bacterial colony foraging algorithm: combining chemotaxis, cell-to-cell communication, and self-adaptive strategy. Inf. Sci. **273**, 73–100 (2014)

7. Cai, W., Yang, W., Chen, X.: A global optimization algorithm based on plant growth theory: plant growth optimization. In: 2008 International Conference on Intelligent Computation Technology and Automation (ICICTA), pp. 1194–1199, IEEE Press, New York (2008)

8. Zhang, H., Zhu, Y., Chen, H.: Root growth model: a novel approach to numerical function optimization and simulation of plant root system. Soft. Comput. **18**, 521–537 (2014)

9. He, X., Chen, H., Niu, B., Wang, J.: Root growth optimizer with self-similar propagation. Mathematical Problems in Engineering. Article in Press. http://www.hindawi.com/journals/mpe/aip/498626/ (2015)

10. Trewavas, A.: Green plants as intelligent organisms. Trends Plant Sci. **10**, 413–419 (2005)

11. Struik, P.C., Yin, X., Meinke, H.: Plant neurobiology and green plant intelligence: science, metaphors and nonsense. J. Sci. Food Agric. **88**, 363–370 (2008)

12. Chandra, M., Rani, M.: Categorization of fractal plants. Chaos, Solitons Fractals **41**(3), 1442–1447 (2009)

13. Rian, I.M., Sassone, M.: Tree-inspired dendriforms and fractal-like branching structures in architecture: a brief historical overview. Front. Architectural Res. **3**(3), 298–323 (2014)

14. Newson, R.: A canonical model for production and distribution of root mass in space and time. J. Math. Biol. **33**, 477–488 (1995)

15. Friml, J.: Auxin transport-shaping the plant. Curr. Opin. Plant Biol. **6**, 7–12 (2003)

16. Moubayidin, L., Mambro, R.D., Sabatini, S.: Cytokinin-auxin crosstalk. Trends Plant Sci. **14**(10), 557–562 (2009)

Author Index

Printed in the United States
By Bookmasters